AF324840

Innovative Coastal Zone Management: Sustainable Engineering for a Dynamic Coast

Edited by **Alexandra Schofield**

Proceedings of *Innovative Coastal Zone Management: Sustainable Engineering for a Dynamic Coast*, the seventh International Coastal Management Conference held by the Institution of Civil Engineers in Belfast, UK, on 15 – 16 November 2011.

Sponsored by:

Organising Committee

Alexandra Schofield Black & Veatch (Chair)
Alison Baptiste Environment Agency
Adrian Bell RPS Group
Bryan Curtis Worthing Borough Council
Lucy Edwards Institution of Civil Engineers
Gregor Guthrie Royal Haskoning
Ben Hamer Halcrow Group
Bill Langley RWE npower Renewables
Dr Robin McInnes Coastal and Geotechnical Services

Technical Committee

Andy Baldock Black & Veatch
Dr Rhoda Ballinger Cardiff University
Dr Noel Beech ENBE Ltd
Dr David Brook OBE Independent Consultant
Andrew Chadwick University of the West Indies
Andrew Cooper University of Ulster
David Green Independent Consultant
Jennifer Hart Beca Infrastructure
Ian Meadowcroft Environment Agency
Heather Schlosser United States Army Corp of Engineers
Marcel Stive TU Delft

Published by ICE Publishing, 40 Marsh Wall, London E14 9TP.

Full details of ICE Publishing sales representatives and distributors can be found at
http://www.icevirtuallibrary.com/info/printbooksales

www.icevirtuallibrary.com

A catalogue record for this book is available from the British Library

ISBN 978-0-7277-5749-4

© The authors 2012

ICE Publishing is a division of Thomas Telford Ltd, a wholly-owned subsidiary of the Institution of Civil
Engineers (ICE).

All rights, including translation, reserved. Except as permitted by the Copyright, Designs and Patents Act
1988, no part of this publication may be reproduced, stored in a retrieval system or transmitted in any form or
by any means, electronic, mechanical, photocopying or otherwise, without the prior written permission of
the Publisher, ICE Publishing, 40 Marsh Wall, London E14 9TP.

This book is published on the understanding that the author is solely responsible for the statements made
and opinions expressed in it and that its publication does not necessarily imply that such statements and/or
opinions are or reflect the views or opinions of the publishers. Whilst every effort has been made to ensure
that the statements made and the opinions expressed in this publication provide a safe and accurate guide,
no liability or responsibility can be accepted in this respect by the author or publishers.

Printed and bound by CPI Group (UK) Ltd, Croydon, CR0 4YY

Preface

Innovative Coastal Zone Management: Sustainable Engineering for a Dynamic Coast is the seventh International Coastal Management Conference in the Institution of Civil Engineers' (ICE) coastal conference series recognised for its focus on current issues and its balance between research and practical application.

The 2011 conference call for papers was the most successful to date and over 70 peer reviewed papers were presented and 8 posters were displayed in the exhibition room.

Papers and posters presented at the conference in Belfast covered the following topics:

* Innovative coastal planning, design and management, including coastal defence
* Coast and near shore renewable energy systems
* Social, environmental and climatic change
* Coastal policy – Legislation, targets and the future
* Funding and accountability – Investment, opportunities and growth
* Effective estuarine and coastal engineering

I hope you enjoy reading these papers and benefit from the knowledge and experience of the authors; and I hope that readers are able to implement some of the lessons learnt by others in this field.

I would like to personally extend my thanks to the Organising and Technical Committees for the rigorous peer review that they have conducted, as well as all the hard work that they have put into making this conference a success.

I would also like to extend my thanks to those who chaired the conference and the delegates for attending.

Finally I would like to invite you, the reader, to think about submitting an abstract to the next International Coastal Management Conference which will be held in 2015.

Alexandra Schofield

Chair, Conference Organising Committee

Contents

Section 4: Coastal policy – Legislation, targets and the future

SECTION 1:

INNOVATIVE COASTAL PLANNING, DESIGN AND MANAGEMENT, INCLUDING COASTAL DEFENCE

Innovative Coastal Zone Management
ISBN 978-0-7277-5749-4

ICE Publishing: All rights reserved
doi: 10.1680/iczm.57494.003

Establishing a Partnership Approach to Managing Coastal Assets

Dr Anne Thurston Environment Agency, England & Wales
Graham Lymbery Sefton Metropolitan Borough Council, England

Summary

Maritime Authorities and the Environment Agency have been working together to develop consistent standards and approach to inspecting and recording coastal asset condition information around the coastline of England and Wales.

Two pilot studies in the North West and South West of England have tested the compatibility of existing asset inspection training, methodology and reporting tools when applied to coastal assets. The pilots have taken different approaches and have identified improvements to enable national consistency of coastal asset condition data. They have also demonstrated very positive collaborative working between the Environment Agency and Maritime Authorities.

Background

Effective coastal risk management relies on a knowledge of the performance and condition of the defences in place (this includes both structural defences such as seawalls, and natural defences such as sand dunes). The many authorities who have an involvement in coastal assets currently have no nationally consistent way of inspecting and managing them.

Strategic Overview on the coast

As a result of *Making Space for Water* the Government gave the Environment Agency a Strategic Overview for all sources of flood risk and coastal erosion in England in 2008; the Welsh Assembly Government is proposing an enhanced oversight role for the Environment Agency in Wales.

Under its Strategic Overview role, the Environment Agency sets the direction of how flood and coastal erosion risk is managed in England. This means that the Environment Agency will

- take the lead for all sea flooding risk in England, and fund and oversee coastal erosion works undertaken by local authorities.
- ensure that sustainable long-term Shoreline Management Plans are in place for the English coastline.
- work with local authorities to ensure that the resulting flood and coastal erosion works are properly planned, prioritised, procured, completed and maintained to get the best value for the public purse.
- ensure that third party defences are sustainable.

Up to date asset condition information will be essential to demonstrate the need for both capital and maintenance funding on the coast.

Through the Floods and Water Management Act 2010, there are now legal duties placed upon Coastal management Authorities (including the Environment Agency) requiring consistent Asset Management. For example;

- The Environment Agency is producing a National Flood & Coastal Risk Management Strategy for England, (The Welsh Assembly Government is producing a strategy for Wales).

- The Environment Agency will report to Government on Flood and Coastal Erosion Risk Management matters. This will include condition of coastal assets as this is fundamental to understanding the associated flood or erosion risk, and to inform investment decisions for both the ongoing maintenance and improvement of the defences.

- There is a requirement for Lead Authorities to share information relevant to Flood and Coastal Risk Management and to develop & maintain an asset register, which must include details of Flood and Coastal Risk management assets, their ownership, and their current state of repair.

The Act will also give the Regional Flood and Coastal Committees the power to use levy monies for coastal erosion works, rather than just flood defence. Coastal asset condition data will be vital to secure support for smaller, levy-funded coastal erosion schemes promoted by Local Authorities.

Current Inspection Practices

Since 1999 the Government's aim has been for all flood and coastal assets to be recorded on a National Flood & Coastal Defence Database (NFCDD). In practice this has not happened for many reasons, including limited resources, inappropriate coastal terminology and the inability to access the database from Maritime Authority locations. As a result there are currently several approaches to undertaking visual condition inspections of coastal assets.

The Environment Agency has nationally consistent processes for inspecting fluvial and sea flood defence assets. A risk based programme of inspections is carried out by trained and externally accredited inspectors. Inspection data is stored on NFCDD.

Maritime Authorities have a range of practices for the inspection and storage of coastal asset data. Some share resources for inspections within the Coastal Groups, whilst others use bespoke inspection techniques and recording systems.

Inspections carried out by Maritime Authority inspectors are often more detailed than the initial visual inspections carried out by the Environment Agency, and require a greater understanding of the coastal processes acting within the asset system. Anecdotal evidence suggests that it takes over 2 years' experience for an inspector to be able to properly assess the condition of coastal assets.

Working In Partnership

To meet the requirements of the Floods & Water Management Act, both the Environment Agency and Maritime Authorities recognise the need and benefit of standardised asset data. It is much harder and more time consuming to share information if it is in several different formats. It therefore makes sense to work together to develop a common approach to inspecting, recording and storing our coastal asset information, and the pilot studies are the first step to achieving this.

Coastal Groups cover the whole coast of England and Wales. They contain coastal experts from all Authorities with a management interest within a geographic stretch of coast. They advise on coastal issues, lead the development of high level plans, and inform long term investment plans on how best to manage the risk from sea flooding and coastal erosion.
These Coastal Groups are ideally placed to manage a co-ordinated inspection programme on behalf of all the Maritime Authorities through their Coastal Monitoring Programmes. Figure 1 shows the geographic extent of the coastal groups, and the locations of the pilot study areas.

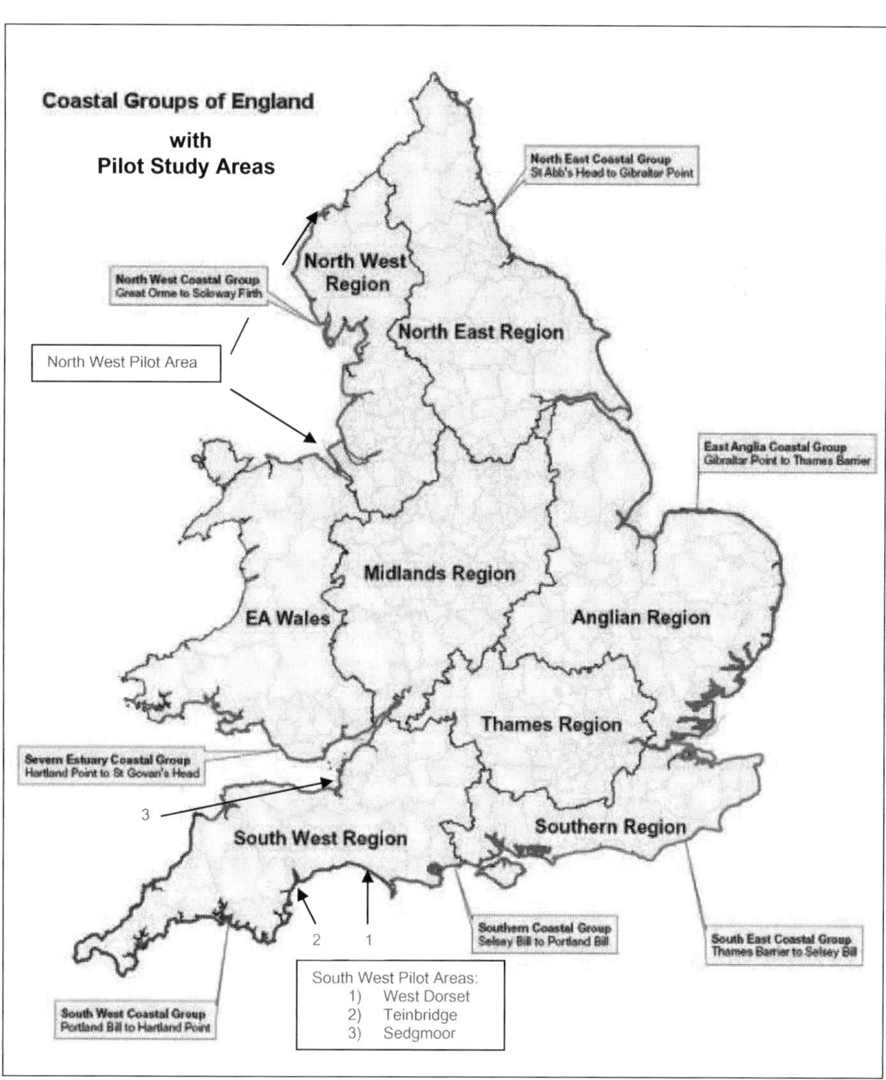

Figure 1. Location of Coastal Groups in England and Pilot Study Areas

Scope of the Pilots

The key aims of the pilots are;

- To investigate use of Regional Coastal Monitoring teams to manage a programme of consistent asset inspections.
- For Maritime Authorities to test appropriateness of the asset inspection accreditation scheme for carrying out inspections on coastal assets
- To examine if NFCDD can be used successfully to record information on coastal assets
- To identify the data required for National level reporting to Government and Operational level asset management.

North West and South West Coastal Groups agreed to lead the 12 month pilots and feedback the successes and constraints of the process, so that lessons learnt could be incorporated into a wider roll out of the inspection methodology to the remaining coastal groups.

Methodology

The first stage was to identify candidates within each coastal group to complete the asset inspection course. The accreditation requires candidates to complete an e-learning module and assessment, followed by a period of on-site asset inspections supervised by an accredited inspector, and concluded with a site assessment.

In the North West, Sefton Metropolitan Borough Council (as Lead Authority for the Regional Coastal Monitoring) identified several candidates, to allow the majority of the coastline to be covered. The South West Region took a different approach, whereby Lead Authority Teignbridge District Council, plus Sedgemoor District Council and West Dorset District Council each identified inspectors to undergo the training. This has allowed 3 micro pilots to be set up in the South West, before widening it to the whole region.

A fundamental part of the pilots was the partnership working between the Environment Agency and Maritime Authorities, so to assist with the accreditation and to learn more about coastal assets, Environment Agency asset inspectors have mentored the Maritime Authorities through the supervision stage of the course, whilst inspecting coastal assets.

Once accredited, the Maritime Authority inspectors have been using the Environment Agency's Condition Assessment Manual (CAM) to assign a condition grade (1-5, where 1 is very good condition, and 5 is very poor) to each coastal asset, and inputting this information into NFCDD. The CAM is a field manual which has photographic examples of all fluvial and coastal defences, at each condition grade, along with key indicators and performance features of each asset type.

Both Maritime Authority and Environment Agency inspectors have been issued with a template so that they can capture their experiences and provide essential feedback on what is working, and what elements of the process need improvement, to allow meaningful and consistent capture of coastal asset condition data.

Results of pilot studies

At time of writing the pilots have been running for 9 months. In the North West, the Maritime Authority inspectors have completed the e-learning, the supervised inspection stage and are awaiting the final site assessment. The coastal asset inspection data has been inputted to NFCDD, either by using ruggedized laptops in the field, and uploading the information in Environment Agency offices, or using computers in Environment Agency offices. Both options are time consuming due to the additional travel time.

In the South West, 5 inspectors are going through the accreditation process, and have just begun the supervised inspections with mentors from the Environment Agency. The inspectors have borrowed laptops from the Environment Agency, again due to complications with using NFCDD in the Maritime Authorities offices. Progress in the South West has been slower than in the North West due to resource constraints within the Maritime Authorities.

Lessons Learnt

The pilots have identified many successes and constraints to achieving a nationally consistent approach to coastal asset inspections, which can be grouped into 3 broad categories, High Level Objectives, Tools and Communications.

High Level Objectives

The early stages of the pilots highlighted several differences in both scale and terminology of asset inspections between the Maritime Authorities and the Environment Agency.

Scale of inspection

The Environment Agency has a staged approach to asset inspections; the field inspector carries out the initial visual inspection, and any assets which are graded below the target condition are reported back to the Asset Management teams who carry out a post inspection process to assess the flood risk and to determine what action should be taken. This may be a further, more detailed inspection, or planned remedial work. The visual inspection is done on individual assets, whereby the post inspection process takes the whole defence system into account when determining the consequence of any single asset failure.

Maritime Authority inspectors generally undertake these steps in one go, hence the longer timescales for becoming proficient at the inspections, as mentioned previously. The different scales of inspection are causing some difficulty in assigning condition grades to individual coastal assets. If this process is to be expanded around the coast, more detailed explanation of the differences in inspections will be required at the early stages to all inspectors.

Terminology

From the outset, the term 'inspection' had different meanings to both the Environment Agency and Maritime Authorities. This largely related to the point above about scale of inspection, i.e. detailed versus purely visual condition indicators. Additionally, whilst the Environment Agency inspectors have separate risk based programmes for asset condition and public safety inspections, Maritime Authorities inspectors simultaneously assess the asset in terms of flood risk, public amenity and safety.

At a more detailed level, there were differences in the terminology of individual asset elements. NFCDD contains asset descriptors which are suitable for fluvial defences, but inappropriate for coastal erosion and sea defence assets. e.g. 'face inner' for seawall elements facing the sea. This can lead to confusion and difficulty in trying to input the correct elements of a coastal asset into the database. This is an area that needs urgent attention if a wider roll out of the process is to be successful.

Use of the Data

The pilots have highlighted that the data collected is required at three levels, Operational, Strategic and National. Each level requires the information in different formats and frequencies.

The success of the long term adoption of a consistent methodology for asset inspection and data recording is to clearly state what information is required and for why. Whilst the idea of collecting data on the condition of coastal assets is well understood by all parties at all levels, what needs to be better defined is what specific information the Environment Agency requires as part of the national reporting requirements. This will help shape how new systems are developed, and data recorded. This will also reduce unnecessary duplication of data gathering for different purposes.

Tools and Technology

NFCDD

One of the main constraints during the pilots was the inability to easily access NFCDD. This meant that data input was very time consuming due to trips to Environment Agency offices, or very slow software performance in Maritime Authority offices. Improvements will be required if this is to be used more widely by Maritime Authorities. Additionally, the Maritime Authority inspectors had difficulty in trying to fit coastal asset elements into the structure of NFCDD and better asset classification is required to suit the asset types commonly inspected by Maritime Authorities.

Asset Inspection training

Maritime Authority inspectors who have completed the course have identified areas of the accreditation scheme which should be improved to cover a wider range of coastal assets. This also included the CAM, which will benefit from a wider section on the assessment of natural defences such as dunes, beaches and saltmarsh. Consideration should also be given to adopting a different approach to assessing the condition of these more dynamic defences, and further work will be required to developing a decision-tree approach, looking at the performance features of the asset, rather than an overall photographic example of each condition grade.

Figures 2 and 3 illustrate the difficulty in visually assessing the condition of sand dunes. The two photographs were taken at the same location, 3 days apart, and show very different features. This may lead to a different condition grade being given, depending on the day of the visit.
The photographs demonstrate the need for a flexible approach to the timing of inspections, fixed frequency of inspections may be suitable for built defences, however natural defences are more prone to change in appearance after storms / tidal events and also seasonal variations, which must be taken into account when carrying out the inspection.

Figure 2. Photograph of sand dune at Freshfield, north of Formby, UK (09/01/2009)

Figure 3. Photograph of same dune, taken 3 days later. (12/01/2009)
(Photographs courtesy of Sefton Metropolitan Borough Council)

Communications

This partnership approach to coastal asset inspections has highlighted the need for good communication. Frequent discussions with the Coastal Group chairs have ensured that the objectives of the pilots are understood, and their support for the process maintained.

One of the really encouraging outcomes from the pilots is the shared learning between Environment Agency and Maritime Authority inspectors. By inspecting different types of assets (coastal and fluvial), they have developed a greater understanding of each environment, and the challenges faced by each authority. Feedback has been positive by all inspectors involved.

The pilots have highlighted the varying demands on resources and priorities of Maritime Authorities which are not always seen by the Environment Agency at a national level. It has confirmed the need to develop a programme for future roll out which is achievable by the remaining Coastal Groups if the process is to be successful in the long term.

Way Forward

The results from the 2 case studies have identified several areas of development required if a consistent approach to asset inspecting around the coast is to be achieved.

Data Reporting Needs

The final report from the pilot studies will set out the specific data needs from both the Environment Agency and Maritime Authorities to enable sufficient data to be collected in one visit. This will satisfy not only the operational needs of the local asset managers, but also to inform investment bids from all parties, and finally to allow accurate National reporting of the State of the Nation's flood and Coastal erosion assets.

Update Inspection Accreditation Course

The existing course was developed for Environment Agency staff to carry out inspections on fluvial and sea defence assets. The pilots have identified that if it is to be applied to all coastal assets there needs to be a wider range of examples in the e-learning module. The CAM also needs to be updated to enable a better condition assessment of dynamic defences such as dunes. This work is underway, and will result in a course and material that is equally applicable on the coast as it is in fluvial environments.

Update NFCDD

This was the most obvious constraint during the pilot studies. A new database is currently being developed which will replace NFCDD. The project team includes representatives from each Coastal Group, who are helping to develop the database with an appropriate data structure for coastal assets, and ensuring that the new system is as suitable for use by Maritime Authorities as it is for the Environment Agency. This is a positive step to enable consistent and accurate data recording, as it will identify the key data fields needed for each level of reporting, i.e. high level national reporting by the Environment Agency, and also operational level asset management by all authorities.

Whilst the database will not be compulsory for Maritime Authorities to use, the Environment Agency are keen to encourage a wider uptake of the new system than was achieved with NFCDD. For this to happen, it needs to demonstrate that it can be successfully used at both levels.

National Roll out to remaining Coastal Groups
Once the issues identified during the pilots and detailed above have been resolved, there is a commitment from the Coastal Groups to widen the methodology out to the remaining coastal groups, with a phased uptake of the process within the next 18 months, resources permitting. (This commitment relates to the inspection process, training, and data structure, not necessarily the use of the new database)

Conclusions

It is recognised that with so many Operating Authorities around the coast, consistent standards of asset condition reporting are essential for effective risk management. The pilots have demonstrated very positive collaborative working between the Environment Agency and Maritime Authorities and whilst alterations to the existing process are required, the methodology can be used successfully by both the Environment Agency and Maritime Authorities.

The results will enable both Maritime Authorities and the Environment Agency to fulfil their duties under the Floods and Water Management Act to develop an asset register and easily share information so that flood and coastal erosion risk can be managed in an integrated way. The asset data is vital to demonstrate the need for investment to maintain and improve the defences which protect our coastal communities.

This approach to solving a particular issue through partnership working has demonstrated the benefits of this manner of working. It has enabled a shared ownership of the problem, mutual understanding of the partners' objectives and the development of a mutually acceptable solution. As such whilst the process itself has delivered positive results we should also consider the value of this approach for future issues.

References

Department for Environment, Food and Rural Affairs, (March 2005); Making Space for Water - First Government response to the autumn 2004 *Making space for water* consultation exercise. Defra publications.

Department for Environment, Food and Rural Affairs, (2008); Coastal Groups of England – The Environment Agency Strategic Overview of sea flooding and coastal erosion risk management. Defra publications.

Innovative Coastal Zone Management
ISBN 978-0-7277-5749-4

ICE Publishing: All rights reserved
doi: 10.1680/iczm.57494.012

HOW HIGH IS HIGH ENOUGH? – THE ART OF ASSESSING EXTREME CONDITIONS FOR EFFECTIVE COASTAL MANAGEMENT

Alastair McMillan, Royal Haskoning, Peterborough, UK
David Worth, Royal Haskoning, Peterborough, UK
Stefan Laeger, Environment Agency, Exeter, UK
Tim Hunt, Environment Agency, Exeter, UK
Angela Scott, Environment Agency, Peterborough, UK
Marc Becker, Scottish Environment Protection Agency, East Kilbride, UK
Dr Kevin Horsburgh, National Oceanography Centre, Liverpool, UK
Dr Mark Lawless, JBA Consulting, Skipton, UK
Crispian Batstone, JBA Consulting, Skipton, UK
Professor Jonathan Tawn, Independent, UK

Introduction

Flooding is caused by extremes in rainfall, river flow, tides or waves (or a combination thereof) which, when they overwhelm flood risk and coastal systems, can lead to severe impacts on people and the environment. Up-to-date and accurate information and statistical methods for assessing these extreme conditions form an essential building block for successful and effective flood risk and coastal management. They help determine which areas are at risk from flooding and influence the required standard of new flood and coastal erosion defences. A consistency in approach to deriving boundary conditions also promotes a level playing field in terms of appraising and allocating funding for schemes nationwide.

This paper outlines major improvements in assessing coastal flood risk and demonstrates the importance of robust information on extremes in flood and coastal risk management. New methods and datasets are presented which provide a nationally consistent evidence base for coastal flood boundary conditions for sea levels and swell waves around the English, Welsh and Scottish coastline. This paper presents new guidance for coastal practitioners describing how these outputs can be used to inform practical coastal management activities. For the first time, information on the uncertainty around extreme values is provided routinely. The paper also introduces a nationally consistent set of storm surge shapes which are recommended for use in coastal flood risk assessment.

Previous Practices

The existing nationally applicable method for deriving extreme sea levels around the coastline is by reference to POL Report 112, Spatial Analyses for the UK Coast[1], published in 1997. This project used tide gauge data up to the early 1990s. In addition, a number of regional and local extreme sea level datasets have been derived for the Environment Agency and SEPA.

These various references offer nationally inconsistent extreme sea level values. The inconsistencies apply both to the method of derivation and to the spatial distribution of the values. Additionally, for much of the coastline the information provided is based on a much shorter tide level record than is now available.

The recently completed project aimed to overcome these shortcomings by using up to date tidal records, by applying a consistent and improved methodology incorporating the currently best available statistical techniques and by providing values at a regular close spacing around the coastline. An additional product, not previously available at a national scale, is the means of generating appropriate storm tide curves.

A similar update was made for offshore swell conditions. Existing nationally available information on swell waves was found in the HR Wallingford report SR409[2]. The information provided in Report SR409 was based on four years of wave data. This project improved on the previous work by using a longer wave record, simultaneously taking advantage of better statistical techniques now available. In addition, this project extended the coverage of swell wave information around the UK coast by including Scotland.

Data Collection

Data used for the extreme sea level analysis and validation of results came from the following sources:

- Records from the UK National Network of Tide Gauges (often referred to as the 'Class A' tide gauge network) run on behalf of United Kingdom Coastal Monitoring and Forecasting by the National Oceanography Centre;
- Gauge data supplied specifically for this project by the Environment Agency;
- Gauge data supplied specifically for this project by SEPA;
- Third party tide gauge data from a number of sources;
- Astronomical tide predictions.

For validation purposes, reference was also made to reports containing information on notable historic events, such as the tidal flooding of 1953[3] along the east coast. Figure 1 shows the locations of all sites for which tide gauge data was obtained.

Data for the swell wave analysis as part of this project was obtained from the UK Met Office and United Kingdom Coastal Monitoring and Forecasting Service (UKCMF); the latter data was sourced via the Cefas Wavenet website[4]. The Met Office data was UK Waters Wave Model, providing a continuous record of modelled wave data around the UK coastline. The UKCMF data was observed wave buoy data from six sites, selected because of their proximity to Met Office model data points. They also provide good coverage of the UK coast.

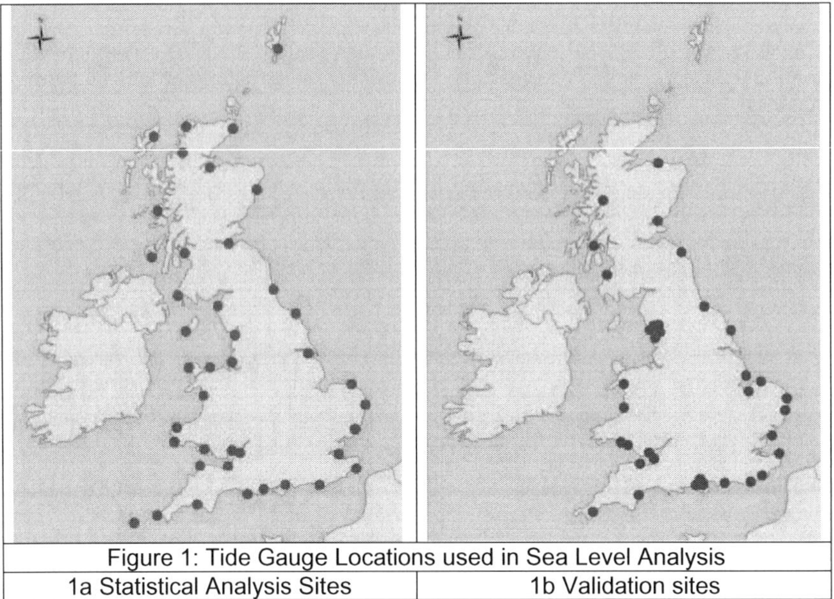

| Figure 1: Tide Gauge Locations used in Sea Level Analysis ||
| 1a Statistical Analysis Sites | 1b Validation sites |

A number of checks were undertaken to ensure the suitability of the data. Daily checks are kept on the performance of the Class A gauges. These are then routinely processed and quality controlled prior to being made available. These checks were replicated for non Class A tide gauge data provided for this project. A further review was undertaken to search for flaws such as missing data, 'spikes' where the gauge is recording erroneously high levels, datum shifts and datum drifts.

The data used for the swell wave analysis involved validation on the Met Office hindcast model data to confirm its acceptability for the project and identify potential weaknesses that might reduce confidence in the ultimate project output.

Method of Deriving Extreme Sea Levels

The challenge of this project was to estimate sea levels which might occur for return periods considerably greater than the length of the dataset on which these estimates are based. Given the relative shortage of data, it was necessary to use statistical analysis to derive estimates of extreme sea levels.

The foundation of the analysis is tide level data as recorded at the 40 Class A gauge sites within the study area, together with equivalent data from five other sites, chosen because the records are of suitable quality and length and also provide data for locations not well represented by the Class A tide gauge coverage.

Statistical analysis was undertaken to generate the joint probabilities of predicted high tide and of storm surge. A full spectrum of theoretical astronomical tides was generated from well-established harmonic constants. These are based on the lunar nodal cycle of high tides which has an approximate period of 18.6 years. Peak tide levels due solely to astronomical tidal forcing are deterministic and have a known absolute maximum. Therefore there is no requirement to use a statistical estimate.

Storm surges are the sea level response to wind stress and atmospheric pressure gradient. Low surface air pressure acts to raise the sea level, high pressure depresses it. Surface winds drive currents that also determine sea levels. The impact on sea level caused by these processes is referred to as storm surge, and these can increase or decrease sea levels. As meteorological processes are independent of tidal forces their influence can occur at any stage of the tide.

Skew Surge Joint Probability Method

The analysis on this project focused on the difference between astronomical high tide and the component introduced by weather. Skew surge is the difference between the predicted astronomical high tide and the nearest experienced high water level, as shown in Figure 2. The use of skew surge as a measure of surge removes any spurious effects due to phase differences (i.e. timing differences) between predicted and observed high water.

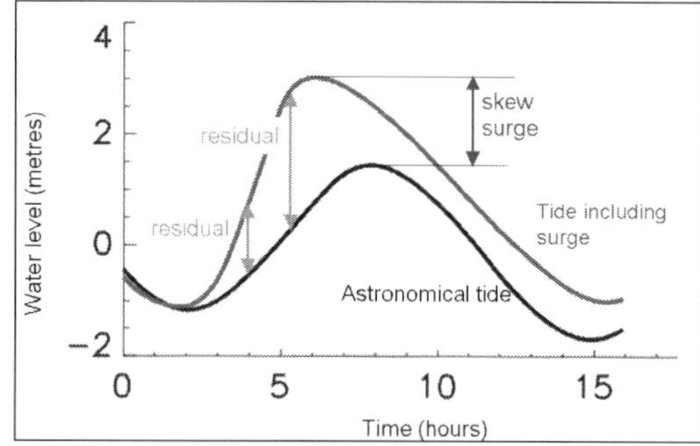

Figure 2: Illustration of the Skew Surge

'Illusory' surge residuals can often occur due to this high water phase difference, especially at the mid-tide stage. This is illustrated by Figure 3 which shows how an 'illusory' surge residual is created merely by the observed tide occurring slightly earlier than predicted, due to meteorological influence or tide gauge timing errors. For this reason, surge residuals seen in the mid-tide range are an unreliable indicator of the peak high water level that will be attained. Similarly, their use in other analyses should be treated with caution.

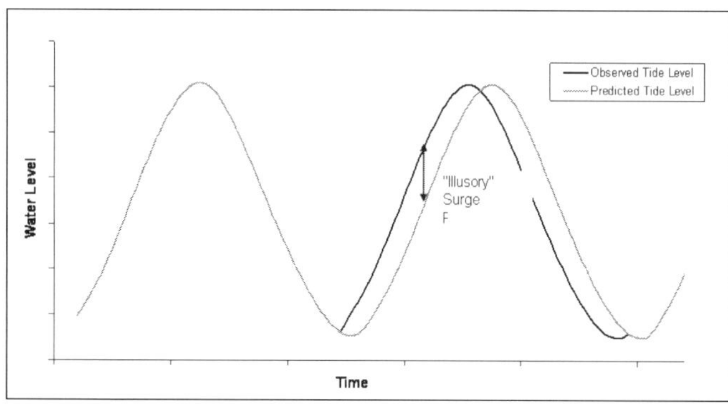

Figure 3: 'Illusory' Surge

The tide gauge data were examined to identify skew surges. This was undertaken for each site for every high tide occurrence, therefore 705 per year. The maximum record length was for Newlyn, with over 90 years of data and the minimum record length was for Bournemouth, with 12 years of data.

Combining the probability of the predicted astronomical high tide and the probability of skew surge gave the overall extreme sea level probabilities, expressed in terms of levels attributed to their respective average return period. This analysis is termed the Skew Surge Joint Probability Method and was developed specifically within study underpinning this paper.

By their nature, observations of extreme values are rare. Consequently the tails of the probability curve for skew surge can be poorly defined. Therefore a statistical model was used to fit a smooth upper tail to the probability distribution, known as the Generalised Pareto Distribution (GPD). The GPD gave the best smoothed fit to the extreme value skew surges above a specific threshold level. Figure 4 shows a schematic of the GPD with all data above a given threshold defined by the statistical distribution.

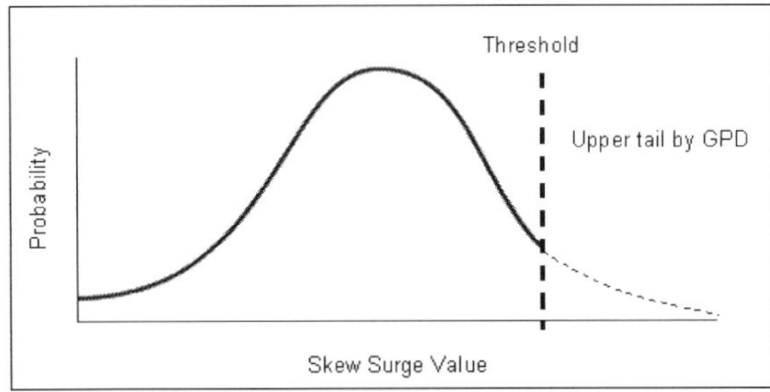

Figure 4: Schematic of Generalised Pareto Distribution

This method allows extrapolation to return periods considerably higher than the data used to generate them. The best information available has been built upon to undertake this extrapolation.

Statistical Smoothing

For some sites it was deemed necessary to apply some statistical smoothing to the growth in sea level with return period. Smoothing was made using data from neighbouring sites, a common approach used in extremes science for balancing spatial variations of estimates. This process added realism to the mathematics by comparing statistically derived estimates with observed data. The decision on the need for smoothing was based on the shape of the growth curve from low to high return period sea levels, aiming to avoid implausibly steep or flat growth compared to other sites. An example to illustrate this is at Hinkley Point (implausibly high growth) in Figure 5.

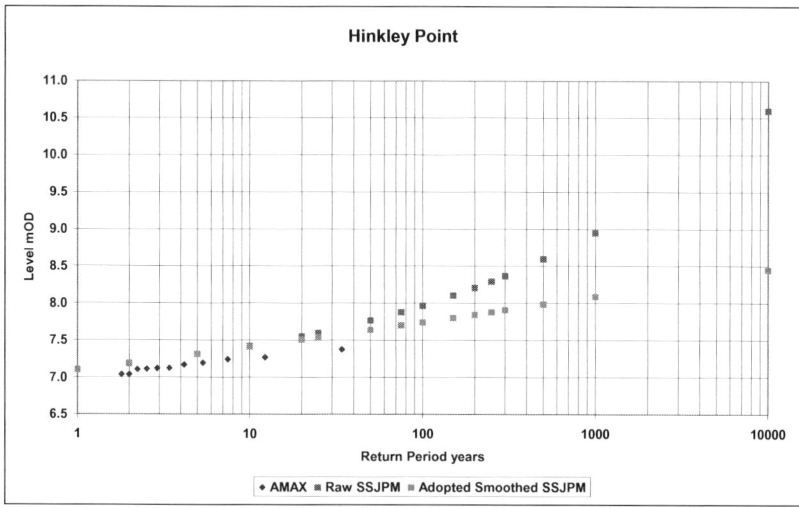

Figure 5: Implausibly high growth curve from 'raw' SSJPM results at Hinkley

Interpolation between Sites

In order to provide a complete coverage of return period sea levels around the coastline, and mindful that using all tide gauge data around the country would be a very lengthy exercise, an interpolation method was used to determine return period sea levels at points between the primary sites. The process was assisted by use of results from the POL Continental Shelf tide-surge model (on a 12km grid) and higher resolution JBA North East Irish Sea model.

The model results could not be used directly, but required a series of corrections before extreme sea levels validated to gauge data were obtained. As a first step, the model results for nodes at the primary gauge data sites were corrected to accord with the values obtained from statistical analysis of the gauge records. Secondly, the models were used as a dynamic interpolator to derive results for intermediate nodes. These results were adjusted in line with the corrections made at the primary data sites. Thirdly, linear interpolation between model nodes was used to give return period sea level values at about 2km spacing along a nominal coastal chainage line. This removed the need for further interpolation by users. Coverage is also given around major islands. This linear interpolation, guided by the concept of sea levels varying smoothly around the coastline, involved smoothing using the relationship between the MHWS and 1-year level and subsequent growth to higher return periods at the model points.

Fourthly, a further check was applied to ensure the return period levels are plausible. This involved comparison of the extreme sea levels along the coastal chainage line with the apparent return period of high levels as indicated by the secondary tide gauge data, being all the data other than that used for the primary analysis. Where inconsistencies arose, the 1-year level was further corrected to give a result more plausibly matching the secondary tide gauge data. The 1-year level values along the chainage line each side of the correction reference point were also adjusted, so avoiding "steps" or "spikes" in the progression of levels around the coast. These adjustments were generally across a distance encompassing four model nodes. The original growth in level to longer return periods was retained.

Confidence Intervals

Extreme sea levels for annual exceedance probability ranging from 100% to 0.01% (1 to 10,000 year return periods) were provided as part of this project. Uncertainty information in the form of confidence intervals is provided for all annual exceedance probabilities as part of this project. Confidence bounds at the statistical analysis sites were used as the starting point to develop empirical confidence intervals. The approach was guided by the need to be precautionary, given the uncertainties, and the need to be consistent around the coast, mindful of the geography.

Derivation of Standard Surge Shapes

This project also provides surge shapes for severe storm conditions to derive total tide curves. For the occurrence of an extreme tide level, the total tide curve will be the combination of underlying astronomical tide with a storm surge. The storm surge can be represented by a shape defining the increase and decrease of surge over time.

In reality, surge shapes are highly variable between different extreme events. However, for practical purposes it is convenient to have a standard surge shape that can be applied and used to generate total design event tide curves in a consistent manner for a particular length of the coast.

In creating standard surge shapes the project was mindful of their probable applications. Shapes were identified which, when used to create total event tide curves, would lead to a precautionary assessment of potential flood risk but without being unduly conservative. The standard surge shapes have been derived by examining the skew surge as recorded at the Class A tide gauge sites. More particularly, for each site the fifteen largest skew surge events were identified, which were then used to determine a standard shape. These shapes have been applied to sections of coastline to provide continuous coverage around the the UK.

How to Generate Design Tide Curves

Associated with this project is practical guidance for coastal managers looking to determine coastal boundary conditions for a site. Three components are required to generate a design tide curve for any given site.

1. Extreme Sea Level
2. Base astronomical tide curve
3. Surge Component

1. Extreme Sea Level

This project produced extreme sea levels at a two kilometre resolution around the coast of the UK. If levels are required which not directly coincide with Class A tide guage locations, the user needs to identify first the site, and obtain the relevant extreme sea level data from the Environment Agency. The chainage is slightly offshore for ease of viewing. Next the user needs to choose a desired return period sea level and select the corresponding confidence intervals. These confidence intervals should be considered when undertaking sensitivity testing in a study or design.

2. Base Astronomical Tide Curve

The method of generating a base tide curve involves using astronomical tide predictions such as from Admiralty Tide Tables or tidal software packages. Further guidance on appropriate tide levels and duration is provided in the Practical Guidance. Essentially the chosen base tide should be high enough to represent a larger than 'normal' event but also reach an appropriate level to reflect an event which occurs every year. It is common practice to base a design tidal graph on a spring tide. However, spring tides vary substantially over time due to the influence of the lunar cycle. Therefore a level between highest astronomical tide (HAT) and mean high water springs (MHWS) is recommended.

The date and time of the selected high tide should be the centre of the base peak astronomical tide. A recommendation for duration is two days prior to the peak tide level and two days after, giving a total duration of approximately 100 hours. This ensures adequate time for a number of high tides. Software to generate tide curves is readily available.

3. Surge Component

Surge shapes are based on analysis of data from the primary sites. Surge shapes and their suggested zone of application is provided as part of this project.

Resultant Curve

Use a spreadsheet to combine the three components (target peak sea level, astronomical tide and surge shape) to produce the resultant tide curve. For practical purposes it is recommended that extreme sea level, astronomical tide level and peak surge height should occur coincidentally. It is also acceptable to place the surge peak either before or after the peak of the base astronomical tide. Sensitivity tests for this timing shift can be undertaken depending on the coastal boundary condition requirements of the study. Figure 6 shows the 3 elements combined to produce a resultant curve.

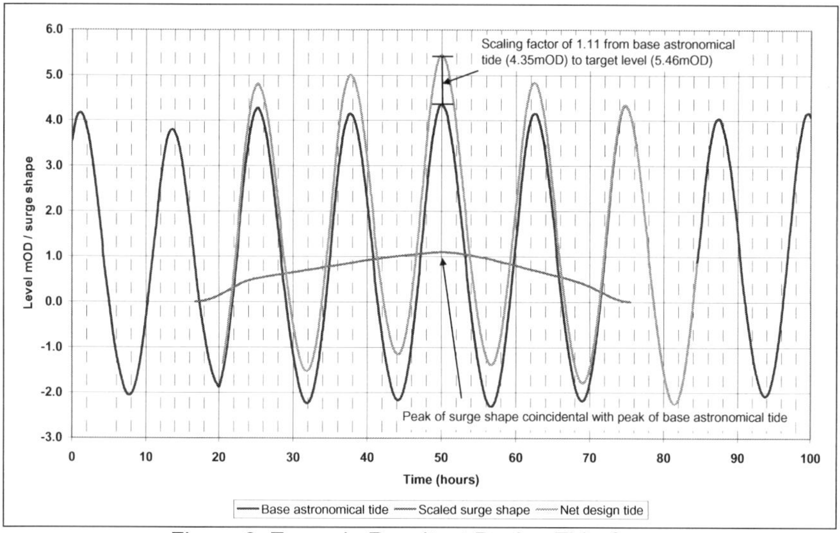

Figure 6: Example Resultant Design Tide Curve

This project did not consider Sea Level Rise due to Climate Change. For the most up-to-date guidance, users are advised to consult the latest climate change guidance for the UK (UKCP09)[6].

Swell Wave Analysis

Another strand of this important work is an update of offshore swell wave parameters around England, Wales, and for the first time, Scotland. The relevance of swell to coastal flood risk and the design of coastal structures arise from their wave period, which is longer than for wind waves. Associated with the longer period is an increase in the power of waves. Thus, swell waves can be very damaging to coastal structures. Their long wave period also means that wave run-up and wave overtopping will be much greater than for wind waves of equivalent wave height.

The Met Office provided hindcast wave model data at 12km resolution for the whole of the UK waters. This provided an opportunity to analyse swell conditions where no observations from wave buoys are available. The swell definition used was the one set out by the Met Office Wave Model. Statistical analysis was undertaken using the same principles as the extreme sea level analysis, using a Generalised Pareto Distribution to fit an upper tail to the probability distribution. The analysis was based on three-hourly values over an eight year record.

The outputs of this work are return period swell wave heights for six directional sectors with outputs at 12km resolution around the coast. Confidence intervals are provided for each swell wave height. An associated wave period is also identified.

Practical guidance has been provided with the swell wave outputs. Mindful that the outputs are for offshore locations, advice is given as to how these waves can be transferred to the coast.

Conclusions

Robust information on sea level and wave conditions is vital for much of the flood and erosion risk management work which is being delivered along the coastline. They inform coastal flood modelling and flood mapping work, design of coastal defences, flood risk assessments and strategic planning for development purposes. The analysis reported here is a major scientific advance to support this important work. It uses improved methods to provide nationally consistent datasets of coastal boundary conditions. The outputs provide continuous coverage around the coast therefore removing the possibility of inconsistent interpolation techniques to derive parameters for specific sites. The associated practical guidance provides some clear guidelines for using the datasets and is written from a practitioner's perspective. The reports can be obtained from the Environment Agency website[5] and the datasets are available under license from the Environment Agency.

Acknowledgements

This work was funded through the Environment Agency/Defra Flood and Coastal Risk Management R&D programme under R&D project SC060064 Improved Information on Coastal Flood Boundary Conditions for UK mainland and islands. The authors would like to express their gratitude to the data suppliers, notably the Environment Agency, Scottish Environment Protection Agency and Met Office.

References

1. DIXON, M.J. AND TAWN, J.A. 1997. Spatial Analyses for the UK Coast, Proudman Oceanographic Laboratory Internal Document No. 112.
2. HAWKES, P.J., BAGENHOLM, C., GOULDBY, B.P. AND EWING, J. 1997. Swell and Bi-modal Wave Climate around the Coast of England and Wales. HR Wallingford Report No. SR409.
3. ENVIRONMENT AGENCY. 2007. Anglian Region, Eastern and Central Areas Report on Extreme Tide Levels
4. CEFAS. 2009. WaveNet [online] Centre for Environment, Fisheries & Aquaculture Science (Cefas). Available from http://www.cefas.co.uk/data/wavenet.aspx.
5. Environment Agency. 2011. R&D Project SC060064: Coastal Boundary Conditions for UK mainland and islands. Available from http://publications.environmentagency.gov.uk/epages/eapublications.storefront?lang=_e.
6. LOWE, J.A., HOWARD, T., PARDAENS, A., TINKER, J., HOLT, J., WAKELIN, S., MILNE, G., LEAKE, J., WOLF, J., HORSBURGH, K., REEDER, T., JENKINS, G., RIDLEY, J., DYE, S. and BRADLEY, S. (2009). UK Climate Projections science report: Marine and coastal projections. Met Office Hadley Centre, Exeter, UK.

Innovative Coastal Zone Management
ISBN 978-0-7277-5749-4

ICE Publishing: All rights reserved
doi: 10.1680/iczm.57494.022

Investigating the Source – Pathway – Receptor – Consequence Framework for Coastal Flood System Analyses

Siddharth Narayan, University of Southampton, Southampton, England
Robert Nicholls, University of Southampton, Southampton, England
Derek Clarke, University of Southampton, Southampton, England
Susan Hanson, University of Southampton, Southampton, England

Introduction

Coastal floods account for a large proportion of the damage to life and property due to natural disasters world-wide. Out of the 15 greatest natural disasters in the world in terms of monetary losses between 2004 and 2008, 14 were coastal (Kron 2008). However, despite the enormity of a particular flood event, people often return to a flood-prone area after a disaster (Cigler 2007; Parker 1995). This necessitates a need for pro-active management of coastal flood systems. A coastal flood system may be thought of as a geographic unit comprising all natural and human elements potentially affected by coastal floods.

Though hard flood defences remain the primary choice for flood protection in many places, the importance of holistic management of coastal flood systems in limiting the consequences of coastal disasters is increasingly being recognised (Thorne et al., 2007 pp 4). However the complexity and size of coastal flood systems, the diverse nature of their sub-systems and the dynamic nature of their interactions makes this a daunting task. Rapid advances have been made in coastal flood modelling and management in attempts to address this challenge (de Moel et al., 2009). However, there is little understanding of how the elements of a coastal flood system interact with and influence one another, in terms of the flood propagation process, or the consequences of a flood event. A review of current flood system studies shows that most studies fall under two categories. These are (1) broad-scale studies (e.g., Kristensen 2004; Penning-Rowsell et al., 2005; Thorne et al., 2007) that consider the state of the system in relation to external drivers, pressures and responses or (2) detailed analyses that focus exclusively on particular parts of the system such as coastal defences or a natural habitat (e.g., Apel et al., 2006; Harvey et al., 2009; Natural England 2007).

System models that fall between these two ends of the spectrum – namely, models that provide an integrated overview of the engineered and natural elements of the flood system and their interactions, are increasingly being recognised as a necessary part of pro-active and strategic flood risk management plans (e.g., Hunt 2002; Pottier et al., 2005; Sayers et al., 2002; Thorne et al., 2007). However, isolated analyses of various sub-systems have been the norm until recently, reflecting administrative fragmentation (Pitt 2008; Parson et al., 2003). The apparent lack of commonality in flood risk analyses in the past has lead to repeated, avoidable failures with severe consequences such as those witnessed during Hurricane Katrina in 2005, and more recently, Cyclone Xynthia in Europe in 2010 (e.g., Seed et al., 2008; Kolen et al., 2010). Therefore, there is a need for a strong, flexible and consistent model that will help map the relative importance and roles of engineered and natural elements within coastal flood systems and facilitate effective integrated risk management policies.

The Source – Pathway – Receptor – Consequence (SPRC) model has been adopted in recent times in coastal flooding in attempts to integrate the different aspects of coastal flood systems. However, a review of the available literature suggests that the use of the model is often poorly defined and unclear, resulting in an ongoing failure to achieve integration of flood system analyses. This paper reviews current flood mapping practices with regard to where and how a conceptual model like the SPRC can prove useful. The manner of implementation of the SPRC model in coastal flood risk assessments is analysed and suggestions are made using a case-study for strengthening the model and placing it at the heart of flood risk analyses to help fill gaps in current understanding of the coastal flood system.

Flood Mapping Practices

Flood mapping has been carried out for many years and various purposes by governments and private institutions around the world. Several recent flood mapping studies have recognised the need for integration of the different elements of a flood system to achieve long-term flood risk reduction and efficient disaster management, and to facilitate effective environment-oriented flood risk reduction policies (e.g., Dawson et al., 2007; Pitt 2008; Thorne et al., 2007; Schmidt-Thomé 2006). Based on existing practices in Europe, flood maps in general may be distinguished as two types: (1) flood event maps that contain information on the probability and magnitude of a flood event and (2) flood risk maps that contain information about the consequences of particular flood events (de Moel et al., 2009). Due to the wide variety of land-use types, consequence calculation methods, objectives and stake-holders, different countries and regions typically have very different flood risk maps. Flood event maps however, follow a general methodology similar to the one outlined in Figure 1, with minor variations between studies, depending on their purpose and the institutions involved with flood mapping.

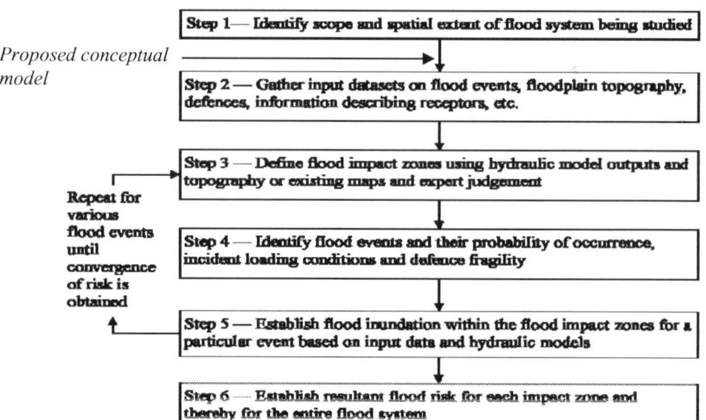

Figure 1: General steps involved in flood risk mapping based on FLOODSite Consortium (2009); DEFRA/EA (2004); de Moel et al., (2009)

While most flood event maps are placed within the context of a broader conceptual framework, it is seen from Figure 1 that the general methodology for flood event mapping does not explicitly include a conceptual model for the physical state of the flood system. The absence of an explicit model may result in inconsistencies in the manner in which the flood system is modelled and makes the collation of information collected from various sources in Step 2 a difficult task. In addition, the absence of a strong conceptual model may result in the exclusion of effects on the overall flood risk, of changes to elements within the flood system. For instance, it is widely recognised that the effects of climate change may include dynamic changes to natural and engineered elements that can have significant knock-on effects on nearby elements with regard to flood propagation and flood risk (e.g., Nicholls et al., 2007; Thorne et al., 2007). Despite this, few flood event maps consider the effects of such changes on future flood events (de Moel et al., 2009). Since the conceptual model used in a study will have a bearing on the nature and scope of its outputs, the inclusion of a conceptual model as a separate step after Step 1 of the methodology will greatly help reduce inconsistencies in the mapping process, provide information on system dynamics and facilitate effective integration of the information gathered in Step 2.

Another important issue in flood risk mapping is that of scale and the inclusion of cross-scale effects. For instance, the development of a port on a coast with extensive wetlands (regional scale change) may affect the wetland distribution, which in turn may affect flood risk in neighbouring towns (local scale effect). While flood event maps that focus on the physical effects of a flood event tend to be applied at local or regional scales (e.g., Horritt & Bates 2002; Gouldby et al., 2008), flood risk maps, that look at the consequences of a flood, are generally prepared at regional to national scales (e.g., FLOODSite Consortium 2009; Thorne et al., 2007). This division in scale can result in key local features within the physical system being overlooked in flood risk maps, especially with regard to their effect on the risk for a particular flood system. A clear conceptual model promotes a system–level understanding of coastal flood systems and provides an overview of temporal and cross-scale changes in flood risk.

Finally, a conceptual model of the state of the coastal flood system is a key step in the comprehension of a large and complex coastal system that comprises many different elements. It facilitates understanding of the system in terms of recognising elements such as sections of coastal defence structures or critical infrastructure that are important with regard to the consequences of a flood event. It also serves as a tool for communication across administrative and institutional boundaries about the aspects of the system that need to be considered in flood management and planning.

It is in an attempt to address the shortcomings in current flood mapping practices and to build a comprehensive model for the state of the flood system that the SPRC conceptual model has been adopted in coastal flooding studies.

The Source – Pathway – Receptor – Consequence Model
Background
The Source – Pathway – Receptor – Consequence (SPRC) model is a common conceptual model in coastal flooding that facilitates integrated flood system analyses by providing a realistic representation of the physical processes of flooding (Thorne et al., 2007). It has been widely adopted in coastal flooding studies in the UK, from its origins in environmental pollution following its use in the Foresight: Future Flooding report in 2004 (Evans et al., 2004). The model is also being used in coastal flooding studies in Europe (FLOODSite Consortium 2009; TU Dresden 2007) and in the USA (North Carolina Division of Emergency Management – Office of Geospatial and Technology Management 2009).

The SPRC model has been in use for more than two decades in the field of environmental pollution, in different forms, to evaluate risks arising from leakage of toxic pollutants. The first detailed definition found in literature is of the 'Source – Pathway – Target' model in Holdgate (1979). It has since been used in a variety of environmental risk assessments from tracing the fate of shotgun pellet pollutants (Sneddon et al., 2009) to classifying the effects of offshore wind farms on their environment (Scottish Government 2010). The UK Environment Agency described the use of the Source – Pathway – Receptor (SPR) approach as being fundamental to good practice in risk assessments for landfill sites (Environment Agency 2004).

SPRC in Coastal Flooding

The main premise of the SPRC model in coastal flooding is that the propagation of a flood event can be schematized in terms of source(s) for the event, receptor(s) and pathway(s) by which the event reaches the receptor(s) (see Figure 2). The 'consequence' term facilitates evaluation of different types of impacts, positive or negative, of a particular event on different types of receptors, both human and natural.

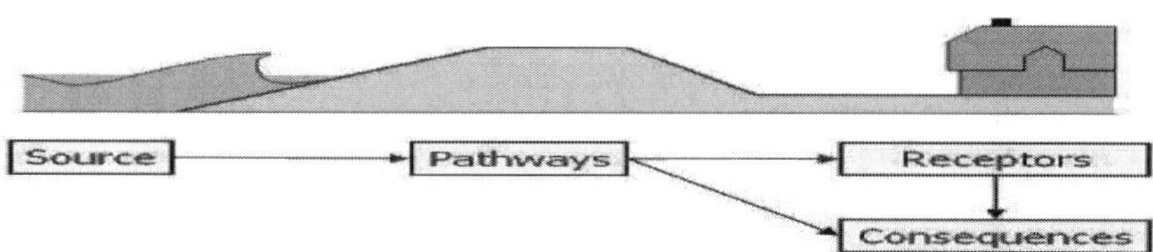

Figure 2: Simplified representation of the SPRC conceptual model for coastal flooding (TU Dresden 2007)

However, despite being cited extensively in coastal flood studies (and flood studies in general), there seems to be little explanation regarding its exact manner of implementation. Compared to other studies that use the SPRC model (e.g., Sayers & Meadowcroft 2005; Reeder et al., 2004; Shrestha 2008) the national-scale Foresight: Future Flooding assessment of the UK (Thorne et al., 2007) provides a comprehensive description of the manner in which the SPRC model has been used in its flood system analyses. Many studies on coastal flooding published after the Foresight report (e.g., Bakewell & Luff 2008; FLOODSite Consortium 2009; North Carolina Division of Emergency Management – Office of Geospatial and Technology Management 2009) use the concept as described therein. The Foresight study however considers a simplified, linear version of the SPRC model that, though useful at a national scale, may overlook several important issues such as element dynamics and the existence of multiple pathways to receptors within the flood system that are relevant to more detailed studies. This gives rise to difficulties in implementing it as a conceptual model for dynamic flood systems with complex, multiple inter-linkages. It is felt that the under-utilisation of the SPRC model is systemic in coastal flooding and strengthening the model to tap its full potential will greatly benefit coastal flood risk studies.

SPRC Implementation and Case – Study
Development of the SPRC Model

The SPRC model in recent applications is not well enough defined to serve as an effective conceptual model for integrated flood systems. However, the model can be developed to create a comprehensive and useful model of the flood system state without unduly sacrificing ease of understanding. The ongoing EU project 'THESEUS: Innovative Coastal Technologies for Safer European Coasts in a Changing Climate' (www.theseusproject.eu) aims to address both the human and ecological consequences of flood events across Europe is using the SPRC model as the basis for integrating these aspects in terms of coastal flood management while providing a system-level overview of the flood propagation process (THESEUS Consortium 2009).

In the example shown in this paper, a system linkage diagram is built, starting from the sources, through the various pathways and ultimately to the intended receptors. The main advantage of this approach is that any element on the system linkage diagram may be picked out by the user as a receptor thereby automatically designating the role of 'pathway' to all elements linking it to the sources. Hence, the definition of pathways and receptors becomes relative, with everything between a receptor and source being a pathway. This will help different users focus on different elements, For instance, in nature conservation, habitats might be regarded as a receptor, while for emergency planners, they are a pathway to people and their assets. This system diagram affords an excellent overview of the various elements and linkages that have been recognised in the study and as such, constitutes an effective and intuitive conceptual model.

Example Implementation in Great Yarmouth, UK

The SPRC system linkage model is applied to Great Yarmouth on the east coast of England. Great Yarmouth is an important regional town and tourist destination, situated between two rivers and the coast with a wide variety of infrastructure, engineering and natural elements. As such, the Great Yarmouth region offers an illustrative example of a site that can benefit greatly from a comprehensive and integrated conceptual model of the physical flood system.

The model is applied to Great Yarmouth following the methodology described below. The methodology is split into two broad stages – analysis of the Source – Pathway – Receptor (SPR) relationships that characterise the state of the physical flood system, followed by an evaluation of the Consequences (C) of a particular flood event. The conceptual model developed in this example focuses on the physical system state and analysis of the SPR relationships within the system. This is most effective when undertaken for the maximum flood extent being modelled without considering internal topographical variations within the boundary of the study. This approach makes it easy to apply the model in the absence of detailed topographical data or advanced hydraulic models.

The site boundaries are defined based on the flood extent of a current 1:100 year flood + 3m in order to account for future sea–level rise. Using standard national (in this case Ordnance Survey) maps broad-scale land-use receptors are identified as polygons on a GIS software as shown in Figure 4, with each receptor given a unique number reference. Coastal management infrastructure is included as a receptor, 'Management' with each type of hard or soft defence being assigned a unique reference. Hard infrastructure such as seawalls and dykes are shown as generally linear features as illustrated in Figure 4 and soft management such as beach replenishment schemes areas, as indicated in Figure 5. Supplementary material such as habitat type, construction material and land use are included as polygon attributes within the GIS software.

Four main sources of flooding are identified for this site: S1 from the open sea, S2 from the downstream confluence of the two tidally-influenced rivers, S3 from River Yare and S4 from River Bure. The site is then divided into three areas A, B and C, based on the general location of the different sources (see Figure 4). A schematic SPR diagram is then built to represent the various elements within the flood system, by considering an individual element as a potential receptor, and identifying for that receptor, all elements through which it may flood. This process is repeated for all the elements identified within the study site. This process offers a succinct illustration of the manner in which a flood may propagate through the system regardless of topography within the boundary flood extent. Since some areas may be flooded by more than one source, the directionality of flood propagation is important. Finally, the specific points at which each source can enter the system are identified. A schematic SPR diagram for Area A is shown in Figure 6 similar diagrams for Areas B and C may be constructed. The SPR diagram will form the basis for subsequent event-based topographical analyses and analyses of consequences.

The conceptual flood system model presented here, based on the SPRC model, provides a strong and robust means of presenting information on the manner of flood propagation through a system as well as the way in which changes to one element may affect the flooding of linked elements. The model thus offers an easy and effective way to qualitatively evaluate the effect of the responses of the flood system to external changes, whether driven by management or climate change factors. Finally, this model presents a simple and effective means of collating and presenting existing knowledge on the different disparate aspects and elements of the physical flood system to the relevant stake-holders. Since the model can be built to any level of detail depending on the data available, the consistency in methodology will ensure that the model provides useful information to the end users at each level.

Conclusions and Further Work

Flood hazard and flood risk mapping, though highly advanced, tend to focus on isolated analyses of elements within the flood system at specific scales and do not consider cross-scale and temporal dynamics between various natural and engineered sub-systems and the defence infrastructure protecting them. Though flood hazard and risk mapping are being increasingly implemented world-wide, flood systems still continue to fail for a variety of reasons, though a lot of the damage can be minimised through effective, integrated flood risk management plans. Policy-makers (e.g., Pitt 2008), managers (e.g., Thorne et al., 2007) and engineers (e.g., Gouldby et al., 2008) are recognising the apparent lack of commonality and the incompleteness of flood mapping practices in this regard. There is therefore an urgent need to develop a dynamic, evolving, rigorous assessment framework for flood risk mapping.

The SPR system model schematic shown in Figure 6 addresses these challenges to consistency and integration in coastal flood system studies and provides a flexible and powerful platform for more detailed assessments of the flood system. Firstly, the model illustrates the complexity of the system being considered. It makes it possible to pick out even at the conceptual stage, sections of coastal defence that may be of particular importance. The schematic identifies and highlights the role of natural habitats in coastal defence and makes it easier to study flood propagation across administrative boundaries. The level of detail of the schematic is representative of the scale of the study, and can be refined or degraded depending on the requirements of the user. This allows for easy and intuitive cross-scale analyses of a particular flood system. Also, this model can easily be developed to include further information on individual elements, such as their vulnerabilities and relative importance, the uncertainties in associated parameters, as well as the effects they may have on linked elements. As part of the THESEUS project, the SPRC system model is being applied in the case study sites and evaluated in the context of decision support for flood risk management. The model will include information on system dynamics, propagation of uncertainty, defence prioritisation, feedback effects of elemental changes and communication of spatial variability in risk during flood events.

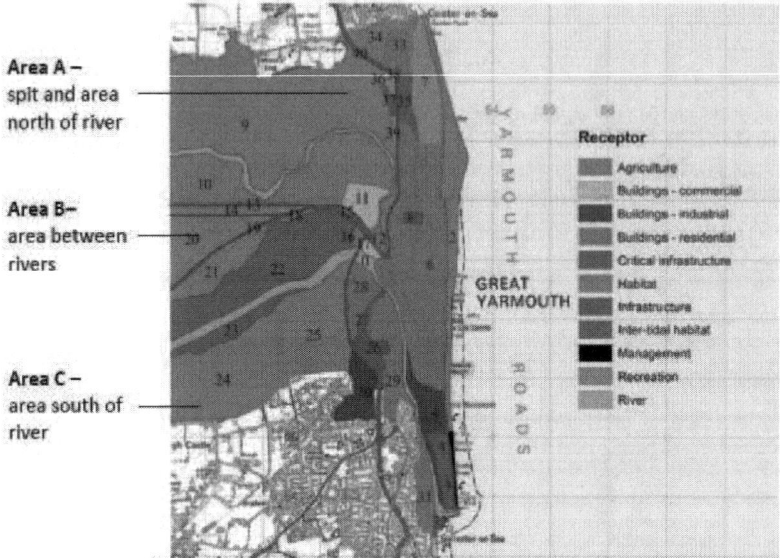

Figure 3: Map of Great Yarmouth study site indicating various receptors and the three compartments for the schematic diagrams based on the sources

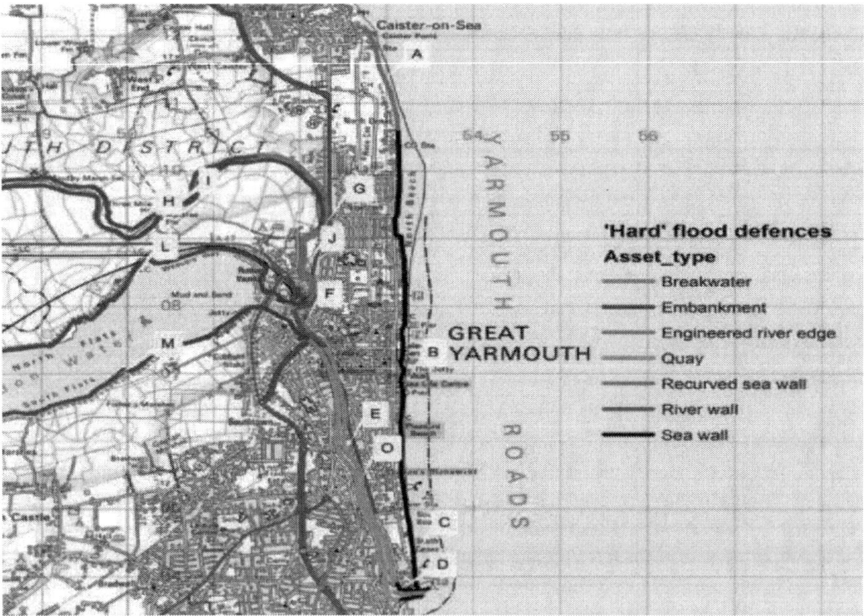

Figure 4: Hard management features in the Yarmouth case-study site

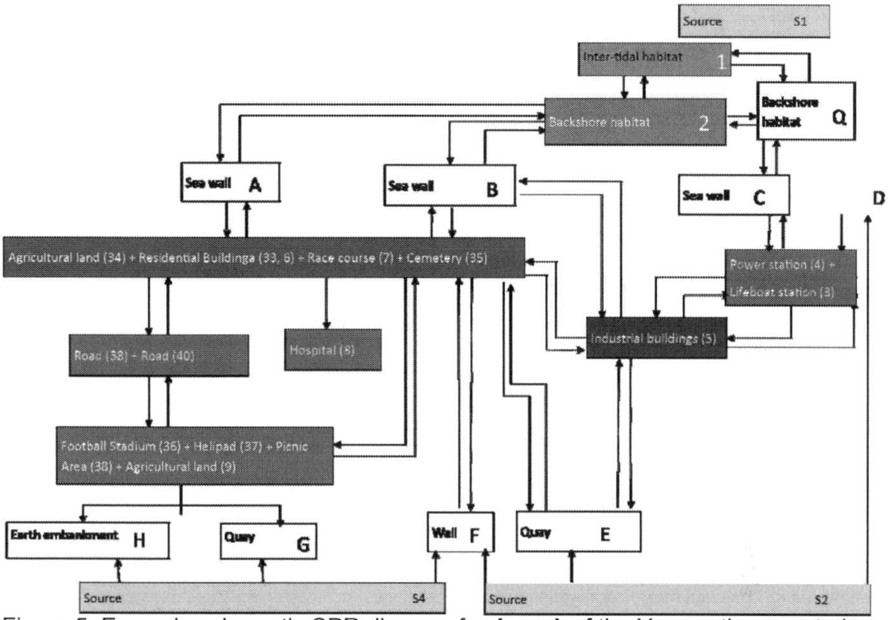

Figure 5: Example schematic SPR diagram for Area A of the Yarmouth case-study site with the sources, pathways and receptors indicated. Unique references have been assigned to receptor classifications (numbers) and hard defence (letters) as indicated in Figures 4 and 5 respectively

Acknowledgements

The support of the European Commission through the project "Innovative Technologies for safer European coasts in a changing climate" (THESEUS), Contract 244104, FP7.2009-1, www.theseusproject.eu, is gratefully acknowledged.

References

ABI, 2006. *Coastal Flood Risk - Thinking for Tomorrow, Acting Today*, Association of British Insurers, London.

Apel, H. Thieken, A.H., Merz, B., Blöschl, G., 2006. A probabilistic modelling system for assessing flood risks. *Natural hazards*, 38(1), p.79–100.

Bakewell, I. & Luff, S., 2008. *North London Strategic Flood Risk Assessment – Final Report*, DEFRA, UK. 158 pages

Cigler, B.A., 2007. The" Big Questions" of Katrina and the 2005 Great Flood of New Orleans. *Public Administration Review*, 67(s1), p.64–76.

Dawson, R., Hall J., Barr S., Batty M., Bristow A., Carney S., Evans S., Ford A., Köhler, J., Tight M., Walsh C., 2007. A blueprint for the integrated assessment of climate change in cities. *Tyndall Centre for Climate Change Research, Working Paper 104*. 154 pages.

de Moel, H., Van Alphen, J. & Aerts, J., 2009. Flood maps in Europe–methods, availability and use. *Nat. Hazards Earth Syst. Sci*, 9, p.289–301.

DEFRA/EA, 2004. *Risk Assessment for Flood and Coastal Defence for Strategic Planning – R&D Technical Report W5B-030/TR*, Bristol, UK. 76 pages.

Environment Agency, 2004. *Guidance on Assessment of Risks from Landfill Sites - External Consultation Report*, Bristol, UK. 73 pages.

Evans, E. Evans, E. Ashley, R., Hall, J., Penning-Rowsell, E., Sayers, P., Thorne, C., Watkinson, A. 2004. *Foresight Future Flooding: Scientific Summary: Volume I - Future Risks and their drivers.* Office of Science and Technology. London, UK.

FLOODSite Consortium, 2009. *Integrated Flood Risk Analysis and Management Methodologies (CD-ROM).* www.floodsite.net.

Gouldby, B. Sayers, P., Mulet-Marti, J., Hassan, M., Benwell, D. 2008. A methodology for regional-scale flood risk assessment. *Water management*, 161(3), p.169–182.

Harvey, H., Hall, J. & Peppé, R., 2009. Outline of a framework for systematic decision analysis in flood risk management. In *18th World IMACS / MODSIM Congress.* Cairns, pp. 4290-4297.

Holdgate, M.W., 1979. *A perspective of environmental pollution*, Cambridge University Press Cambridge, UK.

Horritt, M. & Bates, P., 2002. Evaluation of 1D and 2D numerical models for predicting river flood inundation. *Journal of Hydrology*, 268(1-4), p.87–99.

Hunt, J., 2002. Floods in a changing climate: a review. *Philosophical Transactions of the Royal Society of London. Series A: Mathematical, Physical and Engineering Sciences*, 360(1796), p.1531.

Kolen, B., Slomp, R. & Balen, W. van, 2010. Learning from French experiences with storm Xynthia. *rijksoverheid.nl.*

Kristensen, P., 2004. The DPSIR framework. In *Workshop on a comprehensive/detailed assessment of the vulnerability of water resources to environmental change in Africa using river basin approach.* Nairobi, Kenya.

Kron, W., 2008. COASTS-THE RISKIEST PLACES ON EARTH. In *International Conference on Coastal Engineering (ICCE).* Hamburg, Germany.

Leichtweiß-Institute for Hydraulic Engineering and Water Resources Department of Hydromechanics and Coastal Engineering, 2010. Extreme Storm Surges at Open Coasts and Estuarine Areas Risk Assessment and Mitigation under Climate Change Aspects. Available at: http://www.xtremrisk.de/?page=&lang=en [Accessed November 23, 2010].

Linham, M. & Nicholls, R.J., 2010. Technologies for Climate Change Adaptation: Coastal Erosion and Flooding. In *TNA Guidebook Series.* Roskilde, Denmark: UNEP Risø Centre on Energy, Climate and Sustainable Development, p. 150.

Merz, B., Thieken, A. & Gocht, M., 2007. Flood risk mapping at the local scale: concepts and challenges. *Flood Risk Management in Europe*, p.231–251.

Natural England, 2007. *Planning for biodiversity as climate changes: BRANCH project final report.* www.branchproject.co.uk.

Nicholls, R.J. Wong, P.P., Burkett, V.R., Codignotto, J.O., Hay, J.E., McLean, R.F., Ragoonaden, S., Woodroffe, C.D., 2007. Coastal systems and low-lying areas. Climate Change 2007: Impacts, Adaptation and Vulnerability. Contribution of Working Group II to the Fourth Assessment Report of the Intergovernmental Panel on Climate Change, *Fourth Assessment Report of the Intergovernmental Panel on Climate Change*, p.315–356.

North Carolina Division of Emergency Management – Office of Geospatial and Technology Management, 2009. *North Carolina Sea Level Rise Risk Management Study – Potential Impacts, Risk Assessments and Management Strategies*, North Carolina.

North Norfolk District Council, 2009. *Kelling to Lowestoft Ness Shoreline Management Plan - Final Report*, Cheshire.

Parker, D.J., 1995. Floodplain development policy in England and Wales. *Applied Geography*, 15(4), p.341–363.

Parson, E.A. et al., others, 2003. Understanding climatic impacts, vulnerabilities, and adaptation in the United States: building a capacity for assessment. *Climatic Change*, 57(1), p.9–42.

Penning-Rowsell, E et al., 2005. The benefits of flood and coastal risk management: A manual of assessment techniques (Multicoloured Manual). *Flood Hazard Research Centre.* Enfield, UK.

Pitt, M., 2008. Learning lessons from the 2007 floods: an independent review by Sir Michael Pitt. *London: The Pitt Review, Cabinet Office.* London, UK.

Pottier, N. et al., 2005. Land use and flood protection: contrasting approaches and outcomes in France and in England and Wales. *Applied Geography*, 25(1), p.1–27.

Reeder, T., Donovan, B. & Wicks, J., 2004. Broad Scale Tidal Flood Risk Assessment for London using MDSF and Flood Ranger. In *3rd National CIWEM Conference*. London, UK.

Sayers, P.B., Hall, J.W. & Meadowcroft, I.C., 2002. Towards risk-based flood hazard management in the UK. *Civil Engineering*, 150(5), p.36–42.

Sayers, P.B. & Meadowcroft, I.C., 2005. RASP-A hierarchy of risk-based methods and their application. H.R. Wallingford, UK. Available at: http://eprints.hrwallingford.co.uk/76/ [Accessed January 25, 2011]

Schmidt-Thomé, P., 2006. *Natural and technological hazards and risks affecting the spatial development of European regions*, Geological Survey of Finland. Espoo, Finland. 44 pages.

Scottish Government, 2010. Strategic Environmental Assessment (SEA) of Draft Plan for Offshore Wind Energy in Scottish Territorial Waters: Volume 1: Environmental Report. Available at: http://www.scotland.gov.uk/Publications/2010/05/14155353/6 [Accessed January 25, 2011].

Seed, R.B. Bea, R.G., Abdelmalak, R.I., Athanasopoulos-Zekkos, A., Boutwell, G.P.. Briaud, J.L., Cheung, C., Cobos-Roa, D., Ehrensing, L., Govindasamy, A.V. and others 2008. New Orleans and hurricane Katrina. I: Introduction, overview, and the east flank. *Journal of Geotechnical and Geoenvironmental Engineering*, 134, p.701.

Shrestha, A.B., 2008. Resource Manual on Flash Flood Risk Management.

Sneddon, J. et al., 2009. Source-pathway-receptor investigation of the fate of trace elements derived from shotgun pellets discharged in terrestrial ecosystems managed for game shooting. *Environmental Pollution*, 157(10), p.2663–2669.

Thorne, C.R., Evans, E.P. & Penning-Rowsell, E.C., 2007. *Future flooding and coastal erosion risks*, Thomas Telford Services Ltd. London, UK

THESEUS Consortium, 2009. THESEUS : Innovative technologies for safer European coasts in a changing climate. Available at: www.theseusproject.eu. [Accessed January 25 2011]

TU Dresden, 2007. Flood Protection and Risk Management in the city of Hamburg. Available at: https://floodmaster.hydro.tu-dresden.de/wiki/Saturday_22_September [Accessed January 20 2011]

Innovative Coastal Zone Management
ISBN 978-0-7277-5749-4

ICE Publishing: All rights reserved
doi: 10.1680/iczm.57494.032

Eco-dynamic Development and Design Tested for Coastal Management

Gerard van Raalte Royal Boskalis Westminster nv, Papendrecht (The Netherlands)[1,6]
Mark van Koningsveld Van Oord Dredging and Marine Contractors, Rotterdam (The Netherlands)[2,6]
Jasper Fiselier DHV B.V., Amersfoort, (The Netherlands)[3,6]
Huib de Vriend Delft University of Technology, Delft (The Netherlands)[4,6]
Daan Rijks Royal Boskalis Westminster nv, Papendrecht (The Netherlands)[5,6]

Unlimited Opportunities

Working in coastal planning is nearly always innovative, since no two coasts are the same. Even if they appear to be physically similar, local conditions and governance settings are usually different, thus requiring a different approach. Each project is unique, which makes coastal projects exciting and diverse.

This diversity opens perspectives to a variety of ways to develop a coast, supported by a wide range of guidelines, books and articles on coastal zone management and coastal planning. In general, important factors influencing any given project are:
- functionality (flood protection, recreation & nature, possibly sand or gravel mining),
- physical characteristics (metocean, materials, bathymetry, etc),
- ecosystem characteristics (dune forms and habitat, foreshore and surf zone biota),
- economics (costs and benefits, exploitation, return on investment),
- governance issues (including permitting and politics),
- public opinion.

In most innovative projects the consequences of the interventions are not exactly known, usually less so than in projects taking a traditional approach. This may constitute an obstacle to the implementation of such projects, where a combination of existing 'traditional' designs, measures and materials is often considered a safer choice. This paper claims that this practice should better be changed and demonstrates the feasibility of a more innovative approach.

After an introduction of the philosophies and principles behind 'Building with Nature' and on the steps to be taken within 'Eco-dynamic Design and Development' process, two pilot

---

[1] Expert, Royal Boskalis Westminster, Hydronamic, Papendrecht, (Netherlands), g.h.vanraalte@hydronamic.nl
[2] Senior Engineer, Van Oord Dredging and Marine Contractors, Rotterdam, (Netherlands), mrv@vanoord.com
[3] Senior consultant, DHV, Amersfoort (Netherlands), jasper.fiselier@dhv.com
[4] Professor, Delft University of Technology, Delft (Netherlands), huib.de.vriend@ecoshape.nl
[5] Senior Project Engineer, Royal Boskalis Westminster, Hydronamic, Papendrecht, (Netherlands), d.c.rijks@hydronamic.nl
[6] Also at EcoShape | Building with Nature, Dordrecht (The Netherlands), www.ecoshape.nl

projects are described in which Building with Nature has been tested, both technically and procedurally.

Building with Nature Innovation Programme

The Building with Nature (BwN) innovation programme addresses this issue by bringing together ecologists, engineers and governance experts in a new cooperative setting, enabling them to explore integral solutions, instead of looking for mitigating measures and compensation for environmental impacts after the principal design choices have been made: combining ecology, engineering, economy and decision making to maximise total project value. A unique feature of the programme is that the resulting ideas are then tested in real-life pilot projects in different aquatic environments. BwN is involved in four such projects, one of which, on the Holland coast is highlighted in this paper (other pilot projects focus on bank- and shore stabilisation using natural materials in a way that both. structural stability and ecological development are enhanced). Furthermore, the programme includes 3 generic studies and 19 PhD- studies connected to the pilot projects

---

Building with Nature is a five-year innovation and research programme (2008-2012) carried out by the Foundation EcoShape (www.ecoshape.nl). This 30 million Euro program is initiated by the Dutch dredging industry, while partners represent academia, research institutes, consultancies and public entities.
This approach is created around five program objectives:
1. Develop ecosystem knowledge enabling Building with Nature
2. Develop scientifically sound design rules and norms
3. Develop expertise to apply the BwN concept
4. Make the concept tangible using practical BwN-examples
5. Establish how to bring the BwN-concept forward in society and make it happen

Throughout the program the interaction between disciplines is promoted, involving ecologists, engineers and social scientists.
The work comes together in a work package called 'eco-dynamic development and design', which aims to draft a document with guidelines for eco-dynamic design of hydraulic (mainly coastal) engineering infrastructure. Results will become publicly available in the course of the program, with completion of the design guideline envisaged for December 2012.
Explorations are being made for an extension of the program after 2012, including the extension of international cooperation.

The project teams, which include policy makers, contractors, consultants and scientists, ultimately aim at up-to-date, factual and down-to-earth guidelines for those who wish to implement these ideas in their practice.

The BwN-programme aims at exploiting the potential of this cooperation, looking for ways to combine 'wet' infrastructural engineering with the help of nature, at the same time creating new opportunities for nature. Examples highlighted in this paper, are:

i. Along the coast: re-establishing natural solutions to century-old challenges.

   Examples of ongoing pilot projects are the use of shellfish reefs and other eco-engineers for estuarine shoal stabilisation, ecologically-engineered nourishments, mega nourishments using natural dynamics for their redistribution ('Sand Engine'), etc. The projects are supported by PhD research on key relationships between hydrodynamics, sediment, morphology and ecology, as well as on governance issues. Indicators are sought that show the state of the ecosystem in the vicinity of such projects.

ii. In coastal shelf seas: second life for sand borrowsites:

   Two borrowsites for the extension of Rotterdam harbour are left behind with a sand ridge ($>2Mm^3$) to investigate the potential added value for nature, recreation and fisheries, including the ecological recovery time after sand extraction. The new physical lay-out provides a wider range of water depths, a wider variety of flow conditions and more variation in bed composition, thus probably providing habitat to more species.

iii. In society: governance:
The role of actors and stakeholders in decision making on building-with-nature projects and the effective way to introduce building-with-nature concepts in early stages of the project are investigated and translated into practical guidelines.

The Building with Nature Philosophy: a paradigm shift

As a result of the ongoing demographic concentration in coastal, deltaic and riverine areas, the increasing demand for safety, prosperity and sustainability and the need for action bound to arise from climate change and accelerated sea level rise, there is a persistent demand for large infrastructural projects (Ehrlich and Ehrlich, 1997). Coasts need to provide safety against waves and rising water level, rivers need space to safely accommodate extreme floods (see http://en.wikipedia.org/wiki/Room_for_the_River_(Netherlands)) , capacity of harbours and navigation channels needs to be increased, quality of water bodies and subsoils needs to be maintained or restored, natural resources need to be utilised in a sustainable way, etc.

The past decades have shown that the realisation of infrastructural works can be characterised by the following trends:
– we want more (multifunctional designs, including environmental aspects),
– we know more (knowledge of natural systems has increased enormously), and
– we can do more (increased technological capabilities enable new approaches),

whereas at the same time:
– we have to operate more carefully (care for the environment has increased),
– we have more difficulty getting things done (complexity of society and decision making has increased), and
– we have to meet increasingly complex functional requirements (modern society has high demands).

These developments present hydraulic engineering and environmental protection with challenges as well as opportunities. The only possibility to sustainably reconcile the needs of the environment with the growing demand for infrastructure development and use is to ensure that infrastructure works with nature, rather than against it. This requires a change in thinking, acting and interacting, a paradigm shift in all aspects of project development.

In the past, infrastructural works used to be designed and constructed without paying much attention to the potential environmental impact. Since the 1970's, interest and legislation for the environmental protection have developed and has led to more attention for minimising the environmental impacts caused by infrastructural works: building **in** nature.

Since the 1990's, European environmental legislation adopts the principle of prevention, mitigation and compensation of residual effects. The compensation of any residual environmental loss triggers building **of** nature. Legislation emphasizes conservation of existing nature and puts a lot of emphasis on the precautionary principle. As a consequence, projects are still dominated by impact minimisation, extended with mitigation and compensation measures.

Good engineering complies with nature and makes optimum use of natural forces. In the 80's and 90's (coastal) projects have been implemented that successfully made use of those forces:

building **by** nature. One example is shore nourishment to maintain sandy coasts (http://en.wikipedia.org/wiki/Beach_nourishment; Van Duim et al., 2004).

The next step in the development of hydraulic engineering is to attribute a more active role to natural processes, by utilising these natural forces and at the same time creating opportunities for development of new nature. This concept in now termed building **with** nature (see Figure 1). The credo is no longer 'doing less harm', it now becomes 'doing more good'.

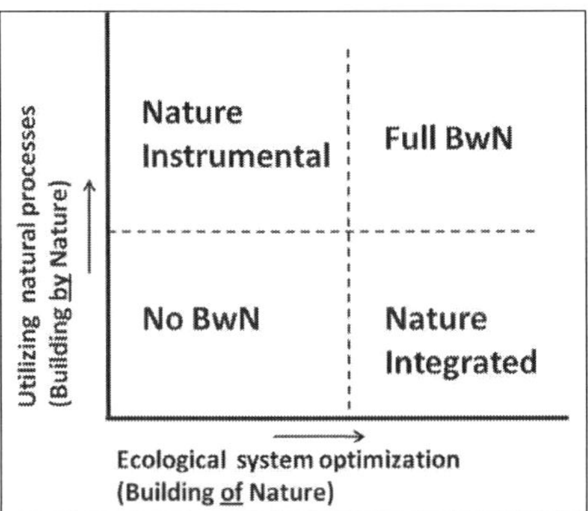

Figure 1: The building-with-nature philosophy in project development

Where the usual approach to large hydraulic projects basically follows the steps:
1. Plan a project or activity, starting from the functionality it has to provide for,
2. describe the effects on the ecosystem,
3. optimize the design to minimize or mitigate detrimental effects,
4. compensate by building of nature, and
5. execute the project in strict adherence to fixed norms and regulations.

the alternative, ecosystem-based approach boils down to:
1. Understand system functioning ('read' the ecosystem, the socio-economic system and the governance system),
2. plan a project or activity taking the system's present and envisaged functions into account (combining functional and ecological specifications),
3. determine how natural processes can be used and stimulated to achieve the project goals and others (using the power of nature),
4. determine how governance processes can be used and stimulated to achieve the project goals (using the power structures in place),
5. monitor the environment during execution, analyse the results, make risk-assessments, and if necessary adapt the monitoring program and/or the project execution (monitoring and adaptive management), and
6. monitor the environment after completion, as to assess the project's performance and to learn for the future (knowledge development).

This alternative approach reflects the notion that from an overall project performance point of view, the best choices are not necessarily the ones that fit best to the individual project phases (project initiation, planning & design, construction, operation & maintenance). Rather we should balance long-term costs and benefits, in monetary and non-monetary terms. The term

Eco-dynamic Development and Design (EDD) is used to refer to this alternative ecosystem-based design approach.

EDD principles

The paradigm shift from Building in Nature, via Building of Nature to Building with Nature, drives the need for an alternative eco-system based design approach: Eco-dynamic Development and Design. In order to achieve this paradigm shift, a different attitude is required, from individuals as well as from social groups. Individuals may need to think, act and interact in another way than they were used to. EDD can only be applied successfully if changes can be achieved in the basic aspects of human behaviour (Edwards, 2009; Kira and Van Eijnatten, 2008):

- ***think differently***: adopt a BwN philosophy, based on a thorough understanding of how the system you are working in functions, and strive for balanced win-win situations;
- ***act differently***: take the entire complex of system functions into account and optimize on the entire project, integrating and evaluating all development phases; cope with the uncertainty that is inherent to nature.
- ***interact differently***: make sure that all parties with a stake in the development can bring in their views and are heard, while being aware that existing societal boundaries might impose limitations to what is feasible; cope with the complexity of multi-party and multi-interest decision making.

EDD steps

A project lifecycle usually has 4 phases: 'initiation', 'planning & design', 'construction' and 'operation & maintenance', with at the end possibly a fifth phase: 'demolition & removal'. Depending on project management and contracting formats, two or more phases may be merged into one. Although each phase offers opportunities for integration of EDD solutions, maximum flexibility and BwN-potential exist in the earliest stages of development. To optimally seize these opportunities, it is recommendable to take a life-time approach and consider the potential of later phases as early as possible (see Figure 2).

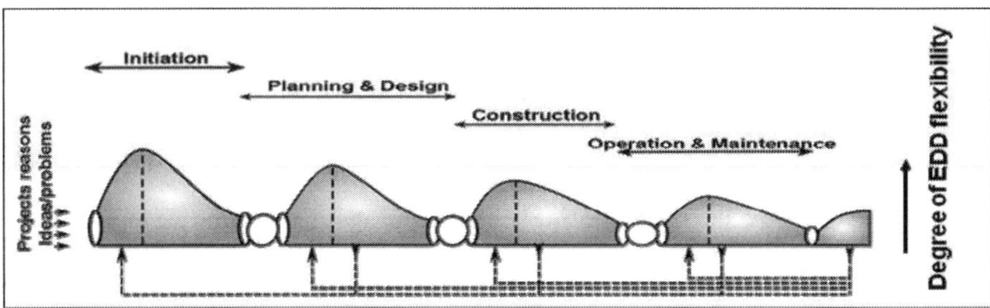

Figure 2: Project phases and flexibility to introduce EDD-principles.

Analysis of practical cases revealed a number of steps that are invariably taken when developing creative Eco-dynamic Designs. Although the process is cyclic, the steps outline a basic creative process that can be followed in any project development phase (Figure 3).

The steps are:
1. ***Gather system understanding.*** Every phase will give the opportunity to amass new information giving more insight into the functioning and the functions of the system one is working in. One shall be clear about primary objectives and realise that finding win-win solutions creates room for flexibility in catering for secondary objectives. Looking at

primary objectives only may lead to a limited definition of the system to consider. Adding secondary objectives entails consideration of other system characteristics: other parameters, other time and spatial scales etc.

2. ***Identify alternatives and options*** that have the potential to meet the project requirements. These alternatives may build upon the options selected in the previous phase, but can also open up for new points of view. This step requires people who can think creatively, beyond the borders of their own profession and expertise.

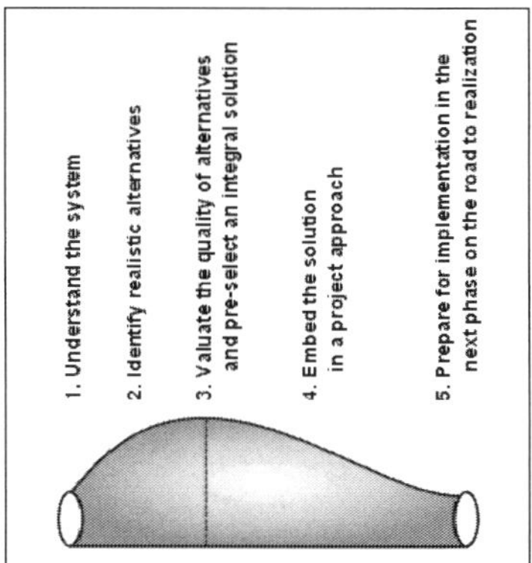

Figure 3: EDD steps in a project phase

3. ***Assess the inherent qualities*** of the alternatives on the basis of a decision framework that includes the BwN sustainability criteria. This may involve an environmental-societal cost benefit analysis and also formal screening against existing rules and regulations, e.g. for nature conservation. Models for estimating ecological values against social and economic values play an important role in this assessment.
4. ***Embed these alternatives*** in the socio-ecological context, notably the allocation of costs, benefits, environmental values, risks and related management.
5. ***Optimise these alternatives*** taking into account multiple goals, costs, dynamics and risks, based on the assessment made. The next project phase will then build upon these optimised alternatives.

The BwN Program is drafting an Eco-dynamic Development and Design Guideline based on the above principles and steps.

Pilot Projects

Sand Engine: Mega nourishments and the "art of letting go"; coastal defence in the hands of nature and natural processes.

A large part of the Dutch coast is sandy, with beaches and dunes forming the first line of sea defence. Part of this coastline is in a state of dynamic equilibrium and hardly requires attention. Other sections, however, require regular maintenance and strengthening, on average 12 million m³ per year of sand nourishment. This implies that certain parts of the coast are disturbed and reworked every so many years, with consequences for the local ecosystem and beach use. Moreover, consistently nourishing the beach and the upper part of the shoreface leads to over-steepening of the lower shoreface (Stive and De Vriend, 1993). These are generally accepted consequences, they can reasonably well be managed, but nevertheless it is of interest to reduce such effects. As a consequence of climate change and sea level rise larger nourishment quantities and/or more frequent campaigns must be expected. This further triggers the interest to explore more sustainable approaches (e.g. http://www.deltacommissie.com/doc/deltareport_full.pdf).

The last decades experiments have been made with beach nourishments following the principle of Building by Nature: rather than placing dredged sand directly on the beach, at the toe of the dunes, sand was placed on the foreshore. There it was left to natural forces, waves, currents, tides and wind, to distribute the material along the coast and within the profile up to the dry beach (for instance, http://www.nottingham.ac.uk/efm/research/shoreface.php). The experiences with these foreshore nourishments are positive: although in total larger volumes may have to be handled, the effectiveness is found to be good, total costs are less and the area is significantly less disturbed. Yet, the overall outcome of each individual nourishment operation is less certain, as one relies on nature doing part of the work.

A next step in this process, expanding from regular small-scale shoreface nourishments, is to place a very large quantity in one operation, at one well selected location (not all along the coast), sufficient to feed a stretch of coastline for 15-20 years, with nature doing the majority of the work. Based on this idea a mega-nourishment, a so-called 'sand engine', was launched. A large volume of sand is put onto the shoreface and left to be re-distributed alongshore and into the dunes, through the continuous action of waves, tides and wind. Besides gradually inducing juvenile dune formation along a larger stretch of coastline over a period of one or more decades, thus contributing to safety against flooding over a longer period of time, the surplus sand volume, before being fully distributed over the coastal system, temporarily creates added value for nature and recreation; amongst others by providing shoals and beach lagoons as rest areas for sea mammals and birds, wide beaches for daily tourism and challenging surf conditions for the local surf community.

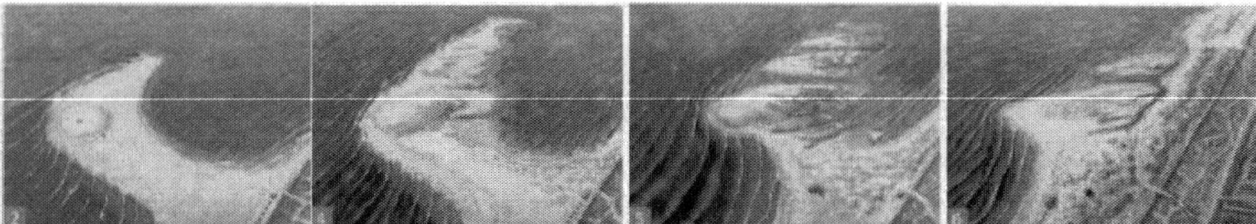

Figure 4: Artist impression of the Sand Engine development on Delfland coast.

This concept has now been adopted for the coast of Delfland, one of the weaker parts of the Holland coast, through an intensive design, governance and communication process with many stakeholders involved.(Aarninkhof et al, 2010; Mulder et al, 2006). In 2011, a volume of more than 20 million m^3 of sand will be placed in this Delfland Sand Engine (Figure 4), in the shape of a sandy hook connected to the beach. Under the BwN program the whole process will be followed intensively through frequent monitoring, both during construction and during the years thereafter. Monitoring covers amongst others climatic conditions, beach- and dune profiles, beach material and water quality, beach and dune vegetation, benthos, fish, sea birds and sea mammals, recreation and swimmers' safety. Monitoring results will enable a concept validation and supports adaptive management actions during the period of natural development. If needed smaller auxiliary (foreshore) nourishments can be considered, should unsafe situations develop.

Results of the Sand Engine pilot are to be reported through the BwN channels and in the open literature when they become available.

Ecological Sand Extraction site: natural hotspots in the sea?

Borrow- or extraction sites, for the purpose of beach nourishment or land reclamation, were traditionally determined in terms of size, shape and location on the basis of geographical, geological, and physical aspects. These conditions were subsequently laid down in permits allowing for little variation. Ecology, with the exception of the expected rehabilitation period, was usually not a main issues. Later on, especially for larger extraction sites, choices / requirements in the permits were based on an EIA. These assessments generally classify extraction sites as having a negative impact on the local environment (De Groot, 1979). The general idea was that during the construction of the site, the top layer of sand and its inhabitants (benthos) were removed, leaving a deeper area with potentially different sediment characteristics. The EIA evaluation usually considered this to require a rehabilitation period of 3-4 years (Boyd, 2004).

The most common mitigation measure was to limit the depth of the extraction to 2m depth, so that re-colonisation could take place under similar hydraulic and morphological conditions as before. However, due to the scale increase in required sand volumes, maintaining a maximum of 2m extraction depth would lead to very large surface areas for the sites becoming affected. After research had shown that deeper extraction is possible, this was allowed at certain sites.

Building with Nature expands on the modern philosophy of searching for opportunities to realise additional quality in a project without compromising its main function. In fact BwN tries to go one step further: in addition to looking for opportunities to enhance the natural value of the project, it seeks ways to utilise nature in the construction and later on operation of the project.

The large-volume sand borrowing sites in the North Sea for Maasvlakte2, the extension of Rotterdam harbour, offered the opportunity for a pilot project. Within the boundaries of the available mining permits and procedures an opportunity was found to create a few sand ridges by leaving an elongated strip of the original bottom untouched (Fig. 5). Two orientations were chosen, one similar to that of the large natural sand waves in the area and the other one similar to that of the tidal ridges in this part of the North Sea. According to ecological models and field observations, the gradients in hydrodynamic and morphological conditions created by this 'ecological landscaping' should lead to a greater biodiversity, a larger biomass production, a faster ecosystem recovery and finally a richer ecology than originally present (Van Dalfsen and Aarninkhof, 2008). This could be achieved without a significant increase of the overall costs of the sand mining project, thus avoiding the question whether project owners would be prepared to pay more for sustainable sand extraction.

The hands-on pilot project also offered all actors and stakeholders, from project initiators and owners to design engineers, permit authorities and contractors, the opportunity to investigate the pros and cons of integrating the ecological function (landscaping) in an existing infrastructure project.

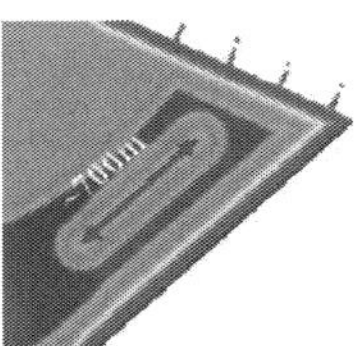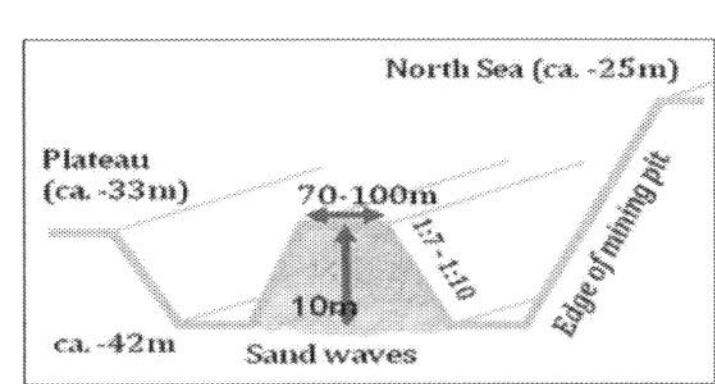

Figure 5: Trailing Suction Hopper Dredgers creating an 'eco-landscape'

Figure 6: Benthos monitoring around ecological sandpit.

Shortly after the sand mining operation the first monitoring results showed that there is already a significant amount of benthic organisms around the ridges, also compared with an undisturbed reference area (Fig. 6). This is considered a first indication of a positive ecological development. In general, the results of the bathymetric- , benthos- and fish surveys will be published when they become available.

Experiences with stakeholder involvement in these projects

Embarking on Building with Nature integrated projects has provided quite some new insight into the governance aspects of such projects. Two 'lessons learned' are highlighted here.

The decision making processes leading to the Sand Engine and the Ecological Mining Pit both included a significant degree of stakeholder involvement. Engaging stakeholders in project planning is nothing new (Hommes, 2008), but here stakeholders were involved all the way from the beginning, and not with the focus on how to address their concerns, but on involving them on how to get the most out of the project. The leading principle in these processes was to be honest, transparent and open, at all levels. Sharing of all information and knowledge sometimes proved to be difficult, but as long as common interests can be defined people and organisations are willing to really cooperate. To avoid deception or frustration later on in the process, it is essential to document all agreements and covenants in a rather detailed, if not contractual format. Nonetheless, the principal lesson learned is that by taking an open, transparent and truly integrated approach win-win situations can really be achieved.

Nature is, by definition, not fully predictable, and the same goes for processes in society. Both are nonlinear dynamic systems driven by partly stochastic inputs (e.g. weather, economy). Working with such systems, like in Building with Nature projects, leads to a higher level of uncertainty in the predicted process outcome and the project effects as compared to a more traditional approach. Willingness, capability and room (also in legislation and regulations) to deal with such uncertainties are crucial to BwN-projects. Uncertainties can for instance be handled by adaptive management procedures (as in the Sand Engine project), or by contingency provisions in the project plan. Obviously, we shall be careful, if not precautionary, when addressing possible negative effects, but potentially positive variations shall be given 'the benefit of the doubt'.

Conclusions

This paper describes the outcome of recent research on 'Building with Nature'. It demonstrates that with proper integration of ecology, economy and societal needs substantial gain can be achieved in development of sustainable hydraulic engineering projects.

Two coastal pilot projects in the Netherlands that have (partly) been developed according to the BwN-philosophy are analysed. The results show that each coastal planning project has its own specific conditions and requirements, so needs a tailor-made solution in order to optimally effective. Where possible, the forces of nature may be used in design, construction and operation in order reduce construction and maintenance costs and to enhance functionality and value.

The analysis of the decision making processes shows that many stakeholders are open to a constructive cooperation, as long as their interests are understood and transparently taken into account as much as realistically possible. Crucial to a successful BwN development is not so much new technology, but much more so a change in attitude: from 'minimising negative impacts' to 'maximising positive development'.

Acknowledgements

The work in this paper is carried out as part of the innovation programme Building with Nature of the EcoShape Foundation (http://www.ecoshape.nl/ecoshape-english/home/). Results and lessons learned from this programme will be published by the end of 2012 in a set of 'Eco-dynamic Development and Design Guidelines'.

References

Aarninkhof, S.G.J., van Dalfsen, J.A., Mulder, J.P.M., and Rijks, D.C. (2010). Sustainable Development of Nourished Shorelines Innovations in project design and realization. In Proc. Of PIANC MMX Congress (2010), Liverpool (UK)

Boyd S.E., K.M. Cooper, D.S. Limpenny, R. Kilbride, H.L. Rees, M.P. Dearnaley, J. Stevenson, W.J. Meadows and C.D. Morris. (2004). Assessment of the re-habilitation of the seabed following marine aggregate dredging. Sci. Ser. Tech. Rep., CEFAS Lowestoft, 121: 154pp.

Cooper, J.A.G. and McKenna, J. 2008. Working with natural processes: the challenge for Coastal Protection Strategies. Geographical Journal 174, 315-331.

Edwards, M. (2009). Through AQAL eyes, part 4: A new theory of holons. http://www.integralworld.net/index.html?edwards8.html

Ehrlich, P.R. and A.H. Ehrlich, (1997). The population explosion: why we should care and what we should do about it. _Environmental Law_, *Winter*, 1997

Hommes, S. (2008). *Conquering complexity,* PhD thesis, University of Twente, 182 pp., ISBN 978-90-365-2742-2

Kira, M. & Eijnatten, F.M. van (2008). Socially sustainability work organizations: A chaordic systems approach. System Research and Behavioral Science, 25 (6), 743-756.

Mulder, J.P.M., Nederbragt, G., Steetzel, H.J., Van Koningsveld, M. and Wang, Z.B. (2006). Different implementation scenarios for the large scale coastal policy of the Netherlands. In: Proc. of Int. Conf. on Coastal Engineering (2006), San Diego (USA)

Stive, M.J.F. and De Vriend, H.J. (1993). Shoreface profile evolution on the time scale of sea level rise. In: J.H. List (Editor), *Large-Scale Coastal Behaviour '93*, USGS Open file Rept. 93-381, p. 189-192.

Van Dalfsen, J.A. and Aarninkhof, S.G.J. (2008). Building with Nature: Mega nourishments and ecological landscaping of extraction areas. In: Proc. of EMSAGG Conference (2009), Rome (Italy).

Van Duim, M.J.P. Wiersma, N.R., Walstra, D,J.R., Van Rijn, L.C. and Stive, M.J.F. (2004). Nourishing the shoreface: observations and hindcasting of the Egmond case, The Netherlands. Coastal Engineering 51(8-9): 813-837.

42

Innovative Coastal Zone Management
ISBN 978-0-7277-5749-4

ICE Publishing: All rights reserved
doi: 10.1680/iczm.57494.043

Threat of coastal inundation in the Solent: real-time forecasting

Matthew Wadey, School of Civil Engineering and the Environment, University of Southampton, UK
Robert Nicholls, School of Civil Engineering and the Environment, University of Southampton, UK
Craig Hutton, Geodata Institute, University of Southampton, UK

Introduction

Flood forecasting and warning can be the most fundamental component of flood risk management (risk here is defined as the product of probability and consequence), because timely coastal flood warnings allow preparation with or without structural defences and provide the opportunity to move possessions or people out of harms way (Floodsite, 2007). The United Kingdom has a well established coastal flood forecasting and warning system; the UK Coastal Monitoring and Forecasting Service (before 2009 known as the Storm Tide Forecasting Service) (see Horsburgh and Flowerdew, 2010). Real-time models generate surge, tide and wave forecasts for coastal flood warnings; with a minimum lead time of 2 hours although forecasts of exceptional surges are usually much further in advance (Horsburgh et al., 2008). This is not the case globally, and existing systems endeavour to provide more accurate and specific flood warnings. At some locations, the spatial gap between offshore forecasts and coastal defences has been reduced by using hydraulic models through the nearshore and surf zones, incorporating ensemble and other probabilistic approaches (Lane et al., 2008; Hawkes et al., 2008; EA, 2009). It has also been demonstrated that models can be assembled into a forecasting chain to assess the probability of flooding and its impacts (Golding, 2009), although these cannot yet run in real-time and doubts over accuracy remain. Research elsewhere includes real-time sensors installed within sea defences to pre-empt structural damage (Urbanflood, 2010), and into social science perspectives relevant to effective warning response (Parker et al., 2007). The importance of this area of research was highlighted on the 28th February 2010 when the well forecast Cyclone Xynthia caused flooding that killed 65 people along parts of the French Atlantic coast (Kolen et al., 2010). Flood-specific dangers were not clearly known beforehand, which meant that warnings did not reach many and authorities did not know where or how to evacuate.

This paper describes research at the University of Southampton using coastal flood forecasting to demonstrate the use of sensor networks and other real-world data for application developers to build systems for environmental management. This system will depict potential coastal flooding rather than provide warnings (a service provided by the Environment Agency). The spatial role of this research within the well known source-pathway-receptor model is defined in figure 1. This has mainly included analysing defence and floodplain (pathway) responses to hydraulic boundary conditions, which in this case are nearshore water levels and waves (sources). A model of these hydraulic boundary conditions is being developed in a parallel piece of research. The Solent in southern England is used as a case study (figure 2).

The structure of this paper is a follows:

- Description of the case study region
- Summary of techniques applied to the model
- An example flood simulation.
- Further work and preliminary conclusions.

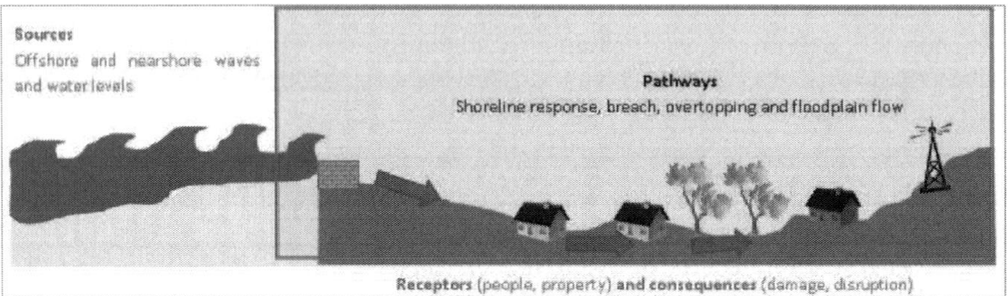

Figure 1: The source-pathway-receptor-consequence model (SPRC), with the main focus of this research outlined by the box.

The Solent

The Solent is a body of water on the south coast of the UK (figure 2) with a history of frequent coastal flooding during the past century (Ruocco et al, 2011; Haigh et al., 2011). This coastline covers over 500 km of tidal rivers, estuaries harbours and open coast. The shoreline includes a variety of natural and structural defences which protect urban and rural landscapes. The tidal regime is complex, with extended high waters and variable tidal range within the region. The maximum meteorological component of total water level is small compared to other areas of the UK with surge heights rarely exceeding 1 m, although this also means that coastal defences are exposed to significant hydraulic loadings across a large proportion of the tidal cycle. Exposure to wave energy is variable due to the shelter of the Isle of Wight and the managed shingle barrier (Hurst Spit) at the western end of the region. Hence fetch-limited wave conditions prevail in most locations, although energetic swell waves can be a cause of floods for open coast areas east of Portsmouth.

Initial estimates of exposure to coastal flooding were generated by intersecting a planar water level (from theoretical extreme sea levels) across a digital elevation model (DEM), not accounting for the effects of defences or dynamics of flood waves. More than 20,000 buildings are on 70 km^2 of land presently situated below a 1 in 200 year sea level, although the uncertainty in the data used to generate this estimate suggests that more than 30,000 buildings could actually be affected by some form of inundation during a hypothetical uniform 1 in 200 year event across the region. The scale and uncertainty of this exposure indicates a challenge for flood warnings and management. Approximately half of the region's exposure is within Portsmouth (cf. ICE, 2009) which is densely populated and with areas below normal high tide levels. Despite this, 25 per cent of the city's sea defences may be below a 1 in 200 water level (Atkins, 2007) whilst development pressure is significant (under

the last government approximately 15,000 new homes were planned for Portsmouth by 2030
– PCC, 2010).

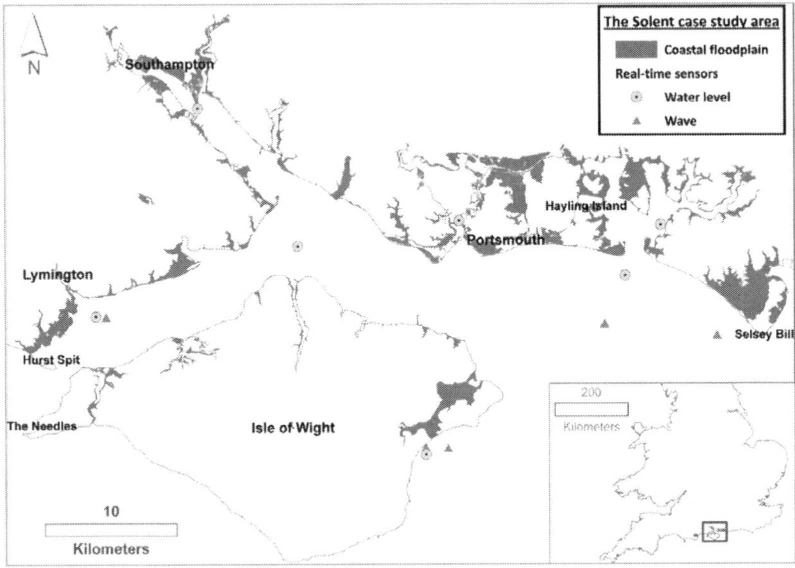

Figure 2. The Solent case study site.

A real-time forecasting model

Figure 3 defines the approach. For a given forecast of wave and water level conditions, the
defences are analysed to define the likelihood of failure and hence the development of a
pathway due to overtopping and/or breaching. Breach failure is unusual, although once other
processes become replaced by significant breaching, the volume of water that passes onto the
floodplain can increase by several orders of magnitude (Muir Wood and Bateman, 2005) and
can vastly increase risk to life and property. Flood inundation modelling is applied to
overtopping and breaching to define the areas that might be flooded (the receptors). These
steps are now discussed in turn, including their integration.

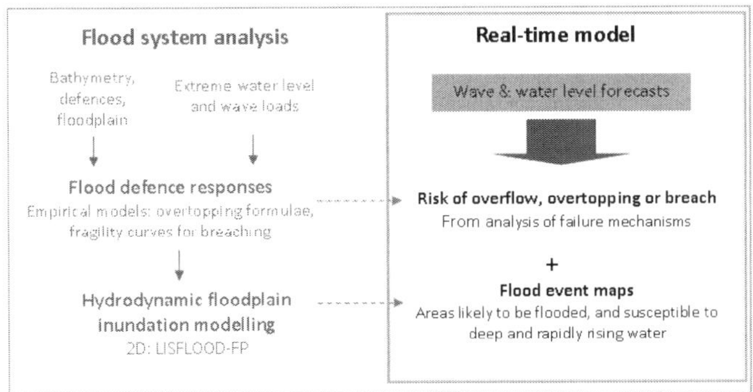

Figure 3. Real-time model components and the analysis within this work.

Predicting pathway responses

Datasets were collated that indicated the capacity of coastal defences to prevent flooding. Crest heights included mainly ground surveys (RTK GPS) for structures, and information on defence type and condition, supplemented by aerial photography and LiDAR (Light Detection and Ranging) data. Wave run-up and overtopping were quantified using methods from the EurOtop Manual (Pullen et al., 2007). Incomplete and uncertain information relating to the likelihood of breach was brought within a probabilistic framework that expresses a conditional relationship between load and strength parameters (Dawson and Hall, 2001) through fragility functions (figure 4) similar to those used in national and regional flood risk assessments (Hall et al., 2003; Gouldby et al., 2008) and widely considered a robust way of understanding flood systems within a system-wide flood risk analysis (e.g. Flikweert and Simm, 2008). These were generated by applying Monte Carlo analysis to selected limit state functions that represented relevant breach failure mechanisms in the Solent region.

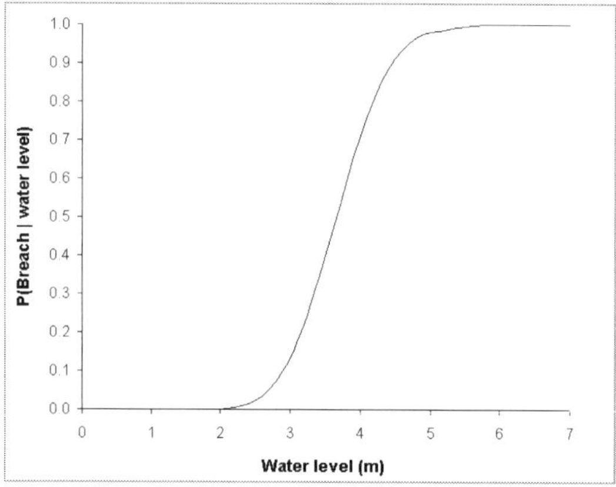

Figure 4. An example fragility curve with loading expressed along the x-axis and probability of failure on the y-axis.

The floodplain surface was represented by a DEM, compiled using LiDAR and Ordnance Survey Landform data. Coastal flood exposure estimates using the planar water level method can be considered to provide the maximum flood outline for the given water level, unlikely to be exceeded by a real flood event except where heavy wave overtopping and limited drainage may combine in some locations. The Solent's exposure to coastal flooding is shown in figure 5. Closer inspection of the DEM and defences revealed 100 flood compartments which appear at the surface as hydraulically discrete when subject to 1 in 1000 year boundary conditions. Each compartment shows varying topographic and defence characteristics, and hence distinct contrasts in the potential likelihood and severity of flooding. The increase in inundated land area and buildings with decreasing probability of water level corresponds to small water level variations, because in the Solent there is only about a 33 cm increase in elevation between a 1 in 10 year and a 1 in 1000 year water level (from analysis of the Portsmouth tide gauge by Haigh, 2009). Hence the predicted flood outline at some locations can be sensitive to the accuracy of the DEM, boundary water levels, overtopping by waves, breaching, and the hydrodynamic characteristics of floodplain inundation.

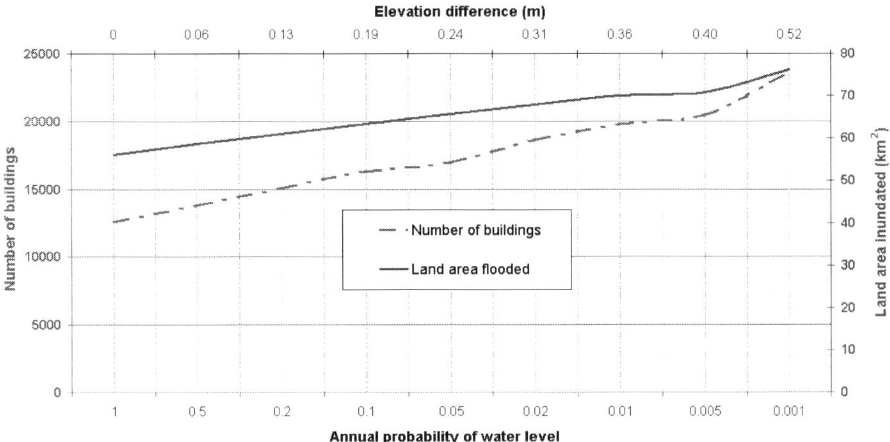

Figure 5. Exposure to flooding in the Solent region – defences not considered.

To set up a hydrodynamic model the DEM was resampled to a 50 m raster grid, with a point placed in each cell at the seaward edge of the DEM in each flood compartment. Collectively these represent the coastal defence line, and each point is considered a potential source of flow into the floodplain. This resulted in approximately 5000 defences spaced uniformly; each assigned data from the fragility and overtopping analysis. Each defence is assumed to behave independently in response to load, although flooding and defence response may be affected by any failure within some flood compartments. A raster-based 2D inundation model, LISFLOOD-FP (see Bates and De Roo, 2000; Bates et al., 2009) was used to relate defence responses to inundation patterns by generating a spatial field of water depths at stages of a simulated flood event. This model has successfully predicted flood extent and depth for coastal applications (Bates et al., 2005). The flood defence line was taken to be the open boundary of the model, with overtopping discharges and/or inflows through breaches simulated at appropriate points in the boundary. Analyses of historic flood events in the Solent are among the techniques used to calibrate and validate the model.

Figure 6 shows results for buildings inundated from combined wave and water level scenarios. Whilst not a formal joint probability analysis this indicates coastal flooding outcomes from a broad range of water level and wave condition combinations. These range from no waves (still-water overflow) to waves of maximum height and period based upon fetch-limited conditions. The 'moderate' wave scenario (figure 6) is approximately an annual extreme mostly gained from data generated by the wave sensor network (figure 2, also see: www.channelcoast.org). Real future flood events are unlikely to be regionally uniform in sea-level extremity, although this is depicted in figures 5 and 6 for demonstrative purposes. It can be seen that even under extreme overtopping conditions; the blocking effects of defences with the effects of mass conservancy and hydraulic connectivity on the floodplain substantially reduce the amount of flooding predicted by simpler exposure estimates. Modelling that has accounted for defence breaching further increased the land area and number of buildings flooded, and significantly increased water depths and rate of inundation at some locations. An example of a regional breach scenario is provided in the following section.

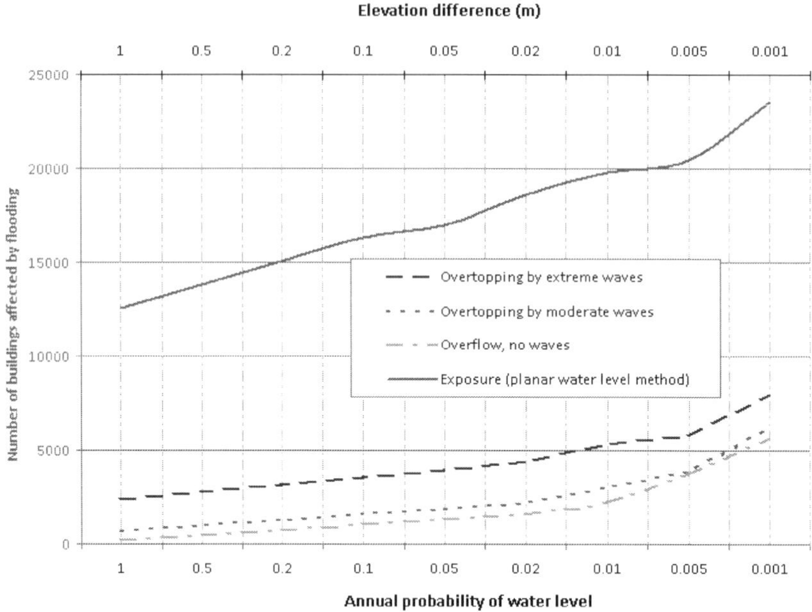

Figure 6. Exposure to flooding in the Solent region compared to results from defence and floodplain (LISFLOOD-FP) modelling

Model application

This model is designed to provide visualisations of defence response accompanied by flood risk maps, which update in response to real-time wave and water level forecasts situated at points in the nearshore zone. Figure 7 is an example of the output for a regional flood simulation that comprised a regional 1 in 200 year water level combined with approximately 1 in 50 year wave conditions. Inflows to the model included overtopping discharges and flow through breaches of defences that had a greater than 10 per cent probability of breach failure under the given loading conditions. This modelled scenario inundated approximately 8000 buildings, which is less than a third of the estimate from the planar water level method (figure 5), but considerably more than by overflow or overtopping (figure 6). This type of output provides quite specific predictions of flood locations and characteristics, whilst not ignoring low probability outcomes. The resulting water depth grids may be interpreted in various ways; for example to indicate where transport could be disrupted or areas unsafe for people.

Further analysis of the results reveals variable failure modes across the region, which may also indicate the timing and nature of flooding onset. There were numerous breaches initiated by overtopping and piping in the western Solent and harbours. Shingle beaches on the east Solent open coast overtopped and breached (these locations have historically flooded during long period swell wave events, notably in November 2005 – Mason et al., 2008; Ruocco et al., 2011). Portsmouth experienced pockets of inundation that accounted for nearly half of the buildings flooded regionally. Defences protecting the most densely populated areas in the south of the city appeared resilient to total failure, although flood risk remains high due to potential consequences of breach that include rapid inundation to depths of more than 2 m.

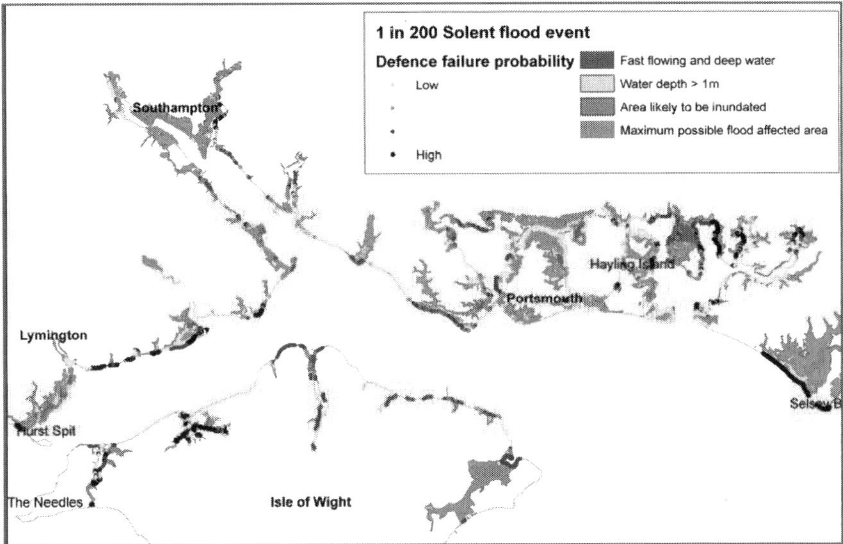

Figure 7. Example of a regional flood simulation using the defences model and LISFLOOD-FP.

Further work and preliminary conclusions

It has been demonstrated how analysis of flood pathways can significantly increase the understanding of a diverse flood system. This can be used to highlight imminent flood risks in response to high resolution real-time source predictions, which may also have the potential to reduce the consequences of coastal flooding. This approach is also relevant to contemporary research elsewhere which aims to improve flood warnings and disaster management.

The model is currently configured to be operable with available data and computing capabilities, but can be extended. For example, to further understand the potential advantages and limitations of these methods for coastal flood forecasting and warning, three sites are being analysed in more detail. Finer resolution floodplain grids may predict flood dynamics and extent more accurately, particularly in urban areas by representation of flow routes and features lost by interpolation of a DEM (Bates and Fewtrell et al, 2008; Syme, 2008). Some defence failures and their impacts upon flood extent may not be well predicted by fragility curves (Flikweert and Simm, 2008) or due to the resolution of the data used to categorise the defences. For example structural weakness or starting points for erosion may exist at transitions between asset types, near point structures, or other features not well collated in defence records (Allsop et al, 2007; Kanning et al, 2007), whilst breach dynamics (including growth rate and final dimensions) may significantly affect flood predictions (Morris et al, 2008). Further modelling of inundation scenarios may contribute to the quantification of these effects on flood predictions in the Solent and elsewhere.

The Solent is a region where flood risk is significant, although so far been overshadowed by flooding on the east and west UK coasts where the scale of storm surges and loss of life have been more prominent (see McRobie et al., 2005; Horsburgh and Horritt, 2006). Under present sea levels coastal defences offer considerable protection to property in the Solent, whilst there is also only quite a small area of floodplain containing receptors that would typically face dangerous flooding. However, the nature of these coastal flood pathways could change significantly with 21st century sea-level rise (Haigh et al, 2011). There is no record of loss of life related to coastal flooding in the Solent region, for the above-mentioned reasons and because defences improvements have kept pace with gradual changes in sea-level, perhaps occasionally prompted by minor flood events (Ruocco et al, 2011). Increased loading conditions accompanied by the potential for faster and deeper floodplain flow reinforces the incentive to continue the development of forecasting techniques and risk-based assessment of coastal defences.

Acknowledgements

This PhD is funded by the 'SemSorGrid4Env' (SG4E) EU research project (http://www.semsorgrid4env.eu/home.jsp) and the School of Civil Engineering and the Environment, University of Southampton. Project partners include the GeoData Institute, EMU Ltd and Niall Quinn. Data has been provided by the Channel Coastal Observatory (www.channelcoast.org), the Environment Agency Southern Region and the Havant Borough Council/Portsmouth City Council Coastal Defence Partnership. Paul Bates, Tim Fewtrell and Jeff Neal from the University of Bristol provided LISLFOOD-FP and expert guidance.

References

Allsop, W., Kortenhaus, A., and Morris, M. (2007). Failure mechanisms for flood defence structures. Floodsite report number: T04-06-01. Lead authors: Task leader: HR Wallingford Ltd.

Atkins (2007). Strategic Flood Risk Assessment for the Partnership for Urban South Hampshire. Document reference: 5049258/72/DG/048.

Bates, P.D., Dawson, R.J., Hall, J.W., Horritt, M., Nicholls, R.J., Wicks, J., Hassan, M.A.A.M. (2005). Simplified two-dimensional numerical modelling of coastal flooding and example applications. Coastal Engineering, 52(9), 793-810.

Bates, P.D. and de Roo, A.P.J. (2000). A simple raster-based model for flood inundation simulation Journal of Hydrology 236 (2000) 54–77.

Bates, P.D., Horritt, M.S., and Fewtrell, T.J. (2009). A simple inertial formulation of the shallow water equations for efficient two-dimensional flood inundation modeling. Journal of Hydrology 387 (2010) 33–45.

Dawson, R.J. and Hall, J.W. (2001). Improved characterisation of coastal defences. 2001: Proceedings of the International Conference on Coastlines, structures and breakwaters, September 26-28, 2001. NWH Allsop (Ed). London: Thomas Telford, 2002, p123-134.

EA (2009). Probabilistic Coastal Flood Forecasting: Forecast Demonstration and Evaluation Science project SC050069/SR2. ISBN: 978-1-84432-974-8.

Fewtrell, T.J., Bates, P.D., de Wit, A., Asselman, N. and Sayers, P. (2009). Comparison of varying complexity numerical models for the prediction of flood inundation in Greenwich, UK, FLOODrisk2008, Keble College, Oxford, Flood Risk Management: Research and Practice, 1, pp. 95-104.

Flikweert, J. and Simm, J. (2008). Improving performance targets for flood defence assets. Journal of Flood Risk Management 1 (2008) 201-212.

Floodsite (2007). Evaluating flood damages: guidance and recommendations on principles and methods. Authors: Messner F, Penning-Rowsell E, Green C, Meyer V, Tunstall S, van der Veen A, Tapsell S, Wilson T, Krywkow J, Logtmeijer C, Fernandez-Bilbao A, Geurts P, Haase D, and Parker P. Floodsite report number: T09-06-01 (January, 2007).

Golding, B.W. (2009). Uncertainty propagation in a London flood simulation. Journal of Flood Risk Management, 2 (2), 213-221.

Gouldby, B., Sayers, P., Mulet-Marti, J., Hassan, M.A.A.M., Benwell, D. (2008). A methodology for regional scale flood risk assessment. Proceedings of the ICE – Water Management, Volume 161, Issue 3, 169-182.

Hall, J.W., Dawson, R.J., Sayers, P.B., Rosu, C., Chatterton, J.B. and Deakin, R. A. (2003. A methodology for national-scale flood risk assessment. (2003).Water and Maritime Engineering, ICE, 156 (3) (2003), 235-247.

Haigh I (2009). Extreme Sea Levels in the English Channel. PhD thesis, School of Civil Engineering and the Environment, University of Southampton.

Haigh I, Nicholls R, Wells N, (2011) Rising sea levels in the English Channel 1900 to 2100. Proceedings of the ICE - Maritime Engineering, Volume 164, Issue 2, 81 –92.

Hawkes, P.J., Tozer, N.P., Scott, A., Flowerdew, J., Mylne, K., and Horsburgh, K. (2008). Probabilistic coastal flood forecasting. In: FLOODrisk 2008, 30 September - 2 October 2008, Keble College, Oxford, UK. (2008)

Horsburgh, K.J. and M. Horritt (2006), The Bristol Channel floods of 1607 – reconstruction and analysis. Weather, 61(10), 272–277

Horsburgh, K. J., Williams, J. A., Flowerdew, J., Mylne, K. (2008) Aspects of operational model predictive skill during an extreme storm surge event. Journal of Flood Risk Management, 1 (4). 213-221.

Horsburgh, K. J., and Flowerdew, J. (2009) Real time coastal flood forecasting. In: Applied uncertainty analysis for flood risk management, Editors K. Beven and J. Hall. World Scientific Press. In press.

ICE (2009). Facing up to rising sea-levels: retreat? Defend? Attack? The future of our coastal and estuarine cities. Building Futures publication, steering group: RIBA.
http://www.buildingfutures.org.uk

Kanning, W., van Baars, S., van Gelder, P.H.A.J.M., and Vrijling, J.K. (2007). Lessons from New Orleans for the design and maintenance of flood defence systems. Risk, Reliability and Societal Safety – Aven & Vinnem (eds) © 2007 Taylor & Francis Group, London, ISBN 978-0-415-44786-7.

Kolen, B., Slomp, R., van Balen, W., Terpstra, T., Bottema, M,, Nieuwenhuis, S. (2010). Learning from French experiences with storm Xynthia – damages after a flood. HKV LIJN IN WATER and Rijkswaterstaat, Waterdienst ISBN 978-90-77051-77-1.

Lane, A., Hu, K., Hedges, T.S., Reis., M.T. (2008). New north east of England tidal flood forecasting system. In: FLOODrisk 2008, 30 September - 2 October 2008, Keble College, Oxford, UK.

McRobie, F.A., Spencer, T., and Gerritsen, H. (2005). The big flood: north sea storm surge Philosophical Transactions of the Royal Society of London Series A: Mathematical, Physical and Engineering Sciences, 363 (1831) 1263-1270.

Mason, T., Bradbury, A., Poate, T., and Newman, R. (2008). Nearshore wave climate of the English Channel – evidence for bi-modal seas. Proceedings of the 31st International Conference on Coastal Engineering, Ed. Jane McKee Smith, World Scientific; 605-616.

Morris, M., Hanson, G., and Hassan, M. (2008) Improving the accuracy of breach modelling: why are we not progressing faster? Journal of Flood Risk Management 1 (2008) 150–161.

Muir Wood, R., and Bateman, W. (2005). Uncertainties and constrains on breaching and their implications for flood loss estimation. Philosophical Transactions of the Royal Society A (2005) 363, 1423-1430.

Parker, D., Tapsell, S., and McCarthy, S. (2007). Enhancing the human benefits of flood warnings. Natural Hazards (2007) 43:397-414.

PCC (2010) Portsmouth City Council – policy statement on flood and coastal defence. Report available at: http://www.portsmouth.gov.uk/media/HCS_coastaldefencepolicystatement.pdf

Pullen, T., Allsop, W., Bruce, T., Kortenhaus, A., Shuttrumpf, H., and van der Meer, J.W. (2007). EurOtop – wave overtopping of sea defences and related structures: assessment manual. Available from : http://www.overtopping-manual.com/.

Ruocco, A., Nicholls, R., Haigh, I., Wadey, M. (2011). Reconstructing coastal flood occurrence combining sea level and media sources: a case study of the Solent since 1935. Natural Hazards. DOI: 10.1007/s11069-011-9868-7

Syme, W.J. (2008). Flooding in Urban Areas - 2D Modelling Approaches for Buildings and Fences. Engineers Australia, 9th National Conference on Hydraulics in Water Engineering Darwin Convention Centre, Australia 23-26 September 2008.

Urbanflood (2010) Joint UrbanFlood & SSG4Env Workshop – Amsterdam, the Netherlands Monitoring and Flood Safety – Proceedings, Thursday, 11 November and Friday, 12 November 2010. See: http://urbanflood.eu/ResearchPublications.aspx.

SECTION 2:

COAST AND NEAR SHORE RENEWABLE ENERGY SYSTEMS

Innovative Coastal Zone Management
ISBN 978-0-7277-5749-4

ICE Publishing: All rights reserved
doi: 10.1680/iczm.57494.055

Preliminary Investigation of a Novel Breakwater Combining Coastal Defence and Energy Generation for Near Shore Environments

Mark Hillmann, Wave Energy Developments, Wigton, Cumbria, UK
Stephen Quayle, Lancaster University, Lancaster, UK
Ming Li, University of Liverpool, Liverpool, UK

Abstract

This paper presents the concept and initial tank testing of a novel low profile floating breakwater incorporating energy generation. The breakwater is intended to be anchored parallel to the coast with a beach at the front of the device to rotate the incoming waves towards collector tubes. The water hammer principle is used to harvest the high water particle velocities, charging a rear mounted high pressure reservoir which is then used to drive a conventional turbine for electrical energy production. The ability to provide breakwater benefits in smaller wave climates should open areas where combined protection and generation give competitive economic returns.

Introduction

The UK Government currently spend around £800 million per annum on flood and coastal defences with future forecasting predicting coastal erosion to increase by three to nine times by 2080 (Foresight, 2010). Coupled to this is the increasing requirement for sustainable energy using low pollution sources. Wave power devices fall into this category, and although many concept and prototype devices exist to convert the energy of sea waves into a useable form, as yet, there is no clear technology winner. To prove attractive, wave power technology devices must survive and provide a reliable source of power. Many have sights set on Atlantic wave climates where systems are remote and operate in vigorous 40 kW/m waves, however starting in smaller more sheltered coastal areas will provide appropriate lessons for further offshore expansion. The added advantage in this instance is that the device provides a robust sea defence.

This breakwater is designed as a floating structure within the near-shore wave zone. As a semi submersible device (figure 1) it remains at wave trough level in larger waves to minimise loadings but fully intercepts smaller design range waves. It takes the waves into surge chambers which capitalise on the water hammer principle to provide a small flow at increased pressure. The surge pressure passes through a one way valve to a spinal collector pipe which gathers the incident energy from several locations along the breakwater length to feed a conventional turbine. The authors are not aware of other devices using this principle.

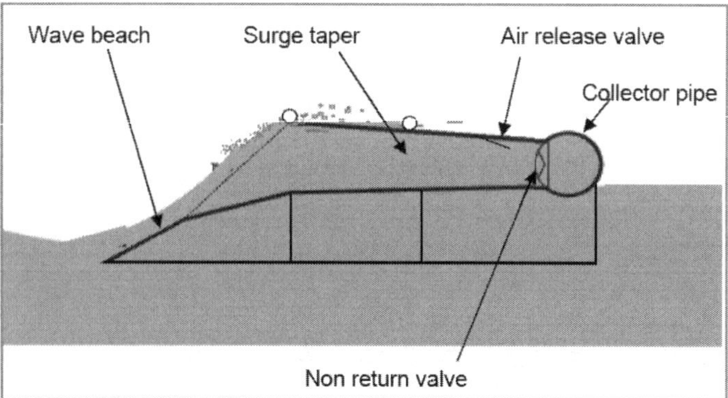

Figure 1 – Cross section through proposed breakwater

Initial wave tank and simulation work have indicated wave attenuation and electricity generation performance. These benefit from the high interception rate of a long device.

A layout similar to that used for segmented shore-parallel breakwaters (SSPBs) (figure 2) is envisaged. These have been used widely to protect coasts by encouraging the accretion of material behind them (Bacon et al, 2007).

Figure 2 – Segmented shore-parallel breakwaters at the Sea Palling, Norfolk

The breakwater can be deployed rapidly to give similar benefits of protection from erosion and flooding as fixed SSPBs and to provide good wave attenuation with minimal visual intrusion. Rapid deployment will also allow their use at marine construction sites to increase the safe working window of sea conditions and limit storm damage during construction.

Floating Breakwater Theory

The breakwater can be regarded as a long horizontal plate at wave trough level. Orientated diagonally to the waves, it is held in place primarily by water inertia and only secondly by the anchors. Waves roll along it, feeding successive inlets and providing a continuous turbine flow.

The wave beach starts by slowing and rotating the waves, improving their approach angle. Upward particle motion is inhibited ahead of the waves and increasing steepness rapidly causes breaking. We have not found a good theoretical treatment for this but used Boussinesq analysis to assess the effect. Wave types and breaking patterns are well described in the literature and the modelling indicates a useful surging break.

Typically for deep water ($h/\lambda \geq \frac{1}{2}$) approximations can be made (McCormick, 2007):

$$\lambda = \frac{gT^2}{2\pi} \quad \text{and} \quad c = \frac{gT}{2\pi} \qquad\qquad \text{Eqns (1) (2)}$$

$$u = \frac{\pi H}{T} e^{kz} \cos(kx - \omega t) \text{ and } w = \frac{\pi H}{T} e^{kz} \sin(kx - \omega t) \qquad \text{Eqns (3) (4)}$$

For shallow water ($h/\lambda < 1/20$):

$$u = \frac{H}{2} \sqrt{\frac{g}{h}} \cos(kx - \omega t) \text{ and } w = \frac{\pi H}{T} \frac{(z+h)}{h} \sin(kx - \omega t) \qquad \text{Eqns (5) (6)}$$

Where:

c = wave phase velocity (m/s)

u, w = horizontal and vertical water particle velocity components (m/s)

h = water depth (m)

λ = wave length (m)

H = wave height (m)

k = wave number $= \frac{2\pi}{\lambda}$

T = wave period (s) $= \frac{2\pi}{\omega}$

This suggests that as the depth decreases, the elliptical paths of particle travel change so that the surface particle velocity increases. The wave phase velocity, c, decreases with decreasing depth until u=c at which point the wave breaks.

Shallow water wave types (Stokes and Cnoidal) have higher horizontal particle velocities than the deep water wave shapes used in the wave tank. The breakwaters are anticipated as clear of the beach but in water shallow enough for the more vigorous wave types to be present. Even the largest waves should be tripped and broken, with the energy dissipated near the breakwater.

Water hammer (or, more generally, fluid hammer) occurs as the air vent on the collector pipes closes, resulting in a pressure surge or wave in the water column due to the sudden change in momentum. Any remaining air within the collector can act as a cushion to this fluid hammer effect, and hence removal of as much air as possible without release of water pressure is required.

Electricity Generation

Hydraulic ram pumps have been used for many years to generate high heads from small stream flows. They click away in the corners of fields using the water hammer principle to provide a reliable water supply with no electricity and very little maintenance.

This principle is used to take wave surge and convert it to a flow with enough head to run a turbine. The principle has been demonstrated with static wave tank models and with computer simulations, but the interaction of a floating device in the usual wide variety of sea conditions has yet to be confirmed.

Survival

Shoreline and buoyant devices are exposed to the full energy of breaking waves and our preferred shallow water, coastal locations have frequent breaking waves. These are not typically the highest energy, long wavelength, deep water waves as these are already broken, reducing peak forces. Our strategy is to allow semi-submergence in severe conditions. Particle velocities at trough level are far below those at crest level, making this the best way to avoid excessive loading. The device will still break these largest waves, but the energy will be dissipated inshore of the device. In locations where storm wave attenuation is not essential, complete submergence can be triggered by excess pressure, with subsequent re-floating by a small compressor.

Comparison with Other Wave Energy Devices

Many wave energy devices have features that we have sought to avoid:

- Waves can arrive every six seconds for long periods, equivalent to 5 million cycles per year. Joints and pistons that will work satisfactorily in seawater for 20 years or 100 million cycles need a high specification.
- High buoyancy exposes large areas to storm wave forces. The large loading from these waves requires high strength and weight.

- Single point moorings are only useful for point absorbers. Anchored ships lie head to tide or wind, seldom head on to the waves. Storms normally have a change of wind direction as they come through, giving a cross sea with waves from an arc of over 90 degrees.
- Energy available is proportional to collection width. Point absorbers may have an enhanced width in long wavelength waves but are fundamentally small.

Considering some other devices:

Pelamis is shown with a single mooring point. Unless it has two additional moorings at the other end with automatic winches, it will often lie broadside to the waves. Our breakwater has no joints and is anchored parallel to the coast. We hide at wave trough level from the large breaking wave forces that the buoyant Pelamis must experience and which broke its joints in Portugal (Copps, 2009).

The Aqua Marine Oyster is hinged to the seabed which may help survival by retreating from storm waves, but maintenance must be done by divers. All our working parts are above water in calm weather and our valves use conveyor belt technology where 100 million cycles is normal. Like Pelamis, towing our breakwater to harbour for maintenance is feasible.

Wavegen Limpet is a land based system using air as the driving medium. This is costly to build and the number of good sites will be limited. Where they have proposed building them into breakwaters we regard their power output per metre as similar to our anticipated output but would expect our costs to be a fraction of theirs. The suggestion of using their air system on a floating rig must expose a large above water structure to the full wave power, with a greater cost penalty.

Wavedragon uses surge on a floating beach like our breakwater, but is an overtopping device with a costly low head turbine and needs a large structure to support the elevated pond. It has wave collection arms which require a single wave approach direction or an automatic mooring adjustment system. Their wave collection efficiency is claimed to be good. We take this as partly because a beach converts both potential and kinetic energy to a useable form, a benefit we also hope to use. Their survival technique allows large waves to pass above the structure, but without our automatic buoyancy reduction in severe conditions their structure is many times larger and heavier.

Single point devices such as buoys, floats and propellers will collect from a small sea area and have a small output. Their moorings normally give high loads in storm conditions requiring robust construction with poor output to cost characteristics. Our proposal is for a semi-submersible design giving fixity in the water allowing slack moorings to minimise loading.

Work to Date

Initial design work was followed by computer simulation using Openfoam software (OpenFOAM, 2011), demonstrating the operating principle. This was followed by initial wave tank model testing at Lancaster University. The model scale of 1:10 for Irish Sea test area conditions is equivalent to about 1:30 scale for Atlantic waves.

It has not been possible to test a long breakwater section within the confines of the Lancaster wave tank, and hence the scale model consists of a short beach section and three collector tubes feeding the pressure manifold. End plates are mounted at either end of the beach to act as artificial boundaries, without which water flow would exit or enter at the sides providing inappropriate conditions for what is essentially a long symmetrical structure. The rear manifold is equipped with a vertical reservoir tube open to atmosphere which acts to record the pressure head. The model is mounted by rigid support beams that span the width of the tank (figure 3a) which somewhat contradicts how it will be anchored at full scale. The intention for these initial tests has been for proof of concept rather than assessment of overall performance and hence future models and tests are to include a better representation of the intended floating structure. The wave direction has primarily been in-line with the collector tubes, and although initial tests at approximately forty five degree angle suggested that the beach turns incoming waves towards the collector tubes, no instrumented data was gathered on this. Instrumentation is through a set of amplified wave gauges, four of which measure the incoming wave height and a fifth records the liquid level in the vertical reservoir tube. Sensors are sampled at 20 Hz.

Subsequent high speed video work showed that the draining of surge tapers between the waves was critical so a revised model was prepared using clear plastic to allow a full visual inspection of operating conditions (figure 3b). This revised model shows the move towards flat plate rather than tubular construction.

Figure 3a – Original scale model mounted in tank with waves approaching from the right
Figure 3b – Revised model using clear plastic sides to aid in visualisation

Results

Power output from the model device is essentially determined by the pressure in the manifold multiplied by the available flow rate. Pressure in the manifold has been recorded by measuring the fluid height above the SWL. Flow rate is more difficult to measure, firstly because the model has a leakage rate, and secondly because the flow rate varies with pressure within the system, this along with vibration of the model makes recording of an absolute liquid level difficult. Sampling at 20 Hz allows the relevant data to be extracted.

During sensor calibration the leakage rate has been monitored and can be approximated by using a spreadsheet trend mapping function. A typical results plot for varying beach depths is shown in figure 4, where the device was equipped with semi-rigid air valves (without allowance for losses). This figure suggests that a deeper beach performs best for energy capture, however, at different wave heights and frequencies this was found not to be the case and a clear definition of beach design was never concluded within the time frame available. Significant work on beach design for overtopping wave power devices has been completed by Aalborg University (Denmark) on the Wave Dragon device (Kofoed et al, 2000).

The most influential factor on device performance was found to be the stiffness of the air valves on each collector tube. Tests were completed with three simple arrangements consisting of a soft flexible rubber hinge, a semi-rigid membrane, and then a third variant with reinforcement. The softer valves were found to be inappropriate, failing to close and form an appropriate seal during surge. This failure to close correctly significantly reduces the fluid hammer effect and it is clear from figure 5 that significantly lower pressures result. The reinforced valve clearly outperformed the other designs in all sizes and frequency of incoming wave. Future work is to mount these valves on flat plate rather than the curved pipe surface to simplify the seal. A limit on the valve travel may also be imposed.

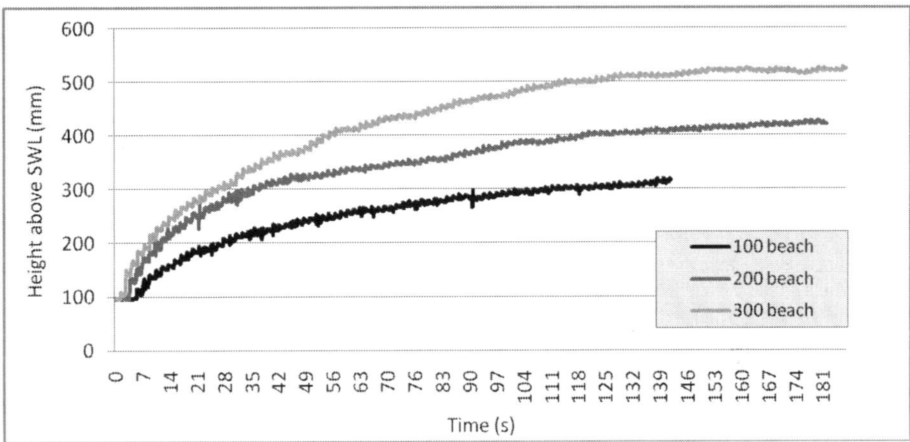

Figure 4 – Effect of beach depth on a 1.5m 5s wave

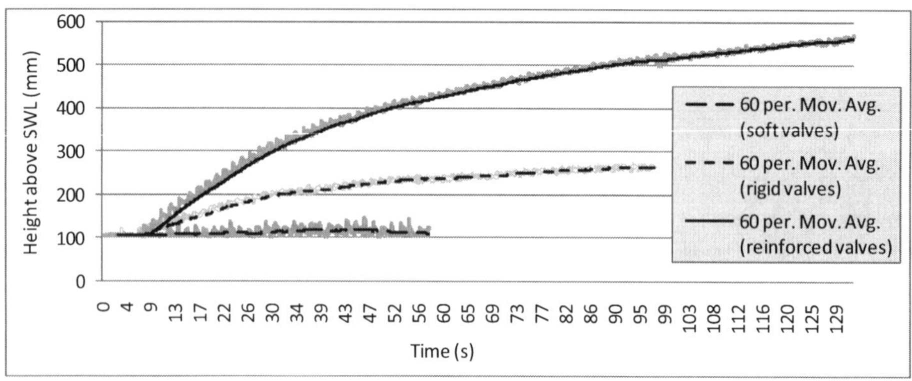

Figure 5 – Influence of air valve material on performance for a 1m 5s wave

Power Capture

The curves shown in figure 6 can be approximated using a second order polynomial expression. By calculating the gradient at a set point, it is possible to calculate an approximate flow rate and hence power output at that instance. The data selected is from the reinforced valves testing with 75mm beach at 10 degrees, using a 5 second 1.5m wave.

The trend line is described by the equation $y = -0.0003x2 + 0.7186x + 114.34$, hence at $y = 300$mm, x= 14.73s, and at $y = 305$mm, x = 15.19s. The flow rate has been calculated to be 72300 mm^3/s, which when added to the leakage rate produces a total flow of 87200 mm^3/s. Model power output at this point is pressure times flow rate, hence at the 300mm height selected power capture is 0.8 w/m, approximately 3% efficient.

A consistent set of operating curves for different positions were obtained. A flat overall output curve indicated that small waves, of a size that initiated water hammer action, had a higher efficiency than larger ones that lost excess energy in turbulence. It is believed with further model modification efficiencies of 5-10% can be readily achieved.

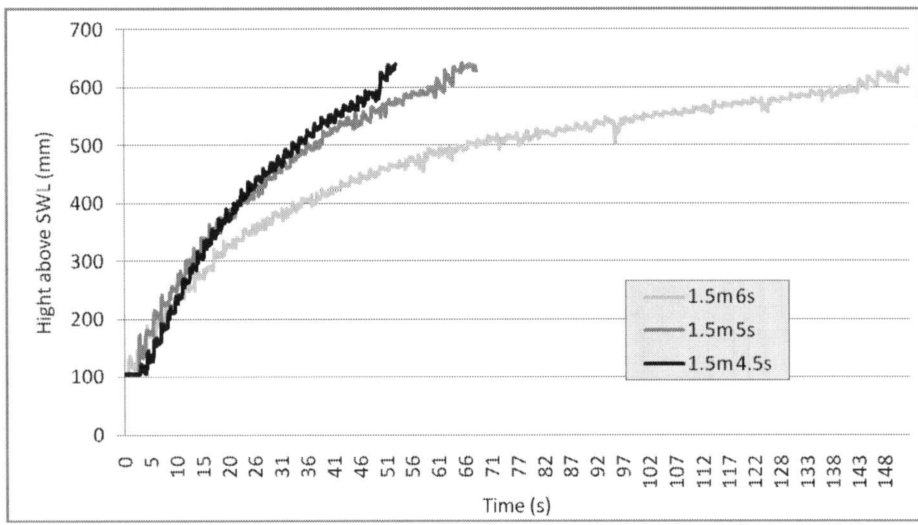

Figure 6 – 75mm beach immersion with 10 degree roof slope, reinforced valves

The initial design assumption of a 1m Irish Sea prototype design wave or 0.1m scaled wave tank design wave was found a little out. The model worked best in the wave range of 0.1 to 0.2m high. Below 0.1m the response dropped off at some wave lengths, although some of these can be attributed to natural frequencies of the tank.

At Liverpool University two studies were done: Offshore wind and wave data from wave buoy and waverider were obtained from CEFAS (CEFAS, 2011) at two potential test sites off the coast of Cumbria. A spectral wave model was used at first to provide the overall wave climate from offshore to the near shore area (Figure 7). This data was then taken into the shallow water sites proposed to give the anticipated shallow water wave energy climate using a Boussinesq Wave module for both monochromatic and random waves. This showed that at these particular locations and using the 6 months of data obtained a 0.7m design wave was more appropriate.

A second computer simulation study used CFD model Fluent (Ansys, 2011) to simulate waves interacting with the structure with various conditions obtained from the first modelling results. The model is based on the volume of fluid approach to compute the instantaneous water surface change and related flow dynamics around the structure and inside the structure. Pressure distribution can be used to evaluate the possible energy generation under the particular wave conditions. Analysis of the model data has shown similar performance as obtained from the laboratory experiments in the Lancaster University wave tank, and will aid in scaling the power capture available.

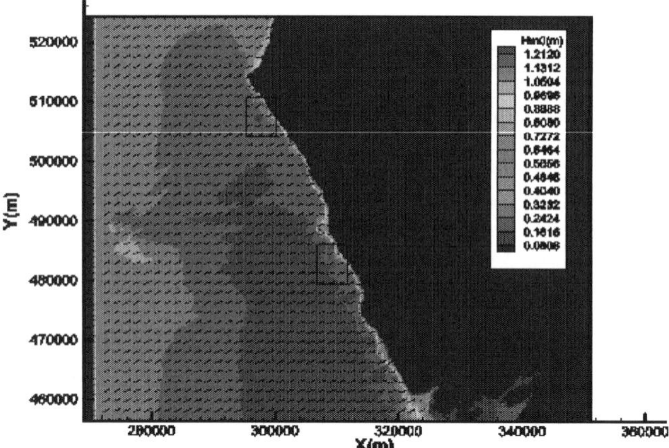

Figure 7 - Spectral wave model results for the wave height distribution near the testing sites off the coast of Cumbria.

Many questions remain:
Froude numbers are used for wave scaling but these do not apply to cavity pressures, particularly at the small laboratory scale experiments (Bredmose & Bullock, 2008). Initial calculations and some simulation work have indicated that this scaling problem should be beneficial to larger scale units.

Mooring loads have still to be established and may be complicated by long wavelength infragravity waves.

Conclusion

Where breakwater benefits are required there will be locations where this may provide a cost effective and more easily installed alternative to conventional solid breakwaters.

The generation performance lies in the interception of all the waves approaching the long device. The Carbon Trust 2006 report (Trust, 2006) suggests a central estimate of 22-25 p/kWh for offshore wave energy converters. It is anticipated that electricity generating costs will be competitive with other devices in large wave energy climates but with the design tailored for the local energy range it should give better returns in smaller climates than existing systems, designed for larger waves but with low interception rates. The ability to provide breakwater benefits in those smaller wave climates should open areas where combined protection and generation give competitive economic returns.

Wave energy economics make areas with a large energy flux more easily viable. The intention is to take the other route and use the breakwater capacity to allow experience to be gained of generation in smaller wave environments. As a new operating principle is used; new problems are anticipated.

Acknowledgements

The authors would like to thank the North West Regional Development Agency, Innovation Voucher Programme for supporting this work. Voucher Ref: 5-036

References

Ansys (2011). Ansys Inc, Retrieved 3 15, 2011 from http://www.ansys.com/

Bacon, J. V. (2007). Shore parallel breakwaters in meso-tidal conditions: Tidal controls on sediment transport and their longer term regional impacts at Sea Palling, UK. *Journal of Coastal Research* , 369-373.

Bredmose, H., Bullock, G.N.(2008). Scaling of wave-impact pressures in trapped air pockets, *International Workshop on Water Waves and Floating Bodies*, Korea

CEFAS (2011). Centre for Environment, Fisheries & Aquaculture Science, Retrieved 3 15, 2011 from http://www.cefas.co.uk/.

Copps, A (2009). Wave power project hits the rocks. *The Sunday Times, 22nd March 2009.*

Foresight Flood and Coastal Defence Project, Future Flooding Executive Summary. Office of Science and Technology. Retrieved 2 11, 2010, from UK Department for Business, Innovation & Skills: http://www.bis.gov.uk/assets/bispartners/foresight/docs/flood-and-coastal-defence/executive_summary.pdf

Kofoed, J., Frigaard, P., Sorenson, H., & Friis-Madsen, E. (2006). Prototype testing of the wave energy converter wave dragon. *Renewable Energy , 31* (2), 181-189.

McCormick, M.E. (2007). *Ocean Wave Energy Conversion,* Dover Publications Inc.

OpenFOAM the open source CFD toolbox. (2011). Retrieved 3 15, 2011, from OpenCFD Ltd : http://www.openfoam.com/

Trust, C. (2006). *Future Marine Energy.* Carbon Trust , CT601.

Innovative Coastal Zone Management
ISBN 978-0-7277-5749-4

ICE Publishing: All rights reserved
doi: 10.1680/iczm.57494.066

The Potential Future Wave Resource Utilisation around Wales

Iain Fairley, College of Engineering, Swansea University
Miles Willis, College of Engineering, Swansea University
Ian Masters, College of Engineering, Swansea University

Abstract

A GIS analysis is used to determine potential deployment areas for wave energy converters in Welsh waters. It is shown that there are substantial areas suitable for wave farm development. A numerical wave model, SWAN, is used as a tool to test the impact and viability of extracting Welsh Assembly Government (WAG) marine renewable energy targets from the GIS defined extraction areas. The effect of wave farms in English waters on Welsh resource is also tested. Wave farms in SW England could cause at most a 10% reduction in wave energy in Welsh deployment areas, but the prime deployment areas around Pembrokeshire would be largely unaffected. This paper suggests that there is a 1.6GW deployment capacity in Welsh waters. However, this magnitude of deployment is unlikely given competing demands on sea space. If it is possible to deploy wave devices of this capacity, under peak generation conditions then coastal reduction in wave height is likely to be noticeable. This may have effects on the south west Gower coast, exposed coasts between Tenby and Milford haven in Pembrokeshire and around Whitesands, based on the tested 1.6GW scenario. It is suggested that numerical modelling of wave resource is crucial to accurately predict technical and practical resources.

Introduction

Marine renewable energy generation is a growth industry due to the rising cost of fossil fuels, the need to reduce carbon dioxide emissions and improve security of energy supply. The United Kingdom has an excellent wave resource; concentrated offshore from South West England and Wales and the North-West of Scotland (ABPmer 2008). Proximity to large ports, industry, associated infrastructure and population centres means that deployment in Wales is attractive in terms of energy demand and supply chain issues.

Welsh Assembly Government (WAG) is committed to producing a significant portion of Welsh energy from marine renewable energy projects. Aspirational targets have been set by WAG for marine renewable energy extraction from Welsh waters of 4GW by 2025 (Davidson 2009). This means that, assuming a 50-50 split between wave and tidal power generation, 2GW of wave energy converters (WECs) will be required. Based on a Wave Dragon (Wave Dragon, 2005) rated at 4MW, this means that 500 devices, with dimensions 260 x 150m, will have to be deployed in Welsh waters. Realisation of such a target will require significant effort on the part of the marine renewable industry, research institutes and WAG. An EU funded Low Carbon Research Institute (LCRI) Marine consortium has been set up to facilitate the successful development of this growth sector. The LCRI Marine consortium (www.lrcimarine.org.uk) involves several Welsh academic institutions and aims to provide independent research, resolve challenges and support the growth of the renewable energy industry in Wales.

Location specific challenges to marine energy deployment include spatial planning with conflicting sea use, reduction of environmental impact and improving public support in stakeholder groups. The extraction of wave energy leads to a reduction in resource in a farm's lee. Therefore far field reduction in wave energy will occur and may detrimentally impact on the 'downstream' environment. This could lead to competition for resource both between individual farm developers and between regions (e.g. England and Wales). Planning and licensing of wave farm developments will fall to the MMO and Crown Estate, it is suggested here that a numerical modelling assessment of resource degradation will be required to adequately plan future developments. The prime wave energy resource in Welsh waters occurs off the coast of Pembrokeshire where there are marine conservation areas and the recreational value of the marine environment is high. Hence proper understanding of the potential environmental impact of energy extraction in these areas is particularly important.

Previous numerical modelling work on the environmental impact of wave energy converters has largely considered the effect of individual demonstration wave energy projects on the marine environment (Palha *et al.* 2010; Millar *et al.* 2007). Clearly, as technologies develop and deployments increase, the cumulative effects of many farms will need to be considered to fully understand environmental impacts. As far as the author is aware no research has been conducted to consider cumulative effects of wave energy extraction from multiple farms, with regional scale work limited to strategic resource assessments at present (PMSS, 2006; PMSS, 2010; RPS Group, 2011). Purely map based assessments can overestimate resource since the effect of removing energy from the system on available resource is not considered.

This paper aims to address this knowledge gap and looks to the future on a regional scale based on consideration of the Welsh resource. The effect of wave farms in South West English waters on the Welsh resource is considered based on locations suggested by the PMSS ORRAD (2010) report. Potential deployment locations around the Welsh coast are determined by inputting existing spatial planning constraints and technological limitations into a GIS framework. A SWAN wave model is then used to investigate a hypothetical scenario based on Wave Dragons deployed in these areas and test the viability and potential effect of extracting WAG energy targets (Davidson, 2009) from the wave resource in Welsh waters. Scenarios are run for summer and winter wave and wind conditions. Tests of the baseline conditions (no wave energy extraction) allow for calculation of the percentage reduction in wave height in the lee of the proposed farms.

Methodology

GIS Analysis

ArcGIS is used to determine the potential sea area available for WEC deployment in Welsh waters. Existing macro scale marine spatial planning constraints including marine conservation areas and military areas are considered. Only macro scale spatial constraints are considered in this analysis. Smaller scale constraints such as marine archaeology sites and existing undersea cables can also constrain wave energy deployment but these constraints affect the specific location of devices rather than general farm areas.

Shipping lanes, military areas, dredging areas and proposed wind farms are considered to exclude the possibility of wave farm deployment. The MoD will consider the deployment of some devices in some MOD areas but that decisions will be on a case by case basis (RPS

Group 2010) and hence the whole areas are excluded in this work.. Shipping lane exclusions are based on the IMO routing locations and on satellite ship tracking (ESA, 2009). Conservation areas are not considered to exclude deployment but are assumed to be less preferable to non-protected areas for farm deployment given the lengthier consenting process associated with environmentally sensitive areas.

Deployment areas are also constrained by technological limitations of devices. These limitations vary depending on device and are expected to be reduced with improvements to technology. Limitations considered here are maximum water depth and minimum wave energy flux (wave power). Minimum wave energy flux is based on the summer conditions. The cut-off level is taken as 15kW/m, in line with the recommended value from a resource assessment of SW England (PMSS, 2010). This study is more stringent than PMSS (2010), taking mean summer (low wave) conditions rather than an annual mean. Wave energy flux (P) is defined as:

$P=EC_g$;   where: E=wave energy; C_g=wave group velocity.

The same study (PMSS, 2010) suggests that by 2050 a likely maximum distance offshore is 50km, based on technological and economic considerations, and this limit is adopted in this work.

Territorial waters, where a country has rights to marine resources, extend to 12 nautical miles; however wave farm developments can potentially be deployed further offshore. Given the devolution of energy supply there may be resource competition between wave farm developers in England and Wales. Equally, competition may arise between Wales and Ireland due to their proximity. In this study it is assumed that the same line as the fishery divisions will be used to demarcate the sea areas available to each nation for energy extraction.

Numerical Modelling using SWAN

The third generation spectral wave model SWAN (Zijlema, 2010; Booij *et al*. 1999) was used in this research. An unstructured triangular grid was created using BatTri (Bilgili *et al*. 2006) and the Triangle mesh generator (Shewchuk, 1996), The domain covers the SW approaches and Irish sea (Figure 1). Wave conditions are implemented at the boundary marked 'incident' in Figure 1, wind conditions are forced over the whole mesh. Given the hypothetical nature of the study, large study region, and lack of data, default settings (SWAN, 2010) are used in the model. A sensitivity test was conducted for the bed friction parameter and only a 0.05m difference noticed for extremes in suggested values.

Two input conditions are tested: summer and winter. Summer conditions are calculated as mean conditions between the months April-September and winter conditions denote mean conditions from October to March. Wave conditions are based on a model output point, at N50.5, W7.5, from a 66 year wave hindcast computed by Dodet *et al* (2010). Summer conditions consist of the following wave parameters: significant wave height H_s= 1.74m; peak period T_p= 9s; and direction θ = 79°. Winter conditions are: H_s=3.3m; T_p=12s and θ =79°. The local wind field is calculated from the NCEP NCAR reanalysis dataset of monthly means from January 1948 – October 2010 (Kalnay *et al*. 1996). Wind speeds at a height of 10m are used. The summer and winter wind fields used are shown in Figure 2.

The wave farms are implemented using the SWAN command 'Obstacle' which allows implementation of a blockage with a wave transmission co-efficient. In these scenarios, the

4MW Wave Dragon devices are used, which have a wave transmission co-efficient (T_r) of 0.68 (Black, 2007). Work on array optimisation conducted for the Wave Dragon device in a north sea wave climate (Beels et al. 2010), suggests that optimal arrays consist of two staggered lines of WECs with separations of two device lengths spacing both lengthwise and depth wise. This array lay-out is used in this study. The gaps between devices increase wave transmission and hence for the whole farm length T_r= 0.78. To investigate the impact of English farms on Welsh resource, a worst case scenario is desired. Therefore, for this part of the study it is assumed that deployed farms could absorb all incident waves (T_r=0).

Potential deployment locations around the coast of south west England were taken from a resource assessment (PMSS, 2010) which gives areas and expected deployment capacities for 2010-2030. The relevant areas, capacities and required deployment lengths are shown in Table 1. Farms were positioned both close to the English 12nm territorial limit line (based on the precedent set by the Wave Hub development) and at the 50km technical limit. Angle of line was based on the PMSS (2010) prescribed areas and on incident wave direction. Since the motivation was to look at any changes to the Welsh wave resource based on English wave farms, the farms for the North Cornwall deployment area were situated as far north in the region as possible. The same approach was not taken for the Lundy area as it is thought that deployments are more likely to be closer to the mainland to minimise access and cabling costs. The locations of the tested farms in English waters are displayed in Figure 3.

The wave arrays in Welsh waters were deployed as farms with a power rating of 100MW, which is a likely array size for 2025 (PMSS, 2010). This corresponds to 25 Wave Dragon's in a 9.75km line when set up in the described (Beels et al. 2010) array. Larger farm sizes are considered unrealistic based on navigation concerns. To reach the WAG 2GW target 20 such farms must be deployed. The model was run with 20 farms distributed through the potential deployment areas with the locations optimised visually to avoid wave shadowing of nearby farms. The wave energy flux was then computed for the summer case with the farms deployed. The same flux constraint of 15kW/m was applied and any farms that were located in regions where the flux was lower than 15kW/m were removed and the model re-run. Percentage reductions in wave height induced by the presence of the wave farms were then calculated for the summer and winter scenarios.

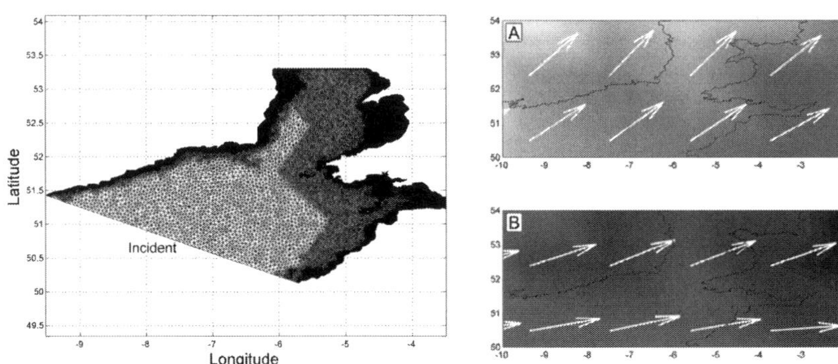

Figure 1: The mesh used in numerical modelling

Figure 2: Wind fields for A) Winter and B) Summer

Table 1: Deployment areas and capacities in English waters		
Deployment area	Estimated capacity (MW)	Length of deployment(km)
Lundy	100	10.9
North Cornwall	540	58.6

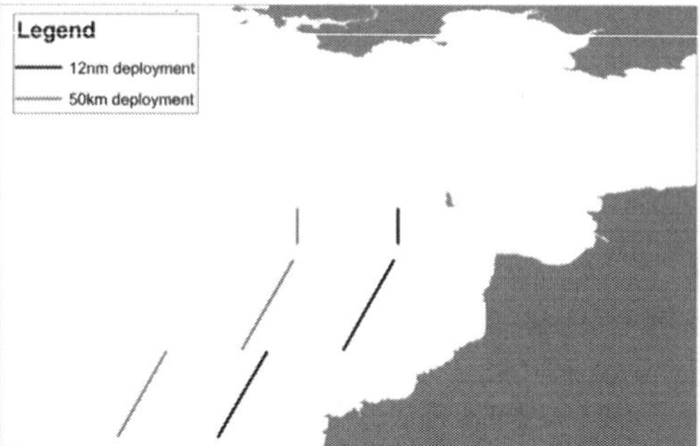

Figure 3: Locations of the wave farms tested to investigate the possible impact of English deployments.

Results

GIS Analysis

The GIS analysis identified three areas that have no constraints prohibitive to wave farm deployment (Figure 4). Figure 4 shows the potential deployment areas in grey and the limits of the conservation areas as thin black lines. Of these deployment areas, the area labelled B in Figure 4 is considered to be the most likely to be developed, despite the presence of conservation areas. Area A is further offshore and the other side of the shipping lane which means that cabling to shore and maintenance visits will become more costly, although the wave resource will be higher in parts. Area C is further into the Bristol Channel and hence has lesser wave exposure and greater tidal modulation of wave heights (reducing wave energy on the ebb tide). This is exacerbated by being behind the Atlantic array which may have a detrimental effect on the incident wave energy. Another consideration is that area C is close to the shore of the Gower Peninsula, an area of great touristic value and Wales's prime surfing resource; hence development may meet strong local opposition.

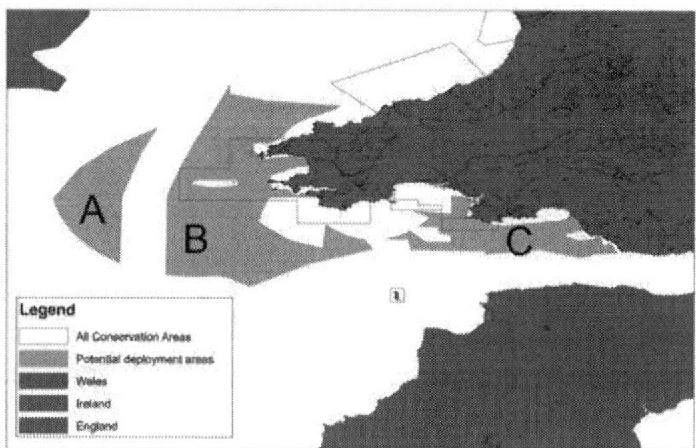

Figure 4: A map showing the potential deployment locations for wave energy converters around Wales.

Swan wave modelling

Effect of English wave farms on Welsh resource.

Despite setting a zero wave transmission co-efficient for the hypothesized farm locations to examine a worst case scenario, only a small effect on the Welsh wave resource is noticeable in the areas available for wave energy extraction. Contour plots of wave energy reduction for the 12nm case are shown in Figure 5. Based on the tested 12nm offshore scenario, the only area affected is deployment area C (figure 4), where wave energy reduction is less than 8% in both winter and summer. The greatest effect is for the near-shore areas around Port Talbot, with offshore wave resources being reduced between 2-6%. The effect of English farms would increase to over 10% reduction in wave energy in deployment area C if farms were deployed to the proposed technological and economical limit of 50km. Conversely, the effect of English farms would decrease to below 2% wave energy reduction if the Lundy deployment area was excluded.

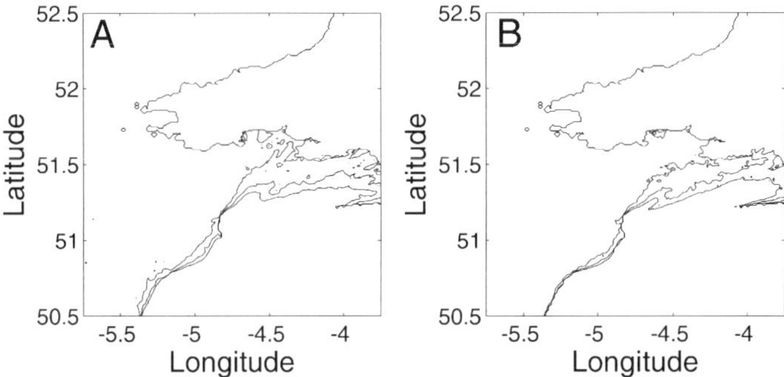

Figure 5: A contour plot of percentage changes to wave energy based on the deployment of wave farms in English waters, for winter conditions (A) and summer conditions (B).

Welsh farms

Constraints based on wave energy flux means that a maximum of 1.6GW could be extracted from the three deployment areas. Figure 6 shows the wave energy flux for the 2GW case with the farms that do not receive sufficient wave power to be commercially viable marked with an asterisk. Deployment area A could hold 5 100MW farms and Area B could potentially hold 10 100MW farms. Area C could only cater for 1 100MW farm, given its size, shape and location in a lesser wave resource. Figure 7 shows the percentage reduction in wave height behind the deployed farms for summer and winter. In both conditions the same regions are most affected by the reduction in wave height. These regions are the south Gower coastline, the Pembrokeshire coastline between Tenby and Milford Haven and around Whitesands, to the west of St Davids. Offshore, the wave reduction is greater, exceeding 20% in the lee of some farms. Reductions in wave height propagate further into the Bristol Channel for the summer condition compared to the winter condition. Conversely, in winter, reduction in wave heights propagate further into Cardigan Bay. These differences are small being around 2%.

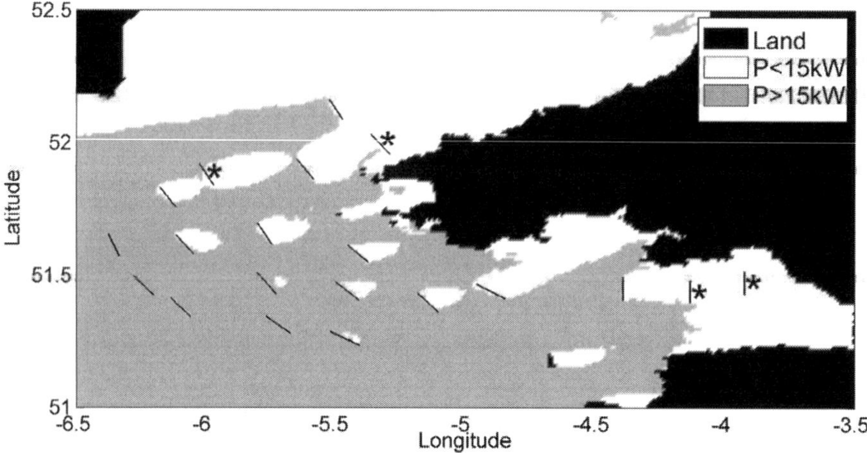

Figure 6: A map of wave energy flux P >15kW/m with the 20 100MW farms in place. The farms which have insufficient power available to be commercially viable are marked with an asterisk.

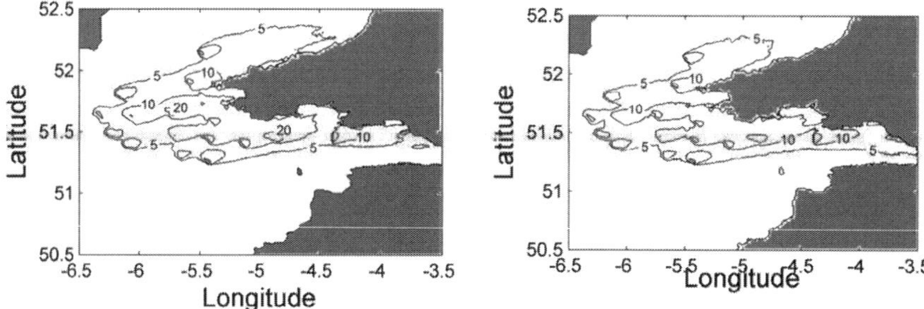

Figure 7: Contours of percentage wave height reduction for a) winter and b) summer.

Discussion

Several assumptions have been made regarding wave array deployment; however, given the strategic level of the work, the authors feel that the assumptions are acceptable. Moreover, until wave energy converter technology is further developed and the first large scale arrays are implemented, no more concrete information is available and assumptions must be made. Only Wave Dragon devices were considered. Other WECs could also have be considered, however this would have generated much more information with little benefit to the study. Wave Dragons have greater power ratings than other devices and hence a lower number of units need to be deployed for the same peak capacity. In reality, it is likely that a range of devices will be installed. The authors believe that device type does not significantly affect the main conclusions of this work. Likewise, differing deployment locations would affect the exact locations of maximum wave height reduction but the regions of wave height reduction will be broadly similar given the constraints on deployment locations and hence testing many different scenarios is not helpful given the illustrative nature of this study. Future work could extend this by testing other device arrays; other locations' and using more sophisticated implementation of wave absorption by farms. WECs are normally tuned to a specific wave height and period. This often means that the long period waves, as favoured by surfers, are not affected to the same extent as shorter period waves. Equally, large waves are not captured to avoid damage to the WEC. To extend the work in this direction, commercially sensitive device specific technical information would be required.

Deployments of wave farms in English waters, based on PMSS (2010), are unlikely to produce a significant effect on the energy in prime Welsh deployment areas. Only the area least likely to be developed, area C, is affected. However, reduction in South Wales wave resource could have impacts on coastal tourism. The affected near-shore areas, Porthcawl and South Gower, are prime tourism locations for surfing, beach going and other marine based recreation. It should also be noted that the Welsh deployments furthest east in the model also have an effect on these regions. Hence there needs to be cross-border discussion over farm placement.

This study shows that there is significant potential for deployment of wave farms around the Welsh coast. The area of sufficient wave energy flux in Welsh waters calculated in this study is larger than in the Renewable Energy Atlas (ABPmer 2008). Mean summer wave conditions rather than annual mean wave conditions (ABPmer 2008) are used here so the described area is not only larger but more commercially viable. The WAG target of 4GW of installed marine renewable devices by 2025, may not be achievable via a direct split of 2GW wave and 2GW tidal. Wave power is likely to be able to supply a significant part, but this work suggests that given one particular type of wave device, a maximum of 1.6GW of wave energy converters could be deployed in Welsh waters. However, tuning of devices to the wave climate in South West Wales may lead to greater potential deployment capacities. Only offshore wave farms are considered in this study. There is some potential to deploy near- and on-shore located wave energy devices, particularly around Port Talbot and Porthcawl, and this could boost wave energy's contribution to WAG targets.

It is probable that the tested 1.6GW deployment levels will not be implemented because an excessive amount of sea space is then utilised for wave farms which would affect other maritime interests. Sufficient sea space must be left free for shipping, recreational boat traffic, fishing and other activities. In deployment areas B and C, there is the added consenting complication of conservation areas. Both positive and negative impacts of wave farms have been described (Inger *et al.* 2009; Boehlert and Gill 2010; Langhamer *et al.* 2010). Positive impacts include the structures acting as fish aggregation devices, by default deployment areas become no-take zones and both these increase bio-diversity. On the negative, there is the potential for the structures, moorings and cabling to act as stepping stones for undesirable invasive species.

Purely map based studies, while valuable to identify potential deployment regions, are not sufficient to accurately predict deployment capacities because the impact of energy extraction cannot be included. A numerical wave model of potential scenarios is required to investigate the impact of adjacent farms on mutual resource; this then allows determination of maximum deployment levels within GIS defined regions. The authors believe an iterative procedure is required to facilitate maximum deployment with minimum impact. Once areas that are likely to be impacted by wave height reductions have been identified, baseline studies can establish maximum permissible wave height reductions. Reductions in wave height at the coast have the potential to affect physical processes, tourism, and ecology. These maximum permissible reductions can then be used to guide the selection of wave farm deployment sites and sizes.

This study suggests that the Welsh coastal regions most likely to be affected by reductions in wave energy are the South Gower coast, the Pembrokeshire coastline between Tenby and Milford Haven and the coastline west of St Davids around Whitesands. While the exact patterns of wave height reduction will depend on the number and type of wave energy devices deployed, due to the spatial constraints on deployments, the affected areas described in this hypothetical study are likely to be correct.

The South Gower coastline comprises of rocky coastlines with limestone platforms and small (<500m) sandy beaches, broken by two larger protected bays: Port Eynon and Oxwich. The protected beaches are less likely to be affected by wave height reduction; however some of the exposed, sandy beaches, such as Slade and Mewslade, which are highly geologically constrained, may display greater sensitivity. These beaches are mainly important as tourist beaches. In Pembrokeshire the coastline consists of sea cliffs and sandy beaches such as Freshwater West, Marloes, Newgale and Whitesands. These beaches are again geologically constrained, but larger than the exposed beaches on South West Gower, with lengths between 600-2700m. Again tourism is the primary function of these beaches. Partners within LCRI Marine has already investigated the baseline processes at Tenby (Thomas et al. 2010; Thomas et al. 2011) and future investigation is planned on beaches in Pembrokeshire to develop understanding of baseline coastal processes. This will enable informed decisions about potential effects of wave height reduction on beach morpho-dynamics and facilitate consenting of future farms.

The predicted affected areas are co-incident with the popular tourism regions of Gower and Pembrokeshire. Reductions in wave height have the potential to negatively affect tourism by reducing the surfing quality of a location. Surf tourism has been boomed in the last decade with specialist shops and surf schools heavily depending on the tourist trade. There is likely to be strong local opposition to deployments which will affect wave resource at popular surfing locations. Conversely, research has shown families to actively seek sheltered beaches and hence this group may benefit from a reduction in wave height (Phillips and House 2010). However this group is unlikely to be pro-actively concerned about wave farm siting. There is concern about reduction of the aesthetic value of the coast caused by offshore installations. The low height of wave energy converters coupled with the distance offshore of deployment means the converters themselves are unlikely to be considered visually obtrusive. However, cable landfalls and associated substations need to be sympathetically designed so that they are not perceived to impact on coastal aesthetics. In order to quantitatively assess potential changes, work is currently being undertaken to provide a financial value for touristic, ecological and aesthetic assets in the Welsh marine environment.

Conclusions

There are three main potential areas for offshore wave energy developments in Wales. Of these, the area offshore from Pembrokeshire is the most likely to be developed. Wave developments in England are unlikely to affect the primary wave resource around Pembrokeshire but could affect wave resource in the Bristol Channel, particularly around Porthcawl. It is estimated that a maximum capacity of 1.6GW could be installed in Welsh waters. Deployment on this scale could lead to reductions in wave height in some areas, primarily South West Gower and Pembrokeshire. This level of wave height reduction has the potential to impact on physical processes, tourism, and ecology, however the level of impact is uncertain. This work suggests geographical areas where focussed research effort into these topics may be fruitful to allow informed consenting and facilitate responsible siting of marine energy devices.

Acknowledgements

The authors would like to acknowledge the use of BatTri, Triangle, and SWAN. Thanks to RPS, Pembrokeshire coastal forum, South Wales Sea Fisheries and Pembrokeshire ranges for data and assistance. The authors wish to acknowledge the financial support of the Welsh Assembly Government, the Higher Education Funding Council for Wales, the Welsh European Funding Office, and the European Regional Development Fund Convergence Programme.

References

ABPmer (2008). Atlas of UK Marine Renewable Energy Resources: Technical Report. , Department for Business, Enterprise & Regulatory Reform.

Beels, C., P. Troch, et al. (2010). "Application of the time-dependent mild-slope equations for the simulation of wake effects in the lee of a farm of Wave Dragon wave energy converters." Renewable Energy 35(8): 1644-1661.

Bilgili, A., K. W. Smith, et al. (2006). "BatTri: A two-dimensional bathymetry-based unstructured triangular grid generator for finite element circulation modeling." Computers & Geosciences 32(5): 632-642.

Black, K. P. (2007). Review of Wave Hub Technical Studies: Impacts on inshore surfing beaches. ASRltd, prepared for SWRDA.

Boehlert, G. W. and A. B. Gill (2010). "Environmental and Ecological Effects of Ocean Renewable Energy Development: A Current Synthesis." Oceanography 23(2): 68-81.

Booij, N., R. C. Ris, et al. (1999). "A third-generation wave model for coastal regions - 1. Model description and validation." Journal of Geophysical Research-Oceans 104(C4): 7649-7666.

Davidson, J. (2009). "Ministerial Policy Statement on Marine Energy in Wales,." from http://wales.gov.uk/topics/environmentcountryside/energy/renewable/marine/marineenergy/;jsessionid=Jm1TL8hdsSwGnfywgTq2N1nS9n0615rk0TnvLDqH19rhhTBGZGTt!2003708271?lang=en.

Dodet, G., X. Bertin, et al. (2010). "Wave climate variability in the North-East Atlantic Ocean over the last six decades." Ocean Modelling 31(3-4): 120-131.

ESA. (2009). Retrieved 10/02/2011, from http://www.esa.int/esaEO/SEMBDI0OWUF_index_0.html.

Inger, R., M. J. Attrill, et al. (2009). "Marine renewable energy: potential benefits to biodiversity? An urgent call for research." Journal of Applied Ecology 46(6): 1145-1153.

Kalnay, E., M. Kanamitsu, et al. (1996). "The NCEP/NCAR 40-year reanalysis project." Bulletin of the American Meteorological Society 77(3): 437-471.

Langhamer, O., K. Haikonen, et al. (2010). "Wave power: Sustainable energy or environmentally costly? A review with special emphasis on linear wave energy converters." Renewable & Sustainable Energy Reviews 14(4): 1329-1335.

Millar, D. L., H. C. M. Smith, et al. (2007). "Modelling analysis of the sensitivity of shoreline change to a wave farm." Ocean Engineering 34(5-6): 884-901.

Palha, A., L. Mendes, et al. "The impact of wave energy farms in the shoreline wave climate: Portuguese pilot zone case study using Pelamis energy wave devices." Renewable Energy 35(1): 62-77.

Phillips, M.R., House, C.(2009)."An evaluation of Priorities for Beach Tourism: Case Studies from South Wales, UK." Tourism Management. 30(2):176-183

PMSS (2006). Wales Marine Energy Site Selection. PMSS, Bath,, Report prepared for Welsh Development Agency.

PMSS (2010). Offshore Renewables Resource Assesment and Development (ORRAD) Project - Technical Report. , Report prepared for South West Regional Development Agency.

RPSgroup (2010). The Potential for Interaction between Wave and Tidal Stream Devices with Military Interests in Welsh Waters, on behalf of Welsh Assembly Government.

RPSgroup (2011). Marine renewable energy strategic framwork for Wales, RPS group.

Shewchuk, J. S. (1996). Triangle: Engineering a 2D Quality Mesh Generator and Delauney Triangulator. Lecture Notes in Computer Science. Berlin, Springer-Verlag. 1148: 203-222.

SWAN (2010), SWAN User Manual, Delft University of Technology, DELFT, The Netherlands

Thomas, T., M. R. Phillips, et al. (2010). "Mesoscale evolution of a headland bay: Beach rotation processes." Geomorphology 123(1-2): 129-141.

Thomas, T., M. R. Phillips, et al. (2011). "Short-term beach rotation, wave climate and the North Atlantic Oscillation (NAO)." Progress in Physical Geography: 1-19.

Wave Dragon. (2005). Retrieved 10/02/2011, from http://www.wavedragon.net/.

Zijlema, M. "Computation of wind-wave spectra in coastal waters with SWAN on unstructured grids." Coastal Engineering 57(3): 267-277.

Innovative Coastal Zone Management
ISBN 978-0-7277-5749-4

ICE Publishing: All rights reserved
doi: 10.1680/iczm.57494.076

Tidal Energy Resource Study for the Copeland Islands

Dr N R Shannon RPS, Belfast, UK
D Laurent AES Electric Ltd, Richmond, Surrey, UK

Introduction

RPS and AES Electric carried out a strategic study to examine the potential for tidal energy generation in the vicinity of the Copeland Islands, located to the south of the mouth of Belfast Lough. This paper presents the case study which was designed to provide information at a strategic level; rather than a detailed academic investigation. The principle aim was to determine if the site was a suitable tidal resource warranting further examination. Numerical modelling was undertaken firstly to develop a representation of the existing tidal flows around the Copeland Islands and subsequently to examine the tidal energy potential with a number of turbine arrangements. In each case the annual tidal energy was determined by extrapolating from a one month simulation of the tides; the month having been specifically chosen to be representative of the mean annual energy potential. The impact of the turbine array was examined in terms of the interaction between neighbouring devices and also the cumulative effect of the overall site development on the tidal regime as a whole at this unconfined site.

Model Overview

The numerical modelling was carried out using the MIKE21 flexible mesh software developed by DHI of Denmark. The flexible mesh allows the size of the cells to vary across the model domain making it computationally efficient. For the Copeland Islands modelling the mesh size around the area of interest was of the order of 30m, this level of detail was also extended into the sounds between the islands to ensure that the bathymetry was depicted in sufficient detail to accurately reproduce the complex tidal flows.

The bathymetric information used to generate the mesh was extracted from digital chart data, provided by CMap of Norway. The north and south boundaries of the detailed Copeland Islands model extended across the North Channel to landfall at both ends. The northern and southern boundaries were defined as flux and level boundaries respectively, with boundary data drawn from the RPS Irish Seas surge model; which is a calibrated model encompassing the entire coast of Ireland and western seaboard of the UK.

Model Calibration

As part of the study field monitoring was carried out to provide data for model calibration. Four locations across the general area of the site were monitored with data collected at hourly intervals throughout the tidal cycle during both spring and neap tides. A fifth site was also monitored using a bed mounted ADCP; which recorded both water level and current data at five minute intervals for a period of one month. The locations of the monitoring sites are shown on Figure 1. In all cases current speed and direction was recorded through the depth of the water column at 1m intervals.

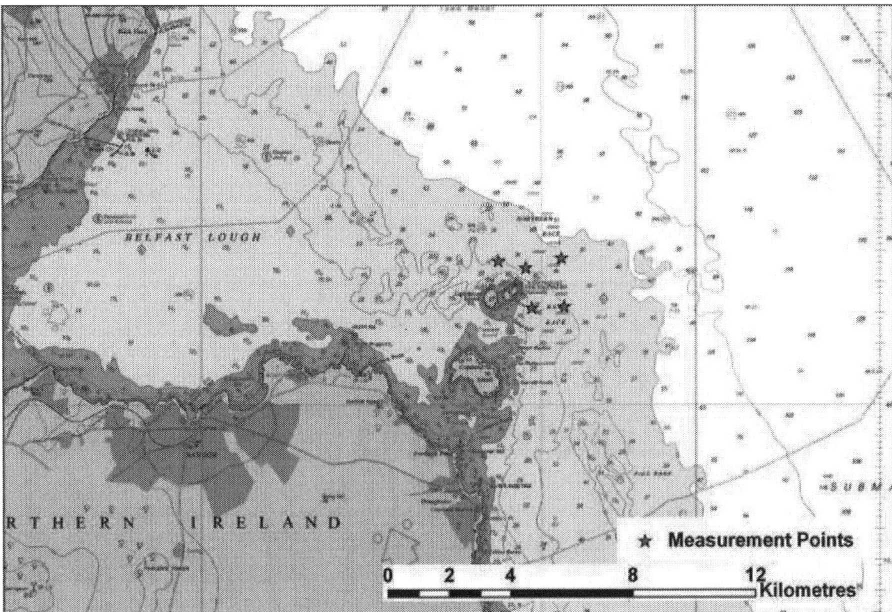

Figure 1: Measurement & data extraction locations

The measured data was compared with simulated data at each of the monitoring sites for a period with a similar tidal range to that recorded over the measurement period. When considering the correlation achieved, it should be noted that the model simulation did not include climatic conditions which may have prevailed during the course of the monitoring programme, i.e. tidal surges. In addition the computational model is also two-dimensional and thus provides a single, depth averaged, velocity for each of the cells at each time-step, in contrast to the monitoring which recorded measurements at up to 35 depths at each location. Thus in order to facilitate comparison, an average measured value was computed and plotted along with the near-bed (minimum) and near-surface (maximum) readings providing data on both the general trend and spread of the recorded data.

Figure 2 shows the correlation achieved with the ADCP data recorded from the bed mounted device during the spring tide, whilst Figure 3 depicts similar information for one of the spring tidal measurements sites. For each of these plots the measurements are discrete points and are shown as such and the simulated value is shown as a continuous line. Comparisons of current direction and velocity from the measured and simulated data are presented in the upper and lower plots respectively.

It was apparent that the flow around the Copeland Islands is complex. Although the ADCP site shows a strongly bi-directional current regime, the sites in the lee of the islands show a more involved regime where eddying takes place. The comparison of the measured and simulated values showed a good correlation between the datasets; the simulated values fall within the range of those recorded for all locations and tidal states. It was therefore concluded that the model was well calibrated and suitable for use in the tidal resource study.

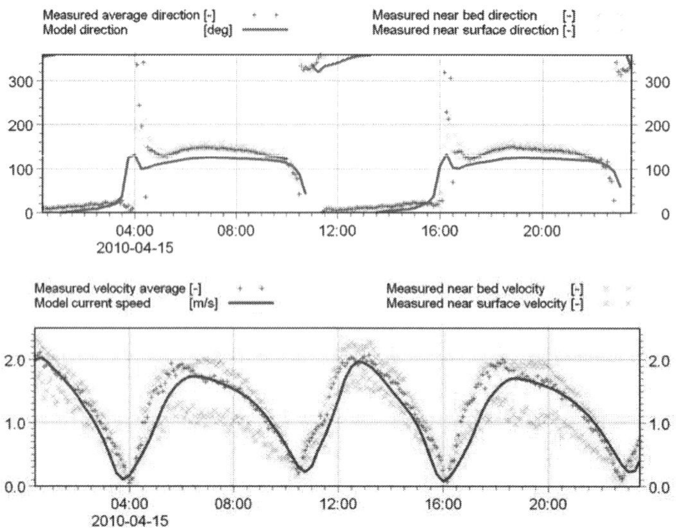

Figure 2: Current direction and speed bed mounted ADCP - spring tide

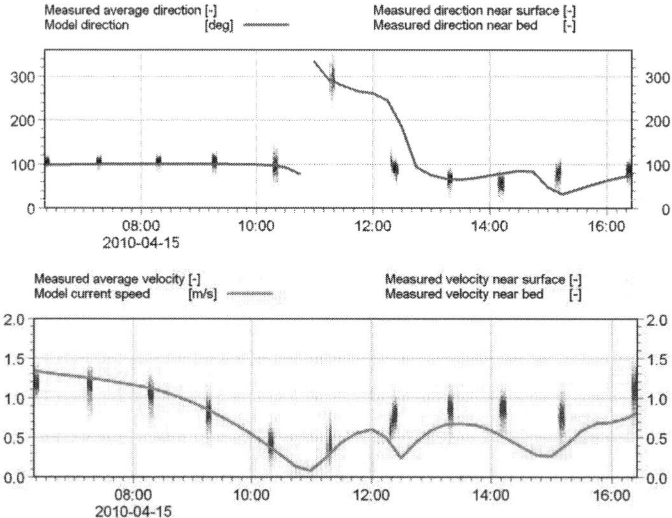

Figure 3: Modelled and measured current direction and speed

Statistical Analysis of Average Month

Following completion of the calibration process the tidal model was used to simulate the tidal flow for a period of one month. The particular month chosen had been shown by statistical analysis to give a good representation of annual tidal energy when scaled to a period of one year. This month was identified by analysis of the tidal amplitude for 30 day periods over a 20 year period. The chosen month contained two medium spring and neap tide cycles; as power output from a device would be significantly higher during a large spring and this is not necessarily balanced by a small neap tide.

Due to the extensive amount of data produced by the model a statistical approach was used to examine and analyse the output. Due to the complexity of the tidal regime the currents are skewed and peak current velocities do not occur at the same time across the site therefore the maximum current speed occurring over the duration of the simulation was extracted for each point in the model, i.e. the maximum value to occur in each cell was extracted from the 15min interval output data for the entire one month period. Figure 4 shows the distribution of maximum tidal current speed over the area surrounding the site. The maximum current speed is a good indication of the tidal potential, however in itself it may be misleading where the tidal flows are 'peaky', i.e. these high flows only persist for short periods. A similar exercise was therefore undertaken to calculate the mean flow over the one month period.

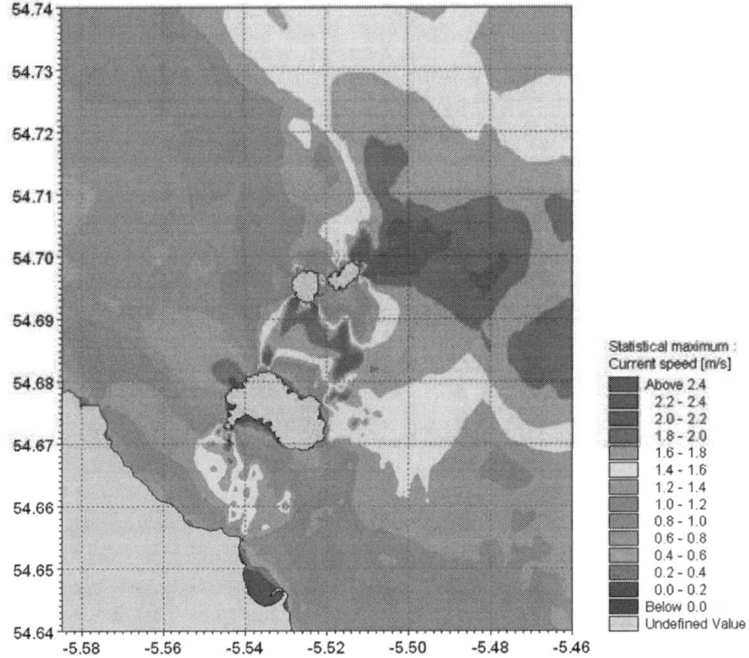

Figure 4: Maximum current speed over average month

Identification of potential array area

Following the initial examination of the general tidal regime throughout the area an investigation of the tidal potential was carried out. It was deemed important to consider the energy potential and impacts of the site as a whole, even if a single/prototype device is to be installed initially. Therefore the first step was to identify a prospective area for locating a turbine array to fully develop the site (rather than a single device). As no decision has been made with regard to the proposed type of tidal energy device to be deployed at the Copelands a range of design parameters for a generic device was established from previous experience of a range of devices and used to identify the optimum site for the array. The potential area was defined as requiring a maximum current speed of greater than 2 m/s and an average current speed greater than 1 m/s. Additionally the water depth was required to be more than 25m for the type of turbine device likely to be suitable; Figure 5 shows the area which was identified using these criteria.

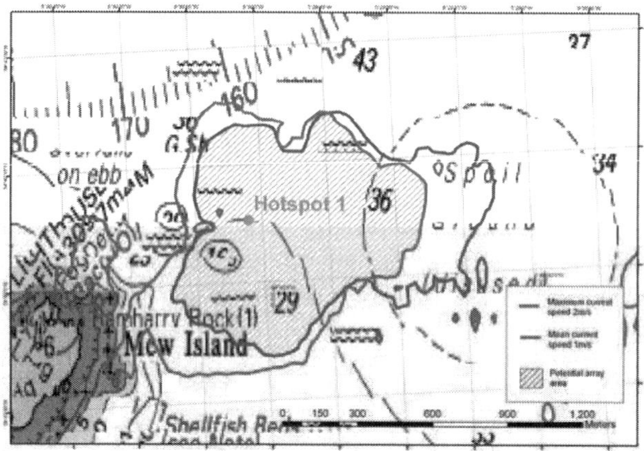

Figure 5: Extent of Potential Array Area

Array Modelling Overview

The modelling was carried out for three scenarios, firstly for a single turbine located at the hotspot selected as it exhibited both the high maximum and average current speeds that were not purely the result of restricted water depth which may be too shallow for a turbine device. Next the modelling was carried out for both an array with a traditional arrangement and for a fence type arrangement. The generic device was assumed to have a diameter of 14m; as this was considered appropriate for the water depth and tidal range across the site. The modelling included the effect on the flow due to each device by evaluating the current induced drag force which was defined as a function of density, effective area and current speed with the generic device having an assumed drag co-efficient of 0.4.

The arrangement of the traditional array used rows of devices normal to the principle current direction, with a spacing of 5 diameters between devices in an individual row and staggered rows at 10 diameters to allow for the wake dissipation. For the fence arrangement, the devices were placed at 2 diameter spacing to maximise the abstraction potential with rows at least 10 diameter spacing. It was assumed for the simulation and subsequent power analysis that each of the devices was oriented normal to the current. Therefore if fixed devices are considered further analysis may be required; as the flood and ebb tides do not run at exactly 180° to one another.

Model Results

The effects of the various turbine arrays were assessed by examining the same time-step for the existing and proposed scenarios incorporating the turbines. For this study, four points in time were investigated, the times that peak current speeds occurred during the ebb and flood tide for the spring and neap conditions defined at the location of the hotspot. The results presented here, Figure 6, relate to the south going flood spring tide. A similar exercise was also carried out for the maximum flood and ebb current speeds at each turbine location although the duration of the northerly and southerly going tides at the hotspot used to define the flood and ebb durations, again using both the spring and neap tides. To quantify the effect of the devices the change from the existing current speed is presented, i.e. the proposed minus the existing current speed, in order to highlight any changes.

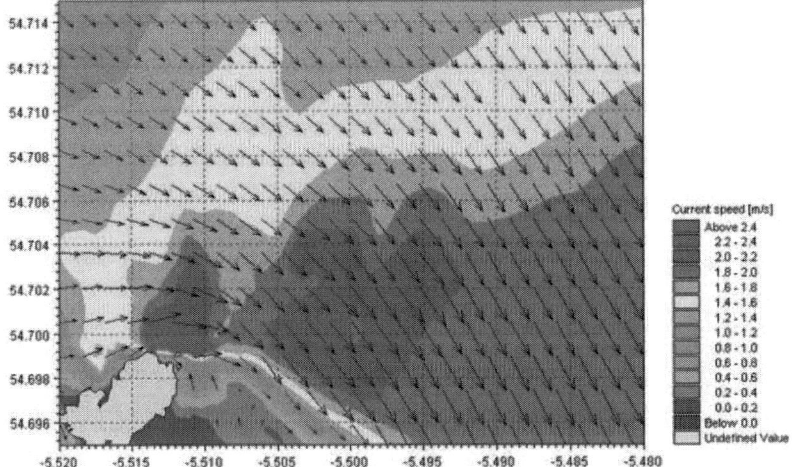

Figure 6: Existing Current Spring South-going Tide (Peak Current at Hotspot)

Single Turbine

The single turbine device is located at the site of the hotspot as illustrated in Figure 1. As anticipated, the effect of a single turbine is larger during spring tides, Figure 7, than during neap. However in all cases the change in current speed is negligible, in the order of mm/s, and would not be noticeable. The direction of flow is also unaltered from the existing flow conditions.

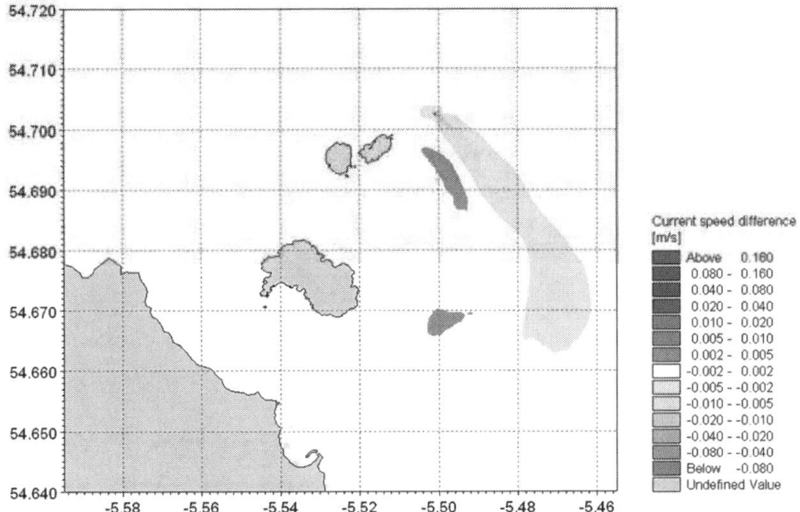

Figure 7: Change in Current Speed Spring South-going Tide - Single Turbine

Traditional Array

For the first turbine array the arrangement as depicted in Figure 8 34, turbines were included in four staggered rows approximately normal to the current direction. The change in current speed is most marked during the spring tide, an example of which is given in Figure 9. It can be seen that the current velocities reduce across the array as the energy is removed from the system, but the velocity also increases on the array boundaries as some of the flow is deflected around the site due to the increased resistance to flow through the array. The much smaller increases in velocity further offshore; are most likely due to small alterations in flow direction rather than a direct effect of the turbine array.

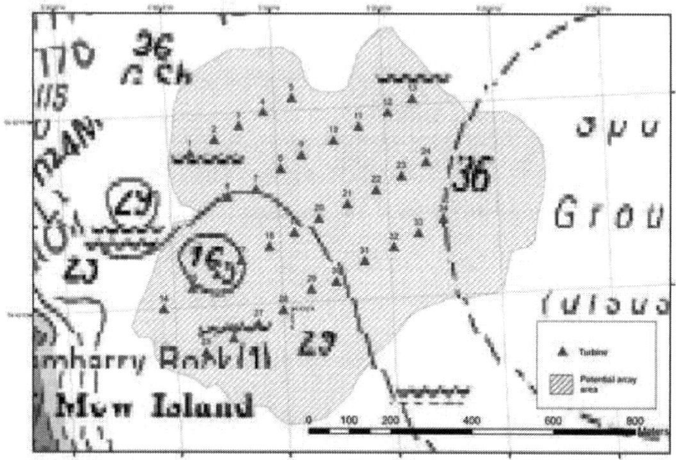

Figure 8: Location of Array Turbines

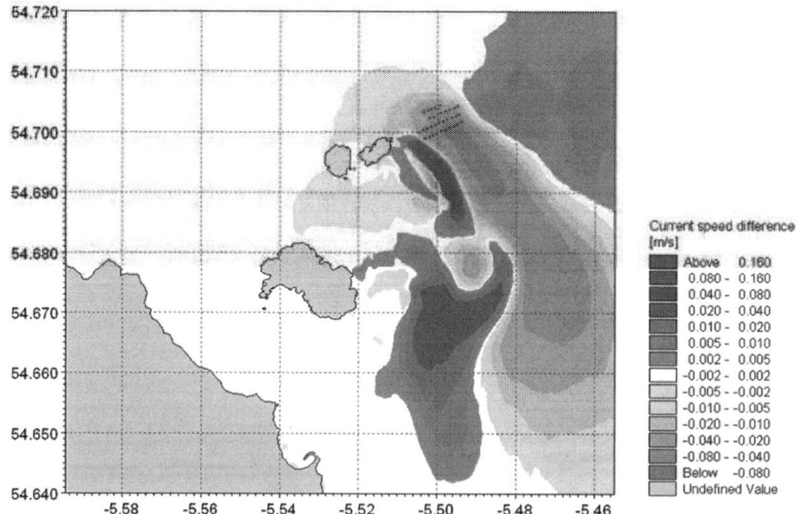

Figure 9: Change in Current Speed Spring South-going Tide – Turbine Array

Fence Array

The fence type turbine array arrangement is designed to extract the maximum amount of energy and is illustrated in Figure 10. In total 76 turbines were included in the array; the rows were increased in spacing from the minimum of 10 diameters as there was insufficient room within the target area for a further line of turbines. The impact on tidal current speeds during the spring south-going tide is shown in Figure 11. This plot illustrates that, as with the previous turbine array the velocity is reduced across the site. These decreases are larger than for the more open traditional array due to the increased number of devices. An increase in velocity, as flow is translated around the site, is also evident. The maximum change in current speed is approaching 0.2m/s and occurs during spring tides. The increases and decreases in current speed in the vicinity of the site are of the same order of magnitude.

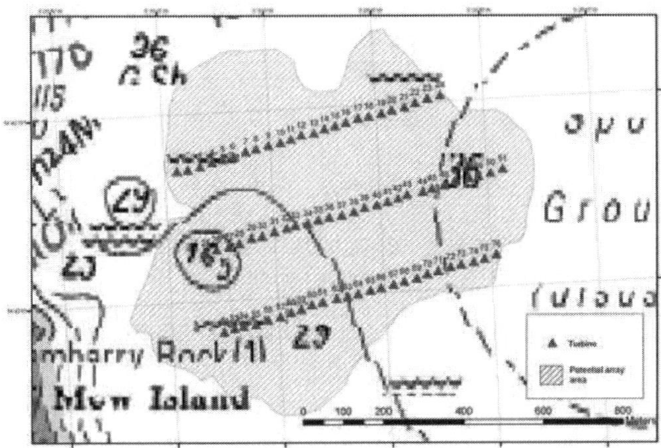

Figure 10: Location of Fence Array Turbines

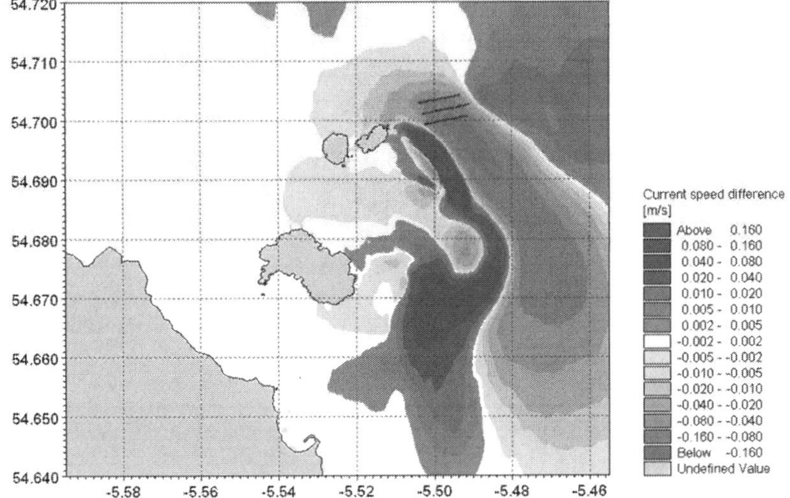

Figure 11: Change in Current Speed Spring South-going Tide – Fence Array

Power Resource

Using various data sources applicable to a range of suitable devices a generic relationship was derived between current speed and power derived by the turbine. Normalisation was carried out to enable this relationship to be applied to a 14m diameter device and the resulting power curve is shown in Figure 12. As, previously stated, it was assumed that variations in current direction did not alter the devices efficiency and the current speed used for power resource assessment is that parallel to the flow direction.

The power curve was applied to the model results from each of the scenarios to derive the potential energy extracted by each turbine under that scenario. In each case the current speed was output for every 15min interval (following model warm-up). The power curve was then applied to calculate the instantaneous power and these values were averaged over the simulation period for each individual turbine. The resulting average instantaneous power was then factored to derive the average annual energy resource in MWh for each device.

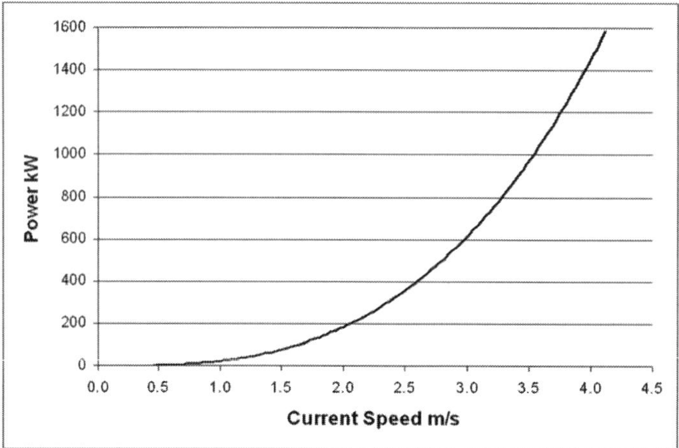

Figure 12: Power Curve for 14m dia Generic Device

The potential output from each site was calculated by summation of the annual energy values for individual turbines in the array. The capacity factor is defined as the ratio of the actual electricity generated per year divided by the electricity it would have generated if it had operated at full capacity all the time. A single turbine (1MW nameplate capacity) was estimated to deliver a capacity factor between 5% and 10%, 34 units in a traditional array would deliver an average capacity factor reduced by 20% when compared with a single turbine and 76 units in a fence arrangement would have an average capacity factor reduced by 30% . This demonstrates the scale of the issues regarding the number of devices and their proximity to one other. This implies that scaling a single resource to the number of devices proposed may result in significant over-prediction of the available resource and illustrates the importance of considering the overall development in any site appraisal.

In order to corroborate this finding, and to ensure that the lower potential is not because the field array simply incorporates a number of lower energy locations, the hotspot location was examined further. For this location alone the capacity factor is between 5% and 10%, the corresponding site in the traditional array has a capacity factor reduced by 7%. The field array does not have a device at this exact location, but the two devices within 100m and in similar tidal flows have an average capacity factor 20% lower than the single device in the hotspot location. Therefore it can be concluded that, as one would expect, the potential of the device will be diminished if turbines located in the vicinity are already reducing the current flow and in order to assess a sites potential the full development scenario should be considered from the outset.

Summary & Conclusions

A calibrated numerical model of the tidal flows around the Copeland Islands was developed utilising monitoring data collected as part of the project. The model was used to assess the tidal potential and impact on flow regimes for three tidal turbine installations; namely, a single turbine, a traditional array of 34 devices and a fence array of 76 machines. A power curve was derived for a generic device, 14m in diameter with a drag co-efficient of 0.4. The model was used to simulate the tides for a period of one month; chosen to be representative of the annual average tidal energy potential. The average annual energy extraction potential was derived for each turbine array as a whole and at a specific location.

It was seen that the presence of the array not only decreased the current flow across the site, as the energy was extracted, but also gave rise to increases in current speed nearby as flow was deflected around the site due to the increased resistance to flow through the site. The magnitude of these impacts were greatest for the fence array during spring tides; which showed changes in current speed of approaching 0.2m/s in the immediate vicinity of the array, reducing with distance from the site. Changes in current direction are also likely to occur, however it was demonstrated that these are liable to be insignificant, even for the fence array.

The modelling of the array scenarios indicated that, when compared with the single turbine, there is a clear impact on the available power resource associated with the number and proximity of further devices. It was therefore concluded that in order to assess a potential site the fully developed site should be considered from the beginning of the project to ensure that the energy potential is not over-predicted.

From a commercial perspective, there are two attributes of the Copeland islands that drive a tidal energy project: (i) their proximity to the Kilroot power plant for a connection to the electricity grid and to the yards of Belfast for operation and maintenance logistics; and (ii) it is a relatively sheltered area with respect to winds and waves which reduce installation and maintenance risks. On the other hand, the tidal flows were known to be slower than in other parts of Northern Ireland; yet this could be offset by the two previous benefits.

The estimation herein of the production potential for one or several 'generic 1 MW' turbines has confirmed the sub-optimal level of the tidal flows: the production per device in an array configuration is equivalent to a capacity factor well below 10%. For a commercial project, even with high level of green certificates, such a capacity factor is too low for the project economics to pay back the initial capital investment. Initial discussion with a few technology developers has confirmed this; at today's stage of development of tidal power technology, no projects are feasible (in a commercial sense) around the Copeland islands. The site is however well suited for demonstration projects at R&D stages. The commercial potential of the site may change with technology development; albeit this is unlikely before the end of the decade. Other sites in Northern Ireland have better (i.e. faster) tidal flows where capacity factors of installed devices would be higher; but those locations are more remote than the Copeland islands (higher costs for grid connection) and also pose more risks for installation and maintenance.

Innovative Coastal Zone Management
ISBN 978-0-7277-5749-4

ICE Publishing: All rights reserved
doi: 10.1680/iczm.57494.086

Reassessing the Severn Barrage: Operational Modes and their Impacts

John Osment, Halcrow Group Limited, Exeter
Peter Halstead, Halcrow Group Limited, Swindon, United Kingdom
Nigel Pontee, Halcrow Group Limited, Swindon, United Kingdom
Ben Hamer, Halcrow Group Limited, Swindon, United Kingdom
Robert Harvey Halcrow Group Limited, Swindon, United Kingdom

Introduction

The Severn Estuary is one of the largest estuaries in the UK, with the second largest tidal range in the world. Plans for a barrage across the Severn have been in existence for over 150 years, starting with Thomas Fulljames in 1849. A number of other studies were undertaken throughout the 1920s, 30s, 70s and 80s. The UK government Department of Energy and Climate Change (DECC) has recently undertaken an assessment of a tidal barrage.

This paper uses the latest numerical models to assess alternative barrage operational modes and make an independent assessment of:

- potential for energy generation,

- hydrodynamic and geomorphological impacts of the Severn Barrage.

There are a number of proposals under consideration for a tidal power scheme in the estuary, but only the full barrage proposed between Cardiff and Weston is discussed in this paper (Figure 1).

Figure 1: Proposed Severn Barrage

Characteristics of the Severn Estuary

The Severn Estuary is one of largest estuaries in the UK. Data on physical parameters of 39 UK estuaries are available from www.estuary-guide.net, including intertidal area. The Severn ranks third in descending order of these estuaries measured by intertidal area. The Severn Estuary has the highest tidal range, and the second highest river flow of UK estuaries, in terms of both mean and maximum flows. It has a number of tributaries including the Severn Usk, Wye and Bristol Avon. Water levels in the estuary can be raised by storm surges. The estuary has high suspended sediment concentrations.

The wave climate of the Bristol Channel is influenced both by swell waves, approaching from the Celtic (Irish) Sea and the Atlantic Ocean, and by locally-generated wind-waves. The average annual wave height decreases upstream along the Bristol Channel and the Severn Estuary. Locally generated wind waves within the Bristol Channel and Severn Estuary are more sporadic in their occurrence and are more variable in direction than swell waves, although they can be significant causes of intertidal erosion during storm events.

Modelling approach

A hydrodynamic/sediment transport model was set up within the DHI Mike Flood shell using the Mike21FM flexible mesh model and the Mike11 river model. The model as presently set up does not include the effects of wave action. Sediment transport has been assessed with the Mud Transport module of the Mike21FM suite.

The model covers the area from Frampton (on the River Severn) to offshore boundaries extending from Cape Cornwall (Cornwall) to Cobh (Eire), and from Ramsey Sound (Wales) to Rosslare (Eire) (Figure 2). Bathymetry and boundary data were obtained from UKHO; tidal data for model calibration were obtained from UKHO and BODC.

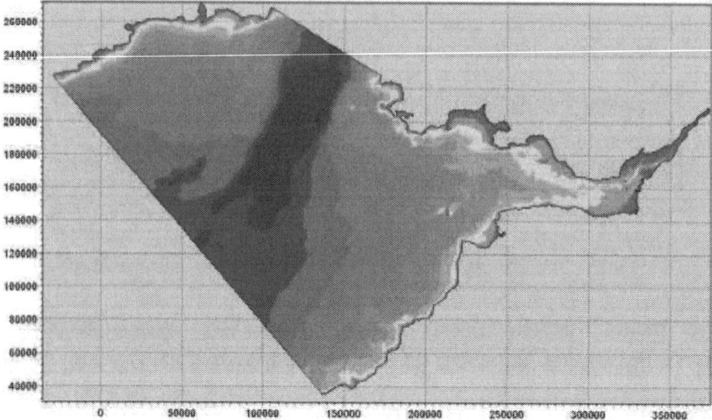

Figure 2: Model Layout

The hydrodynamic model has been set up to represent the existing condition and two barrage operation modes:
- Barrage operating in Ebb-Flood generation mode.
- Barrage operating in Ebb-Only generation mode.

And the sediment model was applied for two specific scenarios:
- Whole bed area covered with erodible sediment layer, no fluvial inputs.
- Fluvial inputs from River Severn, no erosion from bed.

The hydrodynamic model was set up to represent the barrage configuration as proposed by the Severn Tidal Power Group (STPG, 1989). This scheme, commonly referred to as the Cardiff-Weston alignment, spans the estuary from Lavernock Point to Brean Down and represents the largest well-developed barrage design.

The Ebb-Only model was set up with the Severn Tidal Power Group (STPG) configuration, incorporating 216 No. 40MW turbines and 166 No. sluices of two different sizes depending on the depth of water available.

During Ebb-Flood generation, proportionately more flow passes through the turbines, and more turbines are therefore required. However, since the majority of flood tide flows now pass through the turbines, a lower sluice area is required. The Ebb-Flood model was set up with 264 No. 30MW turbines and 132 sluices.

In order to achieve maximum energy yield from the barrage, the various operating parameters relevant to the control of the turbines and sluices must be correctly set as a function of tidal range. This was achieved by creation and application of a zero-dimensional model based on estuary hypsometry, tidal signal and turbine/sluice performance.

Key findings

Hydrodynamics

The impacts of the barrage on tidal levels within the estuary are exemplified in Figure 3. The results presented in Figure 3 are those that result when the barrage is operated to generate maximum power. As noted below, the flood-ebb scheme offers the potential to manipulate these levels by varying the operation of the barrage. The new modelling confirms that the barrage acts to reduce upstream tidal range by reducing high water levels and raising low water levels, and examines the effects of generating mode on these changes.

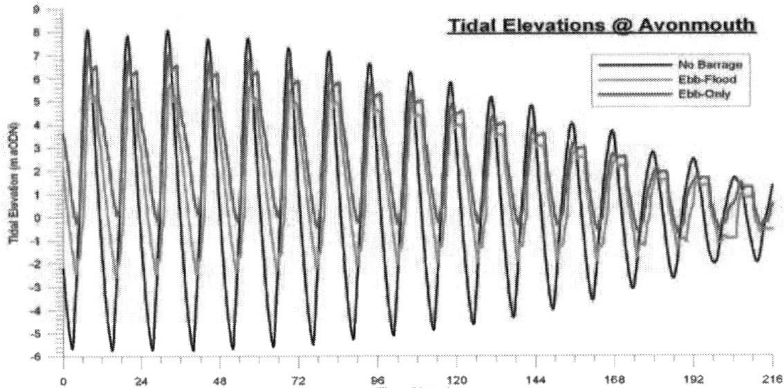

Figure 3: Tidal Elevations at Avonmouth for alternative operating modes

For Ebb-Only generation the model predicts reductions in high water of up to 1m and the raising of low water by up to 6m at Avonmouth. The latter findings are in line with the earlier SDC (2007b) results. The Ebb-Only generation mode will therefore produce some flood risk benefits in terms of reduced upstream high water levels. Raised average and minimum water levels bring benefits for navigation but could cause problems with tide locking of outfalls.

For Ebb-Flood generation the results suggest that high water levels may be reduced by up to 2.5m, whilst low water levels may be raised by up to 2.5m. Under Ebb-Flood generation flood risk benefits may therefore be greater than previous suggestions (SDC, 2007b) based on Ebb-only generation. The reduced average and minimum water levels (in relation to Ebb-Only generation) will reduce the potential for tidal locking of outfalls, but will constitute a major disadvantage for shipping to Avonmouth and other docks in the impounded area.

Early work (SDC, 2007b) noted that there was uncertainty with regard to predicted impacts on water levels downstream of a barrage. The barrage will partially reflect the tidal wave in the estuary, thus modifying the tidal height on the downstream side. The degree of modification will depend on the degree of reflection, so it might be expected that the Ebb-Only scheme (which is more porous on the flood tide than the Ebb-Flood scheme) would produce the lower downstream impacts. This was indeed found to be the case. The new modelling confirms that the downstream impacts of the barrage on maximum water levels depend on its operating mode. During Ebb-Flood generation, maximum water levels are raised by up to 0.4m, whilst for Ebb-Only generation maximum water levels are lowered closer to the barrage and raised by up to 100mm further downstream. Both forms of barrage operation raise low water levels downstream by between 0.5m and 1m. These findings suggest that Ebb-Only generation will produce some benefits for flood risk reduction downstream.

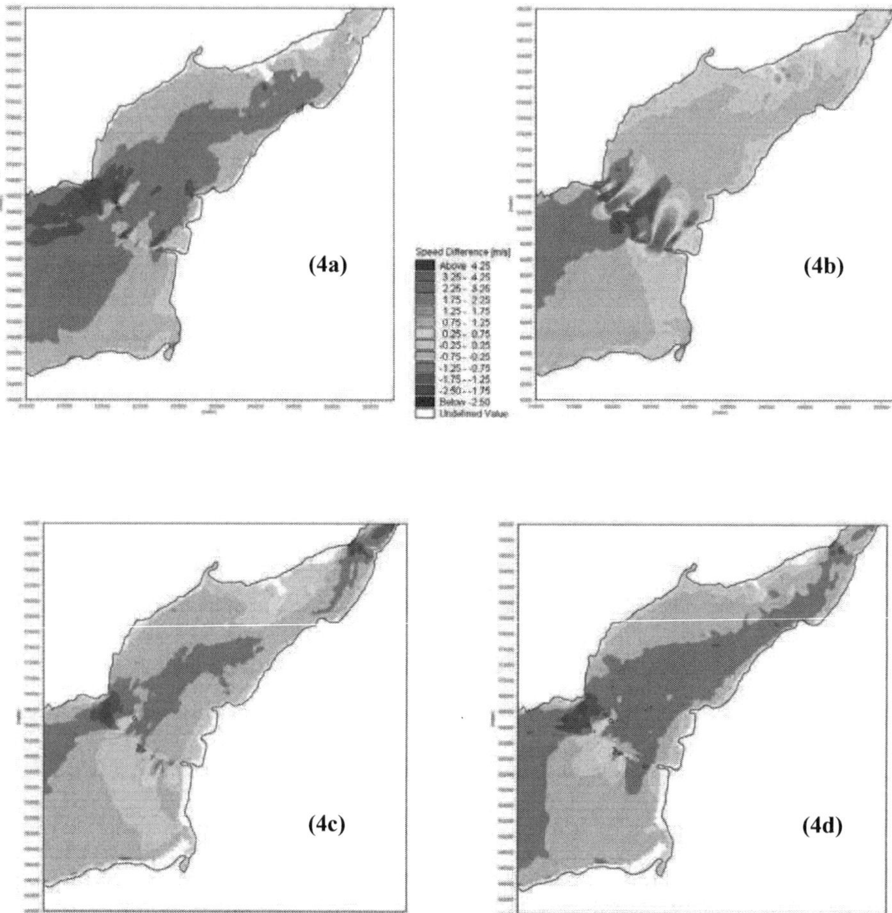

Figures 4a, b, c, d: Difference in peak current speed: (a) spring flood tide, Ebb-Flood generation; (b) spring flood tide, Ebb-Only generation (c) spring ebb tide, Ebb-Flood generation (d) spring ebb tide, Ebb-Only generation.

Current speeds are increased local to the turbines and sluices, but otherwise are reduced both upstream and downstream of the barrage. Again, the degree to which the tidal currents are affected varies according to the generating mode. Flood-Ebb generation results in broadly similar flow rates on both flood and ebb tides, whilst Ebb-Only produces greater flow rates on the flood tide than it does on the ebb tide. The impacts of the barrage on the tidal flows within the estuary are shown in Figures 4a to 4d, and the key impacts of the barrage during a spring tide are summarised as:

- An increase in current speed of up to 4.25m/s in the vicinity of the turbines and sluices.

- A decrease in flow speeds ranging from 0.75m/s to 1.25m/s over wide areas of the estuary upstream and downstream of the barrier.

- A greater decrease of 1.25m/s to 2.5m/s around the western end of the barrage.

- On the flood tide, Ebb-Flood generation produces greater reductions in current speed upstream and downstream of the barrier than Ebb-Only generation because the total flow rate through the turbines on Flood-Ebb is less than the flow rate through the sluices on Ebb-Only.

- On the ebb tide, Ebb-Only generation produces a greater area of reduced velocities upstream and downstream of the barrier because it has fewer turbines than the Flood-Ebb scheme.

Habitat

In the short term, the largest impact on estuary characteristics due to the barrage will be the immediate loss of intertidal area due to the changed tidal regime. Considering the area from Bridgwater Bay to Frampton, the following losses in intertidal area are predicted by the model on a spring tide:

- Baseline: 245km^2 under the existing conditions (Figure 5a).

- 84km^2 of loss under Ebb-Flood generation (34% loss) (Figure 5b).

- 116km^2 of loss under Ebb-Only generation (48% loss) (Figure 5c).

The larger reductions in intertidal habitat occur when the barrage is used in Ebb-Only generation mode because this mode results in higher elevations of low water level. Although the Ebb-Flood generation mode does produce larger reductions in high water level, this has a relatively small impact on the upper limits of intertidal habitats because the edge gradient becomes steeper towards the high water line.

These predictions of habitat extent are smaller than those of the SDC (2007a) which predicted losses of 144km^2 for Ebb-Only generation. These differences may be due to the lower resolution of bathymetry used in the present model as opposed to the LIDAR data used by the SDC (2007a).

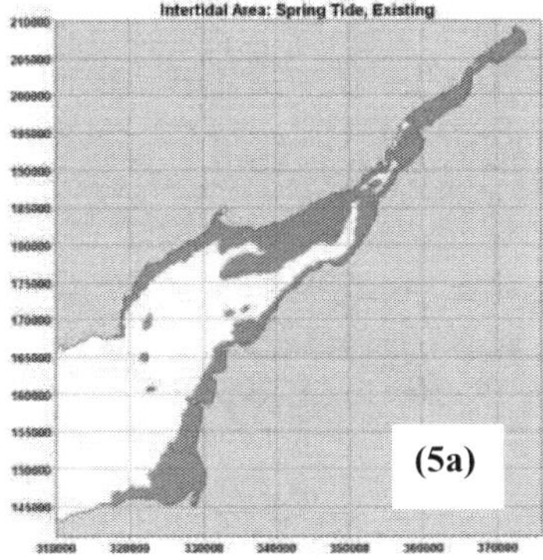

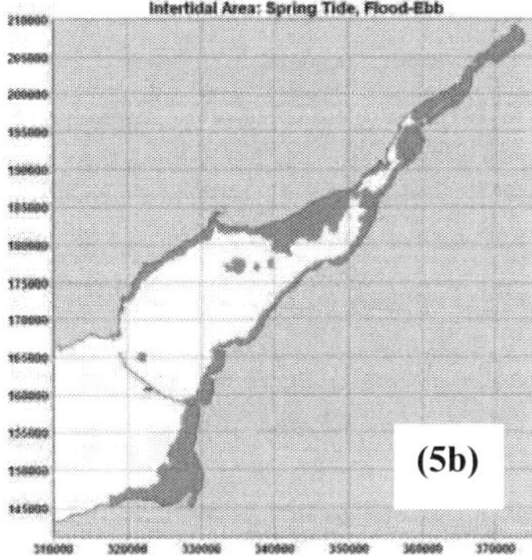

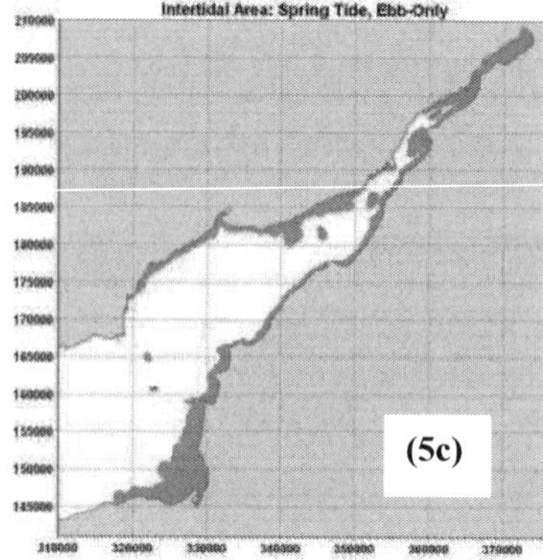

Figures 5a, b, c: Predicted extent of intertidal habitats under different scenarios; a) Existing condition, b) Ebb-Flood, c) Ebb-Only.

Geomorphology

Figures 6a and 6b illustrate the overall magnitudes and extents of erosion/accretion.
Both generation modes show broadly similar impacts, with an overall tendency for erosion of the intertidal and accretion of the subtidal, although there are some local differences.

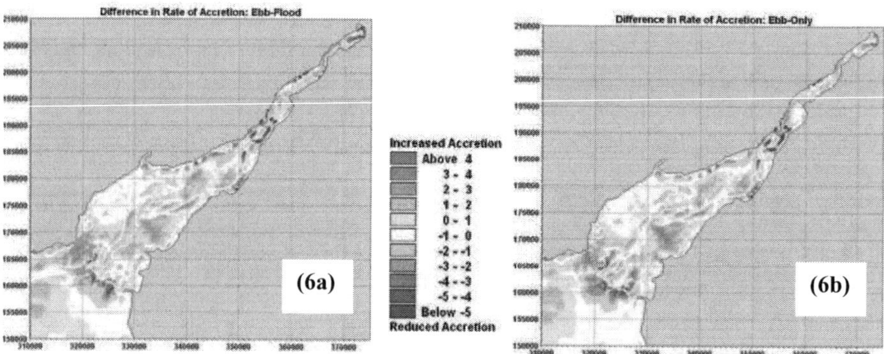

Figure 6a, b: Difference in rate of accretion; a) Ebb-Flood, b) Ebb-Only

A simple investigation of the dispersion of fluvial-derived sediment from the Severn showed that sediment tended to be deposited in the upstream reaches of the estuary and was not transported as far seawards as the barrage. This is consistent with the assessment made by the DECC (2008) report. The volumes of fluvial sediment derived from the River Severn (1M tonnes per year, Posford Duvivier and ABP Research, 2000) are small compared to the 13-30M tonnes (SDC, 2007b) of sediment estimated to be in suspension during spring tides. This suggests that fluvial sediment is not likely to be the main cause of siltation on the intertidal areas upstream of the barrier. In the short term, increased siltation will arise from the deposition of muds currently held in suspension by the high tidal currents. Once the initial pulse of increased sedimentation is complete, then further sedimentation will be governed by the increased wave action on the upper intertidal plus the redistribution of sediments by the new tidal regime.

Pethick *et al.* (2009) concluded that the long term tendency for the area upstream of the barrage, under rising sea levels, will be for erosion of the intertidal saltmarshes and accretion of the subtidal area. This is based on the assumption that there will be insufficient sediment to allow the renewed growth of intertidal marshes at lower levels on the intertidal. The preliminary modelling carried out here supports the assertion that significant amounts of accretion following barrage construction will be in subtidal areas. The recycling of this sediment to carry out intertidal recharge may provide one means of mitigating the loss of intertidal habitats (see 'Mitigation and compensation' section below).

Power generation

In conjunction with zero-dimensional parametric models for definition of operating parameters, the model has been applied to optimise the barrage configuration in terms of layout, number of turbines/sluices and mode of operation. It has been found that the Ebb-Flood mode may offer economic benefits over the Ebb-only mode because the increased flexibility to generate during periods of high energy demand, and thereby deliver higher value electricity, may offset the reduced energy output and increased capital costs associated with an Ebb-Flood barrage.

Variations of the number of turbines and sluices were tested for Ebb-Only operation. Energy output was found to increase with turbine numbers, but with only small improvements above 216 turbines. Sluices were installed to maximise permeability of the barrage to facilitate flood flows. Balancing high energy output and capital costs, a preferred arrangement of 216 No. 40MW turbines and 166 No. sluices was established.

A similar process was undertaken for Ebb-Flood operation. Having modelled a number of layouts a preferred arrangement 264 No. 30MW turbines and 132 sluices was established.

Of the preferred arrangements, Ebb-Only generation delivers marginally more energy than Ebb-Flood. The Ebb-Flood scheme incorporates more turbines of increased complexity, so would be more expensive to construct. Even so, Ebb-Flood operation offers some important benefits.

During Ebb-Flood operation energy is generated over a longer duration at a lower peak output than with Ebb-Only, easing grid load management and potentially reducing the need for grid reinforcement. An Ebb-Flood barrage can be designed to bias towards generating on the ebb tide or the flood tide. Such a scheme could be used to operate Ebb-Only or Flood-Only. This allows flexibility to adjust when energy is generated, so periods of high energy demand can be targeted to maximise the value of energy produced. Similarly, the use of cheaper electricity in times of surplus could be used to store generating capacity for use at times of higher demand by using the turbines as pumps at high or low water.

Ebb-Flood generation operates at a lower average water level, which reduces the amount of intertidal habitat loss. Two-way generation also offers the flexibility to manipulate basin water levels, potentially mitigating environmental impacts, although this would probably incur some loss of energy generation.

An Ebb-Only scheme will deliver more energy at a lower capital cost. However, the potential to manipulate the generation strategy to maximise the value of the energy produced makes Ebb-Flood generation attractive. Environmental impacts may also be reduced with Ebb-Flood generation. To commit to Ebb-Only would significantly limit options for barrage operation. Further work is required to quantify the benefits of Ebb-Flood operation.

Ecological impacts, mitigation and compensation

The modelling results show that the extent of impacts on intertidal habitat will depend on how the barrage is operated. Ebb-only operation offers little scope for mitigation whereas flood/ebb generation enables flexibility in managing water levels and scope to modify the operation of the barrage, partly for environmental mitigation. For example:

- Pumping could be used to intermittently retain higher water levels as a means of pumped storage and to maintain salt marsh in the highest part of the tidal range. Selective timing during periods of low demand could be used to minimise the cost of this strategy.

- Opening sluices and/or switching to flood-only generation during some tidal cycles could reduce low water levels within the impounded area to within 150mm of those outside the barrage. This could conserve *in situ* the majority of existing intertidal habitat, albeit with a modified tidal regime.

- Holding low water levels at low tide for longer than under the existing regime to increase time available for birds to feed, in order to offset a reduction in feeding area. This could be done by delaying the start of flood generation at times of low electricity demand and/or by constructing an intertidal bund around the Welsh Grounds upstream of the barrage to separately manage water levels within the largest intertidal area.

A key concern is the effect of loss of intertidal habitat on bird populations, for which the Severn Estuary is designated as a Special Protection Area and Ramsar Site. Our research shows that the mean peak density of water birds in the Severn is only 4.1 birds/ha, in which respect the Severn ranks 38th out of 39 UK estuaries. This relative impoverishment is due to the aggressive conditions of hypertidal regime, strong currents, high turbidity and associated low primary productivity. Barrage construction will modify the physical characteristics in such a way as to reduce tidal range, currents and turbidity (making the Severn more similar to other UK estuaries), with a likely increase of infaunal biomass. If the post-barrage estuary supported a density of waterbirds comparable to the average for west-coast English and Welsh estuaries (12.5 birds/ha), it could sustain the existing numbers of waterbirds on just 1/3rd of its current intertidal area. This suggests that on-site mitigation may be able to avoid the need for off-site compensatory measures in respect of birds. Further research on the effects of physical changes on biological productivity and carrying capacity for birds is needed.

The Severn Estuary is also designed as a Special Area of Conservation in respect of its migratory fish populations. These may be vulnerable to injury and mortality from passing through the turbines. This is likely to be a greater impact for ebb-flood generation than ebb-only, owing to the higher proportion of time that turbines are operational. However, latest developments in turbine development are reported to be less injurious to fish than traditional bulb turbines. We are exploring approaches to deterring fish from passing through turbines whilst they are operational.

Conclusions

- Greater impacts on intertidal area occur when the barrage is used to generate on the ebb tide only; greater reductions in high water level are seen with generation on both the flood and ebb tides suggesting that the flood risk benefits would be greater than previously thought if ebb and flood generation is adopted.
- The barrage results in a decrease in flow speeds over wide areas of the estuary upstream and downstream of the barrier. Ebb-Flood generation produces a larger reduction of velocities on the flood tide. Ebb-Only generation produces a larger impact on the ebb tide.
- Both generation modes show an overall tendency for erosion of the intertidal and accretion of the subtidal, although there are some local differences. Fluvial derived sediment from the River Severn represents a small percentage of the tidal sediment in suspension within the Severn Estuary. With a barrage in place this sediment is likely to be deposited in the upper reaches of the estuary and is not likely to significantly influence siltation rates or sediment character over the wider area upstream of the barrage.
- The preliminary modelling carried out here supports the assertion that significant amounts of accretion following barrage construction will be in subtidal areas. The recycling of this sediment to carry out intertidal recharge may provide one means of mitigating the loss of intertidal habitats.
- Ebb-flood generation offers significantly more potential to mitigate environmental impacts by management of water levels, especially during periods of low electricity demand.

References

Pethick, J.S., Morris, R.K.A., Evan, D.H., 2009. Nature Conservation Implications of a Severn Tidal Barrage – a preliminary assessment of geomorphological change. Journal for Nature Conservation, Volume 17, Issue 4, December 2009, Pages 183-198.

SDC, 2007a. Turning the tide. Tidal Power in the UK. Report produced by the Sustainable Development Commission. October, 2007. 148pp + references.

SDC, 2007b. Tidal Power in the UK. Research Report 3 – Severn barrage proposals. An evidence based report by Black and Veatch for the Sustainable Development Commission. October 2007.250pp.

UKHO, 2009. 'Admiralty Tide Tables', Vol 1, United Kingdom and Ireland, NP201-10.

DECC, 2008. Severn Tidal Power – Scoping Topic Paper. Hydraulics and Geomorphology. December 2008, 77pp.

Posford Duvivier and ABP Research, 2000. *'Bristol Channel Marine Aggregates: Resources and Constraints Research Project'*. Final Report, Vols. 1 and 2, August.

STPG, 1989. 'The Severn Barrage Project: General Report'. Energy Paper 57. HMSO

Innovative Coastal Zone Management
ISBN 978-0-7277-5749-4

ICE Publishing: All rights reserved
doi: 10.1680/iczm.57494.096

Whitby Pier Power

Nick Cooper, Royal Haskoning, Newcastle upon Tyne, UK
Jamie Grimwade, National Renewable Energy Centre (Narec), Blyth, UK
Stewart Rowe, Scarborough Borough Council, Scarborough, UK
David Langston, Voith Hydro Wavegen, Inverness, UK

Introduction

Whitby Harbour is situated on the North Yorkshire coastline in North East England, at the mouth of the River Esk estuary. It is sheltered against the full force of the North Sea by two piers, each with pier extensions, either side of the harbour mouth (Figure 1). These structures have not only created a 'harbour of refuge' for passing vessels, but have also provided associated benefits to the local economy from the development of quayside facilities and marine trade and industry. In addition, the structures provide a significant contribution to the overall flood and coastal defence system:

- The structures reduce the wave conditions experienced in the downstream reaches of the River Esk estuary, reducing the risk of sea flooding due to overtopping of quay walls.

- The structures help maintain beach levels along the amenity beach to the west of the harbour by trapping eastward drifting sand, protecting the backing sea cliffs against recession due to erosion and landslips.

Whitby Pier Power

Interest in low carbon and renewable energy production has been growing across the North Yorkshire region in recent years and the concept of using some form of wave power device at the harbour become of local interest. Due to this the *Whitby Pier Power Feasibility Study* (Royal Haskoning & Narec, 2010) was undertaken, aimed at investigating the technical feasibility, economic viability, and environmental and social acceptability of generating electricity at Whitby Harbour by harnessing wave energy using renewable technology devices.

Technology Review and Screening

Through an initial desk-based review of 'state-of-the-art' wave energy technologies, more than 60 concepts or devices were identified. These were grouped according to the 'type' of harnessing technology used by the device to extract energy from the waves. It was recognised that there is presently little sign of technology 'type' consolidation because the physical properties of waves are different according to geographical and bathymetric conditions and there remains a general lack of technology maturity across the industry.

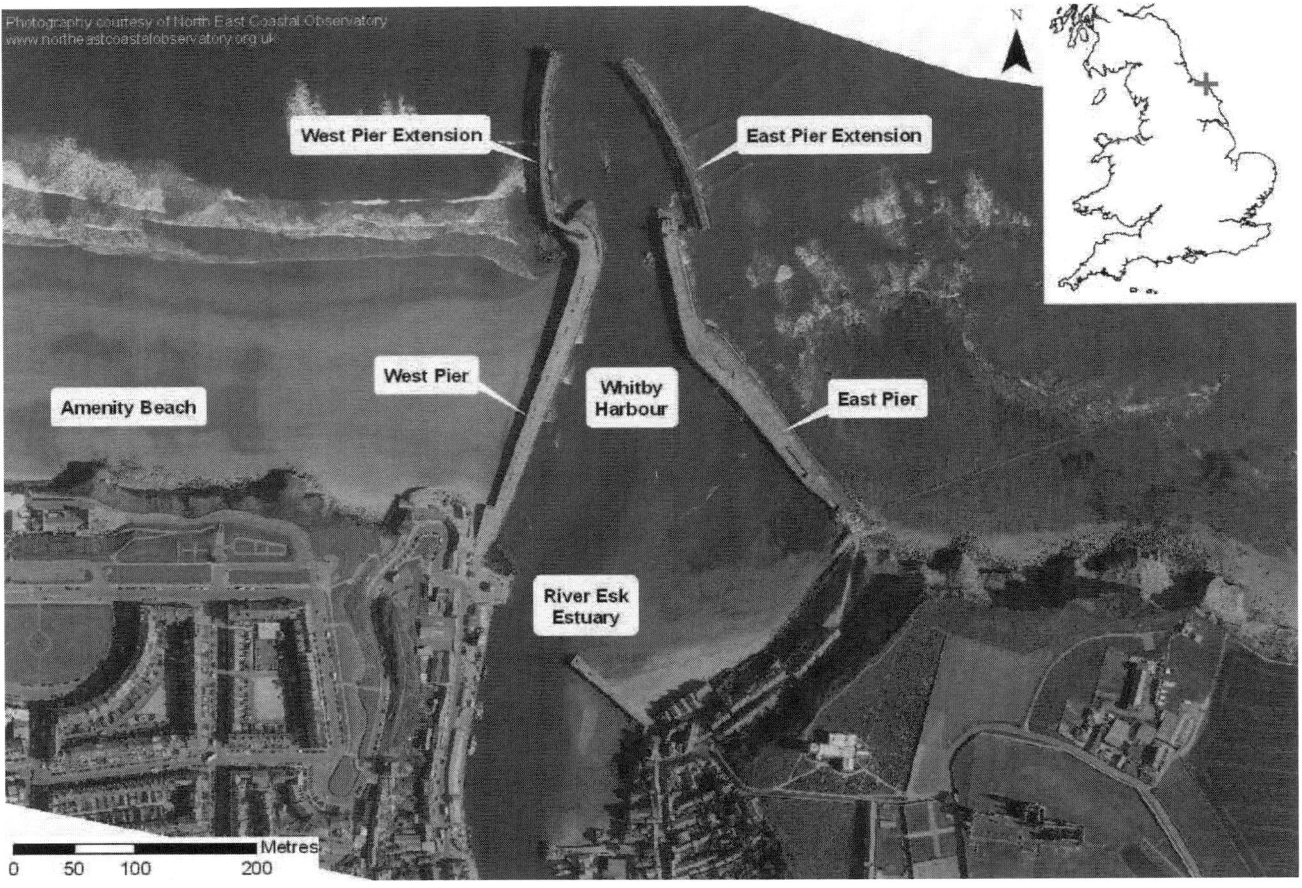

Figure 1 – Whitby Harbour, North Yorkshire, United Kingdom

The device types considered included:

- Overtopping Devices - This type of device relies on the physical capture of water from waves which is held in a reservoir above sea level, before it is returned to the sea through conventional low-head turbines which generate the power. The water capture is achieved by allowing waves to overtop a structure. Both floating and fixed devices of this type exist, with the fixed devices being most suited to potential incorporation into a breakwater, although they remain at an early stage of development.

- Oscillating Water Column (OWC) Devices – An OWC is a partially submerged, hollow structure. It is open to the sea below the water line, enclosing a column of air on top of a column of water. Waves cause the water column to rise and fall, which in turn compresses and decompresses the air column. This trapped air is allowed to flow to and from the atmosphere via a turbine, which usually has the ability to rotate only in one direction regardless of the direction of the airflow. The rotation of the turbine(s) is used to generate electricity. This concept has been implemented in a variety of fixed and floating configurations, with examples of worldwide application including proven and relatively long-running grid-connected operation on the Isle of Islay in Scotland.

- Surge Devices – These devices extract the energy from wave surges and the movement of water particles associated within them. An arm-like member oscillates as a pendulum, mounted on a pivoted joint in response to the movement of the waves. The motion of the arm is used to move hydraulic pistons.

- Attenuator Devices - These are floating devices which work parallel to the incident wave direction and effectively 'ride' the waves. Movements along the length of a device can be selectively constrained to produce power.

- Point Absorber Devices - This type of device is a floating structure which absorbs energy in all directions through its motions at, or near, the water surface. These motions are dominated by vertical (heave) movements. The power take-off system may take a number of forms and overall these devices exploit the relative motion of the float and a reaction body.

- Pressure Differential Devices – With these devices the different pressure between a wave peak and wave tough may be used to cause a relative motion in the buoyant element of a structure which is fully submerged below the water.

Each of these devices types was assessed in terms of their maturity of development, based on their existing scale of deployment and whether their developers were known to have achieved a corporate milestone of raising at least £1m of development funding. In addition, their potential suitability to the specific structural and bathymetric conditions experienced at Whitby Harbour was assessed, particularly focusing on water depths at the structures. This process revealed that OWC technology was best suited to potential application at Whitby Harbour.

Having identified OWC as the preferred device type, technology developers Voith Hydro Wavegen were invited to assist the project through an 'Early Developer Involvement' phase (similar in principal to the more commonly used Early Contractor Involvement (ECI) phase of construction projects). This was principally because the company has proven experience with the grid-connected 'LIMPET' project and is currently working on projects that integrate the OWC technology into breakwaters, including the 'Mutriku' and 'Siadar' projects.

OWC Case Studies

An OWC device was installed on the Isle of Islay, off the west coast of Scotland, as part of the 'LIMPET' (Land Installed Marine Power Energy Transmitter) project in November 2000 and has been operating remotely since that time, supplying energy to the electrical grid in the UK. The device comprises three water columns contained within concrete chambers each measuring internally 6m by 6m and inclined at an angle of 40 degrees relative to the horizontal, giving a total water surface area of $169m^2$. The upper sections of the tubes are inter-connected and power conversion is via a single turbine generator unit connected to the central column. The water columns, with an external width of 21m, are located 17m inland from the natural shoreline in a man-made recess with a water depth of 6m at mean water level. The sides of the recess are virtually parallel and vertical. The power take-off system originally comprised two generators rated at 250kW, giving an installed capacity of 500kW. The plant is now used as a test facility to support the development of new turbines. In 2010 the plant achieved turbine availability of 98% and in February 2011 the plant recorded more that 60,000 grid-connected turbine generating hours. The LIMPET site was selected on the basis of having good wave energy resource, limited tidal range and an easy opportunity to access the local electrical grid.

Figure 2 – LIMPET (Isle of Islay) nearing completion of the construction
(image courtesy of Voith Hydro Wavegen)

A more recent project is the 'Mutriku' project in the Basque Country in northern Spain which involves the integration of OWC technology devices into a new breakwater construction at a small harbour town with a population of 4,800 inhabitants. The new breakwater is an extension and upgrade to an existing construction and therefore has parallels with the Whitby Harbour situation. The justification for the upgrade was principally based upon improving the reliability of navigational access to the Mutriku port, with its associated economic benefits. The integration of OWC into a rock-filled breakwater structure is believed to be a first in Europe. The scheme involves 16 OWC devices housed within concrete caisson units along an overall 'collecting length' of 100m within the breakwater. Each turbine set is rated at 18.5kW which gives a combined installed capacity of 300kW.

The Siadar Wave Energy Project (SWEP) is a 4MW wave farm development, to be built around 300m off the shore of Siadar Bay, on the island of Lewis in Scotland. The project received Scottish Government approval in January 2009. The breakwater structure consists of concrete caissons placed together on the sea bed. Each caisson operates in about 7.5m water depth and will contain 132kW OWC units. The tidal range is in the order of 3m. The breakwater structure may be linked to the shore via the creation of a permanent causeway, which combined with the breakwater, will afford sufficient protection in the Bay to facilitate an increase in leisure and fishing activities.

Wave Conditions at Whitby Harbour

The wave climate at, and around, Whitby Harbour can be severe, particularly during storms, as shown in Figure 3. The waves predominantly approach Whitby from an offshore direction that is just east of northerly, with the most severe offshore waves approaching from due north. Offshore wave heights become notably modified (reduced in power) by transformation processes across the sea bed bathymetry as they propagate towards the shore. One key constraint, however, in terms of wave energy yield is that Whitby Harbour is tidally-influenced. This means that the most energetic wave action along the main piers and the pier extensions is predominantly limited to times coinciding with high water. At times of low water all of the main piers and parts of the pier extensions will often dry out (depending on astronomical and meteorological conditions).

To assess the energy yield potentially attainable from wave power devices, wave climate modelling was undertaken, focusing on three 'zones' of the sea bed around Whitby Harbour, namely: (i) deep water; (ii) intermediate water; and (iii) shallow water depths. Current market-leading wave energy technologies of a type appropriate to each of these zones was considered for use, with primary focus on the potential for OWC use in the shallow water depth. The wave modelling was used to inform assessments of wave power (frequently referred to as wave energy flux) per metre width, using conventional coastal engineering equations.

Figure 3 – Wave climate at Whitby Harbour (note sheltering effect of the harbour piers) (image courtesy of Scarborough Borough Council)

Results showed that none of the water depth zones around Whitby Harbour yield sufficient wave energy to make a scheme commercially attractive at present. However, present state-of-the-art OWC technology deployed below Lowest Astronomical Tide (LAT) along the seaward lengths of the pier extensions would generate an average power output of 12kW or 105MWh per year, equivalent to 21 homes being supplied with electricity. Peak output would be about 5 times the average output value.

Economic Issues

OWC devices in high wave energy environments and facing the incoming wave front could be commercially viable schemes. At Whitby, however, the OWC devices were intended be located along the side of the exiting piers (i.e. not directly facing the incoming waves) and therefore it was not expected to be commercially viable in its own right as a 'stand-alone' renewable energy scheme. There did, however, remain some (limited) potential for OWC technology to be incorporated in the design of a capital coastal defence refurbishment scheme being considered at the time at Whitby Harbour, for example to power the lighting along the piers. In this instance the OWC units would replace rock armour that may otherwise be used to reduce overtopping of the pier extensions by wave action and reduce direct wave loading on the pier structures. The estimated cost of rock armour along the lengths of the pier extensions below LAT is of the order of £0.5M. To act as a replacement to rock armour, the OWC units needed to be of this order of capital expenditure to be broadly economically comparable (notwithstanding issues such as loss of scale economies on construction mobilisation costs, ongoing operation and maintenance (O&M) costs, and revenue incomes/electricity savings, etc).

Estimated capital costs for the construction of the OWC units were made as being of the order of £5.4M, not including design, construction supervision or operational expenditure such as maintenance and annual grid connection fees. Estimated annual operation and maintenance costs for the OWC units and grid connection equipment are in the order of £32k.

Potential revenue from the sale of the electricity generated plus the current value of Renewables Obligation Certificates was estimated as being only of the order of £27k annually under an optimistic economic scenario. This clearly showed that even as a replacement for the rock armour as part of the coastal defence scheme, the OWC units are not a commercially viable proposition at present.

Environmental Issues

Whilst technically feasible to construct the OWC units, the present technology would need to be housed within caisson units that would be of a size that reaches the level of the handrailing along the timber walkway which sits on top of the pier extensions. This would have clear adverse impacts on the landscape and seascape setting and on the heritage value of the harbour structures, as shown in the scheme visualization in Figure 4. Other environmental issues that were identified included potential impacts on nature conservation and earth science heritage features, ornithology and marine mammals. Due to this, despite the over-riding environmental benefit from renewable energy, some of the environmental issues (in particular aesthetics and heritage) were identified as immediate 'show-stoppers'.

Figure 4 – Visualisation of Whitby Pier Power Scheme
(image courtesy of Royal Haskoning)

Conclusions and Recommendations

An investigation was undertaken into the technical feasibility, economic viability, and environmental and social acceptability of using devices to harness wave energy and generate electrical power at Whitby Harbour. This concept was envisioned as a means of incorporating suitable technology into the anticipated capital coastal defence refurbishment of the harbour piers and the pier extensions. A 'Whitby Pier Power' scheme of this nature would contribute to the reduction in greenhouse gas emissions and help improve security of energy supply. It could also potentially be a national exemplar and used as a pilot that was potentially replicable and scaleable to other geographical areas. This would bring with it benefits to the local and regional economy in terms of utilisation of established marine supply chains and development of specialist skills.

The physical setting of Whitby Harbour is such that the wave climate is overall relatively low (notwithstanding the severe conditions that can exist during storm events), the harbour is in shallow water, with much located in the drying inter-tidal zone, and the tidal range is moderate, which is unfavourable for breakwater-mounted OWC type concepts. When these factors are combined the logical conclusion is that whilst it is technically feasible to generate electricity by harnessing energy from the waves using existing wave power technology at Whitby Harbour, the capital and operational costs of the scheme are inordinately in excess of the revenue that would be generated. This situation remains even if an appropriate proportion of the funding came from the savings in cost otherwise incurred in the use of rock as part of the capital refurbishment.

Whilst the situation at present, therefore, is one of lack of economic viability and unacceptable aesthetic and heritage impact, this situation may alter with technological developments over future decades. Indeed these findings set a challenge to renewable energy developers to progress with technological developments over future decades that improve generating efficiencies, reduce device size and reduce cost of schemes. Furthermore, the ongoing discussions with central Government regarding funding for the capital coastal defence scheme, in view of the present economic climate, are leading towards an initial scheme to repair the immediate structural defects, but with a further scheme in future decades to address the issues of wave overtopping. It is therefore anticipated that technological developments, cost reductions due to maturity of the industry, and funding availability may come neatly together in some future decade. Whilst this is a highly optimistic viewpoint, requiring an order of magnitude reduction in capital costs and resolution of complex environmental issues, it would provide an opportunity to put Whitby Harbour forward to research institutions and private developers as a test-case site to encourage technological developments in the future to focus on harnessing energy from lower wave climates and maximizing the opportunities that exist from existing harbour pier structures.

References

Royal Haskoning, 2009. *Whitby Coastal Strategy - Further Investigations at Whitby Harbour.* Report to Scarborough Borough Council, April 2009.

Royal Haskoning & narec, 2010. *Whitby Pier Power – Feasibility Study.* Report to Scarborough Borough Council, July, 2010.

Innovative Coastal Zone Management
ISBN 978-0-7277-5749-4

ICE Publishing: All rights reserved
doi: 10.1680/iczm.57494.104

Developing the 'Atlantic Marine Energy Test Site'

Julie Ascoop, Arup, Dublin, Ireland

Mark Fielding, ESB International, Dublin, Ireland

Brian O'Mahony, Sustainable Energy Authority of Ireland, Dublin, Ireland

Abstract

The development of a full scale, grid connected, wave energy test site by the Irish government is providing an opportunity to learn, for everyone involved in ocean energy. The sharing of knowledge that is at the heart of this project has helped to identify and make known a range of issues that can now be addressed in advance of these becoming bottlenecks to the successful generation of electricity from wave energy. This paper describes the process of designing and consenting to the Atlantic Marine Energy Test Site (AMETS) in Belmullet. It provides and insight into the extensive site investigations required to establish the test areas and cable routes and it describes the particular challenges encountered at this extreme weather location.

The challenges of Ocean Energy

The coastal zone, for centuries the territory of fishermen and seafarers, is now being used by new technologies that seek to harness the energy of the oceans. Offshore oil and gas and wind farms currently extract energy from the marine environment but in doing so they reduce the impact of the large forces associated with extreme seastates by locating their primary energy extraction equipment on raised platforms or piles situated at a safe height above the influence of the extreme waves. Wave and tidal energy converters

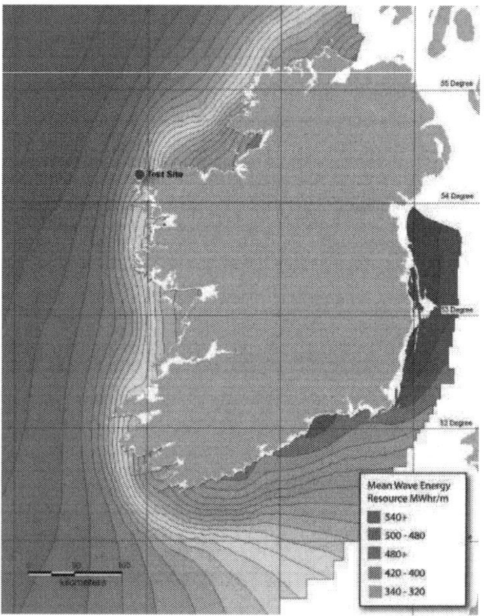

Figure 1: Wave energy resource in Ireland.

actively seek to absorb the energy of the sea. This poses new challenges, not only for the developers of these devices, but also for foreshore legislators, grid operators and other marine users such as surfers and fishermen.

Ireland is excellently placed to harness wave power, and Northern Ireland is well placed to harness tidal power. In the case of Ireland, no country in the Northern hemisphere has a better wave climate, with relatively deep water close to shore. There is competition in this field though with other countries such as Scotland and Portugal also boasting large wave energy resources coupled with good government support systems.

The Irish government is supporting the development of Ocean Energy in several ways. Within the Sustainable Energy Authority of Ireland (SEAI), the Ocean Energy Development Unit (OEDU) has been established to implement the Government's policy decision to accelerate the development of Ocean Energy in Ireland. This unit has, the following objectives:

- The creation, in Ireland, of a centre of excellence in OE technology and the stimulation of a world-class industry cluster.
- The connection of 500MW of ocean energy by 2020

The establishment of a full scale, grid connected, wave energy test site is an important component of the Irish 'national strategy for ocean energy development'.

Introduction to the Atlantic Marine Energy Test Site

Wave energy test sites

There are currently a number of wave energy test sites available around the world. Each site is serving not only a geographical area, but also a specific stage in the development of the devices. Test sites for full scale devices also vary in the size of the wave energy resource available and the type of electrical connection provided.

The Wavehub testsite located off the Cornwall coast for instance, is aiming to provide shared electrical cable facilities for the testing of arrays of devices in wave climate of typical range of 15-25 kW/m, 2x (kilowatts per metre of wave face). It is located16km offshore in 55m of water depth. The EMEC test facility on the Orkneys has a larger wave resource and consists of five cabled berths in up to 70m of water depth approximately 2km offshore.

The proposed AMETS test site in Belmullet will provide dedicated/separate electrical cable connectivity for individual WECs or arrays of WECs at designated test areas between 50m to 100m of water depth, situated between 8-14km offshore in a significantly larger wave resource of approximately 70-75 kw/m.

Site	Resource (kW/m)	Scale	Year opened/ due to open
WaveHub	17	Full Scale	2011
BIMEP	21	Full Scale	2011
EMEC Wave	21	Full Scale	2004
Pilot Zone Portugal	21-25	Full Scale / Pre-commercial	-
Belmullet	55-60 (50m)/70-75(100m)	Full Scale	2012

Figure 2: Extract from Equimar Test Site Catalogue

Development phases

Companies developing Wave Energy Converters (WEC's) generally follow a phased development, testing and evaluation protocol, such as the phases in the HMRC testing and evaluation protocol, as outlined in figure 2. At each one of these phases specific test facilities are required to evaluate the WEC.

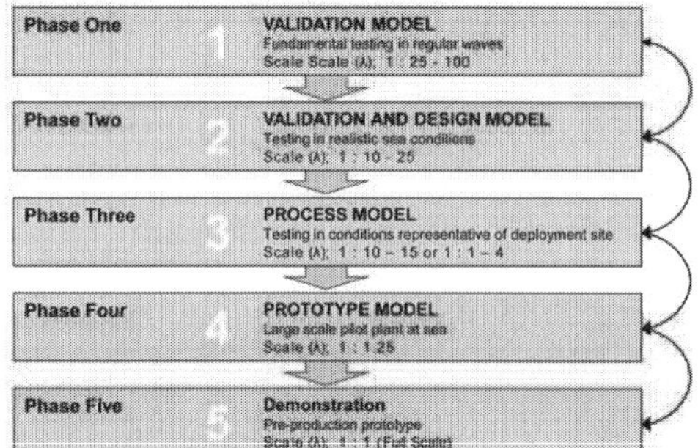

Figure 3: Testing and evaluation protocol (image courtesy of HMRC)

Currently Ireland has facilities for testing in phase one and three. Model scale testing facilities are available at the Hydraulics and Maritime Research Centre (HMRC) in Cork, and a ¼ scale testing facility, operated by SEAI, in Galway Bay. SEAI is now developing a second wave energy test site that is specifically designed for the testing of full scale, pre-commercial WEC's for phase five. In other words: this should be the final test site before the WEC's are considered fully commercial.

The full scale test site will be located off Annagh Head, Co. Mayo in the North West of Ireland, which has one of Europe's best wave climates. In this location WEC's can be tested in a full scale open ocean environment. Their performance can be assessed in terms of their ability to generate electricity over their full envelope of proposed operation and their survivability in such open ocean conditions can be tested.

The test site will provide a facility, not only for testing WEC's, but also for gaining experience on how to develop wave farms for electricity generation, what the environmental impacts are associated with such developments and how electricity generated from WEC's can be integrated and connected to the existing electricity network.

The AMETS Project Team

The Irish full scale wave energy test site has been named: the 'Atlantic Marine Energy Test Site' (AMETS). For the development of this unique project SEAI, has teamed up with a number of project partners with relevant expertise in the field of offshore renewables; ESB International (ESBI) is providing the technical and engineering services with responsibility for all elements of civil, electrical, mechanical and environmental design and development on the project, the Irish Marine Institute (MI) who provide marine survey and wave and tidal data capture services and the users perspective from Tonn Energy: an Irish wave farm development company, with backing and specialist expertise from Vattenfall and Bord Gáis, Arup is providing project management and technical advice services for the development of test site.

The Wave Energy Converters (WEC's)

The development of a full scale, grid connected, wave energy test site is a unique and challenging project. Not only is the test site to be situated in a full scale, open ocean, marine environment off the west coast of Ireland, it will also cater for WEC's that have not yet been developed at full scale. WEC's are still under development and those that succeed in reaching full scale development are not envisaged to be available before 2012/2013. Although the high level functional requirements of the project are well defined, the evolving nature of WEC design and development requires the design approach to incorporate a reasonable degree of flexibility where practicable, to enable the design to adapt to changing requirements. An important part of this project is therefore the interaction with the technology developers and potential suppliers of the equipment that will be used in the test site. To accommodate continuous device development in parallel with the design of the test site, the project team is in regular contact with developers on a whole range of issues including required WEC deployment depths, mooring systems, deployment methodologies and electrical requirements.

Figure 4: The Ocean Energy WEC

AMETS Functional Requirements

The test site should provide an offshore area in the marine environment where arrays of different WEC's can be deployed.

It will be a grid connected test facility and exporting of electricity generated by WEC's will be via submarine electricity cables to the onshore electricity network.

To accommodate for a range of WEC technology requirements two separate test areas will be incorporated in the test site: one at the 50m water depth and one at the 100m water depth.

The maximum export capacity of the test site is set at 10MW.

AMETS Site Selection and Evaluation Study

Until recently, information available about the seabed surrounding the Irish coastline was mainly limited to the UK Hydrographic Office admiralty charts. In recent years however, the Irish government has embarked on a national seabed survey programme called INFOMAR (**IN**tegrated Mapping **FO**r the Sustainable Development of Ireland's **MA**rine **R**esource). This study has made available detailed bathymetry and seabed classification information for large sections of the Irish coastline. This data was analysed by the Marine Institute and ESB International and used to undertake a detailed site selection and evaluation study which was completed in 2008. The study considered a large number of site selection criteria, including the available wave resource, seabed topography and composition, water depth, access to local ports, environmental and planning constraints and the cost of connecting into the local electricity network. All were combined into a weighted scoring matrix comparing the various alternative locations around Ireland.

The results of the site selection and evaluation study ranked the location at Annagh Head, Co. Mayo near the town of Belmullet as the most suitable to develop a full scale wave energy test site.

Development of AMETS

Marine site investigations

The AMETS project team commenced their work in early 2009. The first phase of the project was to develop a detailed concept design in preparation for submission of the various consenting and permitting applications required when developing offshore marine engineering projects. Marine survey and geotechnical investigations were undertaken as part of the concept design phase to provide baseline data for the design and also to inform the consenting and permitting requirements.

Marine hydrographic, geophysical and geotechnical investigations provide essential information for the test site design and tend to dictate critical path in the early stages of a project. Detailed bathymetry, geophysics and geotechnical data are required in order to establish suitable locations for the test areas and identify a suitable cable route corridor.

A licence to carry out offshore site investigations was obtained in the autumn of 2009 allowing for multi-beam echo sounder, side scan sonar, sub bottom profiler, vibro coring and grab sample collection. One of the conditions of the licence to investigate was that a marine mammal observer would be present at all times during the investigations. The site is located in one of the most extreme wave climates in Europe and hence the potential for weather related delays during the offshore investigations was significant. Wave heights of up to 18m have been measured by the wave buoy moored on site in 50m water depth.

Extreme weather conditions were indeed experienced during the site investigation period in autumn 2009, and as a consequence the work was split into two phases and completed in spring 2010. The results from the offshore geotechnical site investigation were used by ESBI to re-assess the preliminary cable route established in 2008. Some re-routing of the cable route corridor was required as part of the final cable route corridor design.

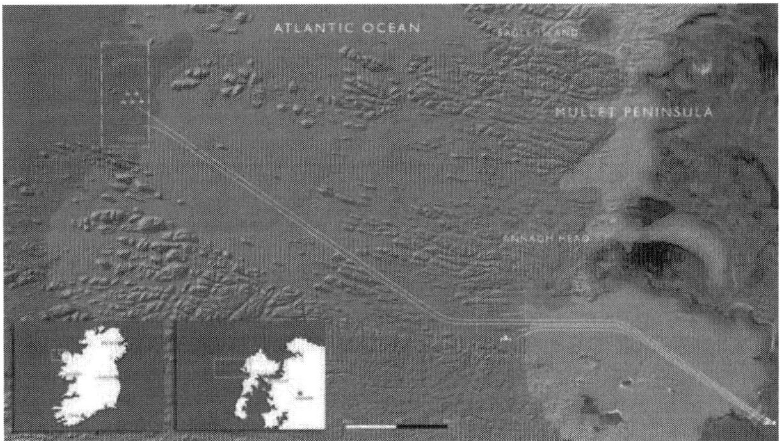

Figure 5: AMETS layout

The design of the test areas

The AMETS project incorporates two offshore test areas: Test Area A at 100m water depth, 14km offshore and Test Area B at 50m water depth, 6km offshore. Both test areas will be connected to the onshore electricity network via submarine electricity cables. Test Area's A and B are designed for surface moored WEC's such as Wavebob, OPT, Ocean Energy and Pelamis.

The size of the test areas allows for two WEC arrays to be deployed at Test Area A in 100m water depth, and one WEC array to be deployed at Test Area B in 50m water depth. A 500m safety zone has been incorporated within the test areas to provide a safety buffer around the deployed WEC's. The size of Test Area A and B will be $3km^2$ and $1.5km^2$ respectively.

The locations and size of the offshore test areas for the wave energy test site were established based upon a number of criteria including:

- Located in vicinity of a feasible cable route
- Located on suitable seabed for mooring of devices.

- Orientated such that the frontage of the area is aligned with the primary wave resource direction
- Suitable water depths for WEC's to deploy
- Avoidance of prime sea fishing (trawling) grounds
- Located within the 12 mile territorial limit
- Located outside any designated military defence zones
- Located outside any designated environmental zones
- Navigation Risk Assessment considered
- Suitable for WEC array configurations (based upon developers current proposed arrangements)

AMETS Electrical Infrastructure

Connectivity between the offshore test areas (A and B) and the onshore electricity network is via installed submarine electricity cables. The final cable route corridor was selected to avoid areas of rocky outcrop on the seabed where possible as was the cable landing location. The cables are designed to be buried approximately one metre below the seabed in order to be adequately protected against damage. In areas where the seabed is rocky or too hard for the burial of cables, alternative protection methods will need to be used such as rock placement or matressing.

Belderra beach was identified as the most suitable cable landing location and an onshore electricity substation will be located at a nearby Greenfield site to export to the electricity network.

Consultation with stakeholders

A vital part of the design process is the consultation with local and statutory stakeholders. Consultation has taken place with statutory bodies such as the Department of Environment, the Coastguard and Commissioners for Irish Lights, with ports and harbours on the West coast of Ireland, and with local stakeholders such as local fishing organisations and surfers.

The extensive consultation process not only served to establish the requirements of other users in the area; it turned out to be a learning process for all involved. Fishermen, surfers and seafarers are now more informed about the workings and requirements of WEC's and the AMETS project team has gained invaluable knowledge about the operations of the former at sea.

Consent applications

The development of the proposed test site is subject to a number of permits and consents which include;

Offshore:

- Foreshore Licence for installation of the submarine electricity cables and deployment of AMETS infrastructure.

- Foreshore Lease for development of the Test Areas.

Onshore:

- Planning permission for the electricity substation

- Grid Connection to export to the electricity network

- Regulator Permits to develop an electricity generation project.

As part of the permitting and consenting process an Environmental Impact Assessment (EIA) has been carried out for AMETS. The EIA establishes a baseline of the environment around the proposed test site, assesses the environmental impact and proposes mitigation measures where impact on the environment cannot be avoided.

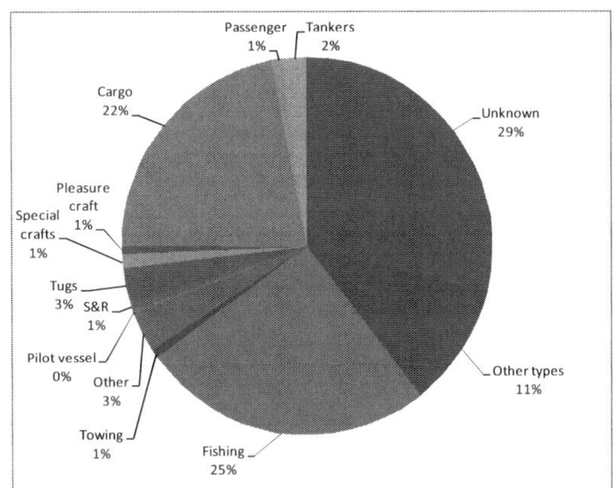

Figure 6: Distribution of vessels using AIS passing in the area of the proposed test site.

The first stage of the EIA process was the preparation of an Environmental Scoping Report which identified the anticipated impacts and outlined how it was proposed to assess these impacts. Extensive consultation with the regulatory and planning authorities, as well as non statutory bodies, was undertaken to refine the scope of the EIA. The data required for the Environmental Impact Assessment was captured and accessed via a large number of surveys and studies of the relevant receptors. Terrestrial and Marine ecological and archaeological surveys were undertaken. The ecological surveys assessed the impacts on local flora and fauna, birds, fish and mammals. The archaeological surveys assessed the potential impacts on potential sites of archaeological significance over the study area. The impact of the proposed AMETS project was further assessed on coastal processes, navigation, visual impact, traffic and noise.

In order to assess the impact of the AMETS project on shipping in the area Arup carried out a navigation risk assessment using anonymised AIS data from the Irish coastguard and data from a radar survey specially commissioned for the test site. AIS signals are transmitted by vessels with a Gross Tonnage (GT) greater than 300tons and display their position and vessel type.

The results of the EIA are incorporated into an Environmental Impact Statement (EIS) prepared by ESBI to accompany a number of the consent applications as required.

Challenges and Programme to completion

The development of this full scale wave energy test site has identified a number of challenges that need to be addressed for the successful completion and operation of the test site. Some of these challenges relate to the establishment of an acceptable test location for all sea users, other challenges relate to technical difficulties with existing electrical connection equipment. An example of the latter is the fact that currently wet-mate electrical connectors are only available up to 6.6kV, and this limits the transmission voltage from WEC's. At a higher voltage power losses during transmission would be reduced. The ocean energy industry and the project team are currently working on solutions for these challenges, which is the ultimate aim of the test site: removing bottlenecks for the further development of ocean energy.

The project team intend to submit the marine consenting application for AMETS to the Department of Environment in 2011 which will consist of a Foreshore Licence and Lease Application and accompanying Environmental Impact Statement. Following successful receipt of a foreshore licence and lease, the cable procurement will be completed. It is envisaged that the test site will be operational for first WEC deployment in 2013. The test site is expected to be in operation for 15 years.

Acknowledgements

The Atlantic Marine Energy Test Site project is being developed by a team of experts from various companies. We would like to thank all of them for their input: Paddy Kavanagh of ESBI, Hannes McNulty of SEAI, Harvey Appelbe of Tonn Energ, James Ryan of Aquavision and Andrea Padovan of Arup.

We wish to thank Eoin Sweeney of SEAI for his kind permission to publish this paper.

References
[1]   Accessible Wave Energy Resource (ESB International 2005)

[2]   An Ocean Energy Strategy for Ireland (Sustainable Energy Ireland and the Marine Institute, 2005)

[3]   Equimar project: Equitable Testing and Evaluation of Marine Energy Extraction Devices in terms of Performance, Cost and Environmental Impact - Test sites catalogue

[4]   Site Selection and Evaluation Report (ESB International and Marine Institute 2008) Mark Fielding, Pat McCullen (ESB International) and James Ryan (Marine Institute)

[5]   Environmental Scoping Report, AMETS Project (ESB International 2009)

[6]   HMRC Testing and evaluation protocol

Innovative Coastal Zone Management
ISBN 978-0-7277-5749-4

ICE Publishing: All rights reserved
doi: 10.1680/iczm.57494.113

Offshore Wind Farms vs. Coastal Erosion

Vicente Negro, Universidad Politécnica de Madrid, Madrid, Spain.
M. Dolores Esteban, Universidad Politécnica de Madrid, Madrid, Spain.
José S. López-Gutiérrez, Universidad Politécnica de Madrid, Madrid, Spain.
J. Javier Diez, Universidad Politécnica de Madrid, Madrid, Spain.

Introduction

The behaviour of maritime facilities (mainly those located close to the coast) as physical obstacles and/or dynamic singularities leads to possible variations of coastal processes and, consequently, sometimes to changes in coastal forms (mainly in the sedimentary ones) relatively near the facilities, but also in some ones sited in farer areas. It is then advisable to carry out coastal processes studies before proceeding to the construction of maritime facilities, trying to avoid problems in the future. The main objective of this type of studies has to be to evaluate the possible coastal affection and, in the case of unacceptable effects, to introduce changes in the projects or, at least, to establish complementary and/or correction measures.

Offshore wind farms have appeared as a new type of maritime facilities. Although the first offshore wind farm experiment took place 20 years ago, facilities like these have continued being constructed since then, most of them being pilot projects. On the other hand, at the beginning of the year 2011, almost 3,000 offshore wind megawatts were in operation, which confirms the current boost happening in the industry in this field, with several countries leading its development (the United Kingdom, Denmark, Holland, Sweden and Germany) (Esteban *et al*, 2011).

However, offshore wind energy can still be considered as an incipient market. As a result of this, the experience is reduced in most of the aspects related to the field, and therefore also in the case of coastal processes analysis for these facilities. In fact, it is necessary to know how the presence of an offshore wind farm can influence on coastal processes and on beach morphology. This is a key aspect due to its possible negative effects for the habitual residents and also for the tourists. Besides, it is important to understand how coastal processes have to be considered for the design of offshore wind farms. In fact, there is a symbiotic relationship between offshore wind farms and coastal processes, and this must be well understood. For that, there are still a lot to research and to advance.

This paper deals with a reflection about the state of the art and the current situation of this type of studies. The content of the paper is: 1) introduction; 2) factors to be known and understood before beginning with the main discussion of this paper; some of these factors are lay-out, foundations, territory, etc.; 3) offshore wind farms versus coastal processes, analyzing how this type of installations influence on coastal processes, and how coastal processes have to be taken into account for the design of offshore wind facilities; and 4) conclusions.

Factors to be considered

Every factor to be taken into account for the design of an offshore wind farm (Figure 1) must be considered in an integral manner. This epigraph only reviews those with a direct and important influence on the analysis "offshore wind farms versus coastal processes". The intrinsic factors, that is the own components of the installation, will be dealt with first. And the rest of factors, the extrinsic and the compound ones, will be dealt later.

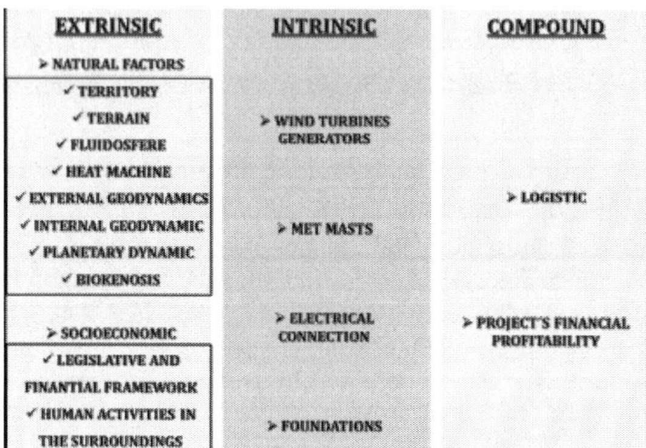

Figure 1. Factors to be considered for the design of an offshore wind farm (Esteban, 2009)

Offshore wind farm's component

Lay-out of wind turbine generators has to follow several criteria. One of these criteria is to make the most of the wind resource, placing the alignments known as rows perpendicular to the predominant wind direction, and the columns parallel to the predominant wind direction. Other criteria is regarding the separation to be respected between adjacent wind turbine generators following the manufacturer specifications; the separation must be around 4-6 times the rotor diameter between the adjacent wind turbine generators placed in the same row, and around 7-9 times the rotor diameter between adjacent wind turbines generators located in the same column.

Transformer substations and electrical cables form part of the electric connection interesting for this study because possible alterations to the sea bottom and scouring processes have to be considered for the protection of the cables.

And regarding the foundations, special attention has to be paid to the description of the different typologies, to their dimensions, to the seabed preparation, if it is necessary, and the tendency to the scouring.

The first offshore wind facilities were constructed in locations near the coast with foundations below 25 metres and with favourable geotechnical properties. Most of them were founded on gravity based structures (also called GBS) foundations and steel monopile foundations. As time passed, these installations were extending to greater depths which, added to an increasing size of wind turbine generators, led to other, more complex foundation solutions appearing, such as jackets and tripods (Figure 2).

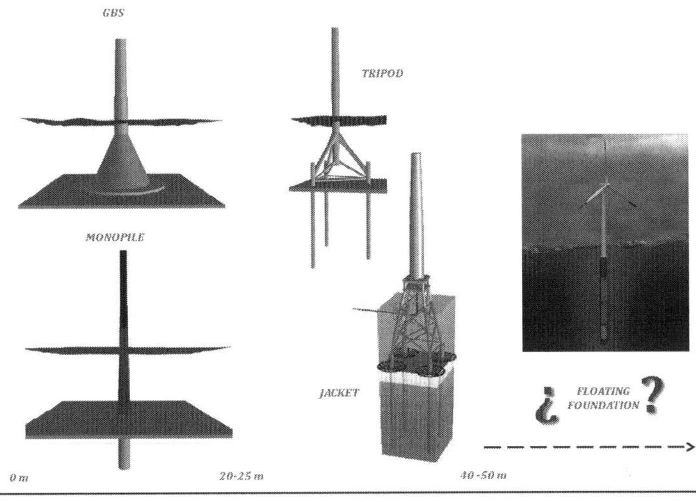

Figure 2. Basic diagram for choosing the type of foundation depending on depth (Esteban *et al*, 2009)

GBS foundations are those able to maintain stability by their own weight. The most usual ones in offshore wind farms concentrate their weight close to the base, achieving a diameter reduction close to the average sea level and, therefore, a reduction also in the hydrodynamic loads. The conical shape has been the most used one up to now, although there are other shaped GBS foundations, generally with diameters in the base around 30-40 metres; anyway, the size is variable and depends on various factors. These foundations usually need prior ground preparation with a berm and a toe bern protecting them from scour phenomena.

Steel monopiles are deep, individual foundations which, through penetrating into the ground, transmit loads to which they are subjected to the latter. Monopiles generally used in offshore wind farms are characterized by being variable in thickness along their whole length, generally in the order of 50 millimetres and about 100 millimetres in the areas carrying most load, having a diameter of about 5 metres and weighing over 600 tonnes. However, these dimensions are variable and depend on various factors. This type of foundation does not call for the ground to be prepared beforehand and possible scouring which may occur in the ground around the monopile has to be considered in its design.

Jacket and tripod foundations are formed with tubular steel inspired on offshore oil rigs. Although they have not been as tested as GBS and monopiles foundations, they will have probably a bright future. The diameters of the tubes forming part of the metal latticework are usually in the order of 1.5-2 metres and foundations directly onto the ground are usually in the form of 1 to 2 metre diameter piers. The dimensions as given may vary depending on several factors. Tripods have a centre column below the wind turbine generator's tower, connected by a tubular structure to three lower legs. And the jackets used in offshore wind farms consist of four legs whilst. Due to the small diameter of their tubes, neither type of foundation is usually exposed to wave and current actions. In general, they do not require ground preparation and point scouring around each of the piers and also that of the overall structure piers must be borne in mind for their design.

Nowadays, it is accepted a technical-economic barrier of around 40-50 metres for foundations of offshore wind farms. Anyway, there are several research projects ongoing with the purpose of finding how to overcome this barrier. Such projects are focused either on the development of new types of direct foundations more effective than those existing or on the development of floating supports.

Direct foundations in a research phase are usually hybrid foundations, which are the result of the various combinations of the types of direct foundations exposed above. Hybrid foundations embrace multiple options since foundations may be completely different and, therefore, so may their impact on littoral dynamics.

Floating supports are not yet technically and economically viable alternatives and the main structural philosophies used up to now have been inherited from the offshore gas and oil industries. Nevertheless, there are certain aspects which have to be specifically considered for offshore wind farms, such as the cost and possible modifications to be undertaken in the wind turbine generator control system to achieve proper operation. Although there are various types of floating supports, they are usually more transparent to wave and current actions than direct foundations. Added to the fact that they are usually further from the coast because of the requirements of greater depths, this would suggest less impact on littoral dynamics, but in any case this have to be checked (Esteban *et al*, 2011).

Other factors
Territory refers to the physical and socioeconomic environment in its entirety and involves a view of the overall surroundings; and also the area's general characteristics, the topography and bathymetry, the geographical location, the environmental aspects, etc.

Bathymetry influences the choice of the type of foundations and its geotechnical properties has an influence on the design of structures, which are aspects influencing the study of littoral dynamics. Also, the properties of the terrain's surface layer including amongst them the possible mobility of the sea floor with the passage of time (see dunes, ripples, etc.), will enable scouring to be expected both at the foundation location and in the surroundings of the undersea electric cable, to be foreseen.

The behaviour of the Earth fluids as a thermal machine gives rise to wind, waves, currents and the meteorological tide. Littoral processes are the result of a combination of some of the above mentioned phenomena with the area's material. And it is the alteration of such phenomena which has to be studied both in the surroundings of foundations and electric cables and also close to the coast.

External geodynamic refers to slowly evolving alterations to the earth's crust occurring in its interface with the contacts of the atmosphere and hydrosphere. This arises from the interaction between the thermal machine and the coastal strip and the continental shelf where offshore wind farms are set up. The thermal machine's operation generates climate forces the effect of which is littoral fluid-dynamics varying a lot in time, which contrasts with the coast's slow morphodynamic response to these fluid-dynamics. The tendency is therefore to establish dynamic equilibrium shaping the landscape. Erosion of the surrounding continental area affected by littorals dynamics, transport of eroded material and sedimentation of the material transported are the fundamental processes forming external geodynamics. The effectiveness of these processes with respect to coastal modelling depends on the intensity and persistence of the different causing agents, as well as the properties of the materials undergoing same. This is therefore one of the key aspects in studying littoral dynamics.

Special attention has to be paid within human activities to human settlements since most coastal towns are very attractive to tourism with beaches being one of the main arguments, though not the only one. Potential users basically value the actual quality of the beaches and the environmental quality of the surroundings. The quality of the beach in itself is linked on the one hand to the users' perception of the actual space available and this is a key point in analysing littoral dynamics and their characteristics and, on the other, to the quality of the services the beach offers. Amongst other aspects, the dry beach width and the area enjoyed by each user clearly influence how users perceive the quality of a certain beach. A change in littoral dynamics could lead to the width of beaches reducing and even their disappearance. This type of analysis not only has to take a specific offshore wind farm into account, but it has to be studied together with other maritime facilities built in the surroundings that may affect littoral processes, such as harbours, other offshore renewable energy using facilities, even other offshore wind farms, etc.

The fact that a prior study of possible impacts on littoral dynamics has been carried out could aid this type of study for selecting sites suitable for offshore wind development included in the regulatory framework, whilst discarding those more problematic sites beforehand.

The logistic project is of interest for littoral dynamics studies because of the possible constructions associated to the building and operating phases to be undertaken, mostly harbour facilities that could influence littoral processes (Esteban *et al*, 2011).

Offshore wind farms vs. Coastal processes

This epigraph has been divided in two different but related parts. The first one exposes how an offshore wind farm can influence on coastal processes. Figure 3 includes the most important ones, differentiating between near and far field effects. And the second part explains how coastal processes have to be taken into account for the design of offshore wind facilities.

Offshore wind farm´s influence on coastal processes

To know if the presence of a specific offshore wind farm will produce some impact on littoral dynamic, firstly it is necessary to analyze the behaviour of foundations being considered them as individual obstacles to the wave. As consequence of this interaction, the following phenomena may occur: 1) wave reflection, depending of the obstacle´s geometry, porosity and roughness; 2) diffraction (in case of being under such regime) (Figure 4a), and 3) energy dissipation due to, for example, turbulence, wave breaking and vortices formation.

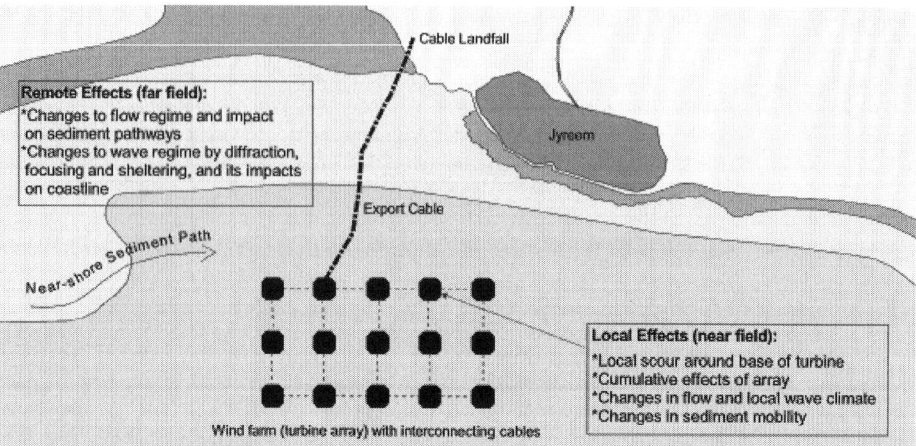

Figure 3. Sketch with the different processes covered in the paper (ABP Marine Environmental Research Ltd., 2002)

According to the bathymetry, the oceanographic variables and the dimensions of the foundations most used up to now in these facilities (monopiles but also in the case of some gravity based structures), it may be claimed that foundations as an isolated element work under Morison's and not a diffraction regime (Figure 4b). Isolated foundations are in general small sized in relation to the wave lengths of waves. If is added to this that new wind farms generally will tend to go deeper, this criterion could be ratified. Nevertheless, it cannot be claimed that they are not in a diffraction regime with shallow depths. This aspect has to be studied in each specific case.

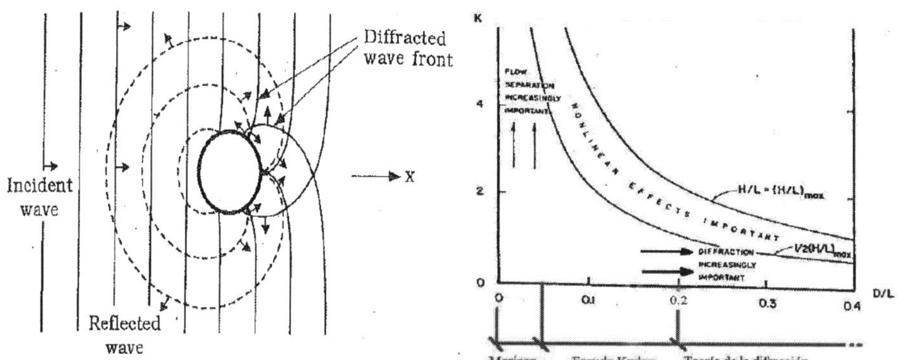

Figure 4.a) Different phenomena around a circular pile (Sumer and Fredsøe, 2002) (left); b) Ranges of use of the different theories (Morison, Froude-Krylov and diffraction) (Isaacson, 1979) (right)

Anyway, it is essential to take into account the effect of the complete offshore wind installation, formed by numerous obstacles. These facilities can provoke modifications in the littoral dynamics existing before the construction of the offshore wind facility, and as a result, changes in beaches close to the location due to variations in wave energy and currents reaching these beaches. This can produce a loss of energy compared to that upstream of the facility. This loss is small when the case of a single obstacle is examined but cannot be ignored when considering the whole offshore wind farm and will increase as the number of rows and columns of wind turbine generators does (Esteban *et al*, 2011). Figure 5 exposes the wave effects of the presence of the offshore wind generators depending on the separation between adjacent turbines, for the same foundation type.

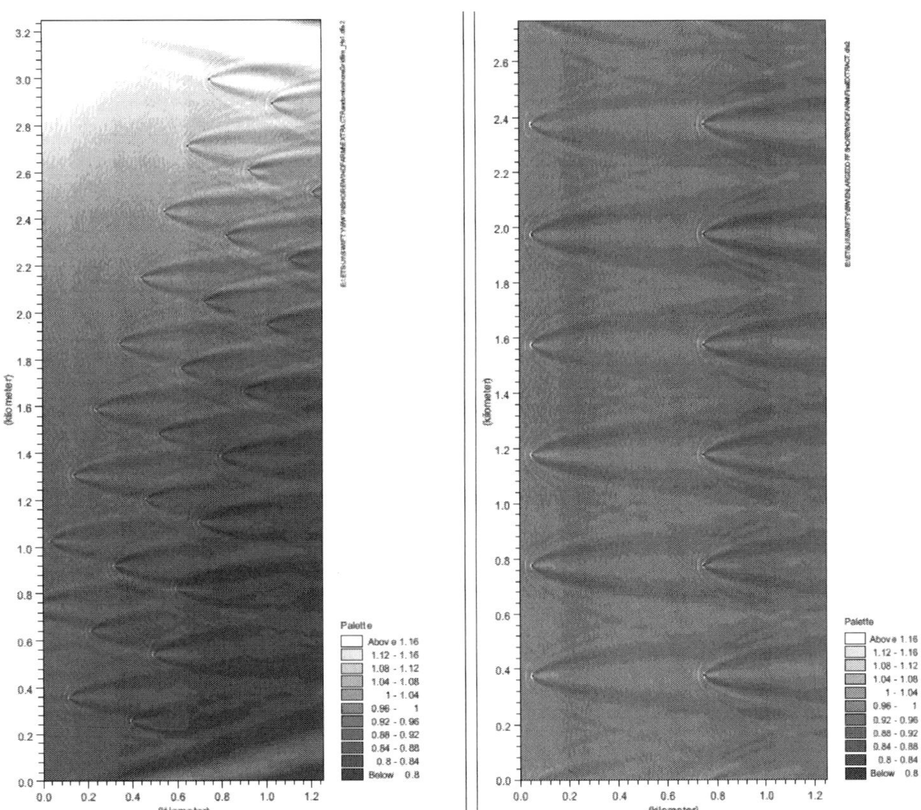

Figure 5. Predicted wave heights for several scenarios. Incoming significant wave height is 1 m and the peak period is 6,5 s (ABP Marine Environmental Research Ltd., 2002)

Considerations will also have to be given to whether the perturbation or interference each element causes in wave transformation processes is re-established before interference with the next structure because of the wind caused wave generator effect, which is a function fundamentally of wind properties and distances between wind turbine generators. There is much still to research in this respect, and the use of physical and numerical models will be highly useful.

Alterations in both the profile and plan of beaches located in the surroundings will have to be analyzed due to variations in wave energy and, therefore, in currents associated to the waves' breaking due to the presence of a new facility (figure 6).

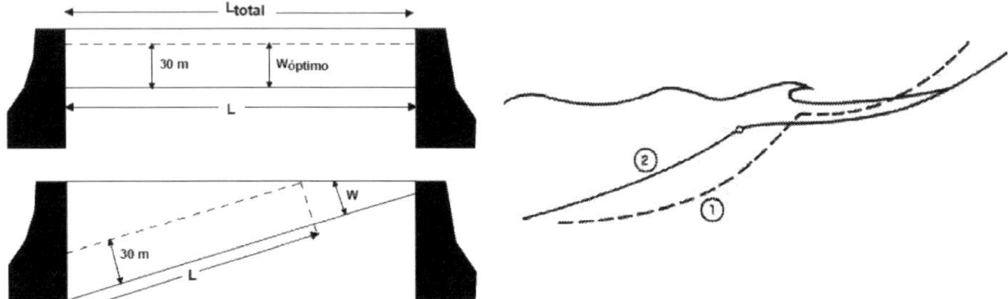

Figure 6. Possible variation of beach plan (left) (Jiménez, 2009) and beach profile (right) (Diez, 2004) depending of the littoral dynamic modification

The effect of an offshore wind farm on coastal processes must not be scorned a priori. It is necessary to analyze specifically each offshore wind farm, at least until some conclusions from the experience can be adopted. It can be stated that coastal erosion mainly depends on dimensions of the foundations, size of the installation (number of rows and columns of wind turbines generators), lay-out of the installation, marine climate (wave regimes), distance between the installation and the coast, coastal forms characteristics, and real sediment transport in the area.

Coastal processes´ influence on offshore wind farm design

Foundations act like obstacles modifying the water flux associated to wave and currents. This provokes an increase of stresses in the seabed, which can be scoured according the characteristics of the soil. The methodology for the estimation of scour dimensions is different depending on the type of foundation (slender piles – tripods, jackets and most of monopiles –, or not slender piles – GBS and some monopiles –). Two different theories can be adopted: to design the foundation with and without scour protection. In the figure 7, in the left picture, a scour phenomenon can be appreciated around the monopile foundation, and the right picture shows a basic scheme of the scour protection designed.

Electrical cables can be damaged due to different causes (fishing, anchoring, and dredging and so on) (Figure 8a). To avoid this and the consequential electrical production losses, it is necessary to protect the cables, either putting over the cables some protection structures (bags, concrete mattresses, etc.) (Figure 8b) or burying the cables under a suitable depth (Figure 8c).

 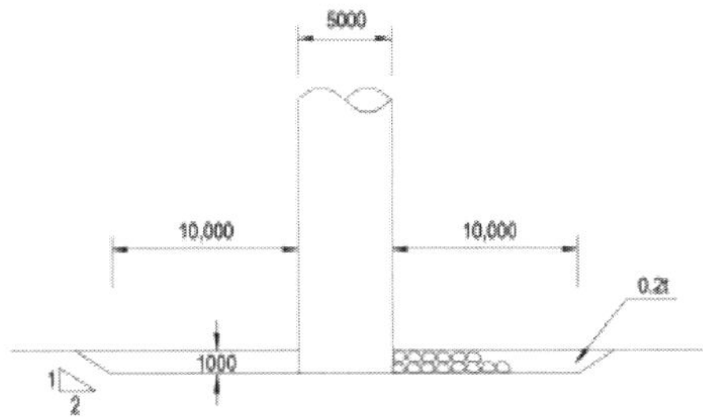

Figure 7. Scour hole around a monopile foundation (left) and scour protection design for a monopile foundation (right) (HR Wallingford *et al*, 2008)

Figure 8. a) Electrical cable damaged; b) Concrete mattress; c) Cable burial plough (Royal Haskoning and BOMEL Ltd, 2008)

Possible sea bottom movements have also to be taken into account because they could uncover part of the foundations. And regarding this, another phenomenon to be considered is the beach profile change due to storm and calm weather periods. These variations involve sandy material from the emerged part (berm-dune) moving to the submerged part (bar) and *vice versa* depending on the incident wave energy and may scour the foundations through loss of material or cause them to silt up, both of which are seasonally varying processes.

Conclusions

Offshore wind energy can be still considered as an incipient market, and therefore the improvement knowledge margin in this technology is still huge in several aspects, being one of them the relationship between offshore wind farms and coastal processes. Regarding this, it is essential to know perfectly the environment characteristics (wave and currents in the area before and after the construction of the installation, soil, coastal morphology, etc.) and the main characteristics of the offshore wind farm (lay-out, type and dimensions of the foundations, etc.) to evaluate the possible coastal erosion due to these facilities and also the influence of coastal processes in the design of the installations..

Firstly, it is important to analyze how an offshore wind farm may affect littoral processes since a consequence may be the reduction of the beach width and even its disappearance; this could have negative effects on coastal towns, because beaches are one of their main attractions. This can have an important negative effect both for tourism and for habitual residents. Therefore, this should be a critical point to be taken into consideration when designing an offshore wind farm, for example in the following factors to decide: the distance to the coast, the number of wind turbine generators, the lay-out of the facility, and the choice of the foundation type. In fact, the effect on coastal processes must not be scorned *a priori*, being necessary to analyze specifically each offshore wind farm, at least until some conclusions from the experience can be adopted.

On the other hand, coastal processes (possible sea bottom movements and scouring around foundations and marine electric cables) have to be considered for the design of the offshore wind farm. For example, the scour happened around foundations influences on their stability; the same effect occurs when sea bottom suffers a downwards movement; and the scour in the surroundings of the marine cables can negatively affects their electrical properties, leading to damages in some case and, as consequence, to electrical production losses.

References

ABP Marine Environmental Research Ltd. (2002). *Potential Effects of Offshore Wind Developments on Coastal Processes*. Technical report, 127 pp.

Diez, J.J (2004). *Dictamen sobre la recuperación de la playa de Nizuc-Cancún y sobre los estudios y documentos consultados al respecto*. Technical report, 75 p. [in Spanish].

DNV (2007). *Offshore Standard DNV-OS-J101. Design of Offshore Wind Turbine Structures*. Technical Reports, 142 pp.

Esteban, M.D.; Diez, J.J.; López, J.S. and Negro, V. (2009). Integral Management applied to Offshore Wind Farms. *Journal of Coastal Research*, SI56, 1204-1208.

Esteban, M.D. (2009). *Propuesta Metodológica para la Implantación de Parques Eólicos Offshore*. Ph.D. Report, Madrid, Spain, 2009 [in Spanish].

Esteban, M.D.; Diez, J.J.; López, J.S. and Negro, V. (2011). Why Offshore Wind Energy? *Renewable Energy*, 36 (2011), 444-450.

Esteban, M.D.; López, J.S.; Diez, J.J. and Negro, V. (2011). Offshore Wind Farms: Foundations and Influence on the Littoral Processes. *Journal of Coastal Research* SI64, 656-660.

HR Wallingford, ABP Marine Environment Research Ltd and the Centre for Environment, Fisheries and Aquaculture Science - CEFAS (2008). *Dynamics of scour pits and scour protection – Synthesis report and recommendations*. Technical Reports, 98 pp.

Isaacson, M. (1979). *Wave induced forces in the diffraction regime*. Mechanics of Wave-Induced Forces on Cylinders. Ed.: T.J: Shaw, Pitman, san Francisco.

Jiménez, J.A. (2009). *Función Recreativa & Morfodinámica*. Máster en Ciencias del Mar. Planeamiento y Gestión del Sistema Costero, 24 pp. [in Spanish].

Royal Haskoning and BOMEL Ltd (2008). *Review of Cabling Techniques and Environmental Effects Applicable to the Offshore Wind Farm Industry*. Technical report, 164 pp.

Sumer, B.M. and Fredsøe (2002). *The Mechanics of Scour in the Marine Environment*. Advanced Series on Ocean Engineering – Volume 17. World Scientific. Singapore.

SECTION 3:
SOCIAL, ENVIRONMENTAL AND CLIMATIC CHANGE

Innovative Coastal Zone Management
ISBN 978-0-7277-5749-4

ICE Publishing: All rights reserved
doi: 10.1680/iczm.57494.125

Coastal Flood Risk – Belfast Case Study

David Porter, Rivers Agency, Belfast, Northern Ireland

Introduction

Rivers Agency is an executive agency within the Department of Agriculture and Rural Development (DARD) and under the Drainage (Northern Ireland) Order 1973 is responsible for flood management from rivers and the sea. The Department is also the Competent Authority for the implementation of the Floods Directive (2007/60/EC) with the day to day delivery by Rivers Agency. In order to provide the public with information on the risk from rivers and the sea, a strategic flood map was developed. This map was launched in November 2008 and it has full coverage of Northern Ireland showing a combined outline of both river and coastal flooding. The map was developed using an approach which enabled this full coverage to be achieved at a reasonable cost so it is not based on close contour surveys or elaborate modelling techniques. This started the process of public flood risk communication which has subsequently been developed as the Flood Directive has been implemented. This directive also requires the identification of significant risk areas (article 5), mapping of flood hazards and risks (article 6) and the development of Flood Risk Management Plans (article 7). This paper will outline the approach taken to produce the strategic flood map, provide a land use policy context and will then compare the current map outputs with those to be produced to satisfy the Floods Directive.

The coastal element of the Strategic Flood map

The methodology for the development of the Northern Ireland coastal inundation map was similar to the strategy used by the Republic of Ireland for the Irish coast. This approach took into consideration both the astronomical tide variation and the surge effect caused by atmospheric pressure, wind and Atlantic Ocean surges. Rivers Agency commissioned one of its framework consultants, RPS, to deliver the model. RPS had previous experience of this work as it had developed a Mike 21 model of the Irish Coastline. Mike 21 is a flexible mesh model which by increasing the mesh density around the coast provides very detailed information on the predicted sea level and therefore its impact on land.

The model extends up to 600km from the Irish coast into the Atlantic to the north and includes an outline of the west coast of Britain.

The model was populated with data derived from historical storm surges recorded in the National Tidal and Sea Level Facility (NTSLF) by the British Oceanographic Data Centre. The tidal boundary conditions were derived from the Kort and Matrikelstyrelsen (KMS) of Denmark, global tidal model. Data was also obtained from Global Sea level Observing System (GLOSS) and the Permanent Service for Mean Sea Level; both from the British Oceanographic Data Centre. Other data held by Rivers Agency was also used including tidal records for the Quoile Barrier near Downpatrick. The meteorological influences were represented in the model using data extracted from the European Centre of Medium Range Weather Forecasts (ECMWF).

The data outputs of the model were then subjected to a rigorous validation including a statistical analysis of the extreme water levels, as it is this output which was subsequently used to generate 48 high water level values at key reference points along the coast. The high water level values were used 'to create a Triangulated Irregular Network (TIN) which is effectively a virtual flood water surface for that area' (Rivers Agency, 2008). The flood water output was then placed over a digital ground model, which was based on the Ortho-DTM with LiDAR data inputs where available, to produce the Strategic Flood Map. In essence the map outputs show the interaction of a coarse ground model with the horizontal projection of a predicted high water level for a 200 year (0.5% AEP) coastal event.

The flood outline accuracy is dependent on the combined accuracy of both the DTM and the water levels and is therefore no more than +/-150mm with 95% confidence interval. The development of the map, its outputs and the resulting accuracy is contained within the 'National Flood Map for Northern Ireland – Methodology report' (Rivers Agency, 2008). The Belfast output from the Strategic Flood Map is shown in Figure 1.

This methodology produces a very precautionary indication of the flood outline but it was considered to be a suitable approach to provide the first generation of the flood map for the entire coastline of Northern Ireland at a reasonable cost. It also enabled a climate change map to be produced looking forward to 2030. This timescale was chosen in recognition that the primary use of the maps is land use planning and this was considered a suitable timescale for development planning taking into account the uncertainty associated with climate science.

Use of the Map

The publicly available Strategic Flood Map shows a combined flood outline for a 1 in 100 year (1% AEP) fluvial (river) event and a 1 in 200 year (0.5% AEP) coastal event. This map provides the public with valuable information which can be used to identify risk areas. This information enables informed land use decisions to be taken, emergency plans to be developed and allows individual citizens to understand the risk posed by flooding.

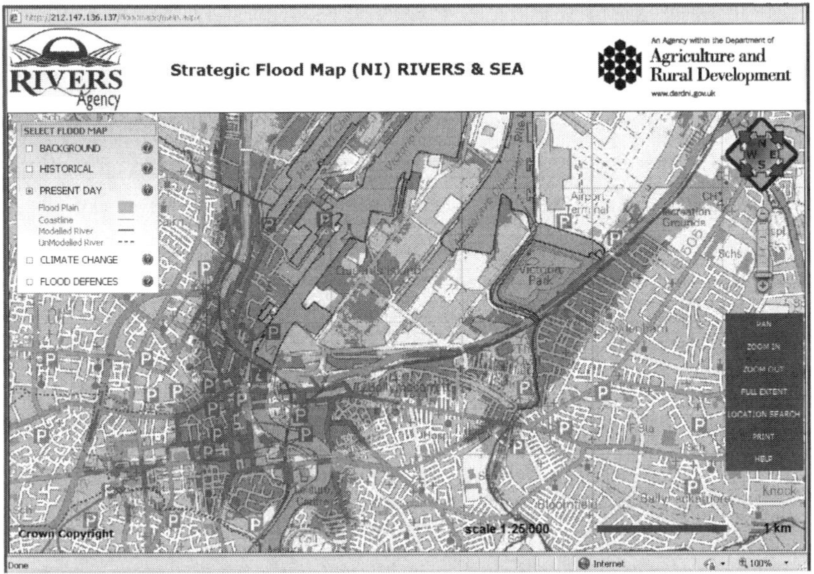

Figure 1 Strategic Flood Map- present day flood outline for Belfast

The Flood Risk of Belfast

The Strategic Flood Map clearly shows that Belfast is at risk from a 1 in 200 year coastal event. The indicated flood extent engulfs the city centre, the Duncrue industrial estate area and numerous residential dwellings in the east of the city. The tide also influences the discharge of a number of significant rivers including the Lagan and Connswater. The combined coastal and river flood outline as indicated by the strategic flood map, impacts on approximately 10,500 properties in the Greater Belfast area.

Planning Policy Position

A key element of flood risk management is land use planning as it makes more sense for flood risk to proactively influence development decisions than to simply react to flood events and construct defences. This approach will become even more critical if the predictions for climate change are realised, as flood risk will increase if sea levels rise, storminess increases and rainfall events become more intensive. In order to address these issues and to have a policy context for the strategic flood map, Rivers Agency worked with the Department of the Environment, Planning Service to develop 'Planning Policy Statement 15: Planning and Flood Risk' (DOE, 2006) or as it is more commonly known, PPS15. There are four policy areas:

FLD1 Development in Flood Plains: which sets out the key importance of restricting development to those areas that are not at risk from flooding unless it satisfies one of the exceptions or the proposal can be considered to be of 'overriding regional importance'.

FLD2 Protection of Existing Flood Defences: which ensures the operational effectiveness of existing defences and their longer term maintenance.

FLD3 Development beyond Flood Plains: is similar in nature to FLD1 but rather than focus on river and coastal flood plains, restricts development in any area with a known flood history; and

FLD4: Flooding and Land Drainage: which restricts the culverting and canalisation of watercourses except in a limited number of exceptional circumstances.

This policy statement enabled the flood maps to be used in both the development of Area Plans, which are strategic level land use documents, and in determining individual planning applications. The conservative approach of the flood map development, particularly that of projecting the high tide level onto the coarse digital ground model as used to determine the coastal flood plain, can however be problematic. This is because the over estimation of the risk that is embedded in the mapping approach, restricts development, as the planning policy requires flood plain avoidance except in a defined set of exceptions. In addition PPS15 in adopting a precautionary approach does not allow infilling of flood plains to enable development. This is a robust policy for river flood plains, as the reduction of the area available for flood water storage will have either a direct impact on water levels, or the impacts will be cumulative over time which may eventually cause a problem on another part of the river system. However the coastal flood plain situation is different and infilling to raise a potential development site provides enhanced protection and will not generally result in increased risk elsewhere as the rise in sea level will be immeasurable. The latter shortcoming is currently being addressed in the review of PPS15 while the accuracy of the coastal flood plain as shown in the strategic flood map will be improved by the development of more detailed hazard and risk maps as required by article 6 of the European Floods Directive (2007/60/EC).

The Floods Directive

In response to over 100 major floods between 1998 and 2004 the European Commission decided to introduce a directive on 'the assessment and management of flood risk' (2007/60/EC). During this period 'floods in Europe have caused some 700 deaths, the displacement of about half a million people and at least £25 billion in insured economic losses' (Europa.eu, 2011) This directive is commonly referred to as the Floods Directive and it is to be implemented in four key stages:

- Transposition into national legislation (Article 17)
- Preliminary flood risk assessment (Article 4)
- The production of flood hazard and risk maps (Article 6)
- Flood Risk Management Plans (Article 7)

For the purposes of this paper it is only necessary to consider the third stage, that of the development of flood hazard and risk maps. The directive requires that flood hazard maps include three return period scenarios; low probability or extreme events; medium probability of about a 100 year event; and high probability events where appropriate. In article 6(6) the directive recognises that where a coastal flood protection structure provides an adequate level of protection, then only the extreme event scenario needs to be mapped. The hazard maps must also indicate extent, water depth or level, as appropriate and velocity. The risk maps are to show the number of inhabitants at risk; the economic activity in the risk area; pollution sources as determined by Directive 96/61/EC and any other information considered necessary by the Member State.

The Belfast Study

In order to progress the implementation of the Floods Directive, the Rivers Agency commissioned a pilot study of Belfast. The outputs from this study were to be used to inform the Flood Risk Management Plan for this area as well as setting the template for the other plan areas. The Belfast area was chosen as it was clearly a significant risk area given that it is the capital city with a large number of both residential and commercial properties at risk from different flood sources such as rivers, the coast, surface water and impoundments or reservoirs.

Part of the commission required the production of a detailed tidal model. This was a 1-dimensional/2-dimensional hydraulic model which included both the Lagan and Connswater rivers and simulated '6 present day (2009) tidal surge events, the Q10, Q50, Q75, Q100, Q200 and Q1000 together with their respective climate change scenarios' (AECOM, 2010). For clarification, Q200 is a 1 in 200 year event (sometimes shortened to a 200 year event), which can also be expressed as having a 0.5% AEP (Annual Exceedance Probability) or indeed a 200 to 1 chance of occurring or being exceeded in any year.

The detailed hydraulic model better represents the tidal cycle as opposed to simply extending a high tide level inland as in the Strategic Flood Map. It therefore better replicates the likely inundation mechanism and therefore enables a better flood outline to be produced. The hydraulic model is also more detailed as it includes an accurate representation of the river channel, bathymetry and the interface between the tidal cycle and the river flows. This more detail approach increases the confidence level in the outputs which subsequently enables a robust policy position to flood risk to be adopted. The Strategic Flood Map was developed to start the communication process to the public and it was decided that it would indicate the flood extent for a 200 year event. The high confidence levels in the detailed model allow a better understanding of how a flood event starts and develops. This is very important as the water level at which damaging flooding occurs, may be achieved in events much lower than a 200 year event. Flood events progress as a tide develops over time and they do not simply arrive with full force on the coastline, except in the case of a tsunami, so all this enhanced understanding will be invaluable when drafting the Flood Risk Management Plan.

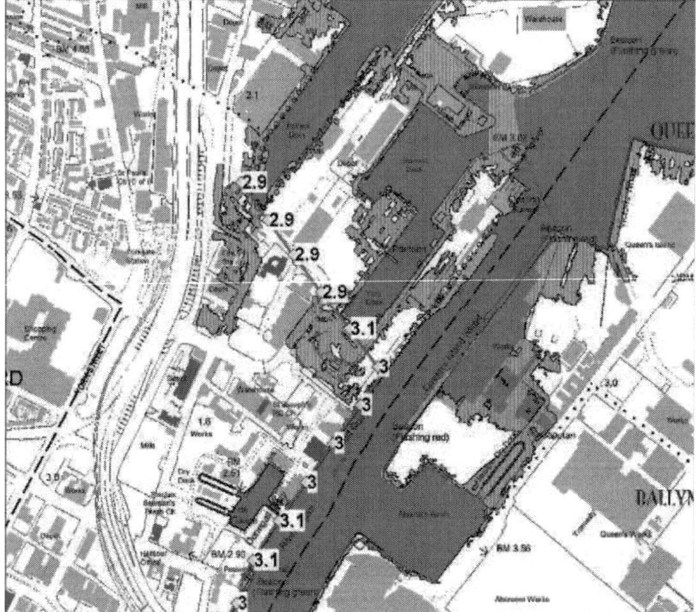

Figure 2 Simulated 2009 Q50 Tidal Inundation Extent with surveyed ground levels (Crown Copyright).

The level of the quay walls in Belfast harbour and the topography of the River Lagan offer protection to the city centre. The model shows that the Q50 event is the point where exceedance of the quay walls occurs with flooding to the industrial areas surrounding the Pollock and York docks (see figure 2). The Q100 and Q200 events show increasing areas of the city being inundated but not to the extent shown in the Strategic Flood Map which was also based on a Q200 event. Even the hydraulic model for the Q1000 tidal inundation does not impact as much of the city as that shown on the Strategic Flood Map, demonstrating the precautionary approach adopted. (See figure 3)

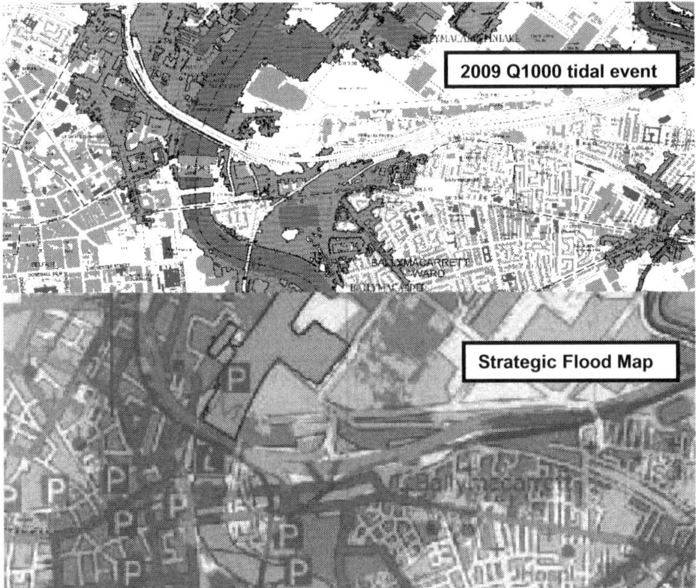

Figure 3 Comparison of Q1000 tidal event from the detailed model with the Q200 Strategic Flood Map (Crown Copyright)

Validation of Coastal Events

The most recent major coastal inundation event of Belfast dates back to early 1900s and the photographic records show flooding in the city centre outside the City Hall (see Figure 4). The information on the event however is very sketchy and the city has developed almost out of all recognition since that time so it is not considered to be suitable for validation. Both the Strategic Flood Map and the Floods Directive maps will suffer from lack of validation data until such times as a real life event occurs!

Lagan Weir

No coastal flood risk assessment of Belfast would be complete without consideration of the Lagan Weir. This structure was constructed in 1994 at a cost of £14 million by the Laganside Corporation and although it looks like a coastal barrage it is actually an impoundment structure. It is designed to control the water levels upstream of the structure to ensure the mud flats are covered. This minimises the decomposition of the mud and therefore reduces the associated odour. This was necessary to encourage waterside redevelopment in this area which today offers office, commercial, recreational, housing and entertainment facilities. 'Additional Measures were also undertaken to improve the water quality, with a programme of river dredging and the installation of an aeration system' (laganside.com, 2011)

Figure 4 Belfast flooding circa 1902

The detailed hydraulic model however indicates that this structure can also perform as a tidal barrage to a limited extent. During a Q50 present day tidal surge the predicted water level is 2.93m OD which is lower than the 3.0m OD top gate level of the weir when it is operational. This means that provided the 'river flows are less than $50m^3s^{-1}$ then the Lagan Weir acts to stop the ingress of the tidal surge and reduce the tidal surge flood risk upstream of the barrier' (AECOM, 2010)

Conclusion

The Strategic Flood Map shows that it is technically possible to provide a coastal inundation outline for a defined scenario at reasonable cost. The approach taken during the development of the map, however, needs to be clearly understood as some of the assumptions will have a direct impact on the accuracy of the outputs. There also needs to be a clear policy position if the flood map is to be a material consideration in the land use planning process.

The EU Floods Directive provides the opportunity for very detailed hydraulic models to be produced for significant risk areas. These will better inform land use decisions, emergency plans and citizens' readiness to react to a flood event. It is important to carefully explain the differences in the mapping process, as the change from the strategic to the detailed looks as if the coastal flood risk to Belfast has reduced. This is incorrect, the risk has in reality not changed but the likely inundation scenario has been better represented by the advances in hydraulic modelling and so the risk can be better understood. The detailed model also provides information on the development of a coastal event and, in the event of a live coastal inundation this could prove invaluable.

References

AECOM, 2010 'Belfast Tidal Modelling – Hydraulic Modelling Technical Report' produced for Rivers Agency

Drainage (Northern Ireland) Order 1973

Council Directive 96/61/EC of 24 September 1996 concerning integrated pollution prevention and control, available at http://eur-lex.europa.eu/LexUriServ/LexUriServ.do?uri=CELEX:31996L0061:en:HTML

Directive 2007/60/EC of the European Parliament and of the Council of 23 October 2007 on the assessment and management of flood risks, available at http://eur-lex.europa.eu/LexUriServ/LexUriServ.do?uri=OJ:L:2007:288:0027:0034:EN:PDF

DoE, 2006, 'Planning Policy Statement 15: Planning and Flood Risk'

Europa.eu, 2011 'A New Floods Directive' available at http://ec.europa.eu/environment/water/flood_risk/index.htm

Laganside.com, 2011, 'Lagan Weir – Why it exists' available at http://www.laganside.com/laganweir.asp

Rivers Agency, 2008, 'National Flood Map for Northern Ireland – Methodology Report' available at http://www.dardni.gov.uk/riversagency/floodmapmethodologyreport.pdf

Innovative Coastal Zone Management
ISBN 978-0-7277-5749-4

ICE Publishing: All rights reserved
doi: 10.1680/iczm.57494.134

Sustainable Development at the Coast: the Lincolnshire Coastal Study

Nikki van Dijk, Atkins Ltd, Peterborough, United Kingdom
Geoff Darch, Atkins Ltd, Peterborough, United Kingdom
Richard Belfield, Lincolnshire County Council, Lincoln, United Kingdom
Alan Freeman, Lincolnshire County Council, Lincoln, United Kingdom
David Hickman, Lincolnshire County Council, Lincoln, United Kingdom

Introduction

Coastal areas are subject to competing and sometimes conflicting land use claims: they are increasingly seen as attractive areas to live, work, retire and recreate as well as being important to the economy and natural environment. However coastal areas are subject to environmental and climatic change, particularly sea level rise and an increase in coastal erosion. The dynamic nature of coastal areas presents decision makers with particular challenges for sustainable development.

Along the Lincolnshire Coast this challenge was highlighted during the review of the East Midlands Regional Spatial Strategy (GOEM 2005) by the Planning Inspectorate in 2007. As a result of the conflict between housing allocation plans and an increasing risk of coastal flooding due to sea level rise, the Planning Inspector placed a moratorium on all new development in three coastal districts of Lincolnshire until a coastal strategy was agreed (Planning Inspectorate 2007). The Lincolnshire Coastal Study Group, consisting of local authorities, the regional planning and development agencies, the Environment Agency, Natural England and local Internal Drainage Boards, was set up to oversee the development of the Strategy. The Lincolnshire Coastal Study was commissioned to inform the development of a Coastal Strategy which would deliver sustainable spatial development over the 21st century.

Through the use of flood hazard maps, socio-economic scenarios and a programme of stakeholder engagement, the Lincolnshire Coastal Study developed a set of Principles for sustainable spatial development in the Study area. The Principles take account of flood risk, local housing requirements, the need for economic regeneration and the natural environment. This paper describes the approaches used in the Lincolnshire Coastal Study to develop an agreed set of Principles for sustainable spatial development.

Methods

The Lincolnshire Coastal Study took a scenario and stakeholder based approach to examining future changes, and to develop Principles for sustainable spatial development. Flood hazard and socio-economic scenarios were developed for the Study Area and presented to a range of technical and non-technical stakeholders whose views were used to develop the Principles.

Flood hazard scenarios

Flood hazard scenarios were developed by mapping the output of Environment Agency flood model data for the Study Area. Flood hazard maps illustrate not only the extent of projected flood events but indicate the degree of hazard associated with the depth, velocity and debris carrying capacity of flood waters. The maps illustrate how flood waters could behave in relation to current defences and the land behind them. The hazard zones shown on the maps produced for the Lincolnshire Coastal Study (red, orange, yellow and green) are based on the Defra and Environment Agency (2008) flood hazard to people classification, see Table 1. The 'little or no hazard' rating is not used in the Defra and Environment Agency classification but was added to illustrate locations in the Study Area which are outside the main hazard zones (Atkins 2010a).

Degree of hazard	Hazard rating	Colour on map	Description of flood water	Description of hazard
None	Little or no hazard	White	Outside of modeled flood extent	Little or no hazard from coastal flooding
Low	Low hazard	Green	Shallow flowing or deep standing water	Caution, low hazard to people
Moderate	Danger to some	Yellow	Fast flowing or deep standing water	Hazard to the vulnerable e.g. children, the elderly and the infirm
Significant	Danger for most	Orange	Fast flowing and deep water with some debris	Hazard to most, including the general public
Extreme	Danger to all	Red	Fast flowing deep water with significant debris	Extreme hazard, danger to all including the emergency services

Table 1 Hazard ratings used on the flood hazard maps in the Lincolnshire Coastal Study (based on Defra and Environment Agency (2008) flood hazard to people classification)

A series of maps were produced which depict the degree of hazard associated with a combination of different types and magnitudes of tidal flood events now and in the future (including sea level rise).

Two different tidal flood mechanisms were modelled: overtopping and breaching. Overtopping occurs when sea defences remain intact but flood water washes over the top. Breaching occurs when sea defences fail and water flows through the gap. The Environment Agency modelled a number of individual breaches at regular intervals along the coast. The width of the breach is dependant on the type and location of defence and the model assumes that 72 hours will lapse before the breach can be sealed. In the absence of data regarding the probability of breach failure and given the seriousness of the consequence of a breach occurring, the precautionary principle was invoked. The consequence of a series of individual breach events along the Lincolnshire coast was combined to form a 'worst case' breach hazard map.

In order to consider the hazard associated with flood events of different magnitudes, maps were prepared based on flood events with a return period of 1 in 200 years and 1 in 1000 years. Flood events were modeled for the present day (2006) and a future scenario (2115).

The future scenario included an allowance for relative sea level rise of 1.13m, based on Defra guidance (Defra 2006). The results of the 2115 scenario were sensitivity tested using lower and higher values of sea level rise (0.55m and 1.60m respectively).

The result of the mapping exercise carried out as part of the Lincolnshire Coastal Study is a series of flood hazard maps which illustrate how tidal flood waters could behave under different scenarios. Figure 1 shows two examples of flood hazard maps.

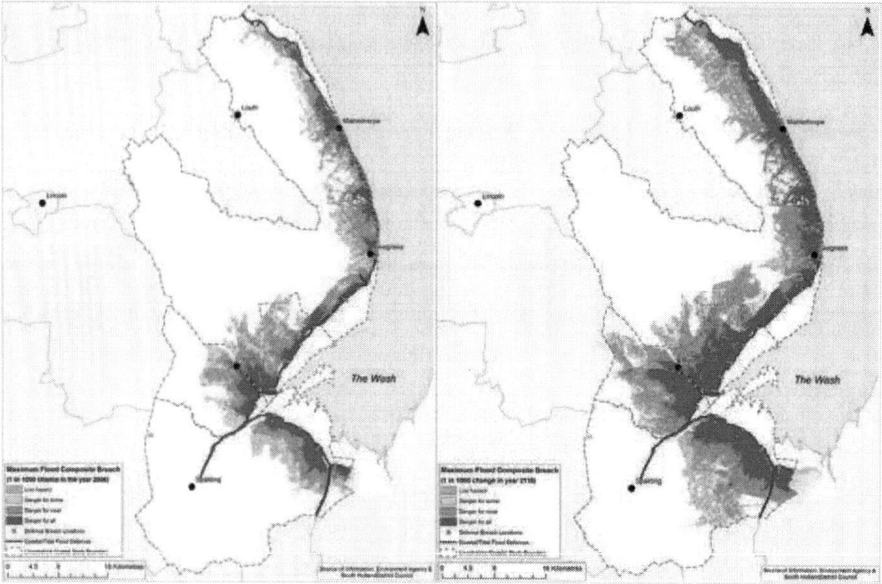

Figure 1 Examples of flood hazard maps produced as part of the Lincolnshire Coastal Study. Maps show flood hazard as a consequence of a 1 in 1000 year breach event in a) 2006 and b) 2115 (with sea level rise of 1.13m) (Atkins 2010a)

Socio-economic scenarios

In addition to flood hazard scenarios, socio-economic scenarios were developed in order to gain better understanding of the potential consequences of flooding on people, property and economic sectors. Based on a review of existing socio-economic scenario sets (Berkhout et al. 1999; UKCIP 2001; OST 2002; Cave et al. 2002), three scenarios for the Study Area were developed. The Conventional Development scenario broadly follows current trends whereas under a National Enterprise focused scenario, Lincolnshire's agriculture and conventional tourism sectors becomes more important. In a third scenario, Green, there is greater concern for the environment of the Study Area (Atkins 2010a).

For each of the three scenarios, qualitative descriptions of the population and major economic sectors in the Study Area were prepared. The descriptions were based on literature review and an understanding of the socio-economic issues facing the Study Area compiled in an early phase of the project. In addition to qualitative narratives, quantitative estimates of population and housing needs in each of the three scenarios were produced. The estimates were based on population projections from the Office of National Statistics extrapolated to the end of the century and assumptions about migration and household size under each of the scenarios. The quantitative projections show that although population under the different scenarios diverges, housing needs stay relatively constant. This is due to assumptions made about household size, see Figure 2.

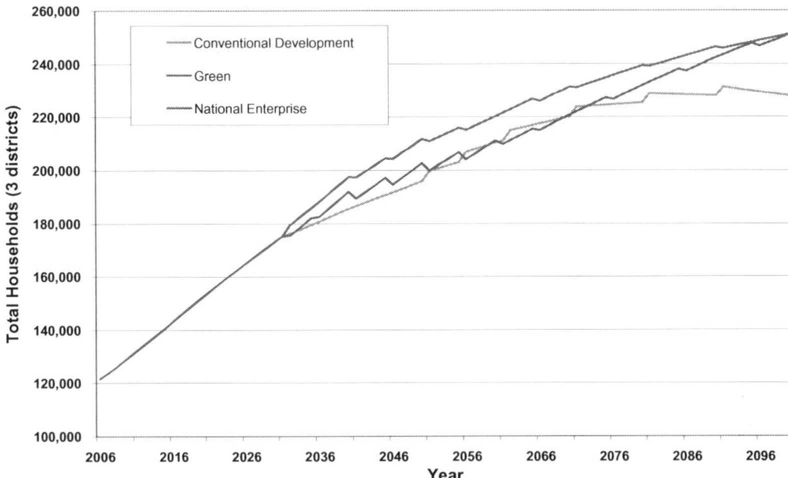

Figure 2 Household projections under the three socio-economic scenarios (Atkins 2010a)

Stakeholder engagement

Technical and non-technical stakeholders were involved throughout the Study and the Principles for sustainable spatial development were developed through this dialogue. Initial consultation was undertaken with the Project Steering Group, made up of Chief Executives from local authorities and representatives from the Environment Agency, Natural England, internal drainage boards and regional government. A Technical Group, made up of flood risk management and planning specialists from the Environment Agency, local authorities and regional planning authorities, was set up to challenge the robustness of the project methodology and emerging findings. In addition to the Project Technical Group, the Study consulted more widely with local experts and specialists in agriculture, emergency planning, economic regeneration and the natural environment. Non-technical stakeholders were also involved in the Study. Elected members in each of the three local authority areas were consulted through a series of workshops and were kept up-to-date with project progress through regular briefing notes.

Workshops held with the Project Steering Group, Technical Group, wider technical stakeholders and elected members followed a similar format. Following a briefing which described the aims and objectives of the project, a selection of the flood hazard maps were presented to stakeholders. A series of base-maps illustrating agricultural interests, biodiversity, deprivation and current flood defences were overlain by flood hazard maps printed on transparencies. These sessions allowed stakeholders to explore the implications of the flood hazard scenarios on the Study Area and provided the background for discussions about how the planning system could contribute to more sustainable development in future.

Private sector interests in the Study Area were represented at two briefing events. A selection of flood hazard maps were presented to representatives from developers, housing associations, architect firms and tourism businesses and their views sought on how future development could be sustainable.

Sustainability assessment of the Principles

An initial set of draft Principles was developed based on the outputs of the stakeholder workshops. These draft Principles were subjected to a sustainability assessment which considered the extent to which they took account of the full range of socio-economic and environmental needs of the Study Area.

Sustainability objectives were developed at the outset of the project through a scoping workshop attended by technical stakeholders. At this event, baseline socio-economic and environmental information was collated and key issues in the Study Area identified (Atkins 2009). These key issues were used to identify objectives for the sustainability assessment (see Table 2). The compatibility of the draft Principles with the sustainability objectives was assessed. Based on the recommendations of the sustainability assessment, the draft Principles were refined to increase their compatibility with the objectives and ensure they deliver sustainable development as far as possible.

As a result of the sustainability assessment, Principle 1 is considered broadly compatible with objectives 4, 10 and 15. There is still a potential conflict between Principle 1 and objective 1 as the Principle will create restrictions on where housing can be located and the options for housing might not be ideal from a traditional housing needs perspective. Principle 2 is broadly compatible with objectives 2 and 10, as the Principle aims to reduce people's vulnerability to flooding through flood resilience, emergency planning and increased awareness. This is likely to have health benefits in terms of reduced anxiety and mental health problems as well as fewer health problems caused as a direct result of flooding of properties and businesses. One of the main recommendations of the initial compatibility assessment was to broaden Principle 3 to encompass a wider range of socio-economic and environmental considerations. As a result of these changes, Principle 3 is now broadly compatible with objectives 3, 4, 5, 6, 7, 8, 9, 13, 14 and 15.

Number	Sustainability objective
Social objectives	
1	To ensure that the existing and future housing stock meets the housing needs of all communities in the study area
2	To improve health and reduce health inequalities by promoting healthy lifestyles, protecting health and providing health services
3	To improve accessibility to jobs and services by increasing the use of public transport, cycling and walking, and reducing traffic growth and congestion
4	To reduce social exclusion and improve equality of opportunity amongst social groups
5	To improve educational achievement and skills levels
Environmental objectives	
6	To protect, enhance and sustain biodiversity and to conserve geodiversity
7	To restore and manage the historic environment and built environment
8	To manage landscape change and enhance landscape quality
9	To encourage sustainable use of resources, specifically land and water
10	To reduce risk of fluvial and coastal flooding by adapting to climate change and coastal erosion
11	To reduce levels of atmospheric GHG by reducing emissions and increasing extent of carbon sinks
12	To promote the utilisation of the local renewable energy potential
Economic objectives	
13	To maximise the area's economic comparative advantages
14	To ensure the competitiveness of agriculture and food processing/packaging industries
15	To promote tourism in the area

Table 2 Sustainability objectives used in the sustainability assessment of the draft Principles (Atkins 2010b)

Outcome of the Study

The outcome of the scenario and stakeholder engagement approach was a series of Principles for sustainable spatial development of the Lincolnshire coast. The overall aim of the Principles is to reduce the number of people at risk of hazard from tidal flooding in the Study Area. In order to achieve this, the Principles indicate the type and scale of development which is appropriate in the different hazard zones. The Principles also address mitigation of the consequences of flooding through design and emergency planning and the role of the planning system in improving socio-economic and environmental conditions. A single flood hazard scenario has been chosen to accompany the Principles: the hazard zones referred to in the Principles are those associated with a breach event with a 1 in 200 year return period in 2115 (with an allowance of 1.13m of sea level rise). This map is shown in Figure 3.

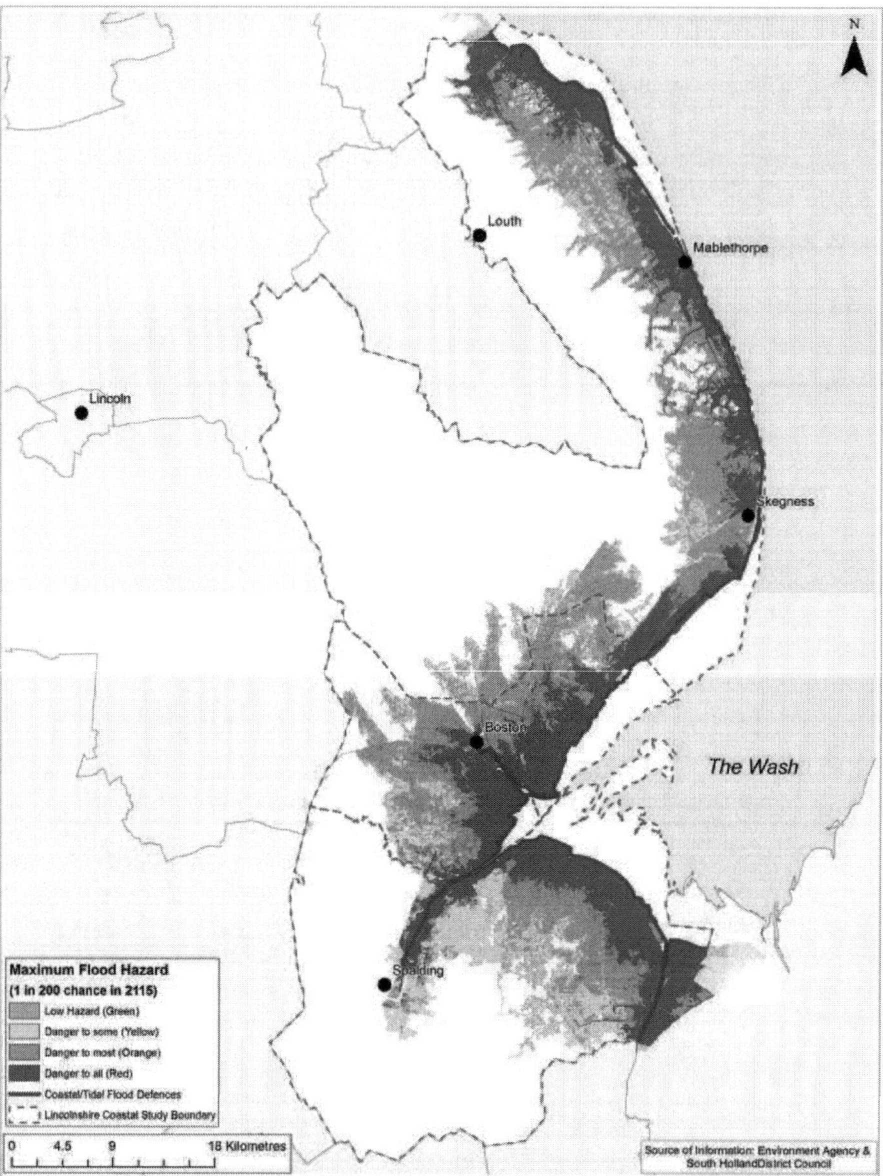

Figure 3 Flood hazard map to be used in conjunction with the Lincolnshire Coastal Study Principles (hazard associated with a breach event with a 1 in 200 year return period in 2115 (1.13m sea level rise). (Atkins 2010b)

The Principles are as follows (Atkins 2010b).

Principle 1
Development will be guided by the level of flood hazard.

With respect to the red, orange and yellow zones:
- Major development[1] will be employment or business related only;
- Exceptionally, development to meet local housing needs[2] may continue subject to the mitigation of flood risk through flood resilient design and emergency planning;
- It will not be appropriate for housing development in the red, orange and yellow zones to contribute to meeting the Region's strategic housing requirements. Rather, any new housing development should be of a level and type designed to keep the population in these zones broadly stable.

With respect to the green zone:
- Exceptionally, major development may be possible so long as flood risk is mitigated through flood resilient design and emergency planning.

With respect to all flood hazard zones:
- New and replacement community buildings may be permitted subject to flood risk being mitigated through flood resilient design and emergency planning;
- New caravan sites or extensions to existing sites may be allowed for short-let tourist use between the months of April and September subject to the mitigation of flood risk through flood resilient design and emergency planning;
- Development of buildings and infrastructure explicitly for use in emergencies may be permitted subject to flood risk being mitigated through flood resilient design.

Principle 2
The consequence of flooding for people in all flood hazard zones will, over time, be reduced by:
- The installation of flood resilience measures in domestic and public buildings, caravan sites and for essential infrastructure;
- Improving emergency planning and emergency response and evacuation arrangements;
- Improving public awareness and understanding of flood risk and responses.

Principle 3
Development decisions will aim to improve social, economic and environmental conditions in existing and new communities by:

---

[1] For the purpose of this study, 'major development' is defined by the Town & Country Planning (General Development Procedure) Order 1995 (as amended). The following are examples of 'major development': provision of more than 10 dwelling-houses; development on a site with an area of 0.5 hectares or more; provision of a building or buildings where the floor space to be created is 1,000 m^2 or more; and development carried out on a site having an area of 1 ha or more.

[2] Local housing needs have been defined as: those required to meet the housing needs of existing communities and should include a mixture of open market and affordable housing (as defined by Annex B of Planning Policy Statement 3).

- Minimising the loss of high quality agricultural land;
- Diversifying the tourism industry;
- Improving green infrastructure;
- Protecting and enhancing water infrastructure;
- Protecting natural, cultural and historic assets;
- Improving transport infrastructure and services;
- Improving the quality of existing housing stock and access to jobs, training and services for local people.

There will be a particular focus on more deprived areas.

The final wording of the Principles was agreed by all members of the project steering group. In order to achieve this consensus, the concerns of steering group members had to be resolved and time was built into the project programme to allow this. There was some initial reluctance to accept the precautionary principle in the absence of data relating to the probability of breaches occurring. It was recognised that variations in the assumptions used to model flood hazard may alter the boundaries of the flood hazard zones and improved information on breach probability may lead to reclassification of the hazard zones in future. However, whilst the extent of the different hazard zones may differ under different assumptions, it was agreed that the nature and level of development that is appropriate in the different zones would not alter.

Conclusion

The Lincolnshire Coastal Study was an innovative project that developed a set of Principles for sustainable spatial development based on the development of a range of scenarios and stakeholder engagement. The Principles formed the basis of a Coastal Strategy which was agreed by the Lincolnshire Coastal Study Group and submitted in draft to the Secretary of State in March 2010.

Following the change of government in March 2010, the intention was announced to revoke the Regional Spatial Strategy. Work on the review of the current Regional Spatial Strategy has therefore ceased. Planning authorities in the Study Area are now using the Lincolnshire Coastal Study as an evidence base, alongside other evidence required by Planning Policy Statements to inform their Local Development Frameworks.

The Principles have been developed specifically for the Lincolnshire coast although are likely to be applicable in locations facing similar pressures (particularly Principle 1 and 2). Bottom-up methodologies, such as the scenario- and stakeholder-based approach used in this Study, are appropriate for developing principles for sustainable spatial development. However, the method used in this Study relies on pre-existing flood hazard and socio-economic information. The applicability of the approach to alternative locations may be constrained by the availability of data. Alternative methodologies could be developed based on what data is available, recognising that there will always be assumptions and uncertainties which should be clearly stated.

References

Atkins 2009. *Lincolnshire Coastal Study Task 1 report: evidence base*. Lincolnshire County Council, Lincoln

Atkins 2010a. *Lincolnshire Coastal Study Task 2 report: scenarios*. Lincolnshire County Council, Lincoln

Atkins 2010b. *Lincolnshire Coastal Study Task 3 and 4 report: Principles and Options*. Lincolnshire County Council, Lincoln

Cave, R.R. Ledoux, L, Turner, K., Jickells, T., Andrews, J.E. and Davies, H. The Humber catchment and its coastal area: from UK to European perspectives. *The Science of The Total Environment*. Volumes 314-316, Pages 31-52

Defra. 2006. *Flood and Coastal Defence Appraisal Guidance: Supplementary Note to Operating Authorities – Climate Change Impacts*, Defra, London, 9pp.

Defra and Environment Agency. 2008. *Supplementary note on flood hazard ratings and thresholds for development planning and control purpose – Clarification of the table 13.1 of FD2320/TR2 and Figure 3.2 of FD2321/TR1 (2008),* Environment Agency and HR Wallingford.

Government Office for the East Midlands, 2005. *East Midlands Regional Plan*. Norwich, The Stationary Office, Norwich

Office of Science and Technology (2002) *Foresight Futures 2002: Revised Scenarios and Guidance*. Office of Science and Technology, London, 2002.

UKCIP, 2001. *Socio-economic scenarios for climate-change assessment. A guide to their use in the UK Climate Impacts Programme*. UK Climate Impact Programme, Oxford

Acknowledgements

The authors would like to thank the following people for their contribution to this paper: Sophie Day (née Nicholson-Cole), Paul Morgalla, Toni Hylton, Rob McSweeney and Miriam Ferrari.

Innovative Coastal Zone Management
ISBN 978-0-7277-5749-4

ICE Publishing: All rights reserved
doi: 10.1680/iczm.57494.144

Understanding Risk to the Community of Spurn Head (Humber Estuary, UK)

Dr Julie Holland, Senior Consultant Geomorphologist, ABP Marine Environmental Research, Southampton, UK
Steve Hunt, PhD Student, University of Waikato, Hamilton, New Zealand
Heidi Roberts, Head of Coastal Processes, ABP Marine Environmental Research, Southampton, UK.

Introduction

Spurn lies at the mouth of the Humber and is one of the most distinctive morphological features on the English East coast. The spit partially encloses the Humber estuary mouth and marks the transition from the rapidly eroding shoreline of Holderness to the north and the estuary itself. The spit is subject to multiple, varying, functions in the form of a Royal National Lifeboat Institution (RNLI) lifeboat station (and uniquely the housing of the families associated with the station) and the ABP Vessel Traffic Services (VTS) for the Humber. In addition the spit is owned by Yorkshire Wildlife Trust (YWT) and managed as a nature reserve. The Environment Agency (EA) (which commissioned the project) also holds a vested interest in the spit due to its potential role in flood protection within the estuary and its surrounding land.

It is becoming increasingly important to understand the current and future risk to coastal communities. Historically the focus for such coastal investigations has been the conceptual understanding of the processes operating and the implications of extreme events. This investigation has taken the next important step in understanding the current and future risk to the community and services provided at Spurn by examining both the less extreme but more frequent events and the implications for loss of access, as well as the more extreme events, and hence the safety of access, to the community of Spurn. This is an important factor in understanding the true nature of risk to coastal communities.

The investigation provides predictions of the frequency of overtopping and overwashing of water onto and across the spit, in combination with understanding the implications for beach profile changes. The project draws upon a large and varied number of datasets and analyses including historical analysis, coastal process studies, a geological review, historical breach records and an innovative application of an overwashing equation. This paper aims to detail the findings of this work and the benefits of this approach for other vulnerable coastal communities.

Study Area

Spurn Head is a narrow sand and gravel spit, extending 5.5km southwards from the cliffs of Holderness, East Yorkshire, at Kilnsea Warren across the mouth of the Humber Estuary (Figure 1). Currently, access to the end of the peninsula is via the single track road which is subject to damage during combinations of storm events and high spring tides, causing damage and restricting access. YWT took over the management of Spurn in 1960 and in general the defences have not been maintained since this time, although some remedial works have been undertaken to maintain the roadway to the tip of Spurn.

Spurn is characterised by a narrow neck in the north, connecting the rest of Spurn to the coast at Kilnsea (Figure 1), and a wider 'spatulate' end to the south. The spit consists of beach sands and gravels overlain by dune sands. The southern end of the spit terminates abruptly at the deep water channel at the mouth of the Humber. Sediment from erosion of cliffs to the north, due to exposure to waves from the predominant north to northeast sector, is transported alongshore by rectilinear tidal flows running approximately parallel to the shoreline of Holderness and onto the seaward side of Spurn (HR Wallingford, 2003), where much of the material continues to be moved towards the south and the tip of the spit.

The foreshore on the seaward side of the neck of Spurn is backed by a narrow area of vegetated dunes. The foreshore is a mixed sand and gravel beach, with occasional blocks of concrete and rubble that are remains of the former seawalls and other structures, as well as timber representing former revetment works. At the narrowest point which is just to the north of High Bents, the neck is approximately 50m wide (Figure 1) and has been subject to washover events that have required regular realignment of the access road which leads from Kilnsea to the VTS and lifeboat services at the end of Spurn.

Towards the southern end of Spurn on the seaward side is The Binks (Figure 1), an area of sand and gravel banks extending north-eastwards from the tip of the peninsula. This generally shallow area dissipates wave energy under certain conditions thereby providing a sheltering effect to Spurn, predominantly from waves approaching from the sector east to southeast. Waves break and refract on The Binks before reaching the southern end of the peninsula. The flow in and out of the estuary constrains the length of the peninsula, and a 'deep pit' that lies immediately south of the tip of the peninsula is thought to be the result of strong tidal scour, which is likely to restrict future prolongation of the spit. To the west of the peninsula, on the estuarine side, Spurn Bight is a relatively flat intertidal area that extends north-westwards into the Humber. The Bight is composed of silts and muds, with isolated patches of coarser material (sand and gravel) including the feature called the Old Den (Figure 1). To the south, along the estuarine side of Spurn from Chalk Bank to the tip, sand waves are present, which are gradually moved northwards towards the Old Den (Ciavola, 1997).

Due to the history of damage and the need for ongoing repair of the access road, Spurn Head has a specific Management Plan for maintenance work. Informal discussions with ABP (April 2009) have identified that the current frequency of overtopping/overwashing is approximately twice a year.

Historical coastal change

To determine the future risk it is necessary to characterise the historic behaviour of the study area. This has been done through a review of other studies and the comparison of recent beach profiles (collected by East Riding Council) and EA LiDAR data.

Since the 14th Century there have been a series of breaches of varying size and differing locations in 1360, 1600, 1789/90, 1850 and 1856 (de Boer, 1981; IECS, 1992), although, only the breach of 1849/50 is thought to have persisted for any length of time. The location of the 1849/50 breach is shown in Figure 1 and was eventually artificially sealed using a chalk bank. Little is known about the initiation of the breaches and whether these breaches followed the terminology used in this study i.e. lowering of crest levels leading to channel development and free flow of water over subsequent tides, or if they were significant overwashing events.

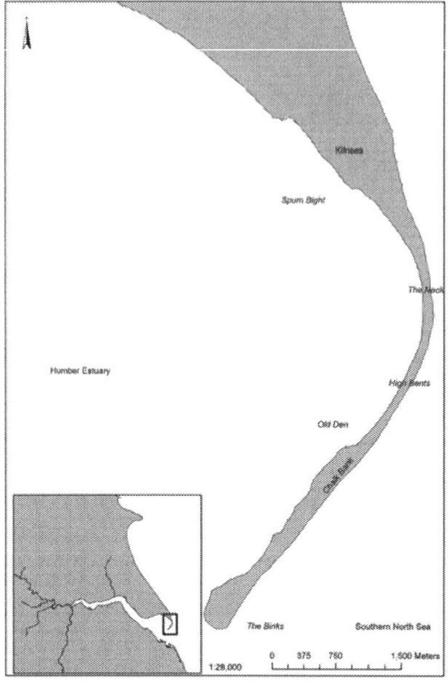
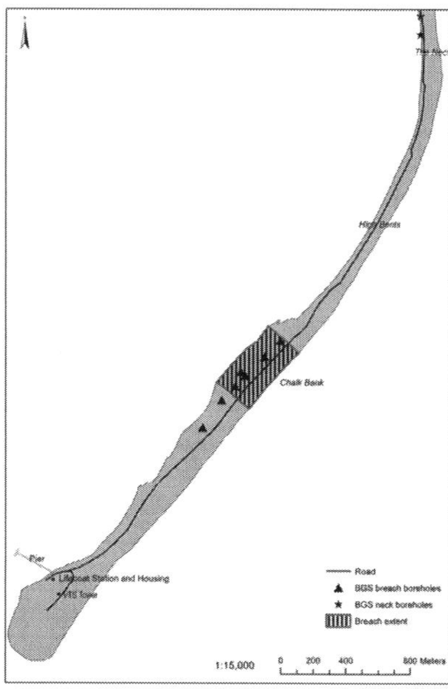

Figure1. Location Maps

At the neck the shoreline changes from a general NNW-SSE orientation in the north to NNE-SSW in the south and wave refraction analysis indicates that wave energy from the north and northeast is focused in this approximate area. To the south, wave refraction diverts energy away from the shoreline for these directions. The main tidal flows, particularly at lower states of tide are directed away from the southern section of the spit by The Binks, thus reducing the potential for supply of sediment in this area compared to the area north of the neck. The historical shoreline record suggests that the shoreline north of Kilnsea has eroded predominantly due to wave attack, moving further away from the main tidal flows and a sediment transport divide has developed in this area. Thus recession has occurred here and the supply of sediment along the shoreline towards the south has reduced over time, thereby slowing the potential for progradation of the spit.

This reduction in supply would have been increased by coastal protection works installed in the 20th Century, and would have probably meant a greater rate of erosion towards the south. As the spit receded, the sub-base of the Binks at about 4-6m below CD became wider, and predominant waves would have been more refracted at greater distance from the shore, thus reducing erosion at the spatulate end of Spurn, even though the supply to the spit from the north continually reduced. Supply of sediment from offshore is likely to have been maintained. Eventually, the width and height of the Binks effectively protected Spurn, south of the neck, from the most severe erosive forces. The change in spit orientation reduces the erosive potential of the north/north-easterly waves.

This hypothesis is generally supported by the observed historic recession of the shoreline, where recession has continued in the area of Kilnsea since 1852. To the south of the neck, however, the seaward shoreline has been, by comparison, essentially stable, and may at certain times and locations have shown progradation. This would indicate that over the last circa 100 years the position of the spit has stabilised, probably aided by the human intervention of construction of coastal defences and groynes, compared with earlier years. The spit is now in approximate equilibrium with prevailing tidal and climatic conditions, with the width of the 'shelf' on which the Binks stands, relative to the offshore deep areas, a significant controlling factor. This possibly suggests the location of the neck will move slowly northwards over time, and therefore the weakest point also moves towards the north, but only whilst the area to the north of Kilnsea continues to erode. The hypothesis put forward, however, would indicate that the erosion would slow in a south to north direction, under the same wave, tide and water level conditions as have occurred over approximately the last 100 years.

Following from the conceptual review summarised above, a number of data sources have been reviewed with the aim of identifying the current vulnerable areas along Spurn, i.e. areas where the spit has a low crest level or where reduction in crest level or retreat is occurring. The datasets used are (i) bi-annual beach profile surveys measured along Spurn between October 1997 and March 2008 by East Riding Council; and (ii) annual EA LiDAR surveys flown between 2000 and 2007 (excluding 2006).

The datasets show a good level of agreement and it has been possible to identify a high risk area in the region of the neck (see Figure 1 for locations). This section has the lowest elevation of between 4 and 6m ODN and is located at the narrowest point on the spit (70m at MHWS in 2007).

Trends of retreat have also been identified in the vicinity of High Bents and the area south of Chalk Bank. At High Bents and immediately south of Chalk Bank the beach profile analysis shows that there is some evidence for a recession of the dune face although as the spit is relatively high at these points (approximately 8 and 9m ODN respectively) it is not considered that these areas have a high risk of overtopping/overwashing.

Along the frontage between Chalk Bank and the VTS tower the retreat identified appears to be restricted to the beach fronting the spit and not the spit itself. Available evidence from this area suggests that the spit has prograded seaward at this point, and in addition as the spit crest is at a height of 10m ODN and above, this region is not considered to be at risk from overtopping.

The location of the 1849/50 breach (at Chalk Bank, Figure 1) shows a steady increase in elevation, indicating that this area is now significantly less vulnerable with a crest level of approximately 9m ODN. The present day stability of the former breach area indicates that the 1849/50 breach could have been significantly influenced by the removal of material for ballast.

Geological analysis

A geological survey and analysis has been undertaken in support of this study. The overall aim was to ascertain whether the underlying geology exerted any control on the location, depth and lateral extent of the 1849/50 breach, once it had formed. Boreholes were drilled at two locations on Spurn: seven boreholes at the site of the 1849/50 breach; and three boreholes on the neck of Spurn (Figure 1). Drilling at the site of the 1849/50 breach revealed a buried surface, believed to represent the floor of the breach channel, lying at around -1.5m ODN, with the greatest depth at -1.88m ODN. The breach floor was underlain by estuarine intertidal deposits and spit deposits, with the glacial till lying several metres lower at between -6.87 and -14.33m ODN. Vetch (1850) reported that by the end of 1851 the breach channel was approximately 4.88m (16ft) deep at high water. MHWS at Spurn is at 3m ODN, which indicates that the base of the channel should lie around -1.88m ODN; this corresponds with the depth of the base of the sands (beach and dune deposits) where they lie on top of the estuarine intertidal deposits. The base of the breach channel was therefore at least 5.4 m above the till surface (BGS, 2008), this indicates that there was no geological control on the depth of the breach channel.

The boreholes at the neck of Spurn showed a very similar sequence to each other with estuarine intertidal silts overlying a glacial till surface, which slopes gently down from north (-4.17m ODN) to the south (-6.23m ODN). The elevation of the glacial till at this location is at a similar elevation to that at the northern end of the 1849/50 breach implying a very low gradient for the surface between these two sites.

Assessment of Access Risk

A review of recent maintenance records held by ABP has identified that damaging events restricting access to Spurn occurred in 1996, 2001, 2004, February 2005 and March 2007. In addition, ongoing maintenance requirements (likely to be a result of damage from smaller scale events) indicate lesser degrees of damage in February 2002, November 2002 and June 2003. This followed major repair work undertaken in 2001, which was supported by the development of a Management Plan, Environmental Impact Assessment and resulting planning consent.

The event in February 2005 causing significant damage to the roadway over a 140m section, was due to a combination of northerly storms and high spring tides. The road over this section was realigned to the west as a result, however in the period before this could be carried out a temporary 4X4 route to the lifeboat station was established and the VTS service was operated from Grimsby. The event in March 2007 was less damaging than 2005 but effects were spread over a distance of 400m. Between 150m and 200m of the tarmac road needed realigning and a further 200m of the concrete mats used to reinforce the road needed lifting, repairing and rebedding.

To understand how access to the infrastructure on Spurn could be affected in the future it is also important to understand how frequently different conditions will result in a health and safety risk to the users. This investigation has focused on understanding the frequency of overtopping (using EurOtop methodologies) and overwashing (using the Ogawa and Shuto methodology). These methods were applied in combination with a number of sea level rise scenarios based on current Defra guidance (Defra, 2006) to evaluate the likely changes in risk from now into the future.

Overtopping analysis

Overtopping is defined as the volume of water likely to be carried over the crest of Spurn, due to waves on top of the tidal water level, i.e. the still water level would be lower than the crest and it is only the wave that causes the overtopping. In such a case the wave does not continue behind the crest and the potential for crest lowering is minimal. The volume of water can then be used as an indicator to determine the frequency of events when it will be too dangerous for vehicles to cross and crest damage could result.

It is possible to relate the overtopping volume, calculated using EurOtop (2007) to crest damage using methods in the US Army Corps of Engineers (USACE) Coastal Engineering Manual (2006). This identifies tolerable limits for overtopping of unprotected embankments and in particular, thresholds for "no damage to crest and rear face of embankment if not protected" (USACE CEM, 2006). The figure given for this threshold and used in this study to denote the onset of breaching (taken to be damage in this case) is 1 l/s/m. Although there are limitations to the application of this method to Spurn, as the method is intended specifically for embankments, it provides a useful indication of the possible crest damage arising from overtopping.

It should be noted that the EurOtop study recommends an overtopping rate of 0.1 l/s/m for the onset of damage on an unprotected embankment. At Spurn overwashing/overtopping generally occurs where some protection exists from the roadway armourflex, therefore a higher value than indicated by EurOtop was considered valid, particularly as this resulted in a current frequency of about twice a year, which is generally considered to the current rate of damage.

Overtopping was calculated using beach profiles collected by the East Riding Council in conjunction with extreme water level data (Black and Veatch, 2006) and sea level rise guidance (Defra, 2006). The results of this analysis show that, as a result of overtopping, damage is most severe along the vulnerable neck of the spit. There is also potential for some damage along some of the lower profiles to the south (between High Bents and the south of Chalk Bank) but there is a lower probability of this occurring. It is interesting to note that at Chalk Bank itself (1849/50 breach location) crest damage is not predicted to occur until approximately the 100 year time horizon due to overtopping.

The analysis has also shown that overtopping increases in frequency with sea level rise; however, more frequent overtopping incidences do not necessarily lead to crest damage. For all locations, the frequency of damage does not become significant until approximately year 50 and beyond. The results of the calculations are summarised in terms of number of incidences of damage per year in Table 1.

Table 1. Indicative incidences of crest damage from overtopping along Spurn Head

Profile No.	Crest Height (mODN)	Incidences of Crest Damage From Overtopping (per year)			
		Present Day	20 Years	50 Years	100 Years
Northern part of neck	4.79	1	2	7	75
Northern part of neck	7.27	2	3	7	19
High Bents	8.77	1	2	4	10
Chalk Bank	8.87	Unlikely*	Unlikely	Unlikely	2
Southern Chalk Bank	9.74	1	2	3	7
VTS	10.43	Unlikely	Unlikely	Unlikely	Unlikely
* Unlikely for crest damage to occur					

Overwashing analysis

Overwashing is defined as when a high tide, plus surge, and wave run-up is higher than the crest of the spit and flows as 'green water' (sometimes referred to as 'weir flow') over the crest. In such a case the overwash depth can be calculated in metres above the crest level. This implies a significant volume of water and sediment can wash over the crest. Overwashing has the potential to move large volumes of sediment over the crest as well as possibly inducing ridge (crest) collapse or lowering, thus potentially initiating a breach.

When conditions are such that there is overwashing of the road, access will certainly need to be restricted. Two methodologies were initially used to assess overwashing, the Bradbury equations (2000) and those by Ogawa & Shuto (1984), to ensure greater robustness in the analysis. The Ogawa and Shuto equations were found to be more applicable to Spurn Head as it is designed specifically for sand features similar to Spurn, whereas Bradbury's method is designed for shingle spits. The results of this analysis are discussed below.

The potential for overwashing has only been identified at the neck of Spurn Head using the Ogawa and Shuto method. This matches the location identified as vulnerable by the beach profile and overtopping analysis. The results of the overwashing analysis are summarised in Table 2 for the neck region of Spurn.

Table 2. Summary of overwashing analysis based on Ogawa & Shuto (1984)

Return Period	Depth of Overwash at the northern part of the neck (m)			
	Present Day	**20 Years**	**50 Years**	**100 Years**
1	None	None	0.31	0.99
2	None	0.03	0.34	1.02
5	None	0.11	0.425	1.11
10	0.17	0.3	0.61	1.3
20	0.33	0.47	0.78	1.46
50	0.45	0.59	0.91	1.6
100	0.54	0.67	0.98	1.67
Timeseries analysis	Overwashing likely 2 times a year	Overwashing likely 4 times a year	Overwashing likely 11 times a year	Overwashing likely 103 times a year

At the southern part of the neck, overwashing is predicted only once a year in 100 years time. The frequency of overwashing at the northern part of the neck is greater because the crest height is more important in this methodology than slope in contrast to the overtopping analysis. Therefore, although, the frequency of overtopping is greater at the southern part of the neck, this has a crest height of 7.27m ODN as opposed to the lower crest height of 4.79m ODN for the northern part of the neck.

The analysis has shown that with 20 years of sea level rise, overwashing is likely to occur about 4 times per year, but when sea level rise accelerates from year 50, this increases to over 100 times (i.e. at moderate high tides) by year 100, resulting in the neck of Spurn becoming extremely vulnerable to increasing water levels and likely breach if maintenance activities are not continued.

All overwashing events have the potential to erode sediments particularly sands from the spit through bed shear stress, thus leading to potential damage. The greater the depth of overwash the longer conditions will persist to do damage. Under present day conditions an event with a statistical return period of between 1:5 and 1:10 years is required before damage to the spit becomes likely. This is reduced to about a 1:2 year return period in 20 years. After 50 years a current 1:20 year return period event can be expected annually. Although there is a significant increase in frequency of overwashing events, the magnitude of the effect is considerably greater. By way of example the excess shear stress created by a 1:20 year return period event in 100 years time will be about 5 times greater than at the present time. For simplicity if we assume this ratio of potential damage per overwash event, damage to the roadway, on average, would be more than 250 times greater than occurs at present in 100 years time, and about 10 times greater than present in 20 years time.

Both the overwashing and crest damage predictions raise the issue of the long-term viability of maintaining regular access along the spit, but not until the latter half of the century as a result of accelerated sea level rise (Defra, 2006).

Assessment of Breach Risk

The assessment of current and future access to Spurn has investigated the frequency of current overtopping and overwashing and predicted changes with sea level rise (Defra, 2006). Concurrent with the present day analysis, the area known as the neck continues to be the most susceptible to breach (Tables 1 and 2). The analysis has shown that the risk of overtopping and overwashing may be acceptable based on the 2007 crest levels but the frequency increases significantly as sea levels rise (or as crest levels lower). Initial analysis suggests that the risk of breach is relatively low for the next 20 years but that the frequency of overwashing accelerates sharply in line with the post 2055 annual sea level predictions (assuming that crest levels do not decrease).

However, of greater importance is the risk of lowering of the crest height along the neck of Spurn, particularly at the southern part of the neck. A reduction in crest height of 10% increases overwashing predictions from twice a year (the current estimate) to 14 times per year. It is likely that this would not leave sufficient time for the profile to recover leading to further degradation and increased overwashing frequency (without maintenance). The sensitivity of reduction of the crest height by 30% (as occurred during the 2005 storm) increases overwashing to over 200 times per year. This frequency is clearly unsustainable and illustrates the need for comprehensive monitoring (and regular maintenance) of this section of the spit to ensure that access can be effectively maintained and the risk of full breach managed.

Future Management

The current management practices for mitigating overwashing can be continued in the immediate term. These methods include:

- Realigning the road to the west where space permits;
- Use of 4WD vehicles for temporary access during repairs; and
- Providing the services from Grimsby during the road access outages, where possible.

The prospect of catastrophic breach in the narrower steeper sections of the peninsula (the neck), where realignment is not practical, means that importation of material to restore the west face (as opposed to re-profiling which only redistributes the same volume of material) could be required in the short-term e.g. within five years. If this is not permitted then in situ repair will be difficult and the risk of repeat failures in the same general area can be expected.

Access to Spurn Head is essential for the services that are provided by the presence of Spurn including the VTS and the Lifeboat Station. The tolerable level of downtime with respect to social needs is probably lower than that for the strategic operational requirements, though the latter implies greater economic damage. Assuming a tolerable level of downtime of 15%, and allowing (qualitatively) for the downtime being greater than the values listed above, then the current management practices could be sustainable for up to 10 to 15 years. Following this more significant engineering works will be required to restore and sustain a viable road connection.

If the policy is to maintain road access in the immediate and near future then further assessment is required to examine viable means of widening the crest at the steep narrow sections (at the neck). This could include recycling or borrowing sediments from areas of surplus to north or south of the problem areas. These measures, once applied, in combination with other current management practices could potentially extend the manageable life of the access to between 10 and 15 years. However, the frequency and cost of repairs, and associated downtime, would continue to increase. If the policy is to extend access beyond 10-15 years, then studies would need to be initiated to examine more robust engineered solutions.

Conclusions

The following conclusions have been reached from this study:

- In terms of spit vulnerability (crest height, width and erosional trends), overtopping and overwashing, the analyses consistently show that the greatest risk to access occurs along the neck of the spit;
- The current probability of overwashing is twice a year and is likely to be manageable in the short term with post-storm maintenance works. However, increased overtopping/ overwashing in the future will further increase the potential for permanent reduction in the crest level, leading to increased risk of overwashing and hence subsequent breach in the future; and
- If the policy continues to be to maintain road access and the presence of the current services provided on the tip of Spurn into the future, then further work is required to examine a viable means of widening the crest at the steep narrow sections (the neck). This could include recycling or borrowing sediments from areas of surplus to the north and south of the neck.

This study has provided the technical analysis of the future risk to access to the full length of the Spurn peninsula. The aim is that the insight gained will be used to plan for the future of the services provided at Spurn. The calculations undertaken are applicable to other locations where overwashing and overtopping may now or in the future present risks to access.

References

BGS, 2008. The geomorphology and underlying geology of the Spurn Head peninsula, East Yorkshire. British Geological Survey, Report CR/07/213, 85pp.

Black & Veatch, 2006. Joint Probability Analysis of Large Waves and High Water Levels in the Outer Humber Estuary. Prepared for the Environment Agency. Final, December 2006.

Bradbury, A.P., 2000. Predicting Breaching of Shingle Barrier Beaches – Recent Advances to Aid Beach Management. Proceedings of the 35th MAFF Conference of River and Coastal Engineers, 05.3.1-05.3.13.

Ciavola, P., 1997. Coastal dynamics and impact of coastal protection work on the Spurn Head spit (UK). Catena 30:369-389 de Boer, G., 1981. Spurn Point: Erosion and Protection after 1849. In: J. Neale and J. Flenley (eds.) The Quaternary in Britain. Pergamon Press, Oxford, pp.206-215.

Defra, 2006. Flood and coastal defence appraisal guidance FCDPAG3 economic appraisal. Supplementary note to operating authorities - climate change impacts.

EurOtop, 2007. Wave Overtopping of Sea Defences and Related Structures: Assessment Manual. www.overtopping-manual.com.

HR Wallingford, 2003. Humber Estuary Shoreline Management Plan, Phase 2. Coastal behaviour from Easington to Mablethorpe: summary report for HESMP2. HR Wallingford and BGS, EX4846, 114pp.

IECS, 1992. Spurn Heritage Coast Study, Final Report. Institute of Estuarine and Coastal Studies, University of Hull, January 1992.

Ogawa, Y. & Shuto, N., 1984. Run-up of periodic waves on beaches of non-uniform slope. Proceedings 19th Coastal Engineering Conference (Houston, Texas, ASCE). 328-344.

USACE CEM, 2006. Coastal Engineering Manual EM 1110-2-1100. US Army Corps of Engineers.

Vetch, J., 1850. Report of Captain James Vetch, Royal Engineers, to the Lords Commissioners of the Admiralty, on the subject of the Breach made by the Sea across the Spurn Point, at the entrance to the River Humber.

Innovative Coastal Zone Management
ISBN 978-0-7277-5749-4

ICE Publishing: All rights reserved
doi: 10.1680/iczm.57494.154

Delivering Sustainable Coastal Management Solutions through Community Engagement in Sensitive Environments: Case Studies from the Western Isles, UK

Daniel Dray, Mott MacDonald Ltd., Croydon, UK
Peter Phipps, Mott MacDonald Ltd., Croydon, UK
Murdo Gray, Comhairle nan Eilean Siar (Western Isles Council), UK

Introduction

On the 11th and 12th of January 2005, an extreme storm event struck the West Coast of Scotland. The storm caused extensive damage to buildings and infrastructure, as well as the natural environment. Large scale coastal erosion was recorded, roads were cut, trees uprooted and widespread flooding was experienced. Communities on the Western Isles (or Outer Hebrides), an Island chain off the west coast of Scotland (Figure 1), were particularly badly affected due to their exposed location. On the Western Isles the storm was manifest by winds of up to 92 knots (106.0 mph) (Angus and Rennie, 2006), which coincided with a high spring tide and a large tidal surge (HRW, 2006). In many areas roads were either eroded or blocked with debris, services were cut and agricultural and semi-urban land was lost.

In response to the resulting damage, the Local Authority (Comhairle nan Eilean Siar) commissioned Mott MacDonald to undertake a review of the twelve most affected sites and where appropriate make applications for funding to the Scottish Executive under the Coast Protection Act 1949. Though the Project Appraisal process, £8m was secured to design and implement various coastal management solutions at 5 of the sites. Due to the sensitive nature of the Western Isles, both environmentally and socially, it was critical that sustainability principles play a key role throughout the entire process from conceptualisation, option appraisal, detailed design development, construction and maintenance.

This paper presents the design and delivery of sustainable coastal management solutions in the sensitive environmental and social setting of the Western Isles.

Physical Setting

Geographical Setting

The Western Isles is an archipelago of islands lying off the west coast of Scotland. The island chain forms part of the Hebrides, separated from the Scottish mainland and from the Inner Hebrides by the waters of the Minch, the Little Minch and the Sea of the Hebrides.

The Western Isles chain runs roughly north to south and has an area of approximately 306,916 ha (289,798 ha excluding freshwater and inter-tidal areas) and approximately 2,700 km of coastline (the result of numerous sea lochs, bays and inlets). The islands extend 210 km in length and are 60 km at the widest point (CnES, 2010).

The larger main islands are Lewis, Harris and North Uist, Benbecula, South Uist and Barra. There are also smaller populated islands, including Vatersay, Esrikay, Baleshare and St. Kilda. Of the 12 most affected sites, the 5 which were ultimately taken forward for funding application were located on the southern Islands of Benbecula, South Uist and Barra (Figure 2).

Figure 1. Location of the Western Isles relative to the United Kingdom

Figure 2. Map of the study area including site locations referred to in the text

Geomorphological Setting

The islands of the Western Isles are predominantly Pre-Cambrian basement rocks, known collectively as 'Lewisian'. Geologists have dated Lewisian Gneiss at nearly 2,800 million years, making it the oldest rock formation in Britain (Moorbath et al 1975).

Much of the Atlantic coast is characterised by a series of blown sand landforms, known collectively as 'machair' as illustrated in Figure 3. They are best developed along the western coasts of the Uists and consist of a mixture of siliceous and calcareous fractions in varying proportions. The British Geological Survey (BSG) estimates that 10% of land area in the Western Isles is composed of these 'low-lying windswept, coastal sandy plains (machairs) together with sand dunes and hillocks'.

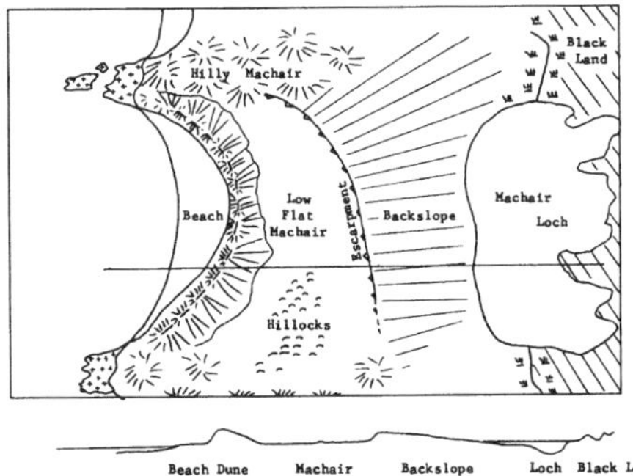

Figure 3. Plan view and cross section through machair land (Ritchie, 1971)

Environmental Setting

The natural environment of the islands is recognised internationally as being of global importance. A large percentage of land, areas of inland water and marine sites are designated for nature conservation purposes. The islands are also home to a number of rare species such as the golden eagle, basking shark, whale, dolphin, otter and corncrake. The importance of the environment extends beyond the land to the seas around the Islands as the coast and coastal waters are also important for a range of species and habitats. The Western Isles has numerous conservation designations including a World Heritage Site (the only site in the UK to have dual natural and cultural status), 15 Special Protection Areas, 14 Special Areas of Conservation (SAC) including marine SACs, 4 RAMSAR sites, 1 Biosphere Reserve (Loch Druidibeg, South Uist), 7 Marine Consultation Areas, 3 National Scenic Area designations covering over one-third of the Western Isles land area, 53 Sites of Special Scientific Interest, 4 National Nature Reserves and 4 Outstanding Conservation Areas (CnES, 2010). A number of these designations had to be considered in the development of focused sustainable management solutions.

Land use

Due to the peaty swamp land, inhospitable mountainous terrain and numerous lochs, the majority of the land on South Uist, Benbecula, Barra and Vatersay is not actively utilised. Transport routes across the Islands tend to be single track with local passing places due to the sparse population spread and distance between population concentrations in towns, villages and hamlets. However, these roads play critical role in maintaining an integrated sustainable community in this isolated island location. The majority of the settlements are concentrated on the western coastline where the land tends to be low and flat. The more rugged hillsides on the east of the islands are less conducive to settlement and therefore remain as agricultural, rough pasture or open land. The low flat inhabited land is still farmed by a traditional crofting system, its fertility a product of centuries of traditional farming practices.

Social Setting

Population

The Western Isles are sparsely populated. The General Register Office for Scotland (GROS) estimated the population of the Western Isles to be 26,200 for June 2008. This figure represents a population decrease of 100 persons since 2007. This corresponds to a larger decrease than in 2006 (-20) and 2007 (-50). Statistics show the long term trend to be one of a declining and ageing population. Figure 4 below illustrates GROS population estimates over the last ten years (1998-2008) for the Outer Hebrides (CnES, 2010).

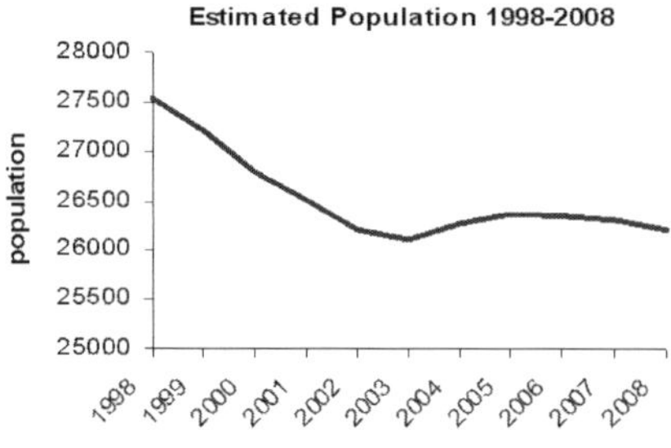

Figure 4. Estimated Population of the Western Isles from 1998 to 2008. (CnES. 2010)

Within the study area (the islands of South Uist, Benbecula and Barra), the most populated island is South Uist. However, the town of Balivanich on the Island of Benbecula is the largest in the study area. The level of population decline for the large Scottish island groups (Western Isles, Shetlands and Orkney) was higher in the 1991 to 2001 period than the Scottish average (CnES 2010). The decline in population of the Outer Hebrides over the last ten years (between 1995 and 2005) was the highest of any Local Authority in Scotland at -8.5% (CnES 2010).

Economy

The biggest employer in the Western Isles is the public sector (administration, education and health) followed by retail and construction. Fishing and farming also employ a large number of the population, although most people who maintain crofts are actively involved in other industries to supplement their income. The contribution of tourism to the local economy is increasing each year (CnES 2010). Due to its economic situation, the whole of the Western Isles, including Stornoway, is recognised in policy terms by Highland and Islands Enterprise as an economically 'Fragile Area'. Therefore, the need for sustainable development opportunities, which combine economic, social, cultural and environmental concerns, is paramount (CnES, 2010).

Sustainable Communities

Over the last 100 years the Western Isles has lost around 40% of its population, with whole communities on the outer isles disappearing. The latest population projections are for further decline until at least 2018. This has an adverse impact on population structure, community well being and social fabric of the islands. There is a need for a positive development framework that can play a part in combating the threat of further population decline and help create sustainable communities (CnES, 2003).

CnES view sustainability as integration and balancing of economic, social and environmental objectives. Sustainability is promoted as more a way of thinking than a product.

In the latest Structure Plan, CnES (2003) characterise a 'Sustainable Community' as one that:
- Has a stable and balanced population structure;
- Has a buoyant, diverse economy;
- Has members who are active and enabled who value tolerance, egalitarianism, education, safety and health;
- Provides access for all to good quality housing, neighbourhood facilities and social activities;
- Values and safeguards its environment, culture and heritage;
- Offers a high quality of life for all.

CnES recognises that the environment, local culture and heritage are of significant value in a sustainable community and that once lost these features may never be restored. They are fully aware that planning policies have an important part to play in ensuring that due weight is given to these aspects of community life when threatened by development and change (CnES 2003).

Sustainable Development

To fulfil CnES's vision of sustainable development and the Scottish Executive's requirement to consider life time sustainability, the incorporation of sustainable principles was vital in achieving desirable coastal management solution.

CnES have a policy of 'managed realignment' in response to rising sea levels, which has been calculated to be 300mm by 2060 in line with the Defra (2006) guidance, and coastal erosion. However, exceptions are considered when one of the following are threatened:

- Important habitats such as the machair; or
- Scheduled monuments or listed buildings: or
- major infrastructure and utilities; or
- Occupied buildings.

Where an exception has been identified, CnES works with other agencies and landowners to put in place appropriate defence measures, subject to the availability of resources and the environmental impact.

Option Appraisal and Design Development Phase
Following on from the storm events of January 2005 CnES commission Mott MacDonald Ltd to evaluate a total of 12 affected locations, in order to ascertain if there could be reasonable grounds to obtain funding support from the Scottish Government to implement pro-active coastal management solutions. Following a risk based analysis of the locations the council instructed that 6 of the locations be evaluated further to identify a preferred option for that location based on technical, economic, environmental, strategic and safety grounds.

The sustainability of potential management solutions was not assessed as a stand alone criterion, but within each of the above mention criteria. In most cases, coastal flooding and recession would not lead to extensive damage to property. Therefore, an economically sustainable solution would be required to obtain a positive benefit/cost ratio. The environmentally sensitive nature of the sites also meant it was vital that the environmental sustainability of potential management solutions be assessed carefully.

In a number of cases, the only asset at risk from coastal erosion and flooding was a single lane coastal road. If the benefit assessment was carried purely based on protecting the road from damage, the costs are unlikely to be justified. However, these roads provide the only link between the sparsely populated areas and are therefore vital for maintaining the fabric of an integrated sustainable community.

Project Appraisal Reports, which constituted an application for funding under the Coast Protection Act 1949, for the 6 locations were submitted to the then Scottish Executive. Consequently, 5 or the schemes received funding. As part of the short listing of options in each location a full multi-criteria approach was implemented and stakeholder engagement with key bodies was undertaken. This included but was not limited to the Scottish Government, Scottish Environmental Protection Agency, Scottish Natural Heritage, Royal Society for Protection of Birds and in particular Local Flood Action Groups.

Local Community Engagement
For many of the local communities directly affected by the storm events of January 2005, one response was to set up Flood Action Groups that could act as a mouthpiece

to the Council and its own coastal managers, and also formulate positions as part of the overall stakeholder engagement.

Through an informed position, the flood action groups were able to contribute significantly to the development of viable management solutions and schemes during the engagement process. It was clear that the groups held a wealth of knowledge on historical events and changes to the coastlines which assisted in assessing the potential impacts of future erosion and flooding scenarios. There were other more uncomfortable discussions held which directly related to 'crofting rights' over the land and the impact that these had on the sedimentary features that were protecting much of the coastline. Significant grazing had taken place on the sensitive vegetated dunes, which was observed to have been leading to increasing damage. In addition, at the case study site Stoneybridge, significant quantities of shingle were removed during the 1960s and 1970s. Crofters have the right to extract material for their own use, but the practice has not officially continued. However, historical abstraction does provide a beach system depleted of sediment and potentially out of equilibrium with its environment (Richards and Phipps, 2007).

The overall consensus that came through engagement with Local Flood Action Groups was that whilst there was a recognised need to protect the local communities, it should be implemented in a sustainable manner that was in keeping with the unique aesthetic, cultural and environmental characteristics of the Western Isles.

Detailed Design Development

Once funding had been secured for the implementation of the 5 coastal protection schemes, the detailed design of each scheme was developed. Although a concept design of the preferred option was submitted with the application for funding, there was still a great deal of flexibility in how the concept could be implemented. It was therefore possible to develop the detailed design with the aim of providing the most sustainable solution.

Generally the most sustainable solution is one with the highest life time benefit/cost ratio and the least environmental impacts, without affecting growth, development and quality of life. With this aim in mind the detailed design was developed. The biggest portion of the life time cost came from the upfront capital cost of construction. Therefore it was fundamental to ensure capital cost was kept as low as possible whist still maintaining the technical requirements of flood and erosion protection.

Economically sustainable designs were developed in a number of different ways. Topographic surveying and site investigations were carried out at each site to assess the existing material on site in terms of its volume and properties. The designs were then developed to utilise as much of the existing material as possible. The result of this was reducing the amount of material required to achieve the design. The majority of the Western Isles is composed of Lewisian Gneiss, an extremely hard and wear resistant rock. Thus designs were developed to utilise this abundant resource from armourstone and underlayer, to stone fill for reno-mattresses and gabion baskets.

Stoneybridge Case Study

At the time of writing the coastal protection scheme at Stoneybridge is one of the 2 sites that have been completed to construction, with the remainder due for completion over 2011. Therefore, Stoneybridge is used as a case study to present in more detail some of the community engagement issues, sensitive environmental matters and sustainable engineering and management approaches, which are also relevant in varying degrees to the other four locations being proactively managed.

The Nature of Stoneybridge
Stoneybridge is located on the Island South Uist (Figure 5). The local coastal road runs west of Loch Altabrug and in places is within 30m of the shoreline. Prior to the 2005 storm, the coastal dune system consisted of a partially vegetated shingle ridge between the sandy beach and seawards of the coastal road (Figure 6).

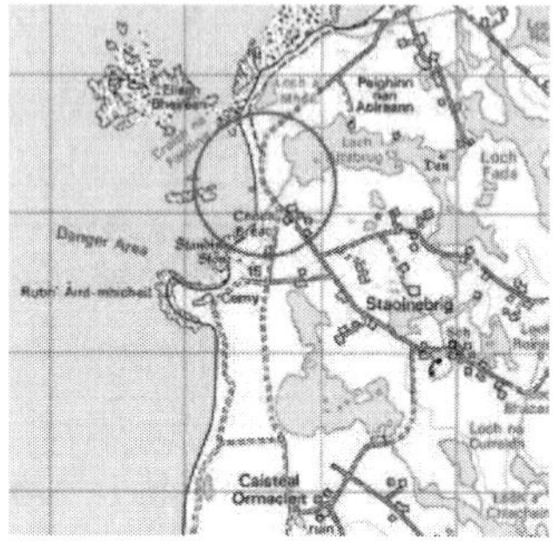

Figure 5. The Study Area at Stoneybridge illustrating location of the coastal road (Phipps and Richards, 2007)

Figure 6. Shingle ridge at Stoneybridge prior to the 2005 storm. Courtesy of Stewart Angus SNH

The high water level and large waves of the 2005 storm led to the shingle ridge being flattened and shingle dispersed up to 100 m inland. The road was covered with shingle debris, leaving it impassable for three days prior to clearance by bulldozer. At the time of field surveys in early 2006 the shingle material had been formed into an unnatural 'knife edge' shape. The resulting ridge was 1-2 m high with a limited crest width of approximately 0.5 m (Figure 7).

Figure 7. 'Knife edge' shaped shingle ridge post 2005 storm.

The coastal road at Stoneybridge forms a vital link in the middle district of South Uist, thus during the storm the community of Stoneybridge was cut off from other communities to the north. This had serious impacts as the communities have a very close relationship with shared facilities. The local school, community hall and cemetery are to the south of Stoneybridge Bay and the two churches and main crofting area to the north. Flood levels in Loch Altabrug and adjacent lochs were such that people were cut off from access routes. Flooding was attributed to a combination of tidal ingress into the lochs, fluvial flooding from inland and over washing by the sea at the depleted shingle bank.

It was established through the option appraisal process and community engagement that the lasting loss of the coastal road at Stoneybridge would not only mean truncation of the community and increased journey times in the Middle District via the A865 inland, but that the associated loss of confidence could lead to suspension of crofting and migration from the area. With the already declining population of the Western Isles, the need to maintain a sustainable community was paramount. The beaches and machair are attractive with a high amenity value, whilst local businesses depend on passing tourist traffic for their livelihood.

The traditional system of low intensity agriculture is integral to the biodiversity of the machair ecosystem and the relationship between the local population and the value of the unique machair landscape could not be ignored. Indeed, when set against overall threat of depopulation in the Western Isles that has led to loss of half the population in 100 years, the importance of sustaining viable communities was a key objective when identifying a preferred option.

Option Appraisal and Design Development
Prior to consultation, the Local Flood Action Group from Stoneybridge had already considered various options for provision of defence to their community from the sea, including an offshore breakwater. However, through the option appraisal process the offshore breakwater option was deemed unsustainable and therefore not developed in detail. Ultimately, the preferred option that satisfied the technical criteria as well proving to be both economically and environmentally sustainable involved the re-distribution and re-profiling of the existing shingle ridge. This would be coupled with the placement of a line of 2-3 tonne rock (local Lewisian Gneiss) up to 2m high, seaward of the coastal road as presented in Figure 8.

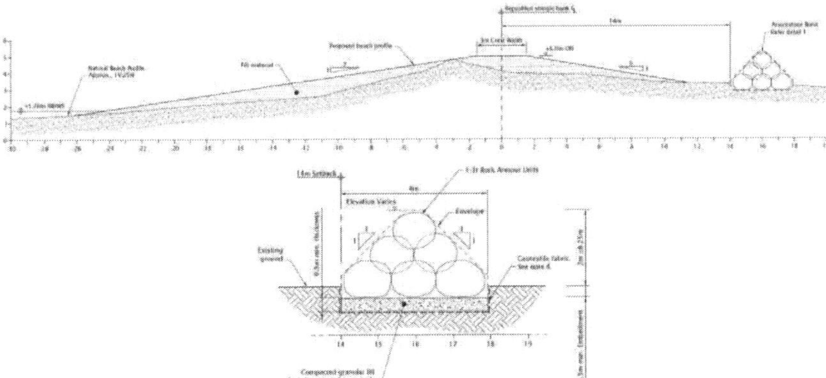

Figure 8. Design beach profile and armourstone bund detail

Topographic surveys indicated that to the south of the site a surplus of shingle material had accumulated, whilst to the north there was a deficit. The design profile was developed to provide a cut/fill balance in combination with the technical requirements of a robust shingle embankment.

The main philosophy behind the development of the solution at Stoneybridge was to maintain the robustness of the shingle embankment to resist wave attack during subsequent storm events and prevent excessive erosion of the underlying machair, whilst still being able to constitute a viable flood defence level from tidal inundation. There was recognition that the shingle ridge would have a tendency to roll back during a storm event, and the rock units were designed to prevent the road being blocked in all but the severest events.

A total length of 500m of works along the coastal frontage was to be implemented. The overall Present Value of the preferred option was £0.53 million including Optimism Bias and contingency with Present Value Benefits of £0.8 million which obtained for the preferred option a Benefit Cost of 1.51.

Construction
A construction contract was developed using the NEC ECC Option A – Lump Sum with Activity Schedule and ultimately a contract awarded to UBC Hebrides. The works commenced with the re-profiling of the shingle embankment (Figure 10) and followed by the construction of the armourstone bund (Figure 11). Substantial completion was declared in January 2010.

Figure 10. Reprofiling the shingle embankment   Figure 11. Construction of the armourstone bund

Sandy organic material removed from beneath the footprint of the armourstone bund was spread over the back face of the shingle embankment to promote vegetation to provide stability.

The out-turn cost of the capital works was £95,000. Whilst there have not been major storm events affecting this frontage, there has been an extreme tidal event on 7 December 2009 but which has not led to any subsequent loss of integrity of the shingle ridge or flooding of the hinterland.

Conclusions

The Western Isles of Scotland are extremely sensitive both environmentally and socially. The low lying western coast is vulnerable to erosion and flooding impacts associated with major storm events. The extreme storm event which hit the Western Isles in 2005 storm led to increased coastal erosion, extensive damage to property and infrastructure and ultimately the loss of life. Management solutions for 5 key locations have been developed following provision of funding support from the Scottish Government, in order to control potential future coastal erosion and maintain a viable level of flood defence. The coastal fringes of the Western Isles are environmentally valuable, comprising rare machair habitats and have major aesthetic and amenity value. The possibility of further de-population of the Western Isles was integral in designing sustainable solutions that strive to maintain and promote important community linkages. Engagement with the affected communities was a fundamental part of the option development, which added value whilst also highlighting needs for the communities to modify their own behaviours to minimize degradation of natural protection structures. The implementation of the scheme at Stoneybridge has provided an example of a sustainable management solution which utilised available resource, augmented the natural environment and ultimately enhanced the aspirations of a sustainable community.

References

Angus, S. and Rennie, A. 2006. The Natural Heritage impact of the storm of 11th January 2005 in the Uists and Barra, Outer Hebrides. Scottish Natural Heritage. (Second Draft).

CnES 2003. Western Isles Structure Plan
www.cne-siar.gov.uk

CnES 2010. Comhairle nan Eilean Siar.
http://www.cne-siar.gov.uk

CnES 2010. Outer Hebrides Fact Card.
http://www.cne-siar.gov.uk/factfile/documents/OHFactCards2010.pdf

CnES 2010. Outer Hebrides Factfile
http://www.cne-siar.gov.uk/factfile/population

CnES 2010. Outer Hebrides Local Development Plan Strategic Environmental Assessment Environmental Report
http://www.w-isles.gov.uk/planningservice/documents/SEA%20Env%20Report%20LDP.pdf

Defra (2006). Flood and Coastal Defence Appraisal Guidance: FCDPAG3 Economic Appraisal: Supplementary note to operating authorities: Climate Change Impacts October 2006.

HR Wallingford 2006. Western Isles Coast Protection Study. Wave and Tidal Conditions on 11th 12th January 2005. HR Wallingford Report TN CBM5582/01. March 2006.

Moorbath, S. Powell, J.L. and Taylor, P.N. 1975. Isotopic evidence for the age and origin of the grey gneiss complex of the Southern Outer Hebrides Scotland. Journal of the Geological Society of London, 131 pg213-222.

Richards, L.A.R. and Phipps, P.J. 2007. Managing the impact of climate change on vulnerable areas: a case study of the Western Isles, UK.

Ritchie, W. 1971. The Beaches of Barra and the Uists. A survey of the beach, dune and machair areas of Barra, South Uist, Benbecula, North Uist and Berneray.

Innovative Coastal Zone Management
ISBN 978-0-7277-5749-4

ICE Publishing: All rights reserved
doi: 10.1680/iczm.57494.164

Stakeholder Engagement and the Severn Estuary Shoreline Management Plan – Lessons Learned for Marine Planning and other Large Scale Plans

Kath Winnard, Atkins Ltd, Swansea, United Kingdom

Introduction

Shoreline Management Plans (SMPs) are high level, strategic and non-statutory plans that set overarching policy on how the shoreline should change over a 100 year time period, divided into three 'epochs' (0-20, 20-50, 50-100 years). Notwithstanding their non-statutory nature, they are considered a vital part of the planning and management of coastal erosion and flooding in England and Wales, and their development has been funded by Defra (via the Environment Agency) and the Welsh Assembly Government (WAG) (via Welsh Local Authorities).

The first round of SMPs was developed during the 1990s. The current round of Shoreline Management Plan reviews are commonly referred to as SMP2s. SMP2s in England have been completed and are at various stages of approval, while the two all-Wales SMP2s are in the later stages of development.

SMP2s aim to help planners and regulators plan for and manage the shoreline over time – by creating, improving or maintaining defences, creating new habitat, enabling natural processes to take a greater role, improve flood warning, manage development in areas of risk and enable people in areas of risk to cope better with that risk, through flood warnings, individual property protection and improving the resilience of buildings and communities to flood events. SMP2s should inform regional and local spatial planning strategies and planning decisions to ensure development does not take place in areas of current or future flood risk and reduce the risk of placing a financial burden on future generations to protect inappropriate development. They should also be developed in collaboration with key stakeholders and the wider public.

However, despite the importance that flood and erosion risk managers place on SMPs, not all land use planners are aware of SMPs, how they should fit into land use planning and consenting and few members of the public are aware of them.

This paper sets out detail about stakeholder engagement in SMP2s in general, some specifics about the stakeholders and challenges of developing the Severn Estuary SMP2 (SE SMP2), the stakeholder engagement that took place, what worked (and what didn't) and some lessons that can be learned for stakeholder engagement in other large scale plans, such as marine plans.

Stakeholder Engagement in SMP2s

Stakeholder engagement is considered a key part of the development of SMP2s. Defra produced guidance documents to steer the development of SMP2s (Defra, 2006a; Defra, 2006b). The guidance was also adopted by WAG with some additional guidance on specific Welsh issues. The guidance defines a number of different stakeholder groups with whom the SMP2 developers should engage:

- Project Management Group (PMG) – responsible for the day to day development of SMP2, makes decisions, works with the consultants appointed to develop the SMP2;
- Coastal Group – has overall responsibility for the development of the SMP – made up of EA regions, Local Authorities, Internal Drainage Boards, statutory nature conservation agency, historic environment agency, etc.
- Elected Members Forum (EMF) – elected members from the local authorities, EA flood management committee(s) – should mirror the membership of the Coastal Group with elected representatives rather than officials;
- Key Stakeholder Group (KSG) – stakeholders with primary interests – industry bodies, parish councils, town councils, nature conservation groups, user groups, etc., and;
- Other stakeholders – everyone else / the public.

The guidance also sets out when in the SMP2 development stakeholder engagement should take place and with which groups (see Figure 1).

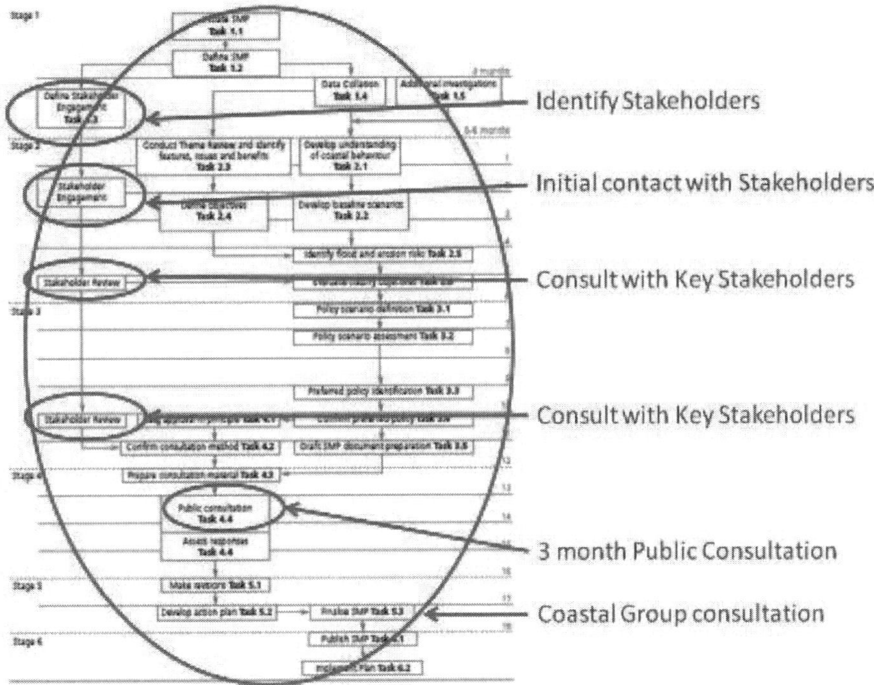

Figure 1.When stakeholder engagement takes place during the development of an SMP2

Stakeholders in the Severn Estuary

The stakeholder groups in the Severn Estuary, for the development of the Severn Estuary SMP2 were:

- Project Management Group (PMG) – a sub-group of the Coastal Group, representing both England and Wales;
- Coastal Group – Severn Estuary Coastal Group (SECG) – comprised of 10 English and Welsh Local Authorities, 3 EA regions, 3 Internal Drainage Boards, 2 Statutory Nature Conservation Bodies (CCW, NE), 2 Historic environment advisors (Cadw, English Heritage), WAG, Defra;
- Elected Members Forum (EMF) – elected members from the Local Authorities, the English Flood Risk Management Committees;
- Key Stakeholder Group (KSG) – over 300 different organisations, and;
- The Public – approximately 500,000 people live within the SMP2 area.

Challenges in the Severn Estuary (and the SMP2)

The Severn Estuary SMP2 is a cross-border plan. The SE SMP2 area runs from Lavernock Point (near Penarth in Wales) to Anchor Head (near Weston Bay in England) via Haw Bridge (near Gloucester). Covering both England and Wales means that as well as including 10 Local Authorities and their policies and plans, it includes two national governments, with their respective guidance, policies and advisors and different roles and responsibilities of key organisations. For example, the EA in England has an overseeing role for coastal defence, while the EA Wales does not; responsibility for coastal defence rests with Local Authorities. This is why WAG funds SMP2s in Wales via Local Authorities and not via the EA Wales (although EAW is a member of both the coastal group and the PMG).

The area is both rural and urban, containing a large expanse of agricultural land and lots of dispersed properties, as well as large urban conurbations, including Cardiff, Newport, Bristol and Gloucester. There is a lot of important utilities and infrastructure very close to the shore or in the flood plain, including the M4, the Severn Crossings and tunnel, mainline railway lines, electricity generation and distribution and water and sewerage. The Severn Estuary has been populated for the last 2,000 years and is, therefore, heavily modified – a large area of land was reclaimed in the past (since the Roman occupation) and there is a huge number of historic environment features.

The area is very low lying and flooding is the main risk to coastal areas, but the sources of flooding are not restricted to tidal flooding – there are several large rivers that pose a fluvial flood risk, rainwater runoff and urban flooding is increasingly posing a problem and tidal surges are a potential risk. The SMP2, however, only deals with the issues associated with tidal flooding.

The entire SE SMP2 plan area is within EU or International conservation sites – the Severn Estuary Special Area of Conservation (SAC), Special Protection Area (SPA) and Ramsar site, not to mention the large number of other EU sites in the region (such as the River Usk SAC), national and local conservation sites.

In addition, while the SMP2 was being developed, the Severn Tidal Energy Feasibility Study was also being undertaken.

What did we do?

A wide range of stakeholder engagement methods were used during the course of the development of the SMP2:

- Severn Estuary Partnership (SEP) – there are a number of existing partnerships and groups in the Severn Estuary for which SEP acts as a facilitator and secretariat[1]. It has a pre-existing website, staff resources, a database of over 2,000 contacts, pre-set timetable of meetings and events, e-mail newsletter, annual conference and quarterly magazine (Severn Tidings);
- Website – via the Severn Estuary Portal – with a specific area for the SMP2;
- Initial mailshot and questionnaire to over 300 organisations;
- Regular updates – monthly email updates using SEP's e-news – this has a circulation of over 1,000 organisations and individuals. Other stakeholders not already on the circulation were added;
- Stakeholder events - advertised on the website, direct e-mail to Key Stakeholders,. Aimed at the KSG, but other stakeholders were not excluded. Stakeholder events were held in January 2009, June 2009 and August 2009. Events were held in locations in England (Bristol, Gloucester, Slimbridge, Clevedon) and Wales (Penarth, Cardiff). In total over 170 people attended key stakeholder events and over 130 responses to questionnaires were received;
- Specific meetings / workshops with important stakeholder groups for key issues – historic environment, natural environment, planning;
- Elected Members Forum – aimed at elected officials from each of the Local Authorities and EA flood committees. Assembly Members, MPs and MEP were also included in the group, and;
- Public Consultation Events – advertised on the website, by e-mail, via SEP e-news, by SECG members, press releases to local papers and radio stations. Events were held in England (Bristol, Gloucester) and Wales (Penarth). Although two events were held in England, Gloucester was seen by the SECG as an important location in order to engage with stakeholders in the upper part of the estuary. Event were held during the afternoon and evening to enable people to attend outside of normal working hours and during their lunch breaks. The consultation questions were available in hard copy at the events, downloadable from the SECG website or could be completed online via the website. A total of 87 responses to the Public Consultation were received.

What worked?

The SECG Project Management Group (PMG) members were asked for their thoughts and opinions on what worked well and what didn't with respect to stakeholder engagement over the course of the development of the SMP2. The comments below are based on the feedback from the PMG. PMG members also passed on comments from colleagues and contacts outside of the SECG, including members of the Environment Agency's Quality Review Group (QRG). The QRG was established to ensure that all SMP2s in England and Wales were produced to a similar standard and followed the Defra SMP2 guidance. No formal

---

[1] ASERA – Association of Severn Estuary Relevant Authorities - represents 32 of the 43 relevant authorities with statutory duties in respect of the nature conservation designations on the Severn Estuary

BCSEG – Bristol Channel Standing Environment Group – made up of agencies with responsibilities for preparedness against marine pollution incidents in the Bristol Channel

SECG – Severn Estuary Coastal Group – made up of local authorities and other organisations with an interest in the management of the coast and erosion and flood risk management

comparison with other SMP2s has been made, although some anecdotal comments have been received via SECG members and coastal group members for other SMP2s.

The website was thought to be a key communication tool, particularly the clickable map of the SMP2 area that allowed people to jump straight to the area they were interested in (see Figure 2) and the clickable electronic version of the SMP2 plan and supporting documents.

However, this did mean that people were able to skip over important supporting information and jumped straight to the 'headline' policy options.

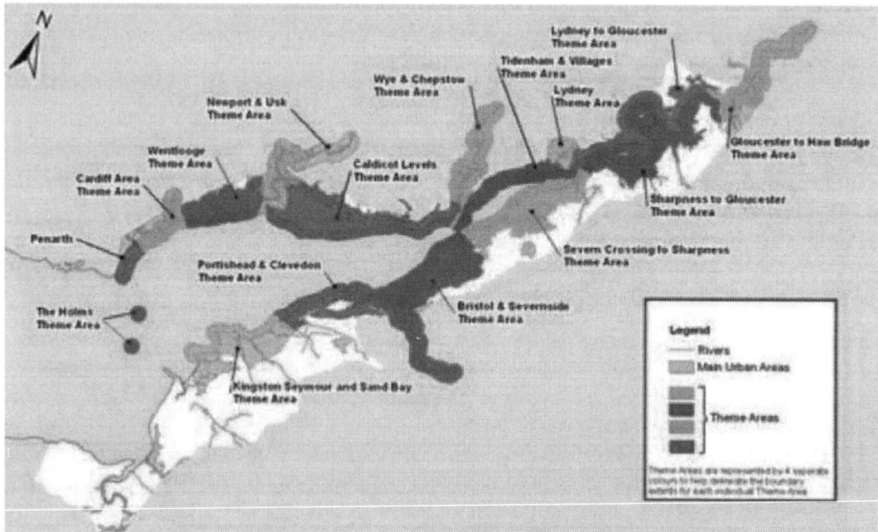

Figure 2. Clickable map of the SMP2 area

Coastal Group members were considered by both the SECG and the consultants to be vital in the engagement process – they were active in their input to the development, involvement in events, providing equipment, information and staff to support events and help disseminate information within and outside their own organisations. It was mainly the SECG members that were part of the SMP2 Project Management Group (PMG) that were particularly active in this respect. The PMG represents approximately 10 different organisations out of approximately 20 organisations on the whole SECG.

Topic-specific meetings, particularly with local authority planners helped to both raise the profile of the SMP2 and the level of understanding about the plan and its role in land use planning. Topic specific meetings were also held with historic environment stakeholders and with CCW, NE and EA advisors on the requirements for the Habitats Directive, IROPI and compensatory habitat creation.

Both SECG members and the consultants felt that the SEP was extremely important to all levels of stakeholder engagement. As a pre-existing Partnership, it had a range of communication tools, knowledge of stakeholders and issues in the area. SEP also had practical knowledge about venues, their accessibility by transport and for less able stakeholders and resources that could be used such as display boards, projectors, screens and staff that were used to organising and executing events.

What didn't work....and why?

Other approaches worked less well, including the Elected Members Forum (EMF). Turnout at EMF meetings was low – at one event only two members attended. Additional EMF events were arranged prior to the public consultation events in an effort to increase engagement with this group. Additional effort by SECG members from local authorities was made in order to raise the profile of the SMP2 and the SECG Chair wrote to local authority Chief Executives several times during the plan development process.

This stakeholder group was not part of the development of earlier SMPs and the lack of involvement of elected local representatives had proven a stumbling block in the adoption of earlier SMPs. However, in the Severn it proved difficult to get politicians engaged – the development of the SMP2 wasn't in general seen as a political issue until policies and the adoption of policies were coming to the fore. The number of elections that took place during the development of the SMP2 was felt by SECG members to 'distract' from the SMP2 process. Over the course of the SMP2 development, local authority elections, Welsh Assembly elections and a general election took place.

Not all Coastal Group members were as involved as others – those members on the PMG tended to be more engaged and involved than other SECG members. It was felt by those that were actively engaged that part of the reason for less involvement from other members was in part due to coastal matters forming only a small part of their overall role. The time commitments given to the SMP2 development process by the PMG members was significant (monthly meetings, large and technical documents to read and comment on, stakeholder events to attend) and not all SECG members can commit to this level of engagement for the time period over which the SMP2 was developed (approx. 18 months).

The complexity of the SMP2, the development process, the high level, strategic nature of the SMP2 and the technical issues involved (the different types of flooding, the terminology, statutory assessments) make it difficult for non-technical and general members of the public to pick out which elements they need to / should be involved with and comment on. This is common with flood and coastal erosion risk management strategies. Comments from SECG members and QRG praised the Severn Estuary SMP2 for being written in an easily accessible and non-technical manner.

Feedback from SECG members based further up the estuary suggested that people particularly towards the top of the estuary did not view themselves as being on the coast and so didn't think the SMP2 was relevant to them.

The problems experienced with some of these techniques in the Severn Estuary are not necessarily the same as for other SMP2s in other locations. Discussions with members on other coastal groups found that in some areas the EMF was highly engaged throughout the process.

What lessons can be learned?

There are a number of key lessons that the development of the SMP2 and the stakeholder engagement process highlight, which are directly relevant to large scale strategies and plans, such as the developing marine plans. Many of these have been discussed in the body of work on stakeholder engagement that is widely available. These findings are, however, the views of this author based on personal and practical experience during the course of the development of the Severn Estuary SMP2 and comments from SECG members.

- Time – it takes time (and resources and effort) to produce documents that are clear, concise and easy for a non-technical audience to understand – much more time than you would think;
- Who are you asking – information needs to be tailored to the audience - not everyone understands all the technical information and background analysis that goes into to developing a plan;
- Who really HAS to be involved – identify the people that really need to know and make sure they are involved. Don't expect people to know that they should be involved.
- What are you asking – make it clear what you are asking about, what influence people can have on the decision and what they can't influence
- Everyone has a role to play in the engagement process – representatives of organisation have a role to play in feeding back information and gathering opinion from their own organisations. They don't necessarily recognise this;
- Make it easy – use clear text, Plain English, easy to navigate, use maps, diagrams and colours that are visible to everyone (some people are colour blind);
- Be flexible and open to newcomers – not everyone will find out about your plan or consultation at the same time. It is important not to close the door to latecomers and this is particularly true for marine planning because it is so new. However, it is important that newcomers don't upset the process or damage relationships between those stakeholders that are already engaged.
- Use a range of different methods – not everything works for everyone or for everything you want to engage on;
- You won't please everyone (or maybe not anyone) – people will always think you could have done more / done it differently

It is, above all, always important to remember why you are engaging with stakeholders. Stakeholder engagement is not marketing, awareness raising, education or capacity building. It may help with these things, but it is not the aim. Stakeholder engagement is about getting people involved in the decision making process and using the information and opinions they provide (often at a cost of time and money to themselves) to inform the decisions you make. Keep this in mind and tailor the process to the stakeholder.

References

Defra, (2006a), Shoreline Management Plan Guidance Volume 1: Aims and requirements, Department for Environment, Food and Rural Affairs, Crown Copyright.

Defra, (2006b), Shoreline Management Plan Guidance Volume 2: Procedures, Department for Environment, Food and Rural Affairs, Crown Copyright.

Innovative Coastal Zone Management
ISBN 978-0-7277-5749-4

ICE Publishing: All rights reserved
doi: 10.1680/iczm.57494.171

Developing a Practical Tool for Coastal Flood Forecasting and Warning

John Ray, Environment Agency. Lincoln, England
Elizabeth Coulling, Environment Agency. Lincoln, England
Sun Yan Evans, Mott MacDonald. Cambridge, England
Kiki Spink, Mott MacDonald. Cambridge, England

Background

The Environment Agency is responsible for issuing flood warnings across England and Wales for flooding from main rivers and the sea. It is continually looking for ways to improve its flood forecasting and warning services to minimise the risk to people and damages to properties.

The Lincolnshire coastline has always been prone to flooding as land levels are typically three to four metres below tidal surge levels. The coastal floodplain, which extends up to ten kilometres inland in places, relies entirely on man made sea defences for protection and is shown in Figure 1.

These defences have been raised and strengthened over the years to reduce the risk of flooding. Nevertheless, surges, high tides and strong winds remain a threat, as complete protection against the most extreme floods can never be guaranteed.

Figure 1: Lincolnshire's Tidal Floodplain

In order to inform the effective planning for responding to tidal flooding and long term land use planning policies, we have undertaken a comprehensive assessment of the current and future flood risk along the coastline from the Humber to the Wash.

This assessment has modelled the likely amount of overtopping of the tidal defences for a range of scenarios, from a 2% (1 in 50) annual chance tide level through to an extreme 0.1% (1 in 1000) annual chance level. Flooding as a consequence of failure (breaching) of the tidal defences has also been modelled and mapped.

This new modelling and mapping has been used to improve the coastal flood forecasting and community based flood warnings along the Lincolnshire coastline. A range of practical tools have been developed to assist with the real time forecasting of tidal overtopping rates and prioritisation of potential breach locations.

The purpose of this paper is four-fold:
(i)   To highlight the limitations with the original flood warning information;
(ii)   To describe the assessment of the Source – Pathway – Receptor components of flood risk;
(iii)   To outline the new tool developed to assist the flood forecasting and warning decision process;
(iv)   To outline future considerations for further performance refinement.

Original Flood Warning Information

Although recent years have seen big improvements in the forecasting of atmospheric and tidal conditions there has been relatively little progress in translating these offshore forecasts into meaningful information for flood warning and emergency response purposes.

Flood warnings were provided based on a simple look up matrix based on forecast tide level, together with wind speed and direction. The matrix simply indicated whether an operational threshold was likely to be exceeded (blue boxes) or if the issuing of a flood alert (green boxes), flood warning (yellow / amber boxes) or severe flood warning (red boxes) should be considered.

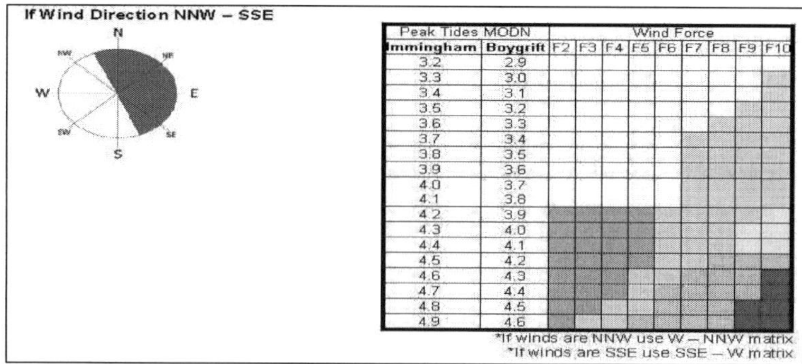

Peak Tides MODN		Wind Force								
Immingham	Boygrift	F2	F3	F4	F5	F6	F7	F8	F9	F10
3.2	2.9									
3.3	3.0									
3.4	3.1									
3.5	3.2									
3.6	3.3									
3.7	3.4									
3.8	3.5									
3.9	3.6									
4.0	3.7									
4.1	3.8									
4.2	3.9									
4.3	4.0									
4.4	4.1									
4.5	4.2									
4.6	4.3									
4.7	4.4									
4.8	4.5									
4.9	4.6									

Figure 2 : Example of the original flood warning matrix

In recent years the performance of the matrix has been monitored after each use, to see if the conditions on the ground actually reflected those expected for the relevant flood warning type.

The origins of the matrix, which has been in use for over 20 years, have been lost during the passage of time. Given the risk to life posed by an east coast tidal surge it was necessary to have a more scientific evidence base to support our flood warning decisions.

Using the matrix in Figure 2 it was possible to determine the flood warning type to issue. However the areas covered by the flood warnings were large, with only four flood warning areas covering the whole of the coastline from the Humber to the Wash. These were:

- The Humber South Bank;
- The East Coast;
- The Wash Frontages;
- Boston.

The only available flood mapping for the coastline was the basic flood map shown in Figure 1. Maps showing the areas at risk of flooding in real tidal events, from overtopping or failure of the defences, were simply not available. This resulted in flood warnings being issued to all properties in the floodplain, with no distinction of the varying levels of risk they faced.

In order to provide the community based flood warnings our customers needed and to do so based on up to date science, it was clear improvements were needed. The opportunity to make these improvements came in late 2010 following the completion of the Northern Tidal Modelling (NTM) project which had modelled and mapped the consequences of flooding for a range of scenarios, including overtopping and breaching of the sea defences.

Source – Pathway - Receptor Mapping

The objective of the project was to update the existing flood warning matrix to make full use of the latest available data and science and data to provide timely, community focused, flood warnings along the Lincolnshire coastline.

In providing an effective flood warning service it is important to understand the SOURCE – PATHWAY – RECEPTOR components of the flooding where:

- SOURCE is the tide and meteorological conditions, such as tide level, wind speed and direction and wave height;
- PATHWAY is the mechanism of flooding, such as overtopping or breaching of the defences;
- RECEPTOR is the floodplain and the things that could be flooded. This may be a residential property, business, school, care home, or critical infrastructure.

Once these were understood it was then possible to develop the new flood warning tool around these components.

Source

The first thing to consider is the source of the flooding, in this case the tide level and wave heights at the defence. As part of the NTM project a new set of extreme sea levels for the Lincolnshire coastline had been calculated, based on available gauged tide level records. Offshore and near shore buoys provided information on recorded wave heights and wind speed. As a starter these were analysed to determine the critical wind speed and direction which produced the mean annual (one in one chance) wave height.

This critical wind speed and direction is of use when considering a 'design event' of a tidal surge with an X% annual chance of occurring, but is of little use in real events where it is necessary to convert available forecast data into a flood warning and potential consequence.

Tidal forecasts from the Met Office provide expected values of sea level, wind speed and direction and offshore wave heights at specific forecast points offshore. What was needed was some way to translate these offshore forecasts to forecasts at the defence, on which overtopping calculations could be applied.

A full analysis of offshore and nearshore buoy records was carried out to consider the relationships between wave height, wind speed and wind direction. Once this was established the relationship between offshore and onshore wave height was also assessed.

The final stage was to carry out a multi-relationship regression analysis to produce a single formula which could be used to estimate nearshore wave heights based on offshore forecasts of sea level, wind speed and direction. The final modelled relationship is shown below and illustrated in Figure 3.

$$Hsn = A + B \times Wspd + (C \times Wspd + D) \times Sin(Wdir + E) + F \times Hso$$

Where:

     Hsn is the nearshore wave height
     Hso is the offshore wave height
     Wspd is the offshore predicted wind speed
     Wdir is the offshore predicted wind direction
     A, B, C, D, E, F and **G** are coefficients to be calculated to fit the data

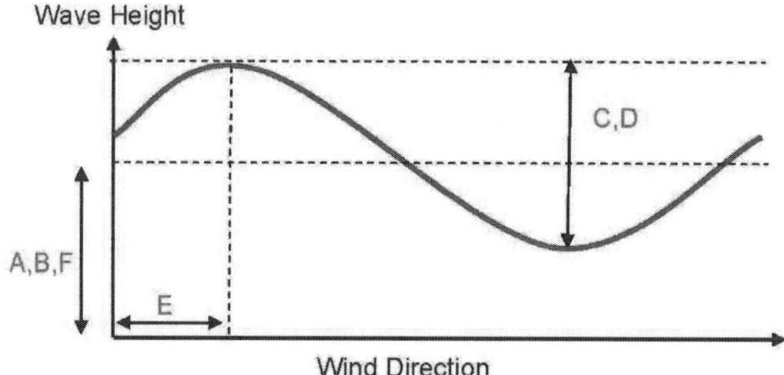

Figure 3 : The Modelled Relationship

Figure 3 illustrates what this model looks like and how the coefficients affect it:
- A, B and F affect the midpoint of the sine cure (i.e. the average wave height over all directions). Therefore the wind speed and offshore wave height effect this.
- C and D affect the amplitude of the sine curve. These coefficients are more influence by wind direction as the wind speed increases.
- E is used to offset the sine curve so the peak direction can be altered.

With the new formula we have the ability to take Met Office offshore forecasts and translate them into forecast conditions at the defence, thereby providing the SOURCE element of the S-P-R relationship.

Pathway

The way in which the source (the sea) gets to the receptor (the floodplain) depends on the tidal conditions and the state of the defences. This would either be from overtopping of the defences or, in extreme cases, following a breach on the defences.

Predicting a defence breach in advance of occurrence is almost impossible and certainly unlikely based on a tidal forecast some days ahead of the actual tidal event. Therefore flood warnings in advance of flooding are only based on predicted overtopping of defences.

In order to calculate the expected overtopping rates for specific tidal conditions the coastline was schematised into over 300 individual overtopping sections (OT Sections) based on defence type, shape, crest level, toe level and foreshore condition. For each OT Section the overtopping rate, in cubic metres per second per metre length of defence (m^3/s /m) was calculated for a range of design event scenarios for the critical wind speed and direction, ranging from a 2% (1 in 50) annual chance to a 0.1% (1 in 1000) annual chance tidal event.

Automating the overtopping calculation and embedding it within an Excel spreadsheet opened up the opportunity to rapidly update the estimated overtopping rates based on the offshore forecast conditions, as outlined in the SOURCE section above.

Overtopping calculations were made based on a modified version of the methodologies within the *EA Wave Overtopping of Seawalls Design and Assessment Manual*. A modification was necessary as the project also considered the potential impacts of climate change and this often led to the scenarios under consideration being outside the applicable range of the formula.

Receptor

The RECEPTOR component of the S-P-R relationship can either simply be taken as the floodplain or the things in the floodplain affected by the flooding. It could be considered that the floodplain is an extension of the PATHWAY component, in that this is the route for spreading of the flooding, but for this project the RECEPTOR has been taken to be either the floodplain or the things in the floodplain. This is partly for simplicity but is also an acknowledgement of the value of the agriculture land that makes up a large proportion of the tidal floodplain in Lincolnshire.

Topographic information of the tidal floodplain, based on a two metre grid obtained from aerial LiDAR surveys (Light Detection and Ranging), was used to construct a 2-dimensional computer model of the floodplain using the TuFlow numerical modelling software. Inflows to the model were the overtopping rates for each OT Section along the coastline

Using TuFlow the extent of flooding from the overtopping of the defences was modelled and mapped. The area affected was colour coded to show maximum values of flood depth, velocity and a hazard rating which is a measure of the danger to people and is a function of the flood depth, velocity and a factor for potential debris in the flood water.

A range of design event scenarios, such as the 2% (1 in 50) annual chance through to the 0.1% (1 in 1000) annual chance tidal event were mapped initially.

Using the National Receptors Dataset we have compiled a list of the key receptors, such as residential properties or critical infrastructure, affected by the overtopping and the likely depths of flooding.

Although we do not have the ability to forecast a breach in advance of it occurring it is important to understand the potential consequences should one occur. Making use of the TuFlow model we were also able to map the area flooded in the unlikely event of a breach in the defences. Simulating breaches at regular intervals along the coastline, typically every one to two kilometres, means that should a breach occur during a real event we would have an assessment within close proximity to it. It is accepted that an actual breach would not occur in exactly the same location, manner or to the same width as we have simulated but the pre-prepared assessments would be of use in understanding the likely consequence.

The flooding from a breach has also been mapped in terms of maximum values of flood depth, velocity and hazard rating. Flood progression maps which indicate the direction and rate of flood spreading have also been produced for time intervals of 2, 4, 6, 8, 12, 24 and 72 hours after the breach. The receptors affected for each time interval have also been determined from the National Receptors Dataset and provided in the form of a simple look up table.

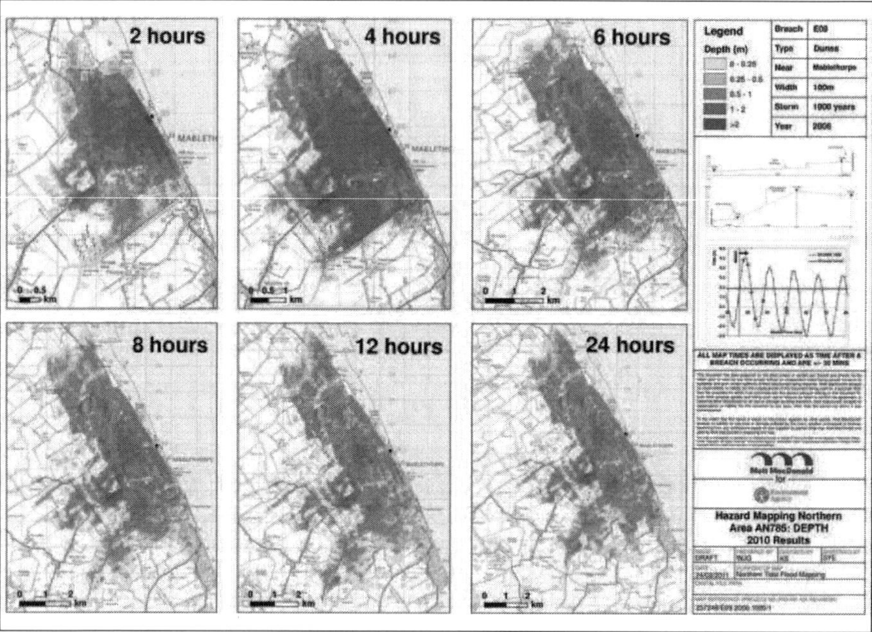

Figure 4 : Example Flood Progression Map

New Tidal Tool Kit

The improved flood warning tool kit brings together a number of individual elements from the work on improved understanding of the SOURCE – PATHWAY – RECEPTOR components of flood risk. These are:

- new community based flood warning areas;
- overtopping rates based on actual forecasts;
- new flood warning thresholds and;
- pre-defined mapping of potential flooding from overtopping or breaching.

The Lincolnshire tidal flood warning areas no longer cover long lengths of the coastline, stretching back as far as the full tidal floodplain. New flood warning areas have been defined based on the extent of flooding from the overtopping or breach of the sea defences.

Along the coastline the flood warnings areas (FWA's) have been made much smaller to cover specific communities at risk. Moving back inland from the sea the FWA's have been divided up with areas immediately behind the defence and at risk from overtopping, being issued a flood warning in advance of those further inland which may only be at risk in the event of a breach in the defence.

Before this project the Lincolnshire coastline was covered by four large flood warning areas. It is now covered by 51 smaller community, risk based, warnings.

This paper has already covered how overtopping rates at the defence have been calculated using offshore forecasts and how this calculation has been automated and embedded in an Excel spreadsheet.

Figures 5, 6 and 7 below illustrate the three key screens from the new tidal flood warning tool and explains how each of these are used.

Map View

The map view in Figure 5 shows the Lincolnshire coastline, with small dots to represent each modelled overtopping section. The user is able to enter the basic forecast information on sea level, wind direction and speed. Once this is entered the macro within the spreadsheet will automatically update all the overtopping calculations based on this information. The remainder of the options on the map are dedicated to how the user wishes to display the information.

It is possible to highlight / colour code the overtopping sections based on:

- exceedance of flood warning thresholds;
- exceedance of overtopping rate thresholds, which can be linked to public safety and potential for defence failure;
- where sea level alone exceeds the defence level and;
- where sea level plus wave height exceed the defence level.

This map provides a very quick and simple way to understand of where the potential problems are likely to be given a particular forecast. It has been designed to pass the 3am test – whereby it can be used effectively by duty staff working in the early hours.

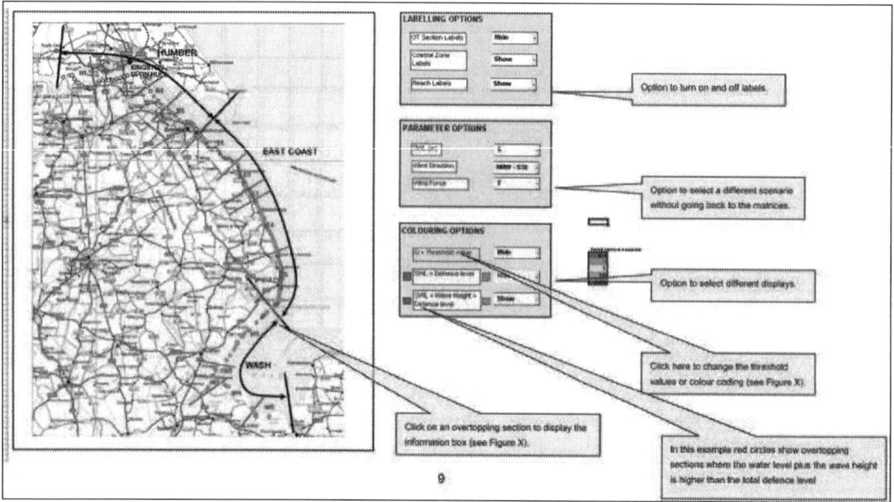

Figure 5 : New Tidal Flood Warning Tool – Map View

Overtopping Rates and Flood Warning Threshold

The main use of the tool is to aid the flood warning decision making process. Figure 6 shows the main tabular view of the tool. This view replaces the original matrix and provides a simple colour coded look up for what thresholds are reached given the forecast information.

Thresholds can be chosen from a drop down of either flood warning thresholds or, in the case of the example shown, by overtopping rates.   The cells within the matrix also show values of maximum and average overtopping rates for that section on coastline.

The overtopping calculations embedded within the tool are based on a credible worst case beach profile, based on recent years survey data. In order to understand the sensitivity to the beach profile it is possible to toggle between a this and a credible best case scenario beach profile. The actual beach profile is likely to be somewhere in between the two though this feature in the tool allows the sensitivity to be easily understood.

Pre-Defined Mapping

The tool is capable of quickly translating a daily tidal forecast into estimated volumes of water overtopping the defences. However current run times for the TuFlow models means that mapping the flooding in real time is not possible.

Therefore the project has developed a library of maps based on a certain number of forecast scenarios, with the flood warning areas affected for each scenario being listed.   On the tabular view it is also possible to colour code the matrix based on which pre-prepared mapped scenario the forecast most closely matches. This is calculated based on the overall volume of flood water within a flood warning area. This allows the tidal forecast to be quickly translated into a list of required flood warnings to be considered.

For example, the volume overtopping the defences for the forecast scenario used in Figure 7 most closely matches the volume mapped for a 1% (1 in 100) annual chance scenario for Reach E4.

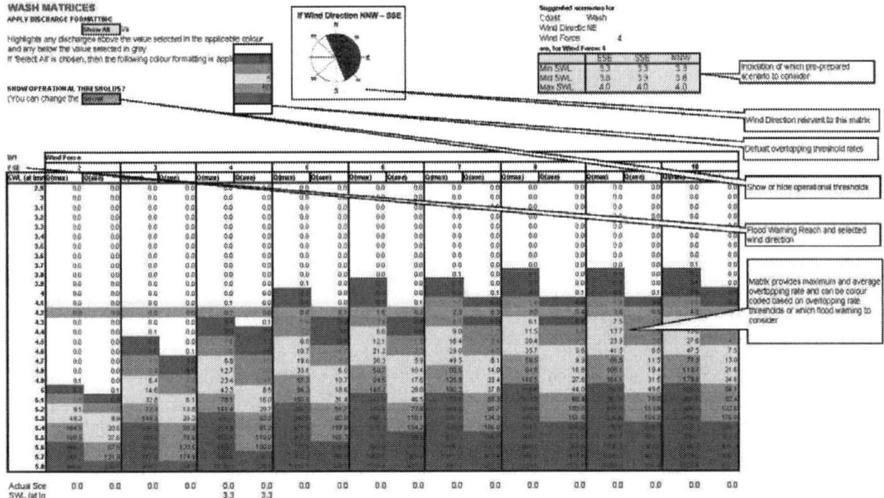

Figure 6 : New Tidal Flood Warning Tool – Tabular View

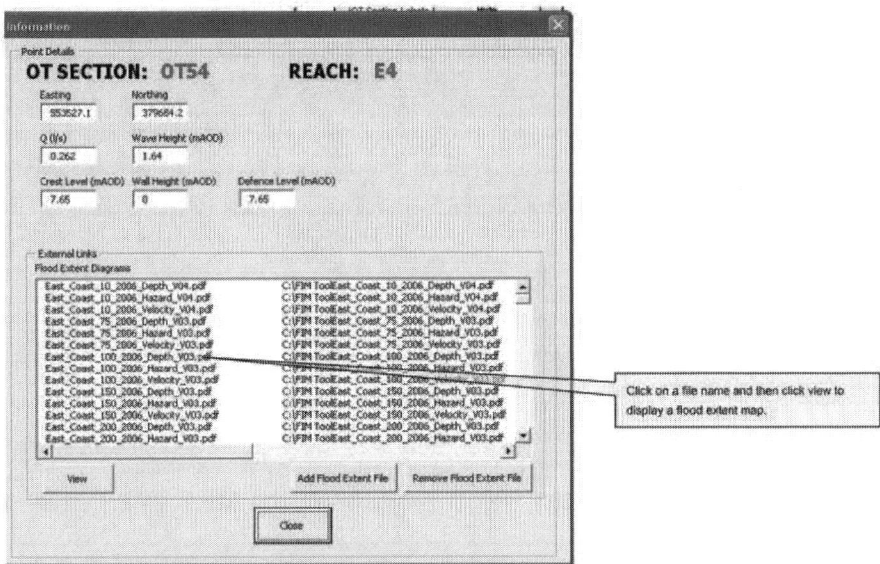

Figure 7 : New Tidal Flood Warning Tool – Pre Defined Mapping

Future Development

It is important to validate the new to check thresholds allocated to the various flood warning types are appropriate. Therefore the new flood warning tool will be used in parallel with previous flood warning matrix for winter 2011/12 to assess its performance.

Although the previous matrix had remained largely unchanged for over 20 years, this is certainly not the plan for the new flood warning tool. No major updates are expected until after the review period, though a number of potential future developments have already been identified.

In particular the ability to import the most recent beach profile survey into the tool is being considered. At present the user of the tool can chose between a credible worst case or credible best case beach profile, based on recent years surveys. This gives a sensitivity range for the overtopping rates but it would be useful to consider the most recent beach survey also, as these are undertaken annually if not more frequently.

Whilst the Northern Tidal Modelling project used the best available information and techniques at that time it is recognised that the science around this subject is continually evolving. Therefore the underlying modelling and mapping within the project is currently being reviewed in light of data and software improvements since the work was commissioned in 2005.

For example, software improvements in 2011 have seen greater use of new rapid flood spreading models. These new 2-dimensional models are capable of mapping flooding from tidal overtopping or breaching in minutes rather than the many hours the current models take to run. This opens up the possibility of being able to map the actual tidal event you are faced with, rather than simply matching it against one of a series of pre-prepared maps.

This review will be completed in early 2012.

Innovative Coastal Zone Management
ISBN 978-0-7277-5749-4

ICE Publishing: All rights reserved
doi: 10.1680/iczm.57494.181

Climate Change Impacts on the Coast of NW England and N Wales

Nigel Pontee   Halcrow Group Limited, Swindon
Daren Price   Halcrow Group Limited, Handforth
Andy Parsons   Halcrow Group Limited, York

Introduction

Climate change guidance for appraisal of flood and coastal erosion risk management projects in England and Wales (Defra 2006) provides allowances to be made for future increases in sea level and gives indicative values for sensitivity tests for future offshore waves. These changes to the hydrodynamic forcing have the potential to change sediment transport and shoreline evolution, thereby necessitating appraisal of impacts when developing shoreline management policies.

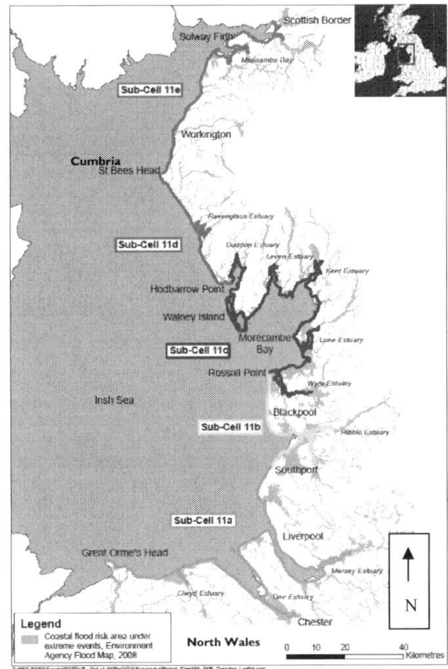

Figure 1. Cell 11 study area stretching from the Great Orme in North Wales to the Scottish Border.

This paper is based on work undertaken within a programme of studies specifically designed to support the development and implementation of long term strategic coastal flood and erosion risk management plans in the coastal Cell 11 region of the UK. The studies which were commissioned by Blackpool Council and funded by the Environment Agency, included: the modelling of waves, tides and sediment transport under contemporary and representative future conditions.

The Cell 11 region extends between the Great Orme in North Wales, along the northwest coast of England to the Scottish Border (Figure 1), encompassing much of the North East Irish Sea. The North East Irish Sea generally has a gently shelving shoreline with maximum depths to the north and south of the Isle of Man of 50m and 90m respectively. The area is macro tidal with the semi-diurnal tidal range for spring tides varying from 6 to 9m. The coast is dominated by the large sandy estuaries of the Dee, Mersey, Ribble, Morecambe Bay, Duddon and the Solway Firth (see Figure 1).

Large areas of land and thousands of properties near this 560km of shoreline are at risk of flooding or erosion. In many locations coastal defence depends upon extensive inter-tidal areas but there is uncertainty over future sedimentary process changes. The Cell 11 area makes an excellent 'natural laboratory' in which to assess climate change responses since it displays a variety of sea level tendencies, wave exposure and sediment supply conditions.

Modelling approach

Regional sediment transport modelling was undertaken as an integrated part of the overall study. The regional model used modules for waves, currents and sand transport for the whole North East Irish Sea. These modules used the MIKE21 FM (flexible mesh) modelling system. The sediment transport modelling was based on developing detailed two-dimensional models of waves and hydrodynamics coupled to the sand transport module in order to develop sediment budgets on both regional and local scales.

An Irish Sea wide hydrodynamic model provided the boundary conditions for the more local and higher resolution North East Irish Sea model used for the sediment transport simulations. The Irish Sea model was driven by water level boundaries to the north and south and was calibrated against water level measurements from a series of class A tide gauges (www.pol.ac.uk/ntslf) throughout the region. Root mean square (RMS) water level differences between observed and simulated water levels for both the Irish Sea model and the North East Irish Sea model were of the order of 0.3m. Current speed RMS errors were of the order of 0.07m/s at two locations in Liverpool Bay where good quality measured data was available.

As part of the Cell 11 Joint Probability Study (Halcrow, 2010a), a broad-scale and a nearshore wave model were set up and calibrated. These wave models were set up primarily to provide wave climate hindcasts at over 50 locations around the Cell 11 coastline for a joint probability analysis of waves and water levels. Additionally, wave timeseries were extracted for an 18 year period at 141 locations (at the closure depth) around the coastline so that littoral transport could be calculated in a consistent manner across the study area.

Following successful calibration of the broad-scale and nearshore wave models for the Joint Probability Study, the same wave parameters were used for a wave model using the same model mesh as the North East Irish Sea hydrodynamic model. This spectral wave model was used to calculate the wave climate (significant wave height, wave period and wave direction) for selected events in the study area.

The aim of the regional sediment modelling was to simulate sediment budgets for a typical year. However, it was impractical to run the North East Irish Sea regional hydrodynamic and sediment transport models for all wave conditions throughout a year due to long simulation times. Therefore an analysis of the 18 year wave data set was undertaken to derive a small number of representative wave conditions which could be used within a probabilistic approach to represent a typical year. These conditions were then used within the hydrodynamic and sediment transport models to provide wave stirring in the sediment transport calculations to allow an approximate representation of the yearly sediment transport potential.

The models were used to investigate the following:

- Sediment transport in subtidal regions for typical annual conditions and a selected surge event;

- Regional variations in littoral transport;

- The impacts of sand banks on waves and sediment transport; and,

- Water and sediment exchanges in two large estuaries.

The potential impacts of future climate change were assessed in terms of:

- Increased sea levels;

- Increases in wave energy;

- Changes in wave direction; and,

- Surge impacts.

Results

Present day conditions

Sediment transport in the subtidal zone

Figure 2 shows the annual transport directions for sand sized sediments throughout the subtidal zone of Cell 11. The sediment transport pathways generally show that there is a potential transport of sediment from the west into the North East Irish Sea and towards the major estuaries. In the southern part of the region there is a general trend for sediment to be transported eastward and onshore. In the northern part of the study area, the transport is onshore towards the Solway Firth, but then southwards towards Morecambe Bay. The Cumbria coast, between St Bees Head and Walney Island, thus represents an exception to the general pattern of onshore transport.

The wave modelling results show that the coast between St Bees Head and Walney Island is the most energetic region being directly exposed to the strong south-westerly storm conditions. The Isle of Man provides some sheltering towards the northern end of the Cumbrian coast for waves from the west, and to the Dee/Mersey/Ribble regions for waves from the northwest. The simulations that were undertaken for surge conditions, which included both the effects of the storm surge, waves and wind, showed some onshore transport towards Walney Island, into the mouth of the Duddon Estuary and towards the shore to the south of the Ravenglass Estuary.

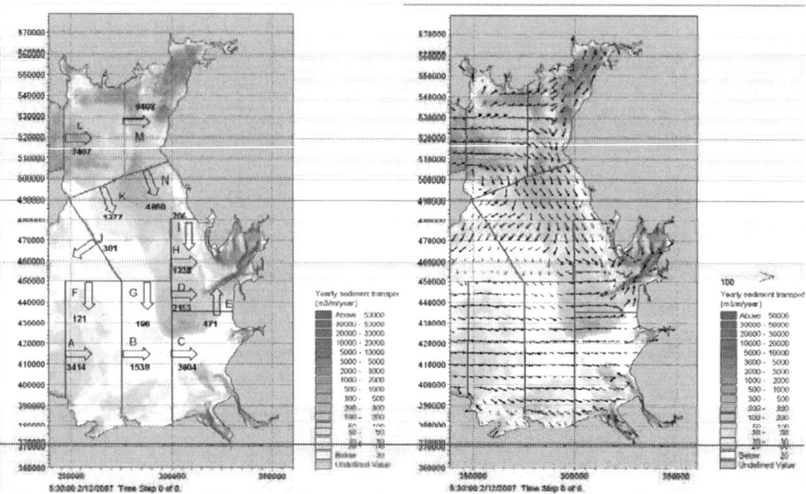

Figure 2. Estimates of annual potential sediment transport across transects in the North East Irish Sea (1000m³). Representative tide simulation with tides and wave forcing. Yearly potential sediment transport vectors are shown as reference on the right.

The modelling results show a general correlation between the mean bed shear stresses and the sea bed sediment grade. The areas of higher mean bed shear stresses were located over areas of coarser sediment, whilst lower stresses occurred where there was finer sediment. These findings suggest that the present day sea bed sediment composition is a reflection of contemporary conditions rather than being controlled by the position of relict deposits.

Some limited modelling of cohesive sediment was undertaken to investigate the re-suspension of fine sediment from the Irish Sea Mudbelt and nearby coastal cliffs during storm conditions. The Irish Sea Mudbelt lies between Cumbria and the Isle of Man and has been shown by sediment cores and dating to be a contemporary deposit (Kershaw *et al.*, 1988). The new modelling confirms that the mudbelt is located in an area of low bed shear stress under normal tidal conditions but under moderate to extreme storm conditions the increase in shear stress is sufficient to cause re-suspension of surface sediment. The results indicated that suspended sediment could be transported over a wide offshore area. The modelling also showed that fine sediment resuspended off the Cumbrian coast could find its way into the Ravenglass Estuary, the Duddon Estuary and Morecambe Bay. These model findings are consistent with previous studies that have studied the dispersal of radioactive pollutants emanating from Sellafield that have been chemically attached to the sediments in the mud belt. Williams *et al.* (1981) reported detection of radioactive isotopes attached to fine sediments in estuaries including Ravenglass Estuary, Duddon Estuary, Morecambe Bay and the Solway Firth.

Littoral transport

Littoral sediment transport calculations were carried out in a consistent way for all wave conditions over the 18 year period at 141 locations around the coastline. These calculations also included the effect of the varying water levels including surges.

The main features of the net littoral transport regime in the Cell 11 region are:

- The littoral drift modelling shows that the Great Orme forms a littoral drift divide which is the reason for the Great Orme was selected as the boundary between coastal Cell 10 and 11.

- Along the coast of North Wales and on the north Wirral coast, between the Dee and Mersey estuaries, the littoral transport is eastward and generally increases with distance eastward.

- On the coast near Southport transport is directed into the Ribble Estuary.

- Just to the south of Blackpool there is a drift divide.

- On the coast near Blackpool net littoral transport is directed northwards towards Morecambe Bay.

- There are drift divergences in the central part of Walney Island and between the Duddon and the Ravenglass estuaries.

- Along the coastline to the north of the Ravenglass Estuary the transport direction varies due to the coastline being aligned close to the mean wave direction, but moving towards St Bees Head the degree of northerly transport increases (although there is a local drift divide at St Bees Head).

- North of St Bees Head the littoral transport is directed northwards all the way to Moricambe Bay in the Solway Firth.

Figure 3 provides an example of the littoral transport modelling results around Morecambe Bay. In this figure the annual average littoral transport directions are superimposed on the sub-tidal sediment transport results from the wave, flow and sediment model.

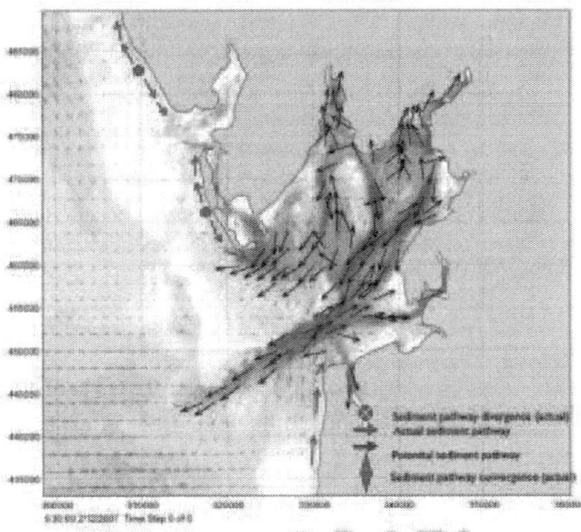

Figure 3. Offshore and littoral transport in the vicinity of Morecambe Bay.

Analysis of littoral transport across the Cell 11 region on a year by year basis showed a large variability in annual rates of transport, but only small changes in transport directions or the locations of drift divides. The standard deviation of annual transport rates is on average 44% of the average annual transport rate. This illustrates the considerable variability of annual net drift rates over the 18 years of data (January 1989 to end of 2006). The year that had the largest transport rates was 1990 (see Figure 4). It is notable that this was the year that there was significant flooding at Towyn on the north Wales coastline following an extended period of significant erosion along the beaches in the vicinity. The year with the lowest transport rates for most of the coast was 2001. An analysis of monthly transport rates showed that the largest littoral transport rates occur during the winter months. The transport rates in the winter rates were on average four times the magnitude of the rates during the summer.

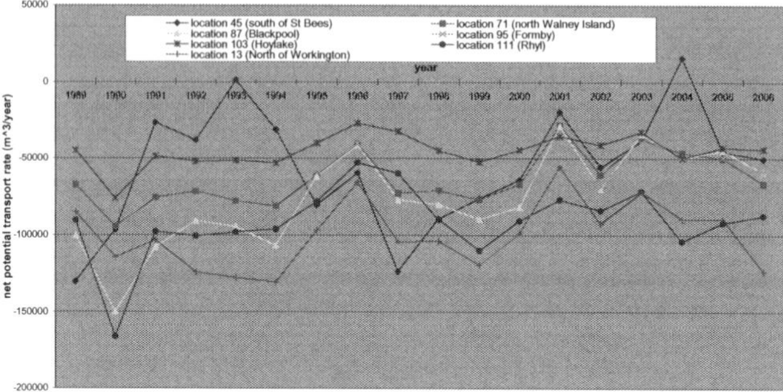

Figure 4. Net annual potential transport rates by year. Negative numbers indicate that the littoral transport is directed to the right when looking offshore.

Potential climate change impacts

Sea level rise

In order to explore the potential impact of a rise in sea level on tidal propagation in the region mean sea level was raised by 0.5 m and the nested north east Irish Sea model run for the same period as for the existing conditions. One key finding was that the increase in mean water level of 0.5m produced an additional increase in high water of up to 0.1m in many of the main estuaries and up to 0.3m in the upper reaches of the Solway Firth. Low water, however, remained almost the same in many of the estuaries, although there were some local decreases of up to approximately 0.3m in the Solway Firth.

In order to explore the potential impact of sea level rise on sediment transport, the model was run with a 0.5m increase in sea level and the same offshore wave conditions as the baseline present day situation. The greater depth of water in the nearshore zone influnced the shoaling characteristics of the incoming waves. The estimated differences in yearly potential non-cohesive sediment transport patterns due to sea level rise are shown in Figure 5. The difference plot on the right of Figure 5 shows small increases between the Isle of Man and Scotland and just north of the Isle of Anglesey. There was a reduction in transport in the approach channel to the Mersey Estuary, possibly due to the increased depth of water. The largest increase in transport rates occurred within the estuaries. This increase was due to several reasons. Firstly, higher sea levels lead to increased tidal prisms in the estuaries and these produced stronger tidal currents. Secondly, the increased depth of water allows larger waves to break further inshore thus leading to higher wave heights in the shallower water within nearshore regions and outer estuaries.

However, the assumption inherent in both of the above simulations was that the bathymetry remained the same under higher sea levels and that no accretion occurred. In Cell 11 the plentiful supply of sediment combined with the general flood dominance (shorter duration flood tide with stronger currents) means that any increase in sea level is likely to be accompanied by a gradual adjustment of nearshore bed levels. The results presented here are therefore likely to exaggerate the changes that may occur under higher sea levels.

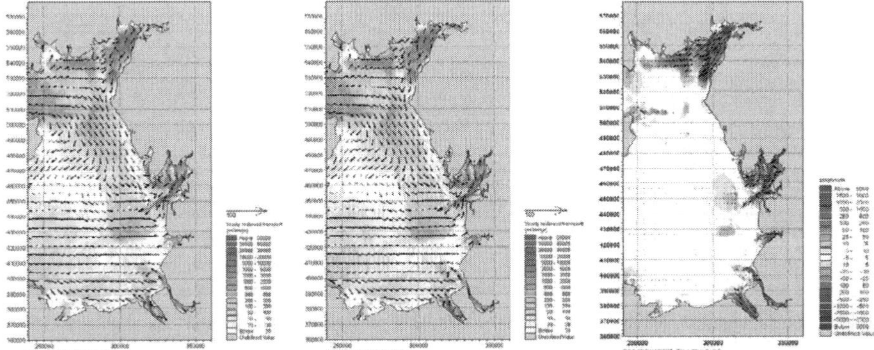

Fig. 5. Estimated potential yearly sand transport (m³/m/year) for tide and wave simulations, left present, middle with sea level rise (0.5m) and right difference in sand transport (positive is increase due to SLR). The vector arrows on the right figure show the direction of change for the sand transport.

Increased wave action

Sensitivity tests on changes to regional sediment transport due to changes in offshore wave heights and directions were undertaken through post processing of the results. This involved changing the frequency of occurrence of the modelled representative wave events. The tests increased offshore wave conditions by 5% and 10% as recommended in the present Defra (2006) guidelines for 2055 and 2085 respectively. The 10% increase was divided into two simulations:

- one assuming 10% increase for waves from the north west; and,

- one assuming 10% increase for waves from south west.

The results showed that the increased wave forcing increased the transport towards the main estuaries e.g. Solway Firth, Morecambe Bay and Ribble and Dee estuaries due to the flood tide dominated sediment transport.

For the littoral transport modelling a sensitivity test was undertaken that considered a general increase in wave heights of 10% all around the whole of Cell 11. This produced an average increase in the littoral annual net transport rates of 19%.

Surges

The hydrodynamic, wave and sediment transport models were also used to examine the effect of a surge event on sediment transport at a regional scale. January 2007 was chosen since a large surge was known to have occurred at this time. The simulation included forcing using measured surge levels and hindcast waves at the models western boundary plus air pressure and hindcast wind over the model domain.

The results from the simulations showed that, compared to the tide only case, there was a tendency for an increased amount of easterly transport offshore of the North Wales coastline. Similarly, for the stretch of coast from the Ribble to Morecambe Bay, there was an increased transport towards the shoreline predicted with higher transport into the northern part of the mouth of Morecambe Bay. Across the mouth of the Ribble, there was an increase in the northerly component of sediment transport. This could potentially provide an additional mechanism for providing sediment movement towards the Blackpool frontage. Along the Cumbrian coast the surge led to increased southerly transport.

For most areas the simulated surge therefore tends to enhance the general onshore transport of sediment compared to the tide only case. This was due to a combination of increased bed shear stresses as well as the larger volume of water entering the estuaries during the surge.

Discussion

The sediment transport modelling shows that the primary mode of sediment transport in the offshore region today is tidal currents, which move sediment as bedload and suspended load. However waves play a greater role in the shallower regions and during storm conditions. Tidal asymmetry is instrumental in driving onshore sediment movement throughout the Cell 11 region. This situation, which is believed to have existed throughout the Holocene period, explains why the majority of coast and estuaries in this region are sediment rich and have shown accretion in the past. These results generally confirm previous coarser scale modelling undertaken for the UK continental shelf area (e.g. Pingree and Griffiths;1979).

The bathymetry for the North East Irish Sea shows a gradual widening of the inshore contours from north to south. This is consistent with the general movement of sediment predicted by the regional offshore sediment modelling, whereby the sediment pathways are in general directed towards the southeast corner, i.e. Liverpool Bay. The general onshore transport of sediment provides a supply to the estuaries in Cell 11. This is especially shown for the Mersey, Dee and Ribble estuaries which have extensive sand banks around their mouths.

The modelling suggests that increased offshore wave heights and raised mean sea level, which may occur under future climate change scenarios, are both likely to increase the potential sediment transport towards the shore and in particular the main estuaries in the south. Storm surges have a similar effect.

While the modelling has calculated potential transport rates and not quantified bed level changes, it is known that there are large areas of sand and mud on the sea bed in Liverpool Bay and the north east Irish Sea available for transport. Preliminary sediment budget calculations (Halcrow, 2010c) indicate that less than 1mm /yr of erosion offshore would be required for intertidal areas to keep pace with the sea level rise allowances given by Defra 2006. The present and expected enhanced tendency in future for onshore transport of sediment suggests that in the Cell 11 region coastal habitats can be maintained in the future. Consequently, in Cell 11 there is likely to be a more limited requirement to recreate habitats to provide compensation for losses to sites designated under the EU Habitats and Birds Directives than in many other areas of the UK, where there are ongoing net losses of coastal habitats that may be worsened due to the combined effects of sea level rise, sediment scarcity and various anthropogenic impacts.

In Cell 11, the maintenance of intertidal areas, sand banks and especially beaches on the open coast should offset some of the impacts of sea level rise. This means that the Shoreline Management Plan for the area (Halcrow, 2010b) can assume that these areas will continue to aid in the dissipation of wave energy in the future. However, in some areas there is no obvious mechanism to move sediment onshore. One such area is the Cumbrian coast south of

St Bees Head, where the lack of onshore supply combines with the higher exposure to storm waves. In this area, the intertidal areas that are already relatively narrow, could be subjected to increased narrowing with increasing mean sea level. This will put increasing pressure on coast protection to infrastructure such as the Cumbrian Coastal Railway. Another such area is the Fylde coastline (north of Blackpool) where both net tidal and littoral transport is northwards creating an erosional environment with no direct onshore feed of sediment. The beaches in this area therefore rely on the supply of sediment from further south.

The littoral transport modelling for this study has provided a consistent assessment of littoral transport rates throughout the region and their variability on a year to year basis. The results confirm earlier work showing drift divergences at Great Orme's Head, south of Blackpool, Walney Island and between the Duddon Estuary and Ravenglass Estuary, and St Bees Head. The exact locations of some of these divergences differs from previous work. The finding that a 10% increase in wave height leads to a 19% increase in littoral transport rate suggests that increases in wave height due to climate change are likely to accelerate coastal change in areas of existing erosion and accretion along the shoreline. This could lead to a number of impacts including:

- Enhanced erosion/accretion;

- Increased variability in beach profiles; and,

- Renewed periods of spit and barrier growth.

The modelling suggests that in the absence of bed level changes, increasing relative mean sea level would lead to tidal amplification in the major estuaries. This results in larger flows in and out of the estuaries than would be the case if the bathymetry kept pace with the sea level increase and, due to the general flood tide dominance throughout the region, provides a mechanism for increased onshore sediment transport into estuaries.

Conclusions

The modelling that has been undertaken has provided a consistent assessment of sediment transport across the whole region, both in terms of the subtidal and littoral regions. The modelling clearly shows the subtidal sediment transport pathways which are key to understanding the large scale sediment budgets for coastal regions. The general onshore directed transport of sand sized sediment provides a supply to the majority of coastal and estuarine areas in Cell 11.

The effect of climate change on the sub-tidal sediment transport due to potential increases in wave height and raised mean sea level was investigated. In general, there was a trend to increase the potential for transport towards the shore. This suggests that in most areas in Cell 11 there is a mechanism for the intertidal areas to respond to sea level rise and many intertidal areas and the sand banks are expected to continue to accrete in the future.

This knowledge will assist more detailed studies and inform estimates of coastal habitat changes required to underpin the Environment Agency's coastal habitat creation programmes that are designed to offset future impacts of coastal defences related to sea level rise.

The continued presence of beaches and wide inter tidal zones will in future continue to assist in dissipating wave energy and reducing the exposure of defences. However, the lack of onshore transport for parts of the Cumbrian coast means that there may be some loss of habitats in these areas in the future and defences may become even more exposed and subject to more frequent damage.

The littoral transport modelling suggests that increases in wave height in the future will increase littoral transport rates and this could lead to more dynamic coastal behaviour in some areas.

The research methodology described in this paper can be applied to other areas to assess the impact of climate change and natural variability on the future evolution of coastal and estuarine areas. The importance of onshore directed tidal residuals is likely to be less import in other areas, however, modelling to help understanding linkages between beaches and potential offshore sediment sources and sinks will be relevant in many locations. The derivation of regional wave models, which can be used to produce consistent estimates of alongshore transport, is also likely to be helpful in many locations. A lesson learned from this particular study is that such modelling approaches require up to date bathymetry and sediment size data in order to derive meaningful results. Such requirements could be usefully built into future regional monitoring programmes.

Acknowledgements

This paper is based on work undertaken as part of the Cell Eleven Tide and Sediment transport Study (CETaSS) led by Blackpool Council (see www.mycoastline.org) and funded by the Environment Agency and Defra.

References

Defra, 2006. FCDPAG3 - *Supplementary Note to Operating Authorities – Climate Change Impacts;* October 2006. Available online at:
http://www.defra.gov.uk/environment/flooding/documents/policy/guidance/fcdpag/fcd3climate.pdf

Halcrow (2010a) *Joint Probability Study. Regional (Broad-scale) and Nearshore Wave Model Calibration.* Report prepared by Halcrow Group Ltd for the North West and North Wales Coastal Group, as part of the Cell Eleven Tide and Sediment transport Study (CETaSS) Stage 2, March 2010. 69pp. Available from http://www.mycoastline.org

Halcrow (2010b). *North West England and North Wales Shoreline Management Plan SMP2 - Main SMP2 Document.* Report prepared by Halcrow Group Ltd for the North West and North Wales Coastal Group as part of the Second Round Shoreline Management Plan for Cell 11. Halcrow, July 2010. 54pp + Annex + Appendices. Available from http://www.mycoastline.org

Halcrow (2010c) Appendix I – Regional assessment of coastal squeeze. Report prepared by Halcrow Group Ltd for the North West and North Wales Coastal Group as part of the Cell Eleven Tide and Sediment transport Study (CETaSS) Stage 2, 2010, 49pp. Available from http://www.mycoastline.org

Pingree, R.D. and Griffiths, D.K. (1979). Sand transport paths around the British Isles resulting from M2 and M4 tidal interactions. *Journal of the Marine Biological Association of the UK*, 59, 497-513.

Bowden, K.F. (1980). Physical and Dynamical Oceanography of the Irish Sea. In: F.T. Banner, M.B. Collins, and K.S. Massie, eds. *The north-west European Shelf Seas: the sea-bed and the sea in motion II) Physical and Chemical Oceanography and Physical Resources.* Amsterdam: Elsevier Press, pp. 391-403.

Kershaw P.J., Swift D.J., and Denoon D.C. (1988). Evidence of recent sedimentation in the eastern Irish Sea. *Marine Geology 85*, 1 - 14.

Williams, S.J., Kirby, R., Smith T.J. and Parker W.R. (1981). Sedimentation studies relevant to low level radioactive effluent dispersal in the Irish Sea. *Institute of Oceanographic Sciences (Taunton) Report No 120.* Report prepared for the Department of the Environment.

Innovative Coastal Zone Management
ISBN 978-0-7277-5749-4

ICE Publishing: All rights reserved
doi: 10.1680/iczm.57494.192

Climate Change Impacts on Disadvantaged Coastal Communities[1]

Amalia Fernández-Bilbao, URS/Scott Wilson, 6 Greencoat Place, London SW1P 1PL, UK
James Allan, URS/Scott Wilson, 6 Greencoat Place, London SW1P 1PL, UK
Mary Zsamboky, URS/Scott Wilson, 6 Greencoat Place, London SW1P 1PL, UK

Introduction and Background

The UK has a long coastline that is relatively accessible to its population and is the focus of important economic, environmental and social activities. Nationally and internationally significant natural habitats and heritage sites are located on the coast as well as concentrations of key economic activities including fishing, ports, offshore energy generation and their land-based supply chains.

However, many coastal communities face a series of important socio-economic challenges including: ageing populations, youth outmigration and inward migration of older people, high proportions of retirees and people receiving benefits, transitory populations, physical isolation, poor-quality housing, an overreliance on tourism, seasonal employment, low income and pressure on services during the summer months (CCA, 2010; CLG, 2007; Centre for Rural Economy, 2006).

This paper is based on a study that explored the vulnerability of disadvantaged coastal communities in the UK to the impacts of climate change. The study aimed to increase knowledge of how climate change effects are likely to be experienced by those communities that are both highly susceptible to the physical impacts of climate change due to their geographical location and are also less likely to have the ability to adapt because of the range of socio-economic issues that affect them.

Therefore, disadvantaged communities in this context have been defined as those which are at risk of the physical impacts of climate change because of their location, and which already suffer from high levels of deprivation or geographic isolation. 'Coastal communities' were defined for the project as those that may be located near to the sea, surrounded by the sea, or experience a close functional relationship to the sea (e.g. seaside resorts, port and estuary towns and island communities).

The research is particularly pertinent in light of the UK government's 'localism agenda' (CLG, 2010, 2011), which puts increasing responsibility on communities and local authorities to lead in service delivery. This paper raises questions about how localism can be successfully implemented in relation to climate change adaptation in disadvantaged coastal

---

[1] The research presented here was funded by the Joseph Rowntree Foundation

areas. These additional responsibilities come at a time of spending cuts for local authorities and flood and coastal defence budgets.[2]

Climate Change on the UK Coast

As described in Zsamboky et al. (2010), climate change is likely to have a severe impact on the UK coast by 2080. The total rise in sea levels off the UK coast may exceed 1 m and could potentially reach 2 m. The frequency of intense storm events is expected to increase and, along with the rise in sea level, to lead to more coastal flooding. Coastal erosion is also expected to increase, partly due to sea level rise. Low-lying and soft-sediment coasts will be most vulnerable (e.g. in the east of England) because they are most easily eroded. The most exposed locations and estuaries may be particularly vulnerable. Temperatures are expected to rise, particularly in the south and east of the UK. Winter precipitation is likely to increase markedly on the northern and western UK coastline.

The impacts of sea level rise and storminess, together with erosion at the coast, present significant threats to coastal communities. These threats will be felt particularly keenly by communities that are highly dependent on coastal areas in terms of where their homes are, transport and communication infrastructure, as well as economic and social activity (Zsamboky et al., 2010).

Social Impacts of Climate Change on the Coast

Climate change is a global problem but the effects will be experienced very differently in different areas and by different groups and communities. Ionescu et al. (2005) provide a useful explanation of the three reasons why this is the case, which will affect the relative vulnerability of coastal communities:

- First, climate change effects are likely to be felt at the coast before they impact elsewhere inland, through increases in the frequency and seriousness of flooding compounded by increasing coastal erosion and land instability (Fletcher and Potts, 2008).
- Second, Ionescu et al. (2005) note that there will be differential impacts between regions and between groups in society. For instance, heavy rainfall can lead to flooding in some areas but not in other areas; heatwaves may be fatal for some elderly people but not affect others and so on.
- The third factor is related to the extent to which regions, communities, groups and individuals are able to prepare for, respond to and address the effects of climate change. Indeed, (lack of) 'adaptive capacity' is recognised as one of the key aspects that determines vulnerability.

Very little research has been carried out to date on the potential social impacts of climate change on the UK coast. Existing literature suggests that climate change is likely to negatively affect people's health, particularly through a greater occurrence of extreme events, such as flooding and heatwaves (Department of Health and Health Protection Agency, 2008). A good understanding of the social impacts of flooding (a type of event that is expected to increase with climate change) has emerged following a number of high-profile flooding events, particularly over the past decade (e.g. Boscastle [2004], Carlisle [2005], Yorkshire, Gloucestershire and other areas [2007], Cumbria [2009], Aberdeenshire [2009] and Cornwall [2010] [LWEC, 2010]). It is now widely understood that while flooding in the UK has

---

[2] See for instance: http://www.broking.co.uk/insurance-age/news/1800081/spending-review-2010-abi-disappointed-flood-defence-cuts

historically had very low mortality rates, it can cause physical and particularly severe mental health effects (Parket et al., 1983; Green et al., 1985; Tapsell et al., 1999, 2001; Ohl & Tapsell, 2000; Tapsell et al., 2002). However, a review undertaken on behalf of Defra (Twigger-Ross, 2005) found that, within the flooding research, while analytical attention was given to the individual and household level, little work in the UK has focused on the social impacts at the community level. In addressing this gap, Walker et al. (2006) examine a wide range of social impacts in terms of how different groups within populations and different kinds of neighbourhoods may be differentially flood risks and flooding experiences. A link was found between high deprivation and high risk of coastal flooding. It is likely that the advancement of work in this field will increasingly widen our understanding of the social impacts of climate change. At the same time, the wide 'array of climate related changes' (Cooper, 2009, p. 4) demands a greater appreciation of differential impacts within the context of communities, households and individuals.

Climate change is also likely to affect coastal livelihoods, particularly for those who depend on coastal and marine environments for employment (e.g. in tourism and fishing) (see Cooper, 2009). The Marine Climate Change Impacts Partnership (MCCIP), in considering how climate change may impact on people and economies, identified the following areas of activity that will experience these impacts: conservation, tourism, infrastructure, transport (ports, shipping, railways, roads), and fisheries (Cooper, 2009). For example, extreme events associated with climate change (such as storms and flooding) are likely to affect key public infrastructure such as health and emergency services, and public transport (e.g. rail lines) along the coast (Health Protection Agency, 2008). Another area that could be affected is agriculture; agricultural land near the coast may also be affected by flooding and erosion (including the loss of land due to managed realignment schemes). This in turn raises questions about potential effects on national food security and the sustainability of agriculture in coastal areas. Additionally, for coastal areas that are already isolated, including island communities, the impacts on transport and infrastructure (e.g. if roads are damaged during and event and the community is cut off) could be particularly serious.

Climate change is also likely to have longer-term impacts on particular neighbourhoods and towns affected by extreme flooding events or that are considered to be at high flood risk (see Cooper, 2009). These areas may be affected by blight and a reduction in housing values, development and investment as investors and house buyers are deterred by the risk of flooding.

Methodology

This study was undertaken according to the following methodology.

- A literature review and climate and coastal change hotspot analysis: the literature review covered existing knowledge of climate change impacts on the UK coast. The hotspot analysis consisted of an identification of areas along the UK coast where the impacts of climate change would be particularly severe. The hotspot analysis drew on published data to provide a projection of the likely impacts of climate change on the UK coast and determined the selection of the case study areas (see below). The hotspot analysis focused particularly on sea level change, storminess, temperature and precipitation changes from the present to 2080, and the implications for coastal vulnerability.

- Stakeholder interviews: these were held with 25 national and local policymakers from a range of agencies including central government officials, local authorities (including planners and flood and coastal management officers) and other agencies.

- Four case studies: these focused on coastal communities in Great Yarmouth, Skegness, Llanelli and Benbecula, where focus groups were run and semi-structured interviews were undertaken with community representatives and local authority officers (including planners and flood, emergency services and coastal management officers). The case studies aimed to explore the understanding and perceptions of the impacts of climate change among local residents and public sector service providers.

- A stakeholder workshop: this event was held to present the preliminary research findings to a range of stakeholders and to consider their implications.

Key Findings

The focus groups and interviews with residents explored their perceptions of climate change impacts. Participants were asked who they thought might be most affected by climate change in their community. All of the focus groups identified the elderly as likely to be most significantly affected. Those with disabilities were mentioned as being particularly vulnerable in Skegness and Great Yarmouth. Participants in Great Yarmouth and Llanelli were also the only groups to mention children. Participants in Benbecula did not feel that any particular demographic group would be worst affected and predicted that those who depended on agriculture for a living would be worst affected.

Local authorities and other interviewees suggested that as well as being particularly vulnerable, the elderly were less likely to be aware of the potential impacts of climate change than other groups. It was suggested that this was sometimes linked to a perception that climate change is distant and will not happen in their lifetime. More generally too, for many residents, daily concerns such as low incomes or unemployment were considered more pressing than climate change.

The case studies suggested that people's knowledge and awareness of climate change is increased by experience of severe events such as storms or flooding. Residents in Benbecula showed higher awareness linked to a recent severe storm event that cost five lives on a nearby island. Elsewhere, where no recent events had been experienced residents were less likely to be aware of the risks of climate change. Even where there was awareness, perceptions of climate change risks were often limited to one manifestation (e.g. flooding) and there was a lack of understanding of the wider impacts and potential consequences (e.g. economic impacts, pressure on services).

The experience of a severe storm near Benbecula was also reported to have led to some improvements in preparedness in the local authority but residents were concerned that not enough had been done by authorities to prepare for a future event. In addition, the way authorities work by commissioning research and looking at options before undertaking any measures is sometimes perceived by residents as 'inaction'. Participants also felt that climate change had recently 'fallen off the authorities agenda' in favour of economic concerns.

Poor awareness and understanding of the risks may be due in part to a lack of effective communication of climate change impacts. However, paradoxically, authorities in the case study locations raised the issue that an increased understanding of the risk of flooding and erosion, which should aid preparedness, could lead to 'blighting'. Once an area is identified as being at risk of flooding or erosion, land values can fall dramatically, resulting in residents being unable to sell their homes. This is a particular problem for areas where it may not be affordable to provide or maintain flood and coastal defences.

The research suggests that coastal communities are likely to experience various challenges in their ability to adapt to climate change. Coastal areas in the UK contain some of the most deprived communities in the country. Individual residents may lack the resources to make physical changes to their homes (e.g. to make them flood resilient) or may not be able to afford, or wish to move away. The day-to-day concerns of residents of coastal areas such as low incomes, lack of jobs or the price of fuel are often seen as more pressing than dealing with climate change. In addition, case study participants highlighted actions undertaken by authorities (e.g. granting planning permission on flood plains) that contributed to increase the risk of climate change impacts to their communities.

Coastal local authorities that contain areas of high deprivation may also not prioritise or be able to afford to undertake adaptation activities. Some of the areas that were examined clearly faced other issues that needed attention, such as regeneration and responding to housing needs, which may mean that there are limited resources to be spent on climate change adaptation.

With central government policy encouraging a shift to localism (CLG, 2010, 2011), there is an ever increasing onus on communities to help themselves to become more resilient. However, the way climate change is communicated (e.g. as a future risk), the lack of awareness of likely impacts and of the actions needed to respond to them may be leading to apathy in some local communities. It is likely that this is exacerbated by the range of daily challenges that more disadvantaged residents in particular face and that, in their view and comparatively, climate change is simply not a major issue. These trade-offs are likely to become more problematic following the coalition government's Comprehensive Spending Review in 2010, which set out real-term reductions of 28% in local authority budgets in the period up to 2014 (Local Government Association, 2010).

Implications for Policy and Practice

The coast is likely to suffer substantial impacts from climate change. In addition, some of the communities living along the coast of the UK can be considered particularly vulnerable to the future climate because of their socio-economic characteristics and current limitations in their ability to adapt. Our research findings support recommendations that climate change adaptation policy and practice should specifically aim to reduce the vulnerability of disadvantaged coastal communities to climate change. This should be a policy priority and the vulnerability of these communities should be considered as part of the on-going climate change risk assessment that is being undertaken by Defra. This research also points to the fact that the likely devolution of responsibility that will be a consequence of the localism agenda should be undertaken with care in relation to climate change adaptation. It is likely that many vulnerable communities and their local authorities, on the coast and possible elsewhere, may need high levels of support from central government if they are to successfully adapt to a changing climate and reduce the risks to their population.

Some coastal communities may face significant challenges to adapt successfully. Some are already disadvantaged and may find it difficult to prepare for the risks or respond to specific events. Knock-on effects may include residents in areas at risk being less able to obtain insurance for their properties. In addition, areas at high risk can become blighted by lack of development and investment.

These challenges are likely to increase the vulnerability of disadvantaged coastal communities to climate change. Adaptation needs to be a key policy priority and with better integration of policies, including regeneration, flood and coastal erosion management and emergency planning.

According to our research, 'Successful adaptation' (Figure 1) requires a combination of:

- Consistent policy priority: adapting to the impacts of climate change needs to be a national and local policy priority. Given the severity of the potential impacts of climate change, this statement may seem obvious. However, participants in our research highlighted that climate change has in recent months almost disappeared from the agenda, in favour of economic concerns. If climate change impacts are not seen as a central government priority, this will reduce the incentive for local service providers and communities to consider climate change impacts. The new localism agenda may influence the management of climate change issues with more emphasis on the local authority and on communities to take the lead. This should be sufficiently supported in terms of resources.

- Good communication and engagement with communities: the research found that the way climate change is communicated by different agencies, the government and the media (e.g. without clear messages about impacts, as part of energy communications, etc.) is causing confusion and lack of trust on the messages. Good communication of climate change should contain clear messages and have an emphasis on increasing understanding of the social impacts of climate change and what actions can be taken to respond.

- Adaptive local and national institutions: the research showed that the actions (or inaction) by local authorities are often maladaptive (e.g. building on the flood plain) or perceived to be so by communities (e.g. because of long processes of decision making). This can give communities the impression that adapting to climate change is not a priority. Adaptation needs to be embedded in all policies and activities by national and local authorities. Policies should also focus on the impacts of climate change on disadvantaged communities. Local authorities also need good information on the impacts of climate change and how it will affect their area and communities as a starting point.

- Long-term development and infrastructure planning: it is essential to avoid putting more people at risk of climate change. Impacts should be incorporated into local authorities' spatial planning and the control or management of development because what is built today will still be around in 50 or 100 years.

- Increased capacity building: in order to increase the resilience of their communities, local authorities and other key stakeholders (e.g. voluntary services) need a good understanding of the range of climate impacts, uncertainty related to these impacts, how to communicate information about impacts and how to embed adaptation in all their policies and activities.

- Support for disadvantaged groups and communities: adaptation activities need to target disadvantaged groups and communities because it is likely that they will suffer the worst impacts.

Policy and practice for climate change adaptation needs to specifically aim to reduce disadvantaged coastal communities' vulnerability to climate change. This needs to be a policy priority and included in the government's UK Climate Change Risk Assessment.

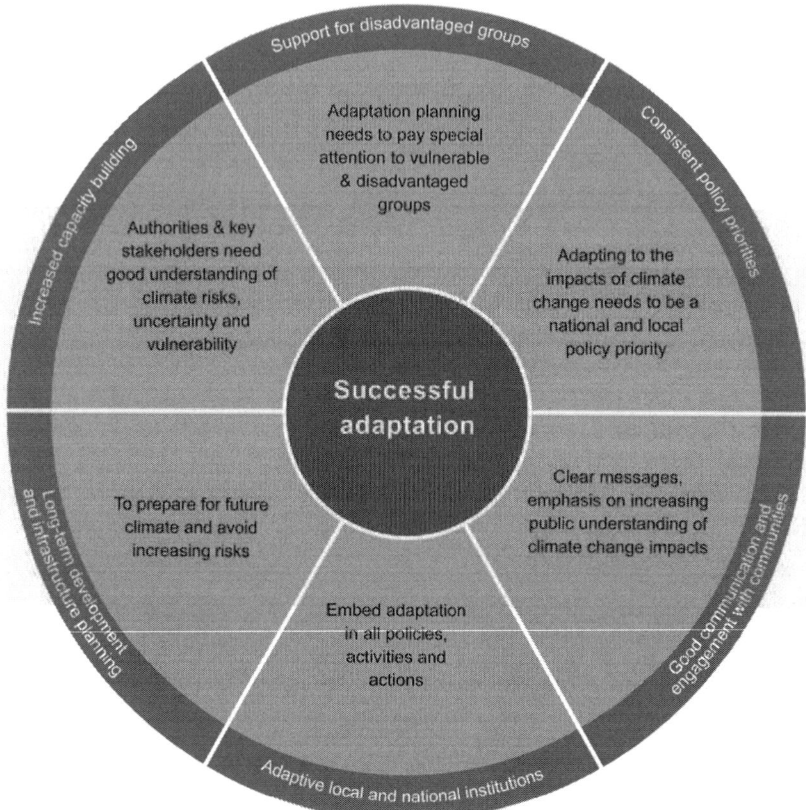

Figure 1. Successful adaptation wheel

References

Cooper, JAG (2009) 'Coastal economies and people' *in* Marine Climate Change Ecosystem Linkages Report Card 2009. (Eds. Baxter JM, Buckley PJ and Frost MT), Online Science reviews, 18 pp.

CCA (2010) Coastal Regeneration Handbook. Available at: http://www.coastalcommunities.co.uk/ (accessed 10 September 2010).

CLG (2007) Committee Report on Coastal Towns.

CLG (2010) Decentralisation and localism bill: an essential guide. Available at: http://www.communities.gov.uk/publications/localgovernment/decentralisationguide (accessed 25 June 2011).

CLG (2011) Localism and the public services revolution. Available at: http://www.communities.gov.uk/speeches/corporate/localism (accessed June 25 2011).

CRE (2006) Ageing and coastal communities. Final report to the Coastal Action Zone.

DoH and HPA (Department of Health and Health Protection Agency) (2008) Health Effects of Climate Change in the UK 2008: An Update of the Department of Health report 2001/2002.

Fletcher, S and Potts, J (2008) 'Coastal and marine governance in the UK: Editorial'. *The Geographical Journal*, 174 (4), 295–298.

HPA (2008) Health Protection Agency: Flooding and Risks to Health. Available at http://www.hpa.org.uk/webw/HPAweb&Page&HPAwebAutoListName/Page/115893460801 1?p=1158934608011 (accessed 10 June 2010).

Green, CH, Emery, PJ, Penning-Rowsell, EC & Parker, DJ (1985) The health effects of flooding: a survey at Uphill, Avon. Flood Hazard Research Centre, Middlesex Polytechnic, Enfield.

Ionescu, C, Klein, RJT, Hinkel, J, Kavi Kumar, KS & Klein, R (2005) Towards a Formal Framework of Vulnerability to Climate Change. NeWater Working Paper 2 and FAVAIA Working Paper 1, Potsdam Institute for Climate Impact Research, Potsdam, Germany.

Living With Environmental Change (2010) UK Flooding Research Strategy: Scope and Aims. Available at: http://www.lwec.org.uk/sites/default/files/Flood_Research_Strategy_Scope_Aims.pdf (accessed 25 June 2011).

Local Government Association (2010) The Spending Review 2010. Available at: http://www.lga.gov.uk/lga/aio/14462985 (accessed 26 February 2011).

Parker, DJ, Green, CH & Penning-Rowsell, EC (1983) Swalecliffe coast protection proposals – evaluation of potential benefit. Flood Hazard Research Centre, Middlesex Polytechnic, Enfield.

Ohl, CA & Tapsell, SM (2000) 'Flooding and human health: the dangers posed are not always obvious'. *Br. Med. J.* 321, 1167–1168.

Tapsell, S (2001) The Hidden Impacts of Flooding: Experiences from Two English Communities. Proceedings of a symposium held at Davis, California, April 2000, Integrated Water Resources Management, IAHS Pub. No. 272, pp. 319–324.

Tapsell, SM, Penning-Rowsell, EC, Tunstall, SM & Wilson, TL (2002) 'Vulnerability to flooding: health and social dimensions'. *Phil. Trans. R. Soc. Lond. A* 360, 1511–1525.

Tapsell, SM, Tunstall, SM, Penning-Rowsell, EC & Handmer, JW (1999) The health effects of the 1998 Easter flooding in Banbury and Kidlington. Report to the Environment Agency, Thames region. Flood Hazard Research Centre, Middlesex University, Enfield.

Twigger-Ross, C (2005) The impact of flooding on urban and rural communities, R&D Technical Report SC040033SR1 (Joint Defra / Environment Agency Flood and Coastal Erosion Risk Management R&D Programme).

Walker G, Burningham K, Fielding J, Smith G, Thrush D and Fay H, (2006) *Addressing Environmental Inequalities: Flood Risk.* Bristol: Environment Agency.

Zsamboky, M, Fernández-Bilbao, A, Smith, D, Knight, J and Allan, J (2010) Impacts of climate change on disadvantaged UK coastal communities. Report to the Joseph Rowntree Foundation.

Innovative Coastal Zone Management
ISBN 978-0-7277-5749-4

ICE Publishing: All rights reserved
doi: 10.1680/iczm.57494.201

Flooding and Eroding Coastal Landfills – Institutions Working Together for Solutions

James Wilkinson BSc(hons), MSc, AMIEMA, Wareham, England

Introduction

Approximately 1,500 of the 22,000 known recorded landfill sites in England and Wales are situated in a coastal flood risk zone (Wilkinson, 2009). Some are already eroding, and a further number of recorded sites in England and Wales are potentially at risk from coastal erosion in the next 100 years. There are numerous current case studies of eroding and flooding landfills in a range of settings, often with their own unique uncertainties and sets of problems. Published Shoreline Management Plans (SMPs) generally recognised the presence of landfills but may not always consider to these in the recommended management options. Consequently, the assessment and management tools which deal with the risks associated with landfill sites, coastal erosion and flooding are not integrated into coastal management.

Responsibility and regulation of sites is complex and falls within different tiers of organisations. It is considered that this is because organisations operate in distinctly separate ways and have not prioritised the resources for the longer term problem. As a result, there has been a haphazard approach to managing the risks posed by landfills and land-based sources of pollution, associated with flooding and coastal erosion.

The situation is not helped since historic landfill records are often incomplete, inaccurate, lost, poorly understood and not disseminated. Organisations are generally unwilling to take responsibility for waste and those who have responsibility have little willingness or incentive to self-report problems before they arise. There is little money available to invest in measures to protect the environment from eroding and flooding landfills.

This is a formidable challenge requiring a long-term vision of sustainable development and joint working with a broad spectrum of professionals. Dealing with the waste problem along our coasts will require multidisciplinary expertise in civil engineering, marine and terrestrial spatial planning, environmental protection, biology, marine conservation, geomorphology. Investment and a long-term assessment of costs and benefits is required. Application to address the growing problem of accumulating landfills is essential and increasing sea-levels and risks from climate change make the issue more problematic and urgent.

The scale of the issue of landfills at risk of coastal flooding and erosion

The precise numbers of landfills are hard to quantify. Generally, landfills are considered to be a site for the deposit of waste onto or into land. There are around 2,000 landfills that are permitted in England and Wales, of which approximately 500 are still operating. About 1,670 sites closed before or during the implementation of the Landfill Directive. The Environment Agency (EA) has a record of around 20,000 historic landfill sites. Records are poor for sites older than 1994 when the enacting Waste Licensing Regulations came in to force. There was no requirement for any records before the Control of Pollution Act 1974. Only landfills satisfying certain data quality are entered onto the landfill database.

Data sets for areas of coastal and tidal flooding were combined with EA records of historic recorded landfills and permitted landfills. This showed that there are around 184 permitted landfills (not including exempt landfill sites based on 2009 data) and that approximately 1,500 of the recorded historic sites are at risk of coastal flooding (Wilkinson 2009). It is not just landfills that can be at risk from flooding but also industrial sites; sea defences (which are often engineered from waste materials) and reclaimed land as evidenced at coastal conurbations such as Poole and Weymouth.

Table 1 Distribution of known coastal landfills in a tidal flood risk zone in each Environment Agency region in 2009 (Wilkinson, 2009)

	National tidal Flood Zone 2 permitted landfill	National tidal Flood Zone 2 historic landfill
Southwest	21	225
Southern	28	331
Thames	8	78
Anglian	42	298
Northeast	27	164
Northwest	23	153
Midlands	10	58
Wales	25	274
Total	**184**	**1561**

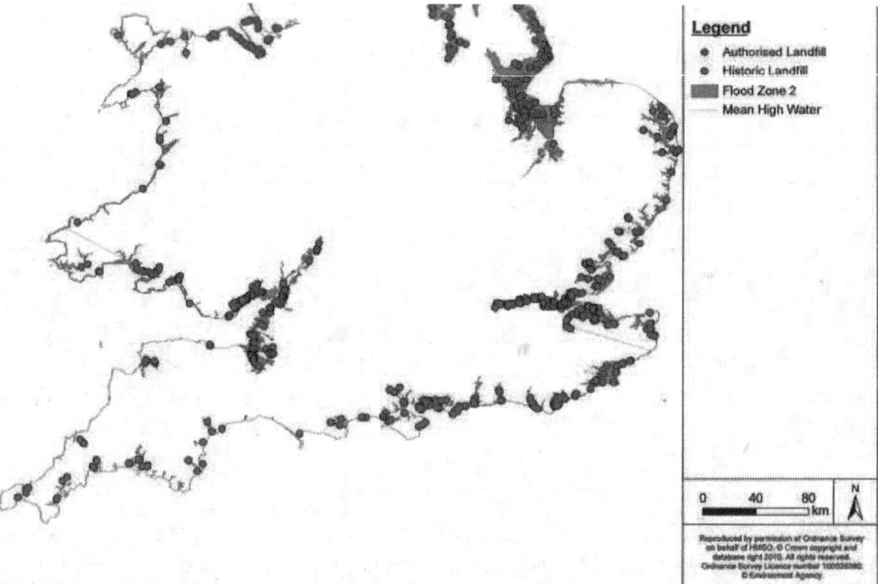

Figure 1. Map showing distribution of recorded landfills at risk of coastal flooding in southern England. Data source Environment Agency reproduced with permission of the Environment Agency.

As you may note from Figure 1 many of the recorded landfills at risk of flooding are in the southeast around Essex and London. Some of the largest are still being filled and are considered to be strategically important.

The average rate of sea level rise is generally accelerating: at the beginning of the twentieth century it was 1mm per year, today the average is 3.5mm per year. In the south of England the landmass is tilting and sinking through crust shifts (Shennan and Horton, 2002). The effects of climate change include more extreme rainfall, storm events, higher fluctuations and rising water table. There will be increased run-off, flooding, erosion, incidents of slope failures in cliffs and potential landslides. These events could potentially increase the frequency and severity of impact on manmade infrastructure including engineered bunds, landfills and industrial sites. This includes industrial and waste sites which generally tend to contain poorly consolidated materials. Any landfill site engineering may not cope with extra forces that were not planned for and so could fail, in some instances catastrophically. Many landfill sites comprise unconsolidated waste materials, making them more vulnerable.

Regulation of waste sites

Since the 1970s legislation has been tightened up and has ensured that some records are kept. Layers of legislation have evolved over time and become more complicated to interpret. As a result, management of the issue has become compartmentalised.

Waste deposit on land is either regulated by the EA under an Environmental Permit (as a disposal or a waste recovery activity), or under a waste exemption from the need of a permit. However, the majority - those which are not subject to EA regulation, and which are historic, are assessed by the Local Authorities and may be considered under Part IIA of the Environmental Protection Act 1990 under the Contaminated Land regime.

A degree of 'stove-pipe' organisational management structures exists in the regulation and management systems that have evolved and to some extent remains entrenched. Flood and coastal risk management (FCRM) seems separate to waste and industry regulation, which seems separate to Local Authorities and contaminated land. Each management hierarchy 'stove pipe' will have their own priorities at any given time. This effect can even be seen within some of the larger public organisations. It may be a contributory factor as to why landfills are largely absent in the actions arising from SMPs.

Permits are granted to the operators of sites perceived to present the greatest risk to the environment to ensure that they take appropriate steps to prevent or reduce pollution. The ongoing regulation of waste deposits on land is risk-based and proportionate to the resources and knowledge. Operational sites accepting non-inert wastes tend to be more closely regulated as the risk to the environment is higher. Closed sites and historic sites are less closely regulated - as the perceived risk from these sites is lower - or have never been regulated because they were active before any controlling legislation was in place. Inspection frequencies of permits are based on a risk score, which is in turn based on a set of risk parameters and the operator's compliance history. Flood risk and erosion are not main components of this system. Nor are they considered within other assessments made under surrender guidance (which considers the risk of pollution at a site that will remain undisturbed) or prioritisation of contamination land forms, which tend to focus on human health. Unpermitted sites or those exempt from needing a permit are generally not inspected, although the site operator or landowner has a responsibility to manage the site to prevent pollution and action will be taken if pollution is identified.

Perceptions of stakeholders

Two on-line surveys were completed to assess perceptions of people with some involvement of landfills (Wilkinson, 2009). One group were members of the Dorset Coast Forum and the other was a sample of EA employees. Many staff from the EA placed the responsibility for regulating un-permitted sites with Local Authorities and Landowners. In the same study of coastal landfills, Dorset Coast Forum (DCF) members with an interest in coastal landfills responded that:

1. A significant proportion considered that the EA was responsible, implying a higher expectation of the EA in dealing with the issue.
2. Organisations may become embroiled in dispute, avoid the issue altogether or react only once a problem emerges.
3. About 75% of relevant DCF respondents considered that one body should be charged with planning and implementing control over historic landfills. General consensus was that this body should be the EA.
4. That it was better to resolve problems of coastal landfills before they occur. A programme of work was needed to manage the risks.
5. It was considered that in the next 100 years a lack of finance would be a significant factor in damage caused by flooding or erosion at coastal landfills sites.

Responsibility for addressing the problem

In dealing with the problems and identifying solutions, including funding sources, responsibility is shared by a number of overlapping agencies.

Land use planning

Planning authorities consider the suitability of the land for landfill sites. In many instances at the coast, swathes of land were deemed suitable for dumping waste. Often it was the Local Authority who granted the permission and operated the site as Waste Disposal Authority. Today Planning Policy Statement (PPS)25 considers the flood risks. Local Authorities are now responsible for considering Annex 1 of the Landfill Directive where it states that 'the location of a landfill must take into consideration requirements relating to the existence of … coastal water, flood risk, landslides, subsidence…'. It appears that existing landfills are very often either not considered by the planning authorities in strategic plans and SMPs for flood and erosion risk, or they are not referred to. This suggests that there is a general low level of awareness or that the issue is not considered of sufficient importance.

When landfills are considered in land-use planning there seems to be little consideration given to geomorphological risks beyond the life of the site. The quantified risks of washed out landfills are little understood. Land-use planning tends to place emphasis on development of new sites and development rather than management of legacy sites. Few studies have been done other than site investigations in a relatively small number of places. This seems to be a missed opportunity for multi-agency management of landfills.

Permitting / environmental regulation

A landfill permit may only be issued if it is possible 'to achieve a high level of protection of the environment'. It also needs to achieve the standards to get an Environmental Permit or to meet exemption criteria under the Waste Framework Directive. Waste deposit on land can be considered to be a recovery operation, where the deposit is seen as beneficial. If a waste deposit on land has a permit, or an exemption then the EA is the relevant authority. In any situation the EA has a duty to act to enforce legislation where there is a threat of pollution or harm to human health from a waste management activity. Whilst there is a permit in place it is possible to vary the permit conditions to require the operator to take actions to protect the environment required by legislation and national policy. In this way a permit can be varied (amended), for example to stop further tipping, to have relevant flood risk plans, and to undertake certain works, monitoring, reporting or other measures. New landfill permit applications consider stability (in terms of landfill liner), impact on groundwater and landfill gas emissions. Planners should also consider rising sea levels and more extreme effects such as erosion, rainfall, drought when deciding where sites should be located. As a statutory consultee on planning applications the EA would advise against a landfill within a flood risk area. Much of the landfill legacy however, has no permit.

Operators' responsibilities

Landfill operators are responsible for ensuring that the landfill does not cause pollution. Records show operators of old sites to be generally unknown, and may be a Local Authority, a company or entity which later ceased to exist. In the absence of the operator or polluter it generally falls to the landowner. Often the Local Authorities have authorised or run the tip on their land. There may be institutional differences between departments in responsibility, role and the authorities themselves.

Contaminated Land and historic sites

Local Authorities have the responsibility to identify potential contaminated land sites as well as being a partner or possibly lead in SMP production. Government departments may find that it is not always possible to work closely with other departments even within the same organisation. This may continue to be the case as staffing decreases and consultants are used. SMP representatives are typically from the FRCM function representing the whole of the EA and coastal engineers from the Local Authority. Other departments may be involved in scoping and decision making, though there is little evidence that waste or contaminated land has been considered to date.

The EA now has the coastal strategic overview role and controls the money for contaminated land capital projects separately. Funding streams are often ring fenced between contaminated land and FCRM work. The EA is also responsible for regulating current landfills and certain installations under Environmental Permitting as well as regulating 'Special Sites' under the part IIA Environmental Protection Act 1990 regime. The guidance for making an application for funding under part IIA tends to be for land based criteria based on scoring and according to EA guidance generally whether contaminants are present or whether contamination poses a certain level of risk.

Where pollution is occurring at historic sites, the EA can undertake the works and recharge the landowner or the Local Authority. The Local Authorities have a responsibility to investigate land that might be a candidate for Contaminated Land status under Part IIA of the Environmental Protection Act. How they perform their responsibilities varies between Local Authorities. Part IIA generally works on a voluntary basis when new planning permission is being applied for. Often, identifying the relevant party is difficult and may implicate the same or another Local Authority, in some instances causing a dispute about who is responsible.

Complicating factors

The matter of land responsibility is more complex around the coast as legal man-made boundaries imposed on the natural environment are less stable. It is here that the legislative functions overlap further and become more complex.

The UK planning system is based on registered parcels of land, with arbitrary manmade boundaries. At the periphery of the coast the convenient coherence of this two dimensional plan breaks down through four dimensional natural processes. For example, what may have been easily defined at the top of a cliff may then be subject to mass movement and be distorted through landslip.

Charities such as the National Trust may buy land between the beach and the cliffs for its natural interest. On the Dorset Coast there is a World Heritage Site designated for its natural interest. It is being allowed to recede naturally to ensure this is maintained. The shoreline belongs to the Crown. If there is a waste site containing hazardous material at the top of a cliff - for example the Spittles landfill, Lyme Regis - then in theory the waste and the problem can pass from the landowner at the top, to the charity owning the cliff, to the Crown Estate at the bottom. In this case hazardous materials and fridges have been involved.

In theory to deposit waste below the water line, a new Marine Consent licence is needed from the Marine Management Organisation and there may be some overlap with marine planning jurisdiction. If permission is sought to dredge coastal areas, sometimes these sediments, once tested, can be considered hazardous and cannot easily be redeposited on land. This again makes the point about the flow of waste materials in the environment. Wastes from eroded landfills might be considered to contribute to sediments of the future.

Further complications arise where assets are managed and policy decisions taken following a SMP without taking into account the presence of landfills. Even if identified, adverse public perception of spending money to defend or move landfill may be difficult to overcome where homes and property are at risk. In areas where coastal squeeze becomes more of an issue there is often a desire to adopt managed realignment and 'making space for water' strategies. One risk for the decision maker and landowners is that if the landfills are not identified or assessed then the risks have not been taken into account. Often both the EA and Local Authorities are responsible for identifying such sites and the risks where present. If these risks are not considered and should have been known about then this could have consequences. The implications of this are that pollution could occur in a way that has been directly permitted in contravention of the Water Resources Act 1991, Section 85 Pollution Offences which states that:-
'A person contravenes this section if he knowingly permits any poisonous, noxious or polluting matter or any solid waste matter to enter any controlled waters.' It could be argued that the responsible institutions should work together to ensure they are aware of the risks involved in not considering this in any plans.

Improving communication
Consideration tends only to be given to sites once they are problematic. Usually it is a case of monitoring the situation and, where possible, the Local Authority clearing the beach of the wastes that pose the greatest hazards to humans. Organisations do work together when an emergency incident arises such as at Lyme Regis and there is a clear structure in place.

There appear to be some landfills that are already eroding. It has been found that not all published second generation SMPs had identified the problem of eroding / flooding landfills and industrial sites. Some case studies are being gathered and best practice is starting to be shared. However, this is work in progress. How one would quantify risk and factors into flood defence funding is unclear at present.

CIRIA announced in July 2009 that a draft proposal would be developed for future guidance and is seeking sponsors. The aim is to address such sites in order 'to minimise environmental impacts, comply with legislation and ensure that the engineering approach will be sustainable in the long-term'. Further funded research projects are known to have been developed in recent months such as a recently NERC funded project with the London University. It is hoped that the waste and environmental management institutions, practitioners, academics and ICE will engage with the FCRM professionals in any future projects and research opportunities.

Impacts

In considering life cycle of products there is often the assumption that materials can be considered from 'cradle to grave', with landfill being considered as the 'grave'. The deposited material is still within the environment and can be released if disturbed or weathered. There may be further impacts post landfill to consider.

A landfill that is being eroded directly into the sea could be considered as a point source of contamination or marine littering. In future a significant and increasing proportion of the sediments around the coast could comprise eroded wastes. The impacts in the sea are largely unknown, but to rely on dilution of pollutants from so many sites seems unwise.

The few studies carried out have shown that there are effects on marine biology in the sediments e.g. at Lyme Regis (Hutchinson *et al*, 2009). Ager and Oakley (2006) give a good account of the literature of impacts from marine and coastal litter.

The Marine Conservation Society (MCS) published studies into marine litter. One such study showed that a large proportion of marine litter is unsourced but is increasing each year, some of this may be coming from rivers. Materials that come from landfill are subject to the same natural processes as any other sediment. There are also potential health risks to people who gain access to landfills on the foreshore from broken glass and other hazardous materials. Because we do not know what is in the waste, the risks are uncertain, although there are case studies that would be worthy of further evaluation.

It is often considered that after a long period of time wastes 'stabilise' in the ground. This may be true of some of the organic material within wastes or some readily leachable material. However none of it just disappears. Some areas erode, whilst others will still accrete along the coast. Lost waste will be subject to all normal coastal and marine processes, and will form part of the of marine sediments in the future. It is likely that some materials will be in the environment for a very long time to come and will be recorded in the rock record.

Finance / cost benefit

The economic costs associated with marine litter are substantial and landfills could be implicated. In 2000 a study found that marine debris accounted for known costs in excess of £5,500,000 per year in the Shetland Isles (Hall, 2000). There were costs to the fishing industry, blockages to power station inlets, beach cleaning expenses, costs to farmers and also lifeboat launches for litter related problems e.g. fouled propellers. The true cost was expected to be higher. 60 – 80% of marine litter is estimated to have come from land based sources. This includes legal and illegal shore-based solid waste disposal and processes (OSPAR, 2007). The MCS's annual Beachwatch Report, which collates data year on year of all the methodically collected litter around many of Britain's beaches, shows increasing quantities of litter.

Whilst a landfill permit is in force, there is a subsistence fee and this funds the EA's regulation against the permit conditions. When a landfill site closes the risk to the environment is considered to decrease. Wastes are contained by engineered structures and the by-products managed by the site operator. The EA regulatory interventions are therefore lower. It is worth considering that the engineering would need to cope with changing physical conditions. Risk may actually increase with time from changing environmental conditions and ageing. That said, the site continues to be regulated, and the conditions of the permit can be varied if required or enforcement action taken if pollution occurs.

In 2008 the EA was given the responsibility from DEFRA for holding the money for Local Authorities who have a responsibility for defending their coast from erosion under the Coastal Protection Act 1949. Capital funding is currently being reduced. The total budget for Contaminated Land Capital Projects for 2010-2011 was £10m. This is for all contaminated land investigation and remediation projects in England and Wales. This year there were just 75 sites that received funding of £3.6m across England and Wales, for the period between July 2010 and August 2011. The FCRM budget is similarly reducing so assessing risks is also a key part of making a case for funding.

Once a landfill is deposited the options are limited. It costs a great deal of money to remove and re-tip the wastes elsewhere. The waste has to be tested and classified as hazardous and non-hazardous before tipping elsewhere. Hazardous wastes being much more expensive. There are two rates of landfill tax, the higher rate at time of publication being £56 / tonne. This do not include other fees associated with haulage or disposal charges. The very act of disturbing landfill in many instances may also present other risks and further costs.

There could be cost benefits by considering protection or dealing with landfills in combination with other management options or scheme. For example, defending property or other infrastructure. Generally it is not economically viable to mine landfills just yet, but the economics might change in the near future. Some landfills may eventually become a useful resource for materials and energy. In the future mining might be beneficial to decrease the footprint of landfills (Horth, 2006).

A strategic risk assessment exercise

It will be necessary to perform a complex risk assessment of landfill sites and to attempt to factor in uncertainties. The risk factors will have to be normalised for comparison across the country. Closed, permitted sites have been undergoing a risk screening exercise for gas and leachate. It may be necessary to re-evaluate the data that is held, since data for historic sites has not been reviewed. Uncertainty is an inherent part of risk and so the absence of information with older records may require a worst-case assessment. This could then more reliably inform the SMPs and following strategies and schemes.

Ideally older records of landfills should not be destroyed, but protected so that they can be shared and re-evaluated. Whilst there can be good reasons for destroying some data, no public body should operate in isolation. Original paper records tend to be more informative and less malleable.

The type of information needs to be general such as type of landfill, wastes, etc. This information does exist for many of the closed and historic sites. This needs to be linked to SMPs, their sediment cells and suggested management options. Many of the closed landfill records of monitoring and risk assessments the EA holds relate to landfill gas and groundwater for protecting human health. It is suggested that the data might in future be linked to areas at risk of landslip, erosion and flooding and perhaps linked to any assessments and records held by Local Authorities. It would be helpful to identify those sites which have the benefits of justifying or contributing to coastal defence and ideally those which form part of defences - what assets protect them, who is responsible for managing them and so on.

In this way it would be possible to assess, prioritise and strategically manage the risks and make informed decisions that could support the SMP process.

Future opportunities

In future the economics may make it more feasible to mine existing sites and recover energy and materials from landfills. Inevitably there will still be some risk involved. Some of the costs of removal and disposal may be prohibitively expensive in the short term. However the longer term benefits could outweigh the costs for future generations. It could be considered to be an environmental deficit that can be dealt with. The impact of climate change will compound the problem if we do not adapt and attempt to control the situation. It may be that the risks involved in control and remediation are worth taking.

The UK has a long industrial legacy and many associated waste sites. It is also one of the most densely populated regions of the world. As such it is ideally placed to share best practice with the world in adapting and managing its coastal waste reserves.

Recommendations

The main recommendations are to:
1. Improve, keep and share data. Even limited primary or anecdotal data may be useful and better than its absence.
2. Share responsibility and outcomes. This will pool expertise and money
3. Stop adding to the problem. Consider extending life cycle assessments of products from cradle to beyond grave, as landfills and their matter are subject to natural geomorphological processes.
4. Act to manage coastal landfills in situ. Landfills are cheaper and more effective to deal with in situ waste than once it is released.
5. Assess risks, prioritise the greatest and then undertake further study.
6. Ensure that organisations are not knowingly permitting the pollution of controlled water (up to 3 miles on the coast). This could even be as an indirect result of schemes using waste for coastal re-alignment, or works that results in the release of contaminants. To resolve this it is necessary to seek advice and to work together.
7. Share best practice and case studies
8. Better cross-working to avoid the risks of stove-pipe hierarchies, especially in public organisations.

Integrated Coastal Zone Management is used more in Europe as way of resolving all these issues. In the UK the SMP process is the way of delivering something similar The roles of coastal forums and groups are very good at providing a neutral space to discuss the issues and this may be a way of addressing a national strategy through localism.

Conclusion

Having identified that there is a serious and growing problem associated with coastal landfills, we can now start to act together to address it. A number of landfills around the coast have already started to erode and may be at risk of causing environmental damage. If we do not invest sufficiently in this issue now then the economic, social and environmental costs and consequences will multiply. Greater general awareness of the issue, and effective and open collaboration between organisations are required to avert a growing potential toxic disaster.

The views and opinions are those of the author's and are not necessarily representative of those views or the work of the Environment Agency.

Acknowledgements

Although not representing the views of the Environment Agency, I would like to thank them, the Dorset Coast Forum and the many contributors for supporting my independent research.

References

Ager, O. and Oakley J. (2006). *Marine and coastal litter: The Marine Biological Association of the United Kingdom.*

Hall, K (2000). *Impacts of Marine Debris and Oil; Economic and Social Costs to Coastal Communities.* Technical Report. Kommunenes Internasjonale Miløoganisasjon (KIMO): Shetland, Scotland.

Horth, H.M. (2006). *Assessment of the Feasibility of Landfill Mining in Norfolk.* MSc dissertation. University of East Anglia.

Hutchinson, T.H., Langstone, W.J., Frickers, P.E., Harris, C., Imamura, M., O'Hara, S., Pope, N.D., Readman, J.W. and Shaw, J.P. (2009) *Assessment of Contaminants from the Lyme Bay Landfill – Preliminary Results of an integrated Chemical & Biological Study.* NERC ref NE/G00773X/1.

OSPAR Commission (2007). *OSPAR Pilot Project 2000-2006 on Monitoring Marine Beach Litter. Final Report.*

Shennan, I. and Horton, B. (2002) Holocene Land and Sea-level Changes in Great Britain. *Journal of Quaterary Science.* **17**(5-6) 511-526. Wiley InterScience.

Wilkinson. J.W.H. (2009) *Climate Change and the Control of Coastal Landfill Sites.* MSc dissertation. University of Bath.

Websites accessed

www.ciria.org
www.environment-agency.gov.uk
http://eur-lex.europa.eu/LexUriServ/LexUriServ.do?uri=CELEX:31999L0031:EN:HTML
http://en.wikipedia.org/wiki/Stovepipe_(organisation)

Innovative Coastal Zone Management
ISBN 978-0-7277-5749-4

ICE Publishing: All rights reserved
doi: 10.1680/iczm.57494.211

Climate Change Effects on Mixing Characteristics of a Large Embayment

Dougal Greer ASR Ltd, P.O. Box 67, Raglan, New Zealand,
Randall Lee, Environment Protection Authority, Victoria, Australia
John Oldman, ASR Ltd, P.O. Box 67, Raglan, New Zealand.

Introduction

Port Phillip Bay (PPB) in Victoria, Southern Australia is a large semi-enclosed bay. It is constrained by a relatively narrow (2.8 km at its narrowest) and deep (maximum depth =70 m) entrance that experiences fast tidal currents up to 2 m s^{-1} (Black et al., 1993) and features extensive sub-tidal sand banks known as the Great Sands (Figure 1). These physical features result in relatively long residence times of 1- 2 years. As a consequence of this, the average salinity of the bay is strongly affected by seasonally and inter-annually varying catchment inflows and also by within-bay evaporation. A major source of fresh water to the bay is the Yarra River in the North resulting in a salinity gradient across the bay becoming more saline towards the entrance. This is augmented by a consistent anthropogenic flow from the Western Treatment Plant (WTP) on the west coast.

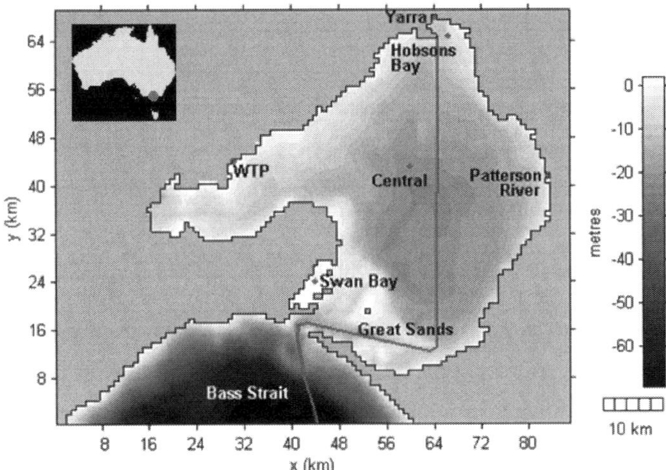

Figure 1 Model bathymetry and domain of PPB with relevant place names. The red line shows the path of the Spirit of Tasmania I passenger ferry.

With a majority of catchment loading at the northern end of the bay, a linear density gradient typically progresses to the open waters at the southern end (the hyposaline condition). Since 1947, monitoring in PPB indicates that during strong and prolonged periods of drought, salinities exceed that of the ocean (the hypersaline condition) and can extend throughout the bay. Unbroken records dating back to 1984 (Figure 2), show that Bay salinity is inversely related to annual rainfall patterns and that until recently there has been a prolonged hypersaline condition since the strong 1997/98 El Niño. With its very long residence time the net evaporation draws more saline water through the entrance from the sea and the salinities correspondingly rise as the evaporation selectively removes the fresh water leaving NaCl behind. This effectively replaces freshwater with saline oceanic water eventually causing the bay to become very salty.

In the mid 1990's the net flow through the entrance was outbound (Harris et al. 1996). This was fed by rainfall and freshwater run-off which exceeded the evaporation from the bay. During the period of reduced rainfall, water recycling and reduced river flows, PPB may have shifted from net outbound to net inbound. This would represent a fundamental transition in the state of the Bay, which may exacerbate increases in salinity within Port Phillip.

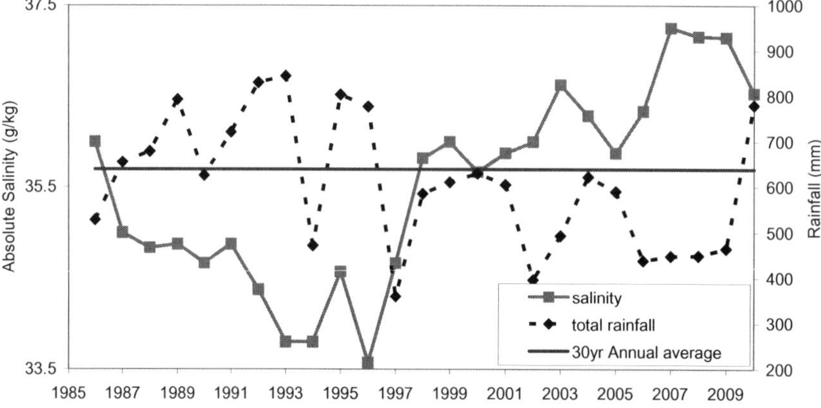

Figure 2 Long-term central bay salinity is inversely related to annual rainfall patterns.

Taking these processes into consideration it is clear that the overall salinity of PPB is greatly affected by precipitation and subsequently river flows. Given that flushing of the bay is governed largely by water movements at the entrance to the bay, salinity variation within the bay has implications for the residence times and circulation. This is particularly pertinent given that climate change forecasts predict reduced rainfall and increased evaporation in the future (Whetton and Power, 2007).

Results presented here summarise modelling efforts undertaken to enhance the understanding of the hydrodynamics of PPB. This work is extended to estimate the effects of predicted climate change scenarios on the Bay in general and on specific locations. The main objectives and methodologies used to develop the suite of hydrodynamic and dispersion models were presented at the 2007 Australasian Coast and Ports conference (Harrison et al., 2007).

Modelling Strategy

Modelling Scope
The models were used to investigate how changes in rainfall, catchment inputs and evaporation lead to changes in residence time within PPB because of the subtle balance between freshwater inflows and bay-wide evaporation.

The model scope included:
1. A dynamic catchment model (defining freshwater, nutrient and mud inputs) (Argent (2006)).
2. 3-dimensional salinity and temperature stratified hydrodynamics (for tidal, wind and density-induced circulation).
3. Dispersal (for effluent, sediment transport and pollutant dispersal).
4. Primary production (for nutrient uptake, phytoplankton and zooplankton).
5. Nested models of priority beaches.

As part of this scope the models simulate the behaviour and dispersal of key water quality indicators. These include nutrients (Total Nitrogen (TN) and Total Phosphorous (TP)), Chlorophyll-a, Total Suspended Solids (TSS), salinity, toxicants, pathogens and litter.

The initial baseline model was used for calibration purposes and simulated the period of 2004-2005. This period is characterised by average rainfall and hypersaline conditions within the bay (**Figure 2**).

Hydrodynamic Model Specification
The hydrodynamic model presented in this paper was produced using MODEL3DD, from the 3DD Suite. The 3DD Suite models of PPB, Western Port and Bass Strait were first developed in the late 1980's and have been utilised many times in Victoria. The current bathymetry grid for PPB (Figure 1) was 800 m x 800 m resolution with 108 x 86 cells and 8 vertical layers and took in the PPB in its entirety and a small section of Bass Strait just outside the bay entrance.

Hourly sea-level measurements from the Lorne tide gauge (located on the open coast approximately 65 km SW from the PPB entrance) were applied uniformly along the southern open boundary of the model. Hourly wind data provided by EPA Victoria from 6 automatic weather stations were applied over the model domain. The data were interpolated by the model using an 'inverse distance squared' algorithm over the wet cells of the bay. Wind contributes heavily to current direction, vertical mixing, and the direction and extent of dispersed plume travel. The model also included full atmospheric heat exchange which included humidity, barometric pressure, air temperature and solar radiation

River inputs were provided by the 'PortsE2' catchment model (Xu and Argent, 2006) which was established for the Port Phillip and neighbouring Western Port as part of the Water Quality Integration Plan (WQIP). The catchment model provided daily flow, TP, TN, TSS, salinity, E. coli, and metals (zinc and lead) loads.

Behind the 3DD model is the unique application of a satellite-derived Sea Surface Temperature (SST) spatial map input over the model domain. The SST is used in the model for 'corrective steering'. This technique involves substituting the model predicted SST (i.e. in the top layer of the 3D model) with 4-km resolution satellite SST observations which have been interpolated and gridded onto every model cell over the PPB region. We employ custom-written MATLAB software (SST_3DD) available with the 3DD Suite to automatically source, scan, filter, and process SST data which is taken from the internet (CSIRO's LAServer). Changing cloud cover leads to intermittent SST observations and so the substitution in the model occurs irregularly, mostly at 1-5 day intervals when a clear image is available. The 2004/05 model run was calibrated by comparing modelled output to measured temperature and salinity at Hobsons Bay (see Figure 3 and Figure 4).

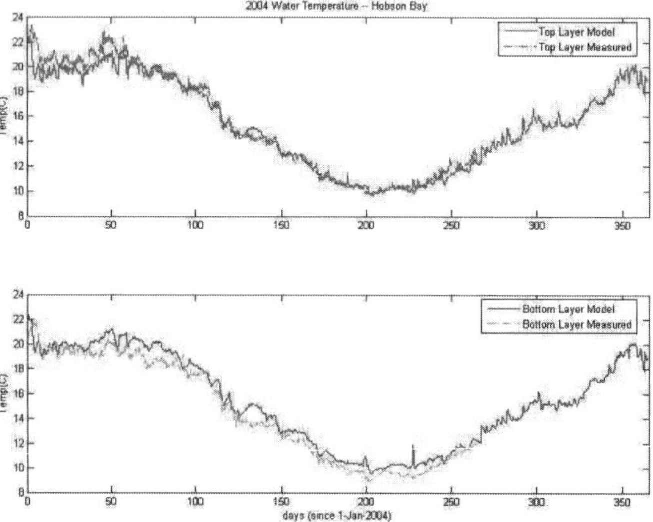

Figure 3   Temperature comparisons of moored observations and model results at the Hobsons Bay site within PPB during 2004.

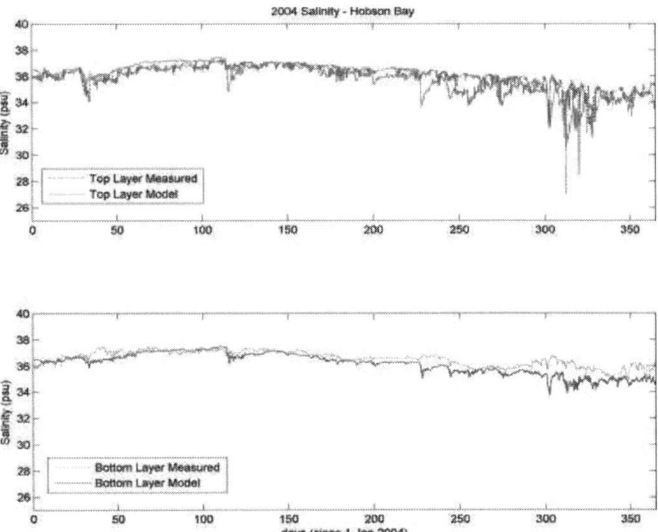

Figure 4 Salinity comparisons of moored observations and model results at the Hobsons Bay site within PPB during 2004-05.

Climate Change Modelling Scenarios

In this paper results from model simulations of PPB are presented under existing conditions and those that are likely occur in 2030 based on the CSIRO Mk3, A1F1 medium sensitivity projection for Melbourne from Whetton and Power (2007). Using baseline conditions (measured conditions in 2005) (Whetton and Power, 2007) blanket changes in evaporation (+4.7%) and rainfall (-2.7%) were adopted which manifested as an average ~10% reduction in flow from the Port Phillip (WBM 2007). Projections for 2070 (evaporation: +8% and rainfall: -18%) were based on the same scenario. In the absence of further climate change scenario runs of the PortsE2 catchment model, projections for 2070 were estimated based on proportional flow reductions for the 2030 scenario with respect to projections for rainfall and evaporation. Models were run by using the same boundary conditions as those in the 2004-2005 case and adjusting them to reflect changes in rainfall and evaporation.

In Situ Observations

Autonomous sampling systems were developed from pilot surveys of the bay in 2007 and were installed on the passenger ferry the Spirit of Tasmania I to collect salinity data (and other parameters), that resolves detailed spatial and temporal dynamics of the bay. These ship-borne measurements are undertaken as part of a national Integrated Marine Observing System (IMOS, 2009). Since 2008, routinely mapped salinity observations, have been available from monitoring from this ferry. The ferry traverses from north to south across the Bay and through Bass Strait each day (Figure 1). The bay-wide transect is the first long term dataset of salinity.

The plots (Figure 5) show consistent hyper-salinity inside the bay regardless of season from 2008 until September 2010 when the bay rapidly became and remained hyposaline due to heavy rainfall through the latter part of 2010. Initially the salinity front at the entrance erodes from intruding Bass Strait water, then by November 2010 the fresh water plume from the Yarra River near Melbourne advances from the north and dilutes the remaining hypersaline water to a fresher state for the first time since 1997.

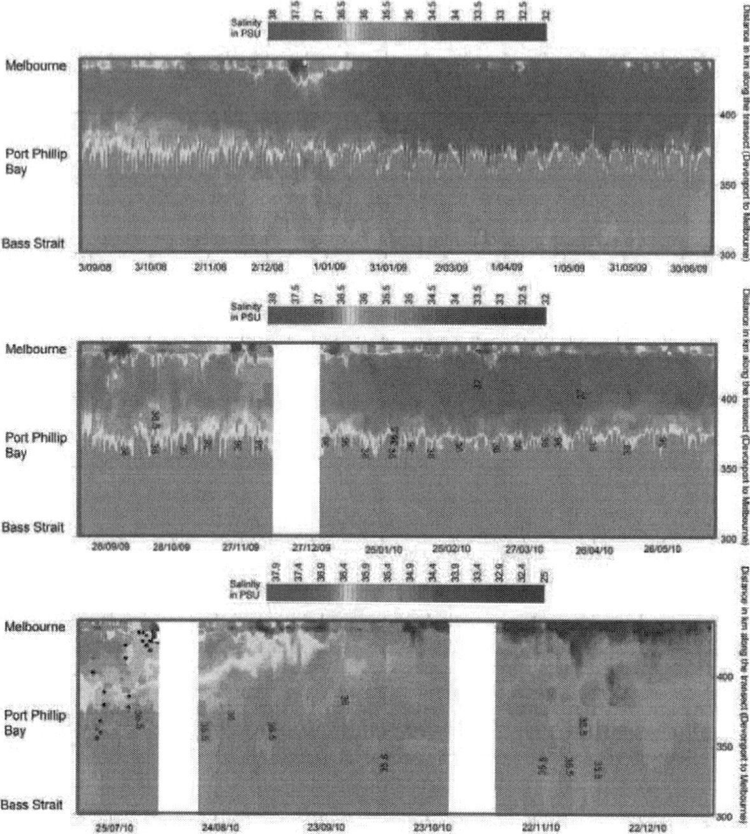

Figure 5 2008-2010 Salinity mapped north-south through the bay by a passenger ferry (ferry track shown in Figure 1).

Principal Outcomes

Residence Times

Residence times were estimated using a Lagrangian model (POL3DD which is part of the 3DD Suite) using the hydrodynamic models described above to drive particle movements. In the model PPB was filled with particles and a particle tracking box was placed at the entrance of PPB. When particles left the bay they were counted and eliminated from the model. T90 times were estimated by keeping track of the number of particles remaining in the bay.

Flushing rates can be expected to vary depending on freshwater inputs and wind conditions. Under future climate change scenarios reduced river input loads and increased evaporation point towards more water entering PPB from Bass Strait (approximately extra 350,000 m³/annum) in the 2030 scenario. With less water leaving PPB it is intuitive to conclude that residence times would be increased however modelling shows that the opposite is true, and the T90 times actually decreased under increased evaporation and decreased catchment loadings and rainfall. T90 values vary depending on the time of year the simulation is started but are typically in the region of 607 days for baseline conditions but around ~1% shorter for the 2030 simulations.

The increased flushing can be understood by considering the increased salinity stratification at the mouth of the Bay which occurs in more hypersaline conditions. On the out-going tide currents near the seabed increase and dense salty Bay water leaves the Bay lower down in the water column. This water is replaced by fresher water from Bass Strait on the surface consequently increasing the exchange between PPB and Bass Strait (Lee et al. 2011, submitted).

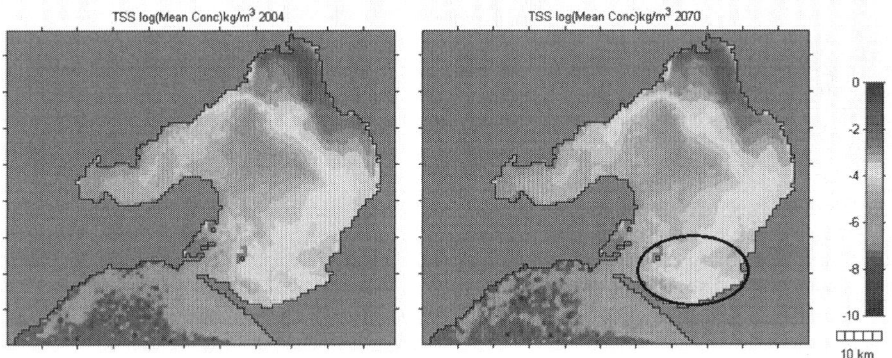

Figure 6 Comparison of suspended sediments in PPB for modelled conditions in 2004 (left) and the climate change scenario for 2070 (right).

TSS and Pollutant Dispersion

Adjusted TSS loads from PortsE2 for a 2070 scenario were driven by the 2070 hydrodynamics to map TSS (mean log concentrations) in the bay in Figure 6. There are overall clear reductions in TSS concentrations for 2070 related to reduced loads predominantly from the Yarra River. This is manifested not only as a reduced footprint in the Hobsons Bay region to the north, but also as significant reductions along the east and south east coast of the bay, that may have ecological consequences, as this region is typically a hotspot for fish larval production. The western side of the bay (near the WTP) shows an increase in TSS for 2070 most likely resulting from reduced dilution and advection of the WTP discharge with increased hypersaline conditions.

Primary Production

Model 3DDLIFE is an Eulerian based, fixed stoichiometry coastal marine ecosystem productivity model. The model solves multiple interactive equations for the state variables in a forward explicit time-stepping scheme, with the variables represented on a regular grid. The model describes nutrient cycling (nitrogen and phosphorus), phytoplankton and zooplankton growth and decay along with the dissolved oxygen conditions within the coastal marine environment, though it's primary concern is phytoplankton and zooplankton dynamics. The model has been described and extensively tested against field measurements in the Bay of Plenty, New Zealand by Longdill (2008) in an assessment of the impacts of mussel farms on primary productivity.

The primary production modelling was undertaken using a new set of boundary conditions which define the mass of nutrients entering the bay from the 3 main sources: The Yarra River, WTP and The Patterson River and was estimated using flows and concentrations from the PORTS E2. Coupling this with the 3DD Hydrodynamic models, nitrogen and phosphorus availability was used to define the growth rate of the phytoplankton. The non-predatory mortality factor was set to 0.01 (1% / day).

Figure 7 shows the phytoplankton response of the 3DDLife model during 2004 and a 2030 scenario extracted for sites at Hobsons Bay and Long Reef (offshore of the WTP). Enhanced nutrient loads (3 fold) for the 2030 were included to test system response and recovery (during major loading events). While the patterns for 2004 and 2030 are similar, the scenario for enhanced nutrient loads doubles growth peaks and extends recovery period back to base line conditions.

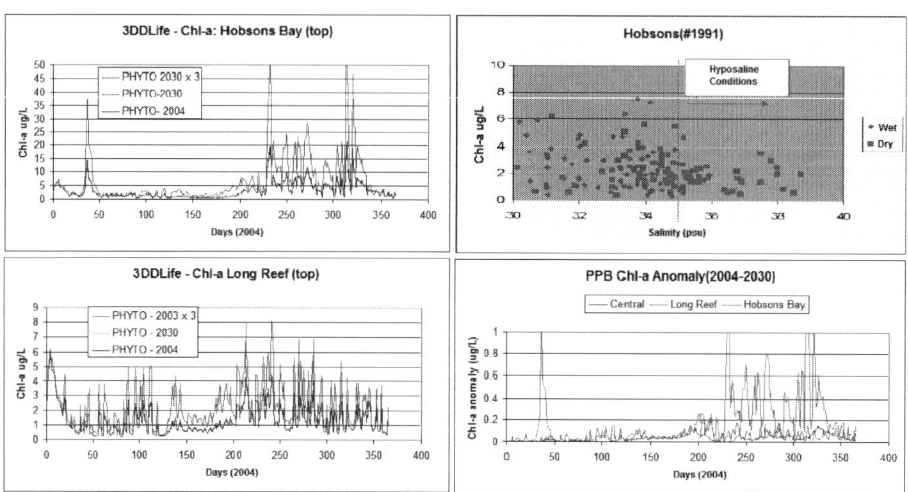

Figure 7 : Modelled phytoplankton response at Hobsons Bay and Long reef (top left and bottom left respectively), the observed effect of salinity conditions on phytoplankton (top right) and Chl-a anomalies for three locations around the bay (bottom right).

The effect of hypersalinity in constraining phytoplankton populations is suggested by observational data from Hobsons Bay (Figure 8). This was tested by comparing the Chl-a anomalies (2004 - 2030 model runs) for the three continuously monitored locations in Port Phillip Bay (Figure 7). All sites indicate constrained production for the enhanced bay salinities in 2030, with the Hobsons Bay site clearly showing the greatest difference, especially when comparing responses to major loading events during the spring (days 230-320).

It was evident from the modelling that increases in nutrients lead to substantial changes in Chl-a levels so it is necessary to know the thresholds above which the bay could be subject to substantial and numerous algal outbreaks, particularly during the favourable spring and summer conditions.

Discussion and Conclusions

Results from this study indicate that predicted climate change scenarios are likely to alter the flushing of PPB and its exchange mechanism with Bass Strait. Increased evaporation and reduced river flow act to increase the amount of water flowing from Bass Strait in PPB. This effect will increase salinity of the bay as fresh water lost by evaporation will be replaced by saline water from Bass Strait. At its most extreme, this mechanism may ultimately lead to the fluxes at the entrance to PPB changing from net outbound to a net inbound. However, this change needs to be viewed in the context of a system which naturally undergoes changes in salinity (switching between hypersaline and hyposaline conditions) which is evident on decadal time scales.

Of particular interest are the recent large rainfall events which have had a noticeable impact on the salinity of the bay. Despite climate change predictions, periods of bay-wide hyposalinity are still to be expected. In situ measurements of the recent transition of the bay from hypersalinity to hyposalinity represent a challenge for further model calibration and an opportunity for enhanced understanding of bay dynamics.

The changes in bay hydrodynamics have implications in a broad range of areas. Altered residence times may change the concentration of pollutants and waste discharged from the catchment throughout the bay. Reduced rainfall will lead to decreased flows and hence increased concentrations at discharge points. The effects of climate change on TSS dispersal are more complex and the altered pattern of dispersal illustrates the effectiveness of using a model which can assimilate the cumulative effects of many forcings.

The models presented here provide a diverse set of tool for developing an understanding of the circulation and associated implications for PPB. They represent a suite of tools which will be used in the future to incorporate updated climate change predictions and also to be calibrated further against on-going data collection.

References

Argent, R. (2006) PortsE2: A decision support system for the Port Phillip Bay and Western Port Water Quality Improvement Plan, CEAH Report 01/06 to MW.

Black, K.; Hatton, D. and Rosenberg, M. (1993) Locally and externally-driven dynamics of a large semi-enclosed bay in southern Australia. Journal of Coastal Research. 9(2):509-538.

Harris, G., Bately, G., Fox, D., Hall, D., Jernakoff, P., Molloy, R., Marray, A., Newell, B., Parslow, J., Skyring, G. and Walker, S. (1996) Port Phillip Bay Environmental Study Final Report. CSIRO. Canberra, Australia.

Harrison S, Lee R.S. & Black, K.P. (2007) An integrated catchment and receiving model for Port Phillip Bay and Western Port to address pressing environmental concerns, proceedings Coasts and Ports, 2007, Melbourne, Australia.

Longdill, P. (2008) Environmentally sustainable aquaculture: An eco-physical perspective. PhD thesis, University of Waikato, NZ.

Whetton, P., and Power, S., (2007) Climate Change in Australia, Technical Report, CSIRO and Bureau of Meteorology, ISBN: 9781921232930, 148 pp.

Xu, M. and Argent, R.M. (2006). Application of a 'Whole-Of-Catchment' Model to the Port Phillip Bay catchment, Australia. www.mssanz.org.au/modsim05/papers/xu_m.pdf

WBM (2007) 2004-05 PortsE2 catchment model outputs for the receiving water quality model, report to EPA Victoria.

Lee, R.S, Black, K.P., Bosserelle, C., Greer, D. (2011) Present and future prolonged drought impacts on a large embayment., Journal of Ocean Dynamics (Submitted)

Innovative Coastal Zone Management
ISBN 978-0-7277-5749-4

ICE Publishing: All rights reserved
doi: 10.1680/iczm.57494.221

Lavernock Point to St Ann's Head SMP2 – Assessing Habitat Changes Under Rising Sea Levels

Marcus Phillips, Senior Coastal Engineer/ Project Manager, Halcrow Group Limited, Swindon, Wiltshire, UK
Helen Jones, Environmental Scientist, Halcrow Group Limited, Cardiff, Wales, UK
Nigel Pontee, Principal Coastal Scientist, Halcrow Group Limited, Swindon, Wiltshire, UK
Phil Williams, Swansea and Carmarthen Bay Coastal Group Chairman, Carmarthenshire County Council, Wales, UK

Introduction

This paper is based on work undertaken during the development of the second generation Shoreline Management Plan (SMP2) for the South Wales coast between Lavernock Point (Vale of Glamorgan, south of Cardiff) and St Ann's Head (Pembrokeshire), see Figure 1.

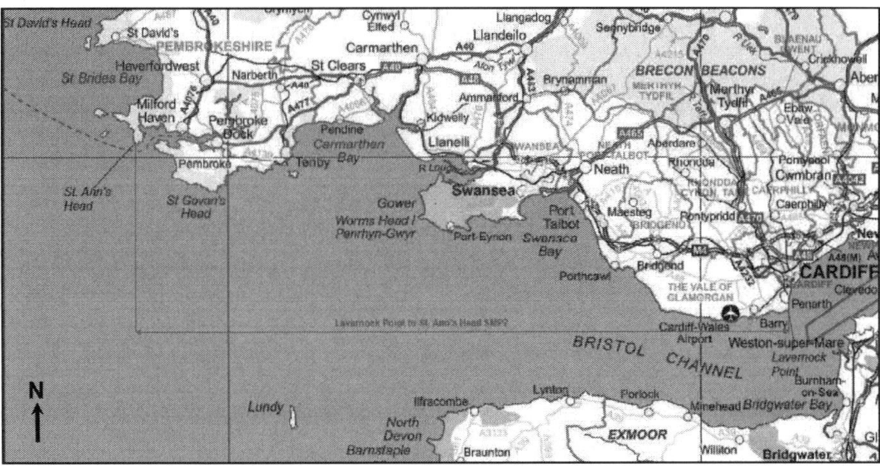

Figure 1: Extent of Lavernock Point to St Ann's Head SMP2

The SMP2 coast is 450km long and comprises thirteen main rivers/ estuaries (including Neath Estuary, Tawe Estuary, Loughor Estuary/ Burry Inlet, Three Rivers Estuarine Complex (Gwendraeth, Towy and Taf) and Milford Haven) as well as a number of smaller estuaries/ outlets and various habitats including: wave cut platforms, cliffs, sand and shingle beaches, dunes, intertidal sandbanks, mudflats and saltmarshes. Along the SMP2 frontage intertidal saltmarshes predominantly occur within Loughor Estuary/ Burry Inlet (see Plate 1), Three Rivers Estuarine Complex and Milford Haven, see Figure 2.

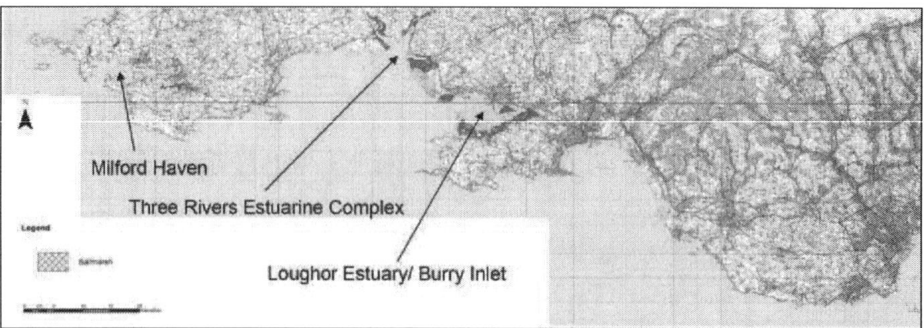

Figure 2: Extent of saltmarshes within the SMP2 study area

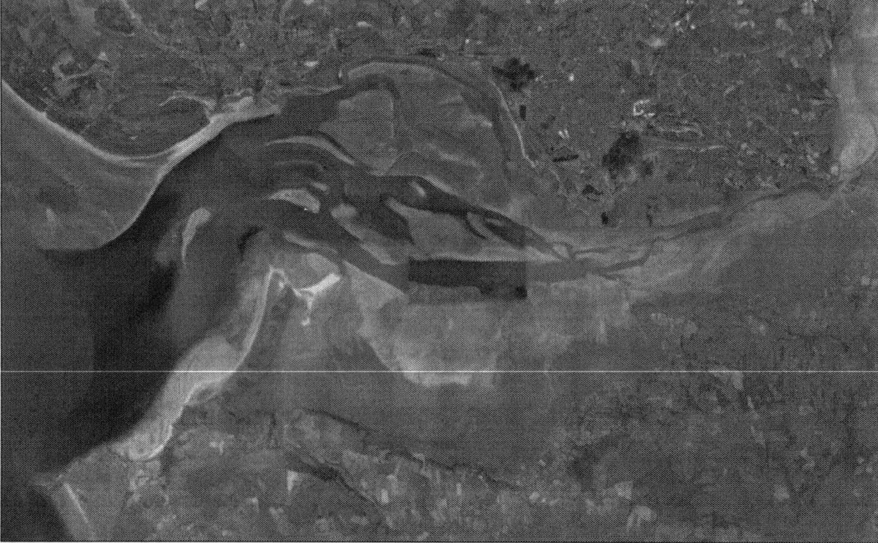

Plate 1: Loughor Estuary/ Burry Inlet aerial photograph (Google Earth Pro, 2009)

The SMP2 area includes nineteen European designated nature conservation sites, twelve of which are Special Areas of Conservation (SAC), four Special Protection Areas (SPA) and three Ramsar sites, see Figure 3. The direct, indirect and in-combination impacts of the SMP2 on these sites were critical to the development of a robust and viable coastal erosion and flood risk management plan. Consequently, the requirements of the European Union Habitats Directive (92/43/EEC) and European Union Birds Directive (79/409/EEC), as implemented in the UK by the Conservation of Habitats and Species Regulations 2010 and the Wildlife and Countryside Act 1981 (as amended), needed to be addressed.

A Habitats Regulations Assessment (HRA) was prepared for the SMP2 in accordance with the Strategic Environmental Assessment Directive (2001/42/EC) and followed the requirements of the Habitats and Birds Directives. This paper explains the approach that was used to undertake the appraisal of habitat gains and losses which was used to inform the development of SMP2 policies.

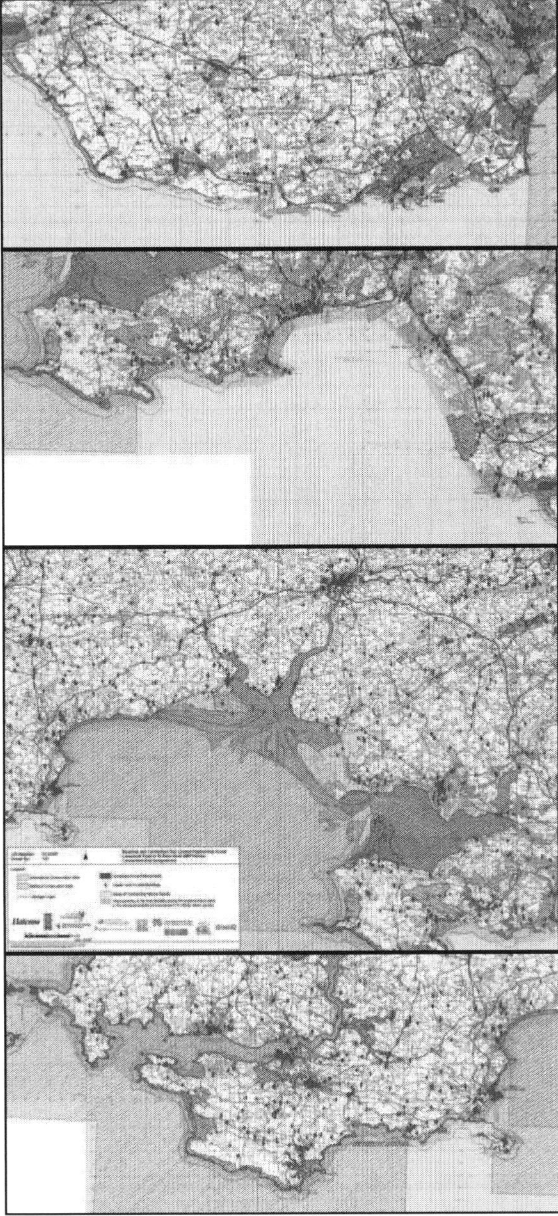

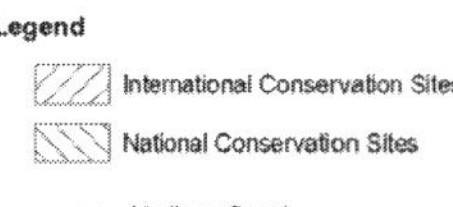

Figure 3: Internationally and nationally designated conservation areas

Methodology

Stage 1

The first stage of the appraisal involved the following:

- defining the extent of intertidal habitats;
- locating areas of high ground inshore, which would constrain inland migration of intertidal habitats; and
- identifying potential low-lying areas where intertidal habitats could be allowed to migrate inshore.

LiDAR (Light Detection and Ranging data) data was obtained from the Environment Agency Wales for the SMP2 study area. Halcrow used a Geographic Information System (GIS) to create a Digital Terrain Model (DTM) from this data and to produce plots of the potential extent of intertidal habitats based on existing sea level and following 1m future sea level rise, see Figure 4 which was produced for the Loughor Estuary/ Burry Inlet. For the purposes of this analysis it was assumed that mudflats occur between mean low water spring (MLWS) tide level and mean high water neap (MHWN) level (shown as light brown and orange on Figure 4), whilst saltmarshes occur between mean high water neap (MWHN) tide level and highest astronomic tide (HAT) level (shown as light green and dark green on Figure 4).

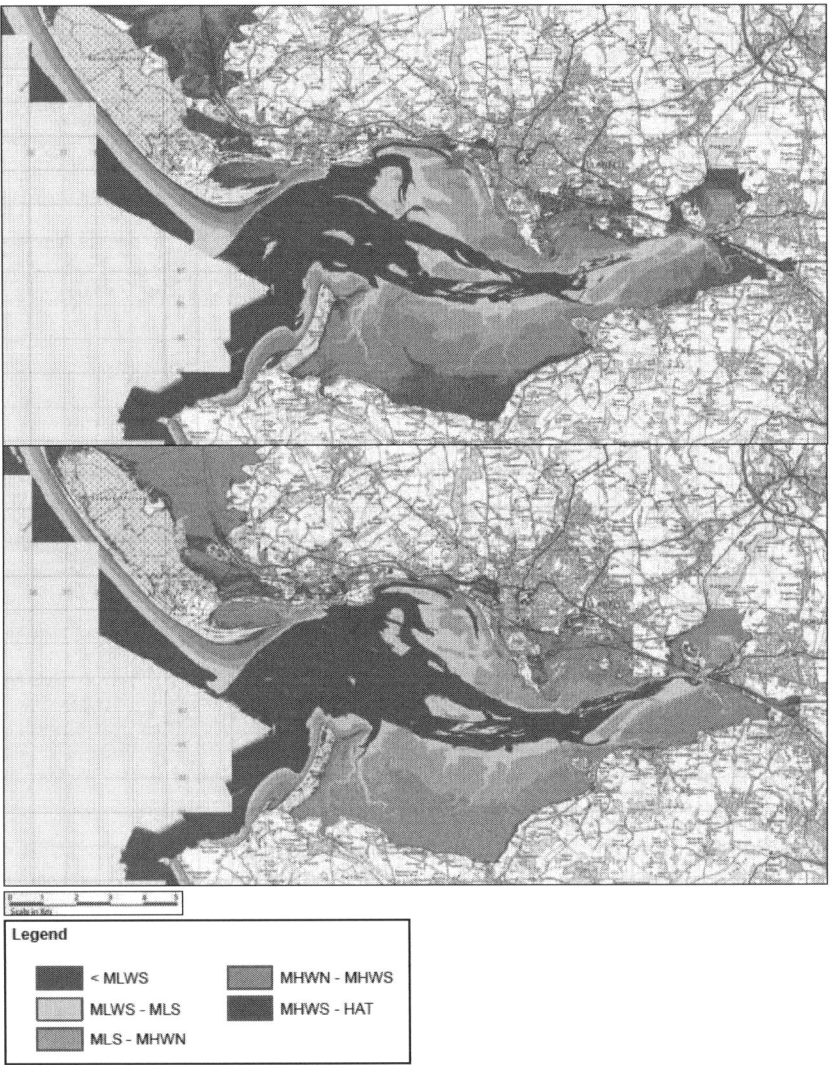

Figure 4: Loughor Estuary/ Burry Inlet - Potential extent of intertidal habitats based on existing sea level (top image) and following 1m sea level rise (bottom image) (mudflats shown as light brown/ orange and saltmarshes shown as light green/ dark green)

Stage 2

The second stage of the appraisal comprised the following tasks:

- quantifying potential intertidal habitat losses over the SMP2 period (100 years) due to sea level rise;
- identifying the impact that the preferred SMP2 policies would have on intertidal habitat losses; and,
- identifying potential sites where managed realignment could be implemented, subject to further detailed studies, investigations and analysis.

The goal was to balance the losses and gains resulting from the preferred SMP2 policies over the SMP2 period (100 years). The methodology was developed by Halcrow, and subsequently refined following discussions with Countryside Council for Wales (CCW) and Environment Agency Wales, to assess the potential losses of intertidal habitat due to coastal squeeze. Coastal squeeze is the term used to describe the loss of intertidal habitats under rising sea levels due to the natural landward migration of intertidal habitats being prevented by a man-made defence, see Figure 5.

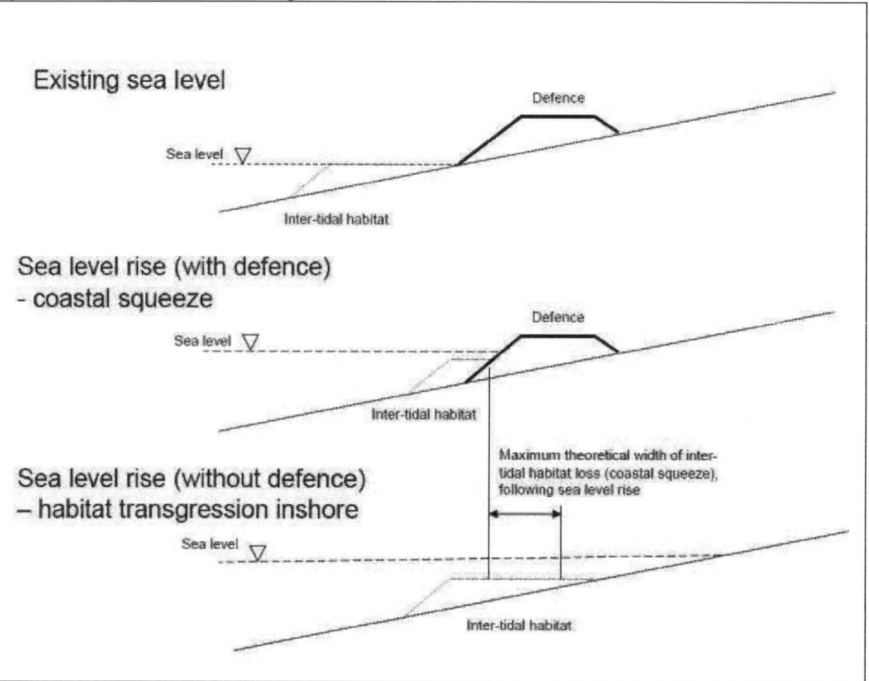

Figure 5: Maximum theoretical width of intertidal habitat loss (coastal squeeze) following sea level rise

The analysis excluded dunes[1], which are not intertidal habitats, and loss of intertidal habitat along defended frontages which are backed by high ground or cliffs, since these losses do not represent coastal squeeze[2].

---

[1] SMP2 polices along dune frontages are Managed Realignment.

[2] SMP2 polices along the majority of undefended frontages are No Active Intervention.

Where the preferred SMP2 policy was to hold the existing line, which may result in coastal squeeze of intertidal habitats, the maximum theoretical width of future intertidal habitat loss was calculated assuming that if the existing defences were removed the existing intertidal habitat would migrate inshore in response to future sea level rise, see Figure 5. This initial appraisal assumed a linear coastal gradient, where the slope landwards of the existing defence was the same as the existing intertidal slope seaward of the defence. The maximum theoretical area of intertidal habitat loss was obtained by multiplying this width by the length of each policy unit.

The initial assessment was refined using GIS to calculate the actual area available for intertidal habitat transgression inshore of the existing defences, following failure or removal of the existing defences and sea level rise. This analysis used the DTM (since the actual coastal gradient is typically not linear landward and seaward of the existing defence), the tidal limits within which intertidal habitats form and future predicted rates of sea level rise. The limitations of this approach are described in the next section.

To provide a precautionary approach the assessment considered a range of future sea level rise projections over the SMP2 period (100 years): UKCP09 (0.2m (low, 5th percentile limit of the range of uncertainty), 0.5m (central estimate) and 0.8m (high, 95th percentile limit of the range of uncertainty)) and Defra/ EA 2006 climate change guidance (1m, for reference), see Figure 6.

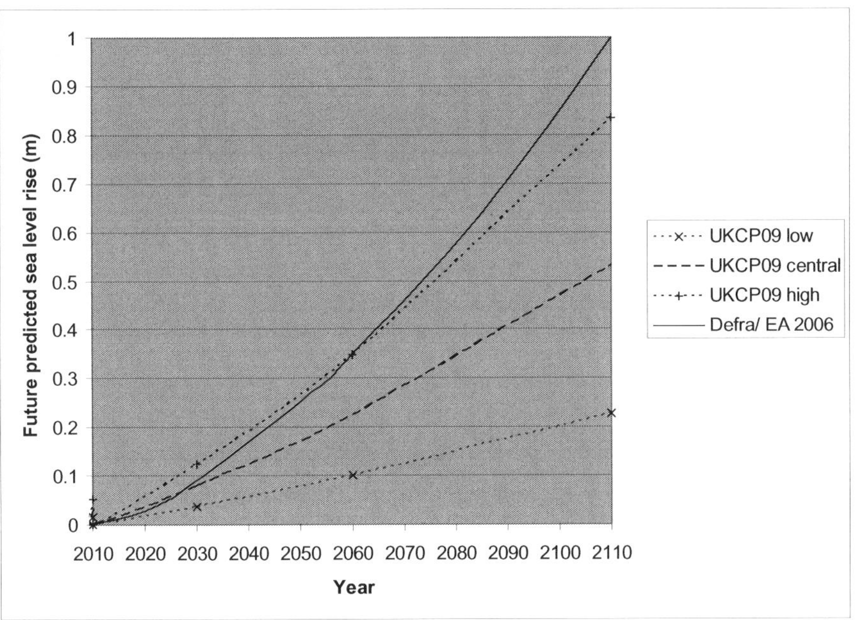

Figure 6: Future predicted rates of sea level rise (Defra/ EA, 2006 and UKCP09)

Losses were calculated over the SMP2 period (100 years) with various adjustments applied to account for the implementation of different policies at different points in time. For example where the policy is hold the existing line in year 0-20, but no active intervention thereafter; or where hold the line only applies along a proportion of the length of the policy unit.

Potential compensatory intertidal habitat areas were identified following review of the 'potential extent of intertidal habitat' plots, see Figure 4. This review sought to identify large sites where the existing topography suggested that it may be viable to create compensatory intertidal habitats, subject to further detailed studies, investigations and analysis. Larger sites are likely to be more sustainable since they are likely to include a range of ground elevations with the potential to provide a range of intertidal habitats and a more flexible response to the uncertainties associated with projected rates of future sea level rise.

Limitations

There are a number of limitations to this high level approach:

- Extent of existing intertidal habitat may vary from the values derived in this analysis for a number of reasons:
    - The analysis was based on a DTM created from data obtained from LiDAR surveys undertaken over a number of years. The DTM for the Loughor Estuary/ Burry Inlet was compiled from a mosaic of LiDAR data from surveys undertaken between 2003 and 2008.
    - The extent of intertidal habitats was based on tide levels predictions provided in the Admiralty tide tables (UKHO, 2009). In reality tide levels may vary from the quoted values, particularly in estuaries. Habitat extent is also dependent on swell waves, locally generated wind waves and/ or tidal currents.
- Future loss of intertidal habitat may vary. This appraisal has assumed that existing habitat will migrate inshore in response to future sea level rise. Such an approach is likely to over-predict the losses in front of structures since: (i) vertical accretion could potentially allow habitats to be maintained to some degree without migrating landward. This is particularly the case in the Loughor Estuary/ Burry Inlet, which has undergone a net import of sediment and an expansion of intertidal habitats over recent decades which may continue in the future; (ii) even if structures were not present the habitats may not migrate landward at a pace to keep up with sea level rise and therefore may experience some losses due to drowning.
- Future extent of intertidal habitats may vary due changes other than sea level rise. For example Whiteford Point, a three kilometre long dune-capped shingle barrier at the mouth of the Loughor Estuary/ Burry Inlet, shelters the estuary from offshore swell waves. Future changes to this feature could lead to significant changes to the wave climate and the extent of intertidal habitats in the estuary. The Loughor Estuary/ Burry Inlet continues to adjust following significant human interventions over the last 200 years including: extensive land reclamation along the northern shore (from the 17th century onwards) and the construction of training walls, to provide a navigable channel to Llanelli Docks (which began in 1883). The tidal channel system is current re-establishing a higher degree of dynamic behaviour following the abandonment and breakdown of the training walls, which could have a significant impact on the future extent of intertidal habitats along the southern shore of the estuary.
- Potential compensatory intertidal habitat areas may have been over-predicted. Other than consideration of topographic levels, this analysis takes no account of the potential suitability of the areas identified. In reality some of the sites may be unavailable or unsuitable for habitat creation.

Despite these limitations the approach provides a high level assessment of the potential gains and losses of intertidal habitat which was appropriate to inform the development of the Lavernock Point to St Ann's Head SMP2.

Results

It was estimated that the SMP2 policies will result in a potential loss of intertidal habitat of 324ha (based on UKCP09 relative sea level rise projection, central estimate), with a range of between 183ha and 460ha (UKCP09 5th and 95th percentile limits of the range of uncertainty) and up to 533ha (Defra/ EA October 2006 sea level rise allowance, which was included for reference) over the SMP2 period (100 years). Potential intertidal habitat losses for each policy unit, calculated using the methodology discussed above, are provided in Table 1. Total intertidal habitat losses over the SMP2 period (100 years) are shown in Figure 7.

A sensitivity test showed that +/- 1° variation in intertidal slope has the potential to lead to a -8% (- 25ha) or +11% (+ 35ha) variation in potential area of habitat loss, based upon the UKCP09 central estimate.

Given that there are a number of uncertainties associated with the future loss of habitat it is recommended that the extent of coastal habitats is monitored in order to update the targets for habitat creation and to inform future management of the estuaries. The SMP2 action plan has recommended improved and regular monitoring of the estuaries, comprising annual LiDAR surveys and vertical aerial photography of entire estuarine systems undertaken over a single period of low spring tides.

The Lavernock Point to St Ann's Head SMP2 is due for review in 5 to 10 years and this will provide an opportunity to review how the intertidal habitats are developing and to revisit habitat creation targets.

A number of sites were identified, for which SMP2 policies of managed realignment were defined, which may be able to provide up to 620ha of compensatory intertidal habitat, subject to further detailed studies and investigations, see Table 2.

Table 1: Areas of potential habitat loss due to coastal squeeze (ha) over the SMP2 period (100 years)

Policy Unit			Areas of potential inter-tidal habitat loss due to coastal squeeze (ha) over 100 years			
			UKCP09 low	UKCP09 central	UKCP09 high	Defra/ EA (2006)
12	1	Whiteford Point to Llanrhidian Marsh	8	20	31	38
12	2	Crofty to Penclawdd	11	11	11	11
12	6	River Loughor West Bank	53	53	53	53
12	7	Morfa Baccas	1	3	5	6
12	8	Wildfowl and Wetlands Centre to Penrhyn Gwyn	20	48	75	90
12	10	Pwll railway frontage	19	25	25	25
14	2	Banc-y-Lord to Commissioner's Bridge	34	80	125	150
14	3	Kidwelly (Commissioner's Bridge to Kidwelly Quay)	23	53	84	100
14	9	Ferryside	8	20	31	38
14	19	South of Laugharne to Ginst Point	4	9	14	16
16	4	Wiseman's Bridge	0	0	0	0
20	6	Gelliswick Bay	0	0	0	0
21	2	Pickleridge (The Gann to Black Rock)	1	2	3	3
21	3	Dale (Black Rock to Dale south)	1	2	4	4
TOTAL			183	324	460	533

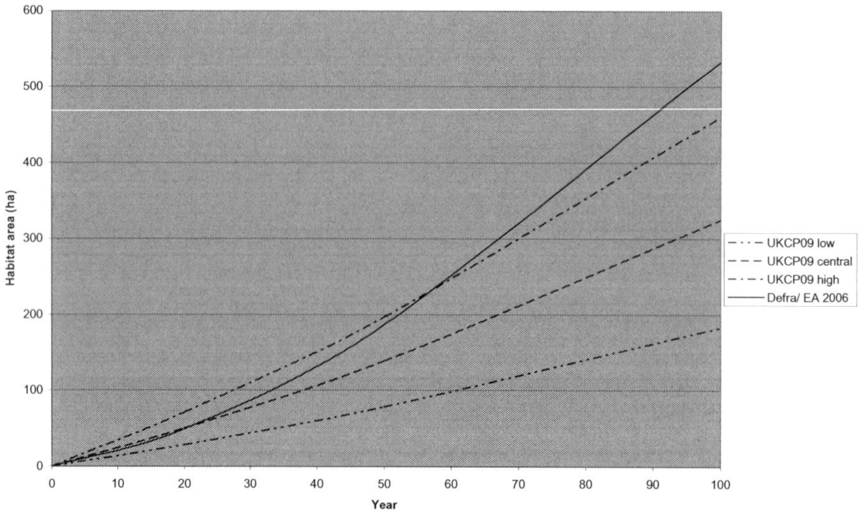

Figure 7: Potential intertidal habitat losses over the SMP2 period (100 years) for a range of future sea level rise projections

Table 2: Areas of potential habitat gain (ha) over the SMP2 period (100 years)

Policy Unit			Potential compensatory sites (ha)
12	3	Gowerton	20
12	7	Morfa Baccas	50
14	15	River Taf East Bank	25
14	19	East Marsh	525
TOTAL			620

Although the SMP policies (assuming that they are implemented) are estimated to produce a net gain in intertidal habitat over the SMP2 period (100 years), the Habitats Regulations Assessment (Halcrow, July 2010) concluded that the SMP2 is still likely to have a significant adverse effect on the following sites:

- Pembrokeshire Marine SAC;

- Carmarthen Bay and Estuaries SAC;

- Burry Inlet SPA; and

- Burry Inlet Ramsar Site.

In respect of these effects, a Statement of Case has been prepared, showing that there are no less damaging, appropriate alternative policies for the frontages concerned and that there are Imperative Reasons of Overriding Public Interest (IROPI) for pursuing the SMP2 policies. The Statement of Case also outlined the compensatory measures that will be delivered to offset the adverse effects of implementing the policies and has been submitted to the Welsh Assembly Government for consideration.

Conclusions

Developing a high level quantitative appraisal of intertidal habitat losses and gains at the outset of the development of SMP2 (which was based on existing available information and made certain assumptions) provided a useful tool to inform the development of the preferred SMP2 policies. If the SMP2 polices are delivered then they would provide a net gain of intertidal habitat within the SMP2 study area over the next 100 years.

The mechanism for delivering intertidal compensatory habitat is currently being developed. It has been suggested that the preferred mechanism for delivering intertidal compensation will be the National Habitat Creation Programme (NHCP) for Wales. This programme is currently being developed by the Environment Agency Wales, supported by the Welsh Assembly Government. It is recommended that detailed studies and investigations are undertaken to determine whether the sites identified are suitable for development to provide compensatory intertidal habitats.

Given that there are a number of uncertainties associated with the future gains and losses of intertidal habitat in the study area it is recommended that the extent of intertidal habitats are regularly monitored in order provide an improved understanding of how they are developing and to update the habitat creation targets outlined in this paper. The SMP2 action plan has therefore recommended improved monitoring of the estuaries. The specific recommendation was for annual LiDAR surveys and vertical aerial photography of entire estuarine systems to be taken over a single period of low spring tides. This will represent a significant improvement over the present coverage which is composed a mosaic of LiDAR surveys undertaken over a number of years. CCW is currently investigating alternative methods of monitoring the extent of intertidal habitats. Proposals for the Wales Coastal Monitoring Centre are currently being developed, which might be similar to the Channel and Plymouth Coastal Observatories and would potentially form part of the National Strategic Regional Coastal Monitoring Programme.

Further investigations are required to establish the suitability of the potential habitat creation sites by considering their specific characteristics in terms of land use and ownership, substrate type, contamination. It should be noted that the Lavernock Point to St Ann's Head SMP2 will itself be due for review in 5 to 10 years and this will provide an opportunity to review how intertidal habitats are developing and to revisit the habitat creation targets.

This paper provides a relatively simple method for quantitative appraisal of intertidal habitat losses and gains which can be used during the development of high level coastal erosion and flood risk management plans. The precautionary approach makes certain assumptions but provides a useful starting point which could identify the requirement for further detailed studies and investigations at sites which may be able to be provide compensatory intertidal habitats. It is recommended that any appraisal undertaken is subject to regular review, informed by regular intertidal habitat monitoring.

Acknowledgements

The views expressed in this paper are those of the authors alone. The authors are grateful to the Swansea and Carmarthen Bay Coastal Group, who commissioned and led the Lavernock Point and St Ann's Head SMP2. The authors would also like to recognise the significant inputs of Helen Jay and Jenny McConkey at Halcrow, who undertook baseline processes assessment, policy development and appraisal, preferred policy scenario testing, development of the policy statements and policy sensitivity analysis. The authors are also grateful for the work carried out by Professor Ken Pye and Simon Blott which informed the development of the SMP2. Their work included a review of coastal processes and shoreline behaviours of estuary-dominated systems in Swansea Bay and Carmarthen Bay and further more detailed analysis of the morphological change in Carmarthen Bay and adjoining estuaries.

References

Bristow, C.R. and Pile, J., 2003, South Wales Estuaries. Carmarthen Bay. Evolution of Estuarine Morphology and Consequences for SAC Management. CCW Contract Science Report No. 528, CCW, Bangor. 116pp.

CIRIA, 2004, Leggett, D.J., Cooper N. and Harvey, R. Coastal and estuarine managed realignment – design issues, Report C628. 217pp.

Defra, 2006, Flood and Coastal Defence Appraisal Guidance FCDPAG3 Economic Appraisal Supplementary Note to Operating Authorities – Climate Change Impacts. October 2006. 9pp.

Halcrow, 2010, Lavernock Point to St Ann's Head Shoreline Management Plan Consultation Draft, Appendix H Habitats Regulations Assessment. July 2010. 1018pp.

HMSO, 1994, The Conservation (Natural Habitats, & c.) Regulations 1994 (SI 1994 No 2716). 61pp.

Environment Agency, 2006, Economics sub-group, 4 December 2006, Economic analysis and climate change. 5pp.

Pye, K. and Blott, S.J., 2009, Coastal Processes and Shoreline Behaviours of Estuary-Dominated Systems in Swansea Bay and Carmarthen Bay. Annex A1 in Halcrow (July 2010), Swansea Bay and Carmarthen Bay Shoreline Management Plan 2, Appendix C: Baseline Processes Understanding. Halcrow Group Ltd., Swindon. 121pp.

Pye, K. and Blott, S.J., 2010, Morphological Change in Carmarthen Bay and Adjoining Estuaries: Further Analysis. Annex A2 in Halcrow (July 2010), Swansea Bay and Carmarthen Bay Shoreline Management Plan 2, Appendix C: Baseline Processes Understanding. Halcrow Group Ltd., Swindon. 103pp.

Robins, P.E., 2009, Development of a morphodynamic model of the Burry Inlet to inform future management decisions, CCW Contract Science Report No 898b, CCW, Bangor. October 2009. 101pp.

Tappin, D.R., Chadwick, R.A., Jackson, A.A., Wingfield, R.T.R. & Smith, N.J.P, 1994, United Kingdom Offshore Regional Report: The Geology of Cardigan Bay and the Bristol Channel. HMSO, London, 107pp.

UKCP09. Marine and coastal projections, June 2009, Jason A. Lowe, Tom Howard, Anne Pardaens, Jonathan Tinker, Geoff Jenkins, Jeff Ridley, Met Office, James Leake, Jason Holt, Sarah Wakelin, Judith Wolf, Kevin Horsburgh, Proudman Oceanic Laboratory, Tim Reeder, Environment Agency, Glenn Milne, Sarah Bradley, University of Durham, Stephen Dye, Marine Climate Change Partnership (MCCIP) ISBN 978-1-906360-03-0. 61pp.

UKHO, 2009, Admiralty Tide Tables, United Kingdom and Ireland. 366pp.

Innovative Coastal Zone Management
ISBN 978-0-7277-5749-4

ICE Publishing: All rights reserved
doi: 10.1680/iczm.57494.233

Lessons for Designing Managed Realignment Sites Along Hyper Tidal Estuaries – A Case Study on The Bristol Port Company's Steart Habitat Creation Scheme

Adrian Wright, ABP Marine Environmental Research (ABPmer), Southampton, England.
Jack Shipton, ABPmer, Southampton, England.
Benjamin Carroll, ABPmer, Southampton, England.
Dr Susanne Armstrong, ABPmer, Southampton, England.

Introduction

The Bristol Port Company (TBPC) is proposing to undertake a managed realignment scheme on the south bank of the hyper-tidal Bristol Channel, along the Steart Peninsula in Bridgwater Bay (Figure 1). Managed realignment can be defined as 'the deliberate breaching, or removal, of existing seawalls, embankments or dikes in order to allow the waters of adjacent coasts, estuaries or rivers to inundate the land behind' (Burd, 1995). To date, more than 100 Managed Realignment schemes have been implemented in Europe. Of these, some 50 schemes have been completed over the last 20 years in the United Kingdom (UK), mainly within estuaries to improve flood risk management, improve biodiversity or provide compensatory habitat (see ABPmer, 2011).

This realignment is being undertaken to create new mudflat and saltmarsh habitats in compensation for the loss of, and impacts on, designated habitat losses resulting from the construction of a new container terminal at Avonmouth (near Bristol) to accommodate the largest deep-sea ships. The Bristol Port Company view this new terminal as a key contributor to the regeneration and modernisation of Avonmouth's ageing infrastructure and to the local and South West regional economy. It is estimated that, once operational, the terminal will generate 480 full time equivalent jobs in the port and contribute over £114 million annually in the local economy (TBPC, 2008). Avonmouth is considered a good location for a new terminal, as the existing port already has good access to national road and rail transport infrastructure, and no upgrade to this is required for the development to become operational. Furthermore, distance to final destination for containers carried by road will be lower than that from many other UK container ports (TBPC, 2008). An application for a Harbour Revision Order, accompanied by an Environmental Statement, was submitted to the Department for Transport (DfT) in July 2008 which was followed by the statutory consultation period during which objections and representations from 35 organisations and individuals were received. TBPC worked with these stakeholders to provide further information and to address outstanding questions and concerns; all objectors subsequently withdrew their objections, in some cases after concluding an agreement with TBPC. The Secretary of State therefore decided that it was not necessary to hold a Public Inquiry. In granting the Order to construct and operate the new container terminal in March 2010, the

Secretary concluded that there were imperative reasons of overriding public interest, of an economic and social nature, as to why the proposals should be permitted.

To compensate for the impacts at Avonmouth, TBPC must create at least 120 hectares (ha) of new 'compensatory' habitat for wetland birds and other wildlife; of which 20ha must be mudflat (which should exist for at least five to ten years). These requirements, together with baseline conditions at the realignment site, has led to a complex scheme design being developed which will entail unprecedented excavation works and include novel design features. The scheme's design and assessment is being underpinned by a robust understanding of the changes that will take place to the hydrodynamics and morphology of the adjacent Bay both in the short and longer term; particularly the wide intertidal flats fronting the site (Stert Flats). Such understanding is vital given the wide range of interests to be found in and around the Bay, ranging from the nuclear power station at Hinkley to European designated nature conservation sites in the Severn Estuary and inner Bristol Channel. A unique set of assessments, drawing on lessons learned from comparable sites, is to be completed by ABPmer and its partners by October 2011.

This paper describes the complex scheme design process currently under way, focusing on how the important hydrodynamic, morphological, and environmental considerations were taken into account and assessed. Results of the numerical modelling exercises are furthermore outlined. First, however, background information regarding the physical setting is provided.

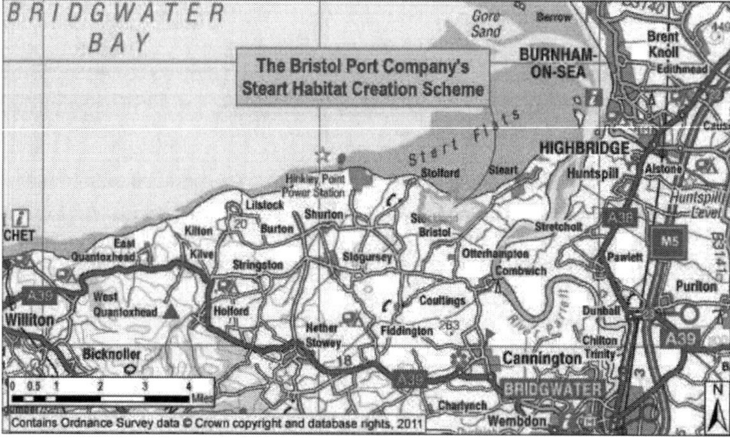

Figure 1. The location of the Bristol Port's Habitat Creation Scheme (Bristol Channel)

Physical Setting and Characteristics

Bridgwater Bay

Bridgwater Bay lies at the mouth of the Severn Estuary, a hyper-tidal estuary system with tidal ranges in the order of 10 to 12m; the spring tidal range at Hinkley Point is 10.7m, compared with a neap tidal range of 4.8m. Tidal currents within the Bay flow approximately parallel to the shoreline, with peak spring velocities reaching 0.7m/s at a distance of 1km offshore from the proposed scheme frontage. Waves within Bridgwater Bay are typically wind driven, dominated by wind arriving from the west-northwest, although the wave regime will also comprise a swell component dominated by the open fetch to the North Atlantic.

The Steart coastline is characterised by the Stert Flats, a large subtidal and intertidal expanse of muddy silts, approximately 18km^2 in size. Sediment distribution in this area typically shows a trend of progressive coarsening in a landwards direction, with mud on the lower intertidal, sand found in the upper intertidal and gravel ridges and sand dunes fronting the Steart Peninsula. Littoral drift is driven by wave induced currents with a predominant easterly direction. Previous studies highlight that the progression of beach material along Stert Flats is extremely slow (Kidson & Carr, 1961; Rendel, Palmer & Tritton, 1986), with most movement confined to the narrow band close to the high water line, and with no significant movement occurring along the shoreline behind Stert Flats.

Within the Bridgwater Bay area, suspended sediment concentrations (SSC) typically vary between 100 and 500mg/l (Environment Agency, 2010), although this is likely to be higher adjacent to the intertidal areas. Generally, SSC are greater on the flood than the ebb tide and during spring tides. The majority of Bridgwater Bay is considered to be a sink for fine sediment (Kirby, 1994), however, the inshore zones appear to be eroding; with mudflat erosion on Stert Flats estimated at 15.6mm per annum (Kirby & Kirby, 2008).

Steart Peninsula

The Steart Peninsula consists of approximately 900ha of low-lying agricultural land, with a small number of scattered settlements. It is made up of a historic sand and gravel spit that was traditionally backed by saltmarsh, the majority of which has been claimed, and forms the current farmland. The peninsula has considerable potential for coastal habitat creation, for which the Environment Agency also currently has plans on land adjacent to the proposed TBPC site (under its 'Steart Coastal Management Project').

The Managed Realignment Site

To achieve the proposed compensatory 'Steart Habitat Creation Project', TBPC has been working closely with the current landowners (11 individual ones), and has secured agreements to purchase approximately 190ha (470 acres) of pasture and arable farmland. One farmstead also forms part of TBPC's holdings; this is to be utilised as a visitor and site management facility once the site is operational. The Royal Society for the Protection of Birds (RSPB) have been chosen as the future site managers, and are assisting in the development of the scheme design to help meet the scheme's ecological objectives.

The land held by TBPC has been shown, through a scoping study, to have the potential to be converted to support new coastal habitats (Royal Haskoning, 2009). The shoreline at the site is fronted by a gravel barrier (approx. height 7.5m to 9.2m Ordnance Datum Newlyn (ODN)), in front of which the intertidal flats are comprised of mud, with well mixed sand in places, as identified through several site visits. Behind the gravel barrier, the land is relatively flat at a height of approximately 4.7m to 5.6mODN. The flat, low lying topography of the site is complimented by numerous small relict creeks and more recent drainage channels. Through modelling, 5.74mODN was identified as the level of Mean High Water Springs (MHWS), and 2.72 as that of Mean High Water Neaps (MHWN) at the TBPC site frontage. As MHWS is the level to which (most) saltmarsh develops, and mudflat can generally be found below the level of MHWN, the site will not 'produce' any mudflat at current elevations.

Design Methodology

In designing this scheme, the project team is building on experience with designing other realignment schemes, including two large-scale schemes on Wallasea Island in Essex (see ABPmer, 2011), and an open-coast scheme at Medmerry in West Sussex (Environment Agency, 2009). The process is being undertaken in an iterative and phased manner in which evidence is collated about the scale of changes and the functioning of the site. The process was started with a site visit, which is an essential element, in particular to assist with the selection of suitable locations for breaches and the identification of alignments of any new coastal defences. Environmental constraints were then determined through numerous ecological surveys. Preliminary design options were developed and assessed as part of the initial studies, however, ABPmer and TBPC are currently working on the final scheme design. Although the scheme depicted in this paper (Figure 2) is not the final scheme, it is expected to be very similar.

Design Criteria

Breach Design

Insights gained from the site visit, as well as from a review of available historic charts and current Light Detection and Ranging (LiDAR) surveys provided a number of possible breach locations, from which one was iteratively selected. The latter approach primarily focused on matching locations to an understanding of sediment transport patterns outside of the site and the impacts these would likely have on the littoral drift and the sustainability of the breach.

The optimum breach and channel dimensions were calculated using recognised equations outlined by Townend (2008). A key input into the breach width equations was the tidal prism (volume of water entering and leaving the site) it needed to accommodate. A breach needs to be sufficiently large with a sill level at the correct height to avoid unwanted stability issues, a lesson learned from implemented realignment schemes, such as Freiston in the Wash (Symonds & Collins, 2007). The breach is ultimately the controlling factor for the scheme design, the dimensions of which controls how the site functions. At present it is estimated that the breach will need to be approximately 200m-wide.

The stability of the breach is sensitive to sediment type, and the designs considered critical flow speeds through the inlet, based on the erosion potential of the *in-situ* sediments. For this project, assumptions have been made in relation to the sediment density (bulk) and material grain size through the breach; these have been supported by site investigation data. Calculations on breach stability finally fed back to the determination of an appropriate breach sill level, which balanced the need to create mudflat below MHWN levels, whilst allowing for some sedimentation and also trying to limit the volume of excavated materials (as it is an objective to re-use all material on site if possible, and anticipated volumes exceed those required for embankment construction). A final breach sill level of 1.9mODN was decided upon, which is approximately 0.9m below MHWN levels, meaning that excavated channels will be inundated on neap tides. The given sill level also requires the excavation of a channel into the foreshore in front of the breach, extending out by approximately 180m.

Channel Design

The primary focus of the scheme design was to create a target area of mudflat of 20ha. Whilst the ideal site preparation approach at any realignment would be one of minimum intervention, the necessity of excavating sections of the site in order to create mudflat was identified early on (due to current site elevations). Based on expert advice and preliminary modelling, it was decided that this habitat was to be provided along the formation of substantial channels (Figure 2), as the relatively high flow speeds through these are expected to help limit sediment accretion, which could otherwise be high due to relatively high SSC in the adjacent estuary system. This design is novel both with respect to the scale of the excavations (up to 2.5m deep) and the use of channels to create mudflat. It builds on lessons learned particularly at the four Humber realignment sites, where large accretion rates of up to 0.25m/annum were observed in the first years following inundation (ABPmer, 2011). The dimensions of the channels were designed to accommodate reasonable flow speeds in order to maintain stability for the local substrate; as a rule of thumb these speeds should not exceed 1m/s. It is currently envisaged that the channels will be 60 to 70m wide near the breach; and will have a combined length of ca. 3.5km. Existing drainage channels and other paleo-creeks in the remainder of the site have been considered to assist as a creek network in promoting the inundation of the entire site on spring tides and in aiding appropriate drainage across the site.

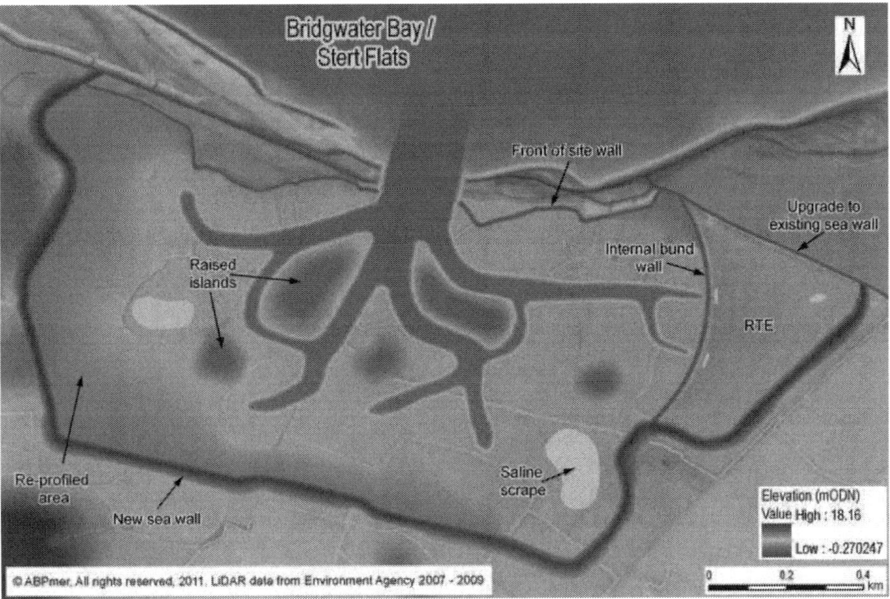

Figure 2. Near-final scheme design for the proposed Steart Managed Realignment Site, designed within a GIS environment.

Environmental Mitigation and Enhancement

Numerous ecological and archaeological surveys were undertaken whilst the preliminary design process was under way. Several species protected under both European and national law, most importantly bats (at least nine species), badger (*Meles meles*), great crested newt (*Triturus cristatus*) and water vole (*Arvicola amphibious*), were found. The archaeological value/potential of the site is currently being assessed. The main mitigation measure which affects the footprint of the realignment was the need to avoid a badger sett along the western edge of the site. The embankment realignment was furthermore influenced by the presence of electricity pylons to the rear area of the site, as well as ownership constraints. Mitigation for water vole and great crested newt is to be provided outside of the realignment footprint. Features to enhance the realignment site for the benefit of the environment and people were also developed, in collaboration with the RSPB. These include:

- o A regulated tidal exchange (RTE) scheme in the eastern section of the site (to provide additional 'wet mudflat' habitat and a high tide roost). The optimum culvert size and its invert level were determined using two models;
- o Raised islands above the level of the highest astronomical tide (HAT), some shingle topping, to provide breeding habitat for birds such as terns;
- o Shallow saline scrapes at the back of the site to further increase habitat diversity; and
- o Landscaping along the back of the site to raise elevations to levels sufficient for the creation of upper and transitional saltmarsh zones (scarce within the estuary).

Design Modelling and Assessment

A numerical model was developed as part of the study, and designed to reproduce water levels, depth averaged flow speeds and directions to an acceptable standard. In addition, the modelling of suspended sediment concentrations was undertaken to examine potential accretion in the site. Wave modelling was also conducted to examine implemented scheme effects on the local wave regime, and provide design criteria for defence structures. An assessment of coastal change that may result from the scheme was also carried out.

Modelling was carried out using the Delft Hydraulics Delft3D modelling software suite, with the model domain comprised of a curvilinear grid, and composed of a total of six nested grids of varying resolution (Figure 3). The highest resolution of 5m extends across the realignment site. A resolution of around 30m to 50m is achieved around Hinkley Point and Burnham-on-Sea, with the resolution becoming coarser elsewhere in the Severn. The boundary of the modal domain is driven by water levels, derived from harmonic constituents, extending across the Severn estuary between Minehead and Aberthaw, some 30km from the managed realignment site. In addition, fluvial inputs into the domain are represented by constant discharges, replicating data on mean river flows from the National River Flow Archive.

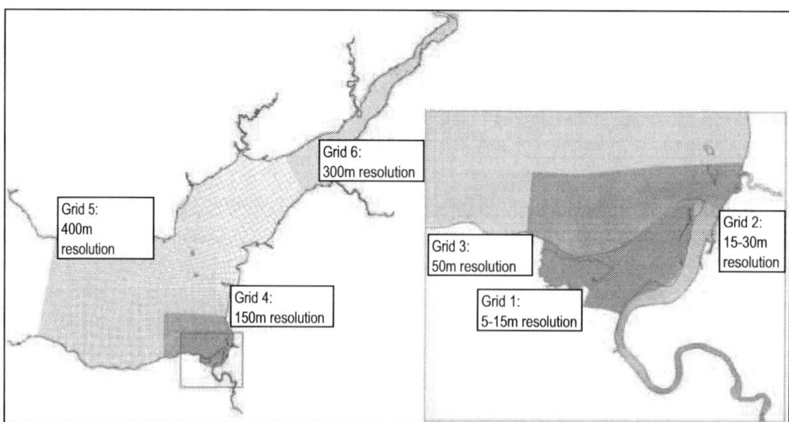

Figure 3. Delft3D nested hydrodynamic model of the Severn Estuary including the proposed Steart Managed Realignment Site

In the scheme model, key hydrodynamic indications as to the stability and longevity of the design were calculated, such as the distribution of flow speeds and bed shear stresses across the site. In addition to the hydrodynamic modelling, the model was used to examine suspended sediment concentrations and any impact the scheme would have on baseline conditions. The sediment model was also used to gain an insight as to how the scheme responded to the large suspended sediment loads locally, and to quantify suspected rates of accretion within the managed realignment site.

Design Consultation

A detailed Communication Strategy is currently being pursued. This represents a fundamental element of the project, and views of statutory consultees and other interested parties have been taken into account with regards to the scheme design. Consultees have to date mostly been concerned with the flood defence and access features of the design (the latter were not presented in Figure 2). In addition to numerous individual meetings, one public exhibition took place in January 2011 to obtain input into the preliminary design. Another is scheduled for September 2011 to discuss the final design. To aid with the community engagement process, photorealistic three-dimensional (3D) Geographical Information System (GIS) visualisations will shortly be produced (by integrating aerial photographs, model outputs and bathymetric data).

Modelling Results

Hydrodynamics

The modelling exercise was undertaken to establish the potential impacts of the proposed scheme on the surrounding area, to aid the scheme design and establish the potential habitats that may be created. For the design process, it is essential that any adverse hydrodynamic impacts such as high flow speeds can be minimised or avoided, and that the functioning of the site is confirmed. The detailed modelling exercise is also used to determine the likely longer term (decadal) changes associated with the scheme, which will be important in calculating the sustainability of the habitats created, and how they may evolve.

Within the site the approximation of the bed roughness has been identified as a key parameter along with the appropriate wetting and drying depth calculations which define the point where the model inactivates a cell depending on the depth of water. For the Steart modelling exercise a model time step of 0.1 minutes was used with a variable bed roughness of 0.02 Mannings 'n' coefficient within the intertidal channels (mudflats) and a higher value of 0.035 on the higher elevations which are more likely to become saltmarsh. A threshold depth of 0.05m was applied within the site compared to a value of 0.1m applied else where within the model.

The results show that the site fully inundates on a mean spring tide, whilst only the tidal creeks are inundated on a mean neap tide. The site's mean spring tidal prism is ca. 1.03×10^6 m^3. Figure 4 shows the maximum high water level within the proposed scheme over a mean high water spring tide. The modelling confirmed that the design performed well, albeit with significant engineering requirements, and that the site should thus be able to support intertidal habitats. The results for maximum flow speeds likely to occur over a mean spring tide are shown in Figure 5. The maximum speed does not generally exceed 1m/s (with the exception of one area, where sensitivity tests will now be undertaken), and peak flow speeds occur on the incoming flood tide. The flows and erosional forces in the breach area are considered likely to be sufficient to maintain the opening of the breach through the shingle ridge.

A SWAN (Simulating Waves Nearshore) wave model was configured to examine the wave conditions at the proposed site, specifically considering how waves propagate into the site. Return period predictions developed by ABPmer for Hinkley Point (ABPmer, 2009) were used, specifically, a wave height of 1.52m was used for the 10 in 1 year return period and 2.77m for 1 in 10 years. The wave investigation indicated that waves will penetrate the scheme entrance for return period events for West and North–East wind directions (Figure 6).

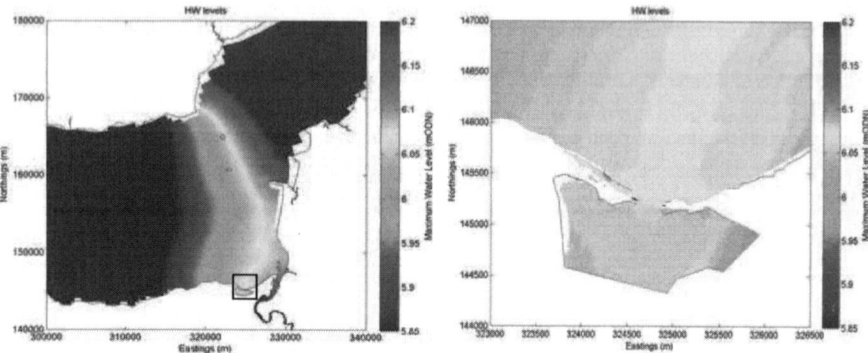

Figure 4. High water levels on a mean spring tide within the proposed Steart Managed Realignment site and corresponding estuary wide levels.

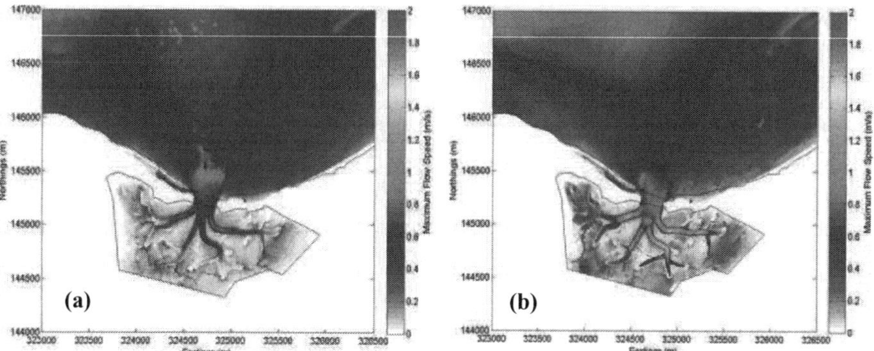

Figure 5. Maximum Ebb (a) and Flood (b) current speeds into and out of the proposed Steart Managed Realignment Scheme.

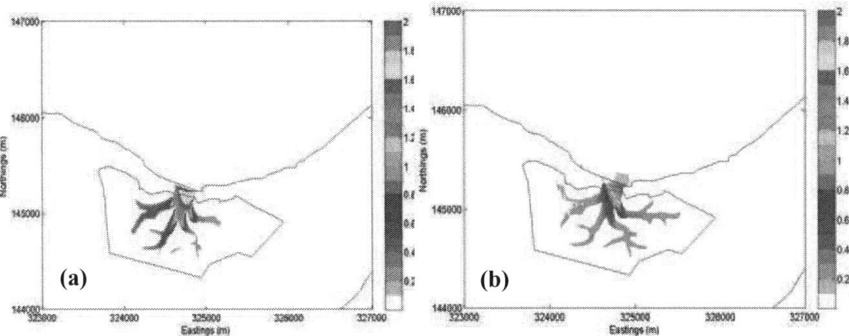

Figure 6. The 1 in 10 year wave height differences (m) at the proposed Steart Managed Realignment site for West (a) and North East (b) wind directions.

Discussion

The successful design and modelling investigation of the proposed managed realignment at Steart have been highlighted along with the scheme findings. The uniqueness of the site lies firstly in its location along the Bristol Channel, which is subject to extreme tidal variations and high sediment loads. Secondly, a novel scheme design has been developed in an iterative manner; this is anticipated to involve significant engineering in order to meet the primary objectives of the project, specifically the creation of mudflat and saltmarsh habitat.

The results of the hydrodynamic and wave modelling undertaken on the (near) final design have shown that the scheme can function, with tidal inundation both on neap and spring tides. Furthermore, the results have provided a mechanism to establish a stable breach size, breach depths and channel dimensions which will be key to the long-term stability of the site. In relation to likely effects of the implementation of the scheme on the morphology of the two adjacent estuaries, it is anticipated that peak water levels throughout the wider study area and close to the realignment site would remain largely unaffected. Flow changes and erosion would generally be restricted to a small area immediately fronting the breach; some changes would also necessarily occur on the fronting intertidal along the drainage path of the site.

Peak wave heights of up to 2m are predicted to occur within the entrance of the site under an extreme event (1 in 10 years) with smaller <1m wave heights under more typical (10 in 1 year) conditions. The wave events are likely to aid in the dispersion and resuspension of material within the scheme (i.e. help lower accretion rates, which would aid in maintaining longer-term mudflat). In engineering terms, the modelling indicates that some armouring may be required around the easterly edge of the breach, and that waves will not need to be taken into account when designing the rear sea walls (as, under all conditions examined, no wave heights were predicted to reach the rear of the proposed managed realignment scheme).

Conclusion

The Steart Managed Realignment site lies in a highly dynamic estuary, the scheme design and modelling investigation has shown that even under these most challenging conditions a scheme that can deliver mudflat and saltmarsh habitat can be created.

The lessons in this case study come in the iterative and methodical approach to the scheme design. By carefully considering the existing environment and undertaken a series of iterative steps starting from a site visit, through to engagement with ecologists and engineers, a successful outcome is possible. In particular, by gaining an in-depth understanding of the uniqueness of the surrounding environment, the design and modelling can be tailored to minimise engineering works and provide a prediction of the site function and form. Furthermore, the results from the modelling investigation can be used to engage stakeholders and the wider public to both help provide an understanding of the proposed options and also provide reassurance against potential negative impacts.

References

ABPmer, 2009. Severn Estuary Flood risk Management Strategy Assessment of Offshore Extreme Conditions. ABPmer report R.1513.

ABPmer, 2011. The Online Managed Realignment Guide. Available at: http://www.abpmer.net/omreg/

Burd, F., 1995. Managed Retreat: a Practical Guide. Peterborough: English Nature, 26p.

Environment Agency, 2009. Medmerry Managed Realignment Environmental Statement Almondsbury, Environment Agency, 166p (excl. app.).

Environment Agency, 2010. Steart Coastal Management Project Scoping Consultation Document. Environment Agency, Exeter, 45p (excl. appendices).

Kidson, C. & Carr, A.P., 1961. Beach drift, experiments at Bridgwater Bay, Somerset. Proc. Bristol Naturalists Soc. 30 (2), 163-180.

Kirby, R., 1994. The evolution of the fine sediment regime of the Severn Estuary and Bristol Channel. Biological Journal of the Linnean Society 51 (1-2), 37-44.

Kirby, J.R. and Kirby, R., 2008. Medium timescale stability of tidal mudflats in Bridgwater Bay, Bristol Channel, UK: Influence of tides, waves and climate. Continental Shelf Research 28 (19), 2615-2629.

Rendel, Palmer & Tritton, 1986. Coastal Physiography Study. Volume 1. Report to central electricity board, system safety strategy branch, Hinkley Point C Power Station pre-application studies.

Symonds, A.M. & Collins, M.B., 2007. The development of artificially created breaches in an embankment as part of a managed realignment. Journal of Coastal Research SI50, 130-134.

Townend I.H., 2008. Breach design for managed realignment sites. Proceedings of the Institution of Civil Engineers 161 (MA1), 9-21.

Royal Haskoning, 2009. Steart Compensation Scheme Environmental Scoping Report. Royal Haskoning, London, 97p.

The Bristol Port Company (TBPC), 2008. Bristol Deep Sea Container Terminal - Environmental Statement. TBPC, Avonmouth, 763p.

Innovative Coastal Zone Management
ISBN 978-0-7277-5749-4

ICE Publishing: All rights reserved
doi: 10.1680/iczm.57494.243

Medmerry Managed Realignment – Sustainable Coastal Management to Gain Multiple Benefits

Joe Pearce, Samina Khan and Pippa Lewis, Environment Agency, Worthing, UK.

Introduction

Construction of the managed realignment scheme at Medmerry is planned to start at the end of 2011. Coastal realignment has been recommended for this area by both the wide scale shoreline management plan and the locally focused coastal flood and erosion risk management strategy. The scheme will both reduce the very significant flood risk for the local area and create new intertidal habitat to offset losses across the shoreline of the northern Solent. The urgency to meet both these requirements has driven a tight programme to design the scheme in outline and gain the major approvals needed for construction.

The proposed scheme at Medmerry will be the first managed realignment on the open coast in the UK. Other schemes have been carried out in more sheltered estuary conditions elsewhere. Innovative approaches were needed to appraise the effects of the scheme and to assure concerned local residents that it will be effective in managing the risk of coastal flooding.

Involvement of local people and groups has been an essential part of designing the realignment scheme. By including representatives of existing community networks and non government organisations from the start of the project, their interests were included in decisions as they were made. This helped gain local support for the Environment Agency's proposals and show the real stake that there was for everyone in meeting all of the objectives of the scheme. Local representatives have accepted the part that the realignment needs to play in adaptation to secure the long term future for their area. Locally driven groups have begun to draw up plans for enhancements beyond the scheme to maximise the opportunities for recreation and access network enhancements in the area.

Location

Medmerry is on the western side of the coastal peninsula approximately 10km south of Chichester on the south coast of England. The coastline borders the Solent facing towards the Isle of Wight. The land area is mainly agricultural, with widely spaced small villages and hamlets. To the west and the east of the site are 2 caravan parks; the Sussex Beach Holiday Village and Bunn Leisure respectively (Figure 1). Low lying land extends towards the sole access road to the town of Selsey and the edge of Pagham Harbour 3km inland. The foreshore and 50 hectares of land adjacent to the defence at Medmerry is designated a Site of Special Scientific Interest (SSSI), owned and maintained as a nature reserve by the RSPB.

Background

The Environment Agency currently manages coastal defences, inland drainage channels and associated structures for the area. Defences are formed of two raised shingle banks, separated by an area of higher ground where the Bunn Leisure holiday park is located.

Maintaining the defences requires constant work throughout each winter to recycle and re-profile the shingle between tides. Despite this work, storms have caused the bank to breach 14 times since 1994 causing flooding to agricultural land, homes and holiday parks. Around 350 homes and businesses are at risk from coastal floods in addition to thousands of fixed caravans and holiday chalets. The sole access road and utilities serving the community of 5,000 households in Selsey are at risk from coastal flooding.

Figure 1 Scheme design outline

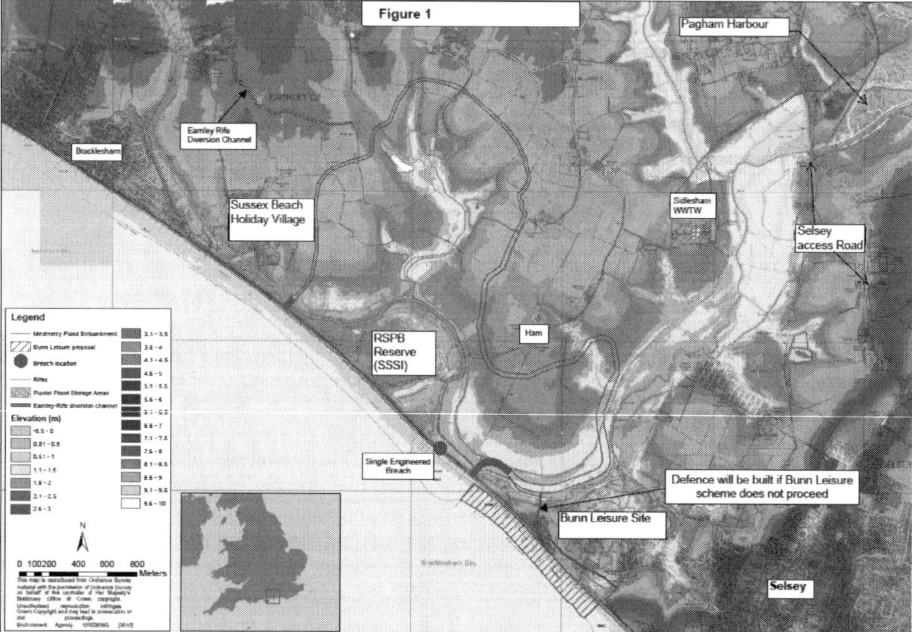

Strategic Issues

Managed realignment at Medmerry is seen as important in long term management of flood and erosion risk by both the wide scale shoreline management plan and the more local flood and erosion risk management strategy.

Coastal flood and erosion risk for the coastal peninsula around Medmerry was considered in the Pagham to East Head Coastal Defence Strategy, (henceforth referred to as the strategy) completed in 2009. Studies concluded that existing shingle defences are unable to prevent flooding now and will not remain effective beyond the short term. Implementation of an alternative way of managing coastal flood risk is needed as soon as possible. Following appraisal of available options, managed realignment was recommended. New inland embankments were proposed to be built largely from locally sourced earth and clay, with a tidal breach allowed to form along the existing coastline.

The strategy recognised the potential for the formation of new intertidal habitat as part of the realignment but did not include plans to actively form new intertidal areas or secure their future. Economic appraisal had included evaluation of environmental benefits of creating new intertidal habitat, contributing approximately 15% to the total economic benefits. Calculations were based on latest available guidance which makes use of research on already completed realignments. The strategy recognised that a realignment scheme based only on managing flood risk for the local area would be unlikely to gain funding and would attract strong opposition.

The North Solent Shoreline Management Plan (SMP) completed in 2010, assessed issues on a wider scale for the coastline stretching some 70km westwards from Medmerry. Natural harbours within the Solent host 4,750 hectares of intertidal habitat that form part of the Natura 2000 network. The Solent Coastal Habitat Management Plan (2003) reported 100 hectares of intertidal habitat loss across the Solent. The Solent Dynamic Coast Project (2008) produced for the SMP, confirmed these losses and also estimated that there will be a loss of 600 hectares over the next 100 years through coastal squeeze caused by rising sea levels. New intertidal habitat is needed to restore the losses already experienced and to offset coastal squeeze effects from maintaining and renewing existing coastal defences across the northern Solent. The SMP was completed after the strategy, confirming the managed realignment recommendation at Medmerry, recognising the importance of the site in creating new intertidal habitat.

To ensure that legal obligations for protection of habitats and species under European Birds and Habitats Directives will be met, while still allowing protection for people and property, the Environment Agency established the Regional Habitat Creation Programme (RHCP). In order to take advantage of the opportunity at Medmerry, following the findings of the Solent Dynamic Coast Project, three farms were purchased by the Environment Agency at Medmerry. This land, together with the SSSI area already owned and run as reserve by RSPB, formed the core of the proposed realignment scheme.

During the production of the strategy, extensive public engagement work was undertaken. Initially, many people would have preferred an option to hold the line of the existing coast. They felt that this would provide a better flood defence than managed realignment could. The Environment Agency worked with local communities to facilitate a greater understanding about how managed realignment works and how it was chosen as the preferred option for managing coastal flood risk at Medmerry. Experience during the development of the Pagham to East Head strategy and from other managed realignment schemes elsewhere showed that ongoing support and involvement of local stakeholders would be essential in promoting the scheme to gain permits and approvals needed for its construction. Many people still view managed realignment as a lesser option for managing flood risk as it brings the sea closer to their properties.

One aspect of the strategy which had given rise to particular concern locally, was the impact on the Bunn Leisure holiday parks situated on low ground stretching inland from the coast at Medmerry. The line of new defences recommended by the strategy at Medmerry extended through part of the holiday park. Unless other action could be taken, this would result in the loss of caravan pitches in areas seaward of the proposed alignment. Bunn Leisure responded by developing plans to safeguard the future of their site by designing and gaining planning permission to build their own coastal defences to protect one kilometre of the coast. The project team has worked with Bunn Leisure in designing the realignment scheme to take account of the construction and operation of both sets of works.

Scheme Objectives

Appraisal of the realignment scheme began immediately after completion of the strategy. There was a priority to move quickly to reduce flood risk for the local area, create the new intertidal habitat and keep the momentum built up by working with local people. Three objectives were defined to address the separate requirements in implementing the managed realignment.

- To provide a sustainable flood and coastal erosion risk management scheme with appropriate standard of protection as cost effectively, quickly and safely as possible.
- Maximise (greatest extent) intertidal habitat and optimise (best use) freshwater/Biodiversity Action Plan (BAP) habitat on the remainder of the Scheme area making best use of existing topography taking climate change into account, and;
- Encourage participation during development of the scheme to deliver a product which can be supported by local people.

The objectives were initially drawn up by the project team and amended following workshops with others to ensure that the concerns of all stakeholders were adequately represented.

Designing the Scheme

In designing the scheme, the challenge was to bring the three project objectives together in order to deliver the habitat and reduce flood risk as soon as possible. A separate habitat creation group was invited to work with the core project team to provide advice on scheme requirements and details. A Medmerry Stakeholder Advisory Group (MStAG) was also formed to provide the focus for engagement with local communities.

The scheme required planning permission, and was subject to an Environmental Impact Assessment (EIA). It was essential that the three groups worked together in order to develop a scheme design that would meet the objectives, and also be acceptable to the planning authority and local people.

The core team included engineers, environmental scientists and engagement specialists from the Environment Agency. Specialist consultants joined the team to complete the environmental and technical appraisal for the scheme and lead the engagement with MStAG. The habitat creation group included Natural England, RSPB, local wildlife groups, academics and local authority officers working with Environment Agency and consultant staff.

The 'Medmerry stakeholder advisory group' (MStAG) was established so that all community groups with an interest in the scheme could be represented in its development. Membership was limited to a workable number of 20 representatives selected by attendees at an initial meeting of a much larger group representing local organisations and interested parties. MStAG meetings were facilitated by specialist engagement consultants who ensured that the members of the group could see how their issues were included in designing the scheme.

All three of the groups played an essential part in the design process, and development of the scheme business case and planning application. Each of them fed into the definition of a number of options to meet project objectives, and helped identify constraints and opportunities associated with each option.

Constraints and opportunities

Design decisions were influenced by key constraints and opportunities at the site which covered all three aspects of the scheme objectives:

- Locating the new inland embankments for cost effective flood risk management *and* to maximise habitat creation opportunities.
- Determining whether the breach should be engineered or allowed to occur naturally. Consideration also included whether one or more breaches would work best and where along the frontage a breach would best be constructed.
- Inclusion of measures to ensure land drainage outside the embankments will continue adequately following the realignment.
- The possible effect on the coastline bordering the realignment. Ensuring they are minimal or can be mitigated.
- The need to mitigate for the effect on fresh-water and terrestrial ecology already present on the realignment site.
- Possibilities for supporting new recreation opportunities as part of the scheme without extra costs.
- Aspirations for additional recreation and other enhancements which could not be included in the scheme.

Technical Details

Numerical modelling and coastal process investigations were undertaken to assess the wider scale impacts on the adjacent coast and to explore how options function in relation to their stability, development, inundation and habitat creation. The modelling also considered the impacts of the Bunn Leisure proposals on the proposed options. The results of these investigations were a key driver in the selection and development of the preferred option.

The history of managed realignment schemes has been summarised by Dixon et. al. (2008) and Rupp-Armstong et. al. (2008). Medmerry differs from previous managed realignment schemes as it is the first to promote managed realignment as a flood risk management measure on the open coast. Previous schemes have largely been promoted for sites adjacent to estuaries. Medmerry has similarities with the site at Porlock in Somerset where a natural realignment was formed by a storm in 1996, but the site at Medmerry is more exposed to prevailing south-westerly winds and waves. This project's numerical modelling and coastal processes studies have applied available tools and knowledge in an innovative way involving an integrated set of discrete models and investigations. These have included breach and tidal inlet stability assessments, coastal process investigations and reviews of existing case studies, hydrodynamic and sediment modelling and habitat mapping. The conclusions of these studies were combined through expert interpretation by the team's consultants and scrutinised and approved by external expert peer reviews.

The importance of sharing major assumptions and outputs from the core project team with all the groups involved in designing the scheme was recognised from the start. For the technical design team, this involved consideration of timing and formats of reports to make information available for comments that could affect the scheme design. Tools for visualising the finished scheme were important to help those involved understand the details.

Environmental Details

An EIA was required for the project as the planning authority considered that there would be a significant impact on the local environment. As discussed, the land taken up by the realignment is mainly used for low-grade agriculture. The EIA highlighted main environmental impacts including the presence of a number of protected species (water voles, reptiles and great crested newts), the potential for disturbing unknown archaeology, impact on public rights of way and the change to the landscape that will take place. The design of the scheme must incorporate measures to mitigate for the impacts to these receptors from building the new embankments and creating new intertidal habitat. Water voles are a particular issue as existing drainage channels provide habitat for a local population stronghold. The strategy for mitigating these impacts is therefore central to the design of the scheme.

In creating the new inter-tidal habitat, a key driver is to maximise the area of new habitat created and the new wildlife opportunities it will offer. Natural development of a stable site was favoured with minimised need for further intervention. However, details of the scheme design were carefully considered in order to create the best possible habitat from the outset. The large material volumes required to build the new embankments will be dug from areas designed specifically to promote the formation of tidal channels and habitat. In consultation with the RSPB, opportunities have also been identified for the creation of high level bird roosting sites for overwintering and breeding birds.

Community involvement

Public acceptance has long been recognised as important in promoting managed realignment schemes. Engagement planning was a high priority in designing and gaining approval for the scheme. The team had the advantage of being able to build on previous engagement for the strategy and SMP in the local area and gain from the experience of implementing schemes elsewhere. As a result of earlier engagement, local communities generally accepted the managed realignment option but significant concerns still remained.

- The ability of the proposed inland embankments to withstand storms.
- The loss of agricultural land and recreation opportunities.
- The perception of wildlife interests being prioritised over properties and land.
- Changes to inland drainage.
- Ease of access to the coast and ability to walk between local villages.
- The effect of changes on holiday parks important to the local economy.

The approach to stakeholder engagement was described by Miller et. al. (2010). It followed the 'building trust with communities' model used within the Environment Agency and applied a stakeholder engagement plan that focused on three core objectives:

- To provide information about the multiple benefits offered.
- To gain widespread support for the implementation of the scheme.
- To get community input into the design.

The plan specifically included engagement methods to ensure that the concerns already voiced could be addressed and where possible, resolved. The issues were explored together with MStAG and were considered in defining the scope of work in planning the scheme. This enabled people's main concerns to be fully integrated and local knowledge to be used in the design.

The potential opportunities of the scheme were considered with MStAG, exploring what people would like to see included to enhance the local area. Ideas mainly focused on improving recreational and local tourism opportunities through inclusion of measures such as footpaths, bridleways and cycle routes linking into the existing network. The views of local residents who would rather not see their privacy compromised by access enhancements were also explored among the group.

Regular meetings and workshops were held with MStAG throughout the design of the scheme where updates were given by the project team, options explored and questions answered. When differing opinions were raised, the group sought to find a compromise that the project team could use in its scheme design. In this way, meetings helped to save project team resource, which might have otherwise been needed to meet with each community group individually. This approach also allowed for more transparent decision-making.

The planned engagement also enabled requirements to be defined and expectations to be managed. As the engineering and habitat creation aspects of the scheme were defined, these were fed into discussions with MStAG together with advice on what was needed to make the scheme acceptable to the planning authority. Many of the enhanced access requirements could be included with little extra cost by making use of maintenance tracks on the site. Other requirements were beyond the scope of the scheme and could not be included.

The project team helped members of MStAG form an 'aspirations group' to explore funding sources and details of enhancements beyond the scope of the scheme such as extending new footpaths to integrate into wider networks. Fully involving local people in the design process and focusing on the opportunities offered was beneficial both in improving the scheme's ability to meet the defined objectives and in gaining local support and enthusiasm.

Outcomes

A separate plan was prepared to address each of the three project objectives in planning and gaining approvals for the realignment (Figures 1, 2 and 3).

The engineering construction shown on Figure 1 for the scheme is relatively straightforward. A total length of 6.7km of new embankments will be constructed stretching up to three kilometres inland. The height of the banks will vary from less than one metre to more than four, making use of the natural rise of the land. Land drainage will be channelled into the realigned area through new outfalls and a new relief channel will be constructed between catchments to the west of the scheme area. After the inland banks are built, a single opening will be cut through the existing coastal shingle bank allowing tidal flows into the site. Construction will begin in autumn 2011 and will be completed in 2013.

One key finding of the technical appraisal was that the future stability of the scheme depends partly on having a large enough tidal exchange to ensure the coastal breach does not close. The larger the volume of water entering and leaving the site with the tide, the less likely it would be unstable. Sharing this information among each of the three groups enabled all to realise that the need for maximising the intertidal area supports both habitat creation and flood risk management objectives.

Reducing Flood Risk

The realigned embankments will reduce the risk of flooding currently faced and the maintenance required. The banks will be built taking into account one metre of sea level rise over the next 100 years in accordance with Defra (2006) guidance. Standard of protection will initially be better than 0.1% (1 in 1000) chance of flooding in any year, falling to 1% (1 in 100) in 100 years time. The Environment Agency will continue to work with Bunn Leisure to ensure works are completed to reduce the risk of coastal flooding across the whole frontage.

Habitat Creation

Up to 300 hectares of new intertidal and transitional habitats will be formed by the realignment to compensate for anticipated coastal squeeze losses across the Solent over the next 100 years. An estimated 163 hectares of new intertidal habitat will be created shortly after construction is completed as illustrated on Figure 2. Medmerry will be an important site for bird life and conservation management among the northern shores of the Solent and will help provide habitat connectivity between Pagham Harbour to the east and Chichester Harbour to the west. In the future, the site will be managed to balance the needs of wildlife and enhanced recreation opportunities. The scheme includes the formation of new habitat areas specifically for protected species including water voles which will be displaced by the realignment. These habitat areas will also be maintained and improved through future conservation management of the site.

Figure 2 Creation of new intertidal habitats

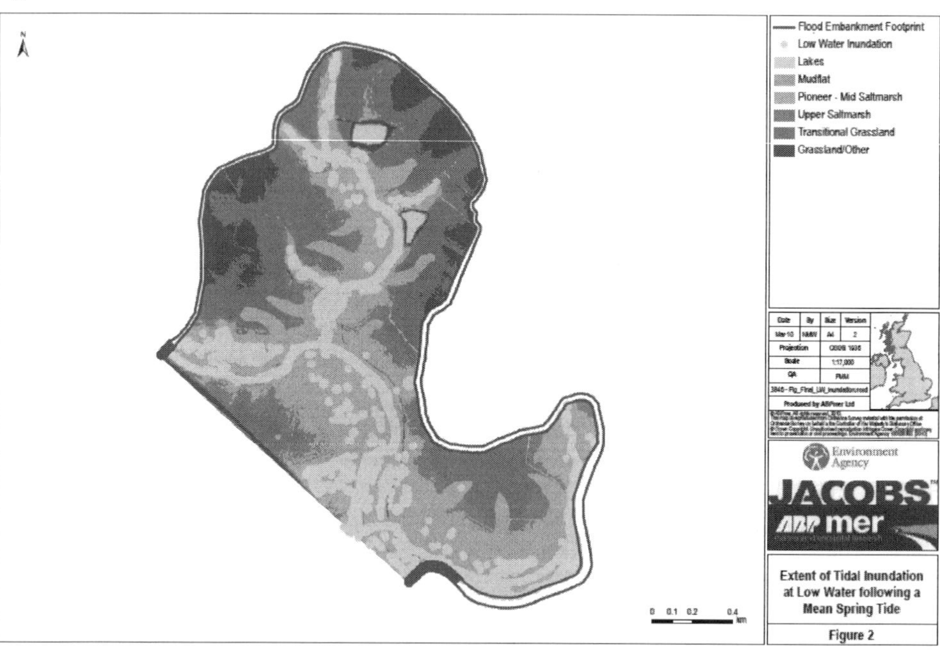

Community Involvement and Recreation Enhancements

Representatives of local communities have helped design the scheme by defining requirements for incorporation in the scheme from the outset. The MStAG group helped the project team resolve the priority of different options where there were conflicts or aspirations could not all be achieved. The Planning Authority was a key participant in MSTAG. Their inputs helped shape the scheme, advising what would be required to gain planning permission for the scheme and ensuring gains for local communities as well as wildlife would be realised through its implementation.

The scheme design has incorporated enhanced access at Medmerry. New footpaths, cycle routes, bridleways, car parks and viewing points will be built as shown on Figure 3. A new access route for emergency vehicles to reach Selsey is included for use when the vulnerable B2145 road is blocked. An aspirations group has formed to pursue the inclusion of extra access enhancements linking the realignment area with the wider path and cycleway network.

Planning permission was gained with support from communities in November 2010 and the business case was approved by the Environment Agency in January 2011. Environmental mitigation actions will commence during spring 2011 and construction will start as programmed in autumn 2011.

Figure 3 Access and enhancement proposals

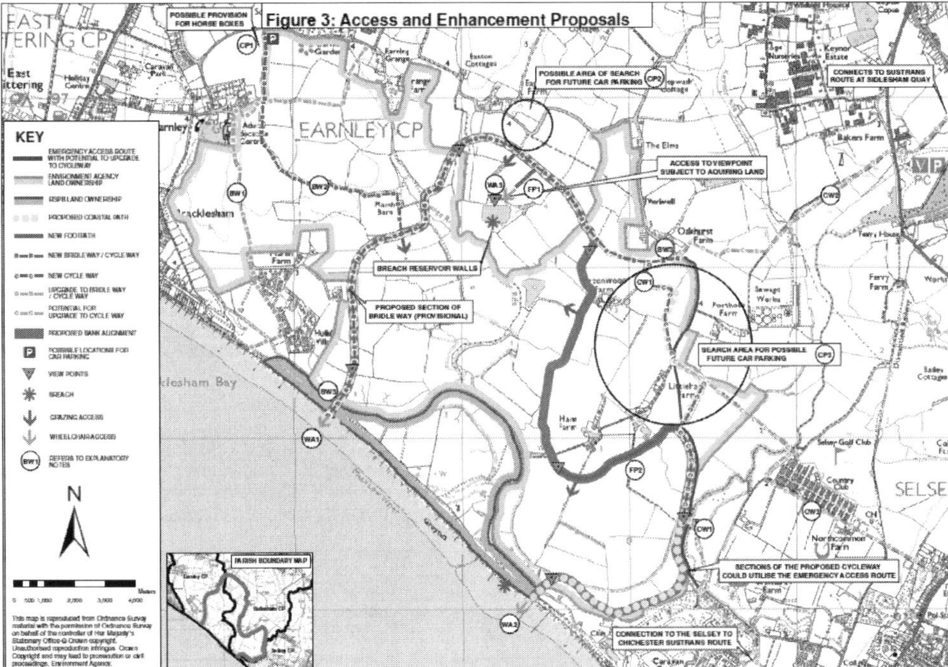

Conclusions

When considered on its own, the engineering design for the managed realignment scheme was relatively straightforward. Working to achieve multiple objectives made the design process more complex. Encouraging the involvement of stakeholders with potentially conflicting interests from the start required time, effort and resources but the dialogue promoted helped those involved to understand each others' positions and reach agreed solutions.

The scheme has been designed as the first planned managed realignment on the open coast in the UK. It will meet the requirements to reduce coastal flood risk, enhance nature conservation by creating new intertidal habitat and incorporate the local priorities for improved recreation and access opportunities. Those involved in designing the scheme have helped explain and promote the scheme to the wider local communities and gain support when its planning application was assessed. The work of the stakeholder group helped ensure that when the scheme was considered by planning authority representatives, it could be fully supported. The ongoing work of the group will also provide a forum for community representatives to help resolve details of outstanding issues such as land drainage.

The vulnerability of the peninsula around Medmerry to coastal flood and erosion risks has been recognised increasingly over the last ten years. Local groups have been considering strategies for adapting to changes that will occur. Work on the managed realignment scheme has provided a focus for local people to be involved in helping shape the future landscape of the area and the opportunities that are presented as it changes.

References

Economic Valuation of Environmental Effects Guidance. (Environment Agency, 2010)

Flood and Coastal Defence Appraisal Guidance FCDPAG3 Economic Appraisal Supplementary Note to Operating Authorities – Climate Change Impacts (Defra, October 2006)

Managed Realignment Electronic platform - www.intertidalmanagement.co.uk

Medmerry Managed Realignment Project Appraisal Report (Environment Agency, 2010)

Miller, S., Giacomelli, J., Hyam, P., Pearce, J., Lewis, P., Pizer, L., and Wilson, G., 'Designing the 'right' scheme through community involvement: lessons learned from community engagement on a coastal protection scheme.' Defra and Environment Agency Flood and Coastal Risk Management Conference (2010)

North Solent Shoreline Management Plan (2010)

Pagham to East Head Coastal Defence Strategy (Environment Agency, 2009)

Rupp-Armstrong, S., Scott, C. and Nicholls, R., 2008. 'Managed realignment and regulated tidal exchange in northern Europe – lessons learned and more.' Defra 43rd Flood and Coastal Management Conference, Manchester July 2008. 9p.

Solent Coastal Habitat Management Plan (2003)

Somerset Coastal Change website - www.somersetcoastalchange.org.uk

Innovative Coastal Zone Management
ISBN 978-0-7277-5749-4

ICE Publishing: All rights reserved
doi: 10.1680/iczm.57494.253

Quantifying the Carbon Footprint of Coastal Construction – A New Tool HRCAT

Rudi Broekens HR Wallingford, Wallingford, United Kingdom
Manuela Escarameia HR Wallingford, Wallingford, United Kingdom
Clemente Cantelmo HR Wallingford, Wallingford, United Kingdom
George Woolhouse HR Wallingford, Wallingford, United Kingdom

Introduction

UK and European regulation aimed at achieving a low carbon society is currently not being sufficiently backed up by tools for the quantification of construction-related carbon emissions. Recent UK Government findings have highlighted that the amount of carbon emitted by construction and maintenance of infrastructure is largely unknown and that consistent carbon accounting is needed (BIS, 2010). It is expected that carbon accounting will become a standard requirement for engineering option appraisal and for any investment justification (be it project specific or at a national scale). Coastal schemes are no exception. Existing tools such as the Environment Agency Carbon Calculator are useful for the UK river and coastal protection market but currently lack the breadth of data and functionality required for the wider range of coastal construction works and for overseas schemes.

This paper explains the process of development of a new carbon accounting tool suitable for coastal construction schemes, illustrating its application on a real breakwater option appraisal.

Development of carbon accounting tool – HRCAT

HR Wallingford has developed HRCAT, a new tool for detailed estimation of carbon emissions of a range of construction schemes including coastal structures such as breakwaters and quay walls, and river and coastal protection (as well as conventional drainage and SuDS – Sustainable Drainage Systems). The development of the tool was underpinned by a thorough assessment of existing regulations and relevant documents, available literature and existing tools (EA, UKWIR, WRAP, Cap$_2$IT). This led to a methodology setting up the required boundaries and steps for the estimation of the carbon footprint of a scheme. Process maps were developed for each of the main subject areas (Figure 1), identifying the individual contributions to the total carbon emissions of the construction materials, transport to site, construction activities, operation and maintenance and disposal at the end of the scheme's design life.

An extensive dataset was gathered from reliable databases such as the Inventory of Carbon & Energy v1.6a (Hammond & Jones, 2008), complemented by manufacturers' own sources when available, the UK Defra/DECC's GHG (Green House Gases) Conversion Factors (Defra/DECC, 2009), as well as data specifically sourced for concrete armour units and rock, and construction and maintenance equipment (land and marine based) including dredgers. A critical assessment of the data was a vitally important aspect of the work to ensure consistency and reliability in the results. In some instances, literature values required some recalculation before they could be used in the tool, for example to avoid double counting of costs.

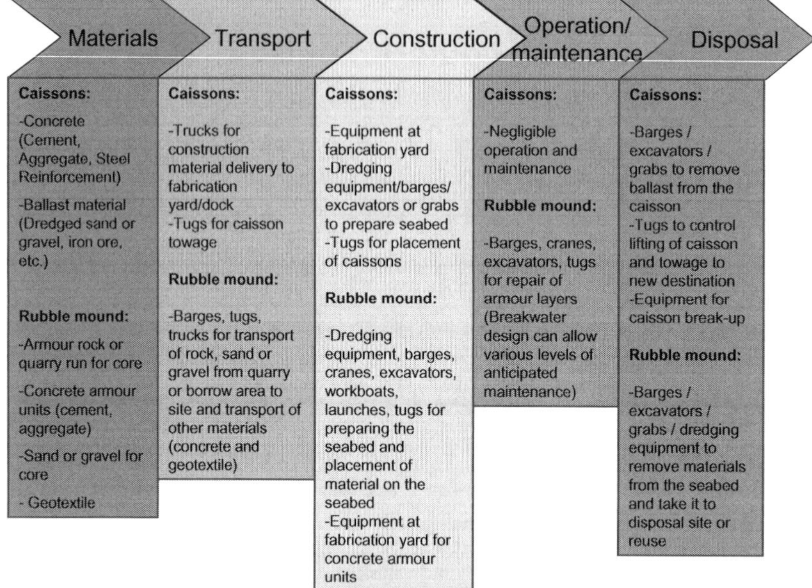

Figure 1 Process map for coastal protection using breakwaters

In the next sections a case study for estimating the carbon footprint of a breakwater construction project during an option appraisal phase is presented. Two different breakwater types are compared and a critical discussion of underlying data on carbon emissions associated with material use and construction activities is made. The paper focuses on a few key items – for an exhaustive discussion see Escarameia *et al* (2011).

Case study – breakwater
Description of scheme
This breakwater forms part of a marine terminal to be built in a remote overseas location with a design life of 50 years. It is detached from the shoreline and provides shelter to berths for ships used for the export of fossil fuels. Its length is approximately 1.4 km and the average seabed level is approximately 14 m below mean sea level. After an initial assessment of various forms of construction of the breakwater, the following breakwater types were selected for a further option appraisal:
1) a rubble mound breakwater using rock for the breakwater core and filter layers and a primary armour formed by concrete armour units in a single layer
2) a concrete caisson breakwater ballasted with sand and placed on a shallow rock foundation mound.

At the time of the option appraisal no suitable quarry for production of rock had been identified and rock was assumed to be produced in quarries at an average distance of about 500 km from the construction site. It was assumed that transport of rock from quarry to site would be carried out over land using trucks with a 32 tonne payload, which would be fully loaded when carrying rock from the quarry to the site and would be empty when returning to the quarry. Concrete caissons were assumed to be fabricated off-site and transported over sea to the site over a distance of about 2500 km. Sand for ballasting of caissons was assumed to be won in a near-by marine borrow area using dredging equipment. Figure 2 shows a schematic of the two breakwater options.

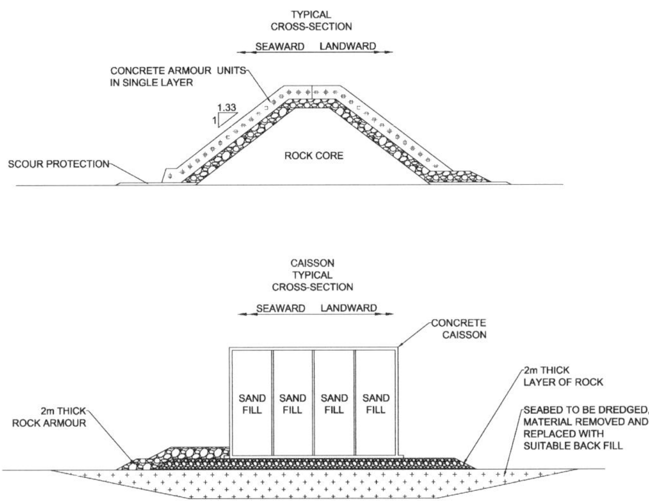

Figure 2 Schematic of breakwater options

HR Wallingford's methodology

A process map for breakwaters has been developed and is presented in Figure 1. As the present case involves an option appraisal at an early design stage, a "cradle to built asset" approach was used (maintenance, operation and disposal were not included in the assessment as these were assumed not to have a significant effect on the outcome of the appraisal). In practice breakwaters, like other large structures such as bridges, are not normally disposed of completely at the end of their design life, but establishing the carbon footprint of disposal (which may not be insignificant) may be relevant in a more detailed option appraisal exercise.

A detailed description of definitions of various parameters associated with carbon footprinting used in available literature is provided by Escarameia *et al* (2011), as well as a description of exclusions (e.g. CO_2 emissions in manufacturing of equipment used for breakwater construction have not been taken into account). It is also noted that both CO_2 and CO_{2e} (equivalent CO_2) are being used in published literature, the latter taking into account green house gases other than CO_2. HRCAT currently does not make a distinction between CO_2 and CO_{2e}, as at present there is insufficient clarity in the available data to make such a distinction.

The carbon emissions associated with the construction of the breakwater have been divided into three main categories:
- production of materials to be used in construction
- transport of materials from where materials are produced to the breakwater site
- construction of the breakwater.

For production of materials the emission is calculated by multiplying the rate of embodied carbon (EC) per material quantity (e.g. $kgCO_2$/tonne) and the estimated total quantity in the scheme. Similarly, the emission associated with transport of materials is calculated by multiplying a rate of EC per quantity of equipment use and the quantity of equipment use. For road transport (trucks) the rate of EC per quantity of equipment use has been defined as $kgCO_2$ per travelled distance. For sea transport this rate has been defined as $kgCO_2$ per energy use ($kgCO_2$/kWh). The latter method has often also been used for calculating emissions related to construction equipment.

Discussion of data values

Material: Rock

A literature search on the carbon emissions associated with production of armour rock or quarry run that are typically used for breakwater construction indicated limited current knowledge. Hammond & Jones (2008) present an embodied carbon (EC) value for "general stone" of 56 $kgCO_2$/tonne. It is not certain whether this value is supposed to be representative of armour rock or of quarry run (or both) and the authors note that data on stone is generally poor. The authors also present values for granite and limestone, both of which can be used for armour rock or quarry run in coastal construction projects. For granite, a wide range of values is presented from 6 to 781 $kgCO_2$/tonne whereas for limestone a value of 17 $kgCO_2$/tonne is given. The upper limit for granite however applies to dimension stone and is based on values presented by Lawson (1996). Therefore this upper limit is most likely not representative of production of armour rock or quarry run. No context for the value for limestone was found.

For a better understanding of the carbon emissions associated with armour rock and quarry run it is necessary to consider the production process in somewhat greater detail. Rock for breakwater construction is typically sourced in either aggregate quarries, dimension stone quarries or dedicated armourstone quarries (CIRIA/CUR/CETMEF, 2007). Carbon emissions for the production of rock may vary significantly depending on the type of quarry that is used. For instance, in a dimension stone quarry, where armour rock is recovered from waste resulting from the production of dimension stone, the EC for production of armour rock is small since most of the energy use of the quarry is taken into account in the production of dimension stone. The EC of rock produced in a dedicated armourstone quarry may be larger than for a dimension stone or aggregate quarry, as there will be energy use associated with setting up the quarry e.g. removal of overburden and there may be energy use associated with the production of waste material, i.e. rock with unsuitable grading. Finally it should be noted that the type of rock (e.g. granite, basalt or limestone) may also affect the EC value.

Some guidance on dimension stone quarries can be derived from EC values of natural dimension stone published by Historic Scotland (2010), indicating that for granite stone the total "cradle to gate" value is 93 $kgCO_{2e}$/tonne, including cutting, polishing, internal transport on site and production of stone waste. The data also indicates that most of the emissions originate from processing the stone (i.e. cutting and polishing) and that only about 20 $kgCO_{2e}$/tonne is associated with extraction at the quarry. It may be argued that if armour rock resulted from waste produced at such a quarry, a value significantly lower than 20 $kgCO_{2e}$/tonne could be achieved. Data provided by STEMA Shipping (2011), distributors of armour and core rock in Northern Europe from a dimension stone quarry in Larvik (Norway) suggest a low value of 0.87 $kgCO_2$/tonne for production of armour rock and 0.99 $kgCO_2$/tonne for production of core rock. A slightly smaller value for armour rock is suggested than for core rock because no crushing is involved in its production.

No specific EC value for rock from an aggregate quarry has been found. For the purpose of the present case study it has been assumed that the EC value for rock from an aggregate quarry is similar to the value of aggregate produced in the same quarry. Hammond & Jones (2008) present a value of 5 $kgCO_2$/tonne for aggregate. Also no data has yet been found on dedicated quarries nor has a clear understanding been developed on the effect of the type of rock on EC estimates. For the present case study a value of 5 $kgCO_2$/tonne was adopted for both armour rock and quarry run, being within the range of values found.

Material: Concrete

As can be seen in data published by Hammond & Jones (2008), the carbon footprint of concrete is strongly dependent on the following factors:

- the compressive strength class of concrete (the underlying reason for this dependency is that cement has a higher EC value than aggregate and that usually a higher cement content is required for concrete with higher compressive strength)
- the amount of cement additions, such as fly ash or ground granulated blast furnace slag (ggbs)
- the amount of steel reinforcement.

Therefore it is important to consider the above factors when developing the EC estimate for concrete material use. Two different structural concrete elements are present in the case study: concrete armour units in the rubble mound option and concrete caissons in the caisson option.

Various types of single-layer concrete armour units may be used for construction of a rubble mound breakwater and their design and construction specification is usually controlled by various licence holders. At the option appraisal stage the required concrete specifications were not available. It has been assumed that a concrete compressive strength class of C25/30 will be required for the armour units. For the amount of cement additions, a low and a high amount of cement additions were selected representing a range suitable for concrete in a marine environment (CIRIA 2010). EC values were calculated using data in Hammond & Jones (2008) - see Table 1. Usually armour units are not reinforced and therefore no additional EC component for steel reinforcement was added.

Table 1 Characteristics of cement for concrete armour units

Cement type	Assumed addition material and percentage	Compressive strength class	EC value unreinforced concrete ($kgCO_2$/tonne)
CEM II B - V	21% fly ash	C25/30	112
CEM III B	80% ggbs	C25/30	51

The Table indicates that the lower limit of the EC range is approximately half the value of the upper limit. The lower limit compares well with the value presented by Van der Horst *et al* (2007) of 110 $kgCO_2$/m^3 (approximately 46 $kgCO_2$/tonne) for the use of CEM III B cement for the fabrication of Xbloc® armour units. Due to the lack of specification, for the present case study an average value of 80 $kgCO_2$/tonne was adopted.

Caissons are usually built using reinforced concrete and a typical concrete strength class of C40/50 was assumed. Again two types of cement were considered in order to establish a range of EC values for reinforced concrete. From experience on similar projects, it was assumed that the amount of steel reinforcement is approximately 0.3 tonne/m^3 concrete. EC values were calculated using data in Hammond & Jones – see Table 2.

Table 2 Characteristics of cement for caissons

Cement type	Addition	Compressive strength class	EC value unreinforced concrete ($kgCO_2$/tonne concrete)	EC value steel reinforcement ($kgCO2$/tonne*)	EC value reinforced concrete ($kgCO_2$/tonne concrete)
CEM II B - V	21% fly ash	C40/50	138	216	354
CEM III B	80% ggbs	C40/50	60	216	276

*EC values are in $kgCO_2$ per tonne reinforced concrete

The Table indicates that the EC value for steel reinforcement is the dominant element for reinforced concrete. For the case study an average value of 300 $kgCO_2$/tonne was adopted for reinforced concrete.

Transport: land transport of rock

The CO_2 emissions associated with truck transport were calculated using data on emission factors for freight transport provided by DEFRA/DECC (2009). This source presents data for carbon dioxide emissions per km for various diesel HGV road freight vehicle classes - see Table 3.

Table 3 CO_2 emissions for road transport

	0% weight laden	100% weight laden	Average
Rigid truck (>17tonne)	0.96 ($kgCO_2$/km)	1.38 ($kgCO_2$/km)	1.17 ($kgCO_2$/km)
Articulated truck (> 33tonne)	0.85 ($kgCO_2$/km)	1.40 ($kgCO_2$/km)	1.13 ($kgCO_2$/km)

Table 3 indicates that the emission rates per km for the two different truck classes are similar. Based on the above numbers a value of 1.1 $kgCO_2$/km was adopted for the case study.

Construction: land based equipment

Land based equipment required for breakwater construction includes wheel-loaders, bulldozers, cranes and excavators. Carbon emission data for land based equipment were taken from the Department of Transport (2004), indicating a fuel consumption rate in terms of grams of diesel per amount of applied power for various types of equipment. It was concluded that the variability of these rates for the various types of equipment is fairly small. The average value was about 260 g diesel/kWh (about 1 $kgCO_2$/kWh). The above reference further provides information on typical loading factors (i.e. the ratio between rated power of equipment and applied power during use of the equipment) for various types of equipment, again indicating a fairly small variability for typical construction equipment and a best estimate of about 30% for most types of plant.

The World Ports Climate Initiative (2009) presents a methodology for estimating greenhouse gas emissions associated with the construction of new port facilities. Their methodology is essentially similar to the one adopted for the present case study as are the EC rates of construction equipment (for example the applied rate for use of a bulldozer is only about 10% higher than the rate used in the present case study).

Construction: Marine equipment

For application of marine equipment the same methodology was used as for land based equipment. Typical loading factors for marine equipment including workboats (45%), tugboats (31%) and ocean tugs (68%) have been presented by the World Ports Climate

Initiative (2009). An approximate emission rate of 0.65 $kgCO_2$/kWh was adopted for these types of marine equipment, also based on the above data.

Results

Using data values as those described in the previous section, estimates of the carbon dioxide emissions were made for the two breakwater options – see Tables 4 and 5.

Table 4 Summary of EC for various construction materials

Material	EC (kgCO$_2$/tonne)	Rubble mound breakwater		Caisson breakwater	
		Quantity (tonne)	EC (million kgCO$_2$)	Quantity (tonne)	EC (million kgCO$_2$)
Rock	5	3,650,000	18.3	510,000	2.6
Concrete (armour units)	80	550,000	44.0		
Concrete (caissons)	300			342,000	91.8
Sand	2*			1,300,000	2.6
Total			**62.3**		**97.0**

* includes transport from borrow area to breakwater

The EC contribution of the use of geotextile was found to be negligible.

Table 5 Carbon footprint of different breakwater types

	Rubble mound with concrete armour units	Caisson
Materials (million kgCO$_2$)	62	97
Transport (million kgCO$_2$)	133	76
Construction (million kgCO$_2$)	17	15
Total (million kgCO$_2$)	**212**	**188**

It is concluded that for the selected case study the caisson breakwater has a smaller carbon footprint than the rubble mound breakwater with concrete armour units, mainly due to a large contribution of transport for the rubble mound breakwater, associated with road transport of rock over a long distance. For the caisson breakwater option the use of reinforced concrete is the dominant element in the carbon dioxide emission estimate. It should however be noted that these results are strongly dependent on specific circumstances for the present case study that affect the design of the breakwaters (e.g. the average water depth) or the breakwater construction (e.g. the distance between rock quarry and breakwater site). Therefore it cannot be concluded that in any option appraisal the caisson option will have a smaller carbon footprint than the rubble mound option. The results also indicate that the contribution of construction related emissions is fairly small. At this point this is not fully understood and a carbon footprint estimate for a case study in the detailed design or construction phase would be helpful to gain a better understanding of these emissions.

Uncertainty

HRCAT enables an estimation of the level of uncertainty in predictions as a qualitative rating of the estimates for each construction scheme element (materials, transport and construction) based on the following criteria:
- knowledge basis (how "good" the data is)
- applicability (how well knowledge can be applied) and
- scheme information (how much detail is available to allow calculations and how much needs to be assumed).

With regard to materials, in the present case study the scheme information is reasonably accurate, as the major quantities have been defined in the conceptual designs of the two breakwater options. On the other hand, the knowledge basis and applicability of materials varies per material: for concrete there small uncertainty but for rock a large variability has been found in published data suggesting a large uncertainty.

With regard to transport the scheme information is quite inaccurate as no quarries had yet been identified and the quarry location will have a major impact on CO_2 emissions. The knowledge basis and applicability of data is thought to be fairly accurate, as variability in data on CO_2 emissions related to road transport is small.

With regard to construction the scheme information is somewhat inaccurate, as no detailed plan has been developed for type, number and duration of use of equipment. The knowledge basis and applicability for construction seems fairly accurate, although more data on loading factors of equipment and emissions associated with temporary works or use of temporary materials such as formwork would be helpful to improve accuracy.

A graphical illustration of uncertainty is presented in Figures 3 and 4.

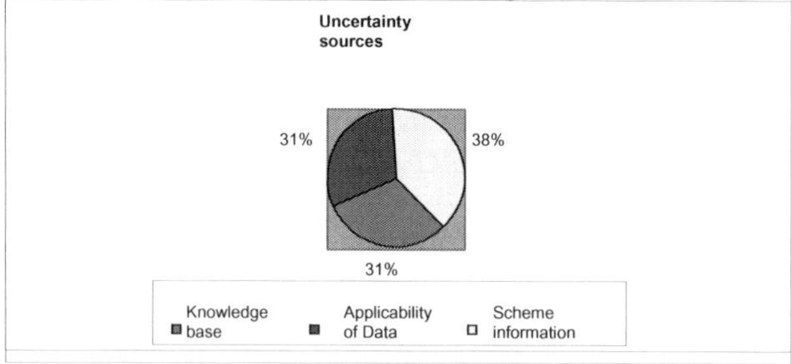

Figure 3 Uncertainty sources per criterion presented by HRCAT

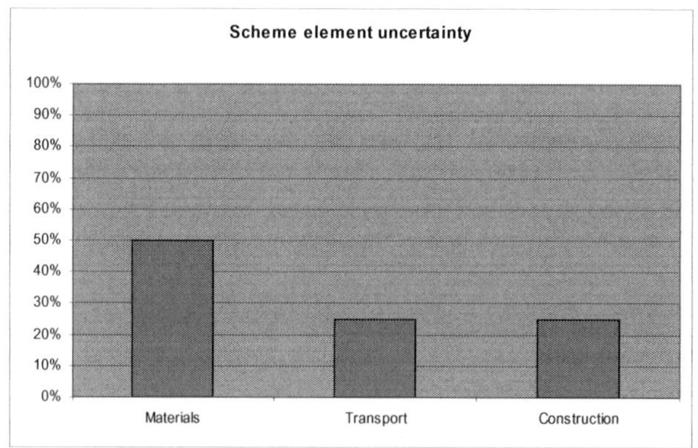

Figure 4 Uncertainty sources per scheme element presented by HRCAT

Discussion and conclusions

Development of HRCAT

HRCAT enables the user to estimate the whole life carbon cost of a coastal construction scheme (as a total or split into its contributing elements), and the level of uncertainty associated with the predictions. The versatility of the tool enables application to various types of coastal construction works and can be extended to include softer coastal management schemes in the future. It can help identify the relative contributions of materials, transport, construction and operation for each construction scheme element.

HRCAT has been tested on a case study for the calculation of the carbon footprint of two different breakwater types. The case study enabled further development of the tool in terms of input/output and its database of embodied carbon (EC) values associated with production of materials and use of construction equipment. HRCAT can support the designer and local authorities in estimating the carbon footprint of various elements of coastal developments, enabling them to focus on the key elements and informing a design with a smaller carbon footprint.

Accuracy and uncertainty

During the development of HRCAT, a critical assessment of published data on EC values related to use of materials and equipment has been carried out. In some key areas of a typical coastal construction project, there can be a wide variability in published values. An example is the large variability in published EC values for the production of rock. This is likely due to inadequate consideration of the processes involved. It was found that different quarry products (e.g. dimension stone, aggregates or rock used in typical coastal construction projects such as armour rock and quarry run) produced in different quarry types (dimension stone quarry, aggregate quarry or dedicated quarry) may have quite different associated EC values. The present variability is illustrated by a comparison of the adopted value for the present case study with the value adopted by Bruce & Chick (2009), who presented a methodology and a worked example for the assessment of breakwater construction. The latter value is 56 $kgCO_2$/tonne (i.e. the value for general stone presented by Hammond & Jones (2008)), whereas the value adopted for the present case study is about 10 times smaller (5 $kgCO_2$/tonne).

It is expected that, as more data becomes available in the coming years on the use of construction materials and equipment, the accuracy of published EC values will improve. It may however be useful to look at the accuracy of carbon footprint estimates in the context of accuracy of cost estimates. At a typical option appraisal stage of a civil engineering construction project, the accuracy of a cost estimate would be about +/- 30%. It may be assumed that, in view of limited experience in the construction industry, the accuracy of a carbon footprint estimate is currently lower than that. Although this could of course improve in the future, carbon footprint estimates may remain somewhat less accurate than cost estimates for some time. Contrary to costs, the exact carbon footprint of completed works are likely to remain unmeasured in many cases, which may limit the build-up of reliable data over the years.

Reducing the carbon footprint of a coastal construction project

Whilst not the focus of the present case study, inevitably some insight has been gained on how to reduce the carbon footprint of a coastal construction project. Not surprisingly, and similar to reducing the construction cost of a breakwater, the primary focus should be on optimising the design and construction process in order to minimise use of materials and equipment. On a secondary level, the case study has showed that when a project includes concrete works, two design options for concrete stand out that seem to be effective in reducing CO_2 emissions:

- using cement with a high percentage of additions (e.g. fly ash, ggbs)
- minimising the amount of steel reinforcement.

Context

Bruce & Chick (2009) provide some context to carbon emissions related to breakwater construction, indicating for example that the construction of 1 km of their example breakwater would have a CO_2 emission equal to approximately 0.035% of the UK's annual emission. Adding to their example, the present case study involves a breakwater for a marine fossil fuel export terminal. The amount of embodied energy in the fossil fuel that could typically be exported through such a terminal during its lifetime has been estimated as approximately 2×10^{10} GJ and the associated CO_2 emissions have been estimated at 1×10^{12} kg, assuming that all fuel will be used for combustion. The CO_2 emissions associated with the construction of the breakwater would therefore be of the order of 0.02% of the emissions eventually resulting from combustion of the fossil fuel exported through the marine terminal of which the breakwater forms a part.

References

BIS (2010). Low Carbon Construction - Emerging findings. Department for Business, Innovation and Skills, HM Government, UK.

Bruce T. and Chick J. (2009). Energy and Carbon Costing of Breakwaters. Coasts, Marine Structures and Breakwaters 2009, 16–18 September 2009, Edinburgh International Conference Centre, UK.

CIRIA (2010). The use of concrete in maritime engineering – a guide to good practice. Construction Industry Research and Information Association, London.

CIRIA/CUR/CETMEF (2007). The Rock Manual – The use of rock in hydraulic engineering. C683, London.

Defra/DECC (2009). 2009 Guidelines to Defra/DECC's GHG Conversion Factors for Company Reporting: Methodology Paper for Emission Factors. Web site accessed in November 2009.

Department for Transport (2004). Non-Road Mobile Machinery. Usage, Life and Correction Factors.

Escarameia M., Woolhouse G., Udale-Clarke H., Woods-Ballard B., Broekens R. (2011). Developing of HR Wallingford carbon footprinting capabilities for hydraulic engineering schemes. April 2011.

Hammond G. and Jones C. (2008). Inventory of Carbon & Energy (ICE), version 1.6a, Sustainable Energy Research Team (SERT), Department of Mechanical Engineering, www.bath.ac.uk.

Historic Scotland (2010). Embodied Carbon in Natural Building Stone in Scotland, Technical Conservation Group, Technical Paper 7 by Chishna N., Goodsir S., Banfill P. and Baker K., March 2010.

Lawson, B. (1996). Building Materials, Energy and the Environment – Towards Ecologically Sustainable Development, School of Architecture The University of New South Wales.

STEMA Shipping (2011), personal communication.

Van der Horst A.Q.C., Van Gaalen J.M. and Van der Vorm-Hoek C.V.A. (2007), "Duurzame toepassing van Xbloc® wereldwijd" (in Dutch).

World Ports Climate Initiative (2009). Carbon Footprinting for Ports, Draft Guidance Document.

Innovative Coastal Zone Management
ISBN 978-0-7277-5749-4

ICE Publishing: All rights reserved
doi: 10.1680/iczm.57494.263

'Will, or Should, the RTPI be Running the ICE Coastal Management Conference in Four Years Time?'
Increasing the Dialogue between Shoreline Management Plans and Spatial Planning

Nick Hardiman, Environment Agency, Bristol, UK
Alison Baptiste, Environment Agency, Bristol, UK
Gregor Guthrie, Royal Haskoning, Peterborough, UK
Japp Flikweert, Royal Haskoning, Peterborough, UK

Introduction

The pattern of historic development along the coast of England[1], and the legacy of coastal management demands that society has inherited as a result, is a familiar story to coastal engineers and spatial planners[2]. The new suite of Shoreline Management Plans (SMPs) completed for England in 2011[3] form a milestone in sustainable shoreline management. They have been developed using the forum provided by Coastal Groups - partnerships between local government, the Environment Agency, and other coastal managers. Together they represent an impressive effort to refine earlier SMPs developed between 1995 and 1999 that were overseen and reviewed by the Ministry of Agriculture Fisheries and Farming (MAFF) as a step towards taking a more strategic management approach to our coasts.

This paper aims to explore the dialogue between Shoreline Management Planning and spatial planning as the key factor both in reducing risk from flooding and erosion at the coast, and in shaping how the developed, historic and natural environment evolves there. It starts with an overview of the roles of SMPs and spatial plans and what they are respectively aiming to achieve. The need for dialogue between 'shoreline planners' and spatial planners in the Local Planning Authority (LPA) in order to achieve these aims is then discussed, with reference to the experience of this dialogue during the development of updated SMPs. A brief appraisal of the degree to which SMP documents should explicitly guide spatial planning decisions follows, before finally looking at how the issue of uncertainty around funding and physical processes affects the relationship between SMPs and spatial plans – and what those affected by this uncertainty can do about it.

Underpinning this discussion is a serious question which is somewhat caricatured in the title of this paper: where does technical risk management decision-making stop and spatial

---

[1] This paper focuses upon England but its principles also extend to the Welsh experience, and many aspects of the discussion are being taken forward in Wales.
[2] See http://www.environment-agency.gov.uk/homeandleisure/107495.aspx for further information.
[3] See http://www.environment-agency.gov.uk/research/planning/105014.aspx for links to each SMP

planning decision-making on responding to that risk begin? This was explored in a report commissioned by the Environment Agency and published by the Planning Co-operative in 2009 called 'Translating Shoreline Management Plans into Spatial Plans', the findings of which provide a platform for this paper.

SMPs and spatial plans – what they are trying to achieve

An SMP is described by Defra as "a large-scale assessment of the risks associated with coastal processes, [which] helps to reduce these risks to people and the developed, historic and natural environment"[4]. SMPs are essentially route maps that manage the transition between the inherited legacy of coastal management and a future of 'balanced sustainability', where competing pressures upon the coast are assessed and prioritised consistently and transparently. Key to achieving this balance is close collaboration and consultation with a diversity of coastal landowners, communities and other interested parties using the Environment Agency's 'Building Trust With Communities' approach and the consultation experience of local authorities.

The SMP is a high level document which assesses coasts at a regional scale, albeit agreeing management policies for local areas. Strategy Plans and, ultimately, individual schemes that focus in more detail upon the nuts and bolts of SMP delivery may prompt the decisions made in SMPs to be re-visited should barriers arise to policy delivery, gradually honing the SMP through an iterative process. Figure 1 summarises the aims and outputs of these iterations.

Stage	SMP	Strategy	Scheme
Aim	To identify policies to manage risks.	To identify appropriate schemes to put the policies into practice.	To identify the type of work to put the preferred scheme into practice.
Delivers	A wide-ranging assessment of risks, opportunities, limits and areas of uncertainty.	Preferred approach, including economic and environmental decisions.	Compare different options for putting the preferred scheme into practice.
Output	Policies.	Type of scheme (such as a seawall).	Design of work.
Outcome	Improved management for the coast over the long-term.	Management measures that will provide the best approach to managing floods and the coast for a specified area.	Reduced risks from floods and coastal erosion to people and assets (see the glossary).

Figure 1: SMPs, Strategies and Schemes – aims and outputs, from Defra, 2006 vol 1

The first Shoreline Management Plans

The original SMPs had only moved part of the way towards balanced sustainability. Subsequent iterations of SMP guidance in 2001 and 2006 noted the progress made towards co-ordination of coastal interests and appraisal of risk beyond administrative boundaries, but called for more ambition. In particular, Defra wanted "more emphasis on improved links with the planning system; more consideration of the effects on the environment; and longer term coastal policies"[5]. There was too much 'business as usual' in the first attempt at strategic planning, providing little impetus to create dialogue between coastal engineers and planners – the message was generally that coastal planners could expect little interference with their aspirations from coastal engineers.

Updated Shoreline Management Plans

---

[4] Defra, 2006 vol 1. p4
[5] Defra, 2006 vol 1. Forword

The updated SMPs have been encouraged to take a different approach, with correspondingly different outcomes and messages for spatial planners. They use a greater range of information to inform the development of preferred management policy options that more pragmatically anticipate the financial, environmental and new legislative barriers to maintaining the status quo over the next 100 years.

In particular, the FutureCoast study that reported in 2002 provided authorities with a baseline Geographical Information System (GIS) of coastal change, and developed a 'behavioural systems approach' that assesses the full range of inputs and outputs in coastal dynamics at different temporal and spatial scales. This approach helps inform the development of strategic objectives in an SMP, and consequently the individual management *policies* that will achieve those objectives. Economic assessments of individual policies are then undertaken to provide a check on their viability – an important factor in delivering the most sustainable option, rather than one purely driven by economics.

Spatial plans

Spatial plans – the Local Development Documents contained within a Local Development Framework (LDF)[6] – share this delineation between an overall objective-led 'vision' and specific enabling policies. Planning Policy Statement 1 (PPS1) stated that spatial planning should "set a clear vision for the future pattern of development, with clear objectives for achieving that vision and strategies for delivery and implementation" – this means more than setting conditions about what can be built where. Unlike traditional land-use planning prior to reforms in 2004, spatial planning frameworks include "the locational and geographic elements of all public policy"[7] Just as SMPs consider needs and solutions in risk management terms, so the LDF should the relate community needs and solutions to the use and development of land, setting out "how social, economic and environmental objectives will be achieved through plan policies."[8]

These parallels between the shoreline and spatial planning systems are bridged by means of a key aim highlighted in Defra's SMP guidance of 2006, namely that . "The SMP must guide and support the planning system in discouraging inappropriate development in areas at risk from flooding and erosion"[9] as well as in potential managed re-alignment sites or where development could upset natural coastal processes. This is also a central tenet of CLG's coastal change supplement to PPS25 published in 2010. This lists SMPs as 'material considerations' in the planning process, specifically calling for local authorities to draw up 'Coastal Change Management Areas' where significant risk is identified, within which development would be restricted.

SMPs and spatial planners - the need for dialogue

In their 2009 report, the Planning Co-operative identified a 'perennial problem' for any organisation charged with producing a plan or strategy that relies upon the planning system to secure its implementation, but which also has its own specific objectives and agenda: "Do you focus attention on your own concerns and hope that the planning system gets the message? Or

---

[6] SMP tie-in with Regional Spatial Strategies (RSS) was also important at the time that they were being developed. The RSS framework is now being dismantled as part of the development of the new Government's National Planning Framework.

[7] Local Government Association Coastal Issues Special Interest Group, 'Local Authority Coastal Risk Management: Information Pack', 2004

[8] CLG, 2005, p13

[9] Defra, 2006, vol 1. p12

do you seek to accommodate [...] the planning system at the risk of diverting focus from your main concerns?"[10] The first is a form of lobbying, the second a form of dialogue. The fact that those drawing together SMPs and LDFs both experience the same pressure to balance competing concerns should provide a sound basis for dialogue between the individuals and organisations drawing them up.

ICZM – the holy grail of maritime planning

The need for dialogue between different coastal managers and interest groups has long been recognised, both in the UK through the findings of Parliamentary Committees and in the EU, which called upon Member States to adopt national strategies for 'Integrated Coastal Zone Management' (ICZM). The principles of ICZM echo those found both in SMPs and in the supplement to CLG's Planning Policy Statement (PPS) 25 on coastal change, such as taking a long term view, working with natural processes where possible, and using clear management policies in an adaptive and flexible way to accommodate change.

ICZM epitomises a successful interface between terrestrial, shoreline and marine planning. In its own strategy for England, one of Defra's key objectives was to "ensure that the future direction of planning activity to address flood and coastal risk management is integrated effectively with other planning and management mechanisms, with full consideration of wider social issues such as regeneration".[11] Yet despite examples of good practice and the hard work of the Coastal Partnerships and others, England's progress towards ICZM has remained mixed, restrained by the long history of sector-based planning at the coast[12]. Why, despite the many possible mechanisms and linkages in place, should this be so?

Part of the reason is no doubt that achieving successful and comprehensive ICZM takes time. However, the Planning Co-operative report suggested that the experience of the three 'pilot' SMP2s (which trialled the more strategic, streamlined versions of original SMPs) highlighted two key obstacles to achieving integration between the objectives of SMPs and spatial plans, which may reflect the broader ICZM experience. First was the lack of direct involvement of spatial planners in drawing up SMP2s, and second – discussed in the next section - was Defra's guidance itself, which limits SMPs to being technical documents unable to explore their own spatial planning implications thoroughly.

The involvement of spatial planners in SMP development

Defra's guidance clearly makes the case for dialogue; in setting up the SMP's 'Client Steering Group' (CSG) "it is imperative that representatives of both planning and engineering disciplines are involved to ensure appropriate 'buy in' and hence uptake of the final Plan"[13]. Planners are therefore not simply 'consultees' but are actively involved in the authorship of the SMP document.

The extent to which this actually happened was variable across the country. An analysis of the CSG membership of the 20 SMPs relating to England show that about two-thirds of SMP CSGs had a representative from the planning department of at least one local authority, and a for small number, such as at The Wash (SMP4) and the Medway & Swale (SMP9), spatial planners were in the majority on either the CSG or the Project Management Board. In some cases, the SMPs which did achieve spatial planner involvement from the start were those with

---

[10] Planning Co-operative, 2009 vol 2. p3
[11] Defra, 2008. p14
[12] Atkins, 2004
[13] Defra, 2006, vol 2. p21

fewer LPAs falling within the SMP's boundaries. The Isle of Wight is a clear example here, where the SMP covered the jurisdiction of a single unitary LPA.

Overall, however, given the clear direction provided by Defra's guidance, it appears that SMP CSGs found it difficult to secure consistent input from spatial planners into SMP development. This is perhaps not surprising given that a local authority coastal engineer or officer usually held the most in-depth understanding of coastal processes and issues, and for many local authorities committing two members of staff to the involved process of SMP development was beyond their resource capacity. Moreover, the presence or absence of a local authority planner on the CSG is a crude measure of the extent to which true 'dialogue' about planning issues raised by the SMP actually occurs; the experience of the controversial 'Humber Estuary Coastal Authorities Group' SMP (Flamborough Head to Gibraltar Point), which has ticked the boxes of planner involvement, shows how this is not a panacea for solving intractable conflicts of interest. What it has prompted, in some areas, is the need for follow-up guidance specifically designed for planners to walk them through an SMP's implications.

Post-adoption efforts to clarify SMP planning implications

An example of a post-adoption effort to guide planners through the implications of SMPs (and of CLG planning policy) is the 'Joint Position Statement' and protocol between King's Lynn & West Norfolk BC and the Environment Agency on planning and coastal flood risk between Wolferton Creek and Hunstanton, setting Defra's stated aim to use SMPs to avoid inappropriate development to the local context. Another is the Environment Agency initiative to produce concise promotional 'flyers' for local authority planners who have not been involved in SMP development – or who have succeeded in office those that have – that summarise the purpose and content of the SMP within their local area, and highlight 'hotspots' where their SMP presents planning issues. An example developed for Scarborough Borough Council[14] is appended to this paper (electronic version).

Other initiatives aim to clear up confusion surrounding landowner rights and responsibilities in relation to the four SMP policy options[15], for example the planning application procedure for those wishing to maintain existing defences in spite of a No Active Intervention policy in the SMP, or the liabilities surrounding third party defences. The Environment Agency 'Supporting Change' landowner's pack initiative in East Anglia is an example, as is the 'Information Note for Landowners, Planners and Developers' for the North Solent SMP. The North Solent note, however, also highlights how spatial planners need to appreciate the localised exceptions or caveats to SMP policies.

Such instances are common in SMPs. The most universal caveat is that policies to hold the existing defence line are subject to availability of funding from Grant In Aid, levies or, increasingly, third parties. Also common are over-arching policies chosen for a stretch of coast with localised exceptions, perhaps stemming from the need to comply with environmental or health and safety legislation. The application and interpretation of the four generic SMP policies as defined in Defra's guidance has been a source of some contention and confusion even in the second round of SMPs despite the guidance being relatively clear, and prompted the Planning Co-operative to suggest that, because they did not easily translate

---

[14] Developed by Royal Haskoning, The Planning Co-operative and Environment Agency with assistance from Scarborough BC, 2010.

[15] Hold the Line, No Active Intervention, Managed Re-alignment and Advance the Line: see Defra, 2006 vol 1 p13-14 for definitions

in spatial planning terms, they should be refined or revised.

Different SMPs have adopted different approaches to this problem in order to gain political acceptance before approval; some adhere to the four policy options and explain the caveats within the Policy Summary, whilst others have added to or changed the policy categories themselves to reflect complexity, such as at The Wash, where a conscious imitation of the more nuanced suite of Catchment Flood Management Plan policies was adopted. Whichever approach is chosen, it is clear that without proper dialogue between SMP authors and spatial planners, these nuances will be lost in translation and either lead to a false sense of 'security' about maintenance of defences, or a focus upon the costs and constraints imposed by chosen policies upon future planning options. In both instances, this often leads to 'surprises' at a later stage among elected members of local authorities, who fear the implications of apparently straightforward policies for future investment in the area.

The experience of updated SMPs indicates that the lesson of involving spatial planners (and elected members) in the SMP development process was only partially learned, and that follow-up work has been necessary to compensate. However, this follow-up work is also a positive sign that SMPs are being kept 'live' instead of stagnating on bookshelves; dialogue with planners should be ongoing, especially as Strategies, Schemes and new information feed back into SMPs and modify their aspirations. Such initiatives also make lengthy and technical SMP documents more easily digestible to spatial planners at a local scale – a feature that will become increasingly important as regional planning is phased out under the 'localism' agenda and more spatial planning decisions hang on the LDF.

Importantly, some, such as the Scarborough example, also go a step further than the SMPs themselves, by suggesting potential planning responses to the issues raised by SMPs. The Planning Co-operative found the lack of – or at best rather vague - discussion with the pilot SMP2s around such planning responses disappointing, and called for them to be better fleshed out in those still under development: "As they stand, the SMP2s are professional pieces or work but they do not realise their potential for communicating to spatial planners [...] what they must do to help deliver the vision of a sustainable coast, i.e. what needs doing, by when, and how this might be achieved."[16]

SMPs and spatial planning – the need for crossover

Should exploring 'how this might be achieved' fall within the remit of the SMP? This is an important question for informing future reviews and ensuring clarity over what kind of documents SMPs are. If SMPs attempt more than coastal risk management decisions using strategic cost:benefit assessments, are they in danger of becoming quite a different document?

Defra's guidance (2006) on where SMPs stop and LDFs begin

Section 2.6 of Defra's guidance (volume 2) on 'Influencing the planning process and how land is used' suggests that SMPs should not try to be spatial planning documents. It discusses how SMPs support planning by putting the LDF in a regional context, informing LDF authors about risk, mitigation measures required for development, and the need for development control. But it also makes it clear that whilst SMPs and LDFs have some shared objectives, they have different responsibilities. The 'issues table' provided in volume 2 of the guidance[17] stops short of including the 'potential planning response' provided in the Scarborough leaflet.

---

[16] Planning Co-operative, 2009 vol 2. p17
[17] Defra, 2006 vol 2. p41

It is also careful to suggest that the testing of policy options against potential 'externalities', such as the building of a harbour structure or a dynamic natural feature, should not extend to speculation about future social attitudes or socio-economic policy[18]. It concludes by stating that *"once the SMP has been adopted* [our emphasis] it is essential that the process for its consideration in the planning framework is progressed"[19]

The extent to which SMPs 'guide' planning decisions

As the North Solent information note highlights, "SMPs are not legally enforceable but are used by Planners and Development Control Officers to assist with decision making for proposed development on or near the coast. Each planning application will be considered on its individual merits on a case-by-case basis against various constraints, opportunities, development plan and policies that apply [...] Planning permissions are not determined solely by the SMP coastal defence policy."[20]. Each LDF has to consider coastal management policy as one of many other issues affecting the trajectory of development; whilst SMP2s do indeed contain important constraints, Defra discourages the more explicit direction called for by the Planning Co-operative within SMPs because they lack the rounded appreciation of other unrelated constraints that impact upon the LDF.

The extent to which the constraints imposed by an SMP outweigh those from other sources are location-dependent. In areas where coastal flooding and erosion are relatively minor issues due to natural geomorphology, the LDF may have the luxury of setting out a vision for community development that responds more directly to socio-economic needs. Indeed, with proper dialogue between planners and engineers, the SMP policy appraisal may be able to accommodate the aspirations of the LDF quite easily. Where flooding and (especially) coastal change present higher and more widespread risk, this flexibility will be reduced and the LDF must be more responsive to the SMP. In such cases, if the SMP includes more explicit planning responses according to their own objectives, it may risk 'jumping ahead' of the democratic process needed for affected communities to decide what their own objectives are in the face of change.

Planning for uncertainty

Even with a successful dialogue between coastal engineers and spatial planners, this dependency of the LDF upon constraints set out in the SMP means that in different places, local communities tend to feel different degrees of 'empowerment' over how much they can 'shape their own destiny'. Likewise local authority planners may feel more or less empowered to respond to the (perceived) needs or wishes of that community in developing the LDF. This is important as it is quite possible that sea level rise will force the 'vision' we have for our local area or region to develop rapidly, even in larger towns and cities.[21] Despite this, local and national political leaders are often keen to promise certainty to their electorate, in order to maintain safe places to live and invest.

The Planning Co-operative report also calls for greater certainty from SMPs, particularly in predicting sea level rise and its effects (such as erosion), so that planners are better placed to respond[22]. Unfortunately this is usually difficult to provide – the behavioural systems approach improves our understanding of the large-scale effects of coastal processes, but also

---

[18] Defra, 2006, vol 2. p57
[19] Defra, 2006, vol 2. p74
[20] Environment Agency, 2010. p2
[21] ICE / Royal Institute of British Architects, 2010
[22] Planning Co-operative, 2009. p24

highlights the complexity inherent in making predictions, especially for individual properties. SMP authors are rightly wary of being 'held to account' when coastal assets are threatened by erosion sooner or flooded more frequently than predicted. New national coastal erosion risk information becoming available on www.direct.gov.uk purposefully displays predictions not as definitive 'lines on maps' as most SMPs do, but as a range denoting what is most likely to be eroded over a given timescale.

This lack of certainty makes SMP outputs inherently difficult to apply to spatial planning even with strong input from planners. However, it is also a common feature of all types of development management: many infrastructure and development projects are subject to funding appraisal and intense competition for resources, as well as dependencies upon third party funding and the uncertainty that entails. In all cases where a LPA must 'plan for uncertainty', the two key responses are the same – find the resources to maintain the status quo, or adapt.

Responding to uncertainty: defence

Uncertainty surrounding future flooding or erosion rates can of course be addressed by the provision of defences (notwithstanding that erosion is often compounded by landslip), but as noted above, even where an SMP intends to 'Hold the Line', the uncertainty for communities lies in the provision of funding. This is the cause of 'blight' highlighted by coastal pressure groups that can hinder it's ability to regenerate or even remain viable. Although SMPs may take account of development plan policies that might influence the pattern, intensity and type of future development and land use, they can only appraise against current situation, using the system of ranking benefits and objectives set out in Defra's guidance.

Instead of the SMP setting out a planning response to its policies, Defra and the Environment Agency are encouraging more flexibility at the Strategy and Scheme level of coastal risk management and planning – namely by formalising a system by which third parties can contribute towards defence management and provision within the appraisal process. This forms the backbone of the Environment Agency's Long Term Investment Strategy and has been the subject of consultation on 'Partnership Funding' by Defra in 2011[23]. A multiple-funder model for risk management provision will not iron out uncertainty completely (it is ever-present when money is involved) but should reduce it in the long term and at least empower both communities and planners to respond to SMPs more pro-actively, instead of feeling 'imprisoned' by them.

Responding to uncertainty: adaptation

Even with funding, defence is not always possible for technical reasons or because it causes further damage to the country's already beleaguered coastal environment. Where community adaptation is required, the dialogue between planners and engineers is particularly crucial both to ensure that the suite impacts are fully appreciated in the SMP's policy appraisal, and to allow adaptation options to be explored properly – and early – so they can be captured in the LDF.

Whilst it is a much more difficult route politically and requires significant engagement effort, innovation and compromise, adaptation can ultimately be the way in which communities feel empowered to 'create opportunities out of losses'. A consistent policy framework is being developed by Defra to help communities to manage this process rather than being destroyed

---

[23] Defra, 2011

by it – currently being tested against lessons learned by a suite of Defra-funded 'coastal change pathfinder projects' that explore a range of adaptation options[24].

Conclusion

The answer to the question posed in the title is no it probably won't, and probably shouldn't, although perhaps there is an argument for a joint hosting so that planners and engineers may be encouraged to consider the implications of their work for each other in person. There has certainly been a call to encourage greater planner participation in the Environment Agency Flood and Coastal Risk Management annual conferences. As far as the broader question of 'should spatial planning be a greater component of SMPs' is concerned, this paper has argued that SMPs should remain focussed upon their remit and not try to 'do the spatial planners' job for them'. It has also emphasised that in reviewing SMPs, spatial planners need to be fully involved so that all parties can understand each others' pressures and goals and influence the objectives of SMPs and LDFs through an iterative process.

This is essentially what is involved in drawing up the Coastal Change Management Areas (CCMAs) proposed by CLG. The approach taken by the Planning Co-operative to engaging people in Scratby, Suffolk, about responding to the local policies of their SMP[25] was to involve them in drawing up their own CCMA. As well as identifying areas of constraint as set out in CLG's coastal change supplement to PPS25, areas of *opportunity* were also defined as part of the CCMA, where the at-risk elements of the community could potentially evolve and improve, ensuring it they were fit to meet the environmental and financial challenges that the 21st century might present them, without damaging the natural environment or the ability of other communities to do the same. This is a response to the SMP, rather than something the SMP should include, but it is perhaps the type of approach that future iterations of SMPs should highlight and draw upon in order to add a 'human face' to otherwise highly technical documents.

References

Atkins 2004, 'ICZM in the UK – a stock take' Final report to Defra at
http://archive.defra.gov.uk/environment/marine/documents/protected/iczm/st-full-report.pdf

Borough of Kings Lynn & West Norfolk Council / Environment Agency 2010, 'Coastal Flood Risk – planning protocol joint position statement', http://www.west-norfolk.gov.uk/pdf/coastal%20protocol.pdf

Client Steering Group, North Solent SMP 2010, 'North Solent Shoreline Management Plan – Information Note for landowners, planners and developers on privately owned coastal defences',
http://www.northsolentsmp.co.uk/media/adobe/j/c/North_Solent_SMP_information_note_for_landowners_and_planners_250810.pdf

Communities and Local Government 2010, 'Planning Policy Statement 25 Supplement: Development and Coastal Change', CLG publications

---

[24] http://ww2.defra.gov.uk/environment/flooding/coastal-change-pathfinders/ for more information
[25] Part of the Great Yarmouth Defra coastal change pathfinder
http://ww2.defra.gov.uk/environment/flooding/coastal-change-pathfinders/east/ for more information.

Defra 2005, 'Making Space for Water: Taking forward a new government strategy for flood and coastal erosion management in England: Government first response' www.defra.gov.uk/environ/fcd/policy/strategy.htm

Defra 2008, 'Coastal Groups in England', Defra publications

Defra and Welsh Office 2001, 'Shoreline management plans: a guide for coastal defence authorities', Defra Publications

Defra 2006, 'Shoreline Management Plan Guidance, Volume 1: Aims and Requirements', Defra publications

Defra 2006, 'Shoreline Management Plan Guidance, Volume 2: Procedures', Defra publications

Defra 2008, 'A strategy for promoting an integrated approach to the management of coastal areas in England', Defra publications

Defra 2011, 'Flood and Coastal Resilience Partnership Funding: Defra policy statement on an outcome-focused, partnership approach to funding flood and coastal erosion risk management', Defra publications

Department of the Environment 1992, 'Planning Policy Guidance Note (PPG) 20: Coastal Planning', The Stationery Office

Environment Agency 2009, 'Working With Others: Building Trust With Communities', Environment Agency internal publication

Environment Agency 2009, 'Investing for the Future: Flood and coastal risk management in England – a long term investment strategy', Environment Agency publications

Environment Agency 2010, 'Landowner Information Pack' produced by Environment Agency Anglian Region

Halcrow 2002, 'Futurecoast' (www.defra.gov.uk/environ/fcd/Futurecoast.htm)

Institution of Civil Engineers / Royal Institute of British Architects (Building Futures) 2010, 'Facing up to rising sea levels – Retreat? Defend? Attack?', ICE publication http://www.buildingfutures.org.uk/assets/downloads/Facing_Up_To_Rising_Sea_Levels.pdf

Local Government Association Coastal Special Interest Group 2004 'Managing Coastal Risk information pack', http://www.coastalsig.lga.gov.uk/info_pack.htm

MAFF and Welsh Office 1995, 'Shoreline management plans: a guide for coastal defence authorities', MAFF publications (Replaced by 2001 SMP Guidance Note.)

MAFF 2000, 'A Review of shoreline management plans, 1996 to 1999', Final Report MAFF Flood and Coastal Defence, March 2000

Office of the Deputy Prime Minister 2005, 'Planning Policy Statement 1: Delivering Sustainable Development', The Stationary Office

The Planning Co-operative 2009, 'Translating Shoreline Management Plans into Spatial Plans (Volumes 1 & 2)', Report for the Environment Agency

A full list of links to Shoreline Management Plans referred to in this paper can be found at http://www.environment-agency.gov.uk/research/planning/105014.aspx

Section 4:

Coastal policy – legislation, targets and the future

Innovative Coastal Zone Management
ISBN 978-0-7277-5749-4

ICE Publishing: All rights reserved
doi: 10.1680/iczm.57494.277

Coastal Flood Risk – A European Legislation Perspective

David Porter, Rivers Agency, Belfast, Northern Ireland

Introduction

The desire for a common legislative approach within the European Union has directed all areas of government policy development. This is clearly evident in the area of the environment which 'achieved more prominence during the 1980s and 1990s than it had in the earlier period of community policy making' (Wallace, 1999). Indeed Haigh (1992) says 'it cannot be repeated too often that it is impossible to understand the environmental policy of any of the EC member states without understanding EC environmental policy'.

Initially the water policy focus was on improving the water quality and the need for change was identified towards the end of the 1980s but it took over the years and 'several interim steps until the Commission finally published its first proposal in February 1997' [Europe.eu, 2011]. This became Directive 2000/60/EC which established a 'Framework for Community action in the field of water policy' (EC, 2000) commonly known as the Water Framework Directive (WFD). This directive requires a survey of water bodies to establish the baseline conditions; introduction of the concept of 'good ecological status'; and the use of River Basin Management Plans to drive improvement.

With its focus on water quality the WFD did not consider flood risk management and in response to a number of widespread flood events across central Europe the Commission introduced Directive 2007/60/EC on the assessment and management of flood risks (EC, 2007) commonly known as the Floods Directive. This Directive requires the consideration of all significant sources of flooding, in order to provide mapped information to the public and the development of Flood Risk Management Plans.

Flood Risk Context

The level of flood risk from different sources varies dramatically across Europe. There are very large river catchments with a significant history of flooding such as that of the Rhine. This river is 1,320 km in length, flows through nine countries, drains an area of 185,000 km^2, has an average flow rate at its outlet of 2,200 m^3/s and a peak recorded flow rate of 13,000 m^3/s (Sprokkereef, 2010). There is ground water flooding in areas of Europe, such as in the UK for example in and around Oxford, where the water table is high 'typically less than 1m below ground in winter and 2m in summer' (Macdonald et al, 2007) and when high river levels coincide with heavy rain residents are often alerted to flooding by the emergence of ground water. Another potential source is surface water or pluvial flooding that is caused as a result of high intensity rainfall which overwhelms natural or engineered drainage systems resulting in water flowing overland and ponding in depressions in the ground. For example on 12 June 2007 high intensity short duration rainfall impacted upon Belfast in Northern Ireland. Detailed analysis of the Knock and Loop river catchments showed that of the 427 properties affected, 46% of these were as a result of surface water flooding (Smyth et al, 2008). Reservoirs or impoundments can also pose a flood risk and finally there is the flood risk from the sea, for example the Netherlands with '52% or some 8.5 million of the inhabitants actually living below sea levels', (McRobie et al, 2005). Some consider that flash flooding is a source but this term really describes the speed of the event rather than the source and it typically emanates from rivers, the ground or surface water.

During the period between 1998 and 2006 'Floods in Europe have caused 700 deaths, the displacement of about half a million people and at least £25 billion in insured economic losses'. (Europa.eu, 2006)

The Floods Directive

The Floods Directive is to be delivered in four key stages.

1. Transposition into national legislation
2. Complete a Preliminary Flood Risk Assessment.
3. Produce Flood Risk and Hazard Maps for significant risk areas; and
4. Produce Flood Risk Management Plans.

Each of these stages will be considered in some detail and where appropriate the specific impact on coastal flood risk management will be highlighted.

Transposition

When a directive enters into force from the European Parliament it must then be transposed into national statue within each Member State. The Floods Directive, in article 17, requires that the necessary 'laws, regulations and administrative provisions' are in place by 25 November 2009. Transposition starts the process of implementing the directive and is in essence a legal administrative process with little technical input. Some member states also used the introduction of new legislation to update other related drainage and flood risk management laws.

Preliminary Flood Risk Assessment (PFRA)

The real start of the process involving technical input is the preparation of the PFRA as detailed in Article 4. This assessment considers both the past flood events and the future predicted risks for all significant flood sources and given the differing levels of flood risk in member states the directive does not attempt to define significant flood, neither does it attempt to define a threshold above which a flood becomes significant. This very flexibility has however been the subject of much debate both at European and national level. The desire to ensure compliance with a common directive would suggest that a similar approach is necessary but the level of risk present and even the sources of flooding are not equal in all countries. For example, not all member states have a coastline so obviously coastal flood risk does not need to be considered under the directive; and those that do have a coastline are not exposed to the same level of risk from this source. The directive does however define in broad terms, the areas of impact which need to be considered as part of the assessment. These are defined in article 4 (2)(b) as 'human health, the environment, cultural heritage and economic activity'. The directive does not state the relative importance or weighing to be applied to each of these four criteria, neither does it define the exact meaning of each of the criteria, again leaving this flexibility to the member state.

Common Implementation Strategy (EU, 2001)

All directives have an element of flexibility in their text to enable national policy setting to be accommodated but the Commission is also very focused on the overall consistency of implementation. At European level this is managed through working groups under the Common Implementation Strategy (CIS). This approach was established for the WFD to ensure that information was exchanged 'between Countries, European Institutions, the various stakeholders and the interested public' (Europa.eu, 2011). Given the close policy fit of the Floods Directive with the WFD, the CIS was extended to assist in the implementation of the Floods Directive. This also ensures that the legislative requirement of co-ordination with WFD, as detailed in article 9, is more likely to be achieved. Working Group 'F' was established to deliver the Floods Directive and this group meets about twice a year and is the forum for discussing and agreeing common approaches to implementation. This group also highlights any conflicts with national policy, legislation or administrative structures particularly those connected to the reporting of progress to the Commission.

Coastal Flood Risk in the Preliminary Flood Risk Assessment

The PFRA considers the risks from all significant sources of flooding. A sample list of sources is included in article 2 where a flood is defined as:

> *'the temporary covering by water of land not normally covered by water. This shall include floods from rivers, mountain torrents, Mediterranean epherneral water courses, and floods from the sea in coastal areas".* (article 2(1))

This legislation requires all member states who consider their coastal flood hazard to be significant to document the risks. The directive also provides some guidance on the types of information that should be part of the assessment including mapping the river basin and coastal areas (article 4(2)(a)), describing significant past flood events which may reoccur (article 4(2)(b)), the impact of climate change (Article 4(2) and 4(2)(d)) and consideration of the potential consequences of flooding in human health, the environment, cultural heritage and economic activity (article 4(2)(d)). To ensure some cost restraint to this process the directive says that the assessment is to be 'based on available or readily derivable information' (article 4 (2)).

The PFRA is the mechanism for identifying areas of potential significant flood risk (Article 5(1)) after consideration of all flood sources. This includes the quantification of coastal flood risk and its likely impact. This assessment is required by the directive to be complete by 22 December 2011 (Article 4(4)) and those areas identified are then subject to further work in the next stage of the process.

'Significant' Flood

No definition of significant is provided within the directive again this is for individual member states to determine, and there has been much debate amongst Woking Group 'F' members on this topic. It does however have to be flexible because the level of risk from any of the sources of flooding is not consistent across Europe.

As already stated, over half of the inhabitants of the Netherlands live below sea level and the whole defended area has a long history of breech including such major events as the 1953 North Sea flood which caused the death of in excess of 1800 people (naturegrid.org.uk.2011). Compare this level of risk to that of Northern Ireland, where a relatively small proportion of the population occupying about '46,000 properties, around one in eighteen, are at risk of flooding from rivers and the sea' (EEA, 2010). The coastal risk element of this figure is just over half at about 2% of the entire building stock.

The Netherlands clearly has a significant coastal risk, given the high percentage of their population living in defended areas, but this does not mean that the lower risk situation in Northern Ireland is therefore not significant. This is because the Floods Directive requires an assessment on a national basis and this is not intended to be a comparative exercise across Europe. In addition the people of the Netherlands have lived with this risk for centuries and it is engrained in their culture. They have developed a very sophisticated system for controlling this risk including monitoring, maintenance and early warning systems. It could be argued that the residual risk is not significant as the controls are in place to mitigate the effects in all except extreme or catastrophic occurrences. The lower risk of coastal inundation in Northern Ireland is not in the consciousness of the people. They do not perceive it to be real as it has not be experienced to any extent in living memory and therefore, in the event of the risk being realised the impact may be uncontrolled. This is clearly a situation that would benefit from a management plan and under the Floods Directive this should be considered in the PFRA.

Flood Hazard Maps and Flood Risk Maps

The significant risk areas, as identified from data analysed in the PFRA, are then mapped in accordance with article 6. The directive is very specific on the mapping requirements as these are not really a function of the actual risk, nor the flood sources present, nor even the perceived attitude to risk in any particular culture or country but are an output based on a detailed technical specification. Any approach variables are already considered in the preliminary flood risk assessment and have therefore influenced the selection of the significant risk area to be mapped.

The flood hazard maps are to include three probability scenarios;
- Low probability or extreme events (generally taken as 1 in 1,000 year events)
- Medium probability (with a likely return period of around 100 years); and where appropriate
- High probability (those with a return period of somewhere between 10 years and 50 years)

Figure 1 shows an example of a hazard map for a river flood extent. Unfortunately no sample coastal maps were complete at the time of writing this paper but the outputs will be in the same format. In addition to the flood extent this map indicates the presence of flood defence structures and the area benefitting from these systems.

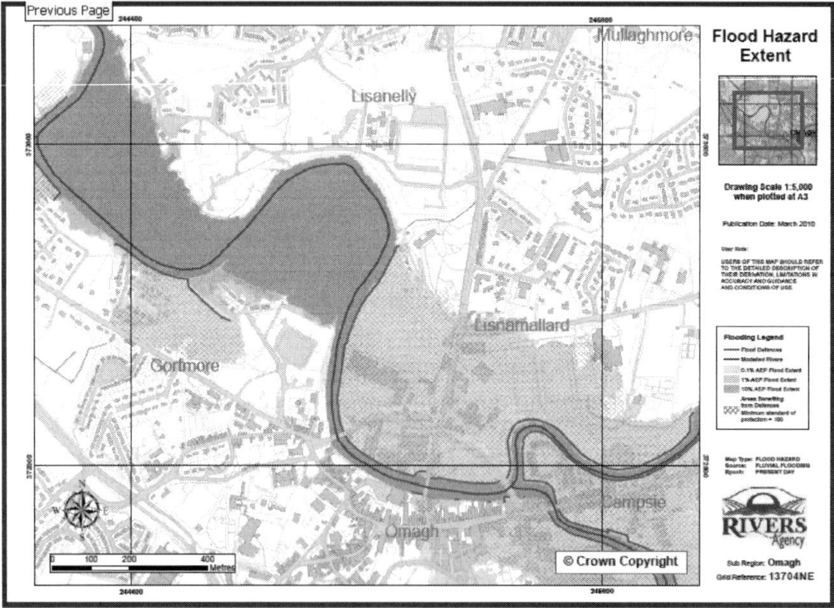

Figure 1: An example of a flood hazard map showing three return period scenarios

The directive also requires the flood hazard maps to include information on the water depth or level and the flow velocity, as appropriate, as shown in Figure 2.

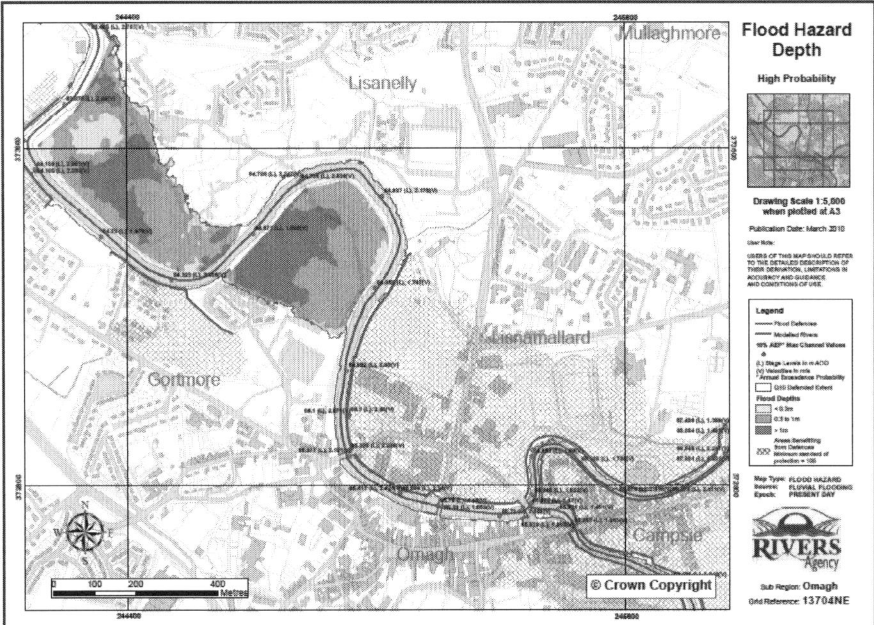

Figure 2: An example of a flood hazard map showing flood levels and velocity

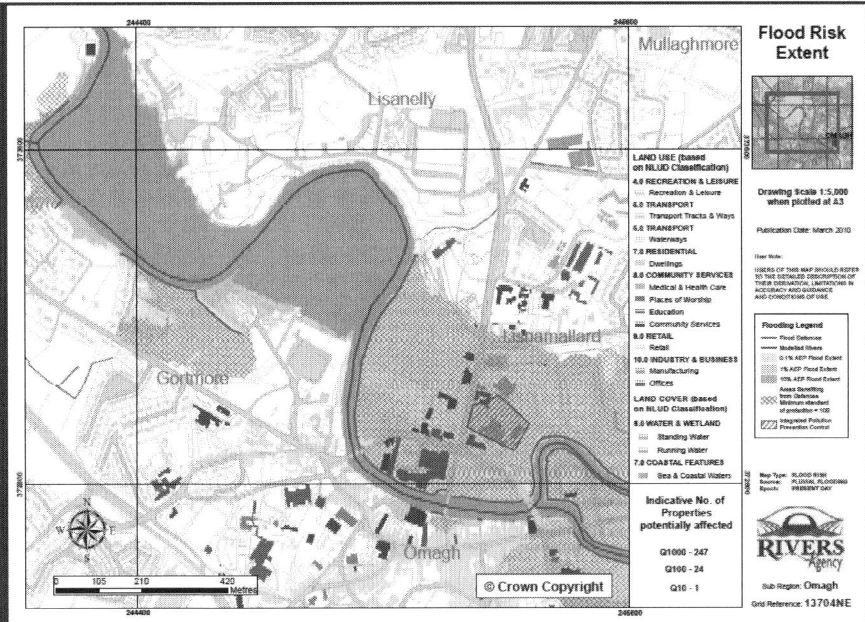

Figure 3: An example of a flood risk map showing economic activity and vulnerable assets

281

The flood risk maps take the hazard information and develop it further to show the consequences of flooding. These maps are to include an indication of the number of inhabitants at risk; the type of economic activity; the location of potentially vulnerable sites which may cause pollution as per Directive 96/61/EC; and if thought useful sediment or debris flow and other sources of pollution. See Figure 3 for an example of a risk map.

Article 6(6) specifically deals with defended coastal areas. In recognition that sea defences typically offer a high level of protection the directive allows the omission of the medium and high probability flood hazard maps for this source but still requires the production of the extreme event scenarios. All maps are to be completed by 22 December 2013 (article 6(8)).

Flood Risk Management Plans

The maps of the significant risk areas, as determined by the PFRA are then used to develop Flood Risk Management Plans. These Plans are to be published by 22 December 2015 (Article 7(5)).

This stage of delivery is focused on the adverse effects of flooding and the same criteria as used in the PFRA are to be used, namely human health, the environment, cultural heritage and economic activity (Article 7(2)). The plans are to be realistic documents which must consider not only the impacts of flooding but the costs and benefits associated with the measures proposed. This is to be achieved by establishing objectives within the plan (Article 7(2)). These are likely to be high level statements which set or describe the flood risk management policy direction of a member state. The Plans are also to include measures which if implemented will assist in achieving the objectives (Article 7(3)).

An example of a coastal flood objective may be 'reduce the number of properties at significant risk from coastal inundation'. In support of this objective there may be a range of both structural and non-structural measures such as 'construct a coastal flood defence structure at a particular location'. This would be a structural measure. An example of a non-structural measure may be 'avoid inappropriate development in coastal flood risk areas by the introduction of planning policy restrictions'.

The directive is clear that the EU wishes the flood defence industry to wholly embrace flood risk management as it requires all aspects of risk management to be considered including 'prevention', 'protection' and 'preparedness'. It explicitly includes such things as 'flood forecasts and early warning systems' (Article 7(3)) in recognition that the flood situation faced across Europe is changing and is one which cannot be tackled by the continued construction of flood defence structures alone. The Flood Risk Management Plans will provide a significant amount of information to the public, commerce, industry and elected representatives which will enable them to plan for flood events in an informed manner.

The Floods Directive Cycle

The dates detailed in the various paragraphs above are seen as starting what will become an ongoing process of assessment, mapping and planning. This is in recognition of the fact that the development of society is not static and therefore vulnerability to flood risk cannot be established once in time. In addition the predicted impacts of climate change and the science that is used in this field is developing and rather than make an ill informed management plan, the directive only explicitly requires climate change to be considered in the PFRA in the first cycle (article 4(2)). In the second cycle, which requires review of the PFRA towards the end of 2018 (article 14(1)), 2019 for the maps (article 14(2)) and 2021 for the Flood Risk Management Plans, climate change is to be considered not only in the PFRA but also in the management plans (article 14(4)). All three elements are then on a six year cycle thereafter.

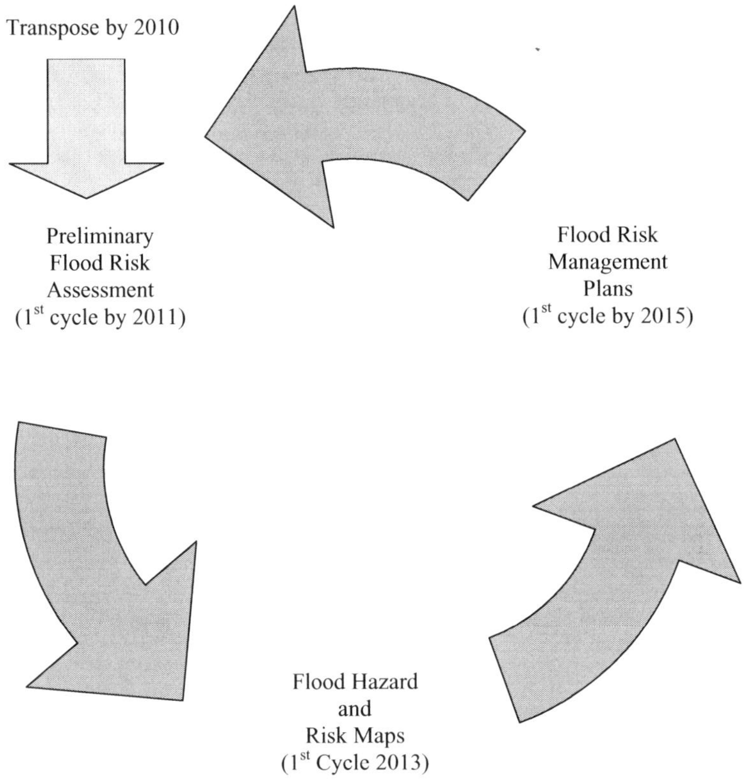

Figure 4: The Floods Directive Cycle

Cross Border Co-operation

Given the application of this directive is EU wide it is very clear that international co-operation is required in both the development of the various stages detailed above and in the implementation of the Flood Risk Management Plans. In addition to this joint working the directive re-enforces the need for solidarity across the EU. It does this by requiring Member States to not only solve their own problems but to think of the adverse consequences on other Member States of their proposed solution and if need be achieve agreement before commencement (Para (13)).

Transitional Measures

As well as the flexibility of the directive outlined in this paper there are further provisions in article 13. This section enables any member state to opt out of the PFRA stage if there is either an existing risk assessment that shows that potential risk exists or that they have decided to produce compliant maps and plans. The decision to use this option had to be taken by 22 December 2010.

Conclusion

The Floods Directive provides a significant opportunity for the European wide flood risk to be understood and subsequently managed. The preliminary flood risk assessment will provide a good base line summary of past flood events and will consider the likelihood and impacts of future floods. The flood hazard and risk maps will provide detailed information to enable informed land use decisions to be taken. They will also allow the further development of emergency plans as the potential flood impact will be identified, as will the receptors. Most importantly the Flood Risk Management Plans will provide both high level objectives and more specific measures which if delivered will improve the flood risk approach.

Interestingly nowhere within the directive does it require problems to be solved as the uncertainty surrounding the issue of climate change is likely to make the job of managing flood risk more complex. However by understanding the hazard and by setting the direction for both government and citizens to work together, the Floods Directive will improve the situation. Not by stopping all flooding but by ensuring the impacts are controlled, that the development mistakes of the past are not repeated without mitigation and that plans are in place to deal with potentially increasing flood risk in the future.

References

Directive 2000/60/EC of the European Parliament and of the Council of 23 October 2000 establishing a framework for Community action in the field of water policy, available at http://eur-lex.europa.eu/LexUriServ/LexUriServ.do?uri=CONSLEG:2000L0060:20011216:EN:PDF

Directive 2007/60/EC of the European Parliament and of the Council of 23 October 2007 on the assessment and management of flood risks, available at http://eur-lex.europa.eu/LexUriServ/LexUriServ.do?uri=OJ:L:2007:288:0027:0034:EN:PDF

EEA, 2010 'The European Environment- State and Outlook 2010' available at http://www.eea.europa.eu/soer/countries/uk/freshwater-state-and-impacts-united-kingdom

EU, 2001, 'Common Implementation Strategy for the Water Framework Directive (2000/60/EC)' available at http://ec.europa.eu/environment/water/water-framework/objectives/pdf/strategy.pdf

Europa.eu (2006) 'Environment: Commission adopts new directive to fight floods' Press Release IP/06/50 issued 18 January 2006

Europa.eu, 2011 'The decision – making process' available at http://ec.europa.eu/environment/water/water-framework/info/decision_en.htm

Europa.eu, 2011 'WFD CIRCA – the Information exchange platform' available at http://ec.europa.eu/environment/water/water-framework/iep/index_en.htm

Haigh N, 1992, 'Manual of Environmental Policy: The EC and Britain' Longman, Harlow

Macdonald D, Hall R, Carden D, Dixon A, Cheetham M, Cornick S, Clegg M, 2007, 'Investigating the interdependencies between surface and groundwater in the Oxford area to help predict the timing and location of groundwater flooding and to optimise flood mitigation measures', available at http://nora.nerc.ac.uk/9884/1/Oxford_Defra_Paper_2007_Final.pdf

McRobie A, Spencer T, Gerritsen H, 2005, 'The Big Flood: North Sea storm surge', 'Philosophical Transactions of the Royal Society, Vol 363 No 1831. available at http://rsta.royalsocietypublishing.org/content/363/1831/1263.full

Smyth P, Falconer R, Dawson S. 2008, 'Fluvial and Pluvial Flood Risk East Belfast Flooding – June 2007' DEFRA Conference, Manchester 1-3 July 2008

Sprokkereef E, 2010, 'The FEWS NL foreceasting stsyem and first experiences with probabilistic predictions in the Netherlands' Centre for Water Management, presentation available at http://www.kcl.ac.uk/content/1/c6/08/12/81/SprokkereefNL.pdf

Wallace H and Wallace W, 1999 (reprint), 'Policy-making in the European Union.' Oxford University Press, England

Innovative Coastal Zone Management
ISBN 978-0-7277-5749-4

ICE Publishing: All rights reserved
doi: 10.1680/iczm.57494.286

A Framework for Coastal and Erosion Risk Management Science

Paul Sayers, Sayers and Partners, Watlington, UK
Owen Tarrant, Environment Agency, Bristol, UK
David S. Brew, Royal Haskoning, Peterborough, UK
Nick Cooper, Royal Haskoning, Peterborough, UK
Michael Wallis, HR Wallingford, Wallingford, UK
Mike Walkden, Royal Haskoning, Peterborough, UK

Introduction

Good decision making at the coast is difficult, reflecting the uncertainty in future climates and demographics, a desire to deliver multi-functionality, the need to maintain a strategic view whilst encouraging local ownership, and the difficulties in understanding the behaviour of coastal systems and impacts of interventions. Often these challenges are set against a background of conflicting and competing management objectives. Although significant progress has been made in the last few decades with regard to the integrated management of our coastline, there remains a broad acceptance that more could and should be done.

Recent initiatives, such as Foresight[i], and its associated Foresight Update[ii], the Pitt Review[iii] and UK Climate Projections (UKCP) 09, have all identified that coastal areas will be at increasing risk of sea flooding and coastal erosion due to climate and socio-economic change. They also highlight the major challenges that persist in our understanding of the coast and how best to manage it. These challenges include gaps in our engineering and physical science understanding and also in our ability to convert good science into good practice. It is these later stages of dissemination and training that have often not been given sufficient priority to make a real difference and break down the barriers between science and uptake in practice.

Robust and practical science and the provision of sound supporting evidence, has long been recognised as a key means of moving practice forward and *much faith has been placed in science, technology and innovation as a means towards a sustainable and sustained economic recovery*[iv].The provision of such science and evidence necessarily requires the co-ordination of a spectrum of activities ranging from basic ('blue skies') and applied research through to development and pilot testing, and then in to implementation. No one single element of this pathway can exist in isolation and if a real difference to our understanding and management of the coast is to be made, a correct balance must be struck between scientific rigour, user relevance and practicality. The Government's white paper on innovation[v] highlights the long and uncertain process that basic research often has to follow to make it to the marketplace, and places an increased emphasis on other sources of innovation including the creative application of tried-and-tested technologies and the role of design in developing innovative products and services.

A vision for the future of coastal research development and dissemination has been recently set out within the **Co**astal **R**esearch **D**evelopment and **Di**ssemination (CoRDDi) Framework. The framework has cross-government buy-in and is being promoted through the joint Environment Agency / Defra Flood and Coastal Erosion Risk Management R&D Programme[vi] in close association with the Living With Environmental Change (LWEC) partnership[vii]. Within the CoRDDi Framework a portfolio of coastal research priorities have been identified to support better delivery of Flood and Coastal Erosion Risk Management (FCERM). Priority areas for RDD over the next five years have been defined together with a clear direction of travel for longer term efforts. The CoRDDi Framework is aimed at all those with an interest in managing flood and coastal erosion risk, including practitioners and the research community. It provides a Framework for all levels of involvement, from researchers and research managers to user organisations who may be called upon to contribute to collaborative funding.

Developing a Shared Vision

The starting point for the development of the CoRDDi Framework was to envision a desired future state for flood and erosion risk management at the coast. The vision (Figure 1) was developed from the following set of criteria/attributes:
- reflecting a bold aspiration for the future of flood and coastal erosion risk management;
- the diverse interests of stakeholders (practitioners, funders and researchers);
- supporting the political context and 'direction of travel';
- a desire to foster collaboration within the coastal community;
- where success can be verified; and
- providing flexibility to ensure continued relevance.

The vision outlines how the timely uptake of user-oriented research can assist with the future management of the coast. This includes managing risk and promoting opportunities, recognizing that a sole focus on risk management alone is limited, fails to maximize return on investment, and is not compatible with a notion of integrated management.

<table>
<tr>
<td valign="top">
"Those with responsibility to manage coastal flood and erosion should have access to useable and relevant tools and techniques that improve their ability to predict change."

The opportunities and constraints of change on all important aspects of the coastal flood and erosion systems are understood and accounted for when making decisions. The decisions taken are fully integrated, nesting UK priorities through to on-the-ground action and maximise opportunities and minimise risks efficiently and effectively.

There is rapid uptake of research, development and dissemination outputs into practice and practical experience and pilot studies routinely refresh research priorities.
</td>
<td valign="top">

</td>
</tr>
</table>

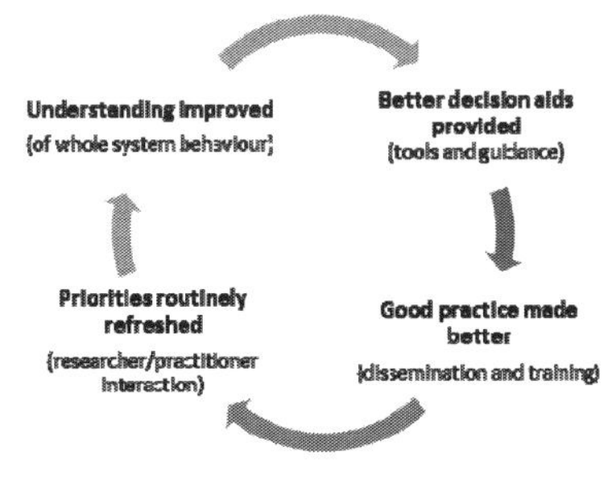

Figure 1. The CoRDDi Vision

Strategic Context of Coastal Research

A cascade of strategies is currently in development to set out the direction of research aimed at supporting flood and coastal erosion risk management. At the top of this cascade is the LWEC initiative; a partnership of 22 major UK public sector funders and users of environmental research, including the research councils and central government departments. Its 10-year programme aims to optimise the coherence and effectiveness of UK environmental research funding and ensure government, business and society have the foresight, knowledge and tools to mitigate, adapt to, and capitalise on environmental change. LWEC is developing a series of strategies to provide high-level direction in the areas of water, health and economy.

Sitting within the LWEC initiative and under the broadly-based UK Water Research and Innovation Framework[viii] is the UK First Flood Research Strategy[ix] (Figure 2). The Flood Strategy continues to evolve but is currently designed around three core themes of i) understanding risk (ii) reducing probability (likelihood), and (iii) reducing consequence (impact).

The delivery of CoRDDi and the Flood Research Strategy will be managed by a team drawn from the LWEC Partner organisations. Their remit will be to deliver the benefits from the portfolio, maintain communications and governance arrangements across LWEC partners, manage risks and issues, facilitate an executive portfolio board and aid framework initiation. The CoRDDi Framework has a pivotal role in setting the future coastal research agenda. It provides a link between the higher-level strategies and their governance structures and specific programmes and projects of the commissioned research (Figure 2).

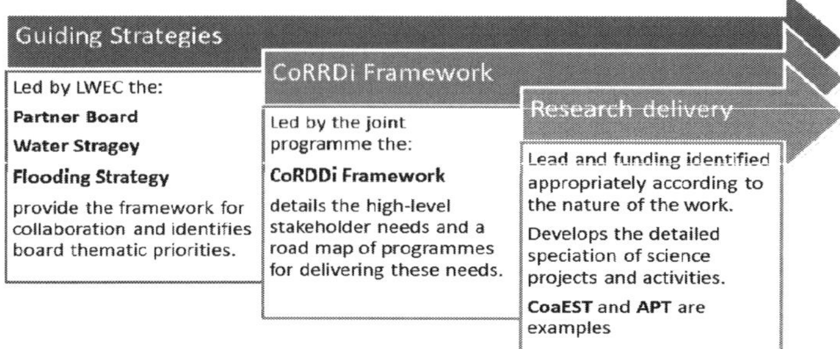

Figure 2. The relationship of the CoRDDi Framework to higher level guiding strategies and research delivery

Aspects of the CoRDDi Framework

Throughout the development of CoRDDi, good science and supporting evidence was recognised as a key attribute for good decision making in relation to the management of coastal flooding and erosion. To support this goal, research, development and dissemination (RDD) extend across a spectrum of activities, from **basic** coastal research (in association with the National Environmental Research Council (NERC), the Engineering and Physical Sciences Research Council (EPSRC) and others), **applied** coastal research (in association with the European Commission), **development** of science into practical tools and guides (in association with the Construction Industry Research and Information Association (CIRIA) and others) and associated **dissemination** and **training** to promote take-up and improve 'on-the-ground' capability (Figure 3).

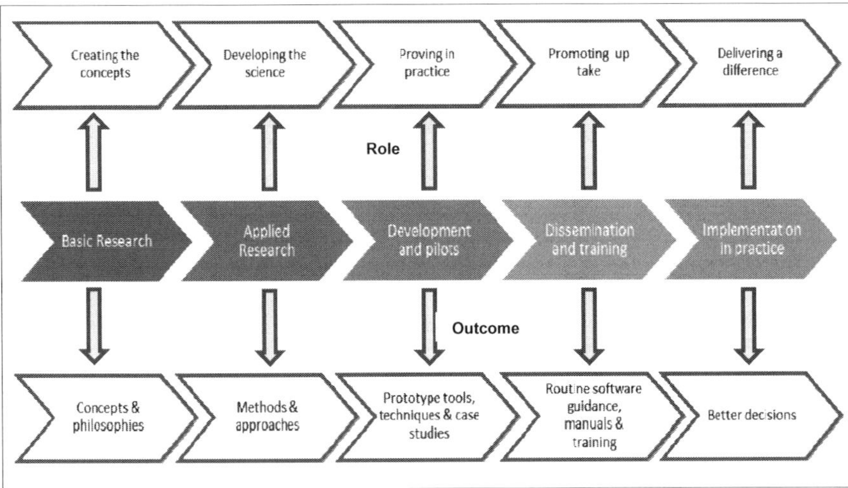

Figure 3. The CoRDDi framework will support all aspects of coastal science

Research Drivers

Various funders are interested and active within the general area of coastal science and many have specific strategic aims and objectives associated with flood and coastal erosion risk management, and more broadly, management of the coast. Achieving better integration of the science initiatives across funders offers significant rewards, including:

- **greater efficiency for each research pound spent:** maximising the use of existing research and multiple funding streams (avoiding duplication);
- **greater effectiveness:** providing a clear and coherent line of sight from basic research through to practice; and
- **better academic innovation and practical applicability:** academic innovation and practicality are not mutually exclusive ideals; often it is simply the lack of understanding of practical requirements by the academic and the lack of understanding of theoretical knowledge by the practitioner that limits advances in either domain. Providing a mechanism by which each of these can come together will enhance both.

Although desirable, delivery of this type of collaborative approach is not trivial. CoRDDi was never envisaged to develop a centralised process of pooling funding and central distribution, but rather to promote integration through:

- **a shared vision:** developing a shared vision for research priorities that can act as a focus for all funders to develop and promote specific initiatives and projects;
- **an active process of update:** evolve this common view through a continuous process of updating and review as the needs and demands of stakeholders change; and
- **common rules of engagement:** often barriers to collaboration result from inequitable sharing of, for example, costs, data, codes, tools and intellectual property rights. Mechanisms to ensure that these become facilitators of collaboration rather than barriers to collaboration will be vital to the success of the CoRDDi Framework. Without this facilitation, individual funders are likely to remain individual.

In lieu of the finalised governance arrangements which will be developed by the National Flood Research Strategy, CoRDDi will operate within the current arrangements in place to run the Joint Programme. These arrangements involve the Coastal Groups, the Joint Programme Technical Advisory Groups (TAGs), and the CoRDDi Project Advisory Group who will all play important roles in delivering the CoRDDi vision. Each group has an essential role in the prioritisation, review and monitoring of the research generated by the framework as it progresses.

To maintain continued relevance it is essential that the Framework should respond to changing user needs and innovation. Key to this is the ability to monitor and verify success and the benefits delivered. If this can be achieved it will enable the Framework to evolve and become self-perpetuating (not self-funding) to incorporate relevant advances in understanding and improved delivery. There will be the need for a continuous discourse between researchers and users as well as scientists, engineers and planners.

Themes

The Framework comprises four themes; 1. Understanding whole system behaviour, 2. Valuing impacts and promoting innovative funding, 3. Decision making and operational practice, and 4. Dissemination, education and training (Figure 4). The first two themes are focused on developing the knowledge base and the third builds on this base to assist the decision-making processes. An overarching strand of the Framework focuses on theme-wide dissemination.

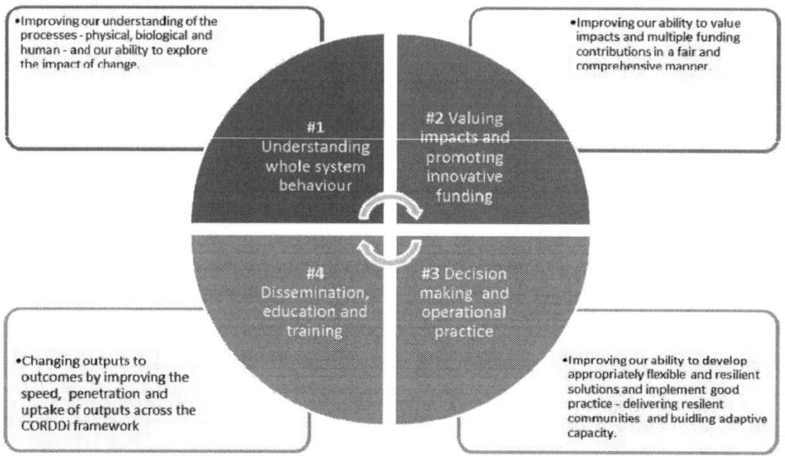

Figure 4. An overview of the four themes within the CoRDDi Framework

Theme 1: Understanding whole system behaviour.

Understanding the behaviour of coastal systems across multiple scales in time and space is now widely acknowledged as a pre-requisite for good decision-making. Shoreline and nearshore hydraulic and sedimentary processes (driven by weather), geomorphological behaviour (driven by these processes), ecological functioning and human intervention at the coast all contribute to a complex system of interacting mechanisms. These linkages present significant challenges in terms of understanding individual components of the system and the system as a whole. However, developing this understanding over multiple temporal and spatial scales is fundamental to providing better management decisions. A significant and important challenge for the CoRDDi Framework is to advance science to develop whole systems understanding in a practical and credible manner. This includes providing new concepts that integrate spatial and temporal scales and enable multi-functional, multi-options approaches to be developed.

Theme 2: Valuing impacts and promoting innovative funding.

The coast is home to many, a work place and business asset for others, as well as a heritage and environment cherished by all. Coastal communities are exposed to potentially wide-spread and life-threatening floods (such as the 1953 storm) as well as erosion losses that can seldom be regained. As such the coast and the risks it faces are unique. Consequently, the way in which the diversity of coastal environments and flood and erosion impacts are valued remains a significant challenge.

The coast is an excellent example of multiple interests demanding multi-functional schemes; promoting opportunities such as harbour use and tourism (for example) whilst reducing risk. Public sector expenditure is likely to be heavily constrained in the coming years, and delivering value for public money will be more important than in previous years. It is also likely that "value for money" will increasingly be scrutinised in terms of the opportunities it promotes for improvement as well as the risk it reduces. This will demand a better understanding of the 'true' risks and opportunities and a more integrated view of the 'value' achieved. Thus, Theme 2 of the CoRDDi Framework will focus on socio-economic and funding issues.

Theme 3: Decision making and operational practice.

Theme 3 focuses on building adaptive capacity within coastal policy, plans and on-the-ground actions. The rates of change in climate, demographics and political setting, and the associated uncertainties, present a major challenge to the decision making processes and operational practice adopted. All levels of decision making are included within this theme, with decisions based on the improved understanding (tools and techniques) developed within CoRDDi Themes 1 and 2. The focus here is how to use this improved understanding to help make better more robust choices in a transparent and participatory manner.

Theme 4: Dissemination, education and training.

The fourth theme within the CoRDDi Framework places emphasis on dissemination to a level not seen in previous studies. It seeks to deliver better dissemination from project inception through to delivery of the final outputs and on to uptake and routine use, whilst keeping research outcomes under review in the light of experience gained through its application. This approach of progressive improvement will provide a continuity of development of new tools and techniques that has occasionally been lacking to date. This will avoid repetition and reinvention, whilst continuing to provide room for real innovation (not repackaged theory). This is not a simple task and one that will demand a wide range of organisations to provide resources and fund activities.

RDD Prioritisation and Project Development

Through a process of consultation with coastal managers, consultants and academics, a wide range of important issues for future research and development were identified. These were then prioritised to distinguish (i) 'essential' and 'desirable' projects; (ii) the scale of impact, i.e. those projects applicable nationally and those applicable to a specific location or region; and (iii) the urgency of the project, i.e. those aimed at supporting current practice and those that will influence future action (for example to inform future Shoreline Management Plans in 5 or 10 years time). The simple scoring matrix used is shown in Figure 5.

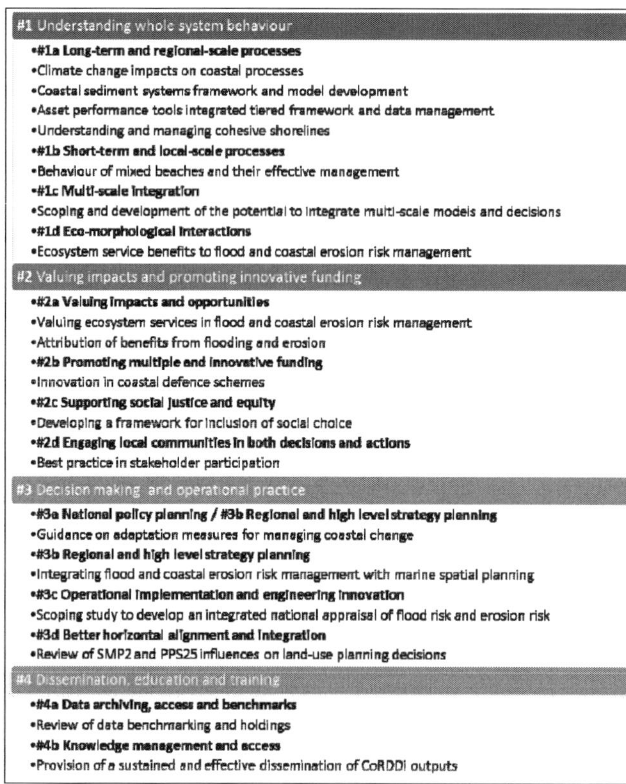

Figure 5. Scoring system for scheduling prioritisation of needs/issues

The projects highlighted as priorities for the coming five years through this process are summarised in Figure 6.

#1 Understanding whole system behaviour

- **#1a Long-term and regional-scale processes**
- Climate change impacts on coastal processes
- Coastal sediment systems framework and model development
- Asset performance tools integrated tiered framework and data management
- Understanding and managing cohesive shorelines
- **#1b Short-term and local-scale processes**
- Behaviour of mixed beaches and their effective management
- **#1c Multi-scale integration**
- Scoping and development of the potential to integrate multi-scale models and decisions
- **#1d Eco-morphological interactions**
- Ecosystem service benefits to flood and coastal erosion risk management

#2 Valuing impacts and promoting innovative funding

- **#2a Valuing impacts and opportunities**
- Valuing ecosystem services in flood and coastal erosion risk management
- Attribution of benefits from flooding and erosion
- **#2b Promoting multiple and innovative funding**
- Innovation in coastal defence schemes
- **#2c Supporting social justice and equity**
- Developing a framework for inclusion of social choice
- **#2d Engaging local communities in both decisions and actions**
- Best practice in stakeholder participation

#3 Decision making and operational practice

- **#3a National policy planning / #3b Regional and high level strategy planning**
- Guidance on adaptation measures for managing coastal change
- **#3b Regional and high level strategy planning**
- Integrating flood and coastal erosion risk management with marine spatial planning
- **#3c Operational implementation and engineering innovation**
- Scoping study to develop an integrated national appraisal of flood risk and erosion risk
- **#3d Better horizontal alignment and integration**
- Review of SMP2 and PPS25 influences on land-use planning decisions

#4 Dissemination, education and training

- **#4a Data archiving, access and benchmarks**
- Review of data benchmarking and holdings
- **#4b Knowledge management and access**
- Provision of a sustained and effective dissemination of CoRDDI outputs

Figure 6. Priority projects across the Themes

Challenges to Overcome and Ownership of Issues

The difficulties and barriers to the successful delivery of the CoRDDi Vision are not simply technical, but rely on an on-going and proactive dialogue with users and researchers alike. Some of these barriers and how they are addressed within CoRDDi are highlighted below in Figure 7.

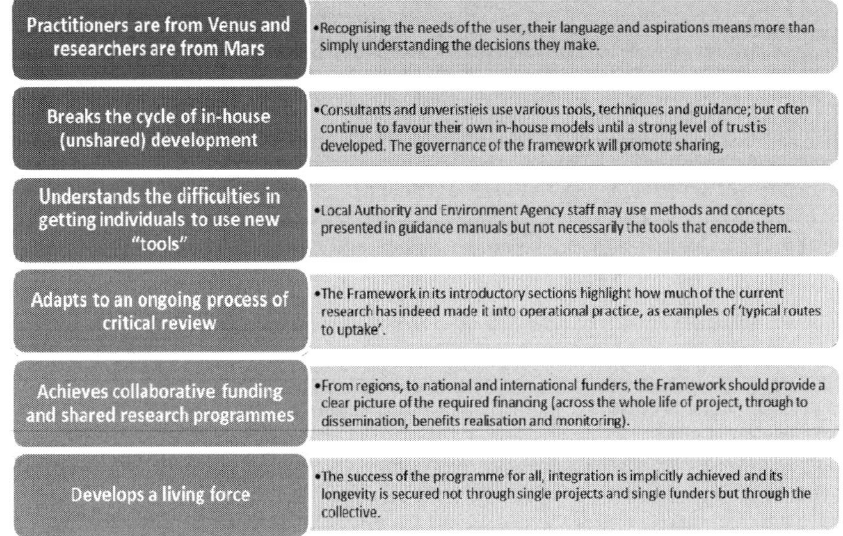

Figure 7. The CoRDDi Framework will provide an on-going process of dialogue with practitioners that has a range of attributes

Some of the more prominent challenges that have been identified whilst developing the CoRDDi Framework that remain unresolved are;

Judging success based on shared criteria: Each funder has a different motivation and hence use different criteria to judge the success or otherwise of research and development projects and programmes. As such, when seeking to be provide collaborative multi-funded research, the success criteria used can either become a facilitator of integration between research and practice or indeed place barriers to successful collaboration.

Evolving shared models, codes and software IP: Developing modelling initiatives into a meaningful community based network of model components capable of being integrated to reflect the demands of a particular decision process will be a crucial challenge facing the coastal community. This will not only demand research but also strong governance to align research with the chosen approach.

Sharing and managing data and data IP: A long held view reinforced in the Environment Agency's data strategies is the idea to collect once and use many times. Coastal observatories are good at sharing the data they have, whilst others are not. Much excellent researcher-collected data is not in the public domain and the governance of CoRDDi needs to try and ensure this data is reused, stored, understood.

Joint funding and/or cost sharing: Who does what, who leads which projects, how the projects are steered and who will sign-off, own and be able to exploit the outputs are all questions to be resolved. The clarity and content of the resolution to this issue will be important facilitators or barriers of successful collaboration.

Maintenance of core capability whilst encouraging innovation and new ideas: Programmes such as the Flood Risk Management Research Consortium (FRMRC) and Environment Agency Frameworks enable like-minded and collaborative thinking to be developed and can, but not always, avoid reinvention. Some argue that such approaches can limit creativity and innovation, so the RDD to be procured needs to ensure it gets the best from the research money spent.

Taking a long term view: It may take some time before concepts become stable methods and a practical reality. For example, many advances have been made in system risk and uncertainty analysis (e.g. RASP[x], Tyndall simulator), but on-going effort is required to ensure such concepts are routinely used in practice and are fully trusted. The governance of CoRDDi should avoid short-termism and recognise the need to continue to evolve tools and techniques.

Although the development of a CoRDDi Framework can go someway to address the issues that are introduced above, the challenges highlighted are, in most cases, cross institutional and are common to many. Given that these issues are often generic it is arguable that they should be effectively tackled at the highest level, by LWEC. The on-going development of the LWEC strategies provides an ideal opportunity to address these and provide clarity to the research community. Without this clarity, collaborative research will remain an unobtainable aspiration. Recent years have seen the move across government to develop a co-ordinated approach to science.

Annual Score Card

The success of the CoRDDi framework will be formally reviewed as part of its on-going management. Attributing societal and environmental benefits directly to investment within CoRDDi will be difficult, as genuine outcomes are likely to occur significantly 'downstream' from the projects. Monitoring benefits will form the basis by which further research funding within CoRDDi will be justified. An Annual Score Card has therefore been developed to form part of the annual review of the CoRDDi programme and the performance of the projects within it.

As each funder has a different motivation, and hence use different criteria to judge the success or otherwise of research and development projects and programmes, a range of criteria (of interest across the broad spectrum of potential funders) has been developed. These success criteria provide important targets for the researchers and therefore have been developed to be facilitators of integration between funders, researchers and practitioners.

Conclusions

This paper details a Vision and Framework for **C**oastal **R**esearch, **D**evelopment and **D**issemination (CoRDDi) that outlines a portfolio of research priorities, including tools and techniques to support better delivery of Flood and Coastal Erosion Risk Management. The vision and framework are of relevance to all those with an interest in managing the coast. Four priority themes for action have been identified; 1. Understanding whole system behaviour, 2. Valuing impacts and promoting innovative funding, 3. Decision making and operational practice, and 4. Dissemination, education and training. The first two are focused on developing the knowledge base and the third builds on this base to assist the decision-making processes. An overarching strand of the Framework focuses on theme-wide dissemination (Theme 4). Within each of these themes, both the longer-term direction of travel and more immediate opportunities for 'quick wins' are outlined. A prioritisation process has been undertaken to distinguish between the 'essential' and 'desirable' needs, between needs of national significance and those of local value, and those needs which are required to influence practice now, and those which will be required at some stage in the future. A set of coherent and fundable projects each with high level objectives has been mapped against the high priority needs.

References

[i] Evans, E P, Ashley, R, Hall, J, Penning-Rowsell, E, Saul, A, Sayers, P, Thorne, C and Watkinson, A (2004). Foresight. Future Flooding. Scientific Summary: Volume I - Future risks and their drivers. DTI/pub 7183/2k/04/04/NP. URN 04/939, Office of Science and Technology, DTI, 1 Victoria Street, London, 366p.

[ii] Evans, E.P., Simm, J.D., Thorne, C.R., Arnell, N.W., Ashley, R.M., Hess, T.M., Lane, S.N., Morris, J., Nicholls, R.J., Penning-Rowsell, E.C., Reynard, N.S., Saul, A.J., Tapsell, S.M., Watkinson, A.R., Wheater, H.S. (2008) An update of the Foresight Future Flooding 2004 qualitative risk analysis. Cabinet Office, London, 159p. Available to download from the Pitt Review website or via:
http://webarchive.nationalarchives.gov.uk/20100807034701/http:/archive.cabinetoffice.gov.uk/pittreview/thepittreview.html

[iii] Pitt, M., 2007. Learning lessons from the 2007 floods: an independent review by Sir Michael Pitt: interim report, London, UK.

[iv] OECD (2010). OECD Science, technology and Industry Outlook. OECD Publishing. ISBN 978-92-08467 (print) or downloadable at http://dx.doi.org/10.1787//sti_outlook-2010-en

[v] Department for Business Industry and Skills, (2008). Innovation Nation. White Paper.

[vi] http://evidence.environment-agency.gov.uk/FCERM/en/Default/HomeAndLeisure/Floods/WhatWereDoing/IntoTheFuture/ScienceProgramme/ResearchAndDevelopment/FCRM.aspx

[vii] http://www.lwec.org.uk/

[viii] http://www.lwec.org.uk/newsletters/2011/01

[ix] http://www.lwec.org.uk/strategic-activities/uk-first-flood-research-strategy

[x] Sayers PB and Meadowcroft IC (2005). RASP - A hierarchy of risk-based methods and their application. Proceedings of the 40th Defra Conf. of River and Coastal Management.

Innovative Coastal Zone Management
ISBN 978-0-7277-5749-4

ICE Publishing: All rights reserved
doi: 10.1680/iczm.57494.296

New century, New Management Approaches – is it Time for Consolidated Legislation for the Coast?

Jennifer Hines, Independent
Jim Hutchison, Independent
Steve Thompsett, Jacobs, London, UK
Jonathan Potts, University of Portsmouth, Portsmouth, UK

Summary

The existing matrix of responsibilities and current legislative structure on the coast of England is, arguably, unnecessarily complex. This paper proposes a consolidated framework for improving coastal risk management which builds on and strengthens existing arrangements under the Environment Agency's strategic overview role for the coast in England. This paper sets out the fundamental components of the current management system which need to be modified and improved to enable more effective future management if we are to truly reduce risk and improve the environment. The focus of these recommendations is based on the need to update, consolidate, and modernise coastal risk management institutional frameworks, making them fit for purpose in a new century and meeting modern day requirements to strengthen management approaches.

Introduction

In the past, it may be argued, that there has been a complex and confusing matrix of responsibilities on the coast. The statutory and organisational framework for the provision of shoreline management is the outcome of a long history of trying to address problems associated with sea flooding and coastal erosion, which has been based on a clear distinction between protection against erosion and defence against flooding (Potts, 1999; Carter, Taussik, Bray, & Hooke, 2000). In 2008, the Department for Environment Food and Rural Affairs (Defra) gave the Environment Agency a 'strategic overview' role for all sea flooding and coastal erosion risk in England with a view to rationalising and clarifying the roles of the key operating authorities on the coast – the Maritime Local Authorities and the Environment Agency.

Since this change in organisational arrangements there has been no formal review of its benefits and shortcomings, and whether it has the potential to be further clarified or improved. In recent years there has also been little research carried out on the statutory arrangements for shoreline management or a reassessment of the institutional frameworks within the context of the present day. This paper, based on a comprehensive literature review and in-depth interviews with expert practitioners at the forefront of delivering the strategic overview approach, begins to address this gap.

For the purposes of this paper, institutional frameworks are defined as the formal and informal rules and supporting organisational structures which are in place to manage the natural environment (Ostrom, 1990). Coastal risk management or shoreline management, as opposed to wider coastal zone management, in this context refers specifically to the statutory provision and management of coastal defence (against flooding) and protection (against erosion) and the associated organisational framework within which it operates.

Historical context

The construction of coastal defence structures has been taking place for many years. The first major developments took place during the Roman period through to the 17th century and were built to protect land claim from inundation by the sea (Charlier, Chaineux, & Morcos, 2005). During the 19th century it became fashionable to visit and take holidays at the coast, which led to the onset of rapid development of coastal areas for tourism (Goodhead & Johnson, 1996). The Victorians were largely responsible for the large scale development of sea walls along the coast of the UK (French, 2004). Built with little regard for, and a poor understanding of coastal processes, they fixed the coastline in place in many locations, preventing it from adapting to changes in natural conditions and often causing increased problems for adjacent coastlines (Fleming, 1992). Solutions to flooding and erosion risks were often reactive and perceived as local problems which required local solutions to resolve, neglecting the larger, regional scale impacts of their existence (Fleming, 1992).

Evolution of statutory arrangements

One of the main reasons for the fragmented organisation of protection and planning and the ad hoc approach to coastal risk management has been the national organisational and institutional structure which governed its operation and management (Potts, Carter and Taussik, 2005). This was and still is based on a distinction between protection against erosion and defence against flooding.

For administrative purposes the coastline of England is generally divided into two sectors, one which provides protection from erosion to land lying above mean high water, and one which defends low lying land, below mean high water, against inundation by the sea. Each sector has its own governing legislation with different Acts of Parliament dealing with flood and erosion risk. The Land Drainage Act 1991 and the Water Resources Act 1994 contain the primary legislation relating to sea defence, while the Coast Protection Act 1949 deals with coast protection against erosion.

Due to a historical development whereby the protection of low-lying coasts from sea flooding was regulated by District Drainage Boards, the Coast Protection Act specifically excluded sea defence from its jurisdiction and as such the distinction between coast protection (erosion) and sea defence (flooding) became entrenched within national legislation (Ricketts, 1986), and has remained so ever since.

The organisational divide

Alongside this divided legislation was a divide within the government departments overseeing activities. Sea defence came under the remit of the Ministry of Agriculture Fisheries and Food (MAFF), while coast protection was the responsibility of the Department of the Environment (DoE). This division continued until 1985 when MAFF, without any changes in legislation, took over responsibility for both flooding and erosion, a role later assumed by Defra in 2001.

This divide was further embedded at the operational level with the split between Local Authorities and Environment Agency functions, and further compounded by the existence of over one hundred maritime Local Authorities and other agencies with statutory responsibilities for providing shoreline management, each with their own political and organisational boundaries, local priorities and drivers. There were, and still are, additional cross border complications with different legal systems and responsible statutory bodies in England, Wales and Scotland, resulting in no one single management structure for the whole of the UK. The system has therefore been characterised by fragmentation, devolution, and the existence of political and administrative boundaries and subsequent inconsistencies, which has prevented coordination and management of the coast within natural process units such as littoral sediment cells.

Operational issues

At the operational level it is often not possible to draw a clear distinction between sea defence and coast protection, with schemes often having a dual purpose and function. The artificial division in statutory legislation therefore creates unnecessary confusion and exposes 'grey areas'. In 2002 the House of Commons Environment Committee concluded that the legislation was too complex and prevented the development of a comprehensive and integrated management approach. Some researchers also argued that the existing legislation constrained shoreline management responses into those biased towards engineering, creating an unsustainable position to maintain with regard to rising sea levels (Pethick, 2001).

Development of a non-statutory approach

It is clear that these institutional frameworks did not provide for an integrated approach to coastal risk management. The failings of this statutory system and the inflexibility of administrative boundaries led to the development of non-statutory measures for coastal defence and protection in the form of regional Coastal Groups and Shoreline Management Plans (SMPs).

Coastal Groups formed at the ground level during the 1980s and 1990s in reaction to the piecemeal, site specific approach to management, and in doing so brought together those responsible for coast protection and sea defence, coordinating activities and promoting a partnership approach to management. In 2008, as part of changes to the organisational framework (see below), the Coastal Groups reformed and became fewer in number and more strategic in operation. This decision was made on an understanding that some previous boundaries were based on a mix of political, administrative and coastal processes boundaries, which overlapped in some cases, and that the new boundaries would be based solely on coastal processes as originally intended (Defra, 2008).

Shoreline Management Plans (SMPs) developed alongside Coastal Groups largely for the same reasons). SMPs form a large-scale assessment of the risks associated with coastal processes and help manage risks in order to reduce the threat to people and the environment. The development of these non-statutory plans was strongly encouraged (and in the main funded) by central Government, being seen to provide a strategic and integrated framework for sustainable decision making and the management of coastal defences.

Current institutional frameworks

Despite the non-statutory efforts to integrate the institutional arrangements for shoreline management there remains in place a complicated legal and organisational system. The continued divide in legislation has been found to aggravate and hinder, rather than support the administrative and operational functions and capabilities of those concerned with defence and protection (Ballinger, *et al*, 2000). In recent years attempts have been made to clarify responsibilities and to simplify the organisational framework, but no similar action has been taken to streamline the legal framework.

Organisational arrangements

In 2005 Making Space for Water, a Defra-led, cross-Government strategy for flood and coastal erosion risk management, comprehensively reviewed and replaced the MAFF Strategy for Flood and Coastal Defence, published in 1993. Making Space for Water made a commitment to adopt a whole catchment and whole shoreline approach to management, thereby attempting to move away from the sectoral approach of the past.

To facilitate the more holistic approach set out in Making Space for Water, the Government decided it would give the Environment Agency an "overarching strategic overview across all flooding and coastal erosion risks" (Defra, 2005). Defra identified that without one single authority having responsibility and a view over all flooding and erosion risks, the confusing roles and responsibilities of the past would remain. This was the first time the need to facilitate a more joined-up approach at the operational, as well as the government level, had been identified by central Government. Ahead of any new legislation, the Environment Agency accepted the coastal strategic overview role in 2008.

Current organisational framework

The following diagram summarises the current position with regard to the organisational arrangements for shoreline management:

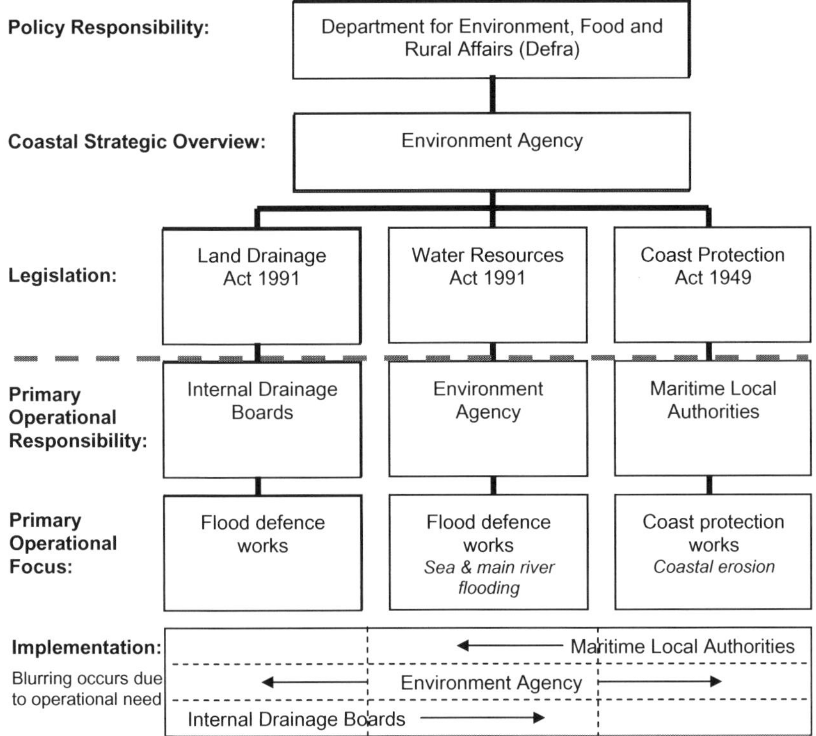

Figure 1. Current framework for the management of coastal defence and protection

This diagram highlights the new overarching role of the Environment Agency, but also the continuing divide and confusion between activities at the operational level. Delivery of coastal defence and protection is often blurred due to operational needs and is not as straightforward as it seems. In reality the strategic overview role remains constrained in its success at the ground level due to the existing legislative framework. The diagram also only covers the operating authorities which have a specific remit to provide defence and protection under legislation. It does not include other statutory consultees such as Natural England and The Crown Estate, and private landowners such as the National Trust and Port and Harbour Authorities, who have interest in managing the coast for their own purposes, but provide coastal defence or protection by default. For the purposes of this paper, consideration of the roles and responsibilities of Internal Drainage Boards have been excluded as they do not generally play a large active role in managing the coast.

Statutory arrangements

The key Acts of Parliament specifically in place for the management of coastal defence and protection activities have already been noted. In addition to these there are a number of Acts and European Directives which both allow for and restrict coastal defence and protection activities in England, for example the Harbours Act 1964, Town and Country Planning Act 1990, Marine and Coastal Access Act 2009, and the Birds and Habitats Directives. The legal framework for shoreline management is neither simple or straightforward. Since the UK became a member of the European Union it has a duty to implement the recommendations from the numerous Directives which relate to the coastal environment. Whilst many of these aim to protect, enhance and conserve the environment, they add a layer of bureaucracy which did not exist in the past.

The Flood and Water Management Act 2010 is the latest piece of legislation to be enacted which affects the way flood and coastal erosion risk is managed in England and Wales. As far as the coast is concerned, this Act legalises the Environment Agency's overview role, amends the Coast Protection Act 1949 to give the Environment Agency the same powers as maritime Local Authorities in relation to coast protection, and extends the remit of the now Regional Flood and Coastal Committees to cover coastal erosion.

The impetus for this new legislation was as a direct result of the severe fluvial and pluvial flooding experienced by much of the country in 2007, and the subsequent Pitt Review, which concluded that there should be a single unifying Act to address flood risk and to clarify responsibilities for flood risk management (Pitt, 2008). Development of this Act provided an opportunity to unite all flood and coastal risk management legislation. However, segregation of sea flooding and erosion risk still remains firmly in place as greater emphasis was placed on joining-up all sources of 'flood risk' and not on joining-up all 'flood and erosion' risk. Despite this, the Flood and Water Management Act 2010 provides the first, albeit small, statutory step forward in attempting to break down the historical division between defence and protection.

Challenges and opportunities

In-depth interviews were held with key individuals representing the views of the front line operating authorities involved in the delivery of shoreline management – the Chairs of the regional Coastal Groups (largely representing maritime Local Authorities), and the Regional Flood and Coastal Risk Managers within the Environment Agency, providing an operational leadership role. Analysis of these interviews lead to the identification of key successes and failings of the current institutional arrangements and of the opportunities for improvement.

Comparing historical approaches to defence and protection measures and those that presently exist, there can be no doubt that the control and management of risks from flooding and erosion has greatly improved. From an organisational perspective, the overview role and the ability and responsibility of the Environment Agency to take a national view over flooding and erosion risks has lead to a more holistic and sustainable management system. However, there are a number of potential areas for improvement.

Organisational arrangements

There is a clear understanding, need and desire for a national, strategic overview role within the organisational framework for shoreline management, and for this role to reside with the Environment Agency. Greater clarity on roles, responsibilities and accountability remains a key area of concern and one which needs to be addressed. However, this clarity is unlikely to be resolved while there remains a division in the legal statutes for sea defence and coast protection, and a subsequent division in the activities undertaken at the operational level by the different operating authorities involved.

Critical partnerships

The success of the strategic overview is largely dependent on partnership working and the continued existence of the (voluntary) Coastal Groups. Whilst amendments to the Coast Protection Act 1949, as set out in the Flood and Water Management Act 2010, provides the Environment Agency with the same legal powers as Local Authorities, operational activities and accountabilities are still divided between the two and therefore successful and integrated shoreline management still remains based on these effective partnerships. It is questionable whether successful and sustainable coastal risk management should be dependent on voluntary partnerships and good collaboration between the individuals concerned, especially against a background of funding cuts, tighter spending restrictions, and increasing environmental challenges exacerbated by climate change.

Uniting the problem, uniting the management

The lack of clarity, tensions between operating authorities not understanding one another, and heavy reliance on informal trust and partnership working, strengthens arguments for one organisation to have sole responsibility and accountability for administering, managing and coordinating shoreline management activities. It remains questionable what benefit is gained from having this continued divide at the operational level, as long as within the one organisation the national overview and local delivery elements were clearly defined and utilised. The Environment Agency, as an existing national organisation with local delivery capabilities and responsibilities, and through extending its overview role would seem a sensible choice. A focus on local delivery and partnership working will be key, particularly with respect to future funding under Defra's proposed Payment for Outcomes approach. However, whether localism can truly be feasible and effective in terms of coastal risk management remains an area for further debate.

Whatever system is implemented in the future it is clear that the technical skills, experience, and local knowledge from both operating authorities is required. If the Environment Agency and Local Authorities are both able to undertake sea defence and coast protection measures, decisions about who is best placed to carry out shoreline management activities could be made locally, based on local skills and resource capabilities. This approach would bring out the strength of those Local Authorities who are more capable than others and who have better knowledge and skills on coastal erosion issues than the Environment Agency. It would also resolve issues where other Local Authorities are under performing, lacking skills or resources

(Cole, 2009), or would prefer to remove the burden of responsibility. Under such an arrangement the role of the Coastal Group would remain, and would continue to provide the technical expertise and a forum for coordinating activities and bringing together the different parties involved, especially those outside of the operating authorities, as they do at present.

Statutory arrangements

The current volume of legislation relevant to coastal risk management is, arguably, too complex. It causes significant confusion for the public through the resulting divided activity at the operational level. Arguably this confusion has been exacerbated by the Environment Agency now having the same powers as Local Authorities for coast protection, as a result of the Flood and Water Management Act 2010. The existing legislation has been enacted over a long period of time, a time in which views, opinions and processes have changed, but one which has not seen any attempts to amalgamate or consolidate the legislation. The Acts themselves were well conceived and they have worked, providing the necessary powers for a considerable time. However, there are improvements which could be gained by creating one Act and improving clarity for all concerned.

Limited powers to act

One of the most noted problems with the legislation in place is that the Coast Protection Act 1949 only provides legal powers to *defend* a coastline, and not to *manage* it in a more sustainable manner. In 1949 nobody had thought of doing anything other than building defences to protect coastal areas. The Acts do not recognise or support adaptation and do not provide the scope for managing the consequences of such actions.

In England, flood and coastal defence legislation is in the main permissive and does not contain any right or statutory duty to provide protection from flooding or erosion. The provision of defences has been based on legal powers to act rather than duties to do so. This permissive rather than mandatory approach to protection has received considerable attack, especially from a social justice and human rights perspective (Howarth, 2002; Cooper & McKenna, 2008). Permissive powers are also considered not strong enough by those who have to use these Acts, and additionally cause unnecessary ambiguity. It is likely that some form of legal mechanism or duty is going to be required to manage coastal risks consistently in the future as we progressively face more significant problems on the coast that will not be resolved by cooperative partnerships.

Technical confusion

The current divide in legislation can cause confusion for experienced practitioners, and not just the public. The Schedule Four boundary, which defines the extent of the Coast Protection Act and notes those inland waters which are excluded, prevents an unnecessary management problem with regard to estuaries, which can be considered a no-mans land between the open coast and main river. These boundaries prevent estuaries being managed as part of the 'coast' and are often the subject of legislative and management confusion.

In the future the dividing line between sea flooding and erosion risk is going to be less clear cut, further adding an element of confusion. In areas where mean high water sea level is close to current ground level, it can be questionable whether the risk should be classed as erosion risk or flood risk as far as the Acts are concerned. If sea levels continue to rise, areas of low lying land could find themselves changing from erosion risk to areas of sea flooding risk. This problem in identifying risk, and for risk sources to change, becomes a greater issue under the current divided arrangements.

Uniting the problem, uniting the statutory legislation

The most comprehensive way to resolve the issues of clarity, divided responsibility and the need to manage rather than just defend the coastline leads to the recommendation to create one comprehensive Act for coastal risk management allowing both coast protection and defence, but facilitating coastal change. If this Act was further to be administered by one organisation, clarity for both the public and professionals would prevail, and would enable the strategic overview role to fully achieve all of its principle objectives.

A new Act could also seek to resolve further issues identified such as the strength and appropriateness of the current permissive powers, the inclusion of tidal estuaries in shoreline management, and problems in identifying whether risk is from flooding or erosion, and whether that risk then changes with the onset of climate change and sea level rise. It could also enable the existing legislation to be updated and consolidated, and to strengthen legal support for those organisations trying to implement adaptation measures. It would also enable a proper review of the ability and strength of the regulatory powers with regard to third party defences.

Time for new legislation for the coast

The Flood and Water Management Act 2010 was developed to integrate and amalgamate legislation relevant for flood risk management. Prior to that the Marine and Coastal Access Act 2009 conducted a similar task for the marine environment. Whilst some elements of shoreline management legislation were considered in both these Acts it could perhaps be considered a missed opportunity that an overhaul, or at least a full review of the legislation relevant to the coast, was not achieved on either of these occasions. It is considered timely to undertake such an activity.

Arguably the existing legislation for the coast has been built upon historical Acts, or developed as a secondary concern after fluvial flood risk, and the political drivers following significant flood events. On a national scale, flood risk is perhaps more apparent, with the consequences more widespread. Coastal erosion impacts are often more localised, but for those involved the risks are the same, if not greater, and once future erosion risk starts to seriously affect large towns and infrastructure, it will become a national scale issue, pushing coastal adaptation at the very top of the political agenda. From a political and legal perspective, coastal risk management needs a robust statutory footing, brought up to date and simplified for a modern century.

A new institutional framework for coastal risk management

From the conclusions drawn, the following diagram presents a proposed new structure for the organisational and statutory arrangements for coastal risk management in England. This diagram is based on current arrangements but provides a significantly more streamlined and simplified framework:

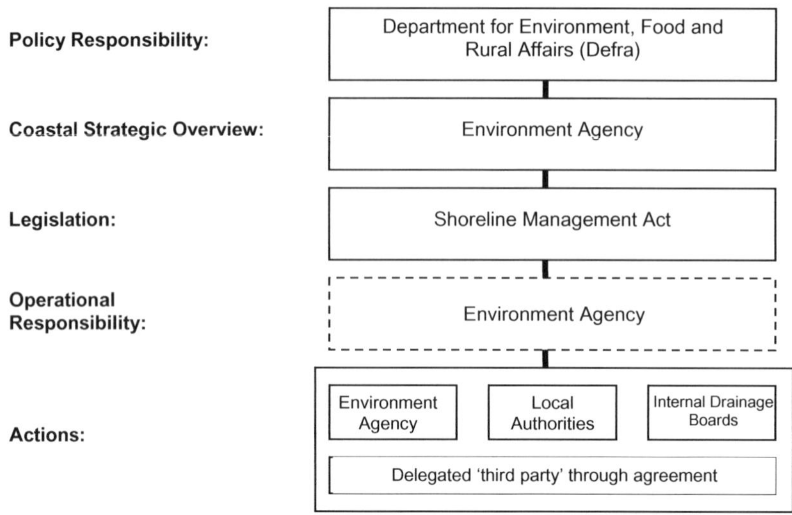

Figure 2. Proposed framework for future management of coastal risk

Under these proposed arrangements, policy responsibility and the role of the strategic overview would remain as it is at present. The current Acts covering shoreline management activities would be updated and amalgamated into one unifying piece of legislation, following a complete review of all existing legislation. It could embrace wider coastal management activities and implications, such as planning and development control, and integrated coastal zone management, but most importantly it could allow the coast to be managed more holistically rather than 'defended' from a legal perspective. There would be one clear organisation with operational responsibility for decision making and coordination of activities, with on the ground actions failing to the operating authority, or third party through agreement, that is locally best placed to carry out shoreline management.

The extent to which localism manifests itself in shoreline management activities will significantly affect political support for an approach which is arguably decreasing direct levels of management and control from local government, and placing it with a national scale organisation. At the time of writing, the future under a localism regime remains unclear.

Whilst the development of a Shoreline Management Act is perhaps not a panacea for resolving all issues, it would certainly be a significant step forward in achieving a fully integrated and clarified position with regard to shoreline management activities and responsibilities. It remains debatable whether the merging of operational organisational activities would yield significant benefits, but what is clear is the need to unite and join-up the existing legislation for coastal risk management. Erosion and flooding are part of the same natural physical system, and as such should be dealt with as one.

Concluding remarks

Over the last 20 years there has been a marked difference in management approach from the historical site specific and ad-hoc of the past, through to a more regional approach based on natural processes with the advent of non-statutory approaches, and finally to a more national and holistic approach with the intervention of the Environment Agency's strategic overview role. Whilst the current institutional framework is working to some degree, there are areas for considerable improvement. It is recommended that a full review of current legislation is undertaken, leading to an updated and amalgamated set of arrangements to create a new Shoreline Management Act, which reflects modern challenges faced at the coast. With restrictions on future spend and the need to consider alternative and adaptive strategies, it is timely that the coast is given greater attention and a new, updated and consolidated Act for coastal risk management is developed, which enables the coast to be *managed* in an integrated way. Since the enactment of the Coast Protection Act 1949, approaches to coastal risk management have advanced significantly, the legal powers to support such activities have not.

References

Ballinger, R. C., Potts, J. S., Bradly, N. J., & Pettit, S. J. (2000). A comparison between coastal hazard planning in New Zealand and the evolving approach in England and Wales. *Ocean & Coastal Management, 43*, 905-925.

Carter, D., Taussik, J., Bray, M., & Hooke, J. (2000). Regional coastal groups in England and Wales: the way ahead. *Periodicum Biologorum, 102(1)*, 215-220.

Charlier, R. H., Chaineux, M. C. P., & Morcos, S. (2005). Panorama of the history of coastal protection. *Journal of Coastal Research, 21*(1), 79-111.

Cole, K. (2009). *Review of Local Authority skills and capacity for coastal risk management functions – a 3-year review.* Coast and Country Projects Limited.

Cooper, J. A. G., & McKenna, J. (2008b). Social justice in coastal erosion management: the temporal and spatial dimensions. *Geoforum, 39*, 294-306.

Department for Environment, Food and Rural Affairs. (2005). *Making space for water: taking forward a new government strategy for flood and coastal erosion risk management in England – first government response to the autumn 2004 making space for water consultation exercise.* London: Defra. Publication PB10516.

Department for Environment, Food and Rural Affairs. (2008). *Coastal groups in England – the Environment Agency strategic overview of sea flooding and coastal erosion risk management.* London. Defra.

Fleming, C. A. (1992). The development of coastal engineering. In M. G. Barrett (Ed.), *Coastal planning and management.* London: Institute of Civil Engineers.

French, P. W. (2004). The changing nature of, and approaches to, UK coastal management at the start of the twenty-first century. *The Geographical Journal, 170*(2), 116-125.

Goodhead, T., & Johnson, D. (1996). *Coastal recreation and management: the sustainable development of maritime leisure.* London: E. and F. N. Spon Publishing.

Howarth, W. (2002). *Flood defence law.* Crayford: Shaw & Sons.

Ostrom, E. (1990). *Governing the commons: the evolution of institutions for collective action.* Cambridge: Cambridge University Press.

Pethick, J. (2001). Coastal management and sea-level rise. *Catena, 42*(2-4), 307-322.

Pitt, M. (2008). *Learning lessons from the 2007 floods.* London. Cabinet Office.

Potts, J.S. (1999) The Non-Statutory Approach to Coastal Defence in England and Wales: Coastal Defence Groups and Shoreline Management Plans. *Marine Policy*, Vol. 23, Nos. 4/5. pp. 479 - 501.

Potts, J.S., Carter, D. and Taussik, J. (2005) Shoreline management: the way ahead. In Smith, H.D. and Potts, J.S (eds.) (2005) *Managing Britain's Marine and Coastal Environment – Towards a sustainable future.* London. Routledge: Advances in Maritime Research Series (Taylor and Francis Group). pp. 239 – 272.

Ricketts, P. J. (1986). National policy and management responses to the hazard of coastal erosion in Britain and the United States. *Applied Geography, 6*(3), 197-221.

Innovative Coastal Zone Management
ISBN 978-0-7277-5749-4

ICE Publishing: All rights reserved
doi: 10.1680/iczm.57494.306

Coastal Policy, Legislation, Targets and the future in Sri Lanka.

V.Mohan, B.Sc (Eng), DIPPCA (SL), M.Sc (PPM for SD, Turin, Italy), Project Director, RDA. Ministry of Ports and Highways, Sri Lanka.

Introduction

Sri Lanka, which is Formerly known as "Ceylon", is in the Indian Ocean separated from the southeastern coast of peninsular India by the Palk Strait and is an island with coastline areas, which are generally low-lying, and landscapes exhibit considerable variety characterized by bays, lagoons, headlands, coastal marshes, peninsulas, spits, bars, and islets and dunes, of 1,585 kilometers and a land area of 64,000 square kilometers (probably more than 2,000 km if the coastlines of lagoons, bays, and inlets are added), situated between the latitudes of 5°55' and 9°51' North and the longitudes of 79°41' and 81°54' East within the tropic of Cancer lying off the southern tip of India. It encompasses a variety of tropical habitats including wetlands (about 120,000 ha); lagoons and estuaries (45 estuaries and 40 lagoons totaling about 42,000 ha); mangroves, salt marshes and sea grass beds (the total extent of mangrove coverage is between 6,000 and 10,000 ha); coral reefs (about 50 linear km of major reefs); and coastal sand dunes, barrier beaches, and spits (sand dunes occur along about 312 km of the coastline).

History

In 1948 Sri Lanka gained its independence after nearly 400 years of continuous colonial rule. After independence, more and more people began to migrate to coastal areas to take advantage of economic and educational opportunities. Today more than half (54.3%) of Sri Lanka's population of 20 million live in coastal districts. Increasing population has led to greater housing densities in coastal areas. The southwestern coastal districts stretching from just north of Colombo to Galle constitute 15% of the total land area of the island, but more than 40% of the island's total population lives in this area. Increasing population has led to greater housing densities in coastal areas; so that temporary shelters are frequently constructed in.

As a result the development of physical, economic and social infrastructure as well as the resource use in the coastal zone has been rapidly increased causing severe environmental degradation especially during the last 25 years. Coastal uses including urban expansion, commercial and fisheries harbor development, river training and out fall schemes, transport and communications, recreational and tourism development, sand and coral mining have all had tremendous impact on the coastal environment. In 1992 tourism accounted for 8% of the total foreign exchange earnings, third after industrial and "major" (agricultural products). With population growth the pressure on the resource base of the coastal zone had also proportionately increased creating new stresses on the coastal environment. Today, plans for sustainable utilization of the island's coastal resources, whether for further settlement, agricultural development, fisheries and aquaculture promotion, tourism or other purposes, underline the need for a deeper awareness and appreciation of the character of this resource base, its potentialities, problems and evolutionary tendencies.

The question that launches this paper's analysis is whether Sri Lanka's   new coastal management strategy of **Special Area Management Planning (SAMP)** is progressive and engaging of local-level participation in planning from beginning to end as suggested by the project literature; or simply a revised discourse that continues to exclude community participation at crucial points in the planning process and therefore continues to exacerbate resource conflicts and ecological problems. These stresses have been further aggravated by the ever present threat of coastal erosion especially occurred in the south and south western part of the country. The policies and the procedures are being successfully implemented, and they have brought credibility and attracted funding for the effort.

Coastal Policy

Coastal management is one of the greatest social, economic, political and environmental challenges facing our generation. Being increasingly aware of global warming, climate change and environmental challenges facing the region, which mainly include sea-level rise, deforestation, soil erosion, siltation, droughts, storms, cyclones, floods, glacier melt and resultant glacial lake outburst floods and urban pollution. Much degradation - coastal erosion, pollution, and species loss-can be attributed to the national economic development strategies pursued since liberalization in 1978, particularly an emphasis on coastal industry and tourism development. The coastal management policy framework born in the late 1980s and early 1990s out of the joint US Agency for International Development (USAID)/Sri Lanka Coast Conservation Department (CCD), Coastal Resources Management Project (CRMP), engages a discourse of sustainable development, particularly sustainable eco-tourism development, that some critics suggest may be inherently unattainable . The idea of coastal management is both beguiling and elusive. Surely, we think, coast should be managed, just as we seek to manage some human activities to protect air and water quality, to promote human health, or to insure the sustainability of renewable resources such as forests and fisheries.

The Major Coastal Issues

The unsustainable manner of resource utilization, lack of planning and management initiatives created serious environmental problems in the coastal zone. These issues are discussed in the subsequent section, can be summarized as follows:

1. Coastal Erosion
2. Loss and degradation of coastal habitats such as coral reefs, mangroves, sand dunes, lagoons and marshes
3. Loss and degradation archaeological, cultural and scenic resources
4. Coastal Pollution

1. Coastal Erosion

Coastal erosion is a process of change that occurs at the land sea interface and is a severe problem in Sri Lanka that results in damage to or loss of houses, hotels and other coastal structures and roads, contributes to the loss of land and disrupts fishing, navigation, recreation and other activities. The impact of coastal erosion is most severe along Sri Lanka's western and south western coasts. It has been estimated that 685 kilometers of coast line in the south, south west and the west coast, about 175,000 - 285,000 square meters of coastal land are lost each year. The principal causes of erosion includes - Natural process due to monsoon generated wave attacks- Man-induced changes occur due to extraction of sand and corals from the coastal zone and improperly cited buildings and maritime structures. While coastal erosion is caused by natural process, human activities such as mining of beach and river sand, mining of corals and construction of ill-planned maritime structures are major factors contributing significantly to coastal erosion.

2. Loss and Degradation of coastal habitats

2.1 Sand Mining

The major source of material supply for the nourishment of beaches in Sri Lanka is from the network of rivers flowing to the sea. River sand is a prime material used in the building industry while beach sand is used for filling purposes. These resources have been traditionally considered as "free" resources and being utilized with no value added for the material itself. In the recent past the rapid growth of the housing and building industry led to increased amount of sand being removed from the rivers and beaches thus affecting the coastal stability and creating environmental problems. Major impacts could be as follows;

a) Coastal erosion
b) Salt water intrusion into the upstream areas and intakes
c) Increase flooding originating from the sea.
d) Water quality problems
e) River bank erosion

2.2 Coral Mining

Mining of coral to obtain Calcium Oxide for building and for other industrial and agricultural purposes. Until recent times it was done on a limited scale and was continued in areas where dead coral deposits in coastal wetlands were found. Within the past three decades, the use of lime has increased tremendously as a rapid growth rate of building industry. A study conducted in 1984 revealed that a quantity in excess 18,000 tons per year of coral is extracted in the coastal reach between Ambalandoga and Dickwella. Due to the controlling measures taken by the Coast Conservation Department this amount has been reduced by 48% just in 1994.

2.3 Construction of maritime structures

Construction of maritime structures such as sea walls, break waters, revetments and jetties has long been undertaken on an ad-hoc basis to deal with the problem of a particular beach stretch. The Master Plan for Coastal Erosion (1989) revealed that more than 49,000 meters of revetment and 6,360 meters of grayness have been constructed. Most of these structures were poorly designed and in some cases led to an increase of coastal erosion. Poorly designed fishery harbours have not only created coast erosion problems but also induced wave refraction and entry of sand bearing currents into the harbours causing siltation.

2.4 Loss and degradation of coastal habitats

The important coastal habitats of Sri Lanka are small and vulnerable to degradation. The areal extent of biologically productive mangrove systems, estuaries, coral reefs, and sea grasses is decreasing. In 1986, it was estimated that 12,000 ha of Mangroves, 23, 000 ha of salt marshes, 7,000 ha of Sand dunes, 158,000 ha of lagoons located in the coastal region. These habitats are being depleted in the recent past due to over exploitation. According to the current information, current rate of depletion will reduce mangrove habitat by up to 50%.

3. Loss and degradation of archaeological historic and scenic sites

Sri Lanka's coastal zone contains many and diverse sites of archaeological, historical and scenic significance. These sites provide valuable evidence of the pattern and the progress of Sri Lanka culture and represent part of Sri Lanka's common heritage. The inventory of places of religious and cultural significance and areas of scenic and recreational value within the coastal zone has been identified 91 sites as high priority for conservation. Today, many of these important sites are threatened with inappropriate development. In some locations, important scenic areas are being degraded and public access to the beach obstructed.

4. Coastal Pollution

Coastal pollution is a direct result of population pressures and misuse of land and water resources. The major sources of coastal water pollution along the south west and south coasts are domestic sewage, industrial waste, solid waste and agricultural chemicals. Growing urban population densities coupled with inadequate housing and lack of water and sewage disposal facilities, has led to fecal contamination of surface and ground water. As a result of contamination of ground water and surface water in the coastal zone major public health problems such as typhoid and hepatitis can be found. About two thirds of Sri Lanka's industrial plants are located in coastal regions, primarily in the Greater Colombo metropolitan region. The main industries contributing to water pollution are textile, paper, tanning, distilleries, paints and chemical production. A study conducted for the Central Environmental Authority (1994) identified 336 industrial facilities in coastal areas as "medium" or "high" pollution potential.

Legislations

The Sri Lanka enacted Coast Conservation Act No 57 in September' 1981 and law went to effect in October 1983 and its amendments, in 1988 provided the legal foundation for activities within the coastal zone, which comprises mainly "the area lying within a limit of three hundred meters landwards of the Mean High Water line and a limit of two kilometers seawards" of the Mean Low Water line. The Act 1981 shifted the emphasis from coast protection to coastal zone management (CZM). The Act decreed the appointment of a **Director of Coast Conservation** with the following responsibilities:

1. Administration and implementation of the provisions of the Act.
2. Formulation and execution of schemes of work for coast conservation within the coastal zone.
3. Conduct of research, in collaboration with other departments, agencies and institutions for the purpose of coast conservation.

The Act states that the Minister in charge of the subject of Coast Conservation "may, having regard to the effect of those development activities on the long term stability, productivity and environmental quality of the Coastal Zone, prescribe the categories of development activity, which may be engaged in within the Coastal Zone without a permit". Such activity should not however include any development activity already prescribed under the CCA. The CCA does not however specify how and when this discretion should be exercised. The CCD interprets this provision as requiring an EIA when the impacts of the project are likely to be significant. In the years since the law was enacted the CCD has conducted a significant amount of research and has prepared a Master Plan for Coast Erosion Management (MPCEM) and a Coastal Zone Management Plan (CZMP). It has also issued 764 permits for development activities, organized seminars and workshops on several aspects of coastal management, and developed effective relationships with several agencies which have management responsibilities in coastal areas.

The Act requires the Director of Coast Conservation, on receiving an EIA Report, to make it available for public inspection and to entertain comments on it. The Act also requires the Director of Coast Conservation to refer the EIA report to the Coast Conservation Advisory Council for comment. The Council is an inter-department, inter-disciplinary advisory body. The Director of Coast Conservation may decide to.

1. Grant approval for the implementation of the proposed project subject to specified conditions, or
2. Refuse approval for the implementation of the project, giving reasons for doing so.

The EIA process is part of the permit procedure mandated in Part II of the CCA for the approval of prescribed development projects and undertakings within the Coastal Zone. The Act has assigned the coast Conservation Department three primary following responsibilities within the designated coastal zone

a) Policy formulation, planning and research
b) Administration of permit procedures regulating coastal development activities
c) Construction and maintenance of shoreline protection works

Coastal Management Targets

As mandated by the Act, The Coastal Conservation Department developed a **Coastal Zone Management Plan (CZMP)** based on a number of specific studies set forth in the Act. The Act mandates the establishment of an advisory council to assist the CCD in the process of plan preparation. In the interim period while the plan is being prepared, the Act stipulates that anyone proposing a development activity in the designated coastal zone must apply to the CCD for a permit.

The landward jurisdiction is somewhat greater for rivers, streams, lagoons, or other bodies of water connected to the sea. Sri Lanka has a strong and vigorous coastal management program. A detailed examination of Sri Lanka's program suggests that its strength and vigor are due in large part to:

1. **The strong coastal orientation of the country;**
2. **The widely shared agreement about what the coastal problems are, what the causes of the problems are, and to a lesser extent, what the appropriate roles of government are in dealing with the problems;**
3. **A law that provides a strong legal basis for management;**
4. **Strong program leadership;**
5. **Adequate political support for planning and management; and**
6. **An adaptive, incremental approach to the development of the planning and management program.**

CZMP set out in 2004, a comprehensive list of interactions required for the management of the coastal zone in an integrated, holistic manner for the conservation and prudent use of its resources while supporting sustainable development. The CZMP highlights the fact that the future approaches for coastal habitat management should be geographically specific and based on clearly understood links between human activities and changes in natural systems. It further states that care has to be taken to ensure that all policies and actions for conservation of coastal habitats comply with the National Physical Development Plan (NPDP), the National Environmental Action Plan (NEAP), the National Biodiversity Conservation Action Plan (NBCAP), and other national planning initiatives. The CZMP addresses key issues related to managing coastal erosion, conserving coastal habitats, controlling coastal water pollution, Special Area Management (SAM), managing sites of special significance and public access, regulatory mechanisms, and integrating coastal fisheries aquaculture with coastal zone management. What emerges from analyses of existing coastal area programs is the recognition that the concept of coastal area management is somewhat elusive. There is no widely accepted blueprint for how to plan a management program. There are no "off-the-shelf" management program models that can be easily adapted and applied. Each country (or other coastal jurisdiction) must carefully tailor its own program to include;

1. An identification of specific coastal problems to be addressed;
2. An identification of priorities among these problems;
3. An analysis of specific processes which cause these problems;
4. An identification of specific management techniques (such as zoning or a permit system) designed to mitigate these problems;
5. A set of organizational arrangements and administrative processes for implementing a management program; and
6. The designation of a geographic area within which management will occur.

As coastal area management program efforts proliferate around the world, these program elements are being addressed in a variety of ways. Among developing countries, however, **few have gone as far as Sri Lanka in developing a coastal management program.**

The coastal zone and management initiatives taken place in the recent past did not cover every e planning coastal problem but focused upon a few well defined issues at a time and to build the program incremental as experience is gained. The policies and the procedures are being successfully implemented, and they have brought credibility and attracted funding for the effort. The first three major coastal issues mentioned above have been tackled during the first generation coastal zone management effort while coastal pollution problem being focused in the second generation effort. Thus to manage the first three major coastal issues two strategies have been adopted by the Coast Conservation Department;

1. Preparation of a National Coastal Zone Management Plan
2. Preparation of a Master Plan for Coastal Erosion
3. Coastal Zone Management Plan (1990)

The Coast Conservation Act also mandated the preparation and implementation of a Coastal Zone Management Plan. Thus the plan was prepared with technical assistance of the University of Rhode Island, USAID and has been implemented since 1990. The Coastal Zone Management Plan (CZMP) 1990 cannot be conceived as a panacea for all the problems and issues resulting from neglect and past mismanagement. The plan addresses only certain critical coastal issues with initial emphasis on problems causing significant economic and social losses which are more amenable to accepted management practices. In particular the plan has attempted to address problems related to coastal erosion, coastal habitats and the loss and degradation of historical and cultural sites and scenic and recreational areas. It has also dealt with the regulatory system including the legal, administrative and fiscal functions of the CCD, while paying some attention to research and education. It has been considered that the plan is an incremental one and coastal zone management planning is a continuing effort. Thus the CZMP 1990 should be considered as a first generation plan and the policies and management strategies outlined therein will have to be tested and their successes and failures evaluated over time. Implementing actions in the coastal Zone Management Plan are of several types: regulations, direct development, research, coordination, education and awareness, plan and policy development. According to the Coast Conservation Act it is mandatory to review the CZMP every four years. Thus the first generation CZMP is under revision now, and the past experience reveals the necessity of inclusion of two new chapters, namely;   on   Pollution   and   Special   Area   Management.

Coast Erosion Management Plan

The Coastal Erosion Management Plan is an integral part of the CZMP and it was prepared in 1986 with assistance given by Danish International Development Agency (DANIDA). It defines the problem of coastal erosion in Sri Lanka within the constraints of available information at that time, and sets out the best possible technical approach towards mitigation and capital investment needed for such action. The CEMP has categorized the coastal erosion problem under "key areas" and "singular cases". The key areas refer to a coastal stretch of several kilometers in length subject to erosion problem which predicts the need for complex solution in view of its morphological complexity. The Singular cases are those isolated problem cases covering limited stretches which would not interfere with the coastal processes to an extent that would require extensive studies prior to litigator actions. The Plan indicated project concepts for structural solutions in erosion areas and also identified investigation needs where there is insufficient data available. Once the Master Plan for Coastal Erosion Management was completed, DANIDA provided further assistance for two stages of coastal protection structures and beach nourishment. Stage 1 (1987 -1989), for the Negombo and Moratuwa coast protection schemes, cost an estimated Rs. 320 million ($US 6,660,000); and Stage 2 (1990 - 1992) for protection of the main road from Beruwela to Weligama, cost approximately Rs. 500 million ($ US 10,400,000). In 1994 the Coastal Erosion Master Plan was updated and presently further donor assistance is being sought for its implementation.

Future Coastal Management

The existing management tools which are being used presently to minimize major coastal issues could be categories as follows;

1. Regulations
2. Direct Development
3. Research
4. Education and Awareness
5. Plan and policy development

The regulatory program as administered by the coast conservation Department is essentially reactive to face the future problems. The program responds to proposals made by other governmental agencies and private developers for construction activities and alteration of the coastline. The CCD attempts to minimize the environmental and social impacts of development projects through its permit procedures, set-back standards, prohibitions and in some cases subjects the development proposals to the Environmental Impact Assessment requirements. It also ensures that environmentally and structurally appropriate decisions are made. A second major type of management tool is being used in the process of coastal zone management is direct development activities undertaken by government. With regard to coastal erosion management the major type of direct development is the construction of shore protection works. A third type of management tool is to identify areas of research and conduct research. Research is necessary because often good management is precluded due to some coastal problems being inadequately understood. During the recent past CCD initiated research on problems related to implementation of the CZM Plan. The CCD has sponsored research on the social and economic aspects of coral mining and sand mining, on how to improve permitting procedures, and how to better educate and involve coastal communities in coastal zone management. Thus the coast conservation Department has implemented vigorous public awareness and education program since 1992 to date. The German Technical Assistance Agency (GTZ) has also been provided assistance for strengthening the coastal Zone Management Program including the aspects of environmental education and awareness. Under plan and policy development, CCD is enabling to focus on Special Are Management as a subsequent planning effort to CZMP. Special Area Management (SAM) planning has emerged as a successful method of managing development in complex coastal settings. It is being tested in two coastal sites in Sri Lanka as an auxiliary coastal zone management tool. Based on the outcome of the present experience of SAM process, an additional twenty one sites have been identified by the coast Conservation Department for future implementation. Also, **1997 revision of the CZM plan sets the strategy for coastal zone management in to the next millennium.** The objectives of the revised plan are to:

1. Identify coastal problems that will be the primary focus of the Coast Conservation Department during the next four years.
2. Indicate why these problems are important.
3. Present the Coast Conservation Department's management
4. Identify what should be done by other Government and non-governmental organizations to reduce the scope and magnitude of coastal problems.
5. Identify research activities to immediate importance to the management of coastal resources.

The revised Coastal Zone Management Plan, like its predecessor of 1990, outlines the interventions to reduce coastal erosion, to minimize degradation of coastal habitats, and to minimize loss and degradation of archeological, cultural and scenic sites. A new focus is the reduction of coastal pollution. A second new focus is Special Area Management, including the recognition of the need for locally based collaborative management. Finally, the plan summarizes the objectives, policies, and actions to be implemented by the Coast Conservation Department and sets forth the priorities for action.

Conclusions

It is too early to declare the Sri Lanka coastal management program a success, but there are enough program and following achievements to make the program worthy of close attention; **1**. A strong permit system has been in place for 3 years and is functioning reasonably well, **2**.A plan has been produced that directs governmental and private development activity in coastal areas, **3**.A substantial investment plan for coastal erosion protection structures has been developed, **4**.Successful in shift the population to coastal areas, **5**.Coastal Area Management in Sri Lanka in the last 5 decades, particularly in the south and southwest, the awareness and attention directed toward coastal management by preventing coastal erosion account for much of the initial interest and the increasing threats to life and property, **6**. The formation of a single government unit to deal with coastal management and the vigorous, professional leadership the agency has received since its inception are important factors in accounting for what has been accomplished, **7**.The development of the MPCEP was also based on more than 10 years experience in balancing political demands with engineering principles and budget realities, **8.** Formal and informal coordination linkages have been developed with some agencies, **9.**Formal linkages are still maintained by means of permit referrals, workshops and meetings and **10.** Three factors, in particular, account for the successes of the incremental learning approach to date. The first is the competence and commitment of CCD's professional staff. CCD's highly energetic and motivated leadership and staff are typical of new agencies with a mission. Second, the credibility of the agency among the public and the political and bureaucratic elite makes it possible for CCD to engage in more experimentation in program development than other agencies might be allowed. CCD's strong record of professionalism in erosion management and its ability to attract international grants and loans to support its projects contributes to that credibility. Finally, the CCD's long record of being able to cope with crises and adjust to changing circumstances gives the staff confidence that they can meet the challenges.

Recommendations

For the successful integrated coastal management;

- The need for environmental awareness and education in support of Sri Lanka's Coastal Zone Management Program was clearly recognized in the Coastal Zone Management Plan.

- The implementation is depending on both understanding and support of the people of Sri Lanka.

- A common framework across coastal planning sectors is essential whereby administrative fragmentation is reduced while inter-departmental and inter-agency cooperation is increased.

- A collaborative effort on the part of several governmental agencies, nongovernmental organizations, and local communities is required and that the geographic area and issues addressed must be expanded.

- The adaptive, learning approach had several features. First, CCD staff chose to focus on a relatively small number of coastal problems rather than the full range of potential concerns. Second, they engaged in an explicit learning-by-doing approach to the implementation of the coastal permit system. They developed an explicit strategy for dealing cooperatively if possible with small landowners and hotel developers and, when cooperation was not possible, they identified minimum conditions that had to be met.

- Major planning events, such as the habitat workshop, help provide the context within which specific interagency agreements and understandings can be developed.

CCD has to cope with the following three very important primary tasks;

1. Phase of program development: the decentralization of the permit system; the development of special area management plans; and the development of specific interagency programs for habitat management.
2. Special area management plans are being considered for some natural habitats.

3. Finally, interagency programs for habitat management, in particular, are beginning to be developed. At present CCD issues permits for activities that affect habitats, but CCD officials hope to work with other agencies to develop a more coordinated effort to conserve mangroves, sea-grass beds, reefs, lagoons, and other habitats.

Reference

Government of Sri Lanka. 1981, Coast Conservation Act No.57

Kem Lowry, University of Hawaii, H.J.M.Wickremaratne Lanka Hydraulic Institute and Coast Conservation Department, Sri lanka "Coastal Area management in Sri lanka"

White, A. (Ed) 1993, Are Coastal Zone Management and Economic Development Complementary in Sri Lanka.

CCD 1997. Revised Coastal Zone Management Plan, Sri Lanka, 1997,

CCD 1990. Coastal Zone Management Plan, Coast Conservation Department, 1990.

CCD 1997. Revised Coastal Zone Management Plan, Sri Lanka, Coast Conservation Olsen, S.D.Sadacharan, J.I.Samarakoon, A.T. White, H.J.M. Wickremaratne, and M.S. Wijeratne, editors. 1992. Coastal 2000: Recommendations for A Resource Management Strategy for Sri Lanka's Coastal Region, Volumes I and II. CRC Technical Report No. 2033, Coast Conservation Department, Coastal Resources Management Project, Sri Lanka And Coastal Resources Center, the University of Rhode Island.

Richardson P, 2001. "Care for the Wild in Sri Lanka". Marine Turtle Newsletter 67:16-19. Samaranayake, R.A.D.B., 2000, Sri Lanka's Agenda for Coastal Zone Management. EEZ Technology. www.sustdev.org/journals/others/iezm/05.d.pdf

Seneviratne, C. Coastal Zone Management in Sri Lanka: Current Issues and Management Strategies. http:www.rabbiteraph.de/edg/p_senev.htm accessed July 20 2002

Innovative Coastal Zone Management
ISBN 978-0-7277-5749-4

ICE Publishing: All rights reserved
doi: 10.1680/iczm.57494.316

Scenarios in Shoreline Management

Emyr Williams B.Sc. (Hons) M.Sc., Pembrokeshire County Council, Haverfordwest, Wales
Gregor Guthrie, B.Eng. (Hons) C. Eng.. M.I.C.E. Royal Haskoning, Peterborough, England.

Introduction

Sea level rise poses a potentially huge threat to the coastal communities of the world over the next century. Whilst it is, technically, entirely possible to provide complete protection for those communities, continued management is likely to be a financial burden that cannot be met, even in a rich western country like Britain. Even if economically justifiable, continued defence may also place communities at unacceptable risk in the future. National policies must therefore encourage communities to adapt if they are to be resilient in the face of such threats. But neither the magnitude nor timescale of the threat is currently accurately known. Planning for such a nebulous threat and getting community engagement is extremely difficult, but vital, if the policies needed to manage the risk are to be accepted by those affected communities in time to be effective. Therefore an approach is needed that allows stakeholders an opportunity to understand the processes used to derive those policies in a manner that is not perceived as threatening.

This paper describes an approach adopted in the preparation of the Shoreline Management Plan 2 (SMP) for the coast of west Wales based upon Scenario Planning (Wack 1985). The SMP process is ideally suited to the application of scenarios and this paper considers the conclusions from real situations in relation to communities in the West of Wales.

Present practice

The Department for Environment, Food and Regional Affairs (DEFRA) guidance (DEFRA 2006) indicated that consideration should be given to policy development for three epochs and implicitly implied that a sea level rise of one meter in a century is to be considered. However, given the uncertainties identified in the United Kingdom Climate Impact Programme Report 2009 (UKCIP 09), basing long term decisions on a single value is felt to limit thought; especially when the focus of the SMP2 is steering management to a position of long term sustainability. The possible maximum of 2m (UKCIP 09 H++) is obviously a very large rise and difficult for anyone without access to the relevant background information to assimilate. Nevertheless, it must be considered as a possibility and allowed for in the options considered, particularly in examining the third epoch and the implications beyond the 100 year outlook of the current SMP 2.

Public policy is also starting to change away from the "Predict and Provide" of the past to a more nuanced approach that explicitly includes increased "resilience" and "adaptation" and "risk management" approaches. In Wales this was explicitly expounded in the "New Approaches" programme which was the subject of a Welsh Audit Office report (Welsh Audit Office 2009) and in England, amongst other documents, Defra draft guidance "Guidance for Community Adaptation Planning and Engagement (CAPE) on the Coast" (Scot Wilson 2009) and "Adapting to Coastal Change: Developing a Policy Framework" (Defra 2010).

These documents outline options for dealing with the management of the risk posed by coastal erosion and tidal flooding and the after effects. Where there is sufficient political and economic justification, such as the protection of London, then resources will be found. However the same cannot be guaranteed for the rural villages in West Wales. With increased rationing of national funds, it may not even be possible to protect locally significant towns. These risks have to be highlighted, so that we develop communities on a sustainable base, allowing planning that builds in a sensible level of resilience.

Application of Scenarios

The many possible options for coastal management presented by the combinations of the range of anticipated sea level rise values, coastal process changes, political pressures and not least an ill defined timeframe for changes both natural and anthropogenic mean that formulation of a coherent definitive plan document for the next century may seem impossible. Developing a "hard" Plan, in the face of such a complexity of issues, can appear daunting, not least in gaining acceptance by elected representatives. However, to allow time for change, we cannot walk away from the issues; we cannot walk away from the possibilities that may drive change, no matter how unpalatable the possible consequences, we cannot hide behind uncertainty. While the change in conditions and the timescale of such change covers a wide range, the decisions that have to be made are really quite narrow, almost to the decision do we continue to defend or do we not. Put in these terms, broad scale scenarios open the opportunity to "ask the unaskable". The Shoreline Management Plan (SMP2) provides the opportunity to think about different possible futures, testing the robustness of a limited number of choices against a much wider range of possibilities.

Scenario Planning was originally used for military planning in the Second World War and adapted for commercial use by Royal Dutch Shell as "a way of rehearsing the future to avoid surprises." (Garvin and Levesque 2006) They consist of carefully researched range of possibilities rather than specific forecasts which were considered bound to be wrong. It does not assume that there is one best answer to a strategic question and involves far more subjective interpretation as well as objective analysis. It is therefore particularly useful in situations where uncertainty and change are high, costly surprises have occurred in the past and the quality of strategic thinking or the supply of new opportunities is low. (Shoemaker 1991)

The situation facing the Royal Dutch Shell Group corporate planning teams in the early 1970's was an analysis of the global oil markets suggesting that the "status quo" was untenable. (Wack 1985) However the policies needed to accommodate the changes anticipated were a radical change from those adopted by the Group senior managers in the past which had tended to be based upon "more of the same". The development of credible scenarios enable those managers to adapt to the rapid and major changes in the global oil market by the price rises in the early 1970's allowing the Shell Group to develop into one of the world's leading oil companies. A similarity exists between the "paradigm shift" experienced by Shell (and the other oil companies) and that facing coastal communities and the authorities responsible for their safety and well being.

Currently there is substantial uncertainty in the degree of sea level rise that will occur over the next century due both to the thermal expansion of the oceans (See Figure 1) to which must be added the contributions made by the melting of glaciers and polar icecaps. The rate of melting has an even greater degree of uncertainty though the consequences are far greater. As an example it has been estimated that the Greenland icecap has enough water to produce a 7m rise in global sea levels – clearly a catastrophic amount – which could arise from a 1.5-2.0°C temperature rise. (Clerk and Huybers 2009). The lack of agreement by governments at the recent climate conference in Cancún does not bode well for succeeding in keeping any anthropogenic temperature rise down to this level.

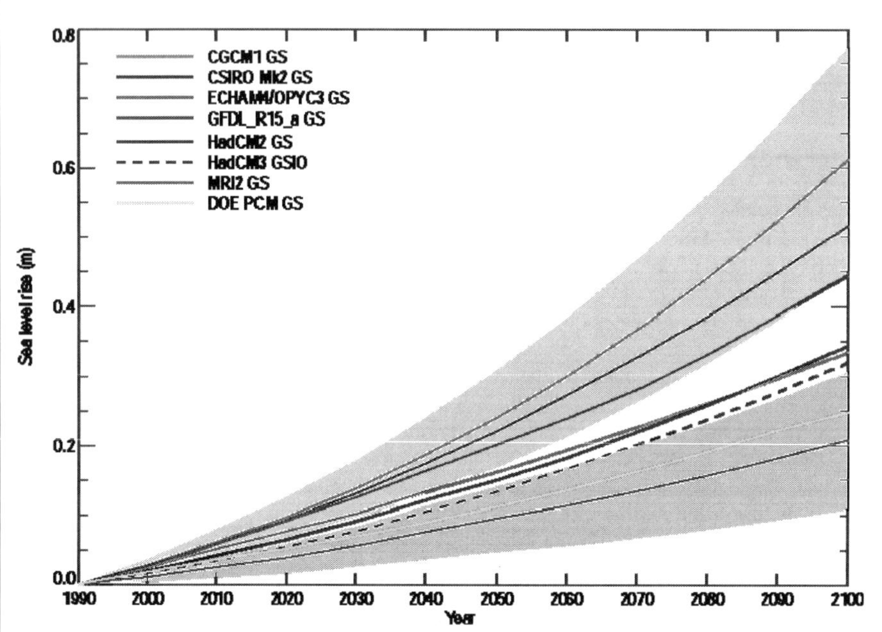

Figure 1: *Global average sea level changes from thermal expansion simulated in AOGCM experiments with historical concentrations of greenhouse gases in the 20th century, then following the IS92a scenario for the 21st century, including the direct effect of sulphate aerosols.* (IPCC 2001)

An example that starkly illustrates the changes in the atmosphere since the Industrial Revolution in the middle of the nineteenth century is the rise in Carbon Dioxide levels. The data from the EPICA ice core indicates that CO_2 levels have not exceeded 300ppm for 800,000years. Starting in approximately 1860 at around 280ppm they have now reached approximately 390ppm, a level not seen for millions of years. (Science Daily 2009). Clearly such levels of a potent greenhouse gas have potentially very significant implications for future global temperatures, but how much and when?

In explaining the science behind the predictions of sea level rise and the full extent of the anticipated consequential inundation an overly complex presentation causes people to lose any sense of understanding and hence acceptance of the risks. The credibility gap opens when values of 3-4° are mentioned. People see this amount of change every day between morning and noon so utterly fail to appreciate the implication for the Earth of such a rise in average temperature.

It is therefore essential that they are presented with a range of potential outcomes starting perhaps from the least threatening and leading up to those with greater implications. It is not only the public and their elected representatives that need persuading. It can be equally problematical for officers of those public bodies such as local authorities and the Environment Agency .

Many key decisions are made at an operating level on default assumptions of individuals. However it is often these individuals who see the first evidence of change and it is important to alert these people to the perhaps subtle effects that indicate significant changes. However, entrenched views can be difficult to overcome to allow acceptance of the conclusions of what are perceived as outlandish prophesies. They can best be dealt with by emphasizing the solid foundations upon which the scenario is based and the logic of the steps followed in its preparation. It helps that other less threatening predictions are available so that they can buy the total package even if elements are unpalatable. (Ringland 2003).

Is 2m really the end of everything?

A key issue in considering scenarios, in any situation, is within the scope of the problems being considered. A 7m sea level rise (melting of the Greenland Ice Cap) would present such catastrophic change to society that the issues considered by Shoreline Management Plans would be trivial. This may be a scenario worth thought but not within the scope of the SMPs.

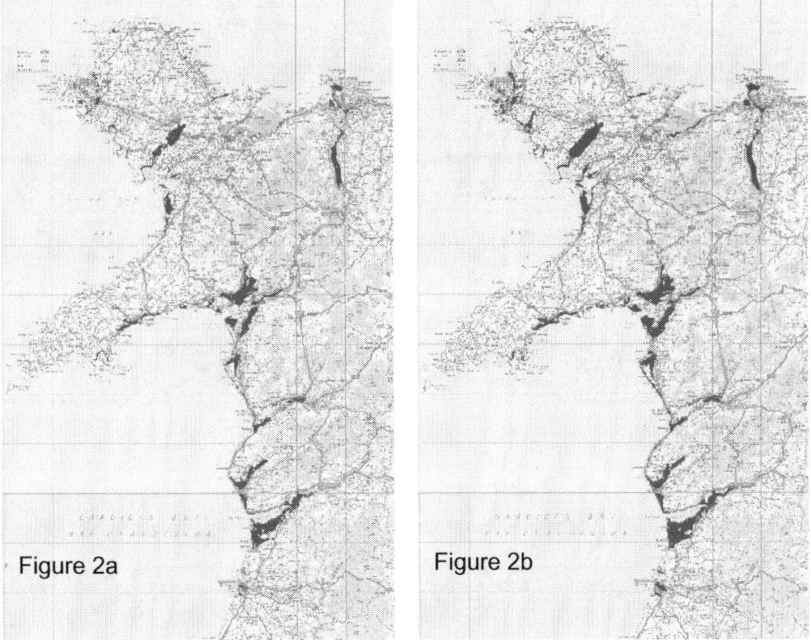

Figure 2. 1:200 year flood risk areas - a) present day, b) +2m sea level rise scenario.

There are many areas around the west Wales SMP coastline at flood risk. It is accepted by communities that this risk will increase with relative sea level rise. However, are we in a different world if the extreme (H++) scenario is considered? Figure 2 compares the present day 1:200 year flood risk probability (figure 2a) with the 1:200 year flood risk with sea level rise of 2m (figure 2b).

Although there are obvious differences, it is primarily a worsening of the existing problem. The +2m scenario extends the range of trend of change, rather than fundamentally changing the scope of issues being considered. We are still considering how we manage the flood and coastal erosion risk. However, by extending the trend we are now not just working within a very fixed set of boundaries and we can consider how we adapt to the expected in a manner that has a view to the future. This has important consequences in terms of management choices, ensuring that we start examining issues as a continuous and continuing process. We can view the 100 years not as "how do we get through the next 100 years?", to a perspective that starts addressing the question "and then what?".

This approach applies to the issues of erosion as much as flood risk but is more obvious in terms of the flood issues discussed below.

Vulnerability and timescale

Even under existing guidance on sea level rise there are significant issues being faced on the west Wales coastline. Despite the length of coast being considered by the SMP (some 1200km.) many of the issues are at a local scale. One of the key considerations is the small communities so characteristic the area; icons of rural Wales, images so beloved of tourism literature.

Figure 3. Indicative high water under a +1m and +2m sea level rise scenario.

The practical issues of sea level rise are highlighted in Figure 3, shown quite graphically in relation to the height of people using the beach area, in relation to the promenade and property.

Taking a purely deterministic 1m scenario, it might be assumed that defences could be raised, even if the beach area were substantially lost. However, then what? +1m rise is about the limit to which an engineering solution could be provided. The expectation of continued defence is built in to the approach, living with and managing the risk of normal tidal levels, accepting the risk and building in resilience to the more extreme events; with no real planning beyond that.

In reality, viewed as a continuum of change, the problem takes on a different perspective, that in the longer term, whether that be 50, 100 or 150 years, we have to accept that this village and the whole character of use of the location will not be sustainable. This situation can no longer be viewed in terms of time-rigid defence policy (20 years Hold the Line, 50 years Hold the Line, 100 years Realignment) but rather has to be considered as planned change over the whole 100 years to deal with an inevitable and radical change.

One aspect drawn out by consideration of the different scenarios is the shift from defence against extremes, to management of the every day occurrence when the current 1:100 flood happens every equinoctial spring tide. This can be seen in relation to the communities of Goodwick and Fairbourne (Figure 4).

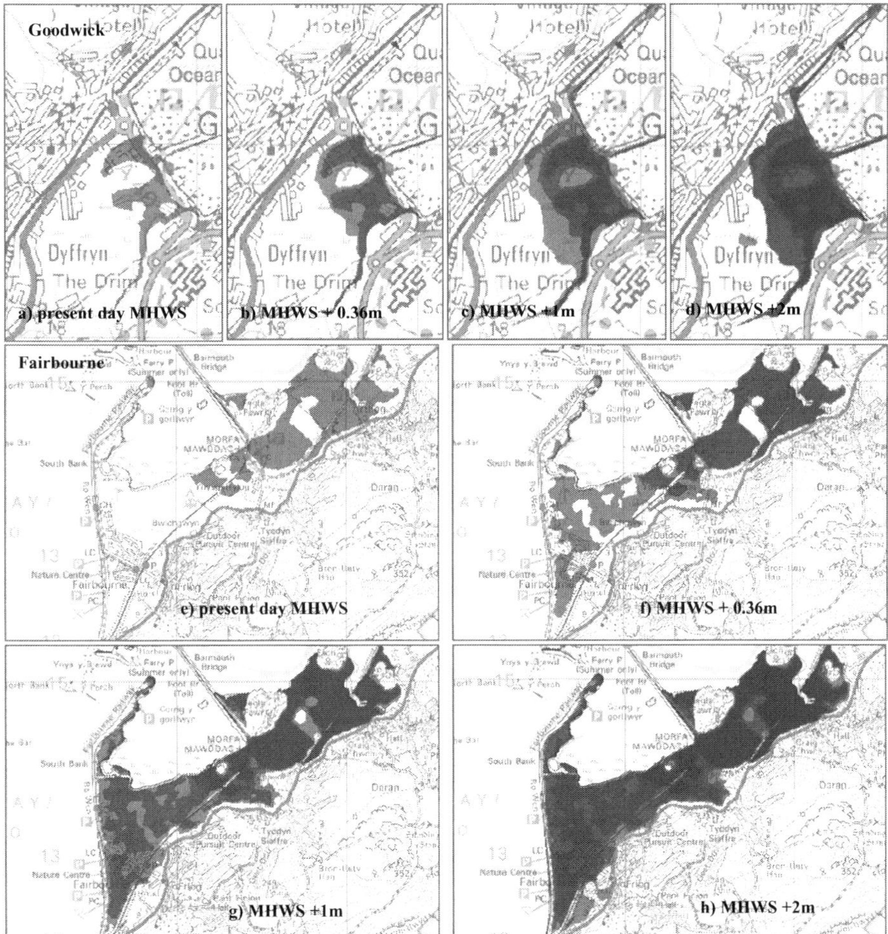

Figure 4. Comparison of MWHS flood risk areas for Goodwick and Fairbourne, with a range of sea level rise scenarios (present day, +0.36m, +1m and +2m).

In both locations the areas are defended. In both locations the general flood risk areas are quite well defined, such that, the more extreme water level flood risk only marginally increases beyond the areas shown in the figure for the 2m SLR scenario for MHWS; it might almost be said that there is a geomorphological memory of past sea level rise.

In the case of Goodwick, in testing a +2m scenario, it may be seen that the main areas of development remain relatively secure against future change in water level. The main issues being identified are in relation to access to the important Port of Fishguard. While limiting further development in the flood plain and addressing the still significant issues of access to the port, the analysis suggests that this area has a sustainable future as a major area of regeneration.

In contrast, figures 4e to 4h clearly demonstrate a worsening condition of flood risk at Fairbourne, not just in terms of managing risk from extremes but in terms of managing ground water levels and fluvial and pluvial events on normal everyday tidal conditions.

At Fairbourne, there would need to be shift in management from one of ensuring the defences are maintained, to one continual pumped flood risk management. Furthermore, while it is economically justifiable to continue to defend some 300 to 400 properties well into the future under either sea level rise scenario, the risk to life increases significantly in the event of defence failure or exceedance of design standards. This has to raise the issue of future sustainability.

If it is accepted that there is a need to move a community, how can this be undertaken, how is this to be planned and over what timescales?

Here again the scenario approach is essential in informing planning decisions. An implicit difference between the 100 year +1m and +2m scenarios is the difference in the rate of sea level rise. Quite crudely, if it is taken that a rise in sea level of 0.5m triggers the need for a threshold change in defence management approach, and the potential timescales needed to plan a potential managed evacuation of a community, this then becomes the criterion by when decisions with respect to the community have to have been implemented; not merely the point at which decisions have to have been made. The time associated with this condition is shown in Figure 5 for the two scenarios of sea level rise.

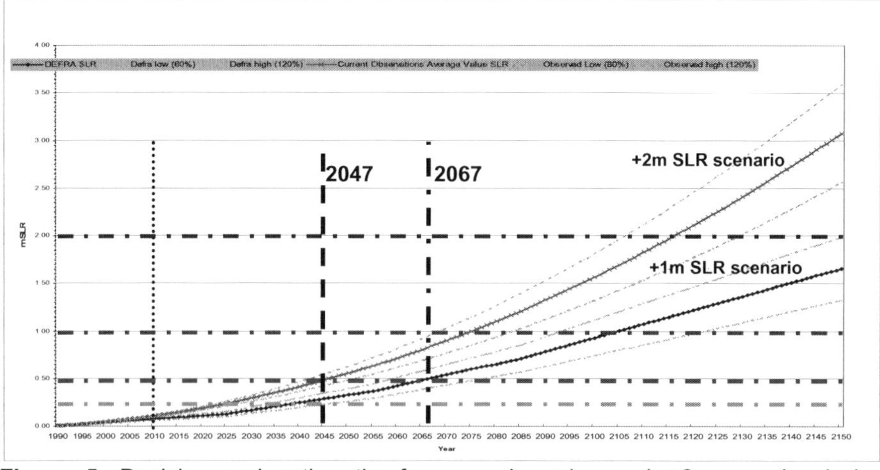

Figure 5. Decision and action timeframe under +1m and +2m sea level rise scenarios.

With this understanding, that actions are driven by conditions not specific timescales, this highlights the need for issues to be considered now and allows time for management of the consequences. Such issues include blight on property, opportunity for future use of the area, integration in assessing future housing needs within the larger area and the fundamental question of how a community may be moved.

Broad scale issues.

The above examples focus on important but local areas. The scenario approach taken within the west Wales SMP2 has also highlighted more broad scale issues with long term management consequences at a regional level.

Many of the major towns, such as Pwllheli, Bangor, Beaumaris, Llandudno, or Aberystwyth have core development within existing flood risk areas. Future developments within these major centres establish trends of community development which may reasonably influence and extend beyond the 100 years, considered specifically by the SMP. This present development establishes the economic core upon which the future of such towns develop and depend into the future. Considering different scenarios provides a different perspective to long term planning.

The same may be said of certain aspects of critical infrastructure, most notably that of the transport network. The main established development is within the coastal zone, with most the major towns of west Wales strung along the shoreline, supporting the rural landuse of the mountainous hinterland. Between the major towns the shoreline is punctuated by smaller communities, interlinked, in many areas, by the road and railways sustaining the character and existence of these communities. In many areas the main railway lines run across the various estuaries, with roads and railways also running along the potentially eroding coastal fringe.

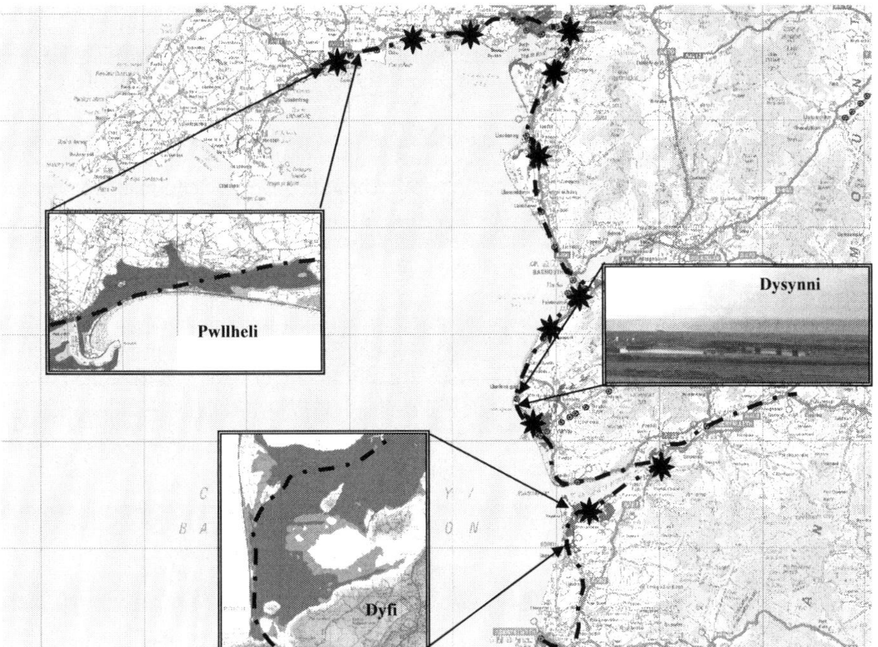

Figure 6. West coast railway line – present and future areas at risk.

Figure 6 shows the west coast of Wales railway line from Aberystwyth through to Pwllheli highlighting potential hotspots for future management.

In many areas the railway is or would in the future constrain appropriate management of a sustainable shoreline. To sustain the railway in the future, providing essential services to many of the major towns and smaller villages, so important to sustaining rural life in west Wales, major works will be required. This may include realignment of the track or major construction of new estuary crossings. The issues cannot therefore be seen as discrete problem sites. In considering this it is critical that all potential risk areas are identified. Only in this way, examining the full range of risk over different possible scenarios can the complete picture be assessed.

Coupled to this is the issue of time scale discussed earlier in relation to Fairbourne.

Conclusion and recommendations

An approach based upon a suite of possible scenarios combining the various possible options and management policies for the coastline can be developed incorporating all the main possible factors, such as sea level rise, the current defenses of a community and the environmental constraints posed by the various European regulations to protect habitats in such a manner as to be acceptable to most stakeholders. Clearly such scenarios would consider a range of, for instance, sea level rises over various timescales. This allows those affected to be able to see that what they may consider to be a reasonable value is considered as part of the wider picture which assists in getting "buy in".

It also allows the consequences of the more extreme values to be discussed and considered in a wider context. It alerts those responsible for developments which have a long "life" such as future development of major towns and transport networks, as well as engaging smaller communities in understanding how their aspirations can or may not be able to be met, to the need to consider the potential consequences of the very extreme changes that might occur.

Considering different scenarios does present problems in developing hard specific policy for risk management. However, this can be seen as being beneficial, within the highly dynamic coastal environment, in reinforcing the principle of shoreline management that we are defining a way ahead. We are defining an approach to risk management not a programme of defence works. It highlights how uncertainty based on key parameters such as sea level rise, must change our thinking from a strict timeline to one where decisions have to be made despite that uncertainty, based on future conditions.

This way of thinking provides an important focus for future monitoring of sea level rise, moving this from the academic to a very practical consideration of social issues. In reality it may take the next 20 years to better distinguish the future trends in the threats. In many areas there is quite possibly not that luxury in planning change for a sustainable future.

References

Clark, Peter U. and Huybers, Peter (2009) **"Interglacial and Future Sea Level"** Nature Vol. 462. Macmillan. London.

DEFRA (2006) **"Shoreline management plan guidance"**. Volumes 1 & 2. London, HMSO

DEFRA (2010) **"Adapting to Coastal Change: Developing a Policy Framework"**. London. DEFRA

Garvin, David A. and Levesque, Lynne C. (2006) **"A note on Scenario Planning"** Harvard Business School. Boston U.S.A.

IPCC (2001) Extract from **"IPCC Climate Change 2001: Working Group I: The Scientific Basis"** View online on January 16th 2011 at http://www.ipcc.ch/ipccreports/tar/wg1/412.htm

IPCC (2007) "**Contribution of Working Group I to the Fourth Assessment Report of the Intergovernmental Panel on Climate Change**" Solomon, S., D. Qin, M. Manning, Z. Chen, M. Marquis, K.B. Averyt, M. Tignor and H.L. Miller (eds.). Cambridge University Press, Cambridge, United Kingdom

The Earth Institute at Columbia University. **"Carbon Dioxide Higher Today Than Last 2.1 Million Years."** Science Daily 21 June 2009. Viewed online on January 17th 2011 at: http://www.sciencedaily.com /releases/2009/06/090618143950.htm

Ringland, Gill. **2002 "Scenarion Planning :Persuading Operating Managers to Take Ownership."** Strategy and Leadership. Vol 31. No6 MCP Ltd. London.

Scot Wilson (2009) "**Guidance for Community Adaptation Planning and Engagement (CAPE) on the Coast**" London. Scott Wilson.

Shoemaker, Paul J. (1991) **"When and How to Use Scenario Planning: A Heuristic Approach with Illustration"**. Journal of Forecasting (10) John Wiley & Sons Ltd. Malden, U.S.A.

United Kingdom Climate Impact Program Report 2009 viewed online on January 14th 2011 at:- http://ukclimateprojections.defra.gov.uk/

Wack P. (1985) **"Scenarios: Uncharted Waters Ahead"** Harvard Business Review, September – October 1985.

Welsh Audit Office (2009) **"Coastal Erosion and Tidal Flooding Risks in Wales"** Cardiff, Welsh Audit Office. Viewed online on 14th January 2011 at:- http://www.wao.gov.uk/assets/englishdocuments/Coastal_flooding_eng.pdf

Innovative Coastal Zone Management
ISBN 978-0-7277-5749-4

ICE Publishing: All rights reserved
doi: 10.1680/iczm.57494.326

Anglian Shoreline Management Plans: Large Scale Processes Driving Local Solutions

Jaap Flikweert, Royal Haskoning, Peterborough, UK
Karen Thomas, Environment Agency, Ipswich, UK
Terry Oakes, Terry Oakes Associates Ltd, Lowestoft, UK
Peter Frew, North Norfolk District Council, Cromer, UK
Mike Dugher, Environment Agency, Lincoln, UK
Greg Guthrie, Royal Haskoning, Peterborough, UK

Introduction

The Anglian coast epitomises all the challenges of coastal management. It is exposed to the fastest net sea level rise in the UK. Erosion of cliffs is threatening villages, but also provides sediment to beaches further down the coast where it prevents erosion of other places. There are large low-lying areas where we rely on flood defences to protect settlements, infrastructure, agricultural land and brackish habitats. However, the presence of these defences may prevent the natural landward movement of the coast in response to sea level rise, which can cause pressure on those same defences and threaten important coastal habitats.

The coastal system usually functions on a large scale; that is the case for coastal geomorphology, but also for society and for the environment. On a local scale however, the mixture of issues and opportunities is different for each location and each community. This is why coastal management is only sensible if it develops local solutions within the context of larger scale thinking. The Government's draft National Flood and Coastal Erosion Risk Management strategy[1] also reflects this approach in its promotion of an approach that helps decision making and action at the appropriate level.

This paper reflects on the second generation of Shoreline Management Plans (SMPs) in Lincolnshire, Norfolk, Suffolk and Essex. It describes the large scale processes and characteristics, using a historical perspective to explain some of the local differences. The paper uses the Anglian SMPs to illustrate that local solutions for coastal management can only be developed in a wider context.

Shoreline Management Plans: partnership working to find the right balance

The second generation Anglian SMPs were finalised in 2010, developed by partnerships consisting of the Environment Agency, local authorities, Natural England, English Heritage and all other organisations with a role in coastal management. For some areas the SMP has been the catalyst that was needed to bring these partnerships together, both for officers and for elected members. The SMPs describe the intent of management for the coast that provides the best balance between all the values and interests, both at a local scale and for UK plc. The SMPs' policies are primarily about the shoreline: does it need active management or not, and if so, should it be held where it is, actively moved or allowed to move. However, by developing the SMPs with an Integrated Coastal Zone Management perspective, through partnerships of coastal defence managers with land use planners, habitat managers and other stakeholders, the management policy for the shoreline is driven by the much wider integrated intent of management for the coast. This approach may not be formal integrated coastal zone management, but it is the best option available within the existing frameworks in the UK. The SMPs are now starting to steer coastal management for the short, medium and long term.

The Anglian coast and its Shoreline Management Plans

Diverse coastal zones with shared challenges

The Anglian coast is characterised by contrasts: from bucket-and-spade Hunstanton to booming bitterns at Blakeney; from the accreting saltmarshes of the Wash to the eroding cliffs at Happisburgh; from the gentility of the Suffolk coast to the international ports of Felixstowe and Harwich; and from to the wild and remote Essex marshes to the cultivated Fens. These contrasts are of course largely determined by geography and morphology, but there is also a significant historical and cultural dimension. The large-scale agricultural landscape around the Wash contrasts with the mosaic of small scale interacting features on the north Norfolk coast. This affects what people want from coastal management, and even how politics work. A common element for most areas is the lasting resonance of the North Sea flood of 1953, which caused the loss of 307 lives in Lincolnshire, Norfolk, Suffolk and Essex and led to the construction of many of the existing defences. The form and current legacy of these defences however varies strongly. Despite these differences, shoreline management has to provide responses to essentially the same broad challenges:

- The coast is changing; do we manage the coast to suit the needs of society and communities, or do we adapt the communities to the natural processes of the coast?
- Money is tight, so who's paying for this ambitious intent of management?
- The future is uncertain; how to deal with this when developing policy up to a 100 years ahead?

The paper illustrates how the SMPs have provided local answers to these generic questions, driven by large scale processes.

The Wash – Gibraltar Point to Hunstanton[2]

This is the SMP with the largest defended flood zone. The embankments around the Wash protect the low-lying land all the way to Lincoln, Peterborough and Cambridge, up to 50km from the shoreline. This flood zone contains some of the most productive agricultural land in the UK, supporting settlements such as Spalding, Boston and King's Lynn and a large number of smaller communities. Many of these are situated in a band around the Wash, some 5km from the shoreline. The sea banks mark a sharp boundary between culture and nature: the Wash itself is characterised by wide expanses of saltmarsh, mudflat and sandflat shaped by the tidal currents. The saltmarsh has been accreting for the last hundreds of years, despite sea level rise. However, this accretional process could reverse over the coming decades. If the saltmarsh eroded, the seabanks would become much more exposed to storm waves; also, the sensitively balanced intertidal habitats would be under threat. With current knowledge, it is impossible to be certain if and when this reversal from accretional to erosional trends could occur. This is a problem, because the impact of this potential change on shoreline management is fundamental. The SMP has made this uncertainty explicit by proposing a conditional policy. There is a clear intent of management that all established communities around this part of the Wash will remain defended. In an accretional future, this will be achieved by holding the current alignment. However, in an erosional future the best way to defend the communities could be to carry out localised and limited landward realignment where needed, which would recreate the saltmarsh bufferzone for the seabanks and replace the intertidal habitats that would be lost in this scenario. See Figure 1.

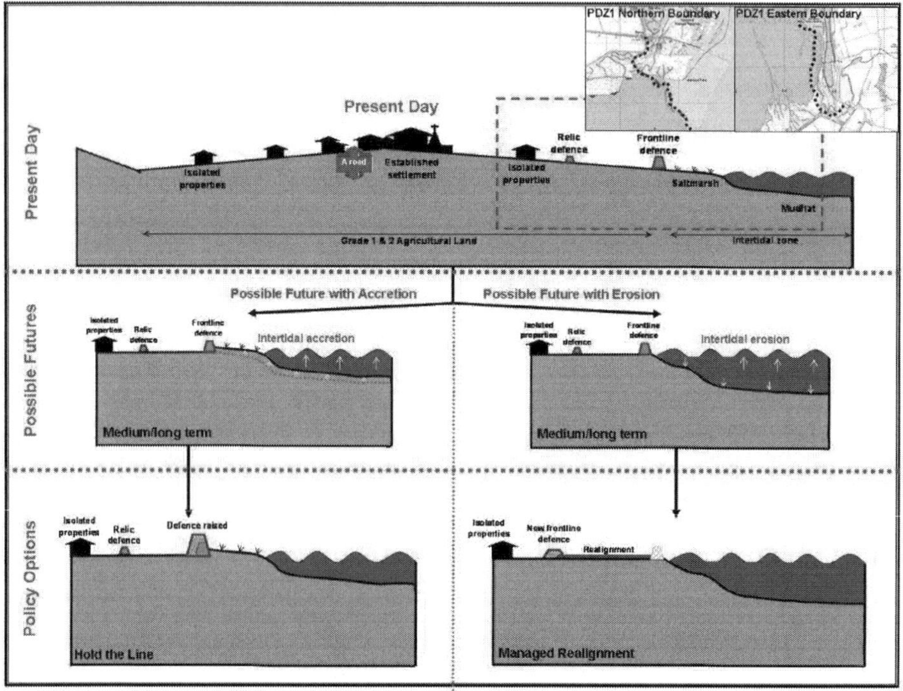

Figure 1: Illustration of SMP policies for west and south coast of The Wash

The east coast of the Wash has a very different character, although there are many environmental, geomorphological and socio-economic interactions with the rest of the area. The cliffs at Hunstanton are partly undefended. The ongoing erosion has given them their unique striped profile and provides some sediment to the system, but is also starting to threaten important parts of the town with a historic value, such as the Lighthouse. The town front's promenade supports the tourist economy, essential for the town and district. South of Hunstanton down to Heacham and Shepherd's Port, there is a low-lying area with a concentration of holiday parks. Further south is the Royal Society for the Protection of Birds' Snettisham reserve, with saline lagoons that attract unique birdlife, which in turn attracts a lot of human visitors. The low-lying area's flood defence largely consists of a managed shingle ridge, needing annual recharge to keep the area habitable. Continuation of this approach is unlikely to be fully funded by national budgets and may have other negative impacts in the long term. The SMP has indicated that the whole of this East Wash frontage, from the cliffs down to the saline lagoons, needs an integrated approach: this means taking account of longshore coastal processes, integrating coastal management planning with land use planning, and involving the businesses and the public in all decisions.

North Norfolk – Old Hunstanton to Kelling[3]

This SMP covers the north facing section of the Norfolk coast, between the cliffs at Hunstanton and those at Weybourne. The shoreline is a mosaic of dunes and embankments that protect grazing marsh, reedbeds and agricultural land, natural barrier islands, spits and shingle ridges, and saltmarsh fronting undefended high ground. There is a string of settlements directly behind the shoreline, such as Thornham, Brancaster, Wells-next-the-Sea and Blakeney; these are all located where the land starts to rise to higher ground. The shoreline is essential for the economy of the area, which is largely based on fisheries and shoreline-related tourism such as boat trips, beach tourism, bird watching and walking, in addition to agriculture (see Figure 2). The shoreline also provides a diverse range of wildlife habitats. The character of the landscape is unique, which is why it is part of the Norfolk Coast Area of Outstanding Natural Beauty. Flood and erosion risk in this SMP is of a very different order than in the neighbouring SMPs. The flood defences protect the low-lying parts of some of the settlements and a caravan park, but most of the defended areas are freshwater habitats and agricultural land. Many of the flood defences are partly natural dunes and shingle ridges, and the embankments are relatively low.

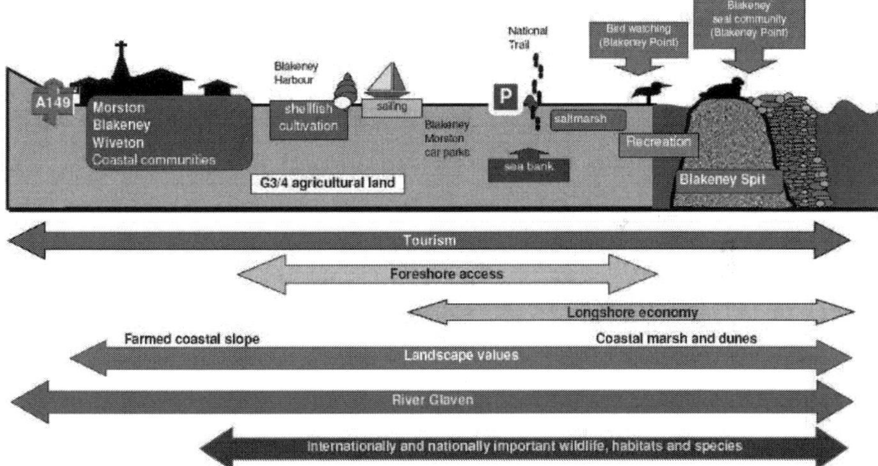

Figure 2: Illustration of the mosaic of features and issues in North Norfolk

This SMP needs to make decisions at a totally different level to The Wash SMP: the challenge of this SMP is to ensure that shoreline management supports the interrelated web of functions and values that characterise this unique stretch of coast. The flood embankments in the area protect mostly freshwater habitats and agricultural land and are not really under pressure from coastal processes. However, their presence does constrain the natural tidal processes that have shaped the area, and some of the channels are suffering from siltation. Opening up some of the reclaimed areas would increase the tidal prism and this could help sustain the channels and hence support navigation and its use for tourism and the economy. The defended freshwater habitats would be lost, to be replaced by intertidal habitats. Again, uncertainty about coastal processes plays an important role: in theory, the increase in tidal prism will have a positive effect, but the local effects are difficult to predict. Broadly speaking, there are two extreme options:

- continue to maintain all defences where they are now, which will support current use of defended land, but may cause the channels to silt up in the course of the next 100 years.
- actively revert to a more natural coastal system by breaching the embankments. This will require adaptation of currently-defended land but is likely to support navigation and can potentially make coastal habitats more resilient to sea level rise.

The SMP concluded that the best solution is different for different frontages. The intention is that all dwellings and all tourist facilities in the floodzone will remain defended, using natural processes as far as possible. There are a number of reclaimed areas where the flood banks are still adequate and currently don't need a lot of maintenance, but they will need refurbishment in about 60 years. For most of these, the SMP identifies that that would be the right time to change the approach and breach or realign them to increase the tidal prism, following land use adaptation. The main exception is around Blakeney, where the problem of channel siltation is particularly urgent. The intent is to breach the defences of Blakeney Freshes in the medium term (before 2055), making the current freshwater habitats intertidal and aiming to use the extra flow in and out of the area to sustain the channels.

East Norfolk –Kelling to Lowestoft[4]

This frontage includes some of the most famous and scenic stretches of coastline in England. To the north the coastline is mostly elevated with clay cliffs dominating the coastline. Sections of this coast are very prone to cliff erosion. The section of the coast in front of the Norfolk Broads is much flatter and is fronted by extensive dunes and broad sandy beaches. This section of the coast is liable to erosion and flooding as the land behind the coastal strip is at or below sea level. South of this the land behind the coast rises again and is less liable to coastal flooding, but there are still areas that are prone to coastal erosion.

The big challenge for this SMP is that coastal processes and sea level rise will lead to increasing pressure on the existing defences, but to keep or improve the defences in the long term could lead to a loss of beaches and sand dunes locally, and have impacts on other parts of the coast, including other communities. However a policy of not defending in the future could have major impacts on local communities and the economy.

This complex situation made clear that the SMP could only develop its policies if it took full account of these social and economic impacts. In addition to shoreline management's traditional focus on coastal processes and the environment, the SMP needed to include the options for social mitigation, to accompany a change in policy. This SMP was one of the first SMPs to start, and this lesson has informed the development of all other SMPs throughout the country.

The final policies have been based on a full understanding of what social mitigation measures are available, including studies to investigate what coastal authorities can do, such as planning roll back, purchase and lease back etc, and what communities can do for themselves, such as adapting local businesses, lobbying government about state insurance/indemnity etc. The resulting policies are a mixture that reflects the varying nature of the coastline:

- Some sections of the coast have never been defended and these will usually remain undefended in the future.
- For major settlements on the coast such as Sheringham, Cromer, Great Yarmouth, Gorleston and Lowestoft North, it is not realistic to stop defending them, so these will continue to be defended. Decisions will need to be made as to how to defend these, and how to alleviate any consequences for other sections of the coast.
- However for a number of sections of the coast the SMP identified that continuing to maintain and raise defences cannot be sustained in the medium or long term, and that a change will be required to a policy of managed realignment or no active intervention. Where a change of policy is proposed within the timeframe of the SMP, it has been essential to stipulate that appropriate social adaptation measures must be identified and implemented (where this is within the power of coastal authorities to do so) well in advance. This will require working closely with the affected communities.
- For one section of the coast, Eccles to Winterton, there are many important issues that need to be weighed up in reaching a policy decision. There will come a time when it will no longer be sustainable to maintain defences, but the timing of this 'tipping point' is very difficult to predict. The 'tipping point' will be reached when the cost, environmental and physical consequences of defending a section of the coast are greater than the social, economic and environmental costs of setting back the defence line. The SMP sets actions to help identify this 'tipping point' and also requires consideration of the social adaptation measures that need to identified before any change to the hold the line policy could be considered. In addition there are many options available, should the coastline ever be realigned, and these will need to be considered years, if not decades in advance of any realignment taking place. This section of the coast is therefore unique in having a conditional hold the line policy in the longer term.

Suffolk – Lowestoft Ness to Landguard Point[5]

The SMP area is from Ness Point in Lowestoft to the mouth of the Stour and Orwell Estuaries at Landguard Point, Felixstowe. This area contains both soft eroding cliffs and lower-lying areas extending inland that are at potential risk from coastal flooding. Three major river estuaries, the Blyth, the Alde/Ore and the Deben, cut the coast and there are several smaller broads or low lying areas, which are closed off by the shingle ridges and beaches which extend over much of the length of the shoreline. There are four major towns - Lowestoft, Southwold, Aldeburgh and Felixstowe - and many smaller towns and villages, all within the coastal strip. The shoreline area is essential for the local, regional and national economy. This relies heavily on shoreline-related tourism, agriculture, two major ports and several smaller harbours. The coastal zone also provides an attractive place to live. This vitality comes from the way in which the coast is used, as well as from its outstanding natural beauty, its largely unspoilt character and the diverse range of wildlife and natural habitats. It has long been an area of human settlement and its cultural and heritage value adds to its overall landscape quality. A historic and central part of that landscape is the changing nature of the coast and this natural change will continue in the future.

The coastal processes here are complex and operate at different scales. The coast is generally made up of soft geology. Much of the sediment has come from erosion of the coast and nearshore area over the last 10,000 years as sea level rose and the East Anglian coast eroded. More recently, the process of erosion has been limited in places by man-made defences, which means that there is now a much smaller supply of sediment. The sediment generally moves from north to south along the shoreline, but its transport rates vary strongly between areas.

Without defences the coast would continue to retreat over its whole length; very slowly in some areas, but in other areas there could be several hundreds of metres of erosion over the next hundred years. The risk of erosion threatens homes, businesses and other assets in all the major towns and several of the smaller villages. It also threatens important historic assets and puts pressure on many of the coastal flood defences. However, erosion in one area also provides sediment that maintains the beaches, provides defence to other areas of the coast and sustains many important nature conservation habitats. There are also many areas of flood risk within SMP area, particularly within the estuaries but also in large areas of Lowestoft and Felixstowe.

There are areas where difficult decisions have had to be made, largely in areas where issues had already been identified in earlier studies. Overall, however, the plan is to continue managing the coast. This does not mean that all sections of the coast can be or should be defended. One of the benefits of the SMP's broader view is in considering how the values associated with the coast may be maintained in a more sustainable manner, achieving a balance over time and adapting the way in which areas are used and defended. One aspect of the coast most affected by continuing erosion will be the historic environment. While many of the individual features within the towns and villages would be protected, as would the Martello towers, there would be a continuing loss of more remote areas where there may be important records of past settlement. This raises the issue of how to manage loss and how to fund investigation to partially mitigate these losses.

All the main towns and many of the smaller villages would continue to be defended. However, in some local areas there will be a loss of individual properties due to erosion. Through adapting the way in which defence is provided, the supply of sediment will continue and beaches would be sustained. The SMP2 approach is about managing change to encourage a more naturally functioning coast. This is a coast of balance, between the essential human activity and the important nature conservation and historic landscape. These are not in conflict but make up a whole coastal area of immense value.

Essex and South Suffolk – Landguard Point to Two Tree Island[6]
The Essex and South Suffolk SMP area covers the Stour and Orwell Estuaries in Suffolk as well as Hamford Water, Colne, Blackwater, Roach and Crouch estuaries. The plan also covers most of the Essex coast as far as Two Tree Island near Southend-on-Sea where it merges with the Thames Estuary 2100 Strategy. Almost all of the 440km of shoreline is made up of embankments that protect low-lying land against flooding. There are also a number of stretches of higher, soft eroding cliffs and these are generally undefended, with the exception of the cliff frontages of Southend and the Tendring Peninsula at Clacton and Holland-on-Sea.

The most prominent economic activity is the commercial shipping that takes place out of the ports at Ipswich, Felixstowe and Harwich. In addition much of the coastal floodplain is used for agriculture. Tourism plays an essential role in the economy of much of the SMP frontage whether beach-based at Frinton, Clacton, Jaywick and Southend or linked to the seafood culture of Maldon and West Mersea. Important recreational activities like walking and sailing also occur throughout the estuaries. Historic towns like Ipswich, Colchester and Chelmsford all have links to the sea. Landscape and wildlife including the southern extent of the Suffolk Coast and Heaths Area of Outstanding Natural Beauty attract many tourists. The low lying areas provide important habitats. At the same time, the coastal zone includes a range of critical infrastructure assets such as sewage treatment works, railway lines and the Bradwell nuclear power station.

Many of the coastal and estuary defences are in good condition, but some locations are under increasing pressure from natural changes (waves, tidal currents, surge tides). These natural processes are already eroding beaches and saltmarshes, leaving defences vulnerable to over-topping by waves and tides or the undermining of defence foundations. At these locations, there is an increased risk of flooding or erosion to local people and property now and in the future. In addition, the loss of beaches and intertidal areas has an impact on local economies and the coastal environment. These are the locations where the SMP needed to consider important management decisions for the future, aiming to find a balance between managing coastal flood and erosion risk and wider socio-economic and environmental needs.

For much of the Essex and South Suffolk coast the defences are in relatively good condition and the intent is to hold them in their current alignment. However, where the coast is under pressure from natural processes, now or in the future, the SMP has proposed managed realignment policies, to be implemented either in the short, medium or long term. At these locations the pressure from coastal processes will make it technically difficult and costly to maintain the defence in its current position, and therefore it makes sense to consider alternative management options. Setting back defences at these locations will create new intertidal areas or beaches which will reduce flood risk locally. There are also a few cases where the cost of continued defence maintenance exceeds the benefits; this can also be a reason to propose managed realignment. The managed realignment policy is mainly proposed where there are no significant communities and in largely rural areas. Significant work was undertaken to engage directly with landowners and their representatives throughout the SMP process as they were seen as the key stakeholder that would be affected by changing management policies. It will be essential to continue to work with landowners and communities in these locations to find the best solution at the appropriate time. Where there are currently no defences, such as soft eroding cliffs in areas of little or no property or infrastructure, the current No active intervention policy will be continued.

Local solutions in a wider context

The stories of these five SMPs already illustrate how policy development needs to take account of different spatial scales. This section picks out some of the most salient examples where it was essential to understand the large scale context (for different aspects) to develop local solutions.

Coastal and estuary processes

The SMPs' boundaries were defined to ensure that the most important coastal process interactions are contained within the SMPs[7]. This means that in many cases, there is an overarching level of coastal processes that governs the whole of the SMP area. In turn, these overarching processes drive more local developments.

Depending on the specific nature of the frontage, the overarching coastal process can lead to a situation where decisions in one location have an impact further along the coast. The classic example is the East Norfolk coast: the presence of local defences to protect one community from cliff erosion can starve neighbouring areas of sediment, which can put another community's defences under pressure or speed up ongoing cliff erosion. The East Norfolk SMP has had to make hard decisions to start helping communities to adapt to coastal change; this was accepted as the best and most sustainable solution at the appropriate scale, even though locally continued defence would have been preferred – see also the paper by Peter Frew for this same conference[8].

Much of the Essex and South Suffolk SMP area consists of estuaries, which are bordered by reclaimed low-lying areas along much of their shoreline. Managed realignment in the upper parts of the estuaries might be a good solution for some cases. However, this would increase the volume of water flowing in and out of the estuary with each tidal cycle (the tidal prism), putting further pressure on the sea banks in the constrained middle and outer parts of the estuaries. This understanding has played an important role in the selection of frontages for a managed realignment policy, in addition to the knowledge whether the defence is or will be under pressure from coastal processes locally.

Socio-economic processes and Communities

In some cases the scale of the socio-economic and community interactions is at least as large as the coastal processes. In The Wash, the seabanks protect a large portion of the best agricultural land in the country: approximately 50% of grade 1 land in England. Any large scale changes in shoreline management could have significant impacts on the local and regional economy, but also on a national scale, particularly against the background of potential future food security concerns. The SMP explicitly acknowledged this importance by determining the impact of policies on the nation's area of grade 1 agricultural land. It could be argued that the farmers and the wider agricultural industry around the Wash form a large-scale community with a large stake in shoreline management, and this was reflected in the NFU's involvement in the SMP's steering group, alongside local authorities, but also Natural England and the RSPB.

Essex and its people have a long history of dynamic interaction of land and sea, which is a vital ingredient of the county's culture and is reflected in its landscape, economy and communities. The Essex and South Suffolk SMP covered a very large area, made up of a number of largely independent coastal and estuary systems. Still, the maritime nature of the county provided a unifying factor to the development of the SMP. This was reflected in the headline principle that guided the development of the SMP, which was to develop policies appropriate to the area's diverse character and its dynamic interaction of land and sea.

Habitats, landscape and historic environment

The Suffolk coast is known and appreciated for its vitality and attractiveness, which comes from its unique mixture of natural beauty, habitats and man's use of the coast throughout history. One of the key features is the mixture of different habitats that make up the coastal landscape, including freshwater habitats in defended areas on the shoreline. If a freshwater habitat is lost at the expense of intertidal habitat as a result of realignment, there is a legal requirement to offset this loss by creating equivalent freshwater habitat elsewhere. However, if this new habitat is created in a different area, then the local balance of features that make up the landscape could be disturbed. These larger scale considerations had to be taken into account in both the Suffolk and the Essex and South Suffolk SMPs.

In the Wash, the whole area seaward of the shoreline is one large interrelated ecosystem, as reflected in its international habitat designation as one unit. This is a much larger scale than what would be typical for flood defence management – this would typically use the four rivers (Witham, Welland, Great Ouse and Nene) to identify units. It is also much larger than the units used for land use planning, because the area covers four districts and two counties. This large scale does however match the uniform nature of the seabanks and their hinterland, as discussed earlier. The SMP started initially by describing the area and its features for smaller units, but it became obvious that the real issues for the SMP to address were at the larger scale, which has then been reflected in the policy development zones and policy units.

Conclusions

The second round of Shoreline Management Plans on the Anglian coast illustrates the diversity of challenges and opportunities for coastal management. Common themes are managing ongoing and future change; balancing widely varying interests and finding opportunities for multiple benefits; dealing with uncertainty.

Decision making for shoreline management has to take account of a range of spatial scales, which can be determined by coastal processes but also by socio-economic structures and communities, by ecological and historic interactions and by landscape-scale connections.

The Government's draft National Flood and Coastal Erosion Risk Management strategy promotes an approach that helps decision making and action at the appropriate level. Given the wide range of issues and the direct impact of shoreline management on communities, it is essential that local authority elected members are actively involved, shaping the process and making decisions. This will require continuation and reinforcement of the partnership approach developed through the second round of Shoreline Management Plans, ensuring the right balance of all the interests in appropriate scale decision making.

References

1 Reducing the threat, building resilience, empowering communities – Consultation on a national flood and coastal erosion risk management strategy for England. Environment Agency, 2010

2 The Wash Shoreline Management Plan 2 – Gibraltar Point to Old Hunstanton. Environment Agency, 2010

3 North Norfolk Shoreline Management Plan – Old Hunstanton to Kelling. Environment Agency, 2010

4 Kelling to Lowestoft Shoreline Management Plan. Halcrow 2006

5 First review of Shoreline Management Plan sub cell 3c - Lowestoft Ness to Landguard Point. Suffolk Coastal District Council / Waveney District Council / Environment Agency, 2010

6 Essex and South Suffolk Shoreline Management Plan. Environment Agency, 2010

7 Shoreline management plan guidance, Defra, 2006

8   Managing coastal change: Use of the Defra Coastal Change Fund in North Norfolk,
        Peter Frew, 2011

Innovative Coastal Zone Management
ISBN 978-0-7277-5749-4

ICE Publishing: All rights reserved
doi: 10.1680/iczm.57494.337

Flamborough Head to Gibraltar Point Shoreline Management Plan – A Consensus-Based Approach

David Dales, URS Scott Wilson, UK
Laura Evans, URS Scott Wilson, UK
John Pos, URS Scott Wilson, UK
Jonathan Short, URS Scott Wilson, UK

Introduction

The Flamborough Head to Gibraltar Point Shoreline Management Plan covers the highly dynamic and varied coastline between Flamborough Head in the East Riding of Yorkshire and Gibraltar Point in Lincolnshire, including the outer Humber Estuary. The extent of the project area is shown in Figure 1. This project was funded by Defra and the Environment Agency and is part of a national framework of Shoreline Management Plans which covers the 6,000 kilometres of coastline in England and Wales. Shoreline Management Plans provide a large-scale assessment of the risks associated with coastal flooding and erosion and present a long-term policy framework to reduce these risks to people and the developed, historic and natural environment in a sustainable manner.

This Shoreline Management Plan, which at 180km in length is one of the largest, has three key issues to address:
- The rapid erosion of the clay cliffs of the Holderness coast threatens towns and villages which in the long-term may not all be defended - the previous Shoreline Management Plan for this length of coast was not adopted because its proposals were not politically acceptable.
- The risk of flooding to the significant economic assets around the Humber and the long-term impacts of flood and erosion management measures on the environmental assets of the Humber.
- The risk of flooding to the extensive low-lying areas of Lincolnshire which are at or just above present day sea level – the protection of which may require funding beyond that available from the Environment Agency who currently manage the defences.

The paper describes how these issues were approached and resolved, using a consensus-based approach.

Project structure

Since the project covers a large area and affects a range of interest groups, multiple organisations were represented on the Client Steering Group (CSG) including four local authorities, the Environment Agency, Natural England, English Heritage and the National Farmers' Union. Each organisation was represented on the CSG by technical officers, and the group was chaired by a technical officer. In addition, an Elected Members' Forum (EMF) was established consisting of councillors from each of the four councils plus officers from the other organisations. The EMF was chaired by a councillor, and was established to ensure that elected members of all the councils were aware of the project, and conversant with the issues, and able to represent the public at large in development of policies.

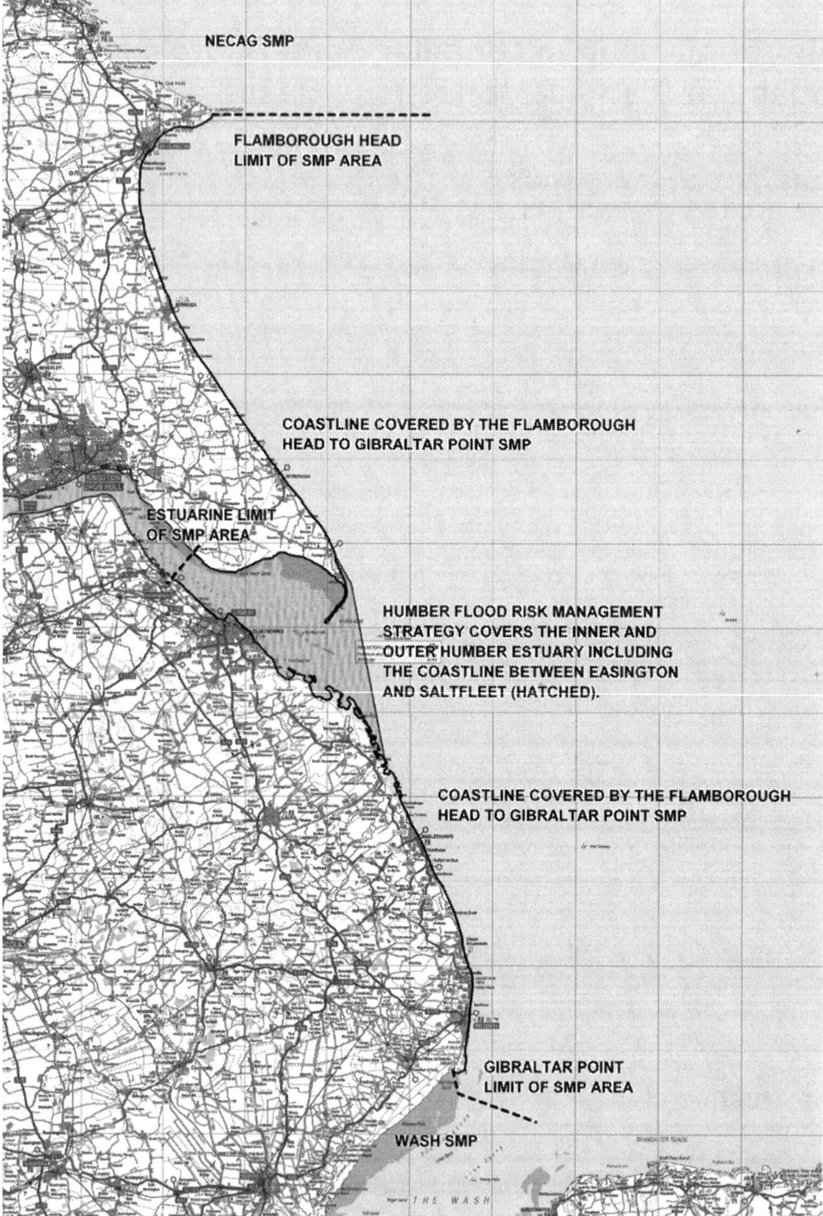

Figure 1: Flamborough Head to Gibraltar Point Shoreline Management Plan area

Policy appraisal process

The appraisal of potential coastal management policies formed the heart of the development of the Shoreline Management Plan. To maintain consistency between adjacent Plans, it was agreed by the CSG and EMF that the appraisal methodology would follow the general approach adopted by the Wash Shoreline Management Plan (Royal Haskoning, 2010), in terms of first establishing over-arching Principles, establishing objectives for a number of Character Areas, and then appraising policies against these objectives.

Principles

The first stage of the process was to develop a set of Project Principles, which effectively set out the aspirations which the shoreline management policies seek to achieve. These Principles were debated at length by the CSG and the EMF until it was agreed that all aspects that would be affected by shoreline management had been adequately covered within the Principles. The Principles agreed for the Flamborough Head to Gibraltar Point Shoreline Management Plan are presented below:

- To balance flood and erosion risk management in a sustainable manner appropriate to the overall value of the features affected
- To ensure that shoreline management policies encompass longer term adaptation options, and give time for communities and individuals to adapt to changing climate conditions and levels of risk
- To develop policies for flood and erosion risk management that will inform spatial planning processes and provide a robust evidence base for Local Development Frameworks
- To support sustainable patterns of development and consider possible effects on communities and their welfare
- To support the nationally, regionally and locally important social and economic assets of the area in a sustainable manner
- To consider the effects of coastal change on local industries, agriculture and employment and provide a secure environment for economic activity and development
- To ensure that local decisions do not have a disproportionately adverse affect on the natural balance of the coastline and shoreline management elsewhere
- To contribute to the positive management and enhancement of environmentally designated sites and protected species, subject to natural change
- To support the conservation and enhancement of biodiversity in the wider coastal zone
- To support the maintenance and enhancement of the character of the coastal landscape
- To support the preservation and enhancement of the historic environment
- To comply with legislative requirements and contribute to a safe and healthy environment.

It was emphasised that all Principles were to be considered in conjunction with one another, and that their order was not significant. The Principles reflect both the generic requirements of any Shoreline Management Plan as set out in the Defra guidance (Defra 2006a, 2006b) but also the concerns of the key stakeholders. Concerns about policies possibly leading to economic blight, loss of agricultural land, loss of habitat or heritage assets are all represented.

Characterisation

At an early stage in the study, an understanding was developed of the important features of the area and issues affecting the frontage. The entire frontage was split into nineteen areas, termed Character Areas. The divisions between the areas were created so that each area has a broadly similar character in terms of land use, geography and coastal character. The characterisation of each area was initially undertaken by URS Scott Wilson, but was then examined, amended and refined in a series of CSG and EMF workshops.

Objective setting

The approach to objective setting was agreed with the CSG and EMF. A set of generic objectives was developed based on the group of project Principles and the understanding of the issues affecting the frontage. The generic objectives are reproduced in Table 1.

Table 1: Generic objectives

Topic	Generic objectives
Flood and erosion risk	Minimise coastal flood and erosion risk to people, property and the environment Make effective use of existing man-made or natural defences
Communities	Protect as many settlements as possible To maintain … as a … (viable town, seaside resort, regional commercial centre etc – pick from list) throughout the plan period
Natural Environment	Maintain natural processes relating to …(relevant biological or geological feature) Maintain and enhance if possible …(relevant biological or geological feature) Ensure that the impact on the UK's intertidal habitat is acceptable
Agriculture and Industry	Protect as much grade 1 and grade 2 land as possible Ensure that the impact on the UK's area of agricultural land is acceptable Maintain and enhance the viability of the area's …(relevant industry inserted from menu) industrial capacity
Tourism	Maintain and enhance the viability of a diverse tourism economy
Infrastructure	Avoid interruption of the functioning of…(relevant infrastructure inserted from menu)
Historic Environment	Protect and where possible, enhance designated and significant historic environment assets
Landscape	To maintain and where possible, improve the quality of the coastal landscape
Coastal Processes	To prevent interruption of coastal processes which supply sediment to other coastlines
Timing of Policy	Provide sufficient time if necessary, for community adaptation, relocation of regional infrastructure etc.

Policy appraisal criteria were then developed for each Character Area using the generic objectives as a 'menu'. The use of a generic set of objectives at the outset allowed policy appraisal criteria to be developed on an equivalent basis for similar issues in different parts of the frontage. A key decision made by the CSG and EMF during the agreement of these objectives was that each would be equal in weight. This was a fundamental decision, based on a desire that everyone's interests would be assessed and considered equally, with the hope that a consensus could be reached.

Policy Packages

For each Character Area, the policies that were deemed sufficiently relevant and viable to justify further appraisal were identified through CSG and EMF workshops. During these workshops, discussions led to the identification of policy options that could definitely be ruled out for full appraisal, and those that were worthy of full appraisal. This process involved systematically assessing the viability of Defra's four shoreline management policy options (described in Table 2) on a Character Area by Character Area basis.

It should be noted that at this stage, the policies selected for appraisal during the workshops were not necessarily considered likely, desirable, feasible or economically viable; rather their evaluation and appraisal was considered to be necessary and in the public interest.

Table 2: Shoreline management policy options available for appraisal

Shoreline management policy	Description of policy
Hold the line	Hold the existing defence line.
Advance the line	Advance the existing defence line by building new defences on the seaward side of the original defences.
Managed realignment	Managed realignment by allowing the shoreline to move backwards, with management to control or limit movement.
No active intervention	A decision not to invest in providing or maintaining defences.
These policies may be applied to any of the three timescales: short term (up to the year 2025), medium term (between 2025 and 2055) and long term (between 2055 and 2105). These three periods are known as 'epochs' within the Shoreline Management Plan.	

Character Areas do not exist in isolation from their neighbours – they are linked by coastal processes and in some cases by a common flood risk cell. In order for policy appraisal to take account of these linkages and common features, 'strings' of policies were formed for stretches of coastline covering multiple Character Areas where issues and processes are largely similar and/or strongly linked. It was considered impractical to appraise every possible policy combination from those selected for each Character Area. Some combinations would also make no logical sense; for example, defending presently undefended rural areas while leaving currently defended urban areas to erode or flood. Therefore, coherent 'strings' of policies were developed representing a particular intent of management across a stretch of coastline. These were termed Policy Packages.

Assessment of policies

Following the development of Policy Packages, an assessment of their impacts was undertaken using policy appraisal criteria derived from the Character Area objectives. Policy Packages were assessed against the appraisal criteria developed for each Character Area and for the short-term, medium-term and long-term. This process was undertaken systematically using an agreed 'traffic light' approach where the results of the appraisal are indicated by a colour (green, amber or red) used to represent the extent to which a policy package fulfilled the individual appraisal criteria. A narrative was also provided to explain the attributed colour and assessment. An integral part of the appraisal process included the assessment of shoreline responses to the different Policy Packages.

In order to make the approach as objective and consistent as possible, within the constraints of a consensus-based system, a set of guidelines was developed by URS Scott Wilson. These fell into two categories: those criteria for which it is possible to quantify a predicted impact (e.g. property loss), and those for which impacts are descriptive (e.g. landscape impacts). An extract from the appraisal guidelines is provided in Table 3. The agreement of the guidelines involved considerable discussion at CSG and EMF.

Table 3: Extract from guidelines used to appraise policies

Topic	Measurement method	Criteria to score a policy as 'green'	Criteria to score a policy as 'amber'	Criteria to score a policy as 'red'
Flood and erosion risk	Appraisal of risk to people and property is undertaken using the order of magnitude of the number of properties predicted to be affected. The projected erosion lines or flood outlines are used to identify the properties at risk in the Character Area for each epoch. The cumulative total of houses lost by the end of each epoch is scored. Flood standard is used as an indicator for scoring in flood areas.	No properties lost to coastal erosion.	No properties lost to coastal erosion. One or more properties with a flood standard between 1 in 50 years and 1 in 20 years.	One or more properties lost to coastal erosion. One or more properties with a flood standard less than 1 in 20 years.
Communities	The project erosion lines are used to identify settlements at risk in the Character Area for each epoch.	No settlements lost or affected.	Properties lost or affected on the periphery of settlements. For areas at risk of flooding, flood standard between 1 in 50 and 1 in 20 years.	Coastal erosion or flooding affects the integrity of one or more settlements. For areas at risk of flooding, flood standard less than 1 in 20 years.
Natural environment	Scoring of damage to natural environment assets is undertaken based on an assessment of the likelihood of impacts and the designation level of the affected site.	None or minimal impact likely on non-designated sites.	Potential for negative impacts on internationally designated sites or significant likely impacts on other sites.	Likely to be negative impacts on internationally designated sites or significant impacts on other sites.
Agriculture and industry	Losses of agricultural land in general are scored according to the order of magnitude of the agricultural land area lost in the Character Area, estimated using the	Less than 100 ha of agricultural land lost and no grade 1 and 2 agricultural	Between 100 – 1,000 ha of agricultural land lost or less than 100 ha of grade 1 and 2	More than 1,000 ha of agricultural land lost or more than 100 ha of grade 1 and 2

	predicted erosion lines. The losses are scored on the cumulative area lost by the end of each epoch. Losses of grade 1 and 2 agricultural land are scored on the basis of the order of magnitude of the grade 1 and 2 combined area lost, estimated using the predicted erosion lines. The losses are scored on the cumulative area lost by the end of each epoch.	land lost. For areas at risk of flooding, flood standard greater than 1 in 50 years.	agricultural land lost. For areas at risk of flooding, flood standard between 1 in 50 and 1 in 20 years.	agricultural land lost. For areas at risk of flooding, flood standard less than 1 in 20 years. Loss of any significant industrial site.
Landscape	The effects of policies on landscape are in comparison to the current landscape condition in the Character Area – on the basis of an expert view.	None or minimal negative impact likely on landscape quality.	Potential for negative impact on landscape quality.	Significant negative impact likely on landscape quality.
Coastal processes	The effect of a policy on coastal processes is compared to the present day baseline and down-drift impacts are considered.	None or minimal negative impact likely on coastal processes.	Potential for negative impact on coastal processes.	Significant negative impact likely on coastal processes.

The initial results of the appraisal were then reviewed by the CSG and EMF at dedicated workshops, and subsequent revisions and fine-tuning of the assessments were undertaken. An extract from the policy appraisal is shown in Table 4.

Table 4: Extract from policy appraisal for Flamborough Head to Sewerby

Policy tested: No Active Intervention for all epochs along the entire frontage, but maintaining access to, and functionality of, the RNLI Station at South Landing						
Objective	Epoch 1 (2025)		Epoch 2 (2055)		Epoch 3 (2105)	
	Score	Explanation	Score	Explanation	Score	Explanation
Landscape						
To maintain and where possible enhance the quality of the coastal landscape	green	A general policy of No Active Intervention would ensure the coastal landscape is maintained.	green	As epoch 1	green	As epoch 1 and 2
Coastal processes						
To prevent interruption of coastal processes which supply sediment to other coastlines	green	A No Active Intervention policy would ensure coastal processes continue and sediment pathways are maintained.	green	As epoch 1	green	As epoch 1 and 2

Historic environment							
Minimise damage to designated and significant historic environment assets (such as Buckden Dyke and Danes Dyke) from coastal erosion	green	This policy would result in the loss of or damage to approximately 7 records noted by RCZAs due to slow erosion of the cliffs	amber	This policy would result in the loss of or damage to approximately 10 records noted by RCZAs due to slow erosion of the cliffs	red	This policy would result in the loss of or damage to approximately 16 records noted by RCZAs due to slow erosion of the cliffs. 2 listed buildings would also be at threat of erosion as well as 2 scheduled monuments.	
Ensure coastal defence works do not threaten designated and significant historic environment assets	green	No new coastal defence works that would threaten designated or significant historic environment assets would be undertaken under this policy	green	As epoch 1	green	As epoch 1 and 2	

Timing objectives: provide sufficient time for:		Explanation
Community adaptation	green	Due to the slow erosion rate in this area, it is considered that there would be sufficient time for communities to adapt.
Relocation/ adaptation of sewage works and other key community services and utilities infrastructure	green	Due to the slow erosion rate in this area, it is considered that there would be sufficient time to adapt or relocate infrastructure.
Research of archaeological features and ecological surveys	green	Due to the slow erosion rate in this area, it is considered that there would be sufficient time for research and surveys.
Provision of recreational access to the foreshore	green	Due to the slow erosion rate in this area, it is considered that there would be sufficient time to provide access to the foreshore at all times.

Following the appraisal, results were represented graphically by category to visualise the balance of outcomes that each policy package achieves (see Figure 2). This representation of the results allowed an efficient evaluation of the relative impacts and effects of the policies, demonstrating those policies which provided greater overall benefits compared to others. In some cases, the balance of outcomes between different policy options was fairly equivocal, meaning further consideration of the two options was required. No attempt was made to quantitatively compare or numerically weight different criteria such as property loss or habitats – instead the overall impacts of each possible policy approach were assessed and the consensus-based approach was used to select a preferred policy for each length of coast. In some cases this led to a future conditional policy being recommended, where the viable alternative could be adopted instead of the currently chosen policy if monitoring and further studies demonstrate that it becomes a more preferable option due to changing circumstances triggering a need for a change in policy.

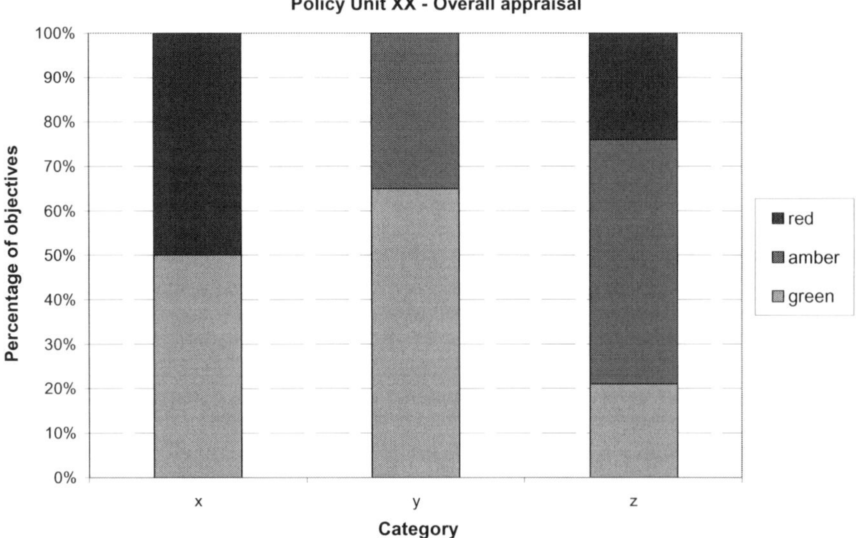

Figure 2: Example of graphical representation of policy appraisal results

In parallel to this 'traffic light' appraisal approach another key part of the policy assessments leading to the selection of the preferred policy was the consideration of various legislative requirements and the wider sediment transport impacts of different policy combinations. Following these iterations, the details of the preferred policies were agreed through CSG and EMF workshops and the draft Plan was confirmed. The draft Shoreline Management Plan was released for public consultation in November 2009 and the final Plan was completed in December 2010.

Consultation

Consultation formed an important and integral part of the policy development process. The public were consulted on the initial policy options and the proposed approach for assessment at an early stage of the process through a number of exhibitions along the frontage. In addition the CSG and EMF were closely involved in the entire process, agreeing the general approach, policy options for testing, appraisal methodology and developing the draft preferred policies. This was an inherently iterative process and one where the preferred policy was fine-tuned and optimised.

Consensus-based approach

The approach necessitated frequent face to face meetings and workshops to debate policy options, policy appraisal and eventually final policy wording. Elected members were involved in meetings from the outset of the project to ensure political concerns were adequately reflected throughout the policy development. This approach meant that progress was steady and iterative rather than dramatic as input was needed from all organisations at each stage of the process before proceeding to the next stage. As a result, more than 130 meetings and workshops were held over the course of the project which took 3 years to complete. However it resulted in a set of policies that all organisations involved signed up to, and which address the difficult and politically sensitive issues which will increasingly be faced in the future.

The policies acknowledge the need to consider managed realignment of defences to ensure they are sustainable. They also acknowledge that in some cases communities will not be protected, and that settlements will have to be abandoned. This is a significant step forward for a region coming to terms with the implications of climate change and sea-level rise. The incorporation of these policies, and their acceptance, is due in no small part to the involvement of elected members in the process of Plan development and the consensus-based approach that has been taken.

A particular outcome of the process, which stems from the consensus-based methodology, has been the adoption of an aspirational approach. In some key areas the ability to provide defences, or to provide a sufficiently high standard of defence, is in doubt. Government funding may not be sufficient. A top-down, expert-led approach may have considered that these policies could not be funded under current arrangement and resulted in a different set of policies. The consensus-based approach has acknowledged this issue, but decided that it should be left to local communities to take this up, and if necessary find local sources of funding. There are examples of private funding of defences in the region, and elsewhere examples of councils setting up tariff-based funding for flood defence. The mechanisms of local funding have not been explored in the development of the Plan, and this is an issue which the Action Plan identifies will need to be addressed in the future.

Conclusion

The approach taken to this Shoreline Management Plan differs from most of the other Plans that have been produced. The large geographic scale, diversity of issues and high significance of the policy outcomes has been addressed by adopting a consensus-based approach to policy appraisal and selection. The process has taken considerable effort by all involved, but the result is a genuinely strategic Plan which has the buy in and backing of key parties and one which can be used to drive sustainable management of a complex coastline.

Acknowledgements

URS Scott Wilson would like to acknowledge the contributions of the following people and organisations over the course of the project: East Riding of Yorkshire Council, particularly the client project manager Mike Ball, CSG chair Jeremy Pickles and EMF chair Councillor Jonathan Owen; Environment Agency, particularly Mike Dugher, Mark Robinson and Philip Winn; Lincolnshire County Council; North East Lincolnshire Council; East Lindsey District Council; English Heritage; Natural England; National Farmers' Union; and Jaap Flikweert of Royal Haskoning.

URS Scott Wilson note that the opinions expressed within this paper are solely the views of the authors and do not necessarily reflect the views of the organisations listed above.

References

1. URS Scott Wilson Ltd, 2010. *Flamborough Head to Gibraltar Point Shoreline Management Plan*
2. Royal Haskoning, 2010. *The Wash Shoreline Management Plan*
3. Defra, 2006a. Shoreline Management Plan Guidance – Volume 1: Aims and requirements
4. Defra, 2006b. Shoreline Management Plan Guidance – Volume 2: Procedures

Innovative Coastal Zone Management
ISBN 978-0-7277-5749-4

ICE Publishing: All rights reserved
doi: 10.1680/iczm.57494.347

"And forward, though I cannot see, I guess and fear." A move from Continual to Continuous Management

Gregor Guthrie, Royal Haskoning, Peterborough, England
Jim Hutchison, Independent – York, England
Jaap Flikweert, Royal Haskoning, Peterborough, England

Introduction

Adaptation lies at the heart of our approach to managing the coast, recognising uncertainty and the need to adapt to the dynamic system that we are managing. Adapting is as relevant to our thinking about how we manage the coast, and in adapting to our changing understanding, as it is to the way in which we actually adapt to the change.

This paper takes a long term perspective, both looking to the past as to where we have come from and looking forward to a future that remains cloaked in many uncertainties; uncertainties that have led some to question whether we can plan for 100 years and beyond. The paper looks specifically at some of the issues through examples and examines how the Shoreline Management Plan process helps by providing a structure and baseline that allows flexible response, rather than proactive reaction. This does, however, require a challenge to our thinking as to how we use and develop upon the work that has been done. We have to manage as a continuous developing process. We have to move from the security of a prescriptive step by step guide into the future to one of at least knowing where we are going and the problems we are trying to avoid.

The second round of Shoreline Management Plans (SMP2) does take this long view forward and has to, if we are to make sensible decisions now that do not drive us down unsustainable dead ends. We see now how, in some places, the old Victorian approach of sea walls presents serious problems for the future. We see the reactive management along large sections of the east coast, following the great storm of 1953, creating issues that will persist well in to the future. Indeed, it might be argued that if the Romans 2000 years ago had had a better eye to the future, they might not have built that bridge over the Thames and we might not now have a large part of our capital city (and headquarters of the Institution of Civil Engineers) within the flood plain.

But our attitude to planning also has to change. We have to accept that we do not fully understand the coast, that our priorities and perceptions may change and that change, in many areas, needs to be through a process of evolution rather than sudden changes in direction. This requires a different sort of planning, one that remains open to new ideas, one that can respond to the unforeseen and one that focuses on the intent of where we are going, less on a prescriptive manner of getting there. This is captured in the Guidance for SMPs[1]:

*"A guiding principle is that the SMP needs to define a long-term sustainable **plan**, even though that may change with time. The SMP will need to provide a 'route map' for decision makers to move from the present situation towards the future."*

Drawing the Line

It is valuable to look back to the discussions that are reported in the Coast Erosion Commission[2] held at Southwold in 1906/07. A significant amount of the evidence presented focussed initially on trying to determine where the coastline actually was. Evidence was given by the Ordnance Survey and various land owners as to how we can measure erosion. There was discussion as to what line defined the shoreline and what was meant by erosion. Classic examples were given trying to understand how, as a coast changes, areas of accretion become areas of erosion and visa versa. Modern equivalents can be seen of this problem in places such as Pembrey, on the south Wales coast. Here a flattening out or erosion of a previously prograding shoreline gives rise to a pattern of lateral accretion either side of the main central section. As erosion continues so the areas of accretion become areas of erosion. Similar processes are seen at the nose of Orford Ness making it difficult to assess when the lighthouse there will be lost.

Far more complex patterns of change are seen elsewhere. On the Teifi Estuary, in the west of Wales, we see the change in growth of the two spits either side of the estuary (Figure 1a).

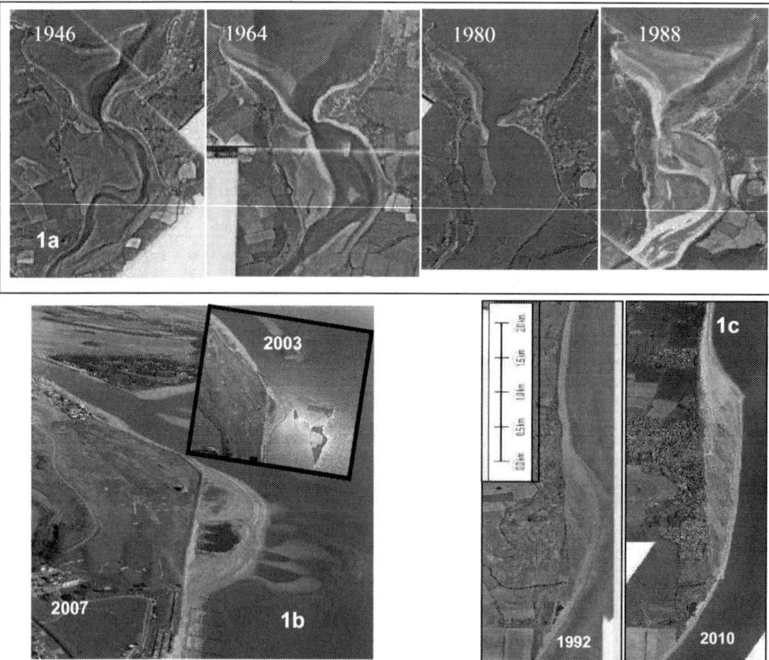

Figure 1. Changes in pattern of behaviour – a) Teifi, b) Deben, c) Benacre

At the Deben (Figure 1b), the changes at the entrance to this estuary results in large pulses of sediment being transferred between the nearshore bank system on one side of the estuary over to the southern shoreline. It also influences the pressure for erosion on the southern shore in an unpredictable manner. These patterns of variability are not confined to the interaction between the coast and estuaries. Just north of Southwold, where the 1907 commission was held, the massive shingle feature of Benacre Ness (Figure 1c) has been moving progressively north possibly for centuries. Typical movement of this feature is measured as being around 20m per year. The Ness covered the outfall to the Kessingland Hundred River valley some 60 years ago and a seawall was built to the village of Kessingland in the 1950s to the north of the Ness at that time. Now the outfall is at risk due to erosion and the sea wall lays several 100s of metres to the back of the active shoreline.

Quite clearly the problems identified 100 years ago, in defining erosion and what is the shoreline, are still with us. This has been the very problem that has taxed minds in developing coastal erosion risk maps over the last few years; how can we define and predict this anamorphous system. It is equally of interest in understanding our change in understanding of the coast.

In evidence to the commission it was also stated that our focus should be on the behaviour of the shoreline, that the nearshore bank systems were not significant in this, in that they had little interaction with the general pattern of sediment movement at the shore. Such an attitude now is hard to perceive but does raise questions of what we now understand to be accepted wisdom.

More recent research (Southern North Sea Sediment Transport Study[3], BLINKS[4], Long Term Geomorphological Change[5]) all highlight the need for a systems understanding of the processes, not replacing the detailed understanding of the mechanisms that drive change but providing a framework within which local change can be assessed. Such good practice does mitigate the chance of the unforeseen but also demonstrates the value of the longer term high level view taken by SMPs.

Headline or Intent

It is against this background that the SMPs are being developed. The flexibility of the SMP2 guidance allows these issues and uncertainties to be discussed, as much as defined.

In the case of the Teifi and the Deben, the respective SMP2s recognise that there is a broad range of interests affected by the way in which the coast and estuaries are managed. In many ways the SMPs take a role of providing an understanding of the processes and history of how the areas have developed, allowing us to look more closely at how changes may affect different interest groups, allowing those interest groups and communities to become involved in a sensible manner. Fundamental to this is the SMP role in defining the risks from coastal change and flooding.

However in the attempt to squeeze such complex issues into the policy "labels", definitions of No Active Intervention, Hold the Line, Managed Realignment or Advance the Line negates the benefit of all the understanding the SMP brings to management.

In the case of the Teifi, the SMP2[6] (still in draft), recommends:

"The management of the area covering Poppit Dunes, The Webley Hotel and Pen yr Ergyd needs to be considered as a whole, with the intent to develop an integrated approach which sustains the various interests while not rigidly fixing the estuary entrance. This approach cannot be defined precisely by the SMP and requires a detailed management plan. This may

require the loss of land to the Caravan Park both to provide width for management and sustainable management of defences to the main area of the Caravan Park. This plan is needed urgently as by default the spit will breach and opportunity for an integrated approach will be lost. The aim would be to maintain the RNLI station." This is supported by a more in-depth discussion within the document. This is a far cry from the more prescriptive approach taken in SMP1, a decade earlier, which considered each of the above frontages section by section.

In the case of Benacre and Kessingland, the SMP highlights the choices that have to be made in terms of the future of the coast. The plan is more definite, that to hold the current line of flood defence is unsustainable. It then goes on to discuss potential management approaches that could still allow management of the key interests. The SMP sets clear boundaries for management but provides the opportunity to develop management at a local level with the communities.

Thorpeness

This setting of boundaries is important in allowing a more continuous process of management to develop. Thorpeness (Figure 2) is situated on the Suffolk coast to the north of the town of Aldeburgh. The ness is an important feature of the coastline and acts to retain sediment to the north, while still allowing sediment to migrate to the south.

Analysis of typical beach face and backshore positions, together with various nearshore features.

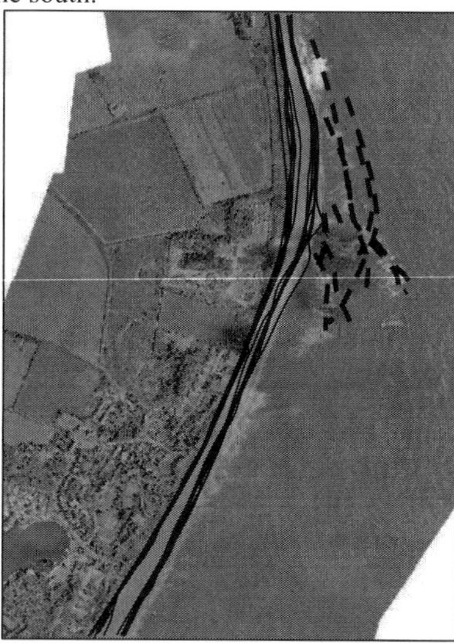

Figure 2. Typical change in coast at Thorpeness.

Collated information for the area within the SMP2 showed that the general line of the coast is retreating slowly at a rate of approximately 0.1m/year. It was identified that due to local behaviour of the ness, there were short periods when the coast, in places, may rapidly alter shape, with erosion increasingly rapidly, locally. This had typically happened every thirty years but the coast readjusts over a period of a few years with accretion back to a more stable condition. Such local erosion may only occur over a length of a few hundred metres. The SMP policy for the general frontage is for No Active Intervention. Under this policy it was estimated that two to three properties would be at risk over the next 50 to 100 years. The policy recognised that there could be local risk and was qualified in allowing local management subject to certain conditions. These being that the long term intent for the frontage was to allow the shoreline to move back gradually, such that management did not interrupt the sediment feed to the south, affecting the sustainable management of defence to Aldeburgh and of the internationally important areas of habitat in the area. The local policy is clearly linked to a suite of policies along the whole section of the coast consistent with management of the whole area.

Within months of agreeing the SMP policy, a particularly severe period of erosion occurred. This threatened some 12 properties in the area, with probable loss of some 5 properties within the year. Using the guidance provided by the SMP a scheme has been developed, with financial support from local residents and the operating authorities, to safeguard these properties in the short term.

Quite naturally, there has been comment that the SMP policy should be changed and, if viewed from a single policy level perspective, there might be some justification for this. However, even within the community, there is support for the approach taken by the SMP. There is recognition that, in the long term, the intent of the plan for the area is sensible, that in the long term, without major intervention that would radically alter the way in which the coast is managed, there needs to be a natural adjustment and adaptation in the way in which we use the coastal area. The plan allows time to adapt without driving management down a dead end.

The high level policy for the area remains the same but, within this, there has been scope to address local short term management issues in an appropriate manner. The further detail developed in understanding these local issues has built upon the information within the SMP and build towards a process of continuous management.

Looking back to SMP1

The shoreline management process has been one of evolution. The first round of SMPs was a major step forward in how we have subsequently looked at the coast and considered what is sustainable. Since the early 1990s there has been a significant revision in the way in which we consider aspects such as sea level rise. Back in the 1990s, climate change was recognised but, within the coastal management community generally, was viewed more deterministically and far less as the major issue that we recognise it to be today.

The changes between SMP1 and SMP2 have in many ways been more subtle. SMP1 guidance was developed from a growing awareness of the need for more holistic thinking. SMP1 was a culmination of on-going discussion developed over time. Comments in this respect, reported in ICE proceedings, from 1949 and 1956 are quite illuminating:

"A country whose leaders were continually urging the necessity of food production and which permitted agricultural land, even of small extent, to be inundated was surely insane.(1949)."

"Post-war housing schemes have brought along their very many problems for the local authority Engineer but it is suggested that one such as the construction of sea defence works of the magnitude now being executed at Aberavon is unique (1956)."

In the first instance it is interesting that, at that time, the issue of food production was considered an issue in much the same way as it is emerging now, given the risk of inundation of major areas of the Wash; even if the perspective of "not to give an inch" is possibly not how this problem might be addressed through current SMPs. The second quote, from the

Deputy Borough Engineer of Port Talbot, highlights the growing concern that the issues of planned development were integral to how we may manage the coastline.

The focus of SMP1 was driven much more by this perspective that the coast could not be managed in isolation from issues relating to land use, the historical and cultural landscape and the natural environment. Taking a 50 year perspective was also a major step forward. This started to think about the long term impact on defences due to coastal change. However, even in taking this broader view, a 50 year perspective typically looked forward over the life of many of the defences. It was only as SMP1s were developed that this raised the understanding that we were not merely considering whether a defence might be managed over its lifetime but should, in fact, be looking beyond that to where this would take us in the future. SMP1 starting the thinking of where do we want to be in the future and how do we get there. The process also awoke the greater complexity of issues relating to how we value the coast and how we need to manage an entire socio-environmental system.

The Wash SMP2[7]
The situation in the Wash provides a good example of how recognising this uncertainty and allowing for it becomes essential to the development of thinking into the future. Although complex in detail, the issues relating to the Wash focus down on two critical aspects. On one side (quite literally in terms of the sea banks) there are the extensive areas of protected low lying land, extending inland up to 50 km from the shoreline, protecting areas of Lincoln, Peterborough and Cambridge as well some of the most productive agricultural land in the UK, including 50% of England's Grade 1 land. The agricultural industry supports settlements such as Spalding, Boston and King's Lynn and a large number of smaller communities, many of which are situated in a band around the Wash, some 5 kilometres from the shoreline. On the other side of the sea banks it's a different world: important expanses of saltmarsh, mudflat and sandflat shaped by the tidal currents. The saltmarsh has been accreting for the last hundreds of years, and is continuing to do so, partly in response to sea level rise. However, this accretional process could reverse over the coming decades. If the saltmarsh eroded, the seabanks would become much more exposed to storm waves and tidal currents; also, the sensitively balanced intertidal habitats would be under threat. With current knowledge, it is impossible to be certain if and when this reversal from accretional to erosional trends could occur. This is a problem, because the impact of this potential change on shoreline management is fundamental. It would be a big mistake, in this case, to base long term policy on effectively a best guess at what will happen. The assumption of an erosional future would set in motion a process toward managed realignment (if only by force of the Habitats Regulations) which might turn out to cause unnecessary loss of good land. An assumption of an accretional or stable future would lead to a firm Hold the Line decision, creating expectations that may be unjustified.

The SMP has dealt with this conundrum and potential political minefield by proposing a conditional policy. There is a clear intent of management that all established communities around this part of the Wash will remain defended. In an accretional future, this will be achieved by holding the current alignment. However, in an erosional future (Figure 2) the best way to defend the communities could be to carry out localised and limited landward realignment where needed, in order to recreate the saltmarsh bufferzone for the seabanks and replace the intertidal habitats that would be lost in this scenario. The wording of the policy confirms the need for monitoring and continuous review of shoreline management, aiming to improve our understanding of intertidal development so that any change in the process can be anticipated and planned for. The conditional policy was the only approach that would enable agreement with all of the SMP's partner authorities and the approach of adaptive management was recognised as the right way to deal with uncertainty.

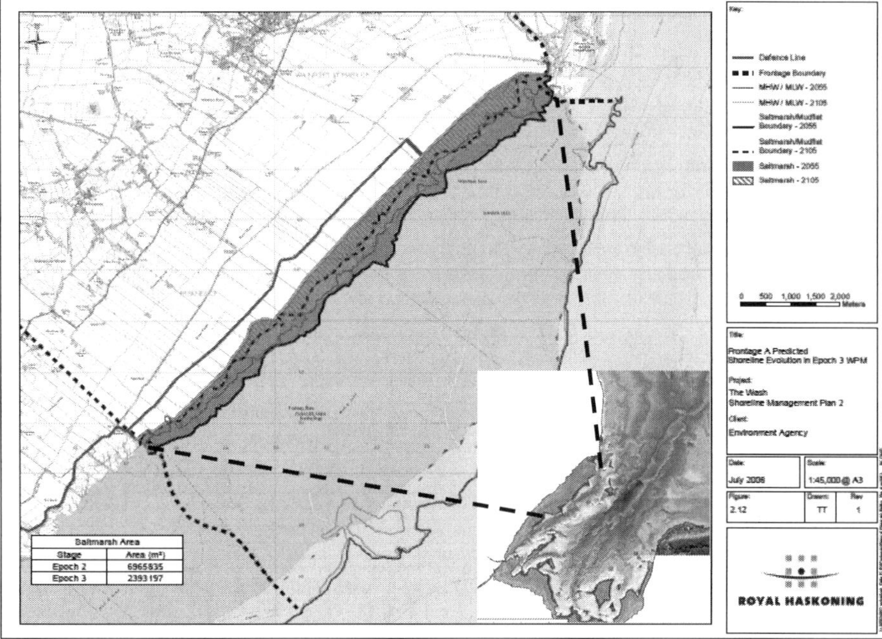

Figure 2. Predicted decrease in saltmarsh in an erosional future.

Understanding the Plan and Policy

Individual polices for local sections of the coast have to be understood as a guide to achieving the plan for the future. Policies are developed over epochs (20 years, 50 years and 100 years) but have to be recognised as not a step in policy but driven by the developing situation and conditions, not by specific time periods. While providing a basis for planning management in detail, informing anticipated investment in the coast, in the broader sense policy cannot be viewed in isolation from the overall intent of management. The focus on policy in isolation has exercised much thought and concern during consultation, not least in terms of how shoreline management relates to the spatial planning process.

Each SMP2 has attempted to address this by adding to the definitions given in the SMP Guidance, trying to reflect, quite correctly, the specific issues being faced in each area of the coast. While these individual definitions are possibly helpful in adding to the understanding of users of the SMPs, there is concern that it introduces a level of inconsistency between SMPs. This in itself highlights a concern that the policy definitions are taking over from the intent of the plan. In reality the policy "headlines" are no more than that, just labels that give an initial flag as to how the coast is seen as being managed. The actual range of policy definition is far more complex and overlapping. The intent of management describes the continuum of ways in which we may manage the shoreline. This is captured in Figure 3, making the distinction between policy and implementation. No Active Intervention may, for example, cover cases where there is an imperative need not to interrupt important sediment drift, it may equally cover cases where there is not felt to be a good case for national funding or it may simply refer to natural sections of the coast where there is no wish to intervene.

Similarly, Hold the Line may have an intent to reinforce the existing defence, it may equally refer to areas where the existing defence approach might sensibly change but where the intent is still to protect assets behind the existing defence. Particularly for flood defence shorelines, there is a range of potential levels of risk management that may be essential for long term shoreline management; for example the headline policies P1 to P5 as used in the Catchment Flood Management Plans. Such distinctions can only really be defined through discussion but that means reading the document. Rather than attempting to more closely define the policy headlines and by implication ending up with a whole complex suite of detailed policy terms, possibly it would have been better to have had just two polices; to manage or not to manage. This may have highlighted the need to understand the complex issues discussed within the SMP documents. It also highlights the need for good engagement with people in delivering that understanding.

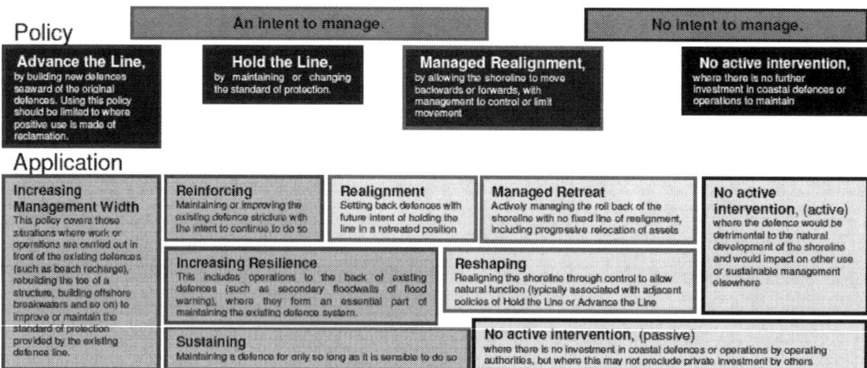

Figure 3. Rationalisation between policy and application of policy.

This difficulty in wanting to clearly fix what each policy means is seen in relation to undertaking the Habitat Regulation Assessment for SMPs, where there can be a perception that a policy of Hold the Line means no movement at all from the line of the existing defence or where a policy of Managed Realignment is seen as either being a managed retreat or retiring a defence to a very specific position. It can equally be misrepresented in planning terms such that in the recent supplementary guidance to PPS25 (Policy Planning Statement – Development and Coastal Change), Coastal Change Management Areas are required for areas other than Hold the Line policy areas.

To Hold the Line may present some of the big issues relating to planning, in looking at sustainable ways to actually hold the line.

Scarbourgh and Pwllheli

Scarborough and Pwllheli are both important regional towns. The policy, unsurprisingly, is to Hold the Line to the central areas of each town. However, this is not the end of the story.

At Scarbourgh, the sea front is an essential part and value of the town. Either of the two bays, North Bay and South Bay, in different ways are integral elements of Scarborough. However, both have problems. In the case of North Bay there is concern over loss of the beach as sea levels rise. There is an opportunity to manage this but ideally associated with the way in which the promenade and backshore areas are developed in the future. Within South Bay, while the beach is relatively stable, there is significant flood risk to the properties, shops and restaurants that line the sea front. The SMP highlighted that while the policy is to Hold the Line there is significant scope, in engineering terms, in landscaping and through adaption to reduce risk and improve the use of the area.

Pwllheli is one of the main towns situated on the south coast of the Llyn Peninsula. The developed centre of the town is recognised to be at significant flood risk with the further risk of erosion along the open coast. With sea level rise these issues increase (Figure 4). These issues are being examined through the Pwllheli Pilot project, involving the local authority, the Environment Agency Wales and CCW, supported by the Welsh Assembly Government. This is not a problem of new development but that of the sustainability of an existing community. Major changes have to be considered, with local studies examining geomorphological change and coastal and fluvial flood risk. The SMP2 has been developed in discussion with this study, providing an important higher level perspective. The issues are complex and it is only through working together, considering the long term view, that a sustainable solution will be possible. Such solutions have to be developed over time with response being planned within the understanding of all the uncertainties. Planning has to happen now, recognising that future actions may have to be adapted to actual outcomes. SMP policy may provide a possible route map but it is really the SMP thinking that will help guide such future management.

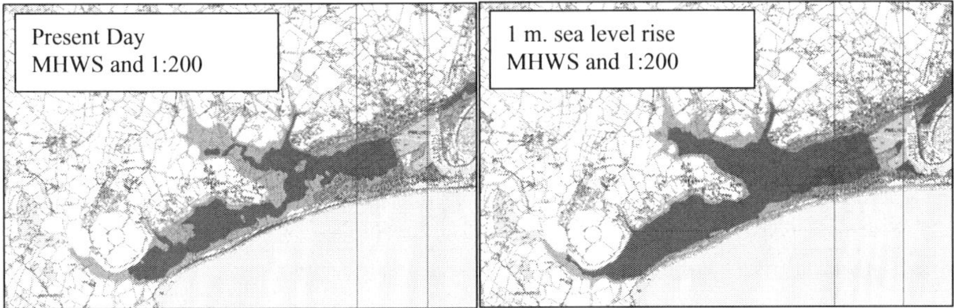

Figure 4. Pwllheli Flood risk – present day and with 1m seal level rise.

Conclusion

We know a lot more than we did in the past and the SMP process, both SMP1 and SMP2, have added immensely to this knowledge and in drawing together the information necessary for sustainable coastal management. This provides a solid base from which to plan how flood and coastal erosion risk management can contribute to this overall process into the future.

However, we also know a lot more, than we did in the past, about what we don't know. Understanding this is equally important. The continuing shoreline management process has to understand this and, we believe, is generally understood. We cannot afford to wait another ten years, and another ten years, undertaking reviews of the plan in discrete steps. We have to develop upon the plans as of now as a continuous process.

Having such broad level strategic plans provides structure and confidence within which local decision making can take place. Far from imposing decisions on communities, the SMPs realise the opportunity for communities to be engaged in decision making. We recognise that there will be change as management is taken forward and will need to adapt to changes in social priorities, funding constraints and our developing understanding of how the coast actually responds. The SMP process recognises this, testing different scenarios and setting out the assumptions that have formed the plan. This provides an understanding of why particular policies have been put in place. We are now in a much stronger position from which we can further test these critical assumptions and adapt the plan accordingly.

The paper has set out some areas of uncertainty and through examples has considered how the SMP process can be used both immediately and in the future. With such threats as sea level rise, present and future threats to existing communities, we possibly do need the courage to both "guess and fear", but from a perspective of taking forward management based on solid foundation with much greater confidence.

References

1.   Shoreline Management Plan Guidance, Defra, 2006

2.   *Coast Erosion Commission Report*, Southwold Commission 1907

3   *Southern North Sea Sediment Transport Study, Phase 2. Report Ex 4526.* HR Wallingford. 2002

4.   *Beach LINKs to Sandbanks, funded by the Natural Environment Research Council, NE/B503917/1.* T.J. Dolphin, C.E. Vincent, C. Coughlan *(*School of Environmental Sciences University of East Anglia, Norwich). 2010

5.   *Characterisation and prediction of large-scale long-term change of coastal geomorphological behaviors. Science Report: SC060074.* Defra / Environment Agency FCERM Research Programme 2009

6   *West of Wales SMP2.* Pembrokeshire County Council (Draft 2011)

7   *The Wash Shoreline Management Plan 2 – Gibraltar Point to Old Hunstanton.* Environment Agency, 2010

Innovative Coastal Zone Management
ISBN 978-0-7277-5749-4

ICE Publishing: All rights reserved
doi: 10.1680/iczm.57494.357

Approaching Future Shoreline Management

Kevin Burgess Halcrow, Swindon, UK
Jim Hutchison Independent, York, UK

Introduction

With the second generation Shoreline Management Plans (SMP) for England and Wales substantially completed, it is timely to reflect on the fundamental reasons for these being produced and consider whether they are fulfilling their potential.

The first round of shoreline management plans were independently reviewed a decade ago and subsequently guidance was produced for plan revision, which included a number of key principles:

- acceptance that current policy may no longer be feasible or acceptable at some time in the future - we should not be tying future generations into inflexible and expensive options for defence;
- the need to recognise longer term coastal evolution and for plans to be more visionary in terms of offering realistic options for the future;
- where there is conflict with natural change then the true consequences of adopting particular policies need to be identified;
- plans and associated policies need to be practical and achievable in the long term, being realistic about how risks can change over time and facing up to these;
- recognition that attitudes, behaviours and thus objectives may well be different in the future to those now
- the plan should provide a route map for change where it will be needed.

Delivering on all of these is perhaps inevitably easier said than done as it remains difficult to reconcile long term planning with current day pressures, and national perspectives with local aspirations. But, of most concern is that short term motivation or fear of change will prevent us from taking the right decisions for the future and we will simply continue to put back the 'hard' decisions until later and allow them to become 'somebody else's problem'.

A changing coastline

Historical records are full of accounts of flooding and disappearing villages and towns, especially along soft and eroding coastlines. In the past whole communities once existed in areas that now lie beneath the sea. One of the most renowned examples is Dunwich, originally the capital of Anglo Saxon East Anglia, now but a single street with almost half a kilometre of the original town having disappeared. The shoreline management plan covering the coast of South Yorkshire includes a map that showed the villages lost since Roman times as a direct consequence of coastal erosion. Elsewhere, shingle barriers have migrated landwards and overrun developments. Today the old town hall in Aldeburgh in Suffolk has an uninterrupted view of the sea; however, when it was built there were six roads between it and the shoreline.

These occurrences and others like them are not a direct consequence of man's intervention, or due to accelerated climate change at the sorts of rates speculated in the media; they are the result of nature at work. This natural change includes long stretches of coast where the entire system is transgressing, retreating and lowering in response to various factors including the consequence of past and present rise in sea level. This is an ongoing process that we should expect to continue into the future irrespective of whether rates of sea level rise accelerate or not. And if it does occur, it will serve only to exacerbate the existing situation, speeding up the rate of this transgression and probably increasing the geographic extent to which it is experienced around the country.

Since the major coastal event in 1953 when several hundred people died, significant public investment has been made to build defences to avoid any such repetition. However, although these defences were constructed to defend us from the sea for all the right reasons at the time, it is only in recent years that we are beginning to learn about the repercussions of attempting to "hold the defence line". In most cases we can defend against the sea, but we cannot resist the inevitability of nature and natural coastal processes. As defences consequently get bigger and bigger to resist these processes, the costs proportionally increase, and this is already becoming unsustainable in many cases. This number will continue to grow unless we consider radically different solutions on our coasts from what we see today.

A technical challenge

Through numerous engagements with the public and various non-technical bodies it is apparent that many people's expectations regarding long-term coastal defence policies are misplaced. There is commonly a call to continue to "hold the existing defence line", but this is coupled with a perception that after 100 years or so the shoreline will continue to look exactly as it does now. This is simply not correct in many instances.

Imagine if the community of Dunwich, Suffolk had been able to successfully hold their original defence line some three to four centuries ago. Even with structures in place to prevent the landward claim of the sea, the naturally transgressive processes would not have prevented natural foreshore lowering from occurring nor continued erosion of the adjacent coast. We would now have a headland extending a half kilometre out to sea, almost certainly though encased in a ring of massive concrete structures. Elsewhere, had Aldeburgh been successfully defended before it lost the six streets seaward of the town hall, it is highly likely that ongoing transgression would have led to coastal 'squeeze' and the complete loss of the foreshore, with a gigantic wall in it place, permanently sitting in deep water and most likely reaching the same height as the rooftops themselves.

This is just an illustration, but we already see coastal squeeze and beach losses in front of many of our older defences. So, whilst we are used to having beaches in front of most of our defences at the present time, as shown in Figure 1, in the future we should expect to see many more towns and villages with frontages similar to that in Figure 2 even at low water, and perhaps undesirable places to reside, with no beach amenity and no tourist trade either.

Figure 1

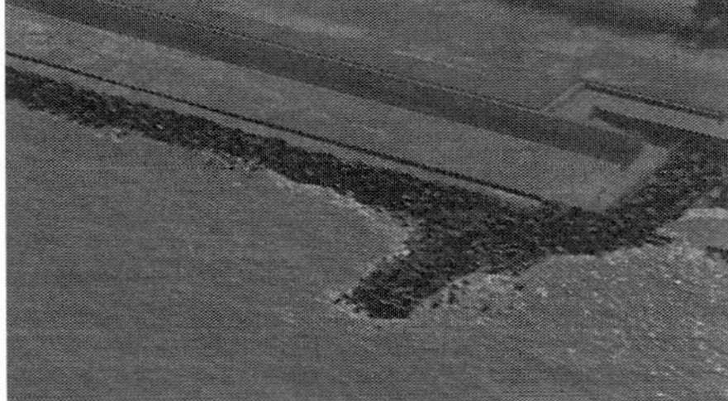

Figure 2

It is evident from policies already presented in the draft second generation shoreline management plans that have gone out to consultation that we intend to continue to hold exactly the same line in 100 years time at almost three-quarters of locations that we currently defend. This perhaps suggests an 'unwillingness', or maybe just a lack of appreciation, to accept the way coastlines are likely to change over future decades, possibly not on the part of those preparing and developing plans but for those in the public and political arenas. Is it simply that the science is not believed? Even without accelerated climate change as forecast, the facts regarding past coastal evolution are indisputable, so why are similar arguments disregarded for future projections?

We certainly live in more affluent and technically advanced times than our predecessors and while we might have the skills, technologies and resources to be able to resist natural processes, we should not expect to be able to sustain that position forever.

The cost of providing defence

With the second generation shoreline management plans potentially promoting little change in policy at a national level, the next question we then have to ask ourselves is, even if the preferred policies are technically acceptable and achievable, are the choices affordable and justifiable?

The majority of our existing defence structures will need replacing or substantial overhaul at some point over the coming decade as they naturally deteriorate, become increasingly vulnerable to structural failure, or simply need to be bolstered to maintain their effectiveness. National level studies to establish the total expenditure necessary to provide flood and coastal defence (e.g. MAFF, 2000; Defra, 2004), have previously concluded that simply replacing those defences we already have when they reach the end of their effective life will require a rolling programme of expenditure every year of approximately double that presently being spent.

Add to this the potential impacts of climate change. Even under the mildest of predicted changes, the estimated cost of replacing our structures to provide similar standards of protection to those enjoyed now will be double that now (Burgess & Townend 2004). At its more extreme this could rise to four times. An increase in funding therefore of at least double, quite likely four times and perhaps even up to eight times that at present, would only occur at the expense of other national needs.

In recent years the Government has provided the highest level of funding we have ever had for flood and coastal defences. Yet the numbers of locations where there is a need for work continues to grow. The cost of future construction to deliver the policies within the second generation shoreline management plans is going to run to several £billion. Unless we believe that this scale of funding is likely, then we have to question whether it is appropriate for these plans to be proposing some of the long term policies stated therein.

It might be argued that our forefathers may have been more pragmatic in 'walking away' from the sea as it claimed the land. The industrial revolution and the subsequent explosion in the size of the urban population offered up both greater incentive and technical knowhow to begin to resist this and have seen society spend more and more money building substantial sea walls and promenades to protect its valuable investment and leisure activities. This is a legacy that we have now inherited, but is it one that we should also leave behind?

It will be argued, quite rightly, that in considering affordability we must be alive to the associated impacts upon those who live, work and use these areas which will come under threat if they are not protected in the future. However, while it is very easy to identify those problems in the current day, are we mistaken in the belief that all of our present objectives will be the same in 100 years time? Social attitudes and behaviours will almost inevitably change over the coming century. One possible example of this is where we make the assumption and thus associated policy decisions around continuing tourism. In the face of a changing coast, and with larger and less attractive coastal defences required to maintain the coast insitu rather than allowing it to evolve, why do we assume people would still come, especially if there is no longer a beach as a result of deeper water at the structure? Also, why do we imagine that people will still be looking for these same types of leisure activities a century from now? If this is the basis upon which long term policy choices are being made, then these should be challenged.

As for communities, again a lot is rightly said about how the full costs of social impacts are not taken into account. But as well as looking at the local costs and impacts, a national perspective must be given to this issue; is it better to spend the taxpayers money on protecting tens or perhaps even hundreds of properties in some existing communities, when for that same level of investment we could bring health, security or infrastructure benefit to tens of thousands? But, are politicians prepared to face the consequences of this approach; and are we actually ready to deal with the losses arising? There is clearly an issue over long term versus short term decision making ability. Localism in decision making for defence provision may do little to improve matters in this respect.

Shoreline management plans

So why might we still have these issues, especially since we have had the shoreline management planning framework since the mid-1990s? We should be concerned that despite our continual improvements in understanding these issues we are not seeing any fundamental shift in thinking about longer term sustainability in terms of shoreline management.

The purpose of the updated Procedural Guidance (Defra, 2006) was to deal with some of the inconsistencies in approach, collect industry best practice and experience, and provide tools to help those developing SMPs. However, the overriding ambition was to help people challenge conventional thinking and encourage informed intellectual debate. We are though beginning to question whether this ambition is being realised and if there is a danger of retreating back into our 'comfort zones', forgetting the reasons why some significant changes to the SMP process were promoted. Or are those in a position to make this necessary change on the coast too driven by local issues?

Unfortunately though, instead of standing back to form well developed direction for future shoreline management, we have again become fixated on 'ticking the boxes' and focussed upon delivering a document rather than delivering ideas. In doing so the creativity on thought and application may have been lost, and as such the true value of the SMP to present a well considered vision of future change is not gained.

The intention of the second generation SMPs was to set out a vision for sustainable management of the shoreline in the future and provide a route map to get there. As such distinction was made between the longer term 'plan' and the more immediate term 'policies'. This is only reasonable as it reflects the uncertainties in both natural change and societal changes that are almost inevitable. However, we remain at a stage where we are, mistakenly, still trying to set long term policy at a local level rather than seeking to understand long term constraints, opportunities and demands, and assess whether present day policy is going to be consistent or contrary to that.

Future approaches

It is clear that there are influences and constraints, external to the current shoreline management planning development structure, that need to be looked at further if we are to achieve our ambition of sustainable shoreline management in the future and being able to provide the route map for change to get there. Decisions that are politically motivated with a local and/or short term perspective is just one of those.

A fundamental driver for future choices will be economics and the need to think about alternative ways to deal with risks that exist on the coast if we are not going to have investment levels many times greater than at present. In reality this means one of three ways for coastal defence in the future. First we can consider ways to better prioritise the available funding, and choose those locations where we believe we shall need facilities to continue to remain at the coast. By default this also means deciding where these features will not continue to be viable in their current form. Or we need to be smarter in ensuring all available funding streams combine to improve the coast. An example of this would be for regeneration funding to be aligned with flood and coastal funding.

Perhaps a better way forward is to consider better ways to deal with the risk once the existing structure reaches the end of its useful life and its ability to provide the level of protection necessary. If we are going to plan for this time frame, which may be in the medium to longer term, we need to use the planning system today to prepare for the necessary radical changes that will be the only way to adapt and avoid increasing costs of defending. In other words, start planning now to allow for a changing coastline in the future.

Even with good planning, there are those outcomes that will be unpalatable for some that have properties so close to edge of a cliff or within a high risk flood area that the decision can only be one of eventually leaving it. But we need to be able to manage this transition in an acceptable fashion. This is why Defra, along with the coastal operating authorities, has been working on an Adaptation Package to consider the changing nature of the coast. Although some of these proposals will undoubtedly provide a wider framework to assist the longer term approach argued by SMP principles, we should ensure that short term issues do not deflect us from taking the right decisions now to ensure future generations have sustainable coasts. To get us there however there are some key questions that we need to pose and consider if we are going to be able to tackle those "tricky" longer term questions.

There is nevertheless a gap between national government policy and what this means 'on the ground'. The expectation perhaps is that the shoreline management plans will provide this translation, but that is not happening. In part the problem is (a) because the structure of local engagement that we are operating in does not adequately enable an objective point of view which takes a national perspective, (b) because there is still some way to go for those in government to actually appreciate the full implications of national policy, and (c) our thinking remains too fixed in the present day and we are not giving ourselves the space or opportunity to be visionary. We therefore need to consider some other ways forward to help us to overcome these obstacles, as outlined below.

1. Have a national coastal vision

Consideration should be given to producing a national government overview on what is possible and where, then having local strategies pick up on the implementation once government policy for each region is set, probably only engaging local public and political input at that stage. This would help to identify the implications of national policies and thus help lead to approaches to deal with them, rather than what we have now where some of these implications are perhaps becoming 'buried' behind inappropriate policies, which is creating a lack of awareness in higher circles of the problems for the future that we are actually storing up.

Through this approach, an SMP would be a high level government piece, not a public consultation document. For example this might be a technical and economic assessment of possible future sustainable options, presenting the risks and opportunities for areas without bias and advising on where investments should and should not be focussed. The development of strategies would then be akin to those at present, but informed by this, and dealing with how to implement a defined plan on the ground and engaging locally in doing so.

2. Take a more holistic approach to coastal management

In planning future approaches we need to do our upmost to bring in all the different factors, not just flood and coastal defence needs but thinking about regional regeneration, recreational demand and suchlike. In doing so the economic benefits and funding streams that each might bring can be used collectively. This is of course a direction that UK government is already taking which is encouraging.

However this introduces new layers of uncertainty that also need to be managed; the coast is a highly dynamic environment and if plans supporting works are withdrawn and policies alter, then the implications for other coastal frontages within a sediment cell can be dramatic. We might consider whether there is scope for a Futurecoast type approach to potential planning scenarios, perhaps tied in with the aforementioned national vision, that could inform this.

3. Accommodate future changes in attitudes and behaviours

Can we really predict how the future will turn out? If a major storm and loss of life was to occur [e.g. 1953] then the public view and government attitude could radically alter. So how can the SMP therefore present a 'realistic' vision? Without considering a range of 'social futures' our policies can perhaps never be considered entirely realistic.

We could adopt a Foresight type approach in producing future shoreline management plans and, considering different social and economic scenarios, determine what each might mean for long term planning. This will enable us to determine where policies should remain the same irrespective of the scenario and where they may differ. The economic and technical implications of each can be established so we can at least move forward with a better understanding of the consequences of the choices we might make. Importantly, we can also establish whether present day policies will conflict or not, and be able to then decide accordingly on the appropriateness of our actions.

4. Focus on solutions not process

There should be nothing to stop us from being visionary and looking beyond conventional approaches, but only if we give ourselves the space and framework to do so. We have brilliant resources within our industry that are capable of developing innovative and even radical solutions. However our decision making structures lead us to look at local schemes based upon piecemeal justifications, sometimes driven by inappropriate short term concerns and bound up by compliance with existing legislation. This can all compounded by bureaucratic demands which take up disproportionate amounts of time from the minds that we might focus upon these tasks. If we can stand back and alter our framework it might enable us to be more innovative in our approaches to finding sustainable management solutions.

Conclusions

The intention of providing guidance for those preparing the second generation SMPs was to promote long term thinking in identifying sustainable shoreline management practices for the future and setting out a plan for change. Nonetheless the authors question whether we are much further forward in developing appropriate plans for the long term management of our shorelines. Some of the questions marks over the plans are:

o Are they technically and physically sustainable, facing up to the environmental changes and the technical and economic challenges this presents?
o Have they embraced flexibility and appreciate that future objectives could be somewhat different from those today, or are local short term concerns driving the long term thinking and generating future outcomes on the belief that these objectives and current legislation will remain the same?
o Are they economically achievable, whether this is through government funding or even with alternative funding sources contributing?

Some of the reasons these question marks exist might be because we are in danger of remaining within our comfort zones, or that we have forgotten the reasons why some significant changes to the shoreline management planning process were promoted. This paper reminds us of some of these.

The second generation shoreline management plans were intended to be a route map for the future; they were not expected to advocate wholesale change now. But policies therein need to recognise that decisions taken now will continue to have implications for decades to come. If we fail to recognise this then we will not be facing up to the real coastal issues. We must make sure that we still stand back and examine what these plans have produced before embarking on wholesale and unchecked implementation. Making the wrong decisions now will inevitably lead to both an increase in risk and ultimately, increasing costs too.

There are some recognised limitations with existing decision-making structures and central direction which constrain the ability of shoreline management plans in their current form to provide the long term sustainable solutions that are strived for. We live at a time where we cannot accurately predict the future but believe we will face considerable technical and economical challenges going forward. We do though have the ability to look ahead at different possibilities and make some informed judgements on whether we are making suitable choices now. The questions posed within this paper may help us to achieve this and the readers of this paper are encouraged to consider and debate whether the suggested ways forward presented within this paper can help to deliver that vision.

References

MAFF, 2000, Assessment of Economic Value of National Assets at Risk from flooding and Coastal Erosion, Defra, London, Report No: Commission FD1702.

Defra, 2004, National Appraisal of Defence Needs and Costs for Flood and Coastal Erosion Management.

Burgess KA, Townend IH, 2004, Guidance and a brief assessment of the impact of climate change upon UK coastal defences, In: 39th Defra Conference of River and Coastal Engineers, Defra, London.

Defra 2006. Shoreline Management Plan Guidance

Innovative Coastal Zone Management
ISBN 978-0-7277-5749-4

ICE Publishing: All rights reserved
doi: 10.1680/iczm.57494.365

Lessons Learned from 20 Years of Managed Realignment and Regulated Tidal Exchange in the UK

Colin Scott, ABP Marine Environmental Research (ABPmer), Southampton, England.
Dr Susanne Armstrong, ABPmer, Southampton, England.
Prof. Ian Townend, HR Wallingford Ltd, Wallingford, England.
Mark Dixon (MBE), self-employed consultant, Mersea Island, England.
Dr Mark Everard, University of the West of England, Bristol, England.

Introduction

In certain coastal and estuarine locations, the best and most sustainable way to enhance flood protection is to realign the primary sea defences in a landward direction. This typically involves building new sea walls at the back of a site and then either breaching the old wall to fully open up the land to tidal waters ('managed realignment' (MR)) or inserting tidal exchange structures such as sluices into the old wall to enable greater control of the new tidal flows ('regulated tidal exchange' (RTE)). Alongside sediment recharge, such 'soft' engineering measures can be used to respond to sea level rise, improve the cost effectiveness of coastal defences and create new intertidal habitat. Over the last 20 years, some 50 individual MR and RTE schemes have been implemented in the UK. Cumulatively, these have created/restored over 1,300ha of coastal habitat. Therefore, there is now a significant (and expanding) amount of accumulated knowledge about how best to implement these schemes, how habitats develop in these sites and how they provide socio-economic functions beyond their core objectives. This increasingly large evidence base can be used to supplement earlier, fairly generic, design and assessment guidance which was based on relatively few implemented schemes (e.g. Leggett *et al.*, 2004). Ensuring that the practical lessons are communicated and disseminated is essential for underpinning the effective implementation of future schemes. This paper therefore summarises these lessons across six topics, namely: scheme implementation costs, project management and communication, key issues in MR design and assessment, ecological development of the schemes (with particular consideration given to fish and shellfish), and concluding with some consideration of the socio-economic benefits. The paper is informed by the authors' practical experiences, as well as consultation and literature used for the creation and ongoing updating of the Online Managed Realignment Guide (OMReG) database (www.abpmer.net/omreg). The paper also draws upon the main issues raised by delegates at a bespoke MR conference in November 2010 hosted by ABPmer.

Costs of implementation

Frequently, one of the main hurdles to undertaking MR/RTE projects is the cost of their implementation as well as the risk of these costs increasing where obstacles are encountered during the various phases (i.e. scheme design, impact assessment, planning and construction).

Lessons from the past

A review of the implementation cost for 35 of the completed UK schemes has shown that the average cost of a scheme is just under £30,000/ha (2010 prices) (see OMReG). However, these costs have ranged greatly from £6,950/ha for the Pillmouth scheme (Torridge estuary) to just over £100,000/ha for the Trimley Marsh (Orwell) and Paull Holme Strays (Humber) schemes. Such variability is to be expected given the distinct challenges and constraints faced at the individual schemes. In general there has been a clear shift over the course of two decades from initial low-cost, small-scale, and relatively inexpensive trial projects to high-cost, larger, projects that were designed to meet specific targets for habitat creation and flood alleviation. This change is not unexpected, but what is much less intuitive is that many of the recent larger projects were not accompanied by improved unit costs (i.e. 'economies of scale'), and thus did not secure enhanced efficiencies in the light of the lessons learned from previous projects. This is illustrated in Figure 1, which demonstrates that schemes implemented after 2000 had higher unit costs. A contributory factor here will be increasing land prices, but also greater costs are being incurred for licensing, assessment, engineering and mitigation requirements. Project objectives are also a factor, with compensatory scheme costs (e.g. those undertaken to offset impacts from port developments such as Welwick (Humber)) being typically much higher, at £70,000/ha on average, than others. It is furthermore clear that the amount and scale of set-back defences is also a critical consideration; where large new defences needed to be constructed these accounted for a large percentage (c. 44%) of the total cost. Therefore, it is unsurprising that five of the six most expensive MR schemes required extensive new defence construction.

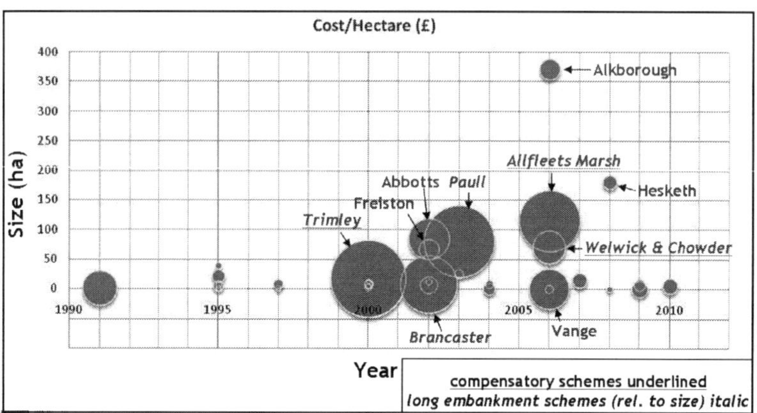

Figure 1: Unit costs of implemented realignments plotted against size and year

The future – need to review and identify cost efficiencies

At this stage there is no sign that the trend of larger projects incurring higher unit costs is likely to change. A number of large-scale projects are in the pipeline at present (e.g. Medmerry (near Selsey) and the Wallasea Island Wild Coast project (Crouch)), which will be relatively high cost projects due to their large size and/or novel complexities associated with their design and build. On an ongoing basis however, there is a need to identify where cost savings can be made with reference to the fees incurred in the past. This could include taking account of the flood protection benefits of intertidal morphology, applying past assessment, mitigation and monitoring regimes to new MRs and designing out the need for future maintenance.

Project management and communication

In general, MR projects should be viewed in the same way as any other engineering development with their implementation requiring careful planning and project management. However, it needs to be noted that, unlike typical infrastructure developments, MRs need to integrate into changing, sensitive, coastal zones and must themselves become adaptive rather than fixed systems.

Lessons from the past

As noted above, there has been a shift from relatively straightforward, smaller MRs to larger, more complex and costly sites that require significant project management to bring them to fruition. For example, the successful realisation of the Allfleet's Marsh site (Crouch; formerly Wallasea) can be attributed to good project management, the competency of the contractors and a clear understanding of the process to be followed (Dixon *et al.*, 2008, p.68). Key documents produced for this project included a clear business case as well as a detailed business plan (setting clear achievable objectives and time lines, outlining robust procurement and cash control). A review of the lessons learned provided for implemented schemes on OMReG also revealed the importance of having committed and enthusiastic implementers on board who learn from previous experience and ensure good cooperation with regulators and wider stakeholders. It is important to note that, once underway, a myriad of issues can cause significant project delays if not completed to specific timeframes. Obtaining planning consent can be a long and complicated process, especially for larger scale projects, and a variety of additional consents/licences are required. Furthermore, construction can be constrained by many factors (weather, tides and protected species windows), and embankments and mitigation habitats may need time to settle and develop. Significant contingencies should be incorporated to allow for aspects such as landowner negotiations, unexpected concerns and issues becoming unexpectedly complex (e.g. the presence of footpaths caused delays at several projects, including at Walborough (Axe)).

Having an effective, clear, honest and early, stakeholder communication strategy emerged as the most important lesson learned from schemes contained in OMReG. This is especially the case as MR is relatively complex and entails largely irreversible land use and visual change. The issue of change from arable land to intertidal has become a more prominent one over recent years, and in 2010, the Donna Nook MR on the Humber was denied planning consent largely for this reason. Based on experience at recent large-scale projects, including Alkborough Flats (Humber) and Medmerry, stakeholder engagement incorporating liaison groups, public exhibitions and individual meetings with interested parties is highly beneficial. This should enable genuine input into areas the public can actually influence and extol the wider benefits of schemes beyond the immediate objectives (focussing on aspects people can relate to, e.g. flood protection).

The future

The requirements and resources needed for project management and consultation are unlikely to diminish in the future, particularly as in some areas, more straightforward sites may have already been realigned. Also, local communities and authorities increasingly demand significant planning gains from MR implementers (e.g. improved flood protection and public access). Cuts in government spending may also make obtaining funding more difficult, so project managers may need to allocate more time to obtaining resources from more diverse sources. Factors which have proven useful during recent consultation experiences, and which are recommended for future schemes, include the 'marketing' of a scheme by providing information on the multiple benefits of MR, and using photorealistic 3D Geographical Information System (GIS) visualisations to demonstrate how a site will look and function.

Design and assessment

It is important at the very earliest stages of implementing a MR to understand the hydrodynamic functionality of a site and the physical interaction it will have with the adjacent estuary or coastal zone. Indeed this element needs to underpin the process of selecting a site in the first instance before then also forming the cornerstone of the majority of the design and assessment work that follows. There is a need to consider both short-term effects likely to arise from introducing a new inundation area as well as longer-term effects given that estuaries can take decades to centuries to respond. However, short-term considerations often dominate the consulting process, with immediate impacts a prominent issue when seeking the necessary consents.

Lessons from the past

Designing and assessing MR projects is a complex process, and can vary greatly in scope between projects. Two issues are crucial to the successful design of MR sites. Firstly, the hydrology and hydraulics within the site have to be designed to support the target habitats. Secondly, the physical changes which occur along adjacent estuaries or coasts following the introduction of a new inundation area need to be assessed; particularly as these can in turn affect other interests, such as designated habitats or flood protection. Understanding such changes often requires detailed hydrodynamic, sediment and wave modelling/assessment exercises. Listing each of these considerations is beyond the scope of this review. However, it is instructive to focus briefly on three of the best-understood and most accurately-quantifiable aspects of the physical changes which occur. These are:

o the amount by which a scheme increases an estuary's tidal prism (which is a simple surrogate indicator of the scale of a scheme's potential near-field and far-field effects);
o the channel formations that occur in front of a site as an indication of its near-field effects;
o the anticipated rate of accretion of sediments within a site, which influence how a site functions and also the rate at which the tidal prism effect reduces over time.

The available guidance on MRs suggests that changes in tidal prism amounting to more than 5% are likely to have significant impacts on the morphology of the adjacent system (e.g. Leggett *et al.*, 2004). However, this has not actually been tested and any schemes implemented to date have at most had a minor effect on the adjacent estuary's prism. For example, the Allfleet's Marsh MR led to a 2% change to the Crouch estuary prism and evidence from this and other projects with comparable changes (incl. Abbotts Hall, Blackwater/Salcott Creek) indicate that this has not led to significant external hydrodynamic effects. With respect to near-field morphological changes, some changes will always necessarily occur along the drainage channel from a site. However, such effects can be minimised by effective site and breach design, drawing on lessons from sites where the breach design was perhaps sub-optimal and thus led to unwanted (largely temporary) effects on fronting habitats (e.g. Freiston, Wash; Symonds & Collins, 2007). Experience from UK sites is that they tend to accrete, thus leading to a gradual reduction in tidal prism over time. The accretion rates differ greatly depending on site design and location. In the turbid Humber estuary initial lower-elevation accretion has been rapid at more than 10cm per year at all of the four Humber MRs. At Allfleet's Marsh, accretion was less rapid at 3-5cm per year; in the four years since implementation, this accretion has already led to a 10% prism reduction. It is interesting to note that at flood storage RTE projects such as Lippenbroek (Belgium), accretion and the consequent reduction in prism are seen as negative. This demonstrates that it is critical that these sites are not seen as static environments but as evolving and changing from day one.

When approaching the assessment and design of MR projects, an iterative and phased process is recommended, whereby there is a building up of evidence about the scale of changes and the functioning of a site. A site visit should be seen as an essential first step in this process. During the next phase, a preliminary design should be developed and its implications assessed on a high level. At the same time, the Environmental Impact Assessment (EIA) process should be commenced, and relevant ecological surveys undertaken to ensure on-site constraints are known (and mitigated for). The final phase should involve the detailed assessment of the scheme's hydrodynamic effects, which will then inform both the finalisation/enhancement of the design and the assessment of the individual EIA topics. This may need to be supported by wave and sediment transport modelling. Design aspects requiring the most careful consideration include tidal prism, breach design (and breach flow speeds), the role of site morphology in delivering particular habitats, and how future accretion may influence site development. Breach placement should be based on insights gained from a site visit, and a review of historic charts, current elevation maps and estuarine/coastal processes. For example, at Allfleet's Marsh, the breaches were largely placed in locations that minimised the losses of fronting saltmarsh habitat. The optimum breach dimensions can be calculated using recognised equations outlined by Townend *et al.* (2010), with a key input being a site's tidal prism. A breach needs to be sufficiently large and deep to avoid unwanted stability issues, a lesson learned from implemented schemes such as Freiston and Hesketh Out Marsh (Ribble). At Allfleet's Marsh, the breaches were deliberately over-designed to ensure that they were in 'regime' with the volumes of water exchanged and, to date, no morphological changes to the breach channels have been observed. Regarding site morphology, the extent of any landform manipulation must be justified with due consideration to project objectives, the potential gains and the likely cost. It has often been the case that materials for new walls need to be sourced on site (e.g. at Medmerry), which provides valuable opportunities for environmental optimisation (e.g. on-site fish lagoons or landward ditches enhanced for freshwater species). At the majority of the implemented MR sites, internal creeks were excavated to facilitate the effective flooding and draining of the site which, in turn, helps to ensure effective habitat creation. In some instances, field drains were already available for this function (e.g. Alkborough, Allfleet's Marsh) whilst, in others, tidal waters were allowed to create their own creek network (e.g. Tollesbury (Blackwater)).

The future

Future projects need to follow and further develop best practices established in the past, particularly with respect to following an iterative approach in which key questions are addressed early. It is recommended that early feasibility studies are carried out that address the key hydrodynamic issues but also begin the EIA scoping process by identifying the relevant impacts based on past experience. Where possible the guiding principles of scheme design work should be to minimise land manipulation and work with the existing topography. However, interventions should be undertaken to facilitate the efficient drainage of a site, maximise the stability of the breach(es) and optimise a site's ecological value from the opportunities presented by wall build excavations. These design principles were for instance applied to design the consented Medmerry scheme (breach anticipated in 2012).

Ecological development (benthos, saltmarsh and birds)

The ecological development of MRs is well studied, particularly where these were implemented as compensatory measures under the EU Habitats Regulations. For these sites there is a requirement to understand whether the created/restored habitats have offset the impacts of the plan or project which they have been designed to compensate. This monitoring usually focuses on mudflat benthos, marsh vegetation and overwintering birds. Some key findings from this work are set out below.

Lessons from the past

With regards to benthos, mudflat invertebrate monitoring undertaken at several MR sites has shown that, where the tidal elevation and physical conditions are appropriate, benthic invertebrates can colonise the accreting mudflat fairly rapidly (e.g. Tollesbury, Allfleet's Marsh and the Humber sites). Site species composition generally becomes more complex and stable over time. Early colonisers such as ragworm, mud snail and mud shrimp often dominate the biomass over the first few years. For example, rapid colonisation was observed at Allfleet's Marsh where there have been 10,000 to 20,000 organisms/m^2 in each year since its breaching. The species composition, abundance and diversity can vary greatly with differences in site elevation and location, and this makes comparison between schemes very difficult. At Chowder Ness for instance, between 500-15,000 organisms/m^2 have been observed, while at the Welwick site, ca. 37km downstream on the Humber, numbers ranged from 700-7,000 organisms/m^2. However, the species diversity at Welwick is still typically lower when compared to fronting, pre-existing mudflats, whereas at Chowder Ness, these figures are already very similar (three years post realignment). Judging assemblages in the context of fronting habitats provides an interesting context but will not necessarily allow the effectiveness of the schemes to be determined given how different the internal conditions can be from those outside (e.g. Nigg Bay (Cromarthy Firth) and Allfleet's Marsh).

Saltmarsh plant colonisation follows a similar successional pattern as that observed for mudflat invertebrates. Rapid colonisation occurs if the conditions are right, especially in relation to drainage and elevation (e.g. Welwick, Chowder Ness). Pioneer vegetation such as glasswort typically colonises within one year, and it may then take several years or decades to achieve a species composition similar to that of adjacent mature marshes. At Freiston, particularly rapid pioneer colonisation was observed; 70% of the area was covered in vegetation within three years. A similar 'exponential' rate was observed at Allfleet's Marsh, where the percentage plant coverage increased over four years from 1% to 6% to 60% and then 100%. At Freiston, the expectation was that the site's species abundance and community types would be equivalent to those outside the site within 10 years of breaching (Brown et al., 2007).

MR sites can rapidly develop into important roosting and feeding sites for waterbirds. At Freiston, Badley and Allcorn (2006) concluded after four years that the site supported 'large numbers of wintering waterbirds, several species in nationally important numbers' (p. 105).

Some sites (e.g. Welwick, Allfleet's Marsh) may initially mainly be utilised as roost sites but, as prey diversity and biomass increases, so should the proportion of feeding birds. Allfleet's Marsh for example supported very good, increasing, numbers of waterbirds in the first three years of its existence; with some 7,000, 10,000 and 12,000 waterbirds observed respectively. At the Tollesbury and Orplands MRs (Blackwater), communities were found to be largely similar to those of surrounding mudflats within five years of the initial breach (Atkinson *et al.*, 2001).

The future –need to clarify site success criteria

With regards to compensatory MRs, one of the most important future objectives will be to clarify the issues associated with measuring a site's ecological performance and, where relevant, addressing the extent to which it has offset the predicted and actual impacts arising from the project which it has compensated. Clearly, appropriately-designed MRs can deliver high ecological and biodiversity value even in a short space of time (weeks/months) alongside a wide range of other gains. However, the process of achieving full equivalency with mature habitat, saltmarsh in particular, may take much longer. The relevance of this needs to be better understood, agreed and communicated amongst coastal managers. In this context it should be noted that measuring value is not simple and different approaches are often taken to review monitoring data. It is also clear that most sites accrete sediments, often very rapidly, and their habitats are therefore inherently adapting from day one. Hence, quantitatively measuring the value and extent of the created habitats will never be a simple task. Each case will probably need to be judged on its own merits, but some kind of auditable framework that addresses the role that new sites play in maintaining the coherence of the Natura 2000 network may be warranted. The most obvious issue is whether designation requirements are met. However, it may be more useful to evaluate the 'functional equivalence' of the new habitats using ecosystem services analysis, particularly given that judgements are made in a non-stationary climate. In 2011, several major schemes will reach the end of their first five-year monitoring cycle and conclusions on their success will need to be reached in the context of these dynamics (and lessons disseminated).

Fish and Shellfish

In general, intertidal habitats are known to be valuable feeding and nursery grounds for many fish species such flounder, herring and bass (Dixon *et al.*, 2008). Hence, the value of managed MR for fish and shellfish populations, as well as for associated commercial and recreational fishing activities, is an important consideration when seeking to understand the socio-economic and ecological gains/benefits that can be achieved. In the UK to date, this has never been a key motive for habitat creation and is typically no more than a tertiary consideration if it is addressed at all. However, there is increasing recognition of the potential importance of MR sites for commercial fishing and food-production in their own right as well as in mitigation for losses of at-risk arable land and as a means to enhance the recruitment of fish and shellfish stocks.

Lessons from the past

Work undertaken at several MRs has confirmed that they are capable of providing suitable habitat within a relatively short space of time. At Allfleet's Marsh, fish sampling undertaken just 1 and 2 months after breaching showed that even though plants and algae had yet to colonise, the lagoonal scrapes in the developing mudflat had high numbers of crustacea and were refuge and feeding areas for juvenile sea bass and herring (amongst others). Fish were feeding on marine zooplankton, crustacea and polychaetes and were leaving the site with fuller stomachs than when they entered (L. Fonseca PhD, pers comm). Longer term surveys undertaken at Freiston and Paull Holme Strays confirmed the value of MRs as nursery areas for economically important fish. At Freiston, high numbers of bass, sprat and herring were observed (Brown *et al.*, 2007). At Paull, eel, flounder, bass and sand goby were abundant, and species composition and density was judged to be largely similar to that of adjacent areas (Hemingway *et al.*, 2008). Of particular note is that sites are of higher value if they provide fish habitat throughout the entire tidal cycle by including channels and ponds which remain flooded at low water. The value of such deep ponds/lagoons has been demonstrated at Abbotts Hall where up to 2000 herring/sprat were once found in one pool alone (along with 10 other species including bass, flounder and eel) (Colclough *et al.*, 2005). Comparative surveys at Brancaster (Norfolk) and Freiston confirmed the value of both sites for fish with the former site having a greater species diversity that was attributed to more diverse habitats and the presence of deep channels (Deaney, 2010). It is also known that fish use RTE areas. At Goosemoor (Exe), for instance, mullet pass through the control device to graze within the site; at Beltringharder Koog (Germany) fish pass through a culvert where average speeds are 4m/s; and at Lippenbroek, fish mostly utilise the outlet sluice.

Increasing work is also being done to understand the use of MRs by shellfish species and the potential for their commercial exploitation. Environment Agency-funded trials of cockle growth undertaken at Allfleet's Marsh demonstrated that this species grows as well within the site as it does outside (Dirt Consultants, pers. comm.). Subsequent RSPB-funded trials at Allfleet's Marsh concluded that MR sites could be used for the initial growing on of juveniles/spat. Benthic monitoring in this site has also shown that bivalve species are thriving more generally. The breach areas now have large rock oyster aggregations as well as occurrences of native oyster and mussels, while the mudflat supports high numbers of clams and occasional cockles. It seems likely that shellfish are feeding on both autochthonous nutritional inputs (internally generated from marsh and algae) and allochthonous organics (imported from external sources).

The future – the importance of environmental optimisation and commercial trials

There is now evidence that MR s are valuable for fish/shellfish populations and certainly that they are important places for the development of juveniles. However, the commercial potential of these sites has yet to be fully realised. In large part this is because these sites are not designed with this as a core objective. Adopting a process of 'environmental optimisation' when designing projects should allow this deficiency to be addressed. It is clear that even modest design changes, often undertaken at limited (if any) extra cost, enable a site to be enhanced in biodiversity terms. The inclusion of ponded areas retaining water at low tide is particularly valuable and also brings about additional benefits for bird species. It will be important to further investigate the commercial exploitability of MRs and to understand how the fish objective can be elevated as part of future scheme design work. This has particular resonance when dealing with the loss of productive (but often at risk) agricultural land and addressing how this can be offset by opportunities in new coastal habitat creation projects.

Socio-Economic Benefits

As discussed above, the main reasons for MR are to enhance flood defences and/or create new coastal habitat. In addition to these objectives, which need to be clearly laid out, secondary benefits can accrue providing additional socio-economic benefits. These can relate to tourism, recreational and commercial fisheries, carbon sequestration and water quality improvements.

Lessons from the past

Although the potential and theoretical benefits of MR have always been well understood, new lessons are being learned about the socio-economic gains that can realistically arise and also, in the light of new guidance from Defra (2007), about how to value ecosystems generally. The RSPB site at Freiston is a good practical example of a site that has been justified on economic and social grounds, having led (among other aspects) to reduced sea wall maintenance and increased visitor numbers (56,000 in 2003) that have boosted the local economy. Anecdotally, businesses near the site have reported increased trade from the visitors to the site and a guesthouse has opened immediately adjacent to the reserve. Separate, ecosystem services review work has also informed the Alkborough scheme (Everard, 2009) as well as projects that have not yet been completed such as the Medmerry and the Wallasea Island Wild Coast projects during their developmental phases. The Alkborough review identified an approximate aggregate benefit of £23m. The Wallasea Island Wild Coast scheme was predicted to lead to the creation of 16 to 21 full-time equivalent jobs in the local economy, and to flood defence-related cost savings of between £0.5 and 10million over the next 10 years (Eftec, 2008).

The future

Projects such as Freiston indicate the clear economic gains that can arise from a scheme, while ecosystem valuation also regularly demonstrates that these projects can have a sound economic rationale. There is a need to obtain more accurate valuation data on the ecosystem services to better understand and quantify these benefits. There is also a need for clarity on how ecosystem valuation will influence future coastal management decisions, because were it to do so, it could shift the balance in favour of coastal habitat creation. In the near future, however, it seems more likely that project valuation will need to be based primarily on the more obvious market services (e.g. reduced defence cost or greater tourism) that are more easily evaluated and communicated to local communities. However, less tangible services (such as local 'sense of place') must mot be overlooked as they can be a source of friction with local residents. Separately, there is a need to use these findings to seek out new funding sources, for instance from commercial organisations that can use such schemes as part of their carbon budgeting or from other as yet unexploited 'paying for ecosystem services' markets. This, if achieved, would help to address one of the key problems of funding MR implementation.

Conclusion

Our understanding about how to implement schemes has been greatly advanced through practical experience. This includes aspects such as how to design MRs, assess their impacts, secure planning consents and construct them, as well as how to engage stakeholders. Knowledge sharing has yet to be achieved adequately, and some projects still do not build upon well-known design opportunities. Major problems still exist in moving from the strategic level (shoreline management planning), which says what should be done, to ground-level implementation which indicates what can be done. Key difficulties include: securing landowner involvement, obtaining appropriate funding (especially when the costs of schemes increase) and communicating the potential socio-economic benefits (carbon sequestration, commercial fisheries productivity, green tourism, etc.). Ultimately, a long-term vision for scheme implementation is required in which a 'conveyor-belt' of future schemes is identified, some of which may not be implemented for decades. Within such a long-term vision there may well be a need to consider whether very large–scale (>>1000ha) projects can now be implemented to help achieve national habitat creation targets and provide services of economic value. It will, however, continue to be difficult to communicate the rationale for this work and to convey the relevant principles of future site and shoreline evolution, when shorter-term visions inherently prevail.

Key References

Large sections of this review are based on the primary and secondary data contained in the OMReG database (incl. many references). This site can be accessed at: www.abpmer.net/omreg

Atkinson, P.W., Crooks, S., Grant, A. & Rehfish, M.M., 2001. The success of creation and restoration schemes in producing intertidal habitat suitable for waterbirds. English Nature, Peterborough, 167p.

Badley J. & Allcorn R.I., 2006. Changes in bird use following the managed realignment at Freiston Shore RSPB Reserve, Lincolnshire, England. Conservation Evidence (2006) 3, 102-105.

Brown, S.L., Pinder, A., Scott, L., Bass, J., Rispin, E., Brown, S., Garbutt, A., Thomson, A., Spencer, T., Moller, I. & Brooks, S.M., 2007. Wash Banks Flood Defence Scheme - Freiston Environmental Monitoring 2002-2006. Centre for Ecology and Hydrology, Dorchester, 374p.

Colclough, S., Fonseca, L., Astley, T., Thomas, K. & Watts, W., 2005. Fish utilisation of managed realignments. Fisheries Management and Ecology 12, 351-360.

Defra, 2007. An introductory guide to valuing ecosystem services. Department for Environment, Food and Rural Affairs, London, 68p.

Dixon M., Morris R.K.A, Scott C. R. Birchenough A. & Colclough S., 2008. Managed coastal realignment: lessons from Wallasea. ICE Proceedings - Maritime Engineering, 161(2), 61-71.

Eftec, 2008. Wallasea Island Economic Benefits Study Report for the East of England Development Agency 24 October 2008. Eftec, London, 35p.

Everard, M., 2009. Ecosystem services case studies. Environment Agency Science Report SCHO0409BPVM-E-E. Environment Agency, Almondsbury, 101p.

Hemingway, K.L., Cutts, N.C. & R. Pérez-Dominguez., 2008. Managed Realignment in the Humber Estuary, UK. University of Hull, Hull, 44p.

Leggett, D.J., Cooper, N. & Harvey, R., 2004. Coastal and estuarine managed realignment – design issues. Construction Industry Research And Information Association, London, 215p.

Symonds, A.M. & Collins, M.B., 2007. The establishment of a temporary creek system in response to managed realignment. Earth Surface Processes and Landforms, 32(12), 1783-1796.

Townend I.H., Scott C.R. & Dixon M., 2010, Managed Realignment: A Coastal Flood Management Strategy, In: Pender G, et al. (Eds.), Flood Risk Science and Management, Blackwell Publishing Ltd, Oxford, pp. 60-86.

Innovative Coastal Zone Management
ISBN 978-0-7277-5749-4

ICE Publishing: All rights reserved
doi: 10.1680/iczm.57494.375

South East Regional Habitat Creation Programme

Ruth Jolley, Environment Agency, Worthing, UK
Emily Allison, Environment Agency, Worthing, UK

Introduction
Sustainable flood risk management is particularly challenging in South East England due to the coastline being both densely populated, and highly designated for national and international wildlife and habitats. Sea levels are predicted to rise over the next 100 years and the resultant coastal squeeze will result in Operating Authorities being legally required to compensate for habitat loss.

Background Drivers
The South East Regional Habitat Creation Programme (RHCP) was created to support the Flood and Coastal Risk Management (FCRM) programme. It provides a strategic, structured and proactive approach to addressing the Environment Agency's and Local Authorities' legal obligations to create habitat in compensation for habitat being lost through Flood and Coastal Risk Management activities. Provision of habitat through this programme allows flood risk schemes to progress in a timely manner so that people and property can be protected from flooding.

There are several drivers for the RHCP including:
- The Conservation of Habitats and Species Regulations 2010 (more commonly known as The Habitats Regulations) which requires Operating Authorities to compensate for any loss to the Natura 2000 network (including Special Areas of Conservation (SAC) and Special Protection Areas (SPA) habitats) caused by their works and provide compensatory habitat in advance of the loss. Potential SPAs, candidate SACs and listed Ramsar Sites are treated in the same way as Natura 2000 sites when considering and appraising flood and coastal proposals (Defra 2005a).
- The Water Framework Directive which requires all artificial or heavily modified waterbodies to have good ecological potential.
- The Environment Agency's corporate target to ensure there is no net loss of BAP (Biodiversity Action Plan) habitat (Defra 2005b, 2007) and to restore saltmarsh to 1992 levels by 2015.

Methodology and how it works
The RHCP has three continual phases to deliver the habitat requirements of flood and coastal risk management activities:
1) **Assessing Habitat Requirements**: Any FCRM plan or project not connected with or necessary for site management and likely to have a significant effect on a designated site is subject to an appropriate assessment under the Habitats Regulations. Appropriate assessments of Shoreline Management Plans (SMPs), FCRM Strategies and Schemes identify losses

and gains of both internationally designated and BAP habitats, and this information is reviewed by the RHCP on an annual basis.

2) **Finding and Securing Habitat Sites**: Potential areas for habitat creation are identified through GIS searches and discussions and workshops with Operating Authorities, local conservation organisations, landowners and interest groups and incorporated in an indicative programme. Suitable sites are then investigated in more detail to ensure they will meet the habitat requirements and business cases developed.

3) **Creating the Habitat**: This involves securing funding, gaining control over the land and the creation and long-term management of the appropriate habitat.

Figure 1 below illustrates how the RHCP works with the FCRM planning process:

- Impacts on designated habitats are identified by the Appropriate Assessment of Shoreline Management Plan (SMP) and Catchment Flood Management Plans (CFMP). The requirements for compensatory habitat are identified in the RHCP allowing the search for this habitat to commence.
- As the projects enter the strategy and scheme phases, options are decided and the habitat requirements are refined. During this time the RHCP is developing the required habitat.
- Once schemes come online, habitat is in place allowing works to progress without delay.

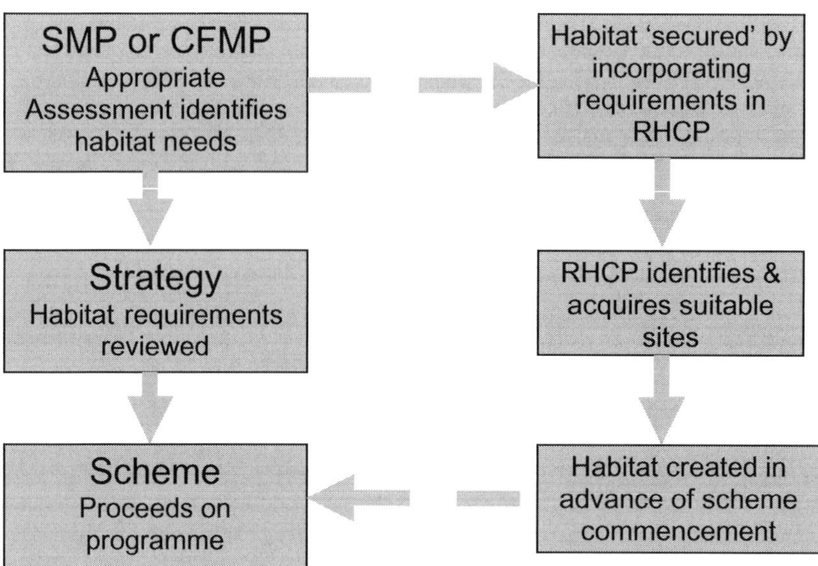

Figure 1 The Regional Habitat Creation Programme process

The RHCP ensures needs are identified at an early stage so that habitat can be developed and delivered prior to flood risk management works being implemented. This process has identified a need for approximately 1800 – 3600 hectares of new habitat in the South East over the next 100 years (Figure 2 and Table 1), along with a potential network of sites that could provide some of the habitat required.

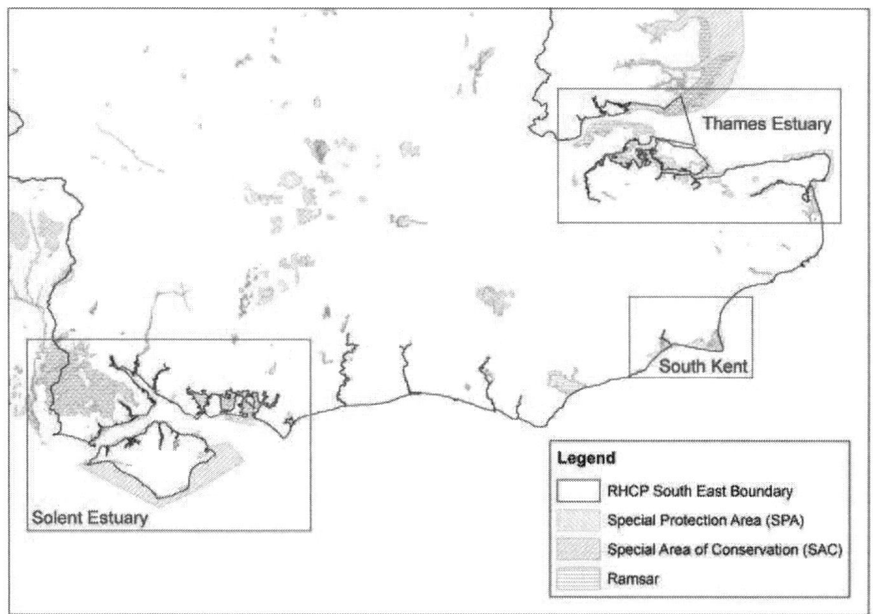

Figure 2 The main estuary complex areas where FCRM activities affect European designated habitat.

Table 1 Compensatory habitat requirements in the south east due to FCRM activities.

	Intertidal Habitat (saltmarsh & mudflat)	Freshwater Habitat (grazing marsh & reedbed)	Vegetated Shingle
Thames Estuary	900 – 1400 ha*	300 – 1500 ha**	0 ha
South Kent	0 ha	0 ha	10ha
Solent Estuary	600 ha	80 ha	0 ha

* Range depends on extent of compensation habitat delivered in Essex
** Range depends on extent of managed realignments

Benefits

The Strategic approach of the programme offers several benefits:

- Clearer understanding of what habitat is being damaged through flood risk management works, when and how;
- Flexible framework which allows pro-active site acquisition at a fair price (Defra 2005c);
- Create habitat in advance of need which avoids delays to protecting people and property from flooding;
- Creating fewer, larger and ecologically more robust areas of habitat to offset a number of small scale losses (Defra 2003, 2005b,c);
- Providing partnership opportunities with other Operating Authorities, Natural England, and conservation organisations.

Timing and location of habitat creation

Different types of designated coastal habitat are affected by FCRM activities in different ways and this can affect the timing of providing any compensatory habitat required. The maintenance and improvement of sea defences on their existing alignment combined with sea level rise results in the gradual loss of intertidal habitat (saltmarsh and mudflat) over time on the seaward side of the defences due to coastal squeeze. The provision of new intertidal habitat needs to keep pace with these losses, but the total estimate of loss does not need to be in place on day one (European Union Commission, 2007). Vegetated shingle is disturbed by beach management activities, but can often re-establish once the disturbance ceases. Where a realignment of defences is planned over a designated freshwater site, such as coastal grazing marsh, the damage occurs as soon as the realignment takes place. In these circumstances, the compensation habitat needs to have been created and be fully functioning before the realignment scheme can commence. This can mean starting work on the compensation habitat 10-20 years in advance of the proposed scheme. Sometimes the only reason a defence is being maintained is because of the duty to protect the designated habitat behind it. There will be circumstances where it will be more cost effective to create new habitat and then cease maintenance of these defences.

As a general rule, compensatory habitat for Natura 2000 sites must be delivered as close to the site of loss as possible. It is important to try and ensure that the network of Natura 2000 sites is maintained, which means considering how any new site will interact with the existing network. However, in the context of climate change, the emphasis of the RHCP is to try and create a smaller number of large sites which will be more ecologically robust than a larger number of small sites that might be in closer proximity to the site of loss (Defra 2003, 2005b,c). The suitability of any proposed site to provide compensation is agreed with Natural England prior to progressing the project.

Defra have suggested that where the RHCP is compensating for coastal squeeze losses, a ratio of 1:1 (loss:replacement) is normally acceptable as these are long-term losses. Where compensation is provided in advance and can be shown to be an effective functional replacement for the habitat, a ratio of 1:1 (replacement:loss) is also appropriate. However, as it is difficult to predict how successful secured compensatory habitat will be in delivering timely compensation, especially in regard to habitats which are dynamic, the area of compensatory habitat secured could be greater than what will be lost (European Union Commission, 2007).

Compensatory habitat has to be secured in perpetuity, so that the habitat cannot be changed or damaged in the future. One of the main ways to achieve this has been through land acquisition, but other approaches are now being explored to avoid the substantial capital investment required up front.

Financial implication of non-compliance with legislation

If a competent authority does not deliver compensatory habitat under The Conservation of Habitats and Species Regulations 2010 they are subject to a fine from the European Court of Justice. A case study in Ireland regarding failure to implement sufficient legislation to address the effect of projects on the environment has resulted in a suggested fine of €33,000 per day for each day after the Court ruling until the infringement ends (European Commission v. Ireland 2007). Although this case study is not directly related to non-delivery of compensatory habitat, it gives an indication of the scale of fine that could be enforced on the Government for non-compliance with The Conservation of Habitats and Species Regulations 2010 .

Issues encountered by the RHCP

The RHCP is a relatively new way of approaching environmental legislation and through its development we have encountered a number of challenges from identifying through to delivering habitat requirements including:

- current financial climate and cost of habitat creation
- land availability, location and timing
- protected status of habitat creation sites
- privately owned defences

Financial Climate

Even though there is a legal requirement to create compensatory habitat to meet legislation, the ability to deliver habitat on the ground has been hindered by the current economic climate. Funding for the RHCP is through the Government's Flood Defence Grant in Aid scheme, and the Environment Agency's budget for the current spending review period has been significant reduced. This reduction does mean that there are likely to be fewer schemes which may directly cause damage to European designated habitats, however there is still an indirect coastal squeeze effect for existing defences, even if they are not being maintained, so the large requirement for compensatory habitat still exists. Securing funding for compensatory habitat sites is challenging, but the RHCP provides a strong evidence base to prioritising habitat development.

Land availability, location and timing

Nationally the Environment Agency uses a benchmark figure of £50,000 per hectare for cost-effective intertidal habitat creation. In the South East we have not yet been able to justify the progression of intertidal habitat schemes purely to address the European legislative requirements, as the cost per hectare has greatly exceeded this benchmark.

The South East is highly developed and there are limited opportunities for habitat creation. This limited availability means there is increased competition from third parties. The RHCP is funded by public money so we can only justify paying market value for land (Defra 2003). The combination of these factors makes the South East more expensive for habitat creation compared to other regions in the United Kingdom. The Government may need to consider increasing the acceptable benchmark cost for delivering compensatory habitat creation, or allow the delivery of more habitat in other regions which may be more cost-effective, but is a greater distance from the site of loss. Under this scenario it is possible that habitat will need to be delivered at a higher ratio than the current 1:1 for coastal squeeze (European Union Commission, 2007).

Protected status of sites

Sites bought for the intention of habitat creation under The Conservation of Habitats and Species Regulations 2010 are not given protected status under planning law until a site is designated as a nature conservation site. This means that sites developed for the purpose of meeting the regulations are at risk of being impacted by future developments.

Ownership of defences

The RHCP includes compensatory requirements of The Environment Agency and Local Authorities for implementing flood and coastal risk management projects for defences they own and maintain. In addition, the RHCP includes habitat requirements for coastal squeeze for existing privately maintained defences, if the standard or footprint of defence is not increased. If a private landowner wishes to improve their defence, they will need to go through the normal planning procedure.

Within the Solent a significant proportion of the defences are in private ownership. In addition there are a number or European designations that are both in front and behind the defences. The RHCP's understanding is that if a private landowner is unable to maintain their defence and this results in damage to freshwater designated habitat behind their defence, they are not liable for compensating for this habitat, this will be the responsibility of the Government. Currently Natural England has a responsibility towards the management of protected sites and each site should have a management agreement with a landowner so this situation can be foreseen. It is anticipated that there is potential for freshwater habitat to be damaged under this scenario, but the RHCP does not currently account for this risk.

Case Studies:
Medmerry

The Medmerry frontage lies between Selsey and Bracklesham on the West Sussex coast (Figure 3). The defence consists of a shingle beach that requires regular maintenance to provide a 1:1 year defence to the low-lying farmland behind. A major breach of the defence could cut off Selsey's only access road. The Pagham to East Head Coastal Defence Strategy (CDS) identified that managed realignment at Medmerry would provide an increased level of protection from flooding for local communities. The RHCP identified that this opportunity could also provide compensation for Natura 2000 losses in the wider Solent Estuary and count towards the Environment Agency's national BAP habitat target.

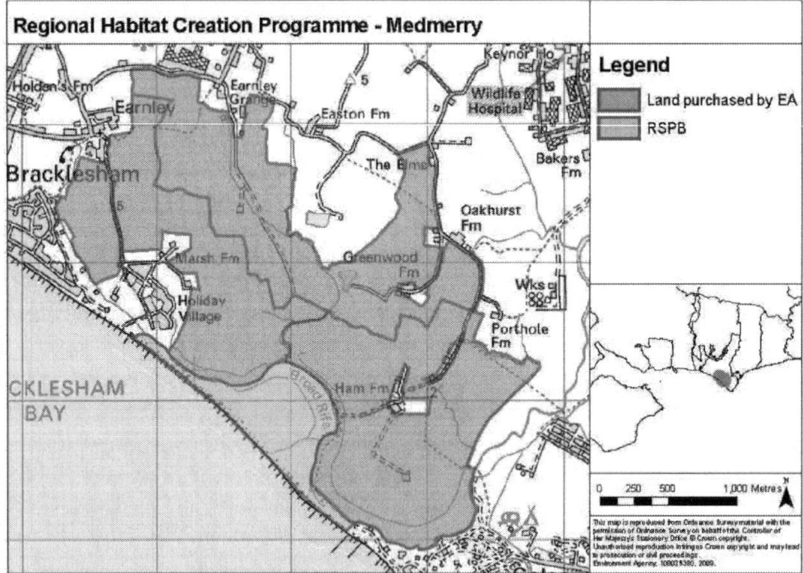

Figure 3 Land ownership at Medmerry

The possibility of realigning the defences at Medmerry had been discussed with the local community for over 10 years, and had always been unpopular. The development of the RHCP in 2008 provided the understanding necessary of the likely requirements for intertidal habitat across the Solent and the limited opportunities to provide it. This enabled a business case to be developed to acquire the land for habitat creation alongside the flood risk management scheme to provide a more sustainable sea defence on a setback alignment. Acquiring the affected land has facilitated acceptance of the realignment scheme by the local community. Five separate landholdings are affected by the proposed realignment. We bought three of the affected farms in Spring 2009, amounting to 498 hectares, and we are in negotiations with the other affected landowners, one of whom is the Royal Society for the Protection of Birds (RSPB, as shown in (Figure 3).

Representatives of the local community had input into the proposed design through a stakeholder group, and the alignment of the new defences was modified to reflect their input (Figure 4). They were particularly interested in the access and amenity benefits that the scheme will offer to the local community. Planning permission was granted in November 2010 and construction is due to commence in autumn 2011 with the mitigation works, followed by the main earthwork construction of the new flood banks in spring 2012.

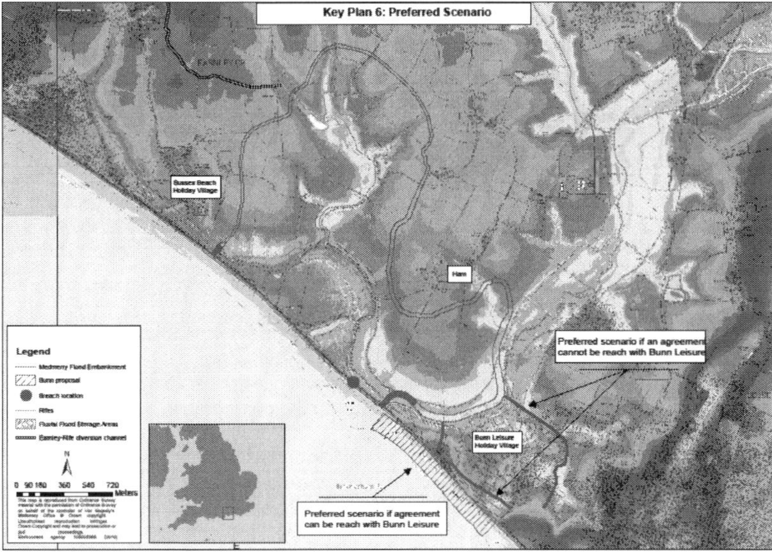

Figure 4 Medmerry managed realignment scheme design.

We estimate that the scheme will create approximately 183 hectares of inter-tidal habitat, predominantly saltmarsh, which will compensate for coastal squeeze losses in the Solent Estuary for the next 30 years. Some of the land behind the defences will be used as mitigation for habitat damaged by the scheme. This project enables important flood and coastal risk management activities to proceed.

Great Bells

Great Bells Farm was put up for sale on the open market and the Environment Agency successfully purchased the site in spring 2009 as the RHCP had identified a strategic need for compensatory habitat in the Thames Estuary (Figure 5). The 193 hectares site has potential to provide high quality grazing marsh habitat to allow managed realignment of unsustainable sea defences identified in the Medway Estuary and Swale and Isle of Grain to South Foreland

Shoreline Management Plans.

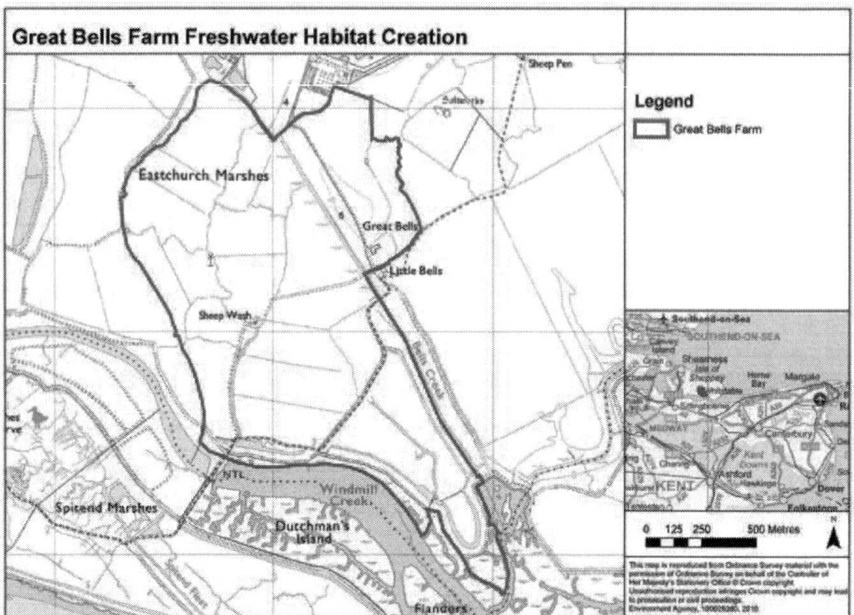

Figure 5 Boundary map and location of Great Bells Farm.

The Environment Agency does not have the experience and resources to design, construct and manage nature conservation sites so a package of works was competitively tendered to experienced conservation partners. The RSPB won this tender, and they have designed the site for the purpose of attracting species identified within the SPA citations in the Thames Estuary area.

Initially this site was also identified as having potential to address coastal squeeze habitat requirements in the Thames Estuary. A consultant was appointed to assess the feasibility of realigning 25 hectares of habitat to the south of the site. This showed that a managed realignment scheme was feasible without significantly affecting the estuary system. However, this scheme was discarded due to the relatively high cost of constructing a new line of defence compared to maintaining the existing defence and maximising the land for freshwater habitat.

Early community engagement highlighted limited support for the managed realignment project. The main reason for the initial negative response appeared to be resistance to change and this is often encountered in managed realignment projects. The freshwater habitat creation element of the project, which is now the objective for the whole site, has not provided such a passionate or negative public response.

It can be argued that much of the works proposed in the design for the project, could be undertaken by a landowner through agricultural works, but in order to be transparent, planning permission for the project will be sought in summer 2011 so that construction can start soon after.

Following best practice within the RSPB, the design will be implemented in three stages over consecutive years. It is anticipated that it will take ten years from construction to develop the habitat and provide evidence that it is of a standard which could be comparable to European designated habitat and classed as compensatory habitat under the habitat regulations.

Conclusion and recommendations

The South East RHCP is a good example of best practice in England and Wales and is supported by Defra and Natural England. The strategic approach of the programme offers several benefits and has allowed the Environment Agency to acquire land at two sites which will deliver some of the compensatory habitat required through European and National legislation. Development of habitat at these sites will allow flood risk management schemes to be implemented in a timely manner which will protect thousands of people and properties from flooding.

References

The Convention on Wetlands of International Importance especially as Waterfowl Habitat 1971 (the 'Ramsar Convention')

The Conservation of Habitats and Species Regulations 2010 (SI No. 2010/490) (The 'Habitats Regulations')

Council Directive 79/409/EEC of 2 April 1979 on the conservation of wild birds ('The Birds Directive')

Council Directive 92/43/EEC of 21 May 1992 on the conservation of natural habitats and of wild fauna and flora ('The Habitats Directive')

Council Directive 2000/60/EC of the European Parliament and of the Council establishing a framework for the Community action in the field of water policy ('The Water Framework Directive')

Defra 2003. Managed Realignment: Land Purchase, Compensation and Payment for Alternative Beneficial Land Use.

Defra Circular, 2005a on Biodiversity and Geological conservation – Statutory obligations and their impacts within the planning system.

Defra 2005b. Coastal Squeeze – Implications for Flood Management. The Requirements of the European Birds and Habitats Directives.

Defra, September 2005c. European Birds and Habitats Directives. Defra Policy Guidance.

Defra, May 2007. Guidance for Public Authorities on Implementing the Biodiversity Duty.

European Union Commission, January 2008. Guidance document on Article 6(4) of the 'Habitats Directive' 92/43/EEC.

The European Union Commission v. Ireland (European Union) C-427/07 (Failure of a Member State to fulfil obligations – Assessment of the effects of projects on the environment – Directive 85/337/EEC) <http://europa.eu/rapid/pressReleasesAction.do?reference=IP/11/168&format=HTML>

Innovative Coastal Zone Management
ISBN 978-0-7277-5749-4

ICE Publishing: All rights reserved
doi: 10.1680/iczm.57494.384

Governance, Planning and the European Ambition for Marine Space

Kate Johnson, Heriot-Watt University, Orkney, UK
Jonathan Side, Heriot Watt University, Orkney, UK
Sandy Kerr, Heriot-Watt University, Orkney, UK

Introduction

Marine space has rarely been at a premium. After all there is so much of it and so few uses to which it could be put. For most of history it has been a vast emptiness steeped in mythology, superstition, ignorance and fear. The two most important traditional uses, transport (getting from A to B) and fishing, occupied tiny areas of marine space only temporarily. They completed their business and were gone. At the margins relatively small areas of space were permanently occupied, land reclaimed from the sea, sand and gravel dredged up, and, most harmfully of all rubbish and waste dumped. Law and jurisdiction was weak with most territorial seas limited to three miles, the range of cannon shot at some distant time in the past.

Gradually at first, the permanent occupation of marine space and the control of traditional activities has become an issue and new governance structures have emerged. It is difficult to point to specific dates of change but one would be the 1970s: the expansion of offshore oil and gas extraction; national ambitions to secure exclusive rights to fishery resources; and mounting concern about the use of the seas as a dump for everything from household rubbish to hazardous waste. Even pirate pop radio created government alarm and played its part in encouraging new jurisdictions and ways to control activities in marine space. The international response was UNCLOS III (United Nations Convention on the Law of the Sea) in 1982 which, although not widely ratified until the 1990s and still not ratified by some including the United States, set and defined internationally recognised boundaries and responsibilities. The twelve mile territorial sea and the two hundred mile Exclusive Economic Zones are universally accepted. The points of measurement and the continental shelf provisions continue to give rise to a myriad of disputes between competing national claims, and the UNCLOS Article 76 Commission charged with ruling on them has a large backlog of cases - over one hundred at the last count including the waters around Rockall where rights are claimed by the United Kingdom, Ireland, Iceland and Denmark (on behalf of the Faroe Islands) [UN 2010]. As the Arctic ice retreats, nations are already lining up to claim resource rights there too. Marine spaces, or more importantly marine resources, are a hotly contested issue.

A second significant date of change in the use and governance of marine space is now, 2011. The underlying cause is the rapid and inexorable growth of technologies which grant access to more of the resources and wealth contained within the oceans and seas. We can drill deeper in more adverse conditions to recover oil and gas; we can farm fish and seafood; we hope to recover more minerals, rare earths and precious metals; we plan biotechnologies; and we can exploit some of the natural energy of the sea - wind, wave and tide. The immediate cause of change is this last activity - the development of renewable energy given political weight by imperatives to limit carbon emissions and mitigate the effects of climate change. The construction of large offshore wind, wave and tidal energy platforms requires the long term occupation of space - 8000km^2 or more in a single development area for some of the larger installations now planned such as a wind farm on the Dogger Bank in the North Sea. New systems of governance, new legislation and new management tools are needed to secure the rights of some and limit the rights of others, to give the confidence which allows these expensive, exposed ventures to proceed while also protecting the environment and the ecosystem as far as is possible, and absolutely in some cases. A key governance tool in the transition from open commons to controlled space is Marine Spatial Planning integrating simultaneous objectives for social, economic and environmental goods. Single sector planning, it is said, is no longer up to the task.

Marine Spatial Planning (MSP) shares most of the features of the land planning system with the important exception that it has to operate in three dimensions with few property rights, underdeveloped law and enforcement, little legal precedent and a natural environment about which knowledge is incomplete and poorly understood. The European Union (EU) has set out its ambitions for marine space including jobs and economic growth from the new marine activities and protection of the natural environment which values ecosystem services. The EU has enshrined many of these objectives into law, such as the Marine Strategy Framework Directive (MSFD); its funded research programmes reveal its thinking for the future, such as multi-use offshore platforms combining several activities (e.g. wind energy and aquaculture) into a single area saving space and sharing services. Such an approach needs forward planning and a master plan predicting interactions between activities and looking well into the future. This paper examines the roll out of MSP in Europe and in Scotland in particular. One pilot area for statutory marine planning in the UK is the busy Pentland Firth and Orkney Waters (PFOW) area which is established as a world centre for wave and tidal energy development. It is a rare example of a marine area in need of planning because of the immediate prospect of many developments which are new in technology and in close proximity to each other and the coast. Several devices are under test and the successful designs will be installed commercially over the next few years encouraged by government policy and public subsidy. The aims of the paper are to identify the critical issues for governance which accompany these changes.

Ambition

Jose Barroso, the President of the European Commission, called in 2009 for boosting new sources of growth, employment and social cohesion in *inter alia* the maritime sector. He stressed that new sustainable sources of growth are to be unlocked, enhancing skills and promoting advanced technologies whilst ensuring protection of Europe's maritime zones. *"Maritime Spatial Planning..."*, he said, *"will be a key instrument for decision making in this respect."* [EC 2010a]. DG MARE (the EU directorate responsible for maritime affairs and fisheries) have gone on to fund research into the economic development of the oceans and seas. In their programme to identify *"Scenarios and drivers for sustainable growth from the oceans, seas and coasts"* they forecast diverse new activities relating especially to food supply and aquaculture, renewable energy, mineral recovery, biotechnologies and coastal land reclamation for housing and industrial development. A subsequent programme *"Oceans of Tomorrow"(FP7 Energy)* [EC 2010b] examines the development of multi-use offshore platforms optimising the use of marine space for a variety of uses but anticipating wind energy and aquaculture in the first instance thereby extending the theme of energy and food security. Of all these possible uses it is the harvesting of marine energy (wind, wave and tide) which is creating the most immediate impact on the use of marine space in Europe and providing the economic drive towards new marine planning and governance measures.

In parallel with the ambition for economic uses of the seas are longstanding concerns about marine ecosystems and environmental issues. Ideas for development are always accompanied by the adjunct 'sustainable' and new regulatory objectives usually include 'the mitigation of the effects of climate change'. The new Marine (Scotland) Act 2010, for example, has as its general duties:

      a) Sustainable development and protection and enhancement of the health of the Scottish marine area; and

      b) Mitigation of, and adaptation to, climate change.

The Scottish Act, like its UK wide counterpart, the Marine and Coastal Access Act 2009, introduces statutory marine planning; a streamlined system of licensing for marine activities; and authority to ministers to establish Marine Protected Areas (MPAs). The legislation is also designed to bring clarity to the wide range of laws and management measures which apply to the marine area. Significant new legislation is the Marine Strategy Framework Directive (MSFD) sponsored by DG ENV (the EU directorate responsible for environmental affairs) and transposed into the domestic law of member countries in 2010. The Directive has the ambition (requirement) for all of Europe's seas to achieve 'Good Environmental Status' (GES) by 2020. In particular the Directive requires Ecosystem Based Management (EBM) and a network approach to impacts which crosses boundaries.

The government in Scotland has uniquely high ambitions for marine renewable energy with the objective to supply all of Scotland's energy needs from renewables by 2020. Located at the centre of one of the highest marine energy environments in the EU and close to potentially viable infrastructure, the government sees the development of marine energy technology as an important part of the industrial and export future of Scotland [Salmond 2010].

Marine Renewables

Marine renewables are the most significant use of marine space to have emerged for some time and possibly the largest so far envisaged. The significance is enhanced by the proximity of many proposed developments to the coast in clear view and in areas where other activities already exist. Offshore wind power is now an established technology moving from research to the mainstream. Single turbine output capacity has gone up from less than 0.5MW to over 5MW in a decade with 7.5MW capacity in prospect. Wind platforms of over 500MW capacity are near to completion and the Crown Estate Round 3 of wind farm leases anticipates possible capacities for a single farm area of up to 9GW capacity occupying an area of sea of over 8000km^2 (Dogger Bank). Statoil have recently commissioned the world's first floating wind turbine off Norway. Offshore wind energy is foreseen as the principal source of renewable energy with the possibility to achieve an output of 520TWh/yr and create 260000 jobs in the UK [Neale 2010].

The wave and tidal power technologies are at a much earlier stage of development and the ambition is more modest. Scotland is a world leader in wet renewables but even here the best estimates are around 1GW of installed capacity by 2020, most of it in the area around Orkney [FREDS 2009]. FREDS (Forum for Renewable Energy Development in Scotland) identifies the need for larger and more accessible funding streams to encourage proof of technologies at scale - *"The most important hurdle for the sector to overcome remains proving the technologies...the sector needs devices to progress from the prototype stage in the near term if it is to make the contribution envisaged"*. The BERR Atlas of UK Marine Renewable Energy Resources [2007] is the most detailed and commonly referenced spatial distribution of wave, tidal and offshore wind resources for UK continental shelf waters. The Atlas overview (Fig 1 and Fig 2) demonstrates the more limited range of feasible wave and tidal sites when compared to those suitable for offshore wind developments.

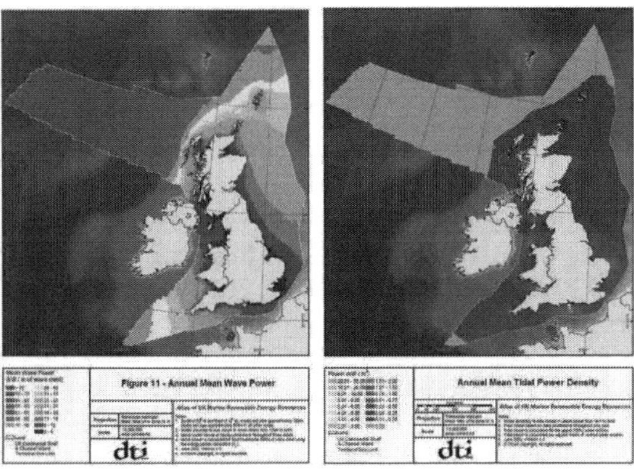

Fig 1 Wave and Fig 2 Tidal Energy Resources in the UK
Source: BERR 2007

The Pentland Firth and Orkney Waters (PFOW) area has become the focal point of wet renewables development because of its location at the heart of one of the few energy rich marine environments where support and grid infrastructure can be made accessible at least cost. In addition to the research and test facilities the Crown Estate have agreed to lease eleven sites to commercial developers in the relatively small area of the PFOW. This high density of potential devices provides a focus for the planning issues which occur with less intensity wherever such development is proposed.

Scottish Environment Link has described the deployment of marine renewables as a double 'U' dilemma of urgency and unknowns [SEL 2010]. The political imperative to build renewable generating capacity quickly is restrained by a lack of knowledge and uncertainty. Wave and tidal devices are so far experimental and there is no clear indication which, if any, of the devices currently under test will progress to commercial application. The generic and specific impacts of these devices on the ecosystem and on other activities are largely unknown; knowledge of the working of the marine environment itself is sketchy. In addition the financial risks of investments in marine renewables are high and dependent on public subsidy. If the political objectives are to be realised a number of value judgements will have to be made in the absence of detailed knowledge. Of many potential environmental and socio-economic impacts, most will fall under one of five headings [Bell and Side 2011]:

1. Physical changes - Changes in velocity of current flow and wave energy; changes in tides and circulation patterns; changes in sedimentary processes (near field and far field).
2. Ecosystem changes - Linked to physical processes; identification of biotopes vulnerable to energy extraction; identification of hydrographic conditions maintaining ecosystem functions.
3. Direct impacts - Displacement of mobile fauna; damage to sensitive habitats; underwater noise; collision and entrapment risks; barriers to free movement.
4. New ecological space - Action as fish aggregation devices (FADs); settlement of marine organisms; stepping stones for exotic and invasive species.
5. Fisheries displacement - Exclusion of fisheries; increases in fishing pressure elsewhere; loss of habitat and spawning areas; possible stock enhancement effects.

Planning

Marine Spatial Planning is in essence a political process of conflict resolution between competing objectives. MSP, as seen now, is a relatively new process and there is little experience or research to call on. Early examples such as the Australian Great Barrier Reef Reserve (1975) and the Galapagos Marine Reserve (1998) are focused on a relatively narrow band of objectives related to nature conservation in huge areas of sea with little human activity beyond fishing and tourism [Douvere and Ehler 2090]. European MSP is evolving to encourage sustainable economic activity and Ecosystem Based Management (EBM) in busy waters already subjected to many uses and abuses. Belgium was one of the first countries to move to multi-objective MSP closely followed by the Netherlands, Germany and Norway all of whom are now moving towards the completion of their second generation plans having implemented and monitored their initial efforts. These countries have tended to use legislation which existed or which could be modified to suit. The UK has opted for new laws. The countries of Europe have a complicated network of boundaries which extend out into the oceans and seas around them. A significant feature of European policy expressed in the Integrated Maritime Policy (IMP) and the Marine Strategy Framework Directive (MSFD) are the needs to consider whole ecosystems in planning. This is also reflected in the OSPAR convention with its emphasis on networks and cross border issues.

MSP aims to create an adaptable master plan for the future of an area or region which identifies interactions and takes an integrated network approach to issues which cross boundaries of several kinds (natural, administrative, international). At the moment it remains largely the preserve of marine scientists and administrators watched anxiously by those directly affected such as marine energy developers and fishermen. The UK government describe MSP as a means to:

"Create and establish a more rational organisation of the use of marine space and the interactions between its uses, to balance demands for development with the need to protect the environment and to achieve social and economic objectives in an open and planned way." [DEFRA 2007]

The goals of MSP are similar many respects to the principles set out for Integrated Coastal Zone Management (ICZM) and recommended by the European Commission in 2002 when seeking to deal with problems of jurisdiction at the land-water interface. However, the results of ICZM have been mixed, possibly due to a failure of governance and little statutory support, and the approaches are different. ICZM has adopted a process approach promoting integration across agencies and sectors. MSP is a spatial and temporal plan approach to effectively allocate space to economic activity and nature conservation. Douvere and Maes [2010] suggest that ICZM principles could benefit and be made operational by the MSP approach. The European Commission directorate responsible for the marine economy in Europe, DG MARE, commissioned a report on the maritime legal framework within which MSP must operate and identified indicators to measure progress in the MSP process [MRAG 2008]. The indicators - *availability and quality of data; supporting laws; efficient bureaucracy; participation; conflict management; trans-boundary cooperation; monitoring, evaluation and adaptation* - are intended to permit a speedy, fair decision taking account of all available information and opinion but adaptable in the light of new information.

Governance

Six of the seven EU MSP indicators relate directly to the governance process with the seventh, the availability and quality of data, being the technical basis for decision making. Governance can be seen as striking a balance between the influences of the state, the market and the community [Jones 2010]. The governance framework emerging for marine space in the UK is weighted towards state control with mainly advisory and consultative roles delegated to local authorities and communities. The influence of the market in marine space is, as yet, immature and underdeveloped. There are few private property rights and all activities are subject to state licence or are unregulated - traditional rights held in common such as the right to fish. The main task and difficulty for new marine governance is to redefine and codify rights between users. Early examples of a developing market are the sale and purchase of aquaculture companies, whose principal assets are leases to marine space and a licence to operate, and the role played by the Crown Estate.

The Crown Estate (TCE) Commissioners administer the Crown ownership of the seabed in the UK territorial sea with an amalgam of public and market responsibilities. The ownership does not extend to the water column or the sea surface above. They have a legal obligation to raise income from the estate but also to care for its wellbeing [Crown Estate Act 1961]. TCE have used this obligation to promote marine renewables and have been proactive in offering agreements to lease areas of seabed to developers subject to them obtaining the necessary permissions and licences from the government. An emerging problem of this approach has been the *de facto* allocation of marine space to renewable energy developments without wider consultation, just in advance of the marine planning regime coming into force, which has antagonised other stakeholders, especially fishermen [Johnson 2011]. TCE invited developers to bid for sites which were, in effect, chosen by the market. The sites have yet to be licensed and a consultation process will follow but the investment which developers have been encouraged to make in these areas is almost bound to be a major consideration. The role of TCE in Scotland is currently the subject of a parliamentary Select Committee inquiry in the run up to the proposed Scotland Bill.

Both the UK and Scottish marine acts assign planning and licensing authority to ministers although the Scottish act goes a bit further in delegating advisory powers to communities in the form of Marine Planning Partnerships (MPPs). Here, the minister will appoint members to the MPPs drawn from local authorities and other stakeholders. They will have the legal duty to draft MSPs for their area but the authority to approve or modify the plan will rest with the minister. The completion of the first plans under this system is unlikely to be before 2015. For areas under the jurisdiction of the UK act the government Marine Management Organisation (MMO) will draw the plans and consult through a statutory 'Statement of Public Participation'. The division of marine jurisdiction between the UK and Scottish parliaments is not straightforward and is made on both an area and activity basis. Renewable energy, for example, is an activity reserved to the UK parliament but executively devolved to Scotland - meaning that Scotland may administer but not change the law. In contrast oil and gas is wholly reserved and fisheries are wholly devolved. Cooperation between the administrations is essential.

The relationship between marine planning, a responsibility of central government, and terrestrial planning, a responsibility of local government, is something that has yet to be worked out in practice in both jurisdictions, but there are requirements for coordination and consultation. Local authorities have few powers to directly control marine energy developments although they are statutory consultees. Even the terrestrial elements of the projects (e.g. cable landfalls and sub-stations) may fall outside the local planning system because every marine energy project over 1MW in Scotland (50MW in England) requires Section 36 consent under the Electricity Act 1989 and the developer is entitled to ask the Minister for deemed planning consent under Section 57(2) of the Town and Country Planning (Scotland) Act 1997. Most, if not all so far, have done so [Forth Energy 2011].

Conclusions

The oceans and seas of the world are at a crucial turning point. The ambition to use them for multifarious purposes, and the technical capability to do so, has never been higher and will keep growing. The use requires a reallocation of rights and new regimes of jurisprudence and planning which convert an open commons into managed and monitored space with closed and private areas and perhaps even a market in marine space. There is already a market established in aquaculture sites as businesses change hands. The ambition to use is balanced by a concern to conserve and to protect essential ecosystem services as well as to mitigate the impacts of climate change. Ecosystem valuation, putting money values to natural processes, is one tool being considered. The principles of national jurisdiction have been established by UNCLOS with basic boundary limits respected by nearly all countries, even those who have not ratified the convention. However, there is still plenty of room for dispute usually related to assessment of the points from which measurements are made and continental shelf limits where data is poor or requires the exercise of judgement.

Marine planning has evolved from purely nature conservation objectives into something which hopes to promote simultaneous objectives for the social, economic and ecological well being of the seas, coasts and communities. MSP is at heart a mechanism for conflict resolution - a political process between competing objectives. In Scotland the SNP government has set its sights, as a national objective, on exploiting its high natural energy environment to create a new renewable energy economy meeting its own needs for electricity and exporting hardware and know how throughout the world. The government is obliged to meet its international and European responsibilities to the environment (e.g. MSFD and OSPAR) but at the same time it is anxious not to scare away investment for development with excessive regulation. The experience of the Pentland Firth and Orkney Waters highlights the need for strategic planning and the scale of the uncertainties in the deployment of new space occupying economic activities such as marine renewables. Events here are running ahead of the pending planning regime revealing a lack of knowledge about impacts and conflicts between 'rights' - for example, the Crown ownership of the sea bed, the right to fish in the water column above and the right to navigate the sea surface above that. In order to expedite the deployment of renewables the Scottish government has adopted a policy of 'deploy and monitor' allowing development to proceed while building knowledge and negotiating to resolve conflicts - although it should be noted that no commercial deployment of wet renewables has yet been licensed. Developers will need more certainty to secure their investment including funds for research and development. Investment in renewables has actually fallen in Europe and the US since 2008 [FT 2010]. Investment in Asia and Oceania has increased rapidly in the same period.

The European ambition for the use of marine space requires planning supported by data about ecological and socio-economic impacts, as comprehensive as possible. It is accepted that this is a long term task, hence the 'deploy and monitor' policy in Scotland although the environmental requirements are increasingly enshrined in law. The ambition also requires a governance structure which is able to negotiate a reallocation of rights between uses which balance the interests of the state, the market and local communities. Key outstanding questions noted in recent meetings with stakeholders in the PFOW [Johnson *et al.* 2011] are the relationships between marine and terrestrial planning in reflecting national and local objectives; and the distribution of benefits from the new activities as part of the negotiation over the reallocation of rights. In the UK and Scotland the government has moved to hold decision making to the centre with only advisory and consultative roles delegated to the local level. Comprehensive local authority powers to control development in the UK territorial sea exist only in the sea around Shetland. These powers are contained within the Zetland County Council Act of 1974 and were negotiated when Shetland faced the arrival of the offshore oil industry in the 1970s [Marshall 1981]. They have been a success and the accompanying 'Disturbance Agreement' with the oil companies has funded a regeneration of the islands. It remains to be seen if the more centrally based powers now planned elsewhere can do any better.

References

Bell M and Side J, 2011, *Tidal Technology Development and Deployment in the UK: Environmental Impacts of Tidal and Wave Power Developments and Key Issues for Consideration by Environment Agencies.* Background Report No. 2 for SNIFFER ER20, Heriot-Watt University, Orkney.

BERR (2007). *Atlas of UK Marine Renewable Energy Resources: a Strategic Environmental Assessment Report.* Produced by ABPmer, the Met. Office and Proudman Oceanographic Laboratory.

Crown Estate, 2010, www.thecrownestate.co.uk/round3

DEFRA, 2007, *A Sea Change*, Marine Bill White Paper, HM Government, London

Douvere F and Ehler C, 2010, *New perspectives on sea use management: Initial findings from the European experience with marine spatial planning,* in: Journal of Environmental Management 90:77-88

Douvere F and Maes F, 2010, *The contribution of marine spatial planning to implementing integrated coastal zone management,* in Green D (ed.) Coastal Zone Management, Thomas Telford, London

EC, 2010a, *Scenarios and drivers for sustainable growth from the oceans, seas and coasts,* European Commission Tender MARE/2010/1

EU, 2010b, *The Ocean of Tomorrow - Multi-use offshore platforms*, EU FP7 Energy Call OCEAN.2011-1:

Forth Energy, 2011, Environmental Impact Assessment, Chapter 2, The Statutory Context

FREDS, 2009, Forum for Renewable Energy Development in Scotland, Marine Energy Group (MEG) Report 2004, *Harnessing Scotland's Marine Energy Potential.* Published by the Scottish Executive

Johnson K, 2011, *Wave and Tidal Energy in the Pentland Firth and Orkney Waters, Diary of a stakeholder meeting - the industry meets the fishers,* Report of a meeting held on 9 March 2011 in Kirkwall, Heriot-Watt University, Orkney

Johnson K, Kerr S, Side J, Jackson A, 2011, *Wave and Tidal Energy in the Pentland Firth Area - stakeholders, who needs them?* Report of the SRDG/MESMA Workshop on 9 February 2011, Heriot-Watt University, Orkney

Jones P, 2010, *Lecture notes on 'Governance',* University College London

Marine (Scotland) Act, 2010, Part 1A Sections 2A and 2B

Marine Scotland, 2010, *Pentland Firth and Orkney Waters Marine Spatial Plan Framework and Regional Locational Guidance for Marine Energy,* Draft Report, Scottish Government, Edinburgh

Marshall E, 1981, *Shetland's Oil Era,* Shetland Islands Council, Lerwick

MRAG, 2008, *Legal Aspects of Maritime Spatial Planning,* Report to the European Commission, Framework Service Contract FISH/2006/09 - LOT2

Neale J, 2010, *One million climate jobs - solutions to the economic and environmental crises,* Campaign against Climate Change, London. www.climate-change-jobs.org

Salmond A, 2010, Announcement by Scottish Government First Minister on 23 September 2010

SEL, 2010, *Avoiding Conflicts in the Marine Environment, Effective planning for marine renewable energy in Scotland,* Scottish Environment Link, Edinburgh

UN, 2010, *Table of claims to maritime jurisdiction (as at 31 July 2010),* Division for Ocean Affairs and the Law of the Sea, Office of Legal Affairs of the United Nations, Division, Room DC2-0460, United Nations, New York, NY 10017, doalos@un.org.

FT, 2011, *World financial investment in clean energy,* UNEP and Bloomberg New Energy Finance, in Financial Times 17 January 2011 Innovation in Energy Special Report.

WEC (1993). *Renewable Energy Resources: Opportunities and Constraints 1990-2020,* World Energy Council, London, June, 1993.

Acknowledgement

This work is part of the ongoing research within the EU FP7 programme "Monitoring and Evaluation of Spatially Managed Areas" (MESMA; Grant number 226661; www.mesma.org). We wish to thank all 21 institutions participating in this research across Europe.

Institution of Civil Engineers

publishing

SECTION 5:

FUNDING AND ACCOUNTABILITY – INVESTMENT, OPPORTUNITIES AND GROWTH

Innovative Coastal Zone Management
ISBN 978-0-7277-5749-4

ICE Publishing: All rights reserved
doi: 10.1680/iczm.57494.395

Community-led Flood and Coastal Management Projects Supporting Adaptation and Localism in Suffolk and Essex

Trazar Astley-Reid, Suffolk Coast and Heaths AONB, Suffolk, UK
Jane Burch, Suffolk County Council, Suffolk, UK
Karen Thomas, Environment Agency, Suffolk, UK

Introduction

In recent years the development of Shoreline Management Plans (SMPs)[1] /Catchment Flood Management Plans (CFMPs) and Flood and Coastal Risk Management (FCRM) Strategies has been the initial driver for dialogue about flooding and erosion between communities and the Environment Agency (EA) and Local Authorities(LAs).

These plans require decisions to be taken about long term flood and erosion risk management.

Continuing to keep defences in their current position where natural processes are impacting on areas of coast and rivers may not be sustainable longer term. In addition, the economics of maintaining defences in these locations may become prohibitive. In some locations the cost of maintaining a defence outweighs the value of what is protected. Finally in areas that are currently undefended building new defences may not be appropriate.

In all these scenarios alternative approaches to managing flood and coastal risk are required. In more vulnerable locations where natural processes are the issue alternative solutions to holding the line will be needed. However, where the issue is about economics and availability of funding, communities and the main beneficiaries of the defences may well need or wish to contribute to their upkeep in future.

Nationally, through flood and coastal management plans, reduction of maintenance and 'no active intervention' policies are being discussed with communities and landowners. The reduction or cessation of maintenance activity by the EA or LAs has been an unpopular move in many communities who risk losing assets without compensation. In other cases the stimulus for local activity has been a lack of any organisational responsibility for the situation. This was certainly the case in many instances of surface water flooding – at least until the Flood and Water Management Act (FWMA),[2] 2010 came into force.

This has resulted in a number of landowners and community groups requesting the opportunity to undertake their own maintenance and, in some cases upgrading, of defences. On the coast these discussions have resulted in the EA, under its Coastal Strategic Overview[3], developing a National Protocol for the Maintenance of Sea Defences[4], 2010 which has been supported by the National Farmers Union, Country Land and Business Association and Natural England (NE). This national protocol has led to the EA producing a new information pack[5] (Supporting Change - Involving landowners in flood and coastal risk management) as well as a more streamlined approach to managing flood defences with the landowning community which is being trialled in the Anglian region and is due to be rolled out nationally.

The budgets for FCRM have been steadily increasing but the challenges of managing flood and coastal risk are also increasing, in particular with increased surface water flood events following periods of heavy rainfall. The emphasis for flood and coastal risk funding must be to protect the most people and property but this leads to rural areas being less of a priority for national funds.

However, this lack of national funding has been the impetus for a number of local, community-driven, flood and coastal erosion projects. The well-documented example of coast protection at East Lane[6], (Bawdsey, Suffolk), funded by enabling local housing development, is one of the best examples of Localism in action and demonstrates how funding can be raised for defence work outside of traditional routes.

The Coalition Government's Big Society/Localism[7] approach will undoubtedly increase the trend of community-led actions in relation to a wide range of issues, including flood and coastal management. Cuts in national funding are likely to increase the need for local contributions to flood and coastal projects, and thus local input into their design and delivery.

However, locally-led projects need to be set in the context of broader strategic objectives in order to ensure that coastal and river environments and natural processes are not negatively affected, at the expense of communities and habitats in adjacent areas. In particular environmental designations are often one of the most challenging issues that need to be addressed by any plan or project. We need to make it easier for local communities to take forward projects whilst balancing wider socio-economic and environmental outcomes within the Localism agenda.

This paper aims to demonstrate, through some local case study examples, some of the challenges communities face in taking forward their own projects. It also aims to highlight some of the opportunities that have arisen for partnership working with key organisations and lessons that have been learnt to make similar projects easier to undertake at other locations.

Case Studies

Shotley Gate, Suffolk is a currently undefended eroding cliff in a unique estuarine location which falls outside the Coast Protection Act [8](Schedule 4) boundary, where neither the EA nor the LA has legal responsibility or access to normal coast protection funds to undertake the work needed. The community approached the AONB's Suffolk Estuaries Officer (SEO) to

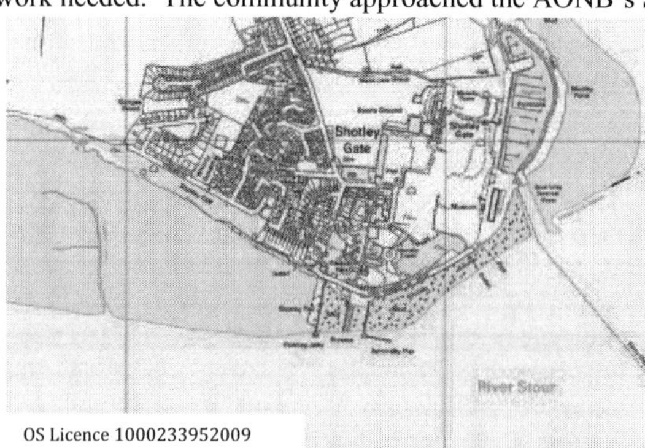

OS Licence 1000233952009

Shotley Gate Stour Estuary, Suffolk

ask for assistance in brokering a way forward, through the Area of Outstanding Natural Beauty (AONB) partnership, with statutory partners. However the community group had already undertaken some inappropriate works, due to the lack of help from previous discussions. This was followed by a realisation from both the community and statutory organisations that the only way forward was one of to an erosion control project acceptable to all parties. The partnership was able to find funding together. However the majority of the project was funded by statutory bodies e.g. Suffolk County Council (SCC), AONB, EA, Haven Gateway Partnership, Port of Felixstowe as well as a from local fundraising. Phase 1 of the project has been completed with a further phase planned for 2011, subject to further funding being raised.

Beyton, Suffolk is a surface water and ordinary water course issue where no individual or organisation is responsible for the problem. The initial impetus came from a frustrated

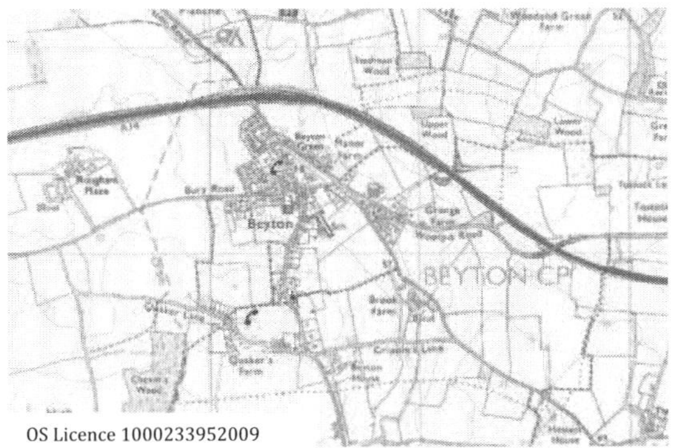

Community invoking the help of their local MP who insisted on all organisations, councils and local landowners getting together to discuss the issue. With SCC taking the leadership role (following the introduction of the FWMA), all parties worked together to develop, fund and deliver a flood alleviation scheme.

Beyton, Suffolk

Deben Estuary, Suffolk - The Deben estuary has 40% of Suffolk's saltmarsh. The local community recognised that large areas of saltmarsh were eroding and wanted to get involved in managing the remaining marshes. They recognised the importance of the marsh for wildlife

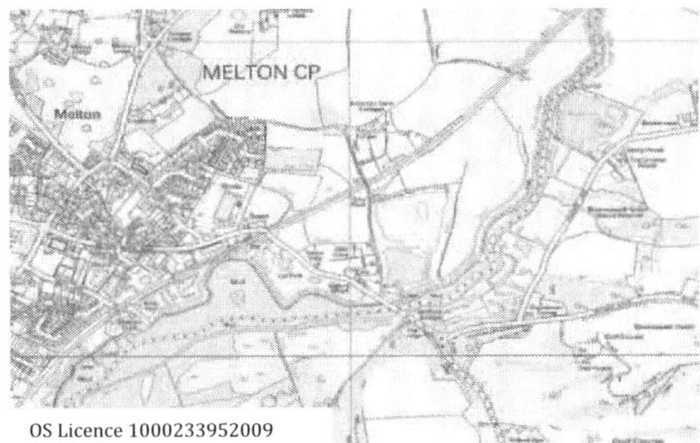

and landscape but also its value in acting as a natural buffer to the estuary flood defences reducing maintenance costs on defences. The Deben Estuary Partnership (DEP) a community initiative supported and facilitated by the AONB's SEO set up a working group with statutory partners to explore management options that would help areas adapt to change. Through this group a number of potential saltmarsh restoration

Sutton Marsh Deben Estuary, Suffolk

projects are in development and the approach has created a basis for future partnership approaches with the community on the upcoming Deben Estuary Flood Risk Management Strategy.

Potton, Essex - An example of a land manager who has taken forward small-scale maintenance works to a defence currently owned by the Ministry of Defence. The land manager is a tenant on Potton Island in the Roach Estuary, South Essex. He raised concerns about the need for maintenance works at locations around the island but was also keen to demonstrate to other landowners that self-help is an important element in managing flood risk, particularly in very isolated rural locations. In addition to under-taking small-scale

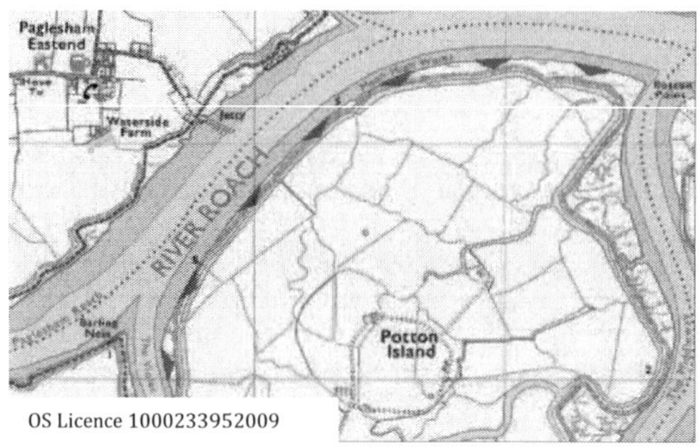

OS Licence 1000233952009

repairs he also invited landowners and agencies to see the repairs, demonstrated how local landowners could manage breaches in defences in the event of a North Sea tidal surge at his own expense. In order to facilitate his proposals the EA provided support through new coastal roles to work with the land manager, Defence Estates and NE to progress a project that could be easily consented through Flood Defence maintenance consent[9].

Potton Island Roach Estuary, Essex

It was clear that Food and Environment Protection Act[10] (FEPA) licensing could have been required for more complicated or extensive work, so partners worked to ensure the project could be delivered through the simplest possible route, to ensure a satisfactory outcome for the land manager and the partners involved.

As more of these community projects are undertaken, lessons can be shared and the background support streamlined. The case studies in this paper were largely ground breaking and as such were a learning process for all involved.

What does a community-led approach mean?

The critical factor for successful projects is a driving force from within the community and a vision that can be supported by organisational partners.

Strong community leadership is a common theme from all our case studies – but this needs to be coupled with strong support and technical advice from other partners. Community frustration with lack of help, lack of a holistic approach or unexplained delays, in spite of seeking help from all available sources (Beyton and Shotley Gate) can be the initial driver that gets a project started.

A more desirable approach is where a consultation process, such as the development of the SMP, (Potton) or a flood risk management strategy (Deben) leads to a realisation that something needs to be done at a local level, but that funding or other constraints mean that it can't be done without input from private individuals and communities.

In the Shotley Gate case, the community leadership was strong, but needed assistance to help them get a technical solution that was acceptable. The community contacted the AONB's SEO who initiated a working group immediately. However due to feeling isolated for some time, the group had already undertaken works that constituted fly-tipping in order to put in place some means of erosion control. The statutory bodies seeing that the community had acted in good faith, came together to deliver an acceptable solution to the erosion issue within the timescale.

In the Deben saltmarsh project the collective aim was agreed at the outset. Support and advice was needed from the AONB estuaries officer and the community to drive forward the impetus to deliver any activity as none of the usual funding sources were available – in spite of the project delivering a key national target. This carefully planned approach with dedicated support has limited the number of unexpected negative issues other projects have faced.

Who can help and support?

A key lesson arising out of these case studies is that all organisations need to be much more alert to community desires and be prepared to work with them at a much earlier stage. In these case studies SCC, EA and the AONB unit have all invested in new roles to focus on these challenging issues.

The Suffolk coast and estuaries are fortunate to have an AONB estuaries officer co-funded by private voluntary and statutory partners who works with local community groups acting as an honest-broker bringing about partnership approaches.

SCC identified a need for additional specialist personnel to work on emerging coastal issues and have now expanded this to cover issues arising from the Flood and Water Management Act 2010.

The EA has additional coastal resource under its Coastal Strategic Overview role (2008) but in Anglian region additional personnel are in post to specifically work with coastal communities and partners.

Having established the necessary partnership which is likely to include relevant partners such as EA, LAs (County, District and Parish), NE, Marine Management Organisation (MMO), AONB and local community, there is an important stage whereby all partners need to understand and accept the desires and constraints of all the other partners and agree a common understanding of the situation.

How to support communities?

Organisations need to establish what are genuine constraints and those which arise from inflexible interpretation of policies. If community projects are to work successfully, inflexible working practices and policies need to be streamlined or removed. In the Beyton case, a solution was found by using existing rules for agricultural and highways work to avoid flood risk management consents and waste licensing.

Any genuine constraints need to be accepted by all, and a position of what is and what is not negotiable needs to be outlined to the community and the project designed around this. Commonly this involves environmental designations, which dictated the timing and design of practical work at Shotley Gate and the Deben thus ultimately dictating the timescale for consents from other partners.

Understanding, empathy and moral support from relevant partners is an important element to the smooth running of the projects. The community needs to know who to turn to for support, especially when things don't go according to plan like the unexpected need to obtain and pay for a Temporary Right of Way Closure at Shotley Gate.

In all our case studies the leader or 'honest-broker' was trusted by the community and became the catalyst to iron out differences and misunderstandings and negotiate the inevitable compromises between partners.

Another key role for relevant partners is to provide data and information. Having one co-ordinator can ensure timely provision of the necessary information and prevent unnecessary duplication of effort. However, this role is time-consuming and thus a costly input into the project. These hidden costs (see appendix) generally do not get considered in initial discussions. The figures outlined in our case studies ably illustrate that this 'hidden' support for community projects often comes at considerable cost to supporting organisations.

How does a community group start a flood or coastal management project?

Having established the community group and wider partners, agreed a common vision and obtained all the necessary data, the time comes to decide what to do and how to do it.

In all the case studies cited, whilst it was often seen as a key driver in the first instance, funding is not the top consideration for such projects.   For all case studies the starting point was 'no funding available', yet practical solutions were found and ultimately funding acquired, albeit after much hard work.

The projects all involved technical and design work in order to establish the scheme details. This work often needs to be undertaken by professional partners (EA or LA) or consultants appointed by the community.

In all the cases studied, the issue of the consents and permissions, needed to be finalised before any work could begin, was a frustrating and tricky one – and involved the community in costs, delays and seemingly unnecessary hurdles. It is important to manage expectations on timescales at this stage and to explain that the outcome will be improved if all partners' considerations and licences are in place i.e. funding and PR opportunities. Even establishing which consents are needed is a task in itself and in the case of the FEPA licence can only be answered by completing a pre-application form. As several of these projects are novel in approach, it took some time to establish whether planning consent was needed or not.

The list below illustrates just some of the consents and related issues that have had to be overcome in local FCERM projects:-

- Coastal Protection Act (1949) for works on coastal erosion frontages (LA)
- Water Resources Act[11] (1991) for works on flood defence frontages (EA)
- Land Drainage Act[12] (1991) may require Flood defence consent from the EA
- Planning permission for new or improved structures (Local planning Authority)
- FEPA licence for works that impact on environment, navigation and fisheries (MMO)
- Habitats and Birds Directives[13] - environmental designations and consents (NE)
- Water Framework Directive[14] - EA
- Environmental Permitting Regulations[15] 2010 for waste permitting (EA)
- Archaeological surveys – often a requirement of the planning consent
- Rights of Way temporary closures/diversions from the County Council
- Highways closures from the Highway Authority or County Highways
- Health and Safety issues during construction and subsequently – planning consent.

Most consents involve completion of complex forms and submission of fees. There is much duplication between the consent processes and the statutory consultees involved, which in the cases studied seemed unnecessary as most of the statutory partners were involved in the development of the projects anyway. The costs and bureaucracy was disproportionate with the scale of the projects – the processes obviously being designed for much larger scale works usually undertaken by the relevant partners.

The schemes at Beyton and Potton were designed to minimise the need for consents by being imaginative within the existing licensing and consents process.

What sort of timescales and resource are needed?

Although not the first concern, funding is always a considerable concern for any community project. Capital funding is much harder to come by, but many organisations can more easily input 'in kind' contributions.

The main lesson is that the right solution, which satisfies the community and relevant partners, is much more likely to attract funding. In all the case studies, funds were acquired from a number of sources as well as raised by the communities themselves (Parish Council contributions, private donations or fund raising activities). The key to success is convincing people and organisations who can help that the project will meet their own objectives. Multiple benefits from a project clearly improve the range of potential funders. Sometimes a creative approach is needed to demonstrate the benefit of a project to a funder. In the case of Shotley Gate the project was clearly designed to control erosion, but in addition it helped to protect amenity woodland and public access, allowing funds to be obtained from a Green Infrastructure fund. In Beyton the local landowner was persuaded that his standing in the community would be enhanced by contributing to the solution (a small strip of land was donated for the watercourse to be widened). At Potton the tenant's willingness to pay for the work led to a contribution of materials from the landowner (Ministry of Defence).

What can prove difficult is to establish realistic costs and timescales. This is an issue for many 'professional' projects and even more difficult for community projects where lack of experience and novel approaches result in many unexpected issues arising along the way. Communities often fail to realise costs involved and the time it takes to get consents, which can be alleviated by managing expectations.

There are a number of practical considerations that may take community groups by surprise – establishing land ownership, deciding who is going to be the applicant for consents, deciding who handles funds and ongoing responsibility for maintenance and monitoring. In addition community volunteers need to input significant amount of personal time. Shotley Parish Council offered to underwrite the project, handle any funding and take ongoing responsibility for the new erosion control measures. However, at a late stage in the project the Parish Council found out that it did not have the necessary 'Power of Well-Being'[16] to undertake this role. With legal advice from SCC it was able to overcome this by use of other (Highways related) powers. This was another critical set-back that dented the confidence and enthusiasm of all concerned and nearly caused the collapse of the project.

The issue of managing expectations and handling media interest is also one that many community groups are poorly equipped to undertake. Here again, relevant partners skills and expertise are essential.

Delivering the project monitoring, ongoing management costs

The case studies illustrate a range of practical works undertaken, all of which required proper supervision and consideration of public safety during construction, as well as the use of specialist contractors. The community group needs to carefully consider the appointment of a suitable supervisor/contractor and factor in the time to get the necessary contracts signed.

Ongoing maintenance and monitoring are a critical requirement for most projects to go forward. Some of the consent forms (e.g. flood defence consent) request a named person/organisation to take ongoing responsibility for the new structure. In the haste to get the project designed and contractors appointed this important issue can easily be forgotten. In Beyton this was a vital consideration as the watercourse will undoubtedly silt up again without ongoing maintenance. An agreement was brokered with the landowner taking on part of the watercourse and Suffolk County Highways taking on the remainder that lay outside the landowner's boundary. In the Deben saltmarsh project ongoing monitoring by National Trust volunteers will be vital in order to assess the effectiveness of the project and thus the suitability of the methodology for use elsewhere.

Key Lessons learnt

1. For local projects to succeed community groups and individuals face a bureaucratic and administrative process that is often disproportionate to the scale of the work they are likely to be carrying out. Current systems are not designed for 'Localism'.

2. There is duplication in the planning and consenting process creating inefficiencies and increased effort for relevant partners and community project groups. In particular planning permission, flood defence consent and FEPA licensing involve broadly the same consultees and require very similar information. This is not efficient use of partner resource and the costs are largely borne by the community group.

3. All the case studies demonstrate that a strong community lead is key to a successful project but in addition a strong and knowledgeable partner lead or "honest broker" in a relevant organisation is also critical to supporting the community and helping them to resolve issues and overcome hurdles. This is in addition to a range of other partners providing time and resource to facilitate all stages of the approach.

4. Localism requires partner support for the community group to be successful this may have hidden costs on both sides. Government departments and relevant partners need to recognise the support communities need to deliver Localism.

5. Financial management and funding requires a formal framework e.g. parish council/Trust need to have ability to raise funds and pay contractors and take ongoing responsibility for the works undertaken.

Recommendations (taking the points above)

Based on the lessons learnt a series of recommendations are made aimed at the key organisations involved in FCRM decision making and community involvement.

Revision of the existing consent processes namely FEPA, Flood Defence consent, planning permission and waste licensing is needed to streamline the approach, or design a new approach that is specifically for small-scale projects within the context of wider environmental social issues. This work is already underway through Suffolk County's Total Environment bid and Central Dredging Association as pilots.

We need to develop specific guidance for community-led projects which will help summarise the process with examples of how to progress and who to involve, how to raise funds and how to formally manage the project as a group, partnership or trust.

We need to establish a definition of 'small-scale' through consultation between relevant partners and organisations, to support the SMP's which have used this term to denote community-led approaches where there is a change in policy or funding issues.

Consideration should be given as to whether a single consenting body is more appropriate for small scale projects as currently several are currently involved.

There should be a review of the consultees for the above processes to ensure that there is an appropriate level of involvement that is proportionate to the scale of likely impacts.

Relevant organisations need to ensure that they are engaging with communities facing flood and coastal management changes and ensure adequate resource to facilitate the adaptation process with the relevant community or group. The success of dedicated or co-funded posts is evident and is recommended.

There should be a more holistic approach allowing greater flexibility and pooling of local authority, EA and other funding streams.

References

1. Shoreline management plan guidance Volume 1: Aims and requirements March 2006 www.defra.gov.uk
2. Flood and Water Management Act 2010 www.defra.gov.uk
3. Environment Agency Strategic Overview for Sea Flooding and coastal erosion 2008 www.defra.gov.uk
4. Asset Maintenance policy protocol for sea defences (for England only), Environment Agency   January 2010 www.environment-agency.gov.uk
5. Supporting Change - Involving landowners in flood and coastal risk management-An information pack for landowners, Environment Agency, July 2010 (contact authors)
6. East Lane (FCRM@2010 paper)   (contact authors)
7. Decentralisation and Localism Bill (Draft-Nov 2010)www.communities.gov.uk
8. Coastal Protection Act, 1949 https://www.og.decc.gov.uk/environment/Guidance_cpa1949.pdf
9. Maintenance consent www.environment-agency.gov.uk
10. Food and Environment Protection Act1985 http://www.marinemanagement.org.uk
11. Water Resources Act 1991 http://www.legislation.gov.uk/ukpga/1991/57/contents
12. Land Drainage Act (1991) http://www.legislation.gov.uk/ukpga/1991/59/contents
13. Conservation of Habitats and Species Regulations 2010 www.defra.gov.uk
14. Water framework Directive October 2010http://ec.europa.eu/environment/water/water - framework/index_en.html
15. Environmental Permitting Regulations (England and Wales) April 2010 www.environment-agency.gov.uk
16. The Well Being Power, Local Government Act 2000 http://www.lga.gov.uk/lga/aio/1293811

Appendices

Costs and Hidden costs
Shotley Gate

Licences		Materials and works		Contributions & Hidden costs
FEPA	£1,500	Design and project supervision	£3,250	**Contributions by** SCC, AONB, Hutchinson Ports, EA, Haven Gateway Partnership, and LA totalled **£34,300**
Planning	£170	Contractor	£24,000	**Community funds**, the Parish Council and fundraising **£6,200**

Land Drainage consent	£50	Materials	£8,500	**Hidden Costs** Staff time circa 300 hours at £18 per hour (a conservative estimate) = £5,400
RoW closure	£300	Waste disposal	£5,000 **Total £42,470**	**Hidden Costs** The community estimated their time at 700 hours

Beyton

Contributions	Activity	**Funding & Hidden costs**
SCC £2500	Minimum 150 hours of staff time for project management & practical works planning and management	**Contributions by:** Highways Agency: £10,000, SCC : £35,000, Mid Suffolk DC: £3000, locality budget: £1000
SCC £500	Environmental investigations & road closure	**Community funds** Beyton PC: £1000,
MSDC £6000	Surveying, technical drawings, planning advice (work undertaken by their contractor, NPS)	**Hidden costs** The full cost of the project, taking into account in-kind and staff time costs would have been another 50% higher. Estimated in-kind contributions £26,000
Landowner £17500+ongoing loss of rent	Loss of land where watercourse widened and saving on waste removal and disposal costs by allowing 400t spoil to be spread on land	**Total cost** of practical works: £50,000.

Deben Estuary

Licences		Materials and works		Funding & Hidden costs
FEPA	£1,800	Design and project supervision	£4,000	SCC, AONB, LSP, EA totalling **£23,000**
Planning	£170	Monitoring	£2,000	Community fund raising totalling **£1,000**
Land Drainage consent	£50	Materials & Construction	£15,000	**Hidden costs** The community has estimated their time at 400 hours. Staff time 220 hours at £18 per hour as a conservative estimate = £3,960.

Potton Island

Licences		Materials and works		Total
Land Drainage consent	£50	30 tonnes of granite rock material	Approx £1,000	
EA Staff time	£1,850	Helicopter hire (landowner)	£10,000	£12,900

Innovative Coastal Zone Management
ISBN 978-0-7277-5749-4

ICE Publishing: All rights reserved
doi: 10.1680/iczm.57494.405

Possible Implications of Localisation and Austerity Measures on Flood and Coastal Risk Management

Nigel Pontee, Halcrow Group Limited, Swindon, UK
Andy Parsons, Halcrow Group Limited, York, UK
Richard Ashby Crane, Halcrow Group Limited, Swindon, UK

Introduction

Over the last 20 years progress towards more sustainable solutions for coastal flood and erosion risk management in England and Wales has been greatly assisted by the adoption of a strategic and long term approach (Defra, 2006). The statutory basis for flood and coastal erosion risk management (FCERM) has recently been confirmed and strengthened. However, proposed changes in the way central government funding for FCERM is prioritised, the amount of funding that is available and plans for the greater involvement of local stakeholders and third party funders, have the potential to impact upon the strategic approach that has been developed.

This paper explores how the current economic climate and the Government's new 'Big Society' programme could have unexpected implications for FCERM. The paper describes:
- Recent and proposed changes to policy and strategy in FCERM;
- The 'Localism Bill' as it relates to FCERM; and,
- The reductions to the FCERM annual budgets.

The discussion section considers the potential implications of these changes on provision of FCERM in the future. The paper demonstrates why it is important to maintain a strategic approach, even in times of austerity.

Existing situation
Strategic management framework

At present central Government plays a key role in setting the policy framework for FCERM in England and Wales. The UK government first stated their commitment to a strategic approach in the 1993 joint strategy for England and Wales (MAFF, 1993): *"new development or the construction of new or improved defence measures should not be considered in isolation. A new flood wall can cause erosion or problems elsewhere; a new development can change flood patterns or create the need for new defence measures. It is therefore important to look at the processes within a related area and the appropriate responses for that area. It is also necessary to consider the affect that any proposed actions, or decisions not to act, will have on the effectiveness of defence measures in adjoining or related areas and the impact these may have on the proposed action."*

The key tenets of the strategic framework that was subsequently developed are that it:
- Takes a long-term, broad scale and holistic view;
- Is forward looking when identifying future problems;

- Seeks to develop innovative, but appropriate, solutions;
- Has a comprehensive regard to impacts;
- Provides for the thorough assessment and reduction of risks;
- Shows balanced decision making when determining the chosen policies; and,
- Allows for the periodic review and updating of management policies in the light of new knowledge.

In England and Wales, the vision for a strategic approach has been implemented through a three level hierarchy of plans and schemes (Table 1).

Table 1: Stages in the strategic planning of flood and coastal erosion risk management.

Stage	Shoreline or Catchment Flood Management Plans	Strategy / Implementation Studies	Scheme Development
Aim	To identify policies to manage risks.	To identify appropriate schemes to put the policies into practice.	To identify the type of work to put the preferred scheme into practice.
Delivers	A wide-ranging assessment of risks, opportunities, limits and areas of uncertainty.	Preferred approach, including economic and environmental decisions.	Compare different options for putting the preferred scheme into practice.
Output	Policies (e.g. hold, retreat, advance) and a broad explanation of how this will be achieved.	Type of scheme (e.g. beach recharge, sea/floodwall, setback embankment, flood storage).	Design of works (e.g. revetment, wall, recycling).
Outcome	Improved management for the coast over the long-term.	Management measures that will provide the best approach to managing floods and the coast for a specified area.	Reduced risks from floods and coastal erosion to people and assets.

The development of this FCERM framework over the last 20 years has promoted the delivery of more sustainable shoreline and catchment management approaches (Pontee and Parsons, 2010). The strategic approach has allowed managers and policy makers to work together and stand back from the local needs and evaluate them in the context of the bigger picture. This involves asking questions about the national and regional impacts of local schemes, the economic justification for one scheme versus another, the prioritisation of schemes at national and Catchment or Shoreline Management Plan (CFMP or SMP) level, and the long-term economic case for schemes.

New draft flood and coastal erosion risk management strategy for England

This strategic approach, which has been adopted under Defra and WAG policy over the years was not backed up by legislation until the Flood and Water Management Act (2010) (FWMA) gave a requirement for national strategies to be developed for England and Wales and gave the Environment Agency a duty to exercise general supervision over all matters relating to FCERM in England..

The national flood and coastal erosion risk management strategy for England (Defra & Environment Agency 2011) was placed before parliament on 23rd May 2011 (EFRA, 2011). This requires the Environment Agency, to set the direction for FCERM through strategic plans that apply government policy to geographic areas and *"ensure that strategic plans, including catchment flood management plans (CFMPs) and shoreline management plans (SMPs) or their equivalents, are in place and monitored to assess progress"*. In Wales, the Welsh Assembly Government (WAG) expects to publish its new National Strategy for flood and Coastal Erosion Risk Management in Wales in the summer of 2011

Proposed changes to the national funding for capital expenditure on flood and coastal erosion risk management in England

At present most FCERM schemes in England and Wales are funded from central government grants. Even before the impact of austerity measures, it was evident that there were not enough funds to carry out all of the proposed schemes in any one year. This led to the development of a centralised prioritisation of funding, firstly the 'priority scoring' and then a revised approach based on contribution to delivery of 'Outcome Measures', which included consideration of the economic benefits; the numbers of households with reduced risk; and the areas of wildlife habitat created or protected.

A public consultation (Defra 2010a) was carried out between November 2010 and February 2011, on a new approach to the allocation of national capital grant-in-aid funding to FCERM schemes. The consultation response, Defra (2011) indicates that a new system will be put in place from April 2012 and that 2012/2013 will be a transitional year. The new system will utilise a new *"payments for outcomes"* approach, whereby Government pays fixed amounts or rates per outcome achieved, such as protecting a household against significant flood risk. the intention is that many schemes will still receive full Government funding, whilst others will qualify for a contribution to the final cost. This differs significantly from the outgoing approach where schemes applying for money were typically either funded in full by Government or not at all. The aim of the proposed changes is to attract more third party funding and to encourage local communities to invest in flood risk management measures, thus achieving more FCERM activity for the same national budget.

Managing risks to areas where defences may no longer be affordable

There are many areas of the coast and inland flood plains where past land use or coastal development took place with less knowledge of natural processes and the risks associated with such development than now available. Natural catchment and coastal processes were sometimes modified in a less than optimum way. There are also examples of past land drainage, flood defence and coast protection schemes that dealt with local problems whilst adversely impacting wider areas of the catchment or cell (e.g. Cooper and Pontee, 2006). This has left a complex and difficult legacy to manage in places, presenting challenges in terms of sustainability, especially in the light of potential future climate change, sea level rise and public opinion.

In the face of a changing climate, limited budgets and the importance of natural processes, it will not be affordable, or appropriate, to continue with "Hold the line" or "sustain level of protection" policies at all currently defended locations. The new national strategy for England (Defra & Environment Agency 2011) builds on previous Defra guidance to the Environment Agency on identifying locations where FCERM intervention is no longer justified. The strategy makes it clear that investment in defences should reflect current circumstances and priorities, and that some areas that once received public investment may no longer do so, and vice versa.

In practice this means that future SMPs, CFMPs and strategies are likely to recommend more polices for Managed Realignment, No Active Intervention, reductions in flood management measures and withdrawal of maintenance. This is particularly expected to be the case around many estuaries, see Figure 1. These changes are expected to lead to a requirement for greater local funding, more local defence maintenance to counter increasing risks, and the potential loss of land and properties which may be currently provided with protection.

Figure 1: The future maintenance of defences protecting low lying agricultural land around many UK estuaries is a key issue in FCERM.

Under current arrangements, those who lose protection, or suffer increasing risk under such policies are not generally eligible for compensation. There is no automatic right to compensation when property or land is lost to flooding or erosion, whether this arises from a change to management policy or not. Not surprisingly, there is often strong local resistance to policies of Managed Realignment, No Active Intervention and withdrawal of maintenance. Local communities have come to expect continued protection for flood and coastal erosion, and are not easily persuaded to give this up. There are also complex 'human rights' issues surrounding the 'legitimate expectation' that historic management will continue. This can be made particularly unpalatable by the knowledge that there is a legal obligation to provide compensatory habitats where designated wildlife habitats are lost or damaged due to FCERM plans or programmes.

Localism Bill

In their Coalition Agreement (May, 2010), the Prime Minister and the Deputy Prime Minister stated that: *"The time has come to disperse power more widely in Britain today."* The Big Society concept and the part played by 'localism' was a key pillar of the Conservative Party Manifesto and has since become central within the Coalition Agreement. As a result the Localism Bill was introduced to Parliament in December 2010.

In essence the Bill will devolve greater powers to councils and neighbourhoods, remove Central Government involvement in decision making in certain areas, and give local communities more control over housing and planning decisions[1]. Key areas covered by the Bill are provisions relating to:

- Local councils and their powers;
- Housing; and,
- Planning and regeneration.

Although there are provisions that might affect the way local funds are allocated to FCERM, this paper concentrates on the potential changes to the planning system that are pertinent to FCERM. The key areas within the planning provisions include:

- Abolishing Regional Spatial Strategies;
- Abolishing the Infrastructure Planning Commission (and a return to a position where the Secretary of State takes the final decision on major infrastructure proposals of national importance);
- Amending the Community Infrastructure Levy, which allows councils to charge developers to pay for infrastructure. Some of the revenue will be available for the local community;
- Introducing 'neighbourhood plans' which would be approved if they received 50% of the votes cast in a referendum;
- Introducing 'neighbourhood development orders' to allow communities to approve development without normal planning consent ('Community Right to Build'); and,
- Introducing new housing and regeneration powers to the Greater London Authority, while abolishing the London Development Agency (a concept which could be rolled out to other cities).

Austerity measures
Background
At present, about 95% of investment in FCERM in England is paid for by central funding from general taxation, which is allocated to the Environment Agency through a block grant, from which the Environment Agency funds most of its own flood risk management activities and provides Grant-in-Aid to other FCERM operating authorities for capital schemes. The routing of most funding in this way has allowed a national prioritisation system, economies in scale and improved certainty in funding (Defra 2010a). The vast majority of capital FCERM schemes are fully funded by Grant-in-Aid.

The present funding system has evolved through a number of steps from a much more complicated system prevalent in the mid to late 1990s. Previously centrally funded grants for between about 45% and 95% of the cost of capital schemes were made available to the Environment Agency and local authorities. Local authorities funded the remainder after grant through revenue funding grants, or "borrowing" funds that were recoverable from the central government in later years. The Environment Agency funding for capital works after grant was through levies on local authorities, and maintenance was funded through central government revenue funding streams.

In response to a number of flood events, including in particular the events of Easter 1998, Autumn 2000 and Summer 2007, there have been significant year on year increases in central

---

[1] Note: The Bill largely relates to England, although there are some provisions relating to Wales and a small number to Scotland. The Bill does not apply to Northern Ireland.

government funding of FCERM, see Figure 2, based on data Defra websites (National Archives 2008).

Long Term Investment Strategy

The most recent of a series of estimates of future funding requirements for the FCERM programme in England was given in the Environment Agency's (2009) 'Long Term Investment Strategy' (LTIS). The LTIS explored the merits of 5 options for future investment scenarios over the next 25 years. The 5 scenarios were estimated to have Benefit to Cost Ratios (BCR) ranging from 4 to 11, with the most favourable scenario having a BCR of 7. This is a good return on pubic funds in comparison to other public sector programmes (Defra 2010a). The favoured Option 4 sought year on year increases in expenditure over the next 25 years to give an 80% increase (over inflation) by 2035. The first ten years of the preferred funding option under the LTIS are shown in Figure 2 below. Note that costs in Figure 2 have not been adjusted for inflation.

The need for continued public investment in FCERM is made very clearly in the LTIS, which indicates that 1 in every 6 properties in England is at flood risk and more than 5 million people live in areas susceptible to flooding by rivers or the sea. In justification of the proposed significant projected increases in FCERM national funding, the LTIS noted that *"Today, around 490,000 properties face a significant risk of flooding. If investment is kept at current levels (in 'cash terms'), there will be 350,000 more properties of which 280,000 will be residential with a significant chance of flooding by 2035. Increasing asset investment in line with inflation would mean that 330,000 more properties would be at significant risk of flooding than now."* Under the most favourable funding option roughly the same numbers of properties were projected to be protected as today, with the increased expenditure being required primarily to replace deteriorating assets and keep pace with climate change and rising sea levels.

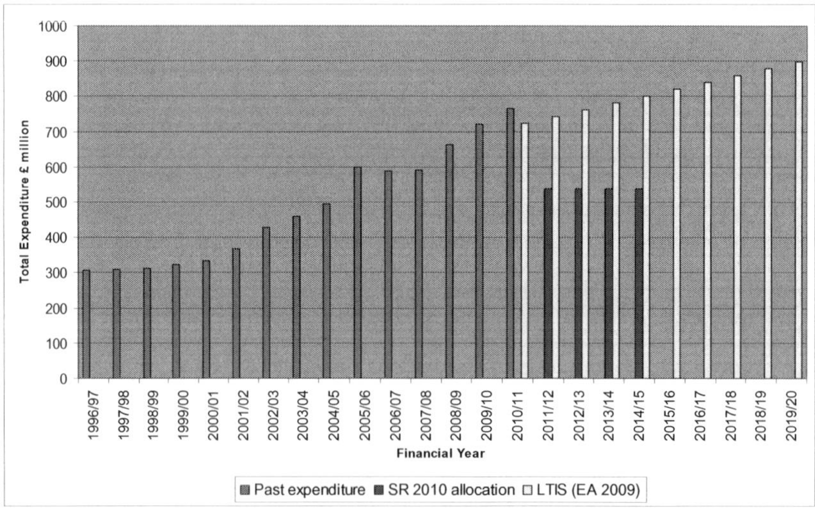

Figure 2: Government Funding for FCERM in England[2]

---

[2] These are cash costs relative to the year in which they were funded or announced and the values are not normalised to 2010. The forward values are at 2010 prices and do not allow for inflation. .

Funding settlement for next 4 years – SR 2010

The current economic crisis has resulted in significant cuts to public expenditure budgets. However, the importance of continued investment in FCERM was specifically recognised in the Government's Autumn 2010 spending review, HM Treasury 2010. It was noted that: *"The Government is also providing funding for adaptation to the impacts of climate change, with continued investment in flood and coastal erosion risk management. This will lead to better protection for 145,000 homes by the end of the Spending Review period"*. It was also noted that Defra expenditure will focus on areas of high economic value, with £2 billion allocated to for FCERM over the four year spending period.

Defra subsequently published its allocations to "Arm's Length Bodies" Defra (2010b), indicating £2.01 billion for Environment Agency's FCERM activities, as shown in Figure 3 below. Defra also clarified that *"Overall for flood and coastal erosion risk management, Defra expects to spend at least £2.1 billion over the next four years. The exact budgets are being finalised, but we expect the final figure to be at least £2.16 billion (an average of £540 million a year). This is approximately 8% less than spend by Defra over the previous four years (an average £590m a year)"*. This can be compared to the £2.15 billion made available over the previous three year spending period. The allocations to the Environment Agency indicate that resource budget will fall year on year, by £10-12 million.

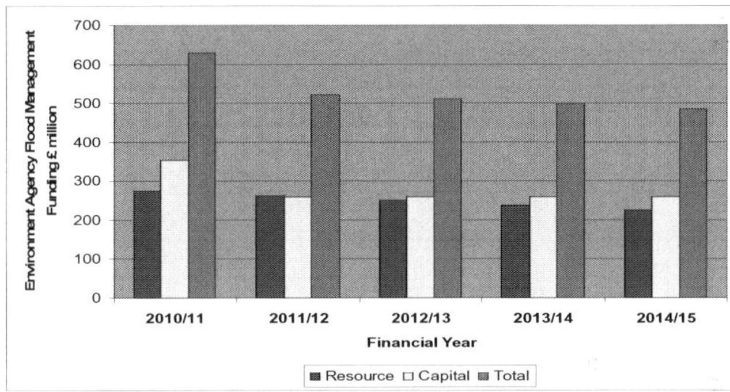

Figure 3: Environment Agency FCERM Funding for 2010/11 to 2014/15 (England)[2]

It appears that the most significant impact of the SR2010 austerity measures will be the significant reduction in the Environment Agency capital budget funding from £354 million in the record year of 2010/11 to £258.8 million in 2011/12, a 27% cut. As indicated in Figure 2 this will return expenditure to 2004/05 levels. Given that there is no adjustment for inflation the relative value will be even less. Commenting on the proposed funding cuts, the Association of British Insurers (ABI) expressed its disappointment and called for a long term plan to tackle the increases in flood risk that will occur over the next 25 years. The ABI also drew attention to the wider economic benefits provided by flood defences to communities and businesses (ABI, 2010).

To help the reduced budget go further HM Treasury and Defra are planning significant efficiency savings of up to 27% across the full Defra budget. Savings are required on capital projects, maintenance and resource costs at Defra and the Environment Agency and restructuring will focus on key priorities that deliver the greatest benefits. This is likely to impact upon the delivery of FCERM and increase requirements for financial contributions to schemes from those that directly benefit.

Discussion
Localism
At first sight the Localism Bill rings a number of 'alarm bells' with respect to the continued provision of strategic planning for FCERM and the implementation of holistic 'sustainable' solutions. The following paragraphs outline some of these concerns.

Community Right to Build (CRtB) & Neighbourhood Plans
It is intended that the CRtB will allow local communities (if they can gain 75% local backing) to carry out certain development, which need not be identified in local development plans, without the need for planning permission. There are obvious concerns here with regard to building in the floodplain and the compatibility of such development with flood risk management plans. Similar concerns arise on open coasts where local erosion prevention measures could lead to downdrift impacts. However, we should also recognise that the 'CRtB' is aimed at 'small' proposals in villages and small communities rather than larger towns and will require a 'sustainability check' as part of the process. Similar concerns surround the concept of Neighbourhood Plans.

It is therefore incumbent upon the Environment Agency and the wider 'flood risk management community' to participate in the development of the CRtB and Neighbourhood Plan concepts, and specifically to ensure the 'sustainability check' includes measures relating to sustainable/strategic FCERM.

Consistency and Locally Empowered Decision Making
Perhaps of more concern is the effect that greater representation of local views (perhaps reflected also by 'local funding') within planning decisions, may have on the way we provide FCERM infrastructure. Potential conflicts could include:

- Increased likelihood of development in the floodplain being consented due to local socio-economic benefits, with potential adverse effects on neighbouring communities;
- Decreased likelihood of strategically important FCERM strategy components being consented, such as flood storage areas or compensatory habitat creation, where these have low local benefit and relatively high impact;
- Increased support for local FCERM schemes that may have adverse effects on downdrift coasts;
- Reduction in the current level of national consistency with respect to the technical requirements of FCERM studies and in the justification (costs and benefits) of schemes for approval. This could potentially lead to poorer quality solutions with less far reaching and holistic benefits; and,
- A risk of inappropriate decisions being made in lower risk areas due to lack of technical rigour/consistency in appraisal, stretched funding and lack of third party financial support. This could result in inadequate maintenance of defences leading to overtopping, breaching and 'unmanaged realignment'.

Localism will inevitably lead to greater dialogue with local stakeholders and there is likely to be an ongoing requirement to inform local stakeholders of the risks that they face from coastal erosion and flooding. The proposed new arrangements for national funding of scheme appraisals and scheme construction costs is complicated and will make this dialogue even more time consuming than before. While dialogue is a good thing, a possible unexpected effect may be the time and cost involved, especially for the Environment Agency and other operating authorities, in facilitating the investigations and decision-making surrounding third-party or multiple-stakeholder funding.

The risks to strategic FCERM from increased local community influence must be managed through active engagement by the Environment Agency (and Defra) and the wider FRM community with the developing Localism Bill to ensure that a 'constructive balance' is maintained between local and strategic interests.

Additional funding sources

The proposed new approach to allocation of national funding for schemes takes a mechanistic approach to encouraging operating authorities to obtain contributions to schemes from other funders. As drafted, lesser priority schemes will have to obtain contributions before appraisal can start. In practice this may be likened to a 'chicken and egg' problem - how can investment in contributions be sought if appraisal of potential options and the benefits they may deliver has not been undertaken?

In addition, if local contributions are taken into account in scheme prioritisation, it may be possible to "buy your way to the front of the queue". This could have implications for less wealthy or less well informed segments of society, potentially adversely affecting 'deprived' communities who are by definition less able to 'help themselves'.

Additional / third party funding might come from a range of sources:
- The Community Infrastructure Levy (CIL);
- Increases to existing RFDC local levy schemes; and,
- Private investment from developers, land owners and infrastructure operators.

Whilst there are a number of ways in which local contributions could be sought, it remains to be seen whether third parties are able/willing to make contributions in times of austerity. Even if additional sources of funding are forthcoming, key issue will be:
- To ensure that this does not amount to a backdoor route for funding development in the flood plain;
- To ensure that these schemes fit within an overall strategic framework; and,
- To avoid unexpectedly disadvantaging already deprived communities that cannot afford to raise local funds.

Transferring defence ownership

The draft national strategy (Defra & Environment Agency, 2010) is seeking to hand over defences that are no longer economically viable, or affordable from national budgets. Such defences could potentially be handed over to private third parties (e.g. land owners, port authorities or commercial property owners), Internal Drainage Boards, who already raise funds through levies on land owners. Similarly, on the coast, the potential exists for the creation of Coastal Protection Boards formed from groups of local land owners.

From a national perspective defence handovers should be welcomed by tax payers as they will free up more funding to deliver national priorities. However, this will not be without cost and under the new funding proposals, further appraisal work will be required to confirm if defences are viable for some proportion of public investment. Additionally, a real concern is that handing over flood or coastal defences could lead to the promotion of local schemes that are unsustainable in a wider context.

Conclusions

In the UK, over the last 20 years, progress towards more sustainable solutions for coastal FCERM has been greatly assisted by the adoption of a strategic approach. This framework was developed after the limitations of pursuing local piecemeal schemes was realised. On coasts the strategic approach has not always been popular with local stakeholders. However, this does not mean that the decisions have been the wrong ones or that there are easier more palatable responses to climate change on our coasts. Importantly, this strategic approach helps deliver against the demands of international legislation such as the Birds and Habitats Directives, the Water Framework Directive, and the Flood Directive.

Proposed changes to the funding of FCERM schemes, austerity measures and the Localism Bill present a number of threats to this strategic approach. A key concern is that the greater involvement of local stakeholders and third party funders may lead to demands for schemes that meet local needs but are unsustainable in the long term or have adverse affects further afield. Significant work will be required to ensure that a strategic approach is maintained and that local empowerment does not lead to the development of unsustainable piecemeal schemes. In the long term, the promotion of piecemeal schemes could lead to the requirement for additional schemes further afield to counter there impacts, potentially increasing costs overall. These increased costs could arise from the requirement for additional defences, remedial works and wider societal costs associated with flooding and erosion.

The combination of localism and austerity brings significant risks to the implementation of long term sustainable FCERM plans. These rely on identifying the big picture and deriving policies that meet the needs of a range of strategic drivers including:

- Maintaining the integrity of sites designated for their nature conservation value in the face of sea level rise / coastal squeeze and human development;
- Providing appropriate FCERM for people, properties, infrastructure and commercial operations;
- Delivering economic, social and health benefits, and promoting equality, especially with respect to deprived communities; and,
- Delivering best value for UK tax payers.

This is possible within a framework of reduced central funding and increased local empowerment, but it will require new approaches to ensure the 'bigger picture' is not lost within the local delivery. A key issue will be engaging a greater range of true 'stakeholders'. These stakeholders include those who are actually providing funds and those who are making some perceived local sacrifices, such as changes of land use for coastal realignment areas. It is also likely that long term FCERM plans will need to be more flexible to reflect the uncertainty associated with local promotion and delivery of components where there is perhaps less 'strategic control'. In turn this may make the assessment of long term and in-combination effects of the plans less certain.

References

ABI (2010) Spending Review – ABI comments on flood investment plans. Wednesday, 20 October 2010 Ref: 52/10 Available online at:

http://www.abi.org.uk/Media/Releases/2010/10/Spending_Review_ABI_comments_on_flood_investment_plans.aspx

Cooper, N.J. and Pontee, N.I. (2006) An appraisal of the 'sediment cell' concept in coastal management: experiences from England and Wales. *Ocean & Coastal Management, 49, 498-510.*

Defra (2006) Shoreline management plan guidance; Volume 1: Aims and requirements; March 2006

Defra (2010a) Future funding for flood and coastal erosion risk management. Consultation on the future Capital Grant-In-Aid allocation process in England, November 2010, 22pp. Available online at:

http://archive.defra.gov.uk/corporate/consult/flood-coastal-erosion/101124-flood-coastal-erosion-condoc.pdf

Defra (2010b) Defra Arm's Length Bodies SR10 Allocations, published 20 December 2010. Available online at:

http://www.defra.gov.uk/corporate/about/what/documents/defra-alb-allocations-101220.pdf

Defra, (2011) Consultation on future funding for flood and coastal erosion risk management

Summary of responses, May 2011, 47pp. Available online at:

http://archive.defra.gov.uk/corporate/consult/flood-coastal-erosion/101124-flood-coastal-erosion-condoc-responses.pdf

Defra & Environment Agency (2011) Understanding the risks, empowering communities, building resilience: the national flood and coastal erosion risk management strategy for England, 23 May 2011, 52pp. Available online at:

http://www.official-documents.gov.uk/document/other/9780108510366/9780108510366.asp

EFRA, (2011) Environment, Food and Rural Affairs. Statement by the Secretary of State for Environment, Food and Rural Affairs (Mrs Caroline Spelman). Monday 23 May 2011. Available online at:

http://www.ipoak.org/monday-23-may-2011/

Environment Agency (2009) Investing for the future - Flood and coastal risk management in England. A long-term investment strategy. Available online at:

http://publications.environment-agency.gov.uk/PDF/GEHO0609BQDF-E-E.pdf

HM Treasury (2010) Spending Review 2010. Presented to Parliament October 2010 Document Cm 7942. Available online at:

cdn.hm-treasury.gov.uk/sr2010_completereport.pdf

MAFF, (1993) Strategy for Flood and Coastal Defence in England and Wales. 1993. Flood and Coastal Defence Division, MAFF, London, PB1471, 39pp.

National Archives (2008):
http://webarchive.nationalarchives.gov.uk/20081027092120/http:/www.defra.gov.uk/

Pontee N. I. and Parsons, A. (2010) A review of coastal risk management in the UK. Proceedings of the Institution of Civil Engineers, Maritime Engineering Journal. Volume 163, Issue MA1, pp31-42.

Innovative Coastal Zone Management
ISBN 978-0-7277-5749-4

ICE Publishing: All rights reserved
doi: 10.1680/iczm.57494.416

Managing Coastal Change: Use of the Defra Coastal Change Fund in North Norfolk

Peter Frew, North Norfolk District Council, Cromer, United Kingdom

Introduction

Historical Perspective

Britain has been an island for many thousands of years. As a result, the British people have developed a close relationship with the sea and the coast. Historically the focus of this relationship has been for trade and food. Hence towns and villages developed where there was access to the coast and a safe haven for boats. The more viable havens developed into larger ports and they, and the towns that supported them, often expanded on to reclaimed land.

In the last 200 years our relationship with the sea and the coast has changed. Though large ports still exist, the most significant change has been to regard the coast as a place to retreat to, either temporarily or permanently. The result has been the development of resorts, both large and small. Today the coast plays host to a multitude of activities, both business and leisure and land on the coast has become increasingly valuable. In many ways, therefore, the coast today is more intensively used than ever before.

In parallel with the growth of seaside towns came the increased use of coast defences. Much of this can be attributed to the development of new materials, particularly concrete and steel. The perception was created that erosion or flooding of a fragile coast could be prevented and this became the generally accepted view. The 1953 storm on the east coast reinforced this view. It engendered a response of "Never again" and led to a massive construction programme of new defences. In Norfolk this started in the immediate aftermath of the storm, reached its peak in the late 1950s / early 1960s and was still in progress in the late 1970s. The aim was a continuous line of defences, protecting the whole coast.

Coastal Processes and Shoreline Management Plans

It was about this time that questions started to be raised (e.g. Clayton & Coventry, 1986)[1] over the appropriateness of such a policy. Our understanding of coastal processes began to improve, particularly the interactions between the sea and man made structures. It is this improved understanding that has led us to question the appropriateness and long term sustainability of defence at all costs. It is this very understanding and the consequent shift in our approach that has been the catalyst for changes in our approach to coastal management.

The first generation of Shoreline Management Plans (now known as SMP1)[2,3], produced in the mid to late 1990s, often reflected nothing more than historical defence practice. There were few instances of retreat or do nothing policies being introduced on to previously defended coasts. The second generation of Shoreline Management Plans (SMP2)[4] changed all that and tended towards policies that worked with, rather than against, natural process, though this is not universal and inevitably there are exceptions. As a document though, the SMPs are unable to respond to the consequences of their policies and hence to a large extent they have been the drivers for the recent moves towards coastal adaptation. What this means in practice is largely the subject of this paper showing how, with innovative approaches to planning and finance, coastal adaptation can become a reality that is not totally harmful to communities or individuals.

Links to Planning

Whatever the outcomes of adaptation and the moves towards greater local determination, a structured planning regime is essential. Firstly, unsuitable development should be prevented; but more importantly, the planning system should create a flexible regime that permits suitable new development and permits existing development to change.

The current forward planning system, the Local Development Framework[5] (LDF) has created a suitable vehicle for the introduction of policies to restrict development in zones at risk. Importantly these policies are subject to rigorous examination through the LDF process that gives to the Local Planning Authority the necessary "teeth" to impose appropriate controls. It is important that the designated erosion risk zones can be shown to be technically sound. Without this backing the policies are open to challenge.

While restriction is the most obvious use of the planning process, adaptation through the encouragement of alternative uses or by rolling back development into "safe" areas are equally, if not more, important. Because without such positive policies, the communities affected will suffer blight and quickly become no-go areas.

Use of Resources

At any time, but particularly when resources are limited, making best use of the available finance is of primary importance. However, the concepts of coastal change, and adaptation to it, are new and it takes time to develop the techniques and approaches and the acceptance of them by both public and practitioners. It is not unreasonable therefore to think in terms of short term measures to bolster community confidence while other measures are developed and put in place.

An example of this might be to prolong the life of the defences. This should not take the form of large scale reconstruction, but rather be limited to extending the life of the critical elements of existing structures, providing low level defence that affords partial protection or providing defences with a finite life, say 10 – 20 years.

In 2006 North Norfolk District Council resolved to implement a 10 year programme of additional expenditure on defences to buy time, increasing its annual maintenance budget from £350,000 to £550,000 for 10 years. Examples of this approach can be seen in Overstrand where the toe of the wall was reinforced and in Happisburgh where a low level rock bund was placed on the beach. Presently, the additional expenditure is recovered (in due course) through the Revenue Support Grant. It is acknowledged that access to an initial fund is necessary to start such an approach.

The development of adaptation measures and methods is not easy. North Norfolk District Council (NNDC) has an advantage in that the need for coastal change has been recognised for some years, even though there is not (and perhaps never will be) a total acceptance of such approaches. Nevertheless, it was in this context, with a momentum already achieved, that NNDC was awarded a £3 Million grant from Defra to develop and trial various adaptation methods. The monies were part of an £11 Million fund set up by Defra to support coastal change. Grants were awarded to Local Authorities to develop and trial as Pathfinders methods of coastal adaptation. How this was done, the early results and some lessons is the main topic of this paper.

The North Norfolk Pathfinder Programme[6]
Outline
In developing its proposals to manage coastal change the Council was keen to explore as many as possible of the issues it faces. Thus the bid it submitted covered whole community, business support and community infrastructure issues. While the award fell short of the actual bid, nevertheless the £3 Million grant has been sufficient to enable most of them to be tested.

a) Happisburgh Whole Settlement Package
i)   Acquisition, demolition and relocation of houses at risk in the 1st SMP epoch;
ii)   Acquisition of houses in the 2nd epoch for lease / lease back;
iii)   Relocation of at risk car park and reprovision of toilets and beach access;
iv)   Support for cliff top caravan site to facilitate relocation;
v)   Clearance of derelict defences and beach debris;
vi)   Recording of village heritage.

b) Business Support Package
i)   Individual business health check;
ii)   Vouchers for business planning and other support (IT, accounting, planning);
iii)   Business development grant scheme;
iv)   Business development loan scheme;
v)   Business cluster analysis;
vi)   Marketing analysis and plan for the east of the District;
vii)   Private contributions to sea defences.

c) Community Infrastructure Package
i)   Support for relocation of Trimingham village hall;
ii)   Diversion and reinstatement of strategic cliff top path in Cromer;
iii)   Clearance of beach debris and reinstatement of safe beach pathway in Beeston Regis.

Finance
The award was made over two financial years with an unalterable split between capital and revenue expenditure. This split was the only condition Defra placed on the award, but with ever greater strictures placed on the definition of "capital" it has been the single most arduous element of the management of the whole programme. This aspect is revisited when considering the financial management.

Apart from ensuring proper financial control the aim was to maximise the benefit of each pound spent and where possible to generate an income. This could either be to recycle the monies to extend the scope / duration of the programme or to provide some longer term community benefit. For example, the relocation of the houses in Happisburgh will generate a capital receipt that can be recycled, and the Happisburgh Parish Council will take over the running of the car park to generate an income to fund the management of the public toilets.

Governance

Overall steer to the Programme was by the Project Board. This Board predated the Pathfinder Programme as the Coastal Management Board, and for the purposes of managing the Pathfinder Programme was augmented by representatives from the Local Strategic Partnership and North Norfolk Business Forum. This latter was particularly relevant given the large business support element to the Programme. The Council was represented by councillors from both main political parties, senior management and representatives of the project team of varying disciplines according to need. The management structure is shown in the diagram below.

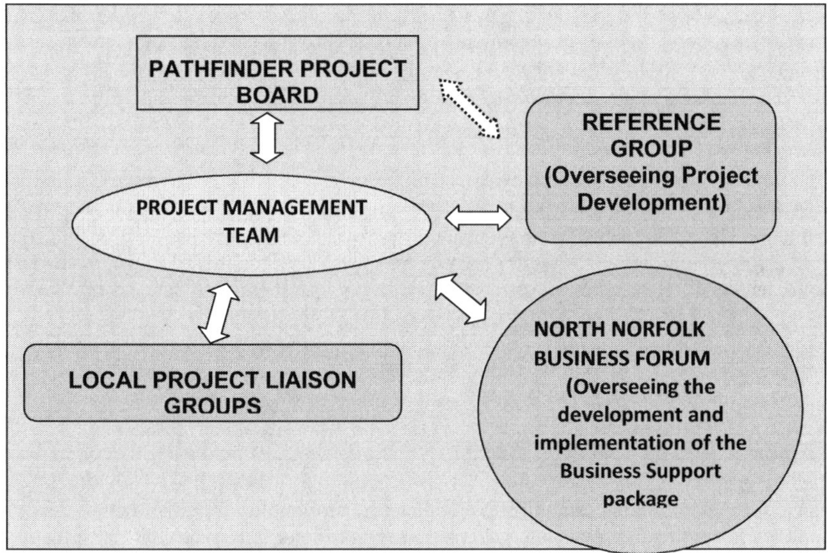

Figure 1. Project Management Structure

Key factors that have contributed to the success of the programme are:-
- Corporate buy in to the programme;
- Community involvement through liaison meetings and the Reference Group;
- Multi disciplinary project team;
- Appointment of specialist property advisers;
- Appointment of a programme manager (by secondment);
- Appointment of a dedicated solicitor;
- Close financial control.

Detailed Discussion of the Programme

Programme Development

With the announcement in the 2007 Comprehensive Spending Review of a Coastal Adaptation Fund of around £30 Million, NNDC always anticipated being able to access a portion of this sum to take forward some of the ideas it had been developing. The impetus for all this had come from the publication in late 2004 of the draft SMP for Kelling to Lowestoft.

The SMP identified the need to consider a process of coastal realignment, some commencing in the first 20 year epoch. The specific erosion risks were translated into the Local Development Framework Core Strategy along with policies to a) prevent development in erosion risk areas and b) permit relocation of properties lost to erosion.

The inclusion in the Council's Corporate Plan of a specific objective of developing a management solution to coastal change enabled some preliminary studies to be undertaken. These focused on determining the full nature of the problem, public awareness and initial thoughts on possible workable solutions. Thus when Defra invited bids in Summer 2009 for a share of an £11 Million Adaptation Fund, the Council was well prepared. The Council submitted a bid for £5.8 Million to undertake the projects previously outlined and was awarded £3 Million. Hence the first task was to trim the projects by something like 50% without losing effectiveness. While most elements suffered reductions certain key elements, particularly business support and housing, were largely preserved. The decision to project manage in house contributed much to the savings on the original budget.

Project Management
Effective project management has been a key factor in the delivery of the programme. With the reduced award the Council felt able to manage the project in house without recourse to external providers. The scale of the work was though beyond the resources of the existing coastal team and so two additional temporary posts were created, a programme manager (by internal secondment) and a solicitor. The latter post was specifically to deal with the significant number of acquisitions and other legal issues that were expected to and did arise. With existing specialist finance, property, business, planning and engineering staff coming together as required the result was a multi-disciplinary team with most of the necessary skills. Despite the numbers involved project overheads have been kept to 9%.

External support
The grant funding was spread over two financial years with a nominal deadline of 31 March 2011. This gave the Council only 16 months to complete the programme from the date of the award. While the Council had many of the skills required, it did not have them all nor did it have the capacity from its own resources. It therefore engaged external consultants to provide specialist property advice, business analysis, business support and tourism / business marketing.

Financial Management
In common with other Pathfinder councils, North Norfolk District Council has found that management of the programme finances has been what can only be described as a "nightmare". In addition to the year on year split Government also split the award into revenue and capital funding based on its own interpretation of the individual bids as to what might qualify as capital and what might qualify as revenue. In fact this was the only rigid stipulation made by Government: in all other respects the monies could be spent in any way the councils chose. The definitions of capital and revenue are laid down by the Chartered Institute of Public Finance and Accountancy (CIPFA) and are rigid and may change at any time. Therefore a great deal of effort has to be expended ensuring that the expenditure remains within the appropriate budgets because while revenue monies can be transferred to capital, the reverse is not true. This is potentially placing limitations on the flexibility of the use of the funds as originally envisaged by Government. The council is currently exploring means through its other budgets to restore this flexibility.

Happisburgh Whole Settlement Package

The main aim of the package of measures put forward for Happisburgh was to introduce the concept of an integrated approach to coastal change within a whole community. The village had gained a high profile, not always for the right reasons, and this package was seen as an opportunity to restore some confidence to the village as well as achieve some tangible objectives. Thus it includes assistance where houses are threatened with loss to erosion, support for the largest economic input to the village (the caravan site) and replacement of infrastructure lost, or at short term risk.

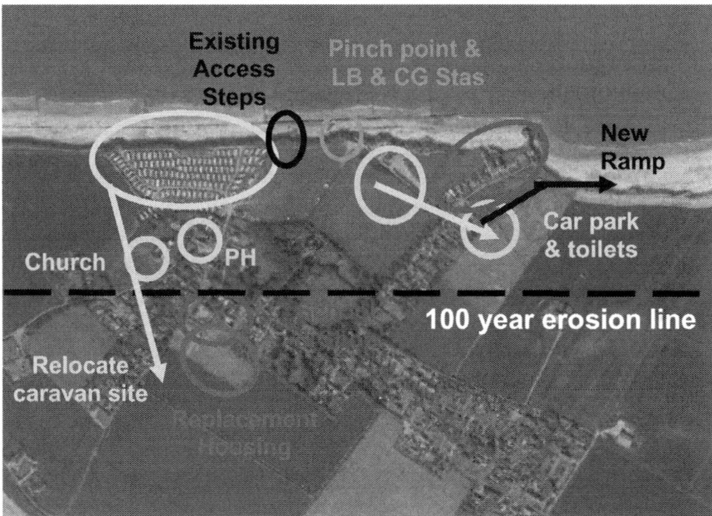

Figure 2. Happisburgh

Housing

Over the previous 20 years nearly 30 properties had been lost to erosion, and those that remained (some still occupied) had become increasingly derelict as there was no incentive to invest in maintenance: the classic downward spiral of blight (See Figs 2 and 3).

Figure 3. Beach Road with some of the houses to be demolished

The highest profile was attached to the proposal for the Council to acquire the 12 remaining cliff top properties and demolish them. Many have referred to this as compensation, but I intend to show that it is far from that and that it is indeed a sound investment that benefits the individuals, the community, the Council and the state through the innovative use of funds.

The basic premise in any Local Authority acquisition is that it should achieve value for money and that for any outlay there should be a demonstrable return. On the face of it acquiring houses (in a poor state of repair) and demolishing them does not meet this criterion. However, because of the existence of the Local Development Framework policy EN12[5] that permits owners of property at risk of loss by erosion to seek consent for a house on otherwise non-development land, such properties carry with them an opportunity value.

The first step towards acquisition was to appoint an independent property valuer / adviser. This was to give some reassurance to owners that the Council was not seeking to influence the outcome. After a competitive tendering process with a significant quality element to the tender appraisal, Bruton Knowles of Gloucester and Nottingham were appointed. Their initial task was to value properties "as seen" with all attendant risks. They also derived a notional "no risk" value as if erosion did not exist and the properties had been maintained. Finally they determined the value that could be attached to each property resulting from the "right" the owner of that property had to obtain planning consent on non-development land.

The offer made to each owner was made up of three elements: the "as seen" value, the planning "right" value and an up lift based upon the degree of risk to the property using a proportion of the "no risk" value. This was set at 10% (owner occupied) or 7½% (second home) discounted to degree of risk using HM Treasury 3½% Test Discount Rate. The end result is a capital payment that can be fully justified in terms of the Council acquiring an asset and which is significantly greater than the "as seen" value.

The next step (currently being explored) is to acquire land for development to replace the lost houses, secure planning consent and develop or dispose of the land at an enhanced value. The Council expects to recover the majority of the outlay for the purchase of the houses and the land. It will then be re-used in similar schemes in other villages. It is not a fully self financing scheme as fees and management costs are not recovered. The initial funds will diminish, but it is expected that the monies can be "recycled" several times.

Clearly there are risks: the initial house purchases may be over valued; there may not be a willing landowner; the land (with consent) may not reach the desired price. However, when compared with either a) the cost of defending the existing properties (if it could be justified) or b) the spiral of blight that would have occurred otherwise, this scheme has advantages. It is cheaper than defence; it permits individuals to move on; it sustains the community.

The houses that have been acquired will be demolished. Of the 12 offers made 9 have accepted. This enables the Council to proceed with its desired cliff top enhancement scheme. This will include a relocation of a car park currently at risk, re-provision of public toilets, re-provision of a ramp to the beach and the opening up of the cliff top as a public space. The Council is also working with the RNLI and HM Coastguard about the relocation of their facilities also at risk. (The lifeboat shed was moved 2 km south in 2003 when the main ramp was lost.)

Caravan Site

As referred to previously the other large project in Happisburgh has been to assist in the relocation of the caravan site (Figure 2). This is an important source of revenue for the village, but it has been losing pitches for a number of years. It cannot retreat further on its own site and therefore it must relocate to another site if it is to remain in business. The Council has made available to the owners grants up to the limit set by the EU State Aid Rules for him to obtain planning and business advice, contribute to other fees such as landscape impact evaluation and towards relocation costs. This project is proceeding well, but no final outcome is known at the time of writing.

Heritage

Finally in terms of the Happisburgh package, funding was secured for a project to record the heritage of the village. In particular it was to enable the village to understand its own heritage, both recent and the more distant past. There is now evidence of human activity in Happisburgh over the last 1 million years, so this project caught the public imagination and has generated a lot of interest and support. It is intended that the concept can be transferred to other similar villages and so represents good value for money. The project was carried out by Norfolk Landscape Archaeology at a coast of £80,000.

Business Support

The eastern coastal wards of North Norfolk all fall within the 35% to 45% most deprived areas in England[7]. In other words, while they don't qualify for additional support, they cannot, nevertheless, be classed as affluent. In recent years the road network has been down graded: what was once an 'A' class road is now 'C' class; many of the minor roads have been designated Quiet Lanes. While this can bring benefits, it also means that the communities of North East Norfolk are more isolated than they once were and businesses have, as a result, tended to struggle. The purpose of this part of the project is to revitalise the economy of the area with, in this case, a focus on businesses in the coastal area, targeting initially, those in the identified erosion risk zone.

Work streams

There were two main work streams: one directed towards individual business and a second more general approach considering the area as a whole.

The support to individual businesses was managed by the North Norfolk Business Forum. After initial launch events businesses that registered were given an initial 'health check' by EEDA's Business Link (at no cost to Pathfinder) to identify obvious needs. Businesses needing support were then offered vouchers of £450 to obtain professional advice, usually business planning as this opened the door to further potential support in the form of grants and loans. The initial target of 80 businesses registering was exceeded with 89 registering and of these 44 have taken the vouchers or are seeking to them take up. At the time of writing the grant arrangements have been set up, but the loan arrangements have yet to be put in place.

In parallel the Council engaged Norfolk and Waveney Enterprise Services to conduct a 'cluster audit' to determine the exact nature, scale and grouping of businesses in the area. The data gathered was used by a second consultant Blue Sail to consider what steps could be taken to improve the marketing of the area. This latter showed that some quite simple, relatively low cost steps could be taken, though in many cases they require the support and commitment of others. Measures identified were better signage (e.g. brown signs and tourist routes), reinstatement of a coastal bus service, separate identification of the area in tourism marketing and emphasising the proximity of other better known tourist draws (e.g. The Broads and the North Norfolk Coast). At the time of writing the findings of this report are being considered.

£860,000 was allocated for support to individual business:-

- Manor Caravan Park, Happisburgh`   £160,000
- Business advice   £151,000
- Further business advice (grants and loans)   £504,000
- Marketing etc   £45,000

Community Infrastructure

Government makes grants available for the provision of coast defences. Likewise Councils and the Environment Agency have budgets for the maintenance of defences. While grants are based on whole life costs that include decommissioning, rarely are funds set aside for the removal of defences and their clearance, largely because it would reduce the funds available for defences and local pressures dictate that it should be not be so used. Therefore the Pathfinder Programme was seen by NNDC as an opportunity to remove derelict defences and debris from them. The aim has been to produce a safer, more attractive beach environment. Two areas were identified, Beeston Regis, near Sheringham, and Happisburgh. The photographs in Figure 4 illustrate the effect at Beeston.

Figure 4. Beeston Regis before and after.

No action has been taken at Happisburgh as yet. That has been awaiting the outcome of negotiations to acquire the houses.   The beach clearance has been closely linked to the relocation of the cliff top car park, the re-provision of public toilets and the re-provision of an access ramp to replace that lost in 2002 (Figure 2). The aim is to improve the facilities for locals and visitors alike and help sustain the economy. These are relatively low cost solutions that have an immediate visual impact and demonstrate more than many of the initiatives that someone cares.

Local Liaison

From the outset the Council has involved local communities in determining the best way to manage the changing coast. This started with village and community workshops well in advance of any announcements from Government and so meant the Council was well placed to submit a bid. After the award local liaison groups were set up through Parish Councils and these have been vital in gauging the local community view and for dissemination of news. A dedicated web site created and all bar the most sensitive documents have been made available on it.

Conclusions

With greater understanding of coastal processes and limitations placed on funding there is a need perhaps greater than ever to devise new an innovative ways of managing our coasts. We will not and should not always seek to defend them. This Pathfinder Programme has been an opportunity to trial some new coastal management tools. Some have been simple and straightforward: realigning a footpath; clearing debris. Others have required more innovative solutions: buying houses, demolishing them and using the planning system to enhance the values and part fund the process. Often it has been the simple relatively cheap solutions that have brought the most benefit: creating an understanding of a community's heritage; clearing past defence debris to create a safer more attractive environment.

However, without the initial funding through the Defra Coastal Change fund none of this would have been possible. Currently most Flood and Coastal Erosion Risk Management legislation (and hence funding) is directed towards constructing and maintaining defences. The Pathfinder Programme has shown that there are alternatives, but that they cannot be implemented without some initial funding. Neither can they be implemented without community support. For many years communities (and individuals) have been led to believe that the only possible solution must be new or upgraded defence. Changing that mindset is very slow process, not helped by (largely) government led aspirations towards regarding homes as investments nor by restrictive legislation and regulations, such as definitions of capital and revenue expenditure.

The North Norfolk Pathfinder Programme has shown that is possible to find a way through; it has needed continuing support from the communities and a dedicated team to deliver it. While it has not yet ended and the full conclusions cannot yet be drawn the Programme has shown that there are alternatives to defences and that they can be effective. What is needed now is Government commitment to take these new initiatives forward into the long term.

References

1   K Clayton and F Coventry, 1986. Report to Nature Conservancy Council: An assessment of the conservation-effectiveness of the modified coast protection work at West Runton SSSI, Norfolk. UEA.

2   Snettisham to Sheringham Shoreline Management Plan. Mouchel 1996.

3   Sheringham to Lowestoft Shoreline Management Plan. Halcrow 1996.

4   Kelling to Lowestoft Shoreline Management Plan. Halcrow 2006.

5   North Norfolk Local Development Framework: Core Strategy. NNDC 2008.

6   Full details at http://www.northnorfolk.org/pathfinder/

7   Indices of Multiple Deprivation. Office of National Statistics

Innovative Coastal Zone Management
ISBN 978-0-7277-5749-4

ICE Publishing: All rights reserved
doi: 10.1680/iczm.57494.426

Cromer Coastal Strategy: An Approach to Prioritising Defence Works

Nick Clarke, URS/Scott Wilson, Basingstoke, England.
David Dales, URS/Scott Wilson, Basingstoke, England.
Peter Frew, North Norfolk District Council, Cromer, England.
Peter Lawton, St La Haye Limited, Cromer, England.
John Pos, URS/Scott Wilson, Basingstoke, England.

Introduction

Cromer is a small Victorian town with a traditional seafront comprising a beach, promenade and pier. The town sits on top of soft cliffs which are defended to prevent erosion. Defence from erosion is provided by a combination of groynes to retain the beach, which lies in front of a seawall. The entire Cromer Coastal Strategy frontage is one mile in length.

Immediately to the east and west of Cromer the high soft cliffs continue. These cliffs adjacent the Cromer frontage are undefended (though some relic defences remain to the east) and are managed under a "No Active Intervention" policy.

Cromer's coastal defences are largely in poor condition and delays in securing funding have exacerbated the risk of defence failure. The Cromer Coastal Strategy reviewed previous studies and prepared a Coastal Defence strategy to sustainably manage the frontage, protecting the town from erosion over the next 100 years.

The Strategy has been prepared during the economic downturn when funding is likely to be limited from the government, with only the highest priority projects going ahead. The emphasis of the Cromer Coastal Strategy has been to combine a long-term sustainable strategy with optimisation of the shorter-term business case in order to gain funding for crucial defence elements.

Doing Nothing

The development of the "Do Nothing" scenario included a Defence Condition Survey. This survey was undertaken in a number of stages, including;

- reviewing previous visual and intrusive condition surveys ;
- conducting a current visual condition survey;
- topographic survey;
- sediment sampling; and
- trial pits.

The Defence Condition Survey report recorded the current condition of the defences in-line with the EA Defence Condition Grading (Environment Agency, 2006). Using all the available information estimates of the residual life for each defence element were made taking into account the potential for a storm to cause failure.

An assessment of sediment transport along the frontage was undertaken. This included physical sampling of sediment along the beach to determine the particle size distribution. Cromer has a dual fraction beach of sand and small shingle with some larger cobble sized fractions; this provided the input to the numerical modelling of sediment transport rates. Importantly Cromer is located adjacent to a drift divide, due to the orientation of the coast and therefore assessments were made to see what deviation in average wave angle would cause the net drift direction to reverse. The change in angle was less than 5 degrees and with the potential for sea level rise to change refraction patterns the future sediment transport direction could not be predicted with any certainty. On this basis an assessment of the current functionality of the groyne field was made. Individual groynes critical to the retention of the beach frontage were identified, these were the groynes that were considered to be most important to maintain a protective beach frontage.

The major risk along the Cromer frontage is that the coastal defences protect the toe of soft clay cliffs some 30 metres in height. Failure of a defence section would likely result in commencement of cliff failure. A singe cliff failure is likely to cause additional damage to defences along the frontage and would begin a sequence of further failures. Under the "Do Nothing" scenario, once cliff recession commenced it would continue over the entire Strategy assessment period. Due to the proximity of Cromer to the cliff edge any failure within the town area would result in loss of assets.

Aerial imagery and historic mapping were used to provide estimates of historic recession rates on the cliffs adjacent to the Cromer frontage. Future projections of cliff recession including allowances for sea level rise were made using a modified Brunn Rule (Lee, 2005). A geotechnic stability assessment of the cliffs in both equilibrium and transient states was undertaken. This identified the zones of instability in the cliff in order to identify which assets along the cliff-top would be at risk in the event of seawall failure and the commencement of cliff erosion. Using both the erosion projection and the instability assessment the envelope affected by the Do Nothing was determined. Table 1 shows the number of key assets that would be lost under the "Do Nothing" scenario and the epoch in which they would be lost.

TABLE 1: Assets Lost under the "Do Nothing" by Strategy Epoch

Asset Type	Short-Term 0-5yrs	Medium-Term 6-50yrs	Long-Term 51-100yrs	Total
Residential	34	279	445	758
Commercial	3	22	21	46
Infrastructure	1	8	2	12
Key Civic Sites	0	5	0	5

Key assets along the frontage include the Anglian Water storm water tank, pump station and outfall, Cromer Pier and Cromer Church as well as a significant number of residential properties. Assets within the envelope were valued by a local estate agency and key infrastructure costs were sought. Table 2 shows the Cash Value (CV) and Present Value (PV) damages accrued per Green Book (Defra, 2003) discount epoch.

TABLE 2: Valuation of "Do Nothing" Damages by Green Book Epoch

Asset Type	Years 0-30		Years 31-75		Years 75-100		Total
	CV (£m)	PV (£m)	CV (£m)	PV (£m)	CV (£m)	PV (£m)	PV (£m)
Residential	34.6	21.3	41.5	8.6	37.8	4.3	34.2
Commercial	16.4	11.7	15.3	3.9	1.1	0.1	15.7
Infrastructure	15.0	12.6	-	-	-	-	12.6
Tourism	33.8	18.0	67.7	15.2	37.6	4.3	37.5
Total	99.8	63.6	124.5	27.7	76.0	8.8	100.16

Option Development and Appraisal

As with most coastal strategies the strategy for Cromer set the objective of implementing the current SMP policies over a 100 year horizon. For Cromer implementation of the "Hold the Line" policy required the continuation of the current defence line for 100 years.

A range of options were proposed, including beach management, seawall replacement, and groyne upgrades in order to provide continued defence. Options were evaluated against four criteria, technical implementation, environmental acceptability, economic performance and residual risk. On this basis it became clear that the approach to continued long-term defence of Cromer would be best achieved through a combination of maintaining a healthy beach frontage as well as a structurally sound seawall.

Costs of the various options were assessed against the "Do Nothing" scenario in order to provide Benefit-Cost Ratios in-line with Defra guidance. It became apparent that any "Do All Now" approach would not be economically justified and an alternative solution would be required.

Whilst some works were required in the immediate to short-term in order to secure the continued function of the defences, other works could be postponed. Therefore, a targeted approach to phasing the works was sought which would ensure the continuation of the defences; achieved over a timescale that provided acceptable risk with the associated economic benefit of delaying capital expenditure.

Prioritisation of Capital Works

The aim of the prioritisation was to give precedence to the works which were urgent and those that would provide maximum benefit to the defence of the frontage. This was to be achieved whilst maximising the economic efficiency of the scheme.

The prioritisation of the capital works was based on a Multi-Criteria Analysis of the Defences. Three criteria were selected for the prioritisation, Urgency, Current Defence Condition, and Importance.

Urgency was based upon the epoch within which the structure was estimated to fail. This information was taken from the Defence Condition Survey where residual life estimates (without maintenance) had been made. Urgency was considered an important criteria as it would differentiate between the mix of defence elements along the frontage and their varying structural lives. For example, a wooden groyne in "Fair Condition" may be considered to only last 5 years without significant maintenance. A concrete seawall in "Fair Condition" may be considered to last 10 years or more due to the differing deterioration rates of the materials involved in the construction. Urgency was scored between 1 and 3 based on the epoch of residual life with those failing first scoring highest. Table 3 shows a summary of the prioritisation criteria the scoring system and definitions.

The Defence Condition was considered important as this would rank the individual defence elements and allow ranking of similar defence types. For example, a wooden groyne in "Poor Condition" would typically require replacement before a wooden groyne in "Good Condition". Current Defence Condition was scored between 1 and 5 according the defence element grade, with structures in the poorest condition scoring highest.

Defence Importance was based on an assessment of how critical the defence element was in the overall continued defence of the frontage. The score is based on the consideration of a number of items including the analysis of sediment transport, coastal processes, position of the defence element along the frontage, vulnerability of the defence element, and the consideration of any strategic or valuable assets directly protected by a defence element. Whilst the Defence Importance criteria were determined qualitatively it was key to ensuring that the elements which are critical to the continued defence of the frontage were prioritised. Through an iterative process it also allowed some refinement in the prioritisation (e.g. consideration could be given to reducing the importance of repairing sections of seawall updrift of groynes that were to be upgraded).

TABLE 3: Defence Prioritisation Criteria

Prioritisation Criteria	Score	Definition
Urgency	1 – Long-term (Years 31-100) 2 – Medium-term (Years 6-30) 3 – Short-term (Years 0-5)	Urgency is selected based on the epoch within which the estimated residual life of the structure falls within.
Current Defence Condition	1 – Very Good 2 – Good 3 – Fair 4 – Poor 5 – Very Poor	Grading system is based on the EA Defence Condition Grading.
Defence Importance	1 – Low Importance 2 – Medium-Low Importance 3 – Medium-High Importance 4 – High Importance	The Importance of a structure is an assessment of how critical the structure is to continued defence of the frontage. The score is related to the value of assets that a defence element directly protects, the coastal processes it controls, the vulnerability of the defence, and the interaction of the defence with other defence elements.

Once the details for each defence element were completed the scores were multiplied together in order to provide a prioritisation score. Elements were then ranked in order of priority. Estimates of the capital costs to refurbish each defence element were added to the ranked list; this permitted the splitting of the list into works within the first phase of capital construction works and those that would be delayed to the second tranch of works.

Once the packages of capital works were determined a number of additional checks were undertaken. These checks ensured that the phasing would result in a robust solution for the continued defence of the frontage. The first check was an "Engineering Sanity" check to ensure that defence elements in the worst condition, known to require urgent attention, were included within the first phase of works. The second check was a "Coverage" check to ensure that suitable coverage of the frontage was achieved. To undertake this, defence elements included in the first phase of works were plotted on GIS; this ensured that works were not concentrated on a single location ensuring that the first phase of investment addressed issues along the entire frontage.

Selection of the works to be included in each phase of works was based on an assessment of the minimum amount of work that could be undertaken to ensure that the defences continued to function with an acceptable level of residual risk.

Impact of Prioritisation

For the Cromer Coastal Strategy a range of options were developed in order to provide continued protection to the cliffs. The options considered covered beach management and repairs to the seawall; two scheme options looked at these individually and the third scheme used a combined approach. Table 4 shows the options tested and the Benefit Cost Ratios.

TABLE 4: Option Benefit Cost Ratios

Option	Description	Benefit Cost Ratio Range
1 - Do Minimum	Repairs to the seawall and groynes, but no significant improvements in structural performance.	4.2 - 36.0
2 - Groyne Refurbishment	Repairs to groynes only to help maintain beach levels.	4.7 - 13.3
3 - Seawall Refurbishment	Repairs to seawall to maintain structural integrity.	4.2 - 8.0
4 - Groyne and Seawall Refurbishment	Repairs to seawall and groynes to maintain beach levels and improve structural integrity of the seawall.	4.2

Options 1 to 3 in Table 4 were given a range of Benefit Cost Ratios. These were the result of sensitivity testing on the likelihood of failure and the cost of intervention, should failure occur. Option 4 had a single Benefit Cost Ratio as it was deemed to provide a holistic solution addressing the beach and seawall (including allowances for future refurbishments and maintenance) and therefore had minimal risk of failure over the appraisal lifetime.

Within the Strategy assessment Options 1 to 3 were rejected as the options were insufficient to reduce the risk of failure of the defence frontage which would result in catastrophic failure and loss of assets.

Option 4 provided a solution that protected the frontage for the entire assessment period with minimal risk of failure. However, the cost of the scheme would have been prohibitively expensive and would not have gained funding. The scheme included all refurbishment and adjustments for sea level rise in a single phase of work.

To improve the Benefit Cost Ratio and reduce Present Value Costs (which would improve the Outcome Measure Score), the prioritisation methodology was applied.

Phasing of the works in to 2 schemes within the first 10 years of the Strategy and providing future schemes for the adaptation of frontage defences to accommodate sea level rise provided the opportunity to delay some of the cash spend and allow for it to be discounted according to Green Book rules (Defra, 2003). The result of the prioritisation is shown in Table 5.

TABLE 5: Prioritised Scheme Benefit Cost Ratio

Option	Description	Benefit Cost Ratio Range
5 – Prioritised Scheme	Phased repairs to the seawall and groynes to maintain beach levels and improve structural integrity of the seawall. Future adaptation to sea level rise.	6.9 - 8.2

The prioritised scheme was still deemed to have some residual risk of failure. Sensitivity analysis demonstrated that the variation in the Benefit Cost Ratio was significantly reduced in comparison with other options.

Conclusions

The development of the Cromer Coastal Strategy highlighted the need for erosion protection schemes to have a targeted approach towards better economic efficiency whilst delivering the SMP policies.

The prioritisation methodology used for Cromer uses a quantitive and qualitative assessment of existing defences in order to prioritise work whilst minimising risk and increasing economic efficiency. The methodology is based on defence characteristics which can be assessed for a wide range of differing defences of varying construction materials. The prioritisation methodology is therefore suitable for application to other coastal projects where increased economic efficiency of a long-term strategy is required to improve the potential for securing funding.

The prioritisation method allows works to be phased within an auditable framework, rather than a piecemeal or ad-hoc selection process. The prioritisation has a number of benefits:

1. Reduces the amount of work to be done in the initial phase. This enables the impact of these works to be monitored allowing for a more bespoke response to issues and conditions along a frontage;
2. Splitting the works into phases and delaying works that are not critical at the time of implementation, improving the economic viability of the Strategy. Capital values of works programmed in the future are discounted, this improves the Benefit-Cost Ratio of the Coastal Strategy;

3. Improving the economic profile of a project not only improves the Benefit-Cost Ratio but will also have a significant impact on improving the Outcome Measures score (Environment Agency, 2008).

References

Defra (2003) Revisions to Economic Appraisal Procedures Arising from the New HM Treasury "Green Book". FCDPAG3 Appraisal: Supplementary Note to Operating Authorities, March 2003.

Environment Agency (2006) Managing Flood Risk: Condition Assessment Manual. Doc Ref 166_03_SD01, October 2006.

Environment Agency (2008) Outcome Measure Prioritisation. (http://publications.environment-agency.gov.uk/pdf/GEHO0210BRXV-e-e.pdf)

Lee (2005) Coastal cliff recession risk: a simple judgement-based model. Quarterly Journal of Engineering Geology and Hydrogeology, vol 38, pg 89-104.

Innovative Coastal Zone Management
ISBN 978-0-7277-5749-4

ICE Publishing: All rights reserved
doi: 10.1680/iczm.57494.433

The Challenges of Delivering Coastal Erosion Risk Management outside Traditional Rules - Thinking Outside the Box

Denner, J.J., Halcrow Group Ltd, Chichester, UK
Fowler, R.E., Independent Consultant, Chichester, UK (formerly Halcrow Group Ltd)
Fenn, T., Risk & Policy Analysts Ltd (RPA), Norfolk, UK
Harris, B., Great Yarmouth Borough Services, Great Yarmouth, UK

Introduction

This paper outlines the challenges faced when trying to develop an erosion risk management scheme that does not meet traditional expectations. The challenges faced have included:

- Non-standard design
- Uncertain funding sources
- Unusual approval requirements
- Need for long term adaptation

This scheme pushes the boundaries of the traditional institutional frameworks in England, and yet is likely to be just an early traveller on the highway to more innovative ways of providing flood and coastal erosion risk management schemes in the face of diminishing public funds.

This paper presents a case study of how a local authority sought to deliver an erosion risk management scheme without reliance on government funding and encompassing adaptation to coastal change. Following the presentation of the scheme background, this paper presents the two interdependent elements of this scheme: 1) the short term engineering solution, and 2) long term adaptation. The adaptation section includes feedback from the empowerment of local communities through Pathfinder[1] studies, showing how local authorities and professionals can engage with and educate the public to develop long term solutions for at-risk communities. The paper shows how this engagement can lead to a better understanding of the reasons for coastal management decisions that affect lives and communities.

Background

Location and the constant threat of coastal erosion

Scratby is the coastal part of the parish of Ormesby St. Margaret in Norfolk (Figure 1). The coastline is orientated north-south and consists of a sandy beach backed by thin band of vegetated sand dunes which front soft cliffs of approximately 15m in height. The Scratby frontage, which extends for approximately 1km, has no coastal protection measures in place and has been subjected to direct coastal erosion for many years. The cliff edge has retreated

---

[1] Defra (2009) Coastal Change Pathfinder funded local authorities to explore new ways of adapting to coastal change through new and innovative approaches to planning for and managing change.

close to the front of the village to a point were the community feels threatened and its future is uncertain.

Halcrow Group Ltd was commissioned by Great Yarmouth Borough Council (GYBC) to identify and develop solutions to protect people and property in this vulnerable village. The appraisal identified a total of 139 residential properties in Scratby that are at risk from coastal erosion within the 100 year appraisal period. Cliff loss is typically episodic and property is expected to be lost in the short term (up to 58 properties over the next 20 years). To the north of the frontage, an advancing dune system protects property.

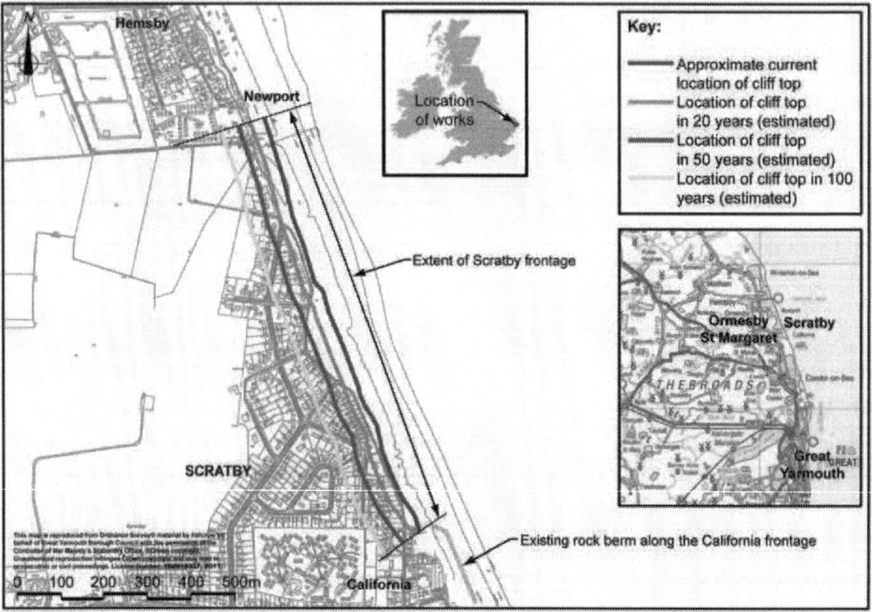

Figure 1: Location Plan

The rock berm at California

In the early 1990s the adjoining community of California was in a similar predicament to that which Scratby faces today. Severe winter storms in 1992 and 1993 saw substantial cliff falls bringing the edge close to a number of properties, but this community's fate was secured for 50 years through a coastal protection scheme comprising a 1.4km rock berm located at the base of the cliff. This rock berm was constructed in 1996 with government funding and has proved successful, considerably extending the life of the homes in California. However, at that time the economic case for extending the California rock berm to cover Scratby was unjustified.

Strategic background and the formation of Scratby Coastal Erosion Group

The policy for the Scratby frontage in the Kelling to Lowestoft Ness Shoreline Management Plan (SMP1) (Halcrow, 1996) was to "hold the (existing) line" of defence for all epochs. The plan was first reviewed in 2006 (SMP2), resulting in the proposed change in policy to "no active intervention". This draft publication raised concerns within the Scratby community, who had felt that it was going to be protected under the previous policy, but not under the new.

In order to promote a change to the SMP2 policy and secure an extension to the successful California rock berm to cover the remaining 1km area of undefended cliff, some of the residents formed the Scratby Coastal Erosion Group. The Scratby community wanted the same level of protection/investment as their neighbouring community and the group continues to exert significant political pressure through the support of their local Member of Parliament, Brandon Lewis (Joint Chairman of the Coastal Erosion All Party Parliamentary Group).

Following extensive public consultation, Great Yarmouth Borough Council promoted and secured a change to the Scratby SMP2 policy for the medium/long term (20 to 100 years) to "managed realignment", which recognises the need to continue to "hold the line" in the short term (0 to 20 years) at Scratby to manage the change in policy from SMP1 to SMP2 (Faber Maunsell, 2010). This change permits short term coastal defence works to be implemented to allow time for the community to develop social adaptation measures to mitigate the long term effects of coastal erosion. The plan acknowledges that the adoption of this policy is likely to lead to the loss of seafront properties in the village of Scratby in the long term.

It is normally expected that, following the development of SMP2 policies, a strategy plan is prepared to identify how the policies are to be implemented. At present there is no coastal strategy in place for the Scratby frontage. However, a decision was made to pursue a short term scheme coupled with a long term adaptation solution that would be initiated through a Defra Pathfinder project (Defra, 2009).

Short term scheme

Halcrow's brief for undertaking the project appraisal for the short term scheme was developed in line with the SMP2 policy and through consultation with Natural England. On submission to the Environment Agency in November 2010 for approval, the scheme presented a 20 year solution to buy time for the Scratby community to develop and implement longer term social adaptation measures in response to coastal change.

Figure 2: Scratby cliff top, with California scheme in the background

Based on the successes of the adjacent scheme, and to minimise alignment discontinuities between the frontages, a simple rock berm scheme was proposed. This low cost, flexible solution incorporates no allowance for maintenance of the structure, such that at the end of its short design life it would become increasingly compromised, thus allowing natural erosion of the cliffs to re-commence in the medium/long term.

Design challenges

As coastal engineers, we aim high, striving to work with the best materials to maximise the design life of our structures. Climate change and development pressures due to population increases provide technical challenges and cost has always been important, but as we face the future with "age of austerity" resonating in our minds, we are forced to reconsider our choice of materials and the appropriateness of our structures.

As we move towards adaptation to coastal change we will be delivering more short term schemes to allow communities to develop and adopt adaptation measures.

Figure 3: Simple rock berm at California

Design Life

British Standards BS6349, Part 1, Section 16 (BSI, 2000) and Eurocode, Part 0 UK National Annex, Section NA.2.1.1 (BSI, 2004) both specify that a design working life of 50 years or greater is normally expected for marine structures. There is a general presumption that coastal defence schemes promoted for government funding are to last 100 years (Environment Agency, 2010a). The requirement to implement a short term scheme at Scratby to allow adaption measures to be adopted provides a clear case for relaxing the design to bring cost savings and an increased chance of attracting funding, especially when it can be combined with adaptation measures that would reduce the risk to people and property in the medium to long term.

Despite a rock berm being a simple solution, the quantity of rock required to protect a 1km stretch of coastline is significant, with supply of rock being the overriding cost component.

Clearly, minimising the overall rock volume and rock size is of key importance in reducing overall costs to meet a short design life of only 20 years. The Rock Manual (CIRIA, 2007) sets out the design basis for rock structures and each of the design criteria were appraised for short design life structures. The following relaxations were accommodated in the design criteria to reduce scheme costs:

1) Allowable overtopping rates were increased. This minimised the cross sectional area of the berm through a reduction in the crest height and width and through a steepening of the structure slopes.
2) The filter layer was removed from the design to minimise overall rock volume and simplify construction.
3) Allowance of high design damage factor for wave attack (Van der Meer).

These measures were considered acceptable as occasional damage of the structure during severe events would be tolerable. The flexible nature of the rock structure means that it will continue to dissipate wave energy, hence erosion of the cliff, even in a failed state. A modest further recession of the cliffs immediately behind the proposed defence is considered tolerable as localised loss of the lower cliffs would not immediately threaten properties. Any such recession would increase the distance of the cliff face from the structure, thus reducing the impact of residual wave energy. However, this tolerance depends on the construction of the rock berm taking place prior to any further significant cliff recession.

Material Selection

The sourcing of lower quality rock was investigated with a view to reducing the unit cost of the rock. Variables considered were rock density, grading, strength and shape specifications.

As no local source of material was identified, there was little opportunity to reduce the overall cost of the rock due to large component of the cost being attributed to transport and placement costs rather than the unit cost of the rock. Lower quality material may prove slower to handle due to the potential for excessive damage during construction. However, another driver for specifying better quality rock was the potential for re-selling the rock at the end of the structure life (or full implementation of the coastal adaptation measures).

With a short term scheme, the future of the materials and the ability to recycle the materials will be of key environmental/economic concern. Good quality rock will preserve the potential resale value and will minimise wastage through handling during deconstruction and re-use.

Approvals challenges

In the UK the Coast Protection Act 1949 (HMSO, 1949) provides for the protection of the coast against erosion and encroachment by the sea, by delegating power to local authorities to undertake coastal protection work as necessary in their area. Such coastal protection work, other than repair or maintenance, is usually supported by central government funding and requires ministerial approval. Privately funded coastal protection schemes can be agreed to by local authorities[2].

The Department for Environment, Food and Rural Affairs (Defra) makes a single block grant in aid to the Environment Agency for capital flood and erosion risk management works in

---

[2] In 1995, at Ulrome on the Holderness coast, the then East Yorkshire Borough Council granted planning permission to a private coastal protection scheme to defend a caravan park. (Select Committee for Agriculture, 1998). In 2009 Suffolk Coastal District Council received private donations to fund a £2.2m coastal protection and flood defence scheme at Bawdsey. (South West Coastal Group, 2011).

England. Defra has also delegated to the Environment Agency the ministerial approval process under the Coast Protection Act 1949 (Environment Agency, 2008). The Environment Agency allocates the grant in aid money to flood defence and coastal erosion risk management projects promoted by the Agency, internal drainage boards and local authorities. The allocation of funding is determined in accordance with agreed procedures, such that maximum return is achieved on the investment of tax payers' money. On a programme-wide basis, this return is measured using a series of Outcome Measures (Defra, 2007), which evaluate the economic, social and environmental aspects of government spending. These Outcome Measures influence the prioritisation of projects for government funding.

At a project level, scheme options are refined to meet objectives set at the start of the project appraisal process. These objectives must reflect government policy, the duties, standards and targets of the operating authorities, and stakeholder engagement (Environment Agency, 2010a). Schemes that are environmentally unacceptable and technically infeasible are discarded. Once the remaining options have been technically refined, decision making as to the preferred option is based on comparison of costs and benefits. Therefore, although technical and environmental issues are considered, the final scheme selection, approval mechanism and ultimately the funding prioritisation is heavily influenced by economic considerations.

At Scratby, a workable solution was identified, which had a positive benefit-cost ratio. However, the Outcome Measures prioritisation score for the scheme was low (below 1.0) and it was expected that this scheme would not go ahead without significant external funding contributions. Accepting this position, GYBC decided to proceed with obtaining ministerial approval, in order to give confidence to potential external funders that the scheme was appropriate and widely acceptable to stakeholders. This approval process therefore focused on the quality of the technical and environmental aspects of this coastal protection scheme.

In accordance with Environment Agency procedures, a Simple Change (Standalone) Project Appraisal Report (PAR) was prepared (Environment Agency, 2010b) and submitted to the Environment Agency's Regional Project Assessment Board (review body for schemes under £10m).

Despite the Council's understanding and the community's acceptance that the scheme was unlikely to attract much, if any, government funding, the Project Assessment Board's comments on the scheme principally focussed on economic and funding aspects. For example, questions were raised about whether the economic appraisal of the scheme should cover a 100 year period, or just the scheme life, i.e. 20 year. The robustness of the cost estimates were also questioned and GYBC were asked to clarify costs for the longer term options that are being considered for beyond the proposed scheme life. Project Approval Board also sought clarification on the Council's exit strategy which largely depends on re-selling the rock at the end of the scheme life.

This focus on economic aspects blurred the original intention of GYBC to seek only ministerial approval for the scheme and not funding approval. Whilst assessment of the economic aspects of the scheme could ultimately prove to be valuable if government contributions were ever secured in the future, it is also recognised that the value for money considerations of private funders, particularly if they are local stakeholders, might be different from the value for money considerations made by government. It is therefore felt that the

Project Assessment Board approvals mechanism needs to accommodate for different perspectives where there are significant external contributions for a scheme.

GYBC's approach could be the first of many occasions when local authorities seek approval for schemes for which funding is not clearly identifiable. Defra's recent consultation on Future Funding of Flood and Coastal Erosion Risk Management in England (Defra, 2010) invited views on potential reforms to the way in which capital grant-in-aid is allocated. This consultation document proposed a new 'payment for outcomes' funding approach for all projects seeking funds from 1 April 2012. It is proposed that flood and coastal erosion risk management projects will be partially or wholly funded by government, depending on the outcomes achieved. Schemes that would otherwise not proceed because they did not have a sufficient Outcome Measures score may in future go ahead if an element of external funding has been secured. This results in those benefitting from such projects contributing more towards them. One of the objectives of this approach is to encourage innovative, cost-effective options to be put forward and it is felt that the Scratby rock berm is a good example of such a scheme.

Medium to Long Term Adaptation: Pathfinder project

Aim of the Pathfinder project

In addition to the short term scheme, GYBC wanted to pursue adaptation measures to address the medium to long term SMP2 policy of "managed realignment". This was initiated when GYBC was successful in applying for Pathfinder funding. The overall aim of the Scratby and California Pathfinder project was to expand the level of detail and depth of engagement with the local community and to empower the local community to explore the range of opportunities that exist for adaptation.

The engagement activities

A range of different engagement activities were undertaken to enable the Pathfinder to explore the views and requirements of the local community. This started with an initial canvass of all residents, undertaken by the Norfolk Rural Community Council (NRCC), to assess understanding and interest in the Pathfinder. This was followed by two questionnaires:

- the first questionnaire asked for people's views on general issues regarding coastal change and was used to explore how much people know about coastal issues;
- the second questionnaire (undertaken by Risk & Policy Analysts, RPA) began to explore options for adaptation and people's preferences for different options and who they thought should pay.

In addition, a number of drop-in sessions were held throughout the project, to obtain additional feedback from the community and to provide an opportunity for the community to keep up-to-date with the project. These activities were supplemented by letters from the Borough Council (since the canvass results showed that contact by letter was preferred). Residents also received summaries of all reports produced during the Pathfinder project.

Views of the local community

The initial engagement activities highlighted that, although some people were well-informed on coastal erosion, coastal change and the SMP2, the level of knowledge for most of the population was variable and inconsistent. Although 93% said they were aware of coastal issues affecting Scratby, only 59% were aware of the SMP. There was also considerable confusion and misunderstanding on the organisations and issues involved. In terms of awareness of the risks for their property, 55% were not aware if coastal change would affect them.

Despite the variability in understanding and awareness of coastal risks, almost all (96%) of those questioned supported the extension of the rock berm. There were some concerns that the Pathfinder project, with its focus on adaptation, might weaken the position when arguing for continued protection. Most people, though, identified the need for longer term planning but recognised that there was a need to 'buy time' to enable adaptation options to be identified, agreed and put into place.

Adaptation options

The Pathfinder project identified and assessed around 35 different ways of providing financial assistance to people living in at-risk properties. Consideration was given to obstacles and barriers that would prevent the approaches being implemented, the likely cost, possible funding sources and the actions needed to remove the barriers. By far the biggest obstacle was the cost. The cost of purchasing all the properties over the next 100 years, at a price that ignored the coastal erosion risk, was estimated at almost £50 million (undiscounted and including borrowing costs). A lower cost option, which fitted well with the wishes of residents to stay in their properties for as long as possible, was to help people pay for maintenance of their home. This was estimated to cost £8 million (undiscounted) where help is available when the at-risk properties have a residual life of 10 years or less. Although this option would help prevent blight on other properties by avoiding gradual dereliction, it would not provide people with financial help once their property was lost.

For comparison, it is possible to estimate what the options might cost if the money was raised through an increase in Council Tax. Table 1 summarises the annual costs over each of the 46,232 households in GYBC and compares them against the amounts people suggested that they would be willing to pay. The results show that people living in Scratby and California may be willing to pay sufficient money to cover the costs of some of the options. However, the willingness to pay of households in the rest of the Borough may be lower (or zero) such that the amount that can be raised may be less than is needed to fund the options locally.

Table 1: Change in Council Tax for each household in Great Yarmouth Borough

Increase in Council Tax to fund options (to nearest £1 to reflect uncertainty)						
	Value if there was no risk of erosion	Purchased at rebuild value	Purchased at value reflecting risk of erosion	Convert to leasehold	Provide alternative property	Subsidised mainten-ance
Total payable per year (on average)	£12	£7	£9	£3	£12	£3 to £11
Amount people said they were willing to pay	£9			£5		£8

Finally, the Pathfinder project looked to establish a Coastal Change Management Area (CCMA) and a set of policies relating to the rollback of development from areas predicted to be at risk. A workshop was held with local residents to identify rollback sites. This was followed by a drop-in session for the wider community.

The results were that the local community wanted as little change as possible, but they acknowledged that some change was inevitable if properties were to be relocated in the village. The community emphasised that any changes should be properly planned and delivered and that the current character of Scratby and California should not be altered. The community also highlighted the importance of maintaining a viable and attractive village, both to live in and as a holiday destination.

Who should pay?

The expectations of the local community were that central Government should provide financial support, whether that is for defences or adaptation. Many of the people who attended the drop-in sessions or completed questionnaires raised questions over the funding of rollback and its cost compared with the costs of extending the rock berm. There was also a strong sense of injustice over the apparent lack of public financial support. Many people feel abandoned under the current process and, although they have not been victims of serious erosion to date, that they are victims of lines drawn on a map.

The outcomes of the Pathfinder project

The Pathfinder project has enabled the views of the local community to be established and for the community to develop their own plan to accommodate roll-back. The community was very clear that they wanted the Pathfinder to come up with ways that could help those who are shown to be at risk in the SMP2, so the negative effects that have occurred since the SMP2 was published could be reduced. At all times, though, the community emphasised that the outcomes of the Pathfinder must not affect the case for an extension of the rock berm.

The plan that has been developed by the community identifies possible rollback areas and includes policies that reflect the requirements for planned, high quality development. The community recognised the need for time to enable those living in property identified as being at risk to take advantage of any potential adaptation options and the proposed roll-back areas. For many, this would only be possible with the extension of the rock berm.

Conclusions

As the demand for flood and coastal risk management funds increases and given the prospect that the government funding will be diminishing, coastal engineers and managers will have to become increasingly innovative in their solutions and how to fund them.

We need to learn how to design for shorter life spans, and to much-reduced budgets, be more creative with materials and their use and re-use. A new approach to government funding that will encourage local contributions is coming. We need to respond to this by becoming much more creative in how we engage with local communities, going beyond our historical role of technical advisors and embracing a much wider leadership role which can help empower these communities to secure own future.

We will need to lead coastal communities in their adaptation to climate change and the effects on their local coastline. It is also important to work with local communities, to help them understand the risks they face and the choices that they will have to make. The Pathfinder project has highlighted the variability in understanding of coastal erosion risk, although it has emphasised people's belief that defences are needed to enable long term adaptation.

Overall, we believe that coastal engineers and managers will increasingly have to 'think outside the box' when delivering erosion risk management solutions to coastal communities.

References

British Standards Institution (BSI), (2000). BS6349 Maritime Structures. Part 1: Code of practice for general criteria, London.

British Standards Institution (BSI), (2004). UK National Annex for Eurocode 0 – Basis of structural design, London.

CIRIA, CUR, CETMEF (2007). The Rock Manual. The use of rock in hydraulic engineering (2nd edition), C683, CIRIA, London.

Defra, (2009), Pathfinder – information from Defra web site on Pathfinder funds (accessed 27th February 2011), London.

Defra, (November 2010). Future funding for flood and coastal erosion risk management, Consultation on the future Capital Grant-In-Aid allocation process in England, London.

Defra, (2007). Comprehensive Spending Review - website summary on outcome measures: http://ww2.defra.gov.uk/environment/flooding/funding-outcomes-insurance/measuring-performance (accessed 27th February 2011), London.

Environment Agency, (January 2008). Guidance Note 3 for Local Authorities and Internal Drainage Boards: Quality Assurance of Flood Risk Management and Coastal Erosion Strategies and Projects, London.

Environment Agency, (March 2010a). Flood and Coastal Erosion Risk Management Appraisal Guidance (FCERM-AG), London.

Environment Agency, (2010b). Project Appraisal Report guidance for a FCRM Simple Change (standalone) Project for Local Authorities & Internal Drainage Boards, London.

Faber Maunsell, (2010). Kelling to Lowestoft Ness Shoreline Management Plan First Review (SMP2), London.

Halcrow Group Ltd, (1996). Kelling to Lowestoft Ness Shoreline Management Plan (SMP1), London.

HMSO, (1949). Coast Protection Act 1949, (Cc. 74 12 13 and 14 Geo 6), London.

Select Committee for Agriculture, (June 1998). Minutes of Evidence, Supplementary Memorandum submitted by The Wildlife Trusts and WWF-UK (F61), (accessed though http://www.publications.parliament.uk/pa/cm199798/cmselect/cmagric/707/8060915.htm), London.

South West Coastal Group, (2011). http://www.southwestcoastalgroup.org.uk/cc_future_finance.html (accessed 27th February 2011), London.

Innovative Coastal Zone Management
ISBN 978-0-7277-5749-4

ICE Publishing: All rights reserved
doi: 10.1680/iczm.57494.443

Sustainable Coastal Management for Clacton and Holland on Sea

Andrew E. Rouse, Environment Agency, Peterborough, England.
Sadia Moeed, Environment Agency, Peterborough, England.

Introduction

The Clacton and Holland Strategy is required to develop a sustainable coastal and flood risk management plan for 9km of the Essex coast. This includes 6.5km of coastal protection fronting Clacton-on-Sea and Holland-on-Sea managed by Tendring District Council (TDC) and 2.3km of coastal flood defences protecting Holland Haven managed by the Environment Agency. The aim is to have one management plan for the entire frontage that is developed by the Environment Agency with support from our framework partners, Royal Haskoning.

There are 3,500 residential properties, including key infrastructure and tourism facilities that are at coastal erosion risk over the next 100 years should the defences fail. This coastal strip also holds significant environmental, ecological and heritage interests including three Sites of Special Scientific Interest (SSSI) and two Martello Towers. The Shoreline Management Plan 2 (SMP2) recommends "hold-the-line" along the whole frontage for the first 50 years with a managed realignment potential of the Holland Haven frontage in year 50.

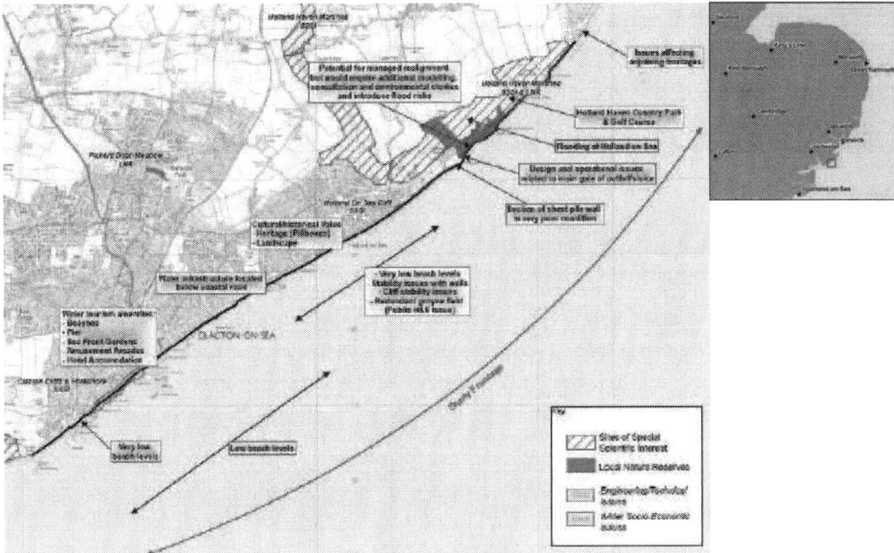

Figure 1 : Study Boundary

A Strategy was developed for this frontage in 2004 which included construction of a breakwater, beach nourishment and groynes refurbishment. The Strategy was agreed by Defra in 2005, but in subsequent years, when TDC applied for funding approval to take forward the first phase of works, they were unsuccessful under the priority scoring system. There is currently no approval to take forward the works recommended in the 2004 Strategy.

Figure 2: The Clacton Frontage (June 2010)

The challenges that the team face are the deterioration of the existing defences and potential failure of the cliff before a management approach can be adopted. Furthermore, in the current economic environment with the focus on localism, it will only be possible to fund any flood risk improvement scheme that is aligned with the local growth and regeneration agenda.

There is an increasing appetite to bring decisions about how we want to see our towns and cities develop economically, socially and environmentally to a local level. There is an opportunity here to join up coastal erosion risk management and the growth and regeneration agenda into an Integrated Coastal Zone Management approach. This falls into line with the Environment Agency's Long Term Investment Strategy which has identified that to be able to deliver coastal and flood risk management in the future, significant proportions of the funding required will need to come from external sources. The key issue for Clacton and Holland is to ensure that the appetite for localism is strong enough to generate funding to implement a coastal management solution before Mother Nature forces our hand and drives the option of retreating inland.

Background

Traditionally Flood and Coastal Erosion Risk Management (FCERM) is undertaken by the Environment Agency through the identification of areas of need. This process is driven by determining the problem and identifying a preferred solution which is in line with government guidance for public expenditure. The Environment Agency undertakes high level planning through SMPs and Strategies to identify where flood and coastal erosion risk could be reduced through the implementation of schemes. These plans result in business case documents which are taken forward to justify funding.

The Clacton and Holland Strategy is required in order to provide a plan and business case for long term investment in this frontage. This Strategy will have to take into account existing policies and drivers but will also need to be deliverable, i.e. fundable/affordable. It is known that deliverability is an issue for this frontage, as the required level of funding needed to implement the 2004 Strategy could not be made available. It is for this reason that the Strategy is being re-visited, but the project team must be careful not to fall into the same trap of producing a Strategy which complies with the guidance but cannot be implemented on the ground. The team will need to look into innovative ways of not only achieving the technical objectives of protecting the frontage, but also identifying solutions which may be attractive to external funders. Any management plan for this frontage will also have to take account of existing national and local policy drivers.

Essex Coastal Policies

A Shoreline Management Plan (SMP) is a high-level policy document in which the organisations that manage the shoreline set out their long term plan. The SMP aims to identify the best ways to manage flood and erosion risk to people and the developed, historic and natural environment and to identify opportunities where shoreline managers can work with others to make improvements. This area falls within the Essex and South Suffolk SMP.

The first plan (SMP1) was created in 1986 which was based on studies carried out for Jaywick frontage adjacent frontage to the south west. Only the first phase of this SMP was implemented (slope protection and Wave Walkers units in the propinquity of the Clacton Pier). The second generation SMP2 is being developed and is planned to be launched in April 2011. It is very important as this document will outline the overall management intent for the frontages which at present is "Hold the Line". Although, there is a dual policy at the Holland Haven frontage, which includes a potential for realignment through management of the coastline inland in Epoch 3 (2055 onwards)[2].

Growth and Regeneration Drivers

Regional Spatial Strategies were used to provide a regional level planning framework, setting out growth and development targets. These were revoked by the UK Parliament in July 2010, following a change in government, however these documents are still a useful reference to outline the regional and local aspirations for growth and regeneration. The Clacton and Holland area is included in the East of England Plan Regional Spatial Strategy.

Clacton and Holland is a part of the Haven Gateway, which is recognised as currently being an area of significant deprivation and therefore a priority area for regeneration. It is also recognised as an area with substantial potential to develop further with a major focus for economic development and growth. The Regional Spatial Strategy also identified a need for the creation of 5,000 jobs in the Haven Gateway, with Clacton Town identified as an area where new sites, premises and infrastructure could create more diverse range of employment[3].

The Coastal Erosion Problem

History of Clacton and Holland Frontage

The entire frontage of Clacton-on-Sea and Holland-on-Sea is presently defended from coastal erosion. These defences were constructed during the time of rapid development of the area following the pier development in the 1870s. Prior to this time the cliffs fronting the Tendring Peninsula were undefended and had a sandy foreshore. However, the pier resulted in the coastal development and the demand to construct defences.

The frontage is defended by stepped and vertical sea walls constructed in the 1950's and 1960's with the exception of a 1,500m length of sloping revetment (east of Clacton Pier) which is armoured with pre-cast concrete Wave Walker units constructed in 1992.

There is a long history of falling beach levels throughout the frontage. Beach loss is most noticeable along the eastern end of the frontage where the underlying clay is now intermittently exposed. In recent years beach levels have also dropped in front of the Wave Walker revetment. The low beach levels mean that the sea walls are exposed to increased wave attack and consequently considerable wear and tear. Overtopping of the walls is also ongoing and worsening. Several repair and emergency work schemes have been completed in 1995 with the construction of rock toe protection, 1999, 2000 and 2002 to stabilise the sea wall.

The condition of the defences remains poor with a significant risk of sea wall failure along the Holland-on-Sea frontage. The condition surveys completed in 2010 demonstrate an ongoing deterioration of the defence structures. The residual life of the defences is dependent upon the life remaining in the structure and foundations of the sea wall. The surveys show that the residual life varies considerably along the frontage, with the most critical conditions along the Holland-on-Sea frontage where lengths of the sea wall have a life of less than 5 years. Certain lengths of frontage have caused great concern over the last 10 years and as a consequence it has been regularly inspected. In some locations the situation has become so bad that for safety reasons, the promenade has been closed or emergency works undertaken to hold the current alignment.

Condition of the Frontage

The problems along the study frontage are directly related to falling beach levels. The low beach levels are a result of a long term and ongoing loss of sediment from the Clacton and Holland frontage. These losses are occurring for the following reasons:

- The protection of the cliffs at Clacton and Holland (and throughout the majority of the Tendring Peninsula) has cut off a significant source of sediment.
- The wave driven longshore movement of sediment away from the frontage to both the north east and south west.
- The cross-shore movement of sediment down the beach during storms.
- Exposure of the clay underlying the beach and the subsequent irreversible erosion of this layer by wave and tidal action.
- Increased reflection of wave energy from the face of the sea wall resulting in localised scour of the beach.[5]

The loss of material from within the system has driven low beach levels. This results in wave energy hitting the frontage with a grater force and drives an ever increasing risk of the sea wall breaching by one or a combination of the following mechanisms:

- Failure of the wall foundations through undermining and/or exposure to larger waves.
- Accelerated "wear and tear" on the sea wall through exposure to larger waves.
- Wash out of the base of the cliff behind the sea wall through excessive wave overtopping.[5]

The first two mechanisms listed above relate to impacts on the residual life of the defences and the third mechanism to the standard of protection provided by the defences. As part of the work done for the Strategy these parameters have been quantified so that the risk of a breach could be assessed and the need for and timing of works established.

Overtopping is known to be a problem along most coastal defences, and this is the case at Clacton and Holland. It is possible that a high level of overtopping would cause failure of the cliff behind the promenade and result in a breach of the defences. An assessment of overtopping quantities during a range of storm events has shown that the most critical conditions again occur along the Holland-on-Sea frontage where lengths of the defences have a standard of protection (against breaching) of less than 10 years.

Project Delivery
Working with Others (ECC – LEP – Community)

As well as the project team members and supply chain partners, this project has a very wide sphere of influence in terms of other stakeholders. It will be critical for the team to engage the right partners at the right time in order to achieve the project objectives. There are two key drivers to working with others:

1. Those that are likely to have a influence on the project i.e. organisations that could aid or block delivery (local planning authority, Essex County Council, Local Enterprise Partnership, major private funders and investors)
2. Those that may not have a big say but will be impacted by any change in coastal management (local community, small businesses, recreational users)

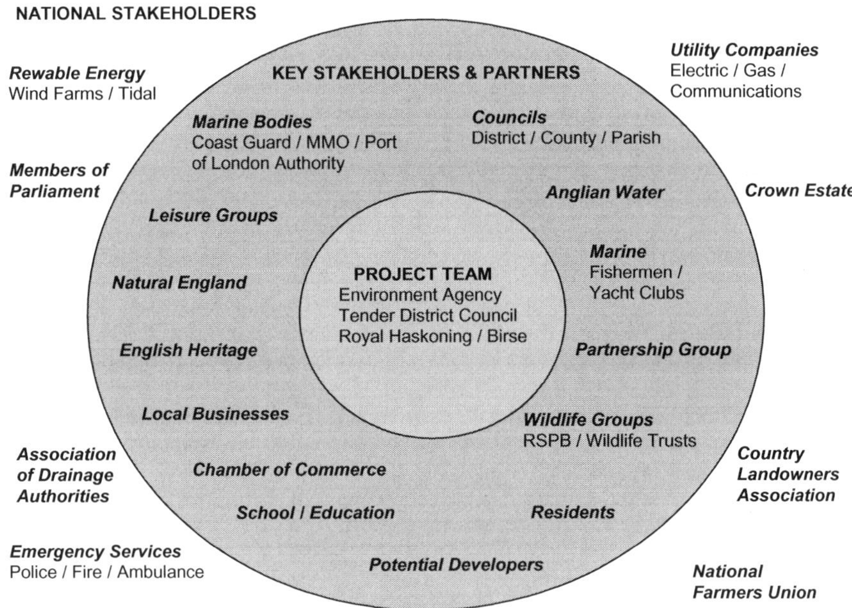

Figure 3 – the team and stakeholders

Key Issues Influencing Project Outcomes
Funding Requirements

The 2004 Strategy gives us a good indication of the level of spend required to manage the frontage over the next 100 years. The Strategy recommended beach nourishment and offshore breakwaters with a present value cost of £58 million, this solution provided a benefit cost ratio of 3.2. Schemes receiving National Flood Defence Grant in Aid (FDGiA) funding in recent years have delivered on average a benefit of £8 for every £1 spent; the target set by central government is to ensure we achieve a £5 return for every £1 spent[6]. The benefit cost ratio for the works identified in the 2004 Strategy fall woefully short of these current national benchmarks for benefit delivery. To ensure that a scheme can be delivered for this frontage it is going to be essential to work with others to identify alternative sources of funding.

In 2011 the Environment Agency and Defra published the new Flood and Coastal Resilience Partnership Funding arrangements. These provide a mechanism for negotiating contributions as well as calculating an appropriate amount of FDGiA which can also be contributed to jointly funded schemes. Prior to the implementation of this, schemes which were unlikely to attract FDGiA funding would have to be 100% externally funded in order to be delivered. With this funding allocation mechanism in place a joint funding approach can be taken with a proportion of the costs coming from FDGiA. This does change how the scheme will need to be developed as not only will it need to meet flood and coastal risk management objectives and deliver Environment Agency targets, it will also need to meet the objectives of external funders such as growth/regeneration or tourism/recreation improvements. The project team will need to engage others to identify opportunities for external funding and find partners who also have a driver to ensure sustainable coastal management in order to achieve their own aims.

The Kent, Essex and East Suffolk Local Enterprise Partnership (LEP) is a recently formed body consisting of business and local government representation. One of the priorities of the LEP is coastal renewal, with Clacton identified as an area for cultural and economic renewal and growth[7]. In addition there is ongoing private investment in Clacton through new developments, which the local authority must ensure contribute to the upkeep of all associated infrastructure. This is done through section 106 agreements in TDC's capacity as the local planning authority. New developments are put forward assuming that Clacton Town continues to be a viable social and economic area for investment, therefore section 106 contributions should be used for infrastructure needs to ensure the ongoing social and economic viability of the town. It is therefore critical that developers, investors and Council members understand that, whilst there are other infrastructure drivers, without sustainable coastal management the entire future of Clacton and Holland as viable settlements is put at risk. The LEP and developers / town planners are examples of potential partners and funding sources, but the coastal scheme would need to consider their vision for the future of Clacton and Holland.

This approach to working with others will result in a very unconventional Flood and Coastal Risk Management project. The team will have to work to understand the aspirations of other organisations and incorporate these in to the appraisal of options. This introduces new risks to the development of a sustainable approach, but also new opportunities in terms of achieving a final outcome which delivers multiple objectives and benefits.

Guidance and Policy Requirements

A coastal management scheme which may be fully or part funded using FDGiA must follow a rigorous set of procedures in terms of ensuring that the final outcome provides the optimum investment of public funds. These procedures are set out in the Flood and Coastal Erosion Risk Management – Appraisal Guidance (FCERM-AG). The investment rules within this guidance are based on the Treasury Greenbook and ensure that all centrally funded schemes apply a consistent approach to assessing damages, valuing benefits and identifying the technically, environmentally and economically preferred solution. Under these rules it is not possible to factor future developments as a part of the benefits of the scheme. This creates an interesting conundrum for the project team as to how to meet the driver to ensure that the final coastal management plan delivers protection to existing properties and also future growth and regeneration opportunities. It is highly likely that the economically optimum scheme identified through the application of FCERM-AG may not be a scheme that is attractive to potential partners. However, by identifying the optimum FCERM-AG approach to managing the frontage, the team will be able to identify the minimum level of cost required for coastal management and also the quantum of FDGiA funding which can be put towards this. This data can then be used as a baseline for working with potential funders to develop an 'ideal' coastal management approach. This ideal approach may be very different to the baseline FCERM-AG solution. However, under Flood and Coastal Resilience Partnership Funding, the same quantum of FDGiA funding can be put towards this as long as it still delivers the required protection to existing properties and provided that all of the remaining funding can be sourced externally.

Therefore, although existing guidance presents constraints, with some innovation and willing partners these should not present a blocker to an integrated management approach which delivers multiple objectives.

Change in Approach – incorporating the localism agenda

All of the opportunities identified for external funding require the involvement of local organisations that have an interest in the future of Clacton Town. This lends itself to a 'bottom-up' approach for developing a sustainable management plan for the Clacton and Holland coastline. Instead of following national government objectives for coastal management, it will be local aspirations that set the drivers and objectives.

Local Regeneration Plan

Figure 4: Opportunity sites identified in the Clacton Town Centre AAP consultation document

The Clacton Town Centre Area Action Plan (AAP) is currently in the process of being completed. The outline plan was published for consultation in 2010, this sets out a vision for Clacton as a vibrant and attractive seaside town. Within this plan, the pier and seafront promenade are identified as one of the key opportunity sites for re-development. The preferred strategy set out in the consultation document for the Area Action Plan is:

'A comprehensive scheme for the seafront, pier, pavilion, and surrounding public realm, that will create a unique, high quality environment, and attract visitors to the town' [8]

Options/Approaches for Coastal Management

Using the team's knowledge of the local aspirations, and working with TDC, the project team have developed a number of potential options for managing the coastline. The options were grouped into key principle approaches which have been listed below:

- Do Nothing – this must be assessed to provide an economic baseline and justification

- Do Minimum (eventually leading to a Do Nothing scenario)
- Managed Realignment (eventually leading to a Do Nothing scenario)
- Hold-the-Line: Asset led investment approach
- Hold-the-Line: Land use led: Protection of Clacton Centre
- Hold-the-Line: Land use led: Protection of Clacton Centre and Residential Areas
- Hold-the-Line: Land use led: Protection of the entire frontage

These options can be sub-grouped into 3 key high level themes; Do Nothing, Asset Led and Land Use Led.

Do Nothing / Do Minimum

Under this scenario no further intervention takes place to manage coastal erosion at Clacton and Holland. The sea defences would fail which would result in immediate loss of the promenade exposing the toe of the cliff to wave attack. This would result in slips / failures in the cliff causing it to recede landwards. Recession rates in the first few years would be high as the beach and cliff try to return to a more natural profile. In physical losses over 100 years it is estimated that 3,250 properties would be lost and the present value economic damages would be approximately £200 million.

'Do Minimum' would involve engineering measures to hold the existing defences either by ad-hoc improvement works as required or by providing a rock revetment to protect the toe of the defences. Whilst this approach may extend the life of the defences, the clay platform on the foreshore will continue to erode and eventually it will not be sustainable or technically possible to maintain the existing line. This option will therefore also eventually revert to the 'Do Nothing' scenario within the 100 year timescale.

By identifying the social, economic and environmental losses which occur in the 'Do Nothing' scenario the team will be able to demonstrate the need for works to implement a sustainable coastal management approach. This will justify that there is an overall benefit to the nation and therefore FDGiA funding should be contributed. It will also present a business case to potential investors by demonstrating that without investment in coastal defence the future viability of any investment in Clacton and Holland becomes questionable.

Asset Led

Under the asset led approach, the optimum solution would be determined using FCERM-AG, and investment would be phased based on the residual life of the existing coastal defence assets. This approach would ensure that the most is made out of any asset investment to date by not upgrading assets until the residual life has expired or is close to expiry. Each phase of the works would form a part of a holistic management approach, so that once all phases had been implemented, a sustainable solution for the whole coastline would be in place. As the key technical issue identified at this frontage is erosion and lowering of the clay foreshore platform, any long term management approach is likely to involve measures on the sea-ward side of the existing defences which will protect the platform. This could be through the provision of a beach and associated structures such as breakwaters and groynes to hold the beach material in place.

Land Use Led

Under the land use led approach, the options appraisal would again seek to determine a sustainable coastal management approach to prevent further lowering of the clay platform. However, the phasing of the works will be led by growth and regeneration opportunities.

Therefore it is likely that the first element of work will take place around the pier area, which is seen as a key strategic site for investors. It is likely that there will be a local drive for investment in the pier and promenade and therefore the first sub-option 'Protection of Clacton Centre' assumes that a long term management solution would be put in place for the pier/promenade/town centre area, and that potentially the rest of the coastal defences will be left to natural processes. The second sub-option assumes that the investment which is brought in by the implementation of protection to the town centre will provide the funds required to then implement the second phase of the long term solution, which will involve protection to the residential areas. In this sub-option the very northern end of the coastline, where the main asset at risk is the Anglian Water sewage treatment works, would be left to natural processes. Under the third sub-option the long term management approach is implemented for the entire frontage with the town centre protection forming the first phase.

The land use led approach is most likely to attract external funding, as it prioritises the local growth and regeneration drivers.

Future of the Clacton and Holland Frontage

Currently the future of this frontage is very uncertain. Much depends on the availability of funding from external sources, partnership organisations and the strength of the local growth and regeneration agenda. One of the critical issues will be generating a local interest in coastal management, and encouraging local partners to take ownership of the problem. This will require a major shift from the current expectation that managing the frontage is the UK Government's problem and therefore will paid for by central funds. With the localism agenda there is an impetus for people to take responsibility for the issues that affect their local community, and this should include flood and coastal risk management.

If the project team is able to strike the right balance, there is a real opportunity for identifying an integrated scheme which not only protects the existing people and property in Clacton and Holland, but also secures a bright social and economic future for the area. It is important to note that the future of this frontage is not just dependent on identifying growth / regeneration and funding opportunities in the short term. This area is an exposed part of the coastline which will always be under pressure from natural processes. Therefore managing this frontage will be an ongoing task with local partners needing to commit to a long term plan of securing funds and implementing works over the next 100 years.

Recommendations / Way Forward

There are three key actions on the project team to ensure successful delivery of this project:

1. Identify the optimum technical, economic and environmental management approach following FCERM-AG and the contributions policy, in order to determine the quantum of FDGiA funding available for coastal management at Clacton and Holland.
2. Identify local aspirations and funding opportunities in order to shape a final solution which is likely to attract a high level of external funding.
3. Engage local partners to develop the final plan.

Conclusion

Undertaking a 'traditional appraisal' which would involve a rigorous application of FCERM-AG, and adherence to national drivers as opposed to local, will result in a Coastal Management Plan (Strategy) document which is unlikely to be implemented. The FCERM-

AG 'preferred solution' could be in the order of £60 million and may not be attractive to potential funders or affordable using government funding alone.

This project is taking forward an innovative new way of undertaking FCERM-AG appraisal in order to develop a long term management plan. Whilst this approach carries risks, there is a great opportunity to deliver multiple benefits. However, all of this must be considered in context with the current physical condition of the defences and risks of failure.

If the team are successful in identifying a deliverable approach for the frontage, this project could pave the way forward for Integrated Coastal Zone Management in other areas. This is becoming more crucial in recent years as the scope for government funding for such schemes is reducing. By working with local partners, and using local drivers and aspirations as priorities for phasing and investment, coastal management is put in the hands of local groups to shape the future of their towns and cities.

References

1   Environment Agency (2009); North Essex Catchment Flood Management Plan
2   Environment Agency (2010); Essex and South Suffolk Shoreline Management Plan (draft for consultation)
3   Government Office for the East of England (May 2008); East of England Plan.
4   Tendring District Council (2004); Coast Protection Scheme Strategy Plan Clacton-on-Sea (Royal Haskoning)
5   Environment Agency (2011); Baseline Processes and Defence Condition Assessment (Royal Haskoning)
6   Environment Agency (2009) Flooding in England – A National Assessment of Flood Risk
7   Environment Agency (January 2011) Kent, Essex and East Sussex Local Enterprise Partnership Briefing Note
8   Tendring District Council (2010); Clacton Town Centre Area Action Plan – Consultation Document,

Innovative Coastal Zone Management
ISBN 978-0-7277-5749-4

ICE Publishing: All rights reserved
doi: 10.1680/iczm.57494.453

Wash East: Integrated Coastal Management

Pippa Lawton, Royal Haskoning, Peterborough, UK.
Mike Dugher, Environment Agency, Anglian, UK.
Mark Robinson, Environment Agency, Anglian, UK.

Context

Coastal strategies have previously followed a standard format; they identified and discussed the problem, developed options to solve the problem and assessed options for a solution that best fits the technical, environmental, social and economic criteria.

In 2007 the European Flood Directive was enforced to assess, map and reduce flood risk, coordinate with the Water Framework Directive and encourage public involvement. In the United Kingdom, this Directive and The Pitt Review, reviewing lessons learnt from the summer 2007 floods, led the development of the 2010 Flood and Water Management Act specifying risk management of flood and coastal erosion.

These changes to legislation and a shifting economy have meant a radical change in the way we look and conduct coastal management. There needs to be more open engagement with people who live and work on the coast and hinterland to understand what is important to them and their aspirations for the future. These priorities and aspirations could, where appropriate, be combined with flood and coastal risk management decisions for a fully integrated, stakeholder driven sustainable future on the coast. This paper demonstrates an example of how this approach is being applied on a part of the United Kingdom coast: The Wash (Figure 1a).

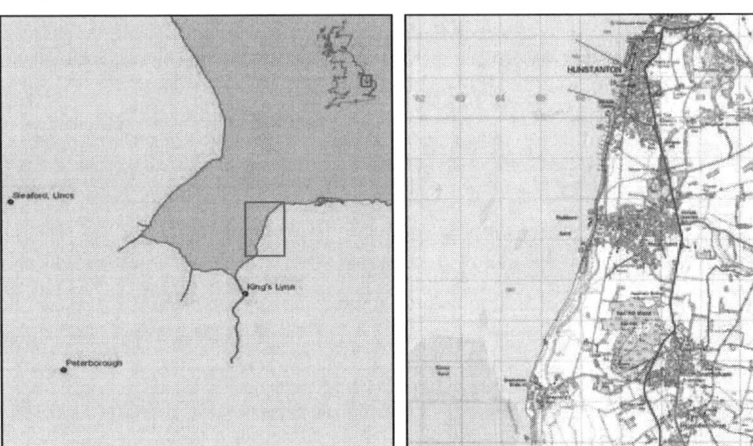

Figure 1: (a) The Wash Embayment and (b) the WECMR Frontage

Introduction

This paper will present the transition from the Shoreline Management Plan (SMP), a high level non-statutory policy document, to the Wash East Coastal Management Review (WECMR), a strategic review of coastal management. The WECMR focuses on some 8.5 kilometres of the frontage from Hunstanton Cliffs in the north to Wolferton Creek (Ingol Outfall) in the south (Figure 1b).

This discusses the stakeholder engagement utilised as part of the SMP and how this has fed into process of developing the WECMR objectives and scope. The paper discusses and touches on the challenges faced in the growing relationship between stakeholders and partners. It tells the start of the story of working with others on innovative ways of managing and funding long term flood and erosion risk along Wash East frontage.

This paper will discuss how the stakeholder engagement funding and accountability, issues were overcome and the lessons that we have learnt insofar from this new approach to coastal management. The paper demonstrates how partnership groups can be drawn together to ensure a sustainable approach to the management of the coast.

Shoreline Management Plan

A SMP is a strategic document which aims to identify the future management intent for a particular length of coastline. Through partnership working it does this by identifying the best ways to manage our coastline, balancing the values of the assets and interests it protects against the risk of flooding and coastal erosion. The plan looks at coastal management issues over the next one hundred years and breaks these down into three time frames covering the short, medium and long term.

The Wash SMP runs from Gibraltar Point in Lincolnshire to Hunstanton Cliffs in North Norfolk and covers around eighty kilometres of shoreline. There are a number of key physical processes occurring in the Wash and an understanding of these processes, combined with defence types, was necessary for development of this plan. The key issues of this plan was understanding of the complex coastal processes within The Wash embayment, economic viability of holding the line beyond the short term and the wide scale impacts of not holding the line on the East of England.

The plan was broken down into four separate Policy Development Zones (PDZs) based on these coastal characteristics and follows the principles and processes set out in the Shoreline Management Plan Guidance issued by Defra in March 2006. The policy development zones are:

> **PDZ 1** - Gibraltar Point to Wolferton Creek
> **PDZ 2** - Wolferton Creek to South Hunstanton
> **PDZ 3** - Hunstanton Town
> **PDZ 4** – Hunstanton Cliffs

The Wash SMP was developed through a strong partnership between the local authorities, Natural England, National Farmers Union, Internal Drainage Boards, English Heritage, Environment Agency, and other organisations. This partnership involved two governing groups, a Client Steering Group and an Elected Members

Forum. This approach was very important as there are strong interactions between shoreline management, coastal land use and the coastal environment when steering strategies and projects that are developed through the SMP Delivery Action Plan.

One of the principal aims of an SMP is to set the intent of management. For coastal flood and erosion management this translated into one of four policies set out by Defra, which best describe the actual management of the shoreline. The policies are:

> **Hold the Line (HtL)** – this involves holding the defence on its existing alignment.
>
> **Advance the Line (AtL)** – this involves constructing new defences seaward of the existing defence line. If relevant, use of this policy is limited to those stretches of coastline where significant land reclamation is considered.
>
> **Managed Realignment (MR)** – this involves allowing the shoreline to move seaward or landward, with associated management to control or limit the impact on land use and environment. This can take various forms, depending on the intent of management to be achieved. All are characterised by managing change, not only technically (by breaching and constructing defences) but also to land use and environment (by facilitating or ensuring adaptation).
>
> **No Active Intervention (NAI)** – this involves no investment in coastal defences or operations.

The above policies do not attribute any particular standard of protection to be provided. In most areas this decision would be beyond the scope of the SMP and left to schemes or strategies to deliver.

Policy appraisal objectives were developed for each specific frontage based on locally-specific features, values and issues. Each policy was appraised against these objectives, which ranged from protecting communities, agriculture, infrastructure, maintaining flood defences and habitat. All relevant options were scored against the objectives, using quantitative and qualitative indicators as appropriate. The results were visualised using a traffic light colour system, which was used to validate the results with the partner organisations to support policy options.

These policy options were then tested through an extensive three month public consultation process whereby the SMP project team engaged with local communities through local media, publications, e-mail and by holding a considerable number of local drop-in events. This proved very beneficial as the team was able to engage with the local community to explain the plan in greater detail and what it meant for them not only in the short-term but also long-term. Also, the team was able to remove areas of uncertainty, invite feedback but more importantly build relationships with the local community. The consultation feedback received helped to deliver a more comprehensive plan in taking the agreed policies forward. The Wash SMP was completed in November 2010.

The Wash embayment relies heavily on raised sea defences to provide flood risk protection to a significant area of low lying land. The south-eastern face is defended by frontline defences consisting of a natural (but maintained) shingle ridge, backed by a secondary line in the form of a sea bank. The Hunstanton coast protection frontage is at low risk of flooding beyond the promenade, but is protected against erosion by a combination of sea walls, promenades, wave return walls and beach control structures (timber and concrete groynes). The only undefended length of the Wash SMP is the Hunstanton Cliffs area, which is allowed to erode naturally, although there is evidence that the base (toe) has been defended in the past.

PDZ 1 of the SMP2 is outside of the WECMR and this paper where we focus on PDZs 2, 3 and 4.

PDZ 2: Wolferton Creek to South Hunstanton

PDZ 2 is very complex and sensitive with a large number of people located behind a relatively low standard flood defence. However, even the existing defence standard will be difficult to sustain into the future because the shingle ridge needs continuous maintenance, with increased costs and detrimental environmental impact. However, the holiday homes and caravan parks that the defence protects are very important for the local and regional economy.

The intent for this area is to jointly develop a sustainable long-term solution based on cooperation between the partner organisations and all people and businesses with an interest in the area. This process commenced in 2010 through forming a strong Key Stakeholder Sub-Group (KSSG) of representatives from the above groups to work together to find alternative sustainable solutions. The KSSG was established encompassing businesses, land owners, local authority and home owners with the aim to support and promote understanding of coastal flood risk and investigate funding mechanisms to facilitate the improvement of the front line defences that protect approximately 3,000 caravans, 600 homes and numerous businesses. The caravan site owners and residents represented on the group indicated early on they are likely to want to contribute to fund holding the line. The Coastal Pathfinder budget, described later, enabled the organisation of a separate project looking at additional opportunities for external funding of coastal defences. These are covered in more detail below.

PDZ 3: Hunstanton Town

The intent of management for PDZ 3 is to sustain the viability of Hunstanton town as a tourist resort and regional commercial centre. The intent is to hold the shoreline defences where they are now. There will be a need to monitor long-term development in relation to the neighbouring frontages.

PDZ 4: Hunstanton Cliffs

The intent of management for this PDZ is to continue to allow the cliffs to erode naturally and provide sediment to help maintain the beaches to the south in PDZ 3, up to the point where the erosion starts to threaten cliff top properties and the B1161; the principal coastal road. It is uncertain when this will be reached, but current knowledge indicates this will be at approximately year 2055. From that time on, the intent is to prevent further cliff erosion to sustain the properties and the road. The WECMR project team held the first community engagement with Hunstanton cliff top community in March 2011 to inform them of the project and of work being done to monitor cliff erosion.

The SMP Delivery Action Plan mentioned previously is key to how we implement policy. A principal action from the SMP was for the Environment Agency to undertake a full review of the Hunstanton cliffs to Snettisham sea defences scheme during 2011 to provide more information on coastal processes, develop and appraise available options and investigate likely funding requirements.

Pathfinder Work

The Coastal Pathfinder funds were gained through an application by North Norfolk District Council (NNDC) and its partners, including the Borough Council of King's Lynn and West Norfolk (BCKL&WN). Some £3 million were allocated to NNDC and its partners to spend on coastal projects. Of this amount, BCKL&WN were given £20,000 to investigate the understanding of flood risk and to investigate alternative funding mechanisms for the flood defence frontage of PDZ 2. The BCKL&WN invested an additional £5,000 into the study. This work initiated from the SMP2 identifying the low economic viability of maintaining the flood defence function of the shingle ridge and embankments into the future.

The project purpose was to be stakeholder led and for the stakeholders to draw their own conclusions of flood risk, how change should be implemented and identify a willingness to pay. The Pathfinder, with support through attendance by the BCKL&WN and the Environment Agency, ran a number of workshops and drop in sessions to understand the local perspective. Stakeholders initially invited were members of the SMP KSSG and as the project gained momentum, additional members of the local community were informed of the sessions, principally on an invite only basis. A drop in session was held in the village of Heacham for the wider community, advertised through posters and leaflets in local shops and community centres. This was to attract the wider audience and give them the opportunity to contribute towards the Pathfinder. However, due to the coldest winter in a number of years and an aging population, community members were deterred from attending. As a result of this bad weather, further opportunity to comment was given through an article distributed in a local newsletter.

As a result of this process, the stakeholders were able to lead the way in suggesting alternative funding mechanisms. There was discussion over the amount payable by communities, the distribution of the expenditure over the local area and if those in higher risk areas should be paying a premium. The overall conclusions have driven support for local funding initiatives and general acceptance that central government will not be able to support flood defence maintenance for this area into the future. With support from the Borough Council, realistic mechanisms were also identified for putting these funding mechanisms in place within the local financial system. At this stage, funding mechanisms still need to be progressed further.

In turn, this work will feed into the upcoming WECMR. The Pathfinder work was limited on programme and budget and unable to progress ideas beyond a local consensus. Through attendance at the Pathfinder workshops, the WECMR team have built relationships and understanding of the perspective of the local community. This ability to empathise with the stakeholders is critically important for the project. The coastal management review has incorporated the findings of the Pathfinder work into the project scope and one of the project objectives focuses on the continuation of the successful Pathfinder work. The project continues to promote and assist the community in developing an understanding of coastal flood risk as well as pursuing a contributions framework. Based on the Defra Funding for Outcomes Consultation document (Jan to March 2011), this frontage would only receive between 8 and 12 % of monies needed to continue maintaining flood defences (RPA Limited, Coastal Pathfinder, 2011). The existing funding for maintenance ceases to exist in 2012 and funding mechanisms need to be sought in the short term as well for long term management.

Coastal Management Review

The existing coastal strategy for the frontage was completed in 2001; this identified the continued need for annual beach recycling at an annual average cost of £60,000 (2011 recycling cost £150,000) and the ten-yearly implementation of beach nourishment, with the last nourishment occurring in 2006. The 2001 strategy also identified the need for new hard defences at South Hunstanton, Heacham and Snettisham; these were implemented in 2001 following a severe tidal surge event.

The flood risk frontage is protected to a one in fifty year standard of protection (2% annual probability of flooding in any one year), although this is now thought locally to be higher as beach recycling, nourishment and defences have allowed the shingle ridge that runs along much of the frontage to develop and strengthen. This shingle ridge is backed by an embankment providing a residual flood defence. There are several properties and businesses located within these two flood defences that are vital to the sustainability of tourism in the local area. The caravan park businesses are a significant part of the tourism trade and significant companies such as Haven Holidays have sites along the frontage.

Located on the cliffs are the lighthouse dating from circa 1830; coastguard station and cottages from circa 1910 and the ruins of St Edmunds Chapel circa 1272 highlighting the importance of heritage. There are concerns over cliff erosion; currently recording between 0.37 and 0.5 metres of erosion annually. No active intervention is the long term policy for the cliffs and there is no existing strategy here. See Figure 1 (a) and (b) for photographic reference.

Figure 2: (a) Hunstanton Cliffs and (b) Heacham South Shingle Ridge.

The coastal strategy is due for a review. The second round of the SMPs for The Wash identified the flood risk frontage from South Hunstanton to Wolferton Creek as an area of low economic viability coastal defence schemes. This project acknowledges the need to manage the standard of protection along the frontage to one that is acceptable to local communities to support the local economy in the future.

There are difficulties to be faced with increasing pressures on funding and affordability of coastal flood protection schemes. As a result of the SMP and through Pathfinder work, the Environment Agency and the BCKL&WN have continually worked with the local community to develop an understanding of flood risk and to recognize funding for further flood defence works will be challenging without external funding initiatives.

Project Process and Development

The project team is working with the KSSG to develop the plan. The team is going to look at a range of different scenarios over the next one hundred years in relation to climate change and social vulnerability. The team will work closely with the BCKL&WN and local community to develop the range of options that will meet planning objectives for the future for both the cliff erosion to the north and flood risk to the south.

There is such enthusiasm and drive amongst the stakeholders for this frontage. The WECMR team initiated a closer knit project team formed as an Advisory Group with the aim to represent key interest groups and statutory consultees. The SMP KSSG Group is represented by approximately twenty members and was deemed too large for the successful development of review. Members were invited to represent interest groups as part of the Advisory Group. A diagrammatic form of how these groups have developed is included in Figure 3.

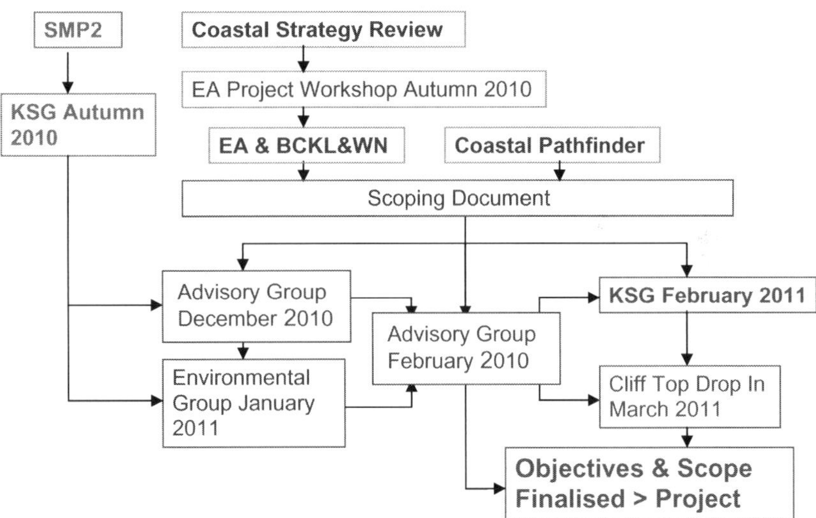

Figure 3: Stakeholder Engagement Development

The purpose of this group is to advise and drive the project forward, with assistance from the Environment Agency and the BCKL&WN. Up front work with the Advisory Group has helped define the project objectives and develop the project scope.

An initial meeting was held in December 2010 with three members from the SMP Group. From this meeting it was clear that the team needed a wider representation for successful development of the project objectives and scope. The response to this was a meeting with members from the environmental community; Natural England, English Heritage and the Royal Society for the Protection of Birds (RSPB). These groups offer great enthusiasm and are optimistic with the early engagement and opportunity to influence the shape of the project.

Further to this, additional stakeholders were identified from the community that would be instrumental to the development of the project. With these additions, the team felt ready to approach the group as a whole, a meeting with the purpose of introducing members to one another and giving them the opportunity to understand the perspectives of all the respective interest groups.

As part of the project development, the team, with the assistance of Royal Haskoning, developed a scoping document highlighting all the issues and constraints associated with the frontages. (It is important to note that this review incorporates Hunstanton town and cliffs as well as the frontage from South Hunstanton to Wolferton Creek). The document identified using a decision-pathway approach to deal with numerous uncertainties such as sea level rise and erosion rates. These triggers will be developed further as part of the project. The stakeholders were engaged throughout the development of this scoping document and helped the team to further understand the issues as perceived by the communities. This understanding and communication has enabled the project team and stakeholders to jointly develop the project objectives. In turn, the objectives outline the project scope.

Challenges Faced / Lessons Learnt

There has been much enthusiasm in developing these stakeholder groups and willingness to take part in all events originating with the SMP and now with the coastal management review. However, one of the principle challenges that the projects have faced and will continue to face into the future is getting wider community engagement to open events. Those currently involved represent their communities and organisations and are entrusted to report back to their groups as the project progresses, but the team also needs to find a way to effectively engage with the wider groups, on an informal basis.

To initiate this, a drop in session was arranged for the community on Hunstanton cliff top in March 2011. Previously there had been limited engagement here although there is representation from the town council on the Advisory Group. In this session the project team presented the project development to a wider audience and gave the opportunity for them to contribute towards the development of the project objectives. The session presented how the project developed from the SMP and to demonstrate how outputs from the Pathfinder study in the flood risk area could be applied to the cliff top frontage, incorporating the frontage wide need for external funding support to effectively manage flood and coastal erosion risk.

As a result of the low number of responses received from the Pathfinder work, the Environment Agency and BCKL&WN have prepared a project with further informative wider stakeholder engagement to understand more perspectives of the frontage community. There will be additional challenges in stakeholder engagement as the WECMR moves on and it is important that the team has the time and the resources in order to engage with the community as far as possible. This will promote ownership of the scheme outcomes.

An additional challenge faced is the willingness by stakeholders to contribute without understanding the underlying processes. A number of stakeholders, early in the process and during the SMP phase, indicated that they would provide the £800,000 (estimated based on SMP economics) per annum required to maintain the flood defences between South Hunstanton and Wolferton Creek (PDZ 2). They were willing to pay if the defences were held. There is a concern that through accepting third party funding there may be increased pressure on the Local Authority and the Environment Agency to allow development in a flood risk area and to Hold the Line where is may not be appropriate or acceptable on other grounds. This expectation needs to be carefully managed as the project moves on.

It has been necessary to inform these stakeholders that money cannot simply be taken from them without further work and engagement. Through the development of the scoping document and discussions, the complexity of the option development for the frontage has since become clearer. The project team have also introduced the concept of adaptation should funding not be made available or the scale of the defences required are deemed unacceptable. Through time working with the stakeholders, it has been made clearer that this may not be sustainable into the future and we cannot discount adaptation, similarly we cannot discount wider management options. The project will develop a process of identifying the benefits of the defences to the wider community. This element of the study has been introduced by the Pathfinder study and needs to be expanded upon for the project development. The stakeholder groups now understand the importance of their engagement in the project development, the importance of ownership of the project and what will be decided into the future.

Working in a partnership between the Environment Agency and the BCKL&WN has allowed the project team to present and develop a fully stakeholder driven project plan. Further work will be required to integrate BCKL&WN planning controls and development framework in looking at options for coastal management in the future. This demonstrates that a partnership approach to coastal management is vital to fully integrate the technical, environmental, economic and social elements of a strategic review.

Conclusions

The WECMR recognises that there have been several studies in the past and this one needs to be different with focus on engagement and funding opportunities integrated with coastal erosion and flood risk management. There is recognition that we are not able to continue in doing the same thing and the output must be developed alongside the wider community, as recognised as part of the EU Floods Directive 2007. The project is focused on the integration of local communities with flood and coastal erosion risk management.

Due to care and diligence in the stakeholder engagement process and time devoted by the project team, the stakeholders have been thoroughly involved from the SMP through to the coastal management review. The Advisory Group has worked closely with the team to feed into the high level scoping document, noting where issues and problems need to be covered by the project. This has subsequently led to the setting of the project objectives and highlighting what elements of the project are important to the wider community, the Environment Agency and BCKL&WN.

The engagement work highlights the community willingness to support the coastal flood and erosion management and how the team has been able to support the local community in their aspirations for the future. Through demonstrating the wish to hold the line, at least into the short term, the stakeholders are showing accountability to hold the line. There is an understanding that the future presents increased risk of inundation and risk to life but there is a feeling that existing flood warning systems are adequate enough to reduce this risk. With a willingness to contribute towards flood and erosion defences, the stakeholders are demonstrating their commitment to provide protection for their businesses, homes and lives.

This project development and work initiating through the SMP illustrates a new and innovative way forward meeting wider legislative requirements. Many stakeholders are accustomed to being informed by the authorities but this project gives them the option to influence how the future on the coast will be shaped. This gives ownership and accountability previously unavailable and is dependent on a strong relationship with stakeholders.

This coastal management review developed through the SMP and the Coastal Pathfinder is expected to be an example of national best practice in the near future.

The project is expected to commence in autumn 2011 when a further update will be available.

Innovative Coastal Zone Management
ISBN 978-0-7277-5749-4

ICE Publishing: All rights reserved
doi: 10.1680/iczm.57494.463

SEATON CAREW COASTAL STRATEGY

John Pos URS Scott Wilson, UK.

Nick Clarke URS Scott Wilson, UK.

David Dales URS Scott Wilson, UK.

Dennis Hancock Hartlepool Borough Council, UK.

Mike Pearson Hartlepool Borough Council, UK.

Introduction

Seaton Carew is situated within the borough of Hartlepool on the North East coast of the UK. URS Scott Wilson was commissioned by Hartlepool Borough Council to undertake the Seaton Carew Coastal Strategy. The study frontage extends from Newburn Bridge in the north to the Tees Estuary in the south, a length of some 10 km. The frontage includes significant residential development as well as major infrastructure assets including Hartlepool Nuclear Power Station. The study objective was to produce a sustainable coastal defence strategy for the frontage for the next 100 years and a robust business case for the implementation of £29M of works needed over this period. A twin-track approach is described in which urgent works were prioritised to prevent catastrophic failure, whilst negotiations with beneficiaries to secure contributions and funding to ensure the future of strategic structures continued after the strategy was presented at National Review Group.

The study frontage comprises the SMP2 Management Areas 12.2, 13.1, 13.2, 13.3 and 13.4 with 13.1 further subdivided as shown in Figure 1. During the study it was felt that the frontage could be better managed by dividing it into the developed, sea wall fronted Northern Management Unit and the more environmentally sensitive Southern Management Unit. The key features and economic and environmental assets for the frontage are also shown in Figure 1. The coastal strategy was produced in the following three stages:

Stage A – Condition and Performance Assessment

Stage A of the study assessed the condition and performance of the existing coastal defences for the study frontage over the next 100 years. The key components of Stage A were as follows: Engagement of Key Stakeholders and Consultation; Existing Defence and Policy Review; Visual Inspection of Defences; Coastal Processes and Geomorphology Desk Study; Topographical Surveys; Aerial Photography; Ground Investigations; Structural Surveys; Numerical Modelling of the Key Coastal Processes; Performance Assessment of the Structural Elements and Baseline Assessment of the Coastline.

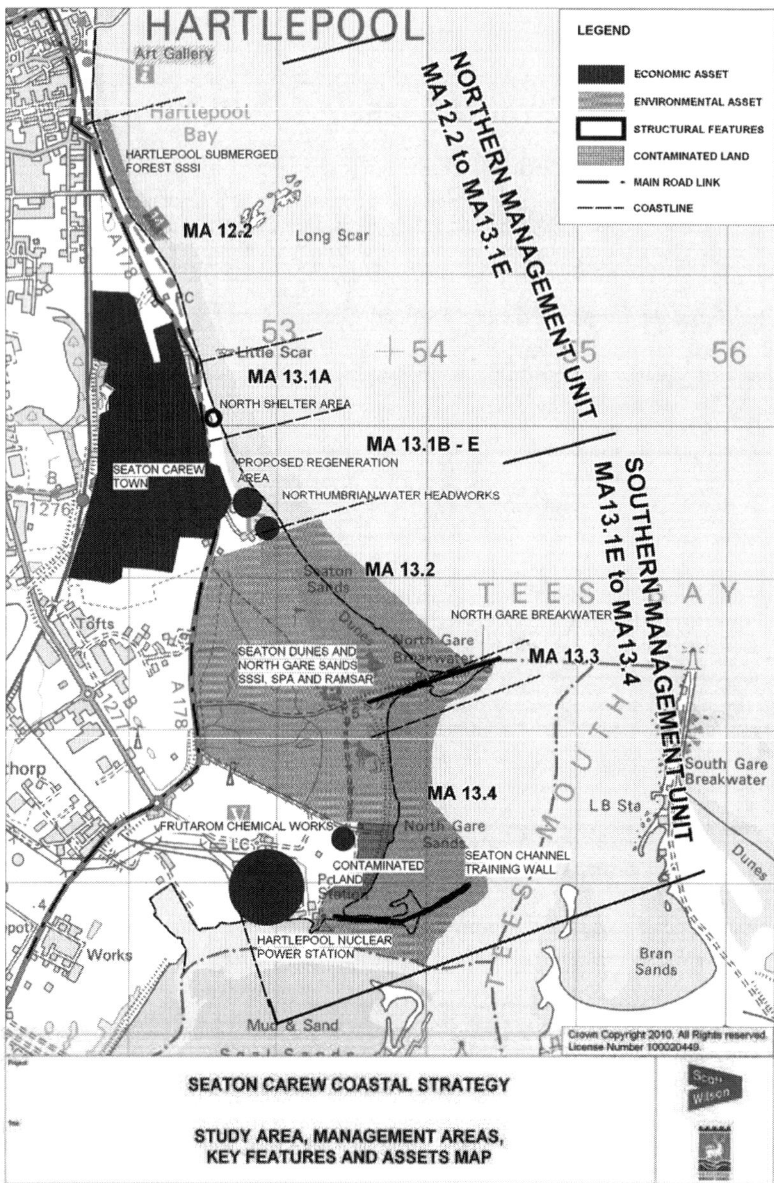

Figure 1. Study area Management Areas, key features and assets.

Stage B – Technical and Environmental Assessment

Stage B of the study developed and appraised the management options for the coastal defences along the study frontage. This included an assessment of the SMP2 management policy, an economic analysis of the preferred options and a strategic environmental analysis. The key components of the Stage B investigations were as follows: Development of Strategic Objectives; Baseline Scenario Analysis; Standard Coastal Defence Option Review; Option Appraisal; Environmental Appraisal (reported fully in the Strategic Environmental Assessment Report); Economic Appraisal and Recommendations for Strategic Management.

Stage C – Strategy Appraisal Report

Stage C culminated in Scott Wilson's recommendation for a preferred long-term strategy to be adopted, in which short-term priorities were highlighted. Stage C was summarised in a Strategic Appraisal Report (StAR) which sets out the business case for the short term (and emergency), medium and long term coastal works.

Northern Management Unit - Outline of the Problem

The most urgent works along this frontage were identified, while negotiations were opened with potential contributing beneficiaries.

Seaton Carew North MA12.2

These rock revetment defences are currently in good condition with a residual life of approximately 30 years without significant maintenance. They currently provide a good standard of protection against overtopping. Although projected sea level rise will reduce the protection provided, the defences are estimated to maintain a good level of overtopping protection over the Strategy duration.

Seaton Carew Town North MA13.1A

The defences along this section were built circa 1938, but are generally in good condition and comprise vertical concrete walls with a recurved parapet. Whilst the walls themselves are in good condition, reduced beach levels have resulted in undermining of the defence toe and loss of fill material from behind the wall. Therefore, the walls have been assessed as having a residual life of approximately 5 years due to the undermined foundations which leave the walls vulnerable to catastrophic failure during significant storm events. Temporary emergency toe protection has been provided to prevent further undermining. However, this does not extend the full length of the wall at risk and is inadequate in its volume.

Within the southern section of this defence length there is a low level area, known as the 'North Shelter', with a significantly lower crest level which is particularly vulnerable to overtopping and breaching. Damage events have occurred at this section in 2006 and 2007. The 2007 event resulted in a near fatal accident when a member of the public fell into a void that opened up in the promenade behind the defence. HBC has since closed this area to public access due to the residual risk of further failures. This incident highlights the urgent need for works to prevent further undermining, wash out of fill material, and failure of the promenade behind the existing walls. Urgent works are required for the 13.1A frontage to prevent undermining and catastrophic failure of the defences.

Seaton Carew Town South MA13.1B-E

Defences along this section were constructed of masonry walls during 1916-1938 with the later addition of mass concrete walls along the promenade. These defences are currently protected by a wide sandy beach but are becoming increasingly exposed to waves.

In places the defences are in a state of disrepair with spalling concrete exposing the reinforcement. Intrusive surveys indicated that the quality of concrete is poor and of low strength. These defences are assessed as having a residual life of 5 to 15 years. They currently rely on the presence of the wide beach, which is essential to reduce the size and frequency of waves reaching the wall. Predicted future sea level rise will reduce beach levels and crest width in front of the defences and will expose the walls to direct wave attack for which they were not designed, resulting in catastrophic failure.

The Northumbrian Water Pumping Station, situated immediately behind the defences, remains vulnerable to inundation.

Southern Management Unit - Outline of the Problem

Two port structures protect this frontage which is of particular environmental interest. Negotiations were opened with the potential contributing beneficiaries.

Seaton Dunes MA13.2

The Seaton Dunes are SPA and Ramsar designated and provide natural flood and erosion protection to MA13.2. The dune system is in a healthy condition, naturally stabilised by Marram Grass, formed on sand which accreted after the construction of the North Gare Breakwater in 1882.

Sediment transport modelling has indicated erosion of the beach below the low water mark resulting in a steepening nearshore profile. This could lead to future erosion of the dune field which would be exacerbated if sediment supply from the north is reduced or prevented. Major erosion and loss of the dunes would result if the North Gare Breakwater was lost.

North Gare Breakwater MA13.3

The North Gare Breakwater provides strategic control to a significant length of the Strategy frontage from MA13.3 up to the southern areas of MA13.1 supporting the beach and Seaton Dunes. Constructed from 1882 to 1892 from a slag waste core and mass concrete, the structure extends approximately 1km. Currently the breakwater is in poor condition and is expected to fail completely within 10 years. A number of failures in recent years have been repaired with reactive maintenance but the structure is at significant risk of catastrophic failure during a storm. Failure would result in substantial loss of beach along MA13.1 and MA13.2. The North Gare Breakwater supports the Seaton Dunes and protects the North Gare Sands, both environmentally designated areas. Contributions from the key beneficiaries will be needed to fund the refurbishment of the breakwater.

North Gare Sands MA13.4

This SPA and Ramsar designated coastline consists of dunes and beach supported by the Seaton Channel Training Wall to the South of the Management Area and protected to the north by the North Gare Breakwater.

The training wall prevents sand migrating into the Seaton Channel and Tees Estuary. Historic accretion of sand has caused the building of a dune system between the North Gare Breakwater and the Training Wall. However, the condition of the Seaton Channel Training Wall is such that it is no longer effective at retaining sediment. Ongoing sand extraction for construction also poses a risk to long-term beach levels. Contributions from the key beneficiaries will be needed to fund the refurbishment of the training wall.

The dunes provide protection to the large reclaimed hinterland which includes Hartlepool Nuclear Power Station and the redundant Leathers Chemical site which contains significantly contaminated land. Failure of the Seaton Channel Training Wall would result in increased risk that the contaminated land would be exposed and leach into the Tees Estuary.

Consequences of Doing Nothing

The "Do Nothing" scenario assumes there is no expenditure on maintaining or improving defences and as a result defences along the frontage would fail depending on their residual life.

Along the Seaton Carew frontage failure of a defence section would lead to commencement of erosion of the hinterland behind the defences and eventual loss of assets. The defences along the 13.1A frontage are the most vulnerable to catastrophic failure due to the undercutting of their toe. Erosion lines were calculated for the entire Strategy period of 100 years. The households lost under "Do Nothing" are summarised in Table 1.

Table 1. Households Lost under "Do Nothing"

Epoch	Households Lost	
	Northern Management Unit (MA12.2 and MA13.1)	Southern Management Unit (MA13.2 / MA13.3 / MA13.4)
0-20	0	0
21-50	153	0
51-100	333	0
Total	486	0

Within the Strategy there are a range of key assets, services and other damages that would occur as a consequence of the "Do Nothing" scenario and these are summarised in Table 2.

Hartlepool Nuclear Power Station has a minimum estimated replacement cost of £2 billion. The inclusion of such losses would significantly skew the economic analysis and result in all options being economically viable. Due to the national importance of the Nuclear Power Station for power production, it would not be allowed to be lost to erosion. Therefore, a substitute loss for providing new intakes/outfalls and coastal defences was included in the economic analysis as the damage value at a notional level of £25 million.

Table 2: Key Damages under "Do Nothing" Scenario

Management Unit	Damage	Value
Northern	Residential and Commercial Assets	£92 million
	General Infrastructure / Services	£17 million
	Northumbrian Water Headworks – Including supporting Infrastructure and Outfall	£40 million
	Recreation and Tourism	£157 million
	Major Health and Safety Risks	Not Costed
	Loss of Coronation Drive - critical transport link.	£2 million
Southern	Loss of Functionality of Teesport Entrance Channel and supported Teeside Industrial Area	£100 million (conservative estimate)
	Loss of Operation / Protection to Hartlepool Power Station	£2 Billion – (Substitute value of providing additional defences £25 million)
	Exposure and Disposal of Contaminated Land	Not Costed
	Loss of Chemical Works	£12 million
	Loss of Environmentally Designated Areas including SPA and Ramsar	Not Costed
	Loss of Golf Course	Not Costed

The area of contaminated land behind the North Gare Sands dunes would be exposed if the dunes were subject to erosion. This issue was considered in the strategic management of the coastal defences along this frontage.

Table 3 provides a summary of the total monetised Present Value losses accrued under the "Do Nothing" scenario. These are the losses that have been used in the economic analysis.

Table 3: Summary of "Do Nothing" Present Value (PV) Losses

Sector	Estimated Present Value Losses (£ millions)	
	Northern Management Unit (MA12.2 and MA13.1)	Southern Management Unit (MA13.2 / MA13.3 / MA13.4)
Residential/Commercial	18.59	0
Industrial	0	82.12
Services	2.41	0
Infrastructure	29.17	0
Recreational	44.35	0
Total	94.52	82.12

Selecting the Preferred Option

Northern Management Unit

Options developed for the Northern Management Unit focused on suitable solutions to implement a "Hold The Line" policy following a review of appropriate policy options. The development of options undertook an early review to reject those unsuitable for use along the frontage due to technical and environmental reasons.

Due to the varying type and condition of the existing defences, the implementation of the short-listed options was tested on individual Management Areas within the Management Unit. Table 4 provides a summary of the options indicating the benefit-cost ratio of the individual schemes, and the key reason for selection as the preferred option.

MA12.2; due to the condition of defences along this section, options other than the "Do Minimum" provided no additional benefit and therefore no further options were investigated. "Do Minimum" maintenance also has the least environmental impact of any potential options. The economics of the option was tested to ensure that the "Do Minimum" remained a sustainable and economic solution over the Strategy period.

MA13.1A; two options were developed to "Hold The Line": Full Height Revetment and Toe Protection (in the form of a low crested revetment). Both options would achieve the same protection to assets as they would prevent failure of the seawall and thus protect against erosion, the key process along this frontage. This frontage required urgent works as the defences were at risk of catastrophic failure.

The benefit of the Full Height Revetment would be significantly reduced overtopping rates. However, overtopping is not a risk to assets but to pedestrian safety along the promenade; the provision of Toe Protection combined with management of access to the promenade area during storms offers a more cost efficient option.

The provision of Toe Protection has a lesser impact on the environment due to its significantly smaller footprint. Toe Protection is also adaptive and could be upgraded to provide a higher level of protection in the future should sea level rise be higher than current estimates. Toe Protection also has a higher benefit-cost ratio due to the reduced structure costs whilst achieving the same protection of assets.

Both options would be combined with the raising of the low area at the "North Shelter" which has previously breached. At this area, the existing seawall would be raised to the level of the adjacent seawall and then protected by either the Full Height Rock Revetment or the Toe Protection. This would address the significant health and safety risks currently posed to pedestrians using this area.

On this basis the Toe Protection option is selected for MA13.1A – this option also includes the raising of the low "North Shelter" area of defence.

MA13.1B-E; two options were developed for this section of frontage to implement the "Hold The Line" policy. A Full Height Revetment or a new Seawall with phased Toe Protection (in the form of a low crested revetment). Both options would achieve the same protection in terms of protection to assets and would also achieve similar overtopping performance.

On the basis that both options provide similar standards of protection and performance and that the benefit-cost ratios are also similar, it was concluded that both options should be carried forward to be investigated in further detail at the PAR stage. However, implementation of either option would be suitable to achieve the objectives of the Strategy.

Table 4: Northern Management Unit - Option Benefit-Cost Assessment

	Benefit-Cost Ratio	Reason for Selection	Selected as Preferred
MA12.2 – Maintenance	7.7	Most economic and least environmental impact to provide continued defence	YES
MA13.1A – Full Height Revetment	7.2	N/A	REJECTED
MA13.1A – Toe Protection	13.0	Most economic and least environmental impact to provide continued defence	YES
MA13.1B-E – Revetment	11.9	To be investigated further at PAR stage	YES *
MA13.1B-E – Seawall	13.7	To be investigated further at PAR stage	YES *
Northern Management Unit TOTAL	12.6	Combined Options	PREFERRED

* Preferred Option to be chosen at PAR stage.

Southern Management Unit

MA13.2; is to be managed under a "No Active Intervention" policy and therefore no options required assessing.

MA13.3; the North Gare Breakwater would be reinstated using the remaining existing structure as a core and providing a layer of protective armour. This solution has been proposed to PD Teesport who own the North Gare Breakwater and has received their support. The North Gare Breakwater is critical to maintain beach widths within the Northern Management Unit.

MA13.4; the Seaton Channel Training Wall would be reinstated using rock or armour units to ensure the structure remains effective at retaining sand. This solution has been proposed to PD Teesport who own the Seaton Channel Training Wall and has received their support.

In order to ensure the continued protection of the SPA and Ramsar sites within the Management Unit, the options for the two structures should be carried out in combination and therefore an assessment of the combined benefit-cost ratio of the Management Scheme has been included in Table 5.

Table 5: Southern Management Unit - Option Benefit-Cost Assessment

	Benefit-Cost Ratio	Reason for Selection	Selected as Preferred
MA13.3 – Reinstate Control Structure	6.5	To protect assets and meet environmental objectives	YES
MA13.4 - Reinstate Control Structure	10.1	To protect assets and meet environmental objectives	YES
Southern Management Unit TOTAL	7.8	Combined Options	PREFERRED

Contributions and Funding

Funding for the Strategy implementation would be sought through the FDGiA process as well as through contributions from the identified beneficiaries. Contributions are critical for the funding of the Southern Management Unit defences and less so for the Northern Management Unit defences.

Contributions are being pursued for the southern section of the Northern Management Unit where there is potential for contributions from Northumbrian Water and regeneration/developers. Initial discussions have taken place between Hartlepool Borough Council and Northumbrian Water. Hartlepool Borough Council has also developed a draft regeneration plan for the development of this frontage. For the strategy, contributions for these elements have been estimated at £1 Million. Formal agreement to funding from third party contributors will be sought during the undertaking of PARs for each section of works in accordance with the Strategy implementation programme. Approval of the PARs for works within the Northern Management Unit will be dependant upon a sufficient level of contributions being received from third parties.

Contributions are being pursued for the two control structures in the Southern Management Unit (North Gare Breakwater and Seaton Channel Training Wall) as these structures are owned by PD Teesport. The structures also provide protection and commercial benefit to PD Teesport, Hartlepool Nuclear Power Station and Frutarom Chemical Works, all of whom have been approached during the development of the Strategy for contributions. Formal commitment from beneficiaries to providing a contribution to the implementation of the Southern Management Area of the Strategy will be sought upon approval of the Strategy. Negotiations with those beneficiaries identified (and any beneficiaries subsequently identified) will take place once the Strategy is approved and prior to the completion of the PAR stages, for each element of works, to confirm total contributions. Contributions are currently estimated at £6.5Million. Approval of the PARs for works within the Southern Management Unit will be dependant upon a sufficient level of contributions being received from third parties.

Summary of Preferred Strategy

Total cash expenditure for the whole strategy is £29.11 Million; this figure includes both capital and non-capital costs as shown in Table 6 below.

Table 6: Summary of Preferred Strategy

	Northern Management Unit	Southern Management Unit	Total
PV Costs (£k)			
Capital	7,240	10,070	17,310
Non-capital	290	440	730
Total PV Costs (£k)	7,530	10,510	18,040
PV Benefits (£k)	94,520	82,120	176,640
Average Benefit/Cost Ratio	12.6	7.8	
Cash Costs (£k)			
Capital	12,504	14,752	27,256
Non-capital	739	1,100	1,849
Total Cash Costs (£k)	13,243	15,862	29,105

Conclusions

The paper describes how the study involved a twin-track approach; urgent works were prioritised to prevent catastrophic failure, whilst negotiations with beneficiaries to secure contributions and funding to ensure the future of strategic structures continued after the strategy was presented at NRG.

The PAR for the most urgent works along the MA13.1A frontage has now been completed and approved. The detailed design has been completed and construction has started.

Progress on the implementation of urgent works and an update on the negotiations with beneficiaries to secure contributions and funding will be reported at the conference.

Innovative Coastal Zone Management
ISBN 978-0-7277-5749-4

ICE Publishing: All rights reserved
doi: 10.1680/iczm.57494.473

Beach Nourishment at the Inner Danish Waters

Christian Helledie, COWI, Parallelvej 2, 2800 Lyngby, Denmark, cel@COWI.com

Introduction

The coasts at the inner Danish waters have been developed over the last century with summer houses and towns, as early planning regulation allowed development close to the sea. The coast is naturally eroding many places, which has led to an increasing demand for coastal protection of private property. The first coastal protection schemes were often uncoordinated and focused on structures such as groynes and revetments, which to a large extent fixed the coastline. Today many beaches at the inner Danish waters are dominated by un-aesthetic and ineffective old shore protection structures. The widespread stabilization of the coast by revetments has reduced natural supply of sand to the beaches and thus extensive beach recession has followed. The eroding beaches provide less protection of the hinterland and access along the coast is often difficult, which reduces the recreational value.

COWI has undertaken a series of successful shore rehabilitation projects at the inner Danish waters focusing on beach nourishment and aiming at providing aesthetic solutions with high recreational value and effective protection, see **Figure 1**. The paper outlines background, objective, financing and technical performance of the projects.

Figure 1 Beach nourishment projects at the inner Danish waters designed by COWI

Beach Nourishment at Funen (1997-2011)
The Site

The site at Funen is situated at the north coast where 2500 summer houses and farms are found along the 12 km low-lying coast, see **Figure 2**. The coastline is convex and the longshore net sediment transport increases from approx. 8,000m³/year at Egebjerggaard to approx. 20,000m³/year at Kristiansminde and approx. 26,000m³/year at Erikshaab resulting in a deficit in the littoral budget estimated to 18,000m³/year. The increasing longshore sediment transport towards east is assessed to be the main reason for the historic beach recession found along the western part of the coast. The erosion problem was smaller east of Kristiansminde

473

due to the input of sand from west and more efficient groynes. Generally, the hinterland is protected from flooding by levees and revetments and the beach is protected from erosion by an old groyne field. The groynes were built with an average spacing of 75m and extend 30m out from the revetments. At some exposed locations breakwaters have been constructed between the groynes. The ineffective and old shore protection structures could not prevent beach recession and storm damage on levees and private plots and the recreational value of the beach was reduced.

Figure 2 Site at the north coast of Funen

Project Objective

In 1998 the local coastal protection association (Det nordfynske kystsikrings-, dige og pumpelag) commissioned COWI to analyse the existing situation and on this basis develop a long term shoreline management plan for the coast. The objective of the project was to re-establish attractive sandy beaches and increase shore and flood protection to reduce risk of damage to private property and at the same time improve the recreational value of the coast.

Beach nourishment was found to be the most feasible and attractive solution considering the available funding. Beach nourishment is the only measure which can re-establish sandy beaches and help mitigating deficits in the sediment budget and hereby eliminate shoreline recession. Due to limited funding, it was decided not to focus on hard coastal protection structures as they would not re-establish sandy beaches. The existing shore protection structures were generally preserved. Levees and revetments were strengthened at selected locations to increase flood protection. The new beaches are designed to be mobile, but new beach breakwaters have been introduced at a few exposed headlands to stabilise the beach.

The beaches are re-nourished at 4 years interval to maintain sufficient beach width and minimise unit cost of sand. The maintenance nourishment is designed to compensate the deficit in the littoral budget estimated to 18,000m^3/year. Due to eastward littoral net drift maintenance nourishments are made along the western beaches. The sand is dispersed along the coast by the littoral drift, thus securing long term input of sand at the eastern beaches.

Financing

The beach nourishment scheme is financed and maintained by the local coastal protection association. The members are private summer house owners and farmers possessing land along the low-lying coast and at the hinterland. Due to severe flooding risk the landowners have had a tradition for working together fighting flooding hazards.

The Municipality is not playing an active role in maintaining the beach scheme, but handles the collection of contributions from the plot owners, which is paid as a separate real estate tax.

Figure 3 and Table 1 show the cost of development and maintenance of coastal protection over the years and the dominant measures implemented. The data has kindly been provided by Mogens Stougaard from the coastal protection association.

The total cost of development and maintenance of coastal and flood protection along the 11.4km coast is estimated to DKK 39 million between 1995 and 2009. The yearly average cost is estimated to DKK 2.6 million, corresponding to DKK 230,000 year/km coastline. The total number of plot owners is around 2500 each paying DKK 1500 year, which is in line with the yearly cost of maintaining the coastal protection scheme when seen over the years.

The maintenance cost varies significantly from year to year. The initial beach nourishment was made in 1999 including 113,000m^3 sand and followed up in 2001 with 70,000m^3 to re-establish the sandy beaches. Hereafter, the cost of the maintenance nourishments is relatively constant including 70,000m^3 every 4 years. At the east side of some headlands, the applied quantities of beach nourishment have not been sufficient to maintain acceptable beach width. Therefore 11 breakwaters were built in 2004 and another 4 in 2009, see Figure 2.

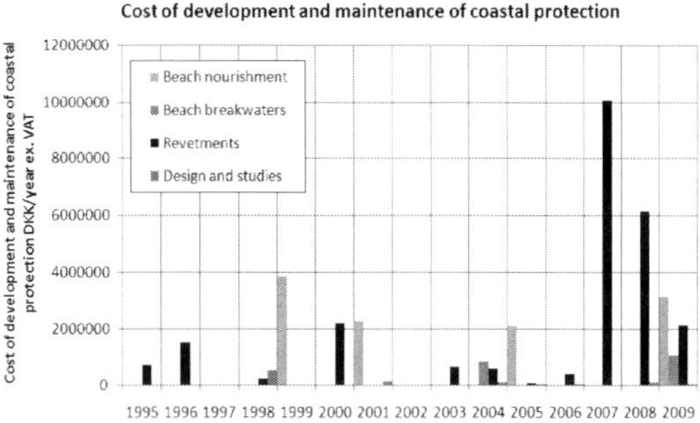

Figure 3 Cost of development and maintenance of coastal protection between 1995-2009

Table 1 Total cost of development and maintenance of coastal protection 1995-2009

Measures	Beach nourishment	Beach breakwaters	Revetments	Design and studies	Total
Cost DKK ex. VAT	11,300,000	1,900,000	24,800,000	1,100,000	39,200,000

In November 2006 the coast was hit by a severe storm with storm surge of up to +1.74m MSL, which is assessed to be a 50 year Return Period event. The storm caused extensive damage to the revetments along the coast. The damages were repaired in 2007, 2008 and 2009 where 90% of the revetments were rebuilt, which increased maintenance cost significantly.

Table 1 shows that the cost of beach nourishment is around half of what has been spent on revetments.

Technical Performance

The shoreline management plan developed by COWI has principally been implemented. However, locations of maintenance nourishment and new breakwaters have been adjusted to optimise the scheme. The extensive rebuilding of revetments was not part of the initial shoreline management plan, but the consequence of the impact of the 50 year Return Period storm in 2006.

The implemented beach nourishment scheme has re-established attractive sandy beaches along the entire coast increasing both recreational value of the beaches and protection of the hinterland, see Figure 4. Figure 5 shows that the beach at the western part of the coast was eroding between 1954 and 1999 due to the convex coastline and the littoral net drift towards east. The beaches towards east were almost stable between 1954 and 1999 due to the more effective groyne field here and the input of sand from the eroding beaches towards west. In 1999 the initial nourishment was made at the western beaches, which significantly increased the average beach width. The severe storm in 2006 caused beach recession along the coast, which can be seen in 2008. However, recently the beach is recovering from the storm as the sand migrates back towards the shore. Additionally, three sections at the western part of the coast were re-nourished in 2009, see Figure 2.

The nourishment has stopped beach recession at the western part of the coast (Groyne 28-112). Positive signs of the beach nourishment can also been found at the eastern part even though nourishments have not been undertaken here (Groyne 112-176).

Groyne 57 Before nourishment 1999 and 2010

Groyne 72 Before nourishment and breakwaters 1999 and 2010

Figure 4 Effect of beach nourishment and coastal protection 1999-2010

Beach nourishment is dynamic and the sand slowly migrates from west towards east thus benefiting the entire coast in the long term. Therefore, beach nourishment can be financed by a large group of land owners along the coast similar to flood protection.

The beach nourishment at Funen must be maintained at intervals to balance the deficit in the littoral budget and avoid beach recession.

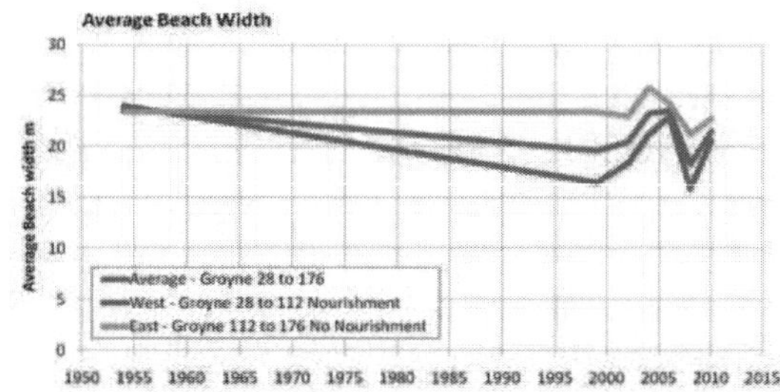

Figure 5 Historic development of beach width

The old groynes have a stabilising effect on the beach as the orientation of the shoreline between the structures generally is oblique. However, the old groyne field is not sufficient to stabilise the beach everywhere. The implemented scheme has not included rehabilitation of the old groyne field. A large part of the shore protection structures are not functioning optimal and reduce the recreational and aesthetic value of the beach. The new breakwaters help stabilising the beach at exposed locations where the old groynes were not sufficient. Beach nourishment is not a stand alone shore protection measure at Funen. Revetments are needed to protect the hinterland against storm impact when houses are located close to the sea.

Rehabilitation of Coastal Protection at Liseleje (1996-2005)
The Site
Liseleje is located at the NW side of Zealand at a long straight coast including a small headland at Hyllingebjerg. The site covers a 2.7km coastline, where high till cliffs have been eroding in the past, see Figure 6. The old Liseleje Breakwater is located at the east end of the

site, which has served to protect boats of the local fishing community over the past century. Towards east a marine foreland is found with dunes and sandy beaches.

Figure 6 Historic development of the beach at Liseleje

The yearly net littoral drift is estimated to 10,000-30,000m^3/year towards NE.
The hinterland at the eroding cliffs has been developed with a large population of summer houses. Cliff erosion has been an increasing problem threatening some villas along the cliff top and the private plot owners independently tried various methods to stabilise the cliffs. As a consequence, the coast in 1996 was dominated by un-aesthetic, un-coordinated and ineffective old coastal protection structures and the beach had disappeared due to erosion, thus reducing the recreational value of the coast, see Figure 6.

Project Objective
In 1996 COWI was commissioned by Frederiksborg County to develop a shoreline management plan for the 2.7km coastline south of the Liseleje Breakwater, see Figure 6. The project was made in corporation with Frederiksvaerk Municipality and the 90 summer house owners along the cliff, who had formed a coastal protection association.
The objective of the project was to provide protection of private property along the cliffs. The shoreline management plan should focus on increasing the width of the beach and access should be provided along the coast. Additionally, the maintenance requirements of the scheme should be minimised.
The shoreline management plan developed by COWI included removal of all existing ineffective shore and coastal protection structures except the Liseleje Breakwater and the two breakwaters towards south. Seven new large, low crested beach breakwaters were constructed at 1.5m water depth to protect the new beaches. The number of breakwaters were minimised to reduce visual impact of the scheme and construction cost. The scheme also included rehabilitation of the revetments along the cliffs. Finally, the sandy beaches were re-established by beach nourishment with a total of 70,000m^3 sand. The shoreline management plan recommended maintenance nourishment at 5-10 years interval.

Financing
The shoreline management plan and the rehabilitation of the shore protection scheme between Hyllingebjerg and Liseleje were financed by Frederiksborg County, Halsnaes Municipality and the private plot owners along the cliff, see Table 2. The new beaches and breakwaters protect the cliffs, but also increase the recreational value of the coast and were therefore partly financed by public funding. The revetments are for stabilising the cliffs only and were therefore financed by the private plot owners along the cliff.

Table 2 Cost of development and maintenance of shore protection at Liseleje 1999-2010. Frits Thaulow, Head of the coastal protection association

Measures	Nourishment and breakwaters DKK ex. VAT	Revetments DKK ex. VAT	Approx. total cost DKK ex. VAT
Initial scheme 1999	~8,800,000	~2,400,000	~11,200,000
Additional breakwater	~1,600,000		~1,600,000
Additional revetments		~1,200,000	~1,200,000
Misc. maintenance	~1,600,000		~1,600,000
Financing	33% Coastal Funding, DCA 33% Municipality 33% Plot owners 1. row	Plot owners 1. row	~15,600,000

The total construction cost of the shore rehabilitation scheme was approx. DKK 11.2 million in 1999 corresponding to DKK 4.1 million per km coastline.
The local coastal protection association is undertaking the maintenance of the scheme at Liseleje, but the Municipality is playing an active role. The total maintenance budget is DKK 420,000 per year split equally between the municipality, the 90 plot owners along the Cliffs and the Coastal Funding administrated by the Danish Coastal Authority, DCA. The contribution of the private plot owners is determined according to the length of the cliff side of the plots equal to DKK 51 per meter, which on average is DKK 1,500 year/plot.
The coastal protection association now focuses on optimising the breakwaters to reduce maintenance cost in the future. Generally, the maintenance budget has been spent on a new breakwater and small adjustments of others. Beach scrapes have been undertaken moving sand from large tombolos to starving sections to optimise the scheme. The yearly maintenance budget is DKK 150,000 km/year, which has been insufficient to maintain the scheme.
In 2005 two additional breakwaters and beach nourishment were constructed north of the Liseleje Breakwater at the popular public beach, which was suffering from leeside erosion. The additional breakwaters were financed by the municipality as they are outside the area administrated by the local coastal protection association.

Technical Performance
The implemented scheme developed by COWI was made with a minimum of breakwaters to reduce visual impact and construction cost. Generally, the pocket beaches between the large breakwaters and the tombolos are relatively stable, which minimises the maintenance requirements. Most of the pocket beaches appear natural and aesthetically attractive with wide sandy beaches as the number of structures is small, see Figure 7.
However, the beaches at Bay 2, 3 and 4 have been eroding and after some years disappeared at the centre of the bays, see Figure 7. Bay 5, 6, 7 and 8 on the other hand have collected more sand over the years, which have increased the width of the beach more than 10 years after the initial nourishment. Dunes have formed at the wide tombolos, which show the efficiency and success of the applied shore protection scheme. From having suffered from

poor recreational value this coast now has some of the most attractive recreational sandy beaches at the north coast of Zealand and the cliff recession has stopped.

Narrow beach at Bay 4 and wide attractive beach and new dunes at Bay 6
Figure 7 Beach and coastal protection at Liseleje, 2010

Large scale maintenance nourishment with imported sand has not been undertaken as initially recommended by COWI. Maintenance nourishment is assessed to be expensive in the long run due to relatively high mobilisation cost of a dredger.

The coastal protection association has chosen to focus on optimisation of the breakwater scheme to preserve the beaches and hereby reduce future loss of sand. However, the effect of the beach scrapes and additional breakwater has not been sufficient to preserve the beaches at the critical sections. Therefore, additional optimisation of the breakwaters is planned.

The shore protection scheme at Liseleje shows that it is possible to re-establish the sandy beaches at the inner Danish waters and to a large extent stabilise the new beaches by use of large beach breakwater even at locations with significant deficit in the littoral budget. It should be stressed that in most cases maintenance nourishment is required even with an efficient breakwater scheme to compensate loss of sand and avoid lee side erosion. Additionally, revetments are required to stabilise the high cliffs against storm erosion.

Large breakwaters are known to form rip currents during periods with rough sea, which has also been observed at Liseleje. Beach users should be made aware of this potential hazard during rough conditions.

Feasibility Study for Beach Nourishment at Zealand (2009-2011)

The Site

The 30km coast in Gribskov Municipality is convex with a series of headlands and bays, see Figure 8. Generally, the headlands are till cliffs. Marine forelands with sand and shingle beaches are found in the bays between. The coast and hinterland is dominated by areas with summerhouses, many places extending to the cliff edge or beach. The littoral net drift is towards NE and E and increases from Tisvilde towards Gilleleje and then declines

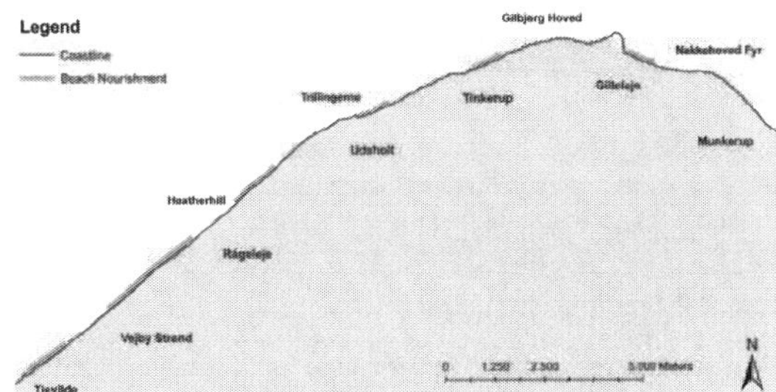

Figure 8 Coast in Gribskov Municipality and nourishment scheme proposed by COWI

east of the harbour. The deficit in the littoral budget is estimated to 40,000-50,000m^3/year between Tisvilde and Gilbjerg Hoved. East of the harbour the deficit is assessed to 15,000-25,000m^3/year, as Gilleleje Harbour blocks most of the littoral drift and sand dredged from the access channel is currently dumped in deep water.

Figure 9 shows the assessment of the historic shoreline recession along the coast in Gribskov

Municipality. Generally, the beaches have receded 10-30m between 1954 and 2009 except at some headlands providing local hard points. The figure also shows that the beaches have accumulated at the west side of Gilleleje Harbour and at Dronningmoelle where the shoreline is concave, thus reducing the littoral drift.

The natural supply of sand from the cliffs at the NW coast

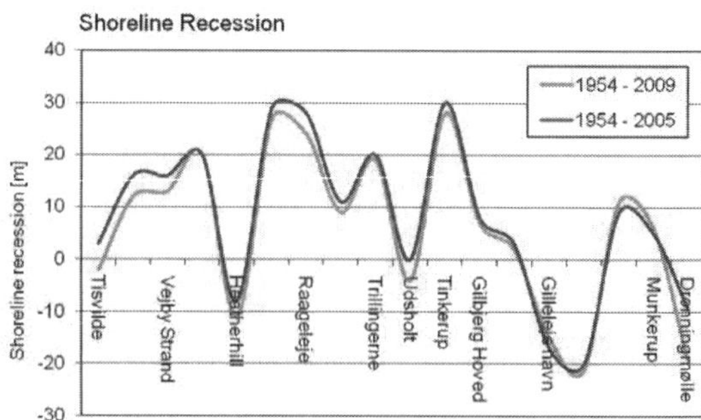

Figure 9 Shoreline recession in Gribskov Municipality

of Zealand has been reduced over the years due to cliff protection by revetments, see Figure 10. Therefore, the sandy beaches have generally receded and disappeared at several locations. The existing protection structures have often been constructed by private plot owners and small coastal protection associations for small sections of coastline. The coastal protection is generally uncoordinated and based on poor design and the structures are of low quality. Furthermore, the structures have generally not been maintained sufficiently probably due to limited funding and lag of coordination. The focus of the plot owners along the cliffs often is to protect the high value properties against erosion typically applying revetments, see Figure 10. The beaches and shore protection structures are secondary. The result is narrow eroding beaches and limited access along the coast thus reducing the recreational value of the coast. Narrow beach in front of revetments furthermore reduces protection of the hinterland and increases risk of storm damage to the properties along the cliffs.

Old coastal protection and narrow beach at Tisvilde and large new revetment at Trillingerne
Figure 10 Existing conditions along the coast in Gribskov Municipality, 2009

Project Objective

Gribskov Municipality has commissioned COWI to prepare a feasibility study for a regionally coordinated beach nourishment scheme for the 30km coastline in Gribskov Municipality. The purpose of the feasibility study is to provide a technical foundation for the political decision process for the nourishment project.

The project has been initiated by private stakeholders headed by Jackob Wandall and Birgit Lund. The initiators, the municipality and other stakeholders have formed a working group with the objective to develop the beach nourishment project and secure regional coordination of interests. The municipality has provided technical and administrative support for the project with Hanne Askholm Larsen as primary contact and political leadership from Mayor Jan Ferdinansen. Ole Juul Jensen and Christian Helledie from COWI and Per Soerensen from DCA have provided technical expertise.

The objective of the nourishment scheme is defined by the working group. The project shall re-establish sandy beaches in Gribskov Municipality and hereby increase protection of the coast and recreational value of the beaches. The beach nourishment should benefit as many citizens as possible and therefore be regionally coordinated within Gribskov Municipality. Ideally, the project should also include the neighbour municipalities Halsnaes and Elsinore, who is part of the regional sediment cell at the north coast of Zealand, see **Figure 1**. Unfortunately, the adjacent municipalities have not found sufficient political support to join the nourishment project at present.

The selected beach nourishment scheme in Gribskov Municipality is seen as a supplement to the existing coastal protection structures that is excluded from the nourishment project. The existing coastal protection structures have been developed and are maintained by private plot owners along the cliffs and primarily focus on protection of the cliffs. However, to obtain the best possible effect of the planned nourishment scheme the local coastal protection associations should clean up their part of the beach and ideally optimise their shore protection structures before the beaches are nourished.

Financing

The selected nourishment scheme is presented in **Figure 8**. The scheme provides a 25m increase in beach width at selected locations, which are concentrated at the SW part of the coast as the need for nourishment is large here. This will also secure the longest possible effect of the nourishments within the municipality, as the littoral net drift disperses the sand towards NE. To obtain an immediate effect along the entire coast suitable sections have also been selected for nourishment along the N and E part of the coast.

The initial beach nourishment is assessed to $625,000m^3$ and the total cost is estimated to DKK 37 million or DKK 1.2 million per km coast. The cost of the initial nourishment is proposed to be split over 3 years to reduce the yearly cost and to optimise the initial nourishment.

The maintenance strategy is to replace the estimated deficit in the littoral budget by beach nourishment at 3 years interval. Hereby, the unit cost of sand can be reduced. The yearly maintenance cost is estimated to DKK 3.2 million or DKK 100,000 km/year. Additionally, sand dredged from the access channel to Gilleleje Harbour is to be placed at the beach east of the harbour on a yearly basis to mitigate part of the deficit in the littoral budget. This is a change in port operation and thus not a part of the nourishment project.

The proposed strategy for the beach nourishment scheme implies that the project can be financed by a large group of citizens, as the project both aims at increasing coastal protection and improves the recreational value of the coast.

A new financing model has been developed by the private landowners headed by Jakob Wandall, see Table 3. The beach nourishment scheme is proposed to be financed by all land owners within a 1km wide zone along the coast through an excise duty. To be consistent with Danish regulations the contribution shall be determined for individual land owners based on utility value here defined as the distance to the beach. The model has been approved by Gribskov Municipality.

The public opinion towards the proposed nourishment scheme has been positive based on experiences from presentations to land owner organisations throughout 2010. However, the financing model has been rejected at a public hearing in June 2011. At present it is undecided if the project will go ahead. Alternatively, the

Table 3 Financing model for beach nourishment DKK ex. VAT

Initail nourishment cost / year / plot owner	Plot owners	DKK excl. VAT
0-200m	2834	1861
200-500m	3079	1241
500-1000m	2378	620
Maintenance cost / year / plot owner		
0-200m	2834	372
200-500m	3079	248
500-1000m	2378	124

project could be financed by all land owners in the municipality through an extra real estate tax. The nourishment project is currently in a political process awaiting the final decision.

Table 4 Comparison of cost of nourishment schemes at the inner Danish waters

	Funen	Liseleje	Gribskov Municipality
Initial cost DKK/km ex. VAT	Included in yearly cost	4,100,000	1,200,000*
Maintenance cost DKK/km/year ex. VAT	230,000	150,000	100,000*

* estimated

Financial Comparison of Nourishment Schemes

Table 4 shows a comparison of the cost of development and maintenance of the beach nourishment schemes at the inner Danish Waters. The coastal protection scheme at Funen has been developed over the years and large initial capital investment has therefore been avoided. Thereby the project economy and the protection scheme have been optimised. The scheme at Liseleje had a large initial cost as hard coastal protection was included initially. Then the coastal protection and the recreational value was optimised form the beginning. However, maintenance is also required for this type of scheme. The cost of the project is relatively large, as the length of the project is small and the coast more exposed. A cost effective scheme can be achieved by focusing on large scale beach nourishment as in Gribskov Municipality. However, this scheme may require additional investment in revetments and breakwaters to protect the hinterland and beach, which is the responsibility of plot owners along the cliffs.

Conclusion and Recommendations

Beach nourishment can re-establish sandy shores at the inner Danish waters and hereby increase protection of the hinterland and create recreational value. However, beach nourishment shall be maintained. Revetments are often required to protect the hinterland against storm impact and beach breakwaters are required at exposed locations to stabilise the sandy beaches. There is a large need for rehabilitation of old ineffective shore protection structures at the inner Danish waters. Small private coastal protection associations are often responsible for maintaining the coastal protection structures, but due to limited funding and lack of coordination shore protection structures are often ineffective and have an un-aesthetic appearance. This has often resulted in beach erosion and thus reduced recreational value of the coast. Accepting the presence of old ineffective shore protection structures may in some cases be necessary in order to implement regional beach nourishment schemes. However, the best option is to coordinate shore rehabilitation projects at regional scale and include the old coastal protection structures as well as large scale beach nourishment initially. Funding is the most critical factor for successful implementation of shore rehabilitation projects. Beach nourishment shall preferably be financed by large group of citizens as it creates recreational value for many as well as increasing protection of the coast. Hereby the costs can be split and thus become acceptable for all. The municipality shall coordinate the regional beach nourishment projects initially, but private coastal protection associations can be responsible for implementation and long term maintenance of large beach nourishment schemes.

Acknowledgement

Thanks to Mogens Stougaard from Funen and Frits Taulow from Liseleje for providing data and background information. Also thanks to Per Sørensen from DCA, Joergen Juhl and Ole Juul Jensen from COWI for reviewing the paper.

SECTION 6:

EFFECTIVE ESTUARINE AND COASTAL ENGINEERING

Innovative Coastal Zone Management
ISBN 978-0-7277-5749-4

ICE Publishing: All rights reserved
doi: 10.1680/iczm.57494.485

Beach Management Plans for Mixed Beaches: Review and Ways Forward

Uwe Dornbusch, Environment Agency, Worthing, UK
Andy Bradbury, New Forest District Council, UK
Bryan Curtis, Worthing Borough Council, UK
Gary Lane, Angmering, UK

Introduction

Over the last two centuries, the predominantly mixed sand and gravel beaches (termed shingle beaches) in Southeast England have become increasingly managed to maintain them in size and position as they often form the only defence against coastal flooding and erosion. Local approaches to management have led to problems along neighbouring frontages leading to current local management practices that take little or no account of those either side or even further along the coast. In most cases management is required to implement the Hold the Line policies of Shoreline Management Plans (SMP2) with the broader management strategy (groyned or open beach) usually defined through more detailed Coastal Defence Strategies. There exists a good understanding of the general beach material movement at all of the frontages, especially as many of these have a long history of management with a basic set of historic annual monitoring data that goes back to the 1970s, and, at a better level of detail and accuracy, since 2003 (CCO 2008).

A review (Dornbusch & Cargo 2011) that extends an earlier assessment (Dornbusch 2008) has been carried out of formal documents (Beach Management Plans) and non-documented practice for >25 management sites covering >130km in Southeast England with both length and number roughly equally divided between Local Authorities and the Environment Agency as the Operating Authority (Figure 1). The distribution of BMP frontages as shown in Figure 1 reflects the distribution of shingle beaches in Southeast England, their role in providing flood and coastal protection and their potential for short term change due to wave and tide conditions. As a consequence the sheltered frontages of the Solent, the Isle of Wight and frontages further to the West do not show many sites.

Compared to sand beaches, volumes required for recycling, bypassing of inlets or maintenance recharge are small for the individual frontage. For the entire frontage shown in Figure 1 >300,000m³ are recycled, >50,000m³ are added through recharge and a significant amount of reprofiling is undertaken; all at a cost in excess of ~ £2.4m on an annual basis.

Beach Management Plans (BMP)

BMPs received a brief mention in the first edition of the Beach Management Manual (Simm et al. 1996). The 2nd Edition of the Beach Management Manual (Rogers et al. 2010) has extended its guidance on BMPs and is devoting an entire chapter to beach management. It contains an exhaustive table of content to be addressed in BMPs as a draft template. While such a comprehensive BMP might be appropriate for frontages with annual management costs of several £100k, for those frontages where only a couple of thousand cubic metres are recycled every few years this seems disproportionate.

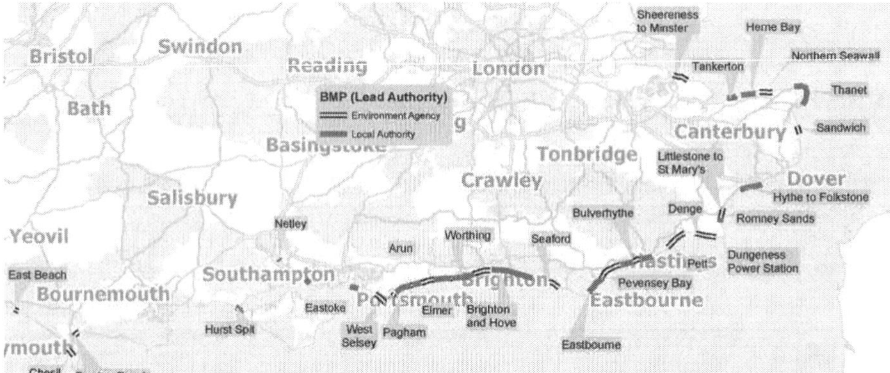

Figure 1 Locations of Beach Management activities in Southeast England.

This paper, based on the analysis of a large number of BMPs and non-documented beach management activities, puts emphasis on the evidence base of BMPs and how this can be assessed and modified in the cycle of design, monitoring and evaluation (Figure 2), which has led to the following definition or vision for BMPs.

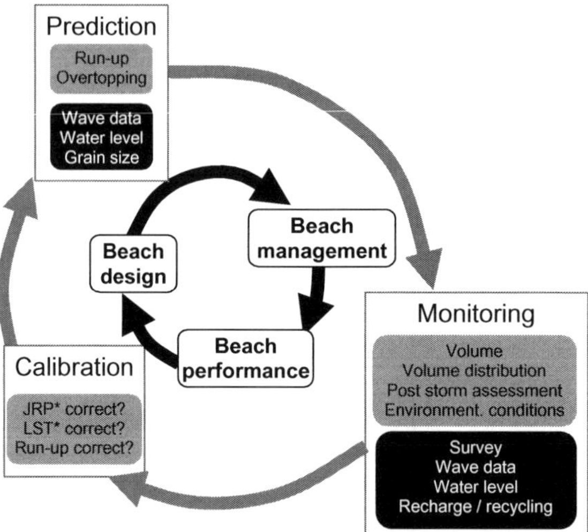

Figure 2 Schematic of the feedback relationship between beach design, management and performance, and prediction, monitoring and calibration of design predictions. Black boxes show examples of primary data that feeds into the process and grey boxes show derived data and data analysis that drives the cycle. * JRP = Joint Return Probability, LST = Longshore Transport.

Beach Management Plans provide an accountable and transparent methodology for managing beaches as coastal defence assets based on risk information that derives from scheme design, monitoring and scientific/research input with the aim of managing the frontage in a sustainable way.

Here, sustainability refers to managing the beach (or the beach plus other defences) as a coastal defence structure in a cost effective way to the right standard of protection to address the risk of flooding or erosion with the least environmental impact so as to enhance vegetated shingle habitat opportunities. While beaches in many locations along the coast also provide a significant amenity, a clear distinction needs to be made between requirements for flood and erosion protection and any additional amenity requirements.

Review Results

The review for the first time collated and summarised any existing beach management documentation together with ad-hoc information at a regional scale allowing for a large scale overview. Table 1 gives a high-level breakdown for the managed frontages, highlighting that the first documents have been produced as early as the mid 1990s; only three of the documents are regularly updated.

Table 1 Breakdown of number or frontages, length of frontages and document year by Operating Authority (* EA = Environment Agency, LA = Local Authority)

			Operators		
			EA*	LA*	Private	Sum
Number of frontages	Ad hoc	6	10	1	17
	Formal Plan	14	7		21
	Sum	**20**	**17**	**1**	**38**
Length of frontages	Ad hoc	17	37		54
	Formal Plan	57	27	2	86
	Sum	**74**	**64**	**0**	**140**
Document production year	1996	1	2		3
	1997	1	1		1
	1999		1		1
	2003	1			1
	2006	2			2
	2007	2			2
	2008	1	1		2
	2009	4			4
	2010	1	2		3
	2011	1			1
	Ad hoc plans or no date available	6	10	1	17
	Sum	**20**	**18**	**1**	**38**

Due to the large variety in format and content of documents an attempt was made to extract information from each document to fit under specific headings including an overview of the risk, the physical parameters driving the design (wave and water level parameters, grain size, longshore transport), design parameters (e.g. crest width & height, slope), trigger levels for intervention, regular and emergency management activities, the recognition of vegetated shingle habitats and costs.

Where possible, information was also collected for the design parameters and for parameters as used in practice. Most of this information was incorporated into a geo-database allowing for the spatial interrogation of the data (an example is shown in Figure 3).

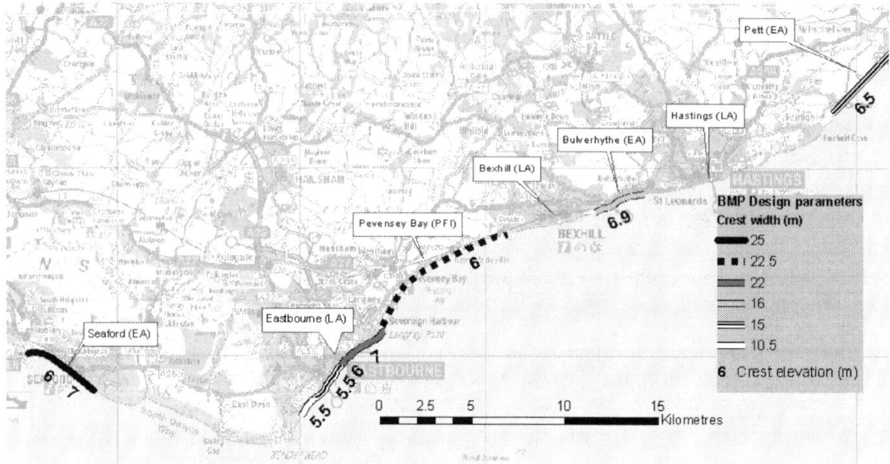

Copyright © Crown Copyright and database right 2011. All rights reserved. Ordnance Survey licence number 100026380.

Figure 3: Example mapping of BMP parameters showing design crest height (numbers) and design crest width (line symbols). Abbreviations in parenthesis are for the Operating Authority: EA = Environment Agency, LA = Local Authority, PFI = Long term contract (2000 – 2025) for contractor to manage for EA un Public Finance Initiative.

While the more recent BMPs in particular provide information within a content framework that resembles that suggested in Rogers et al. (2010) the main shortcomings in the overwhelming majority of documents or undocumented practices reviewed relate to:
- basic modelling, assumptions and data that went into the design,
- environmental considerations in relation to habitats,
- the use of coastal monitoring data,
- lack of considering large frontages beyond administrative or scheme boundaries,
- lack of review following initial project implementation

Basic modelling assumptions and data

From the outset, many BMPs lack in definition by failing to identify whether it is the risk of overtopping, breaching or undermining to which a given standard or protection applies. In the case of overtopping, there is rarely an indication as to what amount of overtopping might be tolerable and why. With rising ground behind the beach, the overtopping for say a 1 in 200 year event should for example be more linked to structural damage than risk to pedestrians as it seems unlikely that a promenade will be used under these conditions.

Overtopping calculations for mixed beaches are inherently uncertain and there does not exist a proven method for its calculation (the EurOtop manual (2007) only refers to shingle beaches and provides a very simple formula that has seen no field testing). Nevertheless, design parameters (crest height, crest width and beach slope) are reported in BMPs and in the more recent ones, reference is linked to EurOtop (2007), however, not giving any more details as to formulae used. This lack of overtopping ro even run-up calculations relates to shingle barriers, shingle beaches in front of vertical seawalls or in front / overlaying sloping concrete or rock revetments. There is also a lack of clarity about the parameters that would be needed for any calculation: such as wave data, water levels and beach characeristics like grain size or elevation of the beach toe. Sometimes, joint return probabilities (JRP) have been used for wave and water level parameters, however, these do not necessarily take wave direction into consideration and bi-modal waves are not considered. parameter Similarly, there is usually uncertainty as to whether wave parameters used are those for an offshore / nearshore location, or relate to the depth limited conditions at the beach toe. This design approach leads to some spatial inconsistencies when for example comparing crest height for beaches of different exposure (Figure 3). The southwest facing beach at Seaford not only has the highest exposure due to the predominant storm waves coming from the southwest but also because the beach toe of the shingle barrier is located at -4 to -5m Ordnance Datum (OD) which is well below Mean Low Water Spring (MLWS at -2.8mOD). Very similar crest elevations are also found for beaches facing less exposed ESE to SSE directions, and where the beach toe is usually well above MLWS. Wave conditions closer inshore are usually transformed from offshore modelling points, or from wind hindcasting for older BMPs. Use of recorded data under the Strategic Regional Coastal Monitoring Programme (CCO) is rare.

Coastal habitat considerations

Especially the more recent BMPs list European and national nature conservation areas within or adjacent to the BMP frontage, however, and assessment or even the recognition of the impact of the advocated beach management activities on existing vegetated shingle or the potential for this habitat to develop (Doody & Randall 2003) is very rare. While some BMPs advocate habitat monitoring amongst the recommendations for monitoring, in general there is no further guidance as to how, when or by whom this might be carried out.

Use of coastal monitoring data

The region considered in this review has been monitored since the 1970s through annual beach profile surveys based on photogrammetry that have been drawn into a more robust programme of beach topography, bathymetry, wave and water level monitoring since 2003. In general, a very limited amount of this data has found its way into BMPs, either in the form of supporting the design of beach parameters or by using it to review and assess design parameters of management activities. Continuation of monitoring is universally recommended, yet many of the BMPs do not take the opportunity to fine tune the regional monitoring programme to meet the specific needs for that frontage

Consideration of longer frontages

Along several stretches of coast, the authorities responsible for managing the beach change every few kilometres (Figure 3). As a consequence, and because many BMPs have been initiated following capital schemes, there is usually no recognition of frontages either side, despite there being an exchange of sediment across boundaries or the potential to combine works.

Review and update of BMPs

Only three BMPs are regularly updated but there is a general absence of evaluating design conditions against monitoring data (for example where do storms experienced fit in with JRP curves) or update predicted longshore transport rates with those measured and with any recycling volumes. The review approach described by Bradbury et al (2009) is uncommon, yet highlights a clear need to review and update design conditions. As BMPs frequently do not include any information about anticipated costs, there is also no review of predicted and actual costs that could inform whether the management practices achieve the assumed benefit cost ratio.

Common large scale issues

The review, together with a previous smaller scale project (Dornbusch 2008) has highlighted some common large scale problems associated with sediment sinks. Major longshore transport pathways are interrupted by rivers and large structures causing accumulation against them. One pathway from Selsey to Brighton is interrupted by the Rivers Arun and Adur with the terminal stop at Brighton Marina. The cell from Eastbourne to Pett is interrupted at Sovereign Harbour with a final stop at Hastings. Finally, the cell on the eastern side of Dungeness has a final stop at Folkestone Harbour. The frontages between these obstacles are too long to carry out land based recycling in a cost-effective and sustainable way. On the other hand, bypassing these obstacles only increases the pressure on the downdrift obstacle to deal with additional beach material accumulations that lead to the blockage of outfalls (Brighton Marina) or has negative impacts on harbour activities (Shoreham Harbour, Hastings Harbour (Hastings Borough Council 2010)).

Recommendations

The 2010 Beach Management Manual (BMM; Rogers et al. 2010) contains "in (Box 8.1) [...] Specific guidance on the preparation of BMPs for England and Wales, prepared in discussion with Defra and the Welsh Assembly Government (WAG)". While this provides an all encompassing framework of what potentially could be included in a BMP the shortcomings identified have led to the following recommendations that should be used as additional guidance that particularly addresses the clarity of the BMP as a single stand-alone document to provide an accountable and transparent methodology for managing beaches as coastal defence assets.

Given the well established needs for beach management, there is a wealth of data contained in the Strategic Regional Coastal Monitoring Programme, SMPs, Coastal Strategies, schemes and beach management activities. BMPs do not therefore have to start from a state of ignorance. Most importantly, there is often some data available to compare the performance with the initial design and with neighbouring or similar frontages in Southeast England.

Unless it is considered to manage a frontage radically differently, for example by adding a groyne field to an open beach or otherwise changing the general principle of management, there seems little merit in starting with the 'Beach Management Scheme Appraisal' outlined in the BMM (Rogers et al. 2010). Instead, the BMPs for the frontages considered should focus on what the BMM calls an 'Operational BMP' but extend it to include the evidence base for why the operations are carried out and using the content suggested below.

The following recommendations are made for the core of an operational BMP which could be supplemented by relevant items from the framework for a BMP listed in the BMM. This should ensure that the management activities and decision making are based on transparent information, based on monitoring and beach performance data; it should also inform and allow for a periodic review of the decision making (Figure 2). In addition, a BMP needs to be in proportion to the average annual costs for the actual management. An all encompassing BMP might be appropriate for frontages with annual management costs of several £100k. For those frontages where only a couple of thousand cubic metres are recycled every few years a more concise, but nevertheless evidence based and transparent BMP may be adequate.

<u>Background</u>: A description and history of the site, including schemes, incidents, general evolution, reference to SMP and Strategies, habitats and environmental constraints and linkages with neighbouring frontages.

<u>Drivers</u>: The drivers are the flood and/or erosion protection, ie the number of properties at risk or any other benefit resulting from protection (e.g. protection of freshwater habitat on the landward side). This will result in the determination of the standard of protection (SoP) required. The SoP should be calculated separately for overtopping and for breach and should take variations of the risk and topography into account. At Seaford (Figure 3) for example, low lying and rising ground alternate several times along the frontage and applying just one SoP for the entire frontage may not be appropriate. The SoP for overtopping should clearly demonstrate the acceptable and unacceptable rates of overtopping depending for example on the back shore type (e.g. promenade, road, wide beach, houses) and its use (e.g. essential road, recreational promenade). This data should be consistent with the Shoreline Management Plan or Coastal Strategy, where these exist.

<u>Physical input</u>: These are the conditions on which design and intervention are based. For better comparison and transparency, these conditions should refer to inshore conditions. Wave heights should consider the breaker wave height rather than offshore conditions that for the frontages considered often bear little relationship to the depth limited or sheltered conditions at the beach. The new national Flood Boundary Conditions data set (McMillan et al. 2011) provides a homogenous and easily accessible data base for water levels. Additional input conditions like grain size description of the beach (and how this was determined), foreshore profile and level of the beach toe need to be included as well. Any modelling or simpler formulae used to derive, for example breaker wave heights and joint probabilities, should be documented in such a way that any results are transparent and the way they were arrived at can be followed and easily duplicated.

<u>Design process and results</u>: The processes required to derive a design profile that satisfies the appropriate SoP identified under point 2 above needs to be clear and transparent and any uncertainties need to be clearly stated. Any modelling needs to be cross checked with coastal monitoring data (e.g. post storm surveys) to provide a reality check – if not a calibration – for the models used. A design that leads to an oversimplified and unrealistic beach cross profile with a horizontal berm of a certain width at a certain elevation and uniform frontal slope needs to be compared with the natural beach profile behaviour under storm conditions. Natural behaviour might involve the creation of a storm berm at an elevation above the design berm; this would have implications for overtopping which should be explored. Beach reprofiling may lead to 1) compaction, 2) habitat destruction and 3) obliterates any sediment sorting that has taken place. 1 and 3 are likely to increase cliffing, when the only reason for reprofiling is often to remove cliffing. Trigger levels need to refer to changed SoPs and in locations where inshore waves are depth limited, foreshore levels should be included as these will have direct impact on input parameters. The result of points 2 to 4 should provide an auditable trail of evidence – that can be easily reviewed at intervals (e.g. how well a BMP performs against physical conditions experienced over a certain interval). This provides the framework for development of a design beach configuration. The consequence of varying this configuration (with respect to changing SoP), or the parameters that went into its calculation, can be easily followed.

<u>Review of management activities</u>: Given that management has happened for many years at most of the sites, and that the existing BMP documents have been rarely reviewed or updated, a review of activities and associated costs should be included. Particular emphasis should be given to the impacts of existing management activity on the beach, habitat (vegetated shingle) and how management activities need to change to improve the habitat. Where beach management is carried out for amenity purposes (for example the common practice of smoothing the beach in spring) this should be put in perspective with damage that this might cause to existing habitats or its impact on the potential for habitat to develop.

<u>Proposed Management activities</u>: Points 2 to 5 provide the background against which the proposed management activities should be based. All activities should be linked to trigger conditions that are themselves linked to SoPs and to monitoring of the beach. Although, for example, regular recycling of a given amount of beach material is easy to plan and budget for, it might not be necessary. Though present funding arrangements seem to favour regular works, if the case is made across many BMPs that more flexibility would be more cost effective, then funding arrangements need to change.

<u>Monitoring activities</u>: Points 2 to 6 should be used to determine monitoring requirements divided into those that can be accommodated within the Strategic Monitoring Programme and those that might be more site specific. Any site specific monitoring should feed back into the Strategic Monitoring Programme data collection. Monitoring should address the individual beach and its problems. For example, some beaches in this review have substantial subtidal portions (e.g. Chesil, Seaford) which cannot be surveyed easily from land. The benefit cost of surveying these differently needs to be assessed.

<u>Costs</u>: This section should include a breakdown of the costs for managing the beach, including any works and time spent by the operating authority for the management and analysis.

<u>Analysis and review</u>: This closes the circle by assessing the performance of all activities (design, management, monitoring and costs) against the risk and the SoP provided. A programme of BMP reviews should be made and published in the Action Plan for the relevant SMP.

Ways forward

A set of hierarchical (Figure 4) and prioritised BMPs has been proposed and will be carried out over the next two years in a project delivered in partnership between all Operating Authorities. Prioritisation for frontages was based on present day annual spend, the existence and date of a BMP, the presence of, and initial assessment of design and trigger parameters and the perceived benefit from combining a frontage with neighbouring ones both in relation to sediment pathways and combining beach works.

However, improving individual BMPs or combining them based on sediment pathways and to achieve economies of scale by combining beach works will only provide limited benefits unless funding and governance structure for beach management is changed at the same time. Funding for maintenance activity from Government Capital Grant (FDGiA) is usually restricted to a short frontage and applied for and granted on an annual basis that is tied to the financial year (April to March). Application for funds has lead-in times often exceeding 12 months. These factors make it impossible to carry out beach management on a 'needs only' basis, both spatially and temporarily, which is likely to lead to funding applications that relate to the 'worst case'. Once the funding is granted, there is no incentive not to spend the funds as this might prejudice future funding. With the financial year ending in March, just short of the highest equinoxial spring tides (usually end of March / beginning of April) and still within the period of the stormy winter season (e.g. severe storm on 11th March 2008), chances are that funds will have been spent unnecessarily prior to an event just to ensure that they are spent within the financial year.

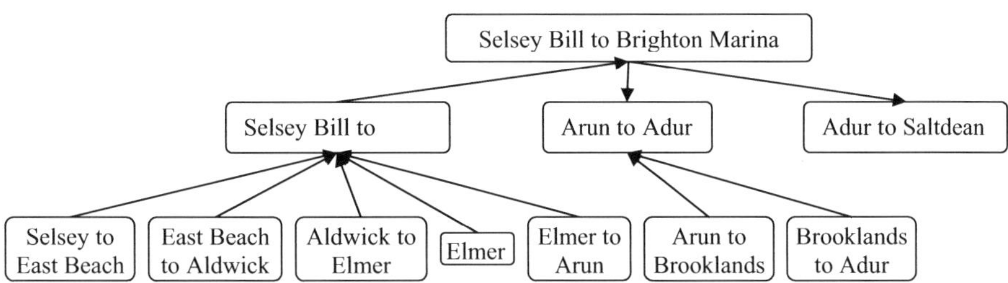

Figure 4: Example of hierarchical subdivision of frontages.

Pool funds in time:
For beach management that is carried out on a needs basis, annual variation can be in the range of between -40 to +25% of the average (pers. com. PCDL 2010). Such savings can only be achieved / higher expenses can only be accommodated, if funds can be carried over between financial years so that savings can be used to build up for years that require higher spend.

Pool funds regionally:
With frontages managed within the Southeast having vastly different storm and surge exposure parts of the annual funding requirement variations could be absorbed through regional variations so that some flexibility could be achieved without the need to pool funds in time. For example, North Kent suffers from northerly to easterly waves and from North Sea surges; there is a quite different wave exposure for Seaford compared to Eastbourne to Hastings.

The largest benefit with regard to flexibility of funding beach management would be to allocate funds on a regional (or Coastal Group) basis and allow for these funds to be carried over into following financial years. This approach would require for the funds to be managed by a group of Operating Authorities that co-ordinates beach management on a needs basis. Such management requires robust BMPs to identify needs and assess risks of activities being carried out or not for individual frontages. Apart from improved management under normal conditions this closer co-operation between Operating Authorities would also benefit emergency planning and activities to recover from extreme events.

Finally, implementing the recommendations will require additional funds for 'yet another plan that does reduce funds available for carrying out work on the ground'. Presently, the answer to a perceived shortage of sediment along a frontage is to recharge (recent examples for 2010-11 are Eastbourne and West Wittering with combined costs of £6m). In both cases recycling across frontage boundaries could have been a more sustainable solution at a fraction of the costs, but either the lack or the narrow focus of a BMP has lead to the recharge option being promoted. Apart from significant post project beach management cost savings, the project itself is designed to deliver BMPs at a fraction of the costs of recent BMPs prepared by engineering consultancies that have cost between ~£8,000 and £10,000 per km. This will be achieved by using Operating Authority staff – many of which are involved in the Regional Coastal Monitoring Project – in working on these BMPs providing direct input of existing information and local knowledge; this will also help maintain and improve staff skills. Comparatively small scale investments into R&D on wave run-up predictions and alternative ways to land-based recycling will benefit all BMPs and achieve disproportionately large benefits. The project budget is < 10% of the annual beach management budget so that the costs of the project to update BMPs and address the present shortcomings is likely to be recovered through savings within a very short period of time.

References

Bradbury, A., Mason, T. & Picksley, D., 2009. A performance based assessment of design tools and design conditions for a beach management scheme. In Coastal, Marine Structure and Breakwater 2009. Edinburgh: ICE.

CCO, 2008. Channel Coastal Observatory. Available at: http://www.channelcoast.org.

Doody, P. & Randall, R., 2003. *A Guide to the Management and Restoration of Coastal Vegetated Shingle*, English Nature.

Dornbusch, U., 2008. Adur to Hastings Integrated Management Plan.

Dornbusch, U. & Cargo, A., 2011. Review of Beach Management Plans and ad-hoc management in Southeast England. Available at: http://www.southerncoastalgroup.org.uk/pdfs/2011-02%20%20BMP%20review_Final.pdf.

EurOtop, 2007. *Wave overtopping of sea defences and related structures: Assessment Manual*, Kuratorium für Forschung im Küsteningenieurwesen. Available at: http://www.overtopping-manual.com/index.html.

Hastings Borough Council, 2010. Hastings Coastal Change Pathfinder. Available at: http://ww2.defra.gov.uk/environment/flooding/coastal-change-pathfinders/south-east/.

McMillan, A., Worth, D. & Lawless, M., 2011. *Coastal flood boundary conditions for UK mainland and islands - Practical guidance design sea levels*, Bristol: Environment Agency. Available at: http://publications.environment-agency.gov.uk/pdf/SCHO0111BTKK-e-e.pdf.

PCDL, 2010. Pevenesey Coastal Defence Ltd. Available at: http://www.pevensey-bay.co.uk/.

Rogers, J. et al., 2010. *Beach Management Manual (second edition)*, London: CIRIA.

Simm, J.D. et al., 1996. *Beach management manual*, Construction Industry Research and Information Association (CIRIA).

Innovative Coastal Zone Management
ISBN 978-0-7277-5749-4

ICE Publishing: All rights reserved
doi: 10.1680/iczm.57494.495

Assessment of Submerged Structures for Coastal Protection in a Low Wave Energy Environment

Jose Borrero, ASR Ltd, PO Box 67, Raglan, New Zealand

John Oldman, ASR Ltd, PO Box 67, Raglan, New Zealand

Laurent Lebreton, ASR Ltd, PO Box 67, Raglan, New Zealand

Shaw Mead, ASR Ltd, PO Box 67, Raglan, New Zealand

Darren James, Department of Sustainability and Environment, Victoria, Australia

Introduction

Beach erosion only becomes a problem when society – perhaps unwittingly – places boundaries or infrastructure in a zone that is experiencing net long term erosion. Since the process may be gradual and/or linked to climatic cycles, the risk to society due to beach erosion may not be revealed until years or decades after particular sites have been established. Coastal structures provide either direct (e.g. seawalls) or indirect (e.g. submerged breakwater) protection of beaches and can be designed to provide varying degrees of overtopping, wave breaking and wave transmission. The changes to local hydrodynamics and sediment transport induced by such structures can result in a shoreline that is wider and in a state of dynamic equilibrium.

Many beaches within Port Phillip Bay ("the Bay"), Victoria, Australia (Figure 1) are artificially managed and require on-going renourishment to provide beach width for recreational use and to protect public infrastructure from the effects of long term and storm induced erosion. Traditional coastal protection works are in use, but have in places led to loss of amenity, difficult beach access and exacerbated down-drift erosion problems. While existing structures are presently serving their intended purpose, their effectiveness in light of global climate change scenarios is uncertain.

The Victorian Coastal Strategy 2008 provides a policy framework for the management of coastal protection. It ensures climate change risks and impacts have been accounted for and that the relative costs and benefits of any future beach protection management options are determined. In this study the feasibility of using detached offshore reefs as a means of coastal protection in the Bay are examined. Such detached offshore reefs can be thought of as providing managed advance consistent with some of the key objectives of the Victorian Coastal Strategy 2008.

In this paper calibrated numerical models are used to quantify the wind driven wave climate, hydrodynamics and sediment transport at two key sites within the Bay (Figure 1). The Mount Martha site faces in the direction of the dominant winter wind direction and has an offshore bar within 100 m of the shoreline. The Sandringham site faces the dominant summer wind direction and the profile is shallower than the Mount Martha site and has a number of smaller offshore bars 100-200m from the shoreline.

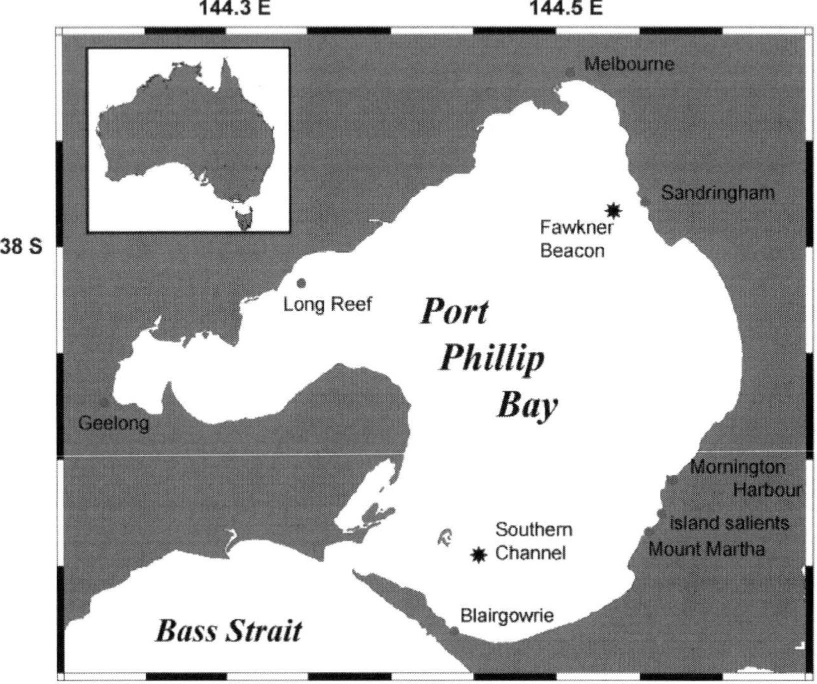

Figure 1. Location diagram showing the study sites of Sandringham and Mount Martha sites plus additional location mentioned in the text.

Port Phillip Bay

Waves within the Bay are driven by local winds, with the exception of the area in the immediate vicinity of the entrance that is exposed to some residual swell waves.. Ten year wind records from the Southern Channel and Fawkner Beacon (Figure 1) automated weather stations were used to derive the wind climate for the Bay. For the period March 1999-March 2009 mean wind speed for the two stations was 7 m/s. The 75th percentile wind speed was 10 m/s and the 90th percentile wind speed was 12 m/s. Winds show a strong seasonal trend with winds tending to be from the north-west in winter and from the south-west in summer. This leads to the observed seasonality in beach dynamics with southerly directed net sediment movement in winter and a northerly directed net sediment movement in summer. Average long term recession of the Bay's beaches is estimated to be 0.5 to 1.0 metre per annum. Net sediment transport is in the order of 1000 to 2000 cubic meters per annum.

The Bay's beaches are highly valued recreational assets for the Victorian community. They protect the coast, generate income through recreational and tourism activities, and enhance land values and business activities. Storm events, coastal development, and the erosion of many Bay beaches, prompted the implementation of extensive renourishment programs in the 1970s and 1980s. Of the Bay's 100 plus beaches, 30 priority beaches have been artificially constructed and maintained through renourishment over the past 40 years. Over the past four years the Victorian government has invested $17 million AUD to renourish 13 of the Bay's beaches.

Shoreline Response to Structures

The shoreline response to emerged (i.e. detached breakwaters) coastal structures has been extensively studied and covered in great detail in the coastal engineering literature. It is generally accepted that offshore structures (including natural features) that reduce and/or redirect incident wave energy can result in net shoreline accretion and salient formation. However there must be an ample supply of sediment within the littoral zone and the structure must be properly designed. Improper design can lead to the formation of a tombolo – where the accreting shoreline reaches the offshore structure, thereby blocking further downstream littoral transport.

Design criteria for emergent structures are relatively well documented (e.g. Dally and Pope (1986), Suh and Dalrymple (1989), Hsu and Silvester 1990) with various empirical formula defining under what conditions a salient may form and what the likely dimensions of the salient may be. Emergent structures essentially reduce wave energy arriving at the beach and weaken the longshore current without deflecting it offshore. This results in the continued transport of sand along shore which does not deprive downdrift beaches (unless a tombolo is formed). The offshore movement of sand during episodic wave conditions is impeded which assists in maintaining the salient. This is particularly important for enclosed bodies of water where there is generally no persistent long-period swell condition to return sand onshore.

For submerged structures, the design guidance is less clear. Pilarczyk (2003) provides an overview of some of the existing literature on the topic. This work is followed up by Burcharth et al. (2007) who provide a full set of guidelines for the design of low crested coastal protection structures. For low crested structures, a fundamental design quantity that differs from an emergent structure is the transmission coefficient - the ratio of the transmitted wave height to the offshore wave height across a structure. The importance of the transmission coefficient is that as wave transmission increases the diffraction effects decrease. This leads to a decrease in size of the salient through direct attack by the transmitted waves and the weakening of the diffraction-current moving sediment into the shadow zone in the lee of the structure (Hanson and Kraus, 1991). Other studies on the shoreline response to submerged structures (i.e. Martin and Smith, 1996; Stauble and Tabar, 2003; and Ransinghe et al., 2006) also note the negative effect of locating submerged structures too close to shore. These studies suggest that such positioning leads to erosive longshore directed currents through two processes (1) the divergence of wave induced current velocities as breaking occurs over the structure and (2) by water level set up that occurs between the structure and the shore line resulting in strong long shore currents generated by radiation stress as the water level attempts to equilibrate. Thus wave transmission is only part of the necessary design criteria important to the design of submerged shore protection structure. Ranasinghe et al. (2006) used a rigorous combination of laboratory and physical modelling of delta shaped reef structures to produce a set of predictive relationships for the shoreline response to a submerged structure.

In summary, properly designed detached breakwaters, whether emerged or submerged can be an effective shore protection strategy even in relatively low energy environments. Examples from the United States can be found at Amelia Island, Florida (Olsen et al., 2006), Lorain Ohio on Lake Erie (Hansen and Kraus, 1991) and at Camellia Shores Beach on Chesapeke Bay (Figure 2). In each of these cases, offshore structures were used in combination with renourishment to create long lived, wide sandy beaches.

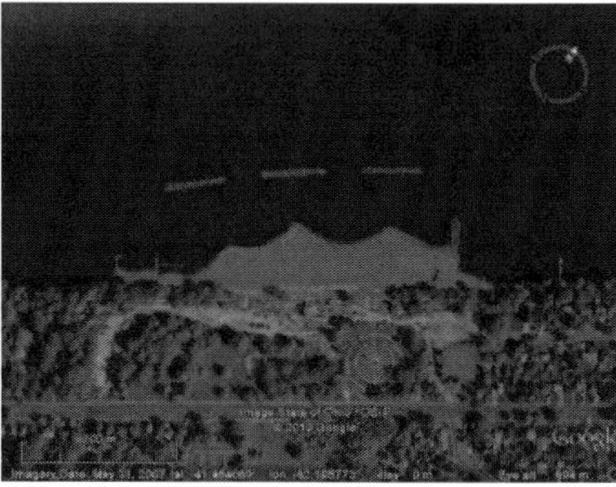

Figure 2. Shore parallel breakwaters, sand renourishment and associated salient response at Amelia Island, Florida (left) and Lorain, Ohio (right), USA.

Wave Climate Assessment

Based on the observed winds within the Bay a wind/wave model was setup using a 200 m finite difference grid. The Windwave GENeration model WGEN was calibrated against wave data collected at Long Reef (Figure 1). WGEN was developed for fetch limited water bodies and applies the JONSWAP equations to simulate wave generation. Figure 3 shows the observed and predicted wave heights, periods and directions for the calibration period. Based on this calibration the model WGEN was used to determine the wave climate at the study sites within the Bay. Figure 4 shows the long term wave distribution plot for the Sandringham site. In general the waves within the Bay are less than 1.5 m with short period waves resulting in relatively steep, low energy waves. The WGEN outputs are representative of the offshore wave conditions (i.e. pre shoaling, pre breaking conditions) and are used to provide boundary conditions for detailed morphological modelling.

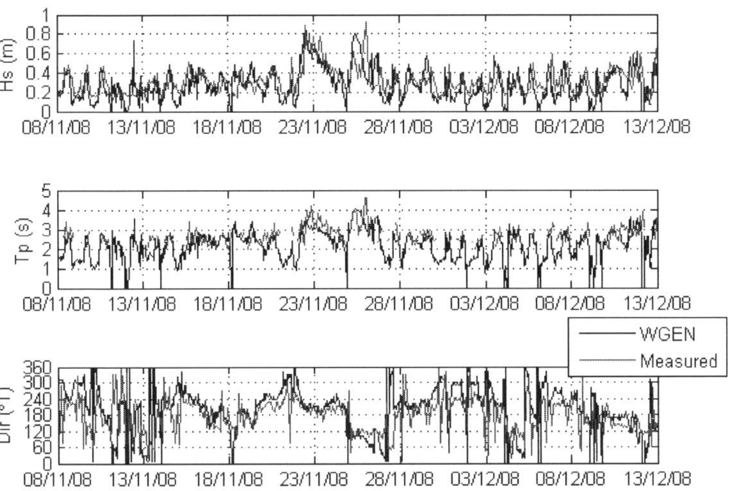

Figure 3. A comparison of WGEN model output and data recorded by an ADCP at Long Reef. Significant height (Hs), Peak period (Tp) and Direction (Dir) at Long Reef are presented in the upper, middle and lower panels respectively.

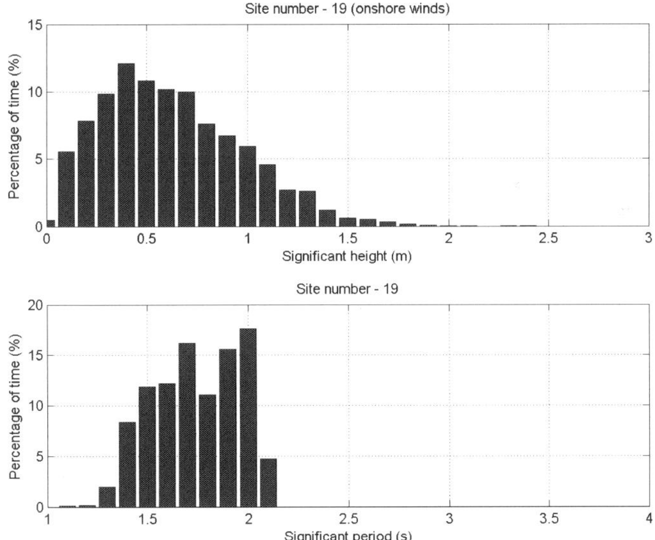

Figure 4. Histogram plots of the significant wave height (m) and significant wave period (s) generated during onshore winds at the Sandringham site.

Empirical Formula Validation

The design methodology started with an examination of existing salient formations in the Bay, both natural and structurally induced, and compared the observed salient formation with those predicted by the empirical formula.

The first such example is the Blairgowrie wave screen/breakwater located in the southern reaches of the Bay (Figure 1). The formation of this salient in response to the structure was discussed in Aitken et al. (2007). In short, the salient response was observed to occur over 12 years following the construction of the offshore wave screen, although it had still not reached dynamic equilibrium. Two additional naturally occurring examples were also found on the eastern side of the Bay south of Mornington Harbour (Figure 1).

Since these salient features are emergent (Figure 5) the salient response was analysed using the empirical relationship described by Hsu and Silvester (1990).

$$\frac{X}{B} = 0.68 \left(\frac{B}{S}\right)^{-1.22}$$

Here B is defined as the structure length, S the distance of the structure offshore and X the distance to the tip of the salient from the structure. The summary of this analysis is given in the following table.

Observed structure and salient dimensions within Port Philip Bay and salient response predicted by Hsu and Silvester (1990).

	B (m)	(S) m	B/S	X (observed)	X (predicted)
Salient 1	75	166	0.45	130	134
Salient 2	40	60	0.67	40	44
Blairgowrie	215	326	0.66	266	242

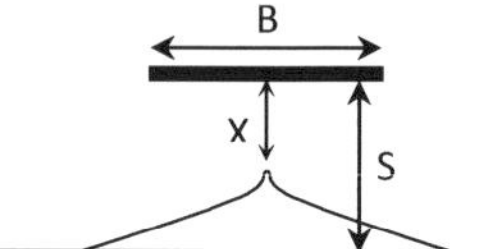

The presence of these features, both naturally occurring and induced by human interaction, are a good indication that the processes leading to salient formation occur within the low energy environment of the Bay. In addition the dimensions of salient within the Bay would appear to conform to existing empirical relationships. However design of such structures should not be limited to the use of an empirical formula but should utilise 2-dimensional numerical modelling tools for hydrodynamics and sediment morphology (e.g. Burcharth 2007).

Figure 5. Dimensions of the salient formed in the lee of the Blairgowrie wave screen (left) and double salient formed in the lee of the two off shore islands on the eastern side of Port Phillip Bay (right).

Morphological Modelling

Having established the wave climate and the applicability of empirical salient response formula for the Bay the model 2DBEACH (Black and Rosenberg, 1992a,b) was used to quantify the local hydrodynamics and shoreline response for proposed reef designs. 2DBEACH uses a non-steady, non-linear hydrodynamic module which is linked to the wave transformation module through a radiation stress terms in the momentum balance equations. The sediment transport model is based on the energetics approach first adopted by Bagnold (1963) and used successfully in a number of studies (e.g. Stive, 1986; Nairn and Southgate, 1993; Ranasinghe et al., 1999). The scheme in 2DBEACH uses a vertically-averaged form of the suspended sediment concentration equations to treat spatial variation in suspended sediment concentration. The model allows for enhanced suspension around the breakpoint (due to turbulence under plunging waves), differential settlement and real time seabed adjustments

Results from the 2DBEACH simulations were in general agreement with established empirical formulae for the reefs with emerged crest or crests at still water level. As found by Ransinghe et al. (2006) a number of combinations of crest height, reef width and wave conditions led to an erosional state on the beach.

At the Sandringham site optimal salient growth for a submerged reef was predicted to occur for a reef located 75 m offshore with the crest at mean sea level. The preliminary design reef was 60 m long, 10 m wide reef and approximately 2000 m^3 in volume. The modelled hydrodynamics for this option show that the velocities induced by the waves passing over the structure form the requisite return cell pattern which induces sediment deposition behind the structure and the formation of a shoreline salient (Figure 6). Predicted salient growth of between 30-50m was predicted under time varying tide and wave conditions. At this site the construction of a reef could be considered as a viable option for beach enhancement especially when considered in combination with the existing beach renourishment regime.

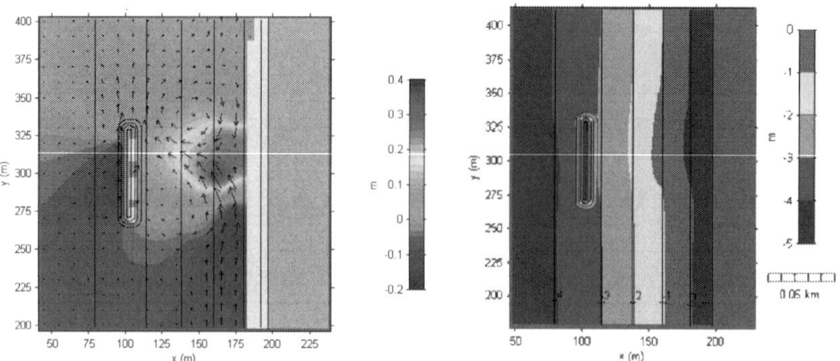

Figure 6. Local hydrodynamics (left) and equilibrium shoreline response (right) for average wave conditions for the design breakwater at Sandringham.

For the Mount Martha site, results from the modelling were not as conclusive in terms of the effectiveness of an offshore reef. The Mount Martha site is characterised by a shore parallel bar situated approximately 150 m offshore. This quasi-permanent bar feature has a crest depth of approximately 1.0 m below mean sea level which dissipates a significant amount of the incident wave energy. Initial results which placed the structure between this bar and the beach did not result in any significant change to the shoreline morphology – neither accretion nor erosion. To achieve any significant shoreline response, the structure had to be moved closer to the bar system itself. The optimal modelled salient response occurred for a reef with a volume of just under 2,400 m^3 (60 m long 20m wide and 120 m from the shore).

While a reef located in the middle of the marine platform does not create a significant change in the shoreline, a wide structure closer to the bar edge allows sand stored in the offshore sandbar to be redistributed behind the reef, leading to salient formation (Figure 7). The model results suggest that the submerged structure works by interacting with the cross-shore sediment transport pathway as sediment is collected inshore of the structure, but is removed from the offshore bar. In this case the salient response is not only due to the structure but also due to feedback between the structure and the bar system. Unlike the Sandringham site where the morphological changes are induced by changes in longshore transport a submerged structure at the Mount Martha site also interacts with the cross-shore sediment transport pathway. Sediment is collected inshore of the structure as well as being removed from the offshore bar system.

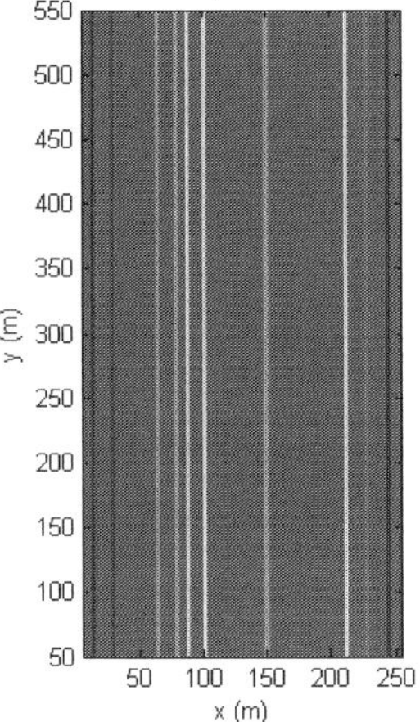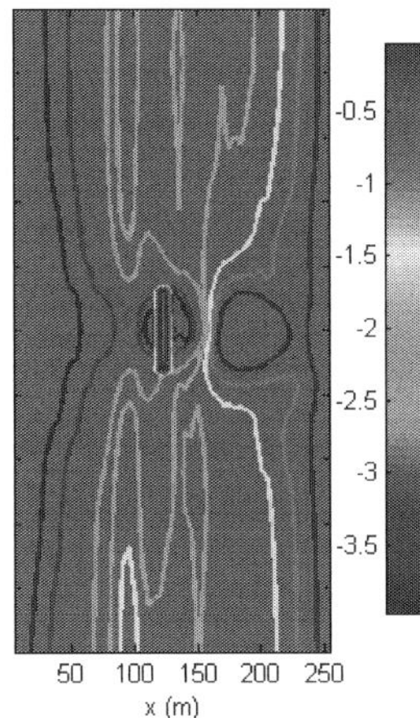

Figure 7. Original schematised isobaths at the Mount Martha site (left) and the resulting bathymetry with a reef placed close to the offshore bar system (right).

Conclusions

The modelling carried out in this study along with the existing body of literature on the topic suggests that constructing offshore structures can provide effective shoreline protection and contribute to a sustainable management plan for Port Phillip Bay.

At the Sandringham site, design criteria derived from empirical formula led to a reef design which was effective in promoting the formation of a salient. A design optimisation study will need to be carried out to minimise construction costs while ensuring maximum benefit for a given reef volume. Once a final design is agreed upon, further modelling would be required to examine long-term morphodynamics of the beach under a wider array of time varying wave and tide conditions than were used within this study. In addition, the costs and benefits of constructing the reef must be quantified in terms of the need for on-going renourishment, risk to infrastructure and loss of amenity.

At the Mount Martha site an empirically based design approach suggested that there would be little salient response under a wide range of reef dimensions, waves and water levels. This highlights the need for the use of more sophisticated models which incorporate beach circulation and sediment transport processes. Such models can be used to quantify not only the local effects of any offshore structures but also help to quantify the likely long term, broader scale effects that introducing any structure in the marine environment may have.

Because of the presence of the offshore bar system at this site, wave attenuation due to the reef placed close to the shoreline was not as significant as that found at the Sandringham site. At this site an effective reef design was achieved by placing the reef further offshore towards the bar system. To finalise the design at this site further studies would be necessary to fully understand the feedback mechanisms between the reef and the natural bar system and to quantify the likely longer term larger scale morphodynamics of the area. This would ensure that any benefits of the reef in terms of local beach dynamics are not offset by any negative consequences.

References

Aitken, T.R., Morris, G.J., Bartle, A. R. (2007) Shoreline Response to an Offshore Wave Screen, Blairgowrie Safe Boat Harbour, Victoria, Australia. Coast and Ports 2007, Melbourne Australia.

Bagnold, R. A., 1963. Mechanics of Marine Sedimentation. In: Hill, M. N (Editor), The Sea: Ideas and Observations on Progress in the Study of Seas.Vol. 3. John Wiley & Sons, Interscience, New York, pp. 507-528.

Black, K. P. and Rosenberg, M. A., 1992(a). Semi-empirical Treatment of Wave Transformation Outside and Inside the Breaker Line. Coastal Engineering, 16: 313-345.

Black, K. P. and Rosenberg, M. A., 1992(b). Natural Stability of Beaches around a Large Bay. J. of Coastal Research, 8(2): 385-397.

Burcharth, H.F., Hawkins, S.J., Zanuttigh, B., Lamberti, A. (2007) Environmental Design Guidelines for Low Crested Coastal Structures.

Dally, W.R. and Pope, J. (1986) Detached breakwaters for shore protection. Technical report CERC-86-1, U.S. Army Engineer WES, Vicksburg, MS.

Hanson, H. and Kraus, N.C. (1991) Numerical simulation of shoreline change at Lorain, Ohio. Journal of Waterway, Port, Coastal and Ocean Engineering, V 117, No 1.

Hsu, J. R.C. and Silvester, R. (1990) Accretion behind single offshore breakwater. Journal of Waterway, Port, Coastal and Ocean Engineering, V. 116, no. 3, May/June 1990, p. 362-380.

Martin, T.R. and Smith, J.B. (1996) Coastal Engineering Technical Note. Analysis of the Performance of the Prefabricated Erosion Prevention (P.E.P.) Reef System Town of Palm Beach, Florida. CETN-II-36.

Nairn, R. B. and Southgate, H. N., 1993. Deterministic profile modeling of nearshore processes. Part 2. Sediment transport and beach profile development. Coastal Engineering, 19: 57-96.

Olsen, E and Bodge, K. (2006). Application of Porous ("Leaky") Rock Terminal Structures, Proceedings of the International Conference of Coastal Engineering, 2006, p 3719 3731.

Pilarczyk, K.W. (2003) Design of low-crested (submerged) structures – an overview. 6th International Conference on Coastal and Port Engineering in Developing Countries, Colombo, Sri Lanka.

Ranasinghe, R., Pattiaratchi, C. and Masselink, G., 1999. A Morphodynamic Model to Simulate the Seasonal Closure of Tidal Inlets. Coastal Engineering, 37: 1-36.

Ranasinghe, R., Turner, I.L., Symonds, G. (2006). Shoreline response to multi-functional artificial surfing reefs: A numerical and physical modelling study. Coastal Engineering 53: 589-611.

Suh, K. and Dalrymple, R.A. (1989) Offshore breakwaters in laboratory and field, Journal of Waterway, Port, Coastal and Ocean Engineering. ASCE, 113 (2), p. 105-121.

Stauble, D.K., and Tabar, J.R. (2003). "The Use of Submerged Narrow-Crested Breakwaters for Shoreline Erosion Control," Journal of Coastal Research 19(3) 684-722.

Innovative Coastal Zone Management
ISBN 978-0-7277-5749-4

ICE Publishing: All rights reserved
doi: 10.1680/iczm.57494.505

Managing Coastal Defence Emergencies (Bawdsey East Lane, Suffolk)

Andrew E. Rouse, Environment Agency, Peterborough, England.
Alec Sleigh, Royal Haskoning, Peterborough, England.
Noel Waterton, Van Oord UK Ltd. Newbury, England.

Introduction

The coast at East Lane, Bawdsey lies on a prominent part of the Suffolk coast between Orfordness and Felixstowe and has been a point of strategic military significance for the last 200 years. In the second world war the promontory was used as gun emplacement and a location where the emerging radar technologies were developed. It is situated at the transition between low lying land to the north and low cliffs to the south.

The Problem

The problem at East Lane related to the shingle beach and ridge north of the East Lane promontory structures. Beach volumes had gradually reduced over recent years leaving the beach and backshore more vulnerable to rapid erosion and loss. Over November 2009 this stretch of coastline was attacked by over four weeks of moderate south-easterly gales and wave action which washed away the beach and exposed the clay layer beneath. This process undermined the concrete apron and exposed Victorian timber groynes. The clay bank which had been uncovered was at risk of imminent failure if exposed to a moderate storm of the type which commonly occur over the winter period. The proposed works were the bare minimum necessary to secure the immediate protection of the floodplain behind the defences.

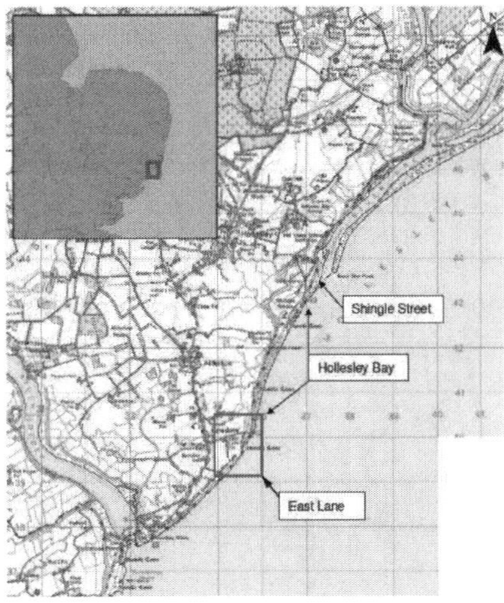

Figure 1 : Location plan

Since the problem occurred, the frontage has experienced relatively benign conditions. For example the spring tides in January 2010 coincided with north westerly winds that did little to threaten this coastline, despite the fact that the tide and surge levels necessitated the Environment Agency issuing flood warnings. If the tides had coincided with more easterly winds then the picture might not have been so good and a breach could have occurred.

Figure 2 :Damage to frontage – Nov 2010

Figure 3 : Exposed Victorian groynes – Nov 2010

The prediction as to whether the defences would have failed is difficult to determine as they are at the mercy of the weather. However, had adverse conditions prevailed, as was quite conceivable over the winter period, then these defences would have failed and properties flooded. It is for this reason that the works needed to be undertaken as quickly as possible. The Environment Agency discussed the issues with Suffolk Coastal District Council (SCDC), Natural England (NE) and the then Marine Fisheries Agency (MFA) now Marine Management Organisation (MMO) and these organisations have supported the Environment Agency in undertaking emergency works.

Project Appraisal Reports (PARs) carried out in 2003 and 2008 demonstrated that the standard of protection should be 1% annual probability of flooding (1 in 100 year event). For a design life of 50 years were present value damages valued at £8.6M (Nov 2009).

Shoreline Management Plan

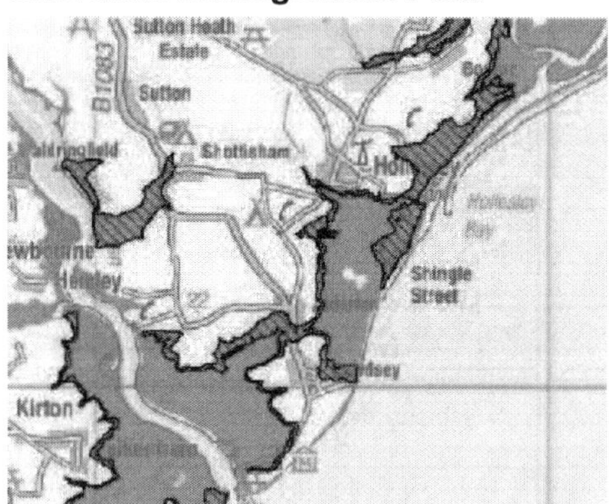

Figure 4: Environment Agency Flood Zone Map

A Shoreline Management Plan (SMP) is a large-scale assessment of the risks associated with coastal processes which seeks to reduce these risks to people and the developed, historic and natural environments. An SMP determines the natural forces which are shaping the shoreline to assess how it is likely to change over the next 100 years, taking account of the condition of existing defences. The SMP develops policies outlining how the shoreline should be managed in the future, balancing the scale of the risks with the social, environmental and financial costs involved, and avoiding adverse impacts on adjacent coastal areas.

The first generation Shoreline Management Plan for the Suffolk coastline, between Lowestoft and Felixstowe, was completed in 1998, covering a length of coastline of approximately 72 km. This SMP is now being reviewed by Royal Haskoning for Suffolk Coastal District

Council as lead authority for the operating authorities. The SMP2 will be launched earlier this year with "Hold-the-Line" as the preferred solution for maintaining the future of this coastal frontage.

The Solution

Following a review of potential options it was concluded that the preferred solution to the problems was the continuation of the existing rock revetment that had been constructed to the south. This option was chosen as it would protect the clay embankment over the next 50 years and have the minimum impact on the natural processes.

The completed works replace a 90m length of the old concrete promenade and extend approximately100m to the north. A typical cross-section through the revetment is shown below.

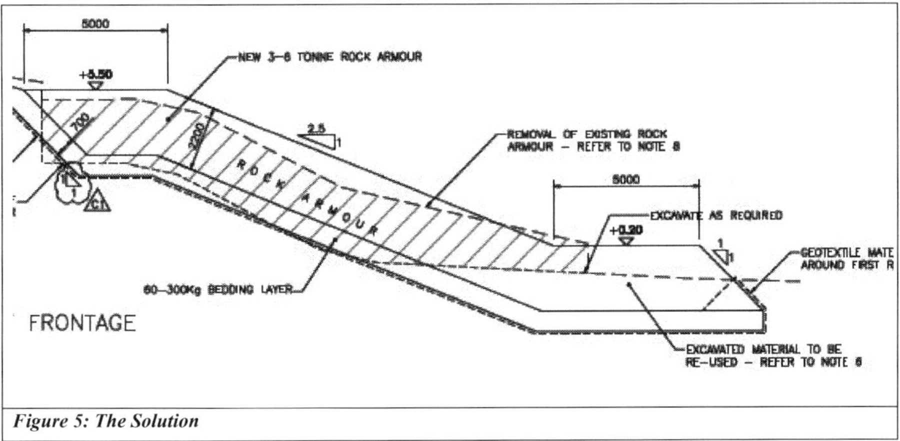

Figure 5: The Solution

Managing the Emergency
Managing the critical path to delivery

The first risk item that required to be managed was gaining the required approvals and legal consents. The Marine Management Organisation (MMO) was contacted immediately to discuss the proposed works and obtain support to move forward as an emergency, using the sea to deliver the work's rock requirement. In addition Natural England were engaged to discuss any concerns they may have with the works being undertaken in close proximity to a Site of Special Scientific Interest. The feedback from the consultees was collated and fed into the Works Information as part of the contract documents (NEC EEC Option E).

Due to the nature of the works, there is a restricted market of suppliers, who can supply and deliver suitable rock for flood risk management works on the highly dynamic North Sea coastline. To secure the supply of rock to coincide with the planned construction start date, an order was placed for rock at an early stage. This required the Environment Bawdt Agency to accept a significant level of risk in placing the order without having the results from surveys or the completed design. A risk based decision was made on the back of historical knowledge from similar schemes within the Area. This highlighted the trust of the Environment Agency in the team selected to undertake the works and the common goal to deliver the works as quickly as possible.

National Frameworks – Suppliers
The Environment Agency's National Frameworks

In accordance with its strategic procurement strategy, the Environment Agency chose to procure the Consultant / Designer and Contractor for this project through their long-term National Frameworks and these provide the basis under which they contract their major flood and coastal defence projects. There are currently six consultants on the second generation of the National Engineering and Environmental Consultancy Agreement (NEECA2) and seven contracting teams signed up to the second generation National Contractors Framework (NCF2). The overarching principle of these frameworks is that all parties will work in partnership using the New Engineering Contract (NEC) as the basis for undertaking individual projects.

Royal Haskoning was commissioned in November 2009 to produce design options to prevent a breach from occurring and to look at a quick design solution to fix the damaged caused to the frontage. Furthermore, Royal Haskoning provided environmental support to the project, including the preparation and obtaining of all necessary consents and approvals. All of Royal Haskoning's input was procured using the NEC Professional Services Contract.

The role of CDM Coordinator for the project under the Construction (Design and Management) Regulations was undertaken by Callsafe Services Limited.

TVO, a Joint Venture of Van Oord UK Ltd, Mackley Construction and May Gurney Ltd, was created specifically to compete for and undertake the Environment Agency's flood and coastal defence works. They were brought onto the project team on the day following the declaration of the emergency when they were directly allocated the role of the Contractor to provide early advice on option costs, buildability issues and construction risks. This involvement was administrated under a construction contract using the Engineering Construction Contract.

As a result of not knowing what the works would be and the requirement to set up contracts quickly to deal with the emergency, all suppliers were procured through Option E, cost reimbursable contracts. This meant that the Environment Agency held the risks but it provided flexibility and speed to operate efficiently. Furthermore, due to the works being classified as Emergency Works is allowed the suppliers to be directly allocated work rather than going through competition as required under the EU directives.

Royal Haskoning were directly appointed based on the following reasons:
- They had historical knowledge of the frontage through their design and supervision work for SDCDC on the frontage to the south of these emergency works;
- They had resource available to assist the Environment Agency in dealing with the problem. In particular, staff who had an expert knowledge of the frontage and who had were familiar with the proposed contractual arrangement of the framework.

TVO were directly appointed based on the following reasons:
- They had the historical knowledge of the frontage through the successful completion of historical schemes at East Lane;
- TVO were able to provide a very competitive rock rate from the supply chain through schemes at Felixstowe and Jaywick (rock represented some 45% of the cost of the works);
- There were further efficiencies from joining/packaging works together through TVO.

Roles and Responsibilities
Partnership Working
The declaration of an emergency lead to the project team being formed quickly. Good project governance was set up to ensure that strong leadership was created, to allow decisions to be made quickly and to enhance the effectiveness of the team in delivering the project. The nature of the project required quick decisions to be made to keep the project moving forward. This led to the empowerment of decision making being between the Area Flood Risk Manager and the Project Executive. Regular briefing were held with the Regional Programme Board, Area and Regional Managers to keep all parties involved in the issues and the scheme status.

To ensure that the project would be delivered in the most effective way, the Environment Agency requested help from their suppliers through the National Frameworks. The good relationship within the supply chain led to companies providing the right people to undertake the scheme and not just who was available.

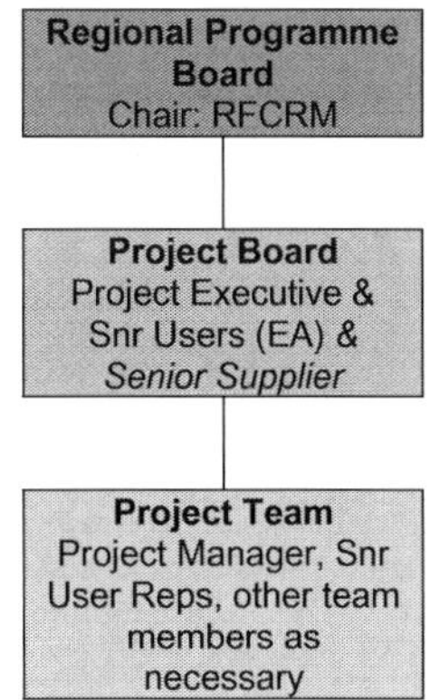

Figure 6 – Governance

To create an effective integrated team, realistic but challenging objectives were set for the team along with clear roles and responsibilities. This allowed the team to work as a professional unit and not to be tripping over each other trying to be helpful.

Effective meetings were held focusing in on the delivery of outputs that concluded with timely focused actions assigned to the right person or people (which ever organisation they came from). The actions were recorded in a Meeting Action Report (MAR). The MAR was the vehicle for quickly informing the team of progress or output from the action. This along with talking to each other, holding regular telecoms and meetings led to excellent communications and team spirit.

Engagement with the local residents
In order to maximise support from the local residents and minimise disruption due to the works, engagement was initiated at an early stage with the residents. Prior to the commencement of the works the project team contacted key stakeholders using contacts developed during previous works at Bawdsey. This previously established rapport again demonstrated the benefit of choosing a project team with experience of the local area.

Prior to the commencement of the works a visit was arranged to the local primary school at Bawdsey. Members of the site team attended a school assembly to inform the pupils and teachers of the works and what to expect so that there were no surprises and they knew what was going to happen and the reasons behind the works.

After the commencement of the works, the site team paid visits to all the local landowners to discuss the works and access requirements. During the first weeks of construction there was a high amount of rainfall, which rutted tracks and created large puddles which made access difficult. In consultation with the residents short term drainage measures were undertaken to minimise the disruption. These visits were maintained at regular intervals throughout the works.

Figure 7: Plant moving material on the beach

Figure 8: The construction of the rock armour

Figure 9: The complete scheme

To maintain good relations with the local community, reinstatement options to repair damage due to construction traffic were identified and discussed with them. The outcome was the reinstatement of an access track to a higher standard together with a turning circle for farm traffic. At a relatively small additional cost to the project this was an excellent public relations measure to maintain the cooperation with the local community.

The engagement throughout this project, together with the relationship developed during previous projects along this frontage, also meant that an area for a strategic stockpile of rock could be agreed. A meeting with the local landowner led to an agreement for rock storage on a field not presently used for agriculture. This would not have been possible without the considerable efforts of the Environment Agency project manager and the project team.

Managing quality
Having built a truly integrated team and strong project governance, the Environment Agency staff were totally engaged, from all internal functions, to help drive and meet the desired outcomes. On the receipt of deliverables (Contract Documents, Drawings, Surveys and the Works), the team ensured that they were fully compliant with the project mandate requirements and their own supplier quality management systems. With this team, the desire to get it right first time ensured that products were produced to a high standard in the office and on site. This approach provided efficiency in time, cost and resources. In addition this meant the gateway reviews were undertaken with confidence by the entire team. The Environment Agency was entirely satisfied with the outcome of the works and feedback since completion has been excellent. The approach adopted on this project is to be used of future schemes within the Region and hopefully will be used to streamline all Environment Agency future schemes.

Maintaining high health and safety and environmental standards

As part of every excellent project team, safety, health and environmental matters are at the top of the agenda. On this scheme, we confirmed our approach was effective through having regular audits undertaken by skilled people, namely NEAS and the CDM-C. All outcomes from the audits were fed back into our planning of the next phase of the works. To streamline the process and produce cost efficiencies, the audits were carried out on the same day as the progress meetings on site to ensure that transfer of information was immediate. This resulted in no SHE related issues on the scheme and all designated areas being fully protected.

Flexible working

As with any coastal scheme, the team needed to be flexible and deal with the tides and the ever changing environment – carrying out emergency works in winter could have raised many problems and issues. Having a professional team that built off each other's drive to succeed led to the team overcoming difficulties and delivering an excellent scheme within three months and within the budget. Largely this could be attributed to careful planning and good risk management. Furthermore, the site supervisor was not on site seven days per week therefore good communication was required with the site manager.

Scheme Cost Estimates

The initial estimate for the works was based on historical schemes undertaken by Team Van Oord (TVO) and the Halcrow design for the adjacent section of the frontage. This led to an estimate at £1.2M for repairs to 100 m of damaged frontage.

Item	Cost (£)	Percentage of Total Cost
Materials	528,750	43.8
Plant & Labour	270,250	22.4
Prelims	218,000	18.0
Other Items	191,196	15.8
Total	**1,208,196**	**100**

Table 1: The breakdown of the cost shows the following.

The costs for assistance with determining the problem, design, site supervision, management of the ECC contract and the formal approval paperwork following the works is estimated to be between £100k and £150k. Furthermore, with Environment Agency time cost and potential risk contingency the works were commenced with a financial year FY09/10 budget of £1.3M. Further monies were also required in financial year FY10/11 to complete the reinstatement works and retrospective environmental approvals.

Risks

There were a number of risks associated with the project, these included:

1. Increases in costs due to changes in the design as a result of site conditions;
2. Compensation claims from fisherman and local land/property owners;
3. Weather delays.

The key to effective risk management is to be able to firstly identify risks and record then within a risk register. Once the register is prepared, the next step is to assess the impact, or consequence should the risk occur and the likelihood or probability of occurrence. The owner of the risk should always be clearly defined along with a common understanding on how the risk will be managed to mitigate the impacts. The final stage is to implement the actions defined in response to the risk. Regular monitoring and updating of progress in managing the risk is important as is escalating issues through reporting through the project governance.

Mitigation

The team worked closely together to provide all the information needed to mitigate the risks, specifically the Environment Agency's Estates Team consulted with fishermen and the local population including the school. SCDC fully supported the scheme by providing some 2,000 tonnes of secondary rock which was used as bedding material to the primary rock.

Added value

A key issue which arose during the works was the presence of the existing earth embankment protruding into the revetment profile. For the proposed revetment profile to continue along the entire frontage this would have resulted in a large volume of clay being excavated and removed from site. This would have required a waste exemption to be registered which would have been a lengthy exercise.

In addition, the high cost of disposing of this clay using an approved waste carrier would have resulted in costs to the Environment Agency of tens of thousands of pounds. To minimise the costs of lost days, a speedy decision was required. The Contractor's project manager recognised the importance of this decision and was on site within hours to discuss the issue. In addition the design team were in constant discussions with the site team regarding a solution. The team worked together, with permission from the Environment Agency project manager to identify the solution which ensured no material was removed from site but also maintained the proposed robustness of the flood defence. The solution was to alter the slope of the revetment to encompass the clay embankment, thus providing further strength to the revetment from the ground conditions below. This was an excellent solution, identified quickly but with due consideration of the risks.

A further value opportunity was the packaging of the rock supply with another scheme within the Region. This lead to the rapid delivery of rock to the site and financial benefits through savings of around £500k.

Lessons Learnt

As part of the Office of Government Commerce (OGC) Gateway 4, it recommended that a lesson's learnt workshop is held. The purpose of these workshops is to understand the parts of a project that were delivered successfully and areas where lessons learnt can be used to help improve project development and delivery in the future. For this scheme the key lessons identified during the workshop were:

- A checklist of requirements for the Environment Agency when declaring Emergencies.
- Collective group of people completed tasks when required, there was no man marking and we all knew what was expected of us and what each other was doing.
- Contractual flexibility (NEC Option E) – Environment Agency was prepared to take a risk based approach.
- Found innovative ways of dealing with Procurement process.
- Quick approvals – as all were bought into reacting to the Emergency.
- Had ownership of the project – knew the scheme would be implemented.
- No SHE issues on site, although the scheme implemented quickly.

Environment Agency guidance for "Declaring Emergencies"

The flow charts below (Figures 10 & 11) captures the findings from the workshop and the teams view on the potential way forward for the Environment Agency when tasked with having to deal with Emergency Works.

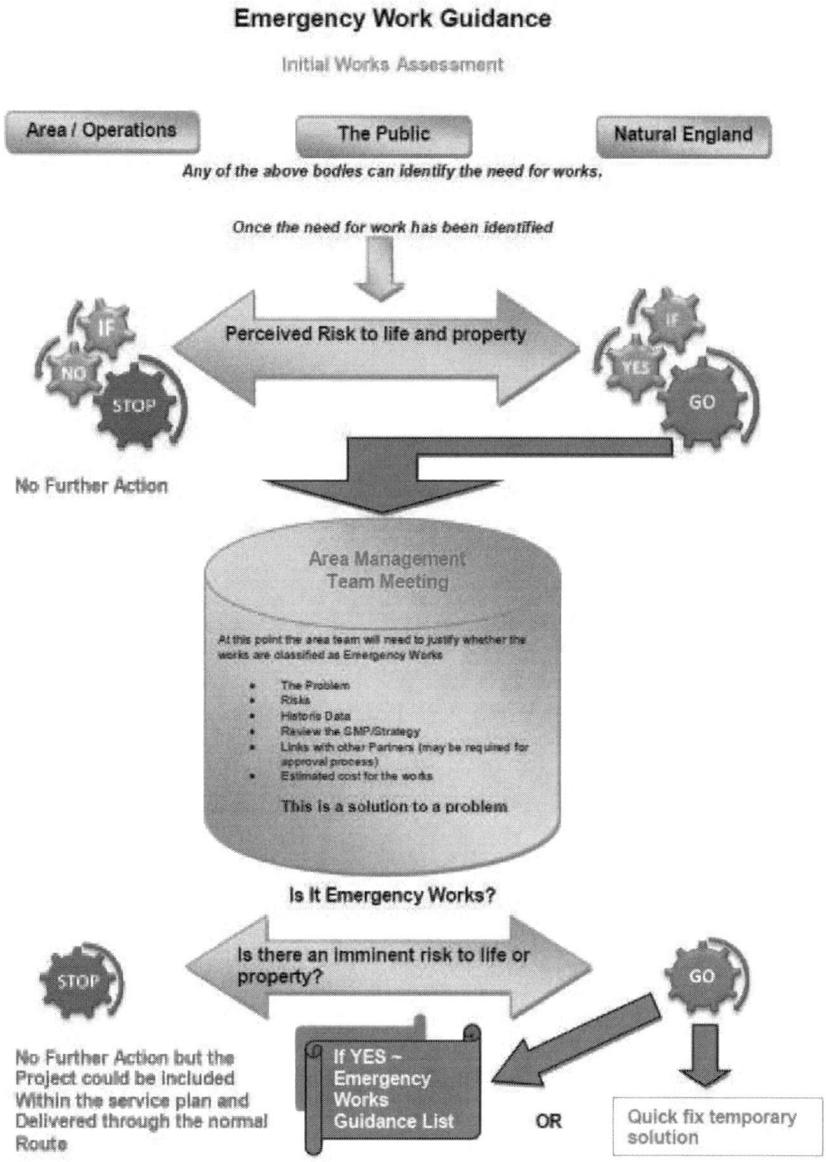

Figure 10: Emergency Work Guidance

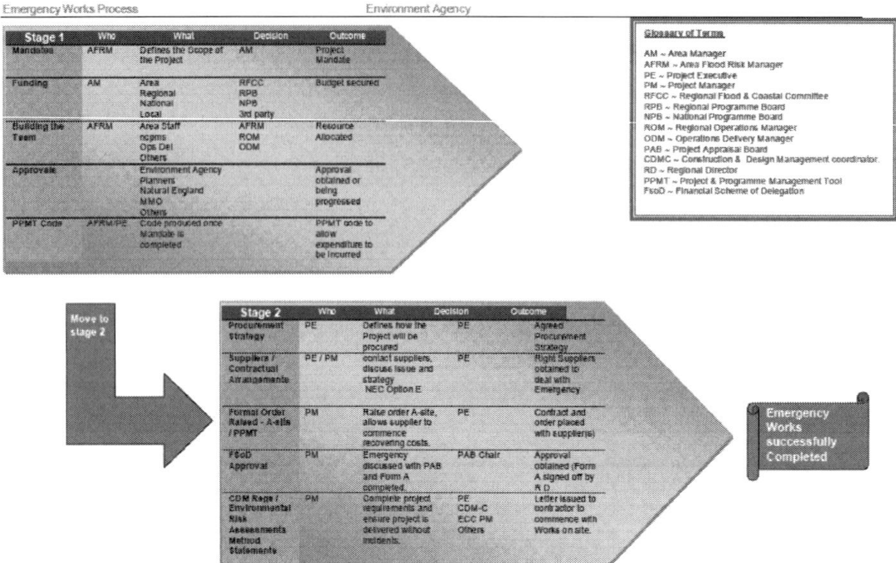

Figure 11: Emergency Work Works Process

Conclusion

The Project Team, the Environment Agency, Royal Haskoning, Team Van Oord and Callsafe Services Limited, has recently completed the £1.3 million project to provide coastal emergency works at East Lane, Suffolk. The completed scheme, which was the result of an intensive period of work, was completed within 4 months.

The Environment Agency considered the scheme to be major success as it dealt with a significant coastal emergency within a short time scale, at a relatively low cost and to an excellent standard. The most significant contributing factor to this success was the project management team and their approach to the delivery of the scheme.

References

1   East Lane Bawdsey – Case for Emergency Works; Environment Agency, December 2009.

2   East Lane Northern Emergency Works, Technical Report, Environment Agency & Royal Haskoning, December 2009.

3   East Lane Bawdsey – Lesson Leant Report; Environment Agency, December 2010.

4   East Lane Bawdsey – Emergency Works Process; Environment Agency, March 2011.

Innovative Coastal Zone Management
ISBN 978-0-7277-5749-4

ICE Publishing: All rights reserved
doi: 10.1680/iczm.57494.515

Innovative Defences from a Precast View

Robert Luther BSc, Birse Coastal, Dartford, UK
Caroline Mottram MBA, Birse Coastal Lancashire, UK
Howard Hibbard BSc CEng MICE, Jacobs, Surrey, UK
Peter Richardson BSc, CEng, MICE, Environment Agency, Worthing, UK

Introduction

The coastline and low lying land within the *Draft Folkestone to Cliff End Flood and Erosion Management Strategy (Halcrow, 2008)* includes one of the largest areas at risk from flooding in southern England. This area, around the Dungeness peninsular on the south coast of Kent, includes over 13,500 homes and 25,000ha of agricultural land. The strategy recommends a 'hold the line' policy for this section of coastline, which comprises a mixture of managed shingle beaches, sand dunes, sea walls, revetments and earth embankments.

The frontage between High Knocke to Dymchurch Redoubt (Dymchurch) lies on the southern Kent coast between Rye and Folkestone (*see fig 1 below*) and forms part of the overall strategy area.

Fig 1: The location of Dymchurch

This location was identified as being at particularly high flooding risk, where a breach of the defences anywhere along this frontage (with flooding risks exacerbated by predicted sea level rise), would pose a risk to life and widespread, potentially irreversible, damage to assets and public safety within a 5km radius of Dymchurch and the whole of the Dungeness peninsula. These assets include:

- Around 2500 residential properties
- Nearly 1000 caravans and 3 holiday camps
- Around 8,000 hectares of low-lying agricultural land
- The A259 trunk road (at risk in a 1 in 10 year flood event)

An overall, two phase scheme was therefore developed and received Defra funding approval in 2004 (ahead of formal strategy approval in recognition of the urgent nature of the works) to achieve the following objectives:

(i) To provide protection against flood risk due to breach of the sea wall against the 0.5% annual chance storm and against overtopping for storms up to the 1% annual chance of occurrence (in any given year), taking account of predicted sea level rise over the next 100 years.

(ii) To provide a scheme which could be implemented in stages to enable funding to be spread, making use of residual asset life.

(iii) To allow the continued use of the beach as a recreational resource and with regard to public and operational safety.

Setting The Scene: The Existing Amenity

Dymchurch is a quintessentially British seaside town. The town itself has an assortment of accommodation types, including hotels, B&B's, caravans and the largest holiday park in the South of England. In the centre of town, there is an amusement park offering the usual family rides. The Romney, Hythe & Dymchurch Railway also has a station in the village providing a unique way of travelling west as far as Dungeness and east as far as the Cinque Port town of Hythe.

The main appeal of Dymchurch as a tourist attraction and holiday destination is its blue-flag, sandy beach. Travel literature describes 'The sandy beach is the jewel in the Dymchurch crown. Stretching for miles toward Dungeness to the west and Folkestone to the east it is washed completely by the tide twice daily to leave a magnificent gently sloping strip of sand almost a 1/4 mile wide at low tide. This expanse of sand is home to donkey rides, kite surfers, sandcastle builders and those who like to relax'. *(www.dymchurchonline.com)*

The Environment Agency recognised that, because the town is predominantly reliant on the tourist trade associated with its sandy beaches, an innovative approach was required.

An Innovative Strategy For Managing And Developing The Coast

The most vulnerable, and thus, first phase of the sea defence refurbishment works (Frontage B) was a 2.3km section just to the north of the town of Dymchurch. This area was not densely populated and had little amenity value or revenue from tourism, thus a practical rock armour revetment scheme was developed, with new wave return wall, improved upper promenade, maintenance ramps and steps connecting promenade level to the beach. The work to Frontage B was completed in 2008. This solution provided a very efficient defence against an aggressive wave climate, including allowance for climate change uncertainty.

The second phase (Frontage A) involved a 2.2km length of the main town seafront and residential areas of Dymchurch, housing approximately 6,000 people plus all the tourist caravan and holiday parks. This meant a radically different solution to the Frontage B works was required. Whilst the primary aim of the second phase was to provide high quality, durable coastal defences, the solution was also heavily influenced by the desire to preserve and enhance the amenity value of the frontage as part of the Environment Agency's '*Creating a better place*' corporate strategy.

The design of the Frontage A defences involved a stepped concrete revetment with wave return wall, new lower promenade, improved upper promenade, raised main flood wall and timber groynes on the foreshore. Additional enhancements included refurbished boat launch slipway, viewing platforms, significantly widened upper promenade at the town centre, improved disabled access and parking.

From the outset, in recognition of the highly public nature of the Frontage A location, plans were made to ensure that that negative impacts of construction in a busy seaside town were minimised and controlled. Considerable efforts were made to confer with and involve the relevant local authorities, stakeholders and residents in all aspects of the scheme. Public liaison through exhibitions, presentations, open days, attendance at local meetings, regular newsletters and public information displays in dedicated buildings continued throughout the works. Significant praise has been received by the project team in this regard.

The construction programme allowed for phasing of elements to avoid the busiest amenity beaches during summer seasons. Initial concerns that construction works in a tourist seaside environment would have a detrimental effect on the town's local economy were unfounded. In fact, watching some of the more significant pre-cast concrete installation processes – from a controlled safe distance – became a tourist attraction in its own right!

Functionality And Value For Money 'versus' Economic And Social Aspirations

The High Knocke to Dymchurch Frontage A Sea Defence Project has been and continues to be a great example of teamwork, bringing together the customer (Environment Agency), consultant (Jacobs), contractor (Birse Coastal) and end user (the public) to develop and implement the best and most cost effective design and construction solutions in challenging conditions.

It was clear from the outset that the design would have to fulfil two fundamental requirements:

(i) to improve the sea defences and increase the level of protection against breach
(ii) to maintain the highly prized amenity value of the beach and to enhance the accessibility of the frontage both physically and visually for the tourist trade upon which the town's economy relies.

These requirements had to be achieved without compromising public safety and without unnecessary expenditure above the need to meet these requirements.

Jacobs and Birse Coastal worked together with the Environment Agency's operations team to develop the outline design, thus ensuring that the designer's aspirations could be met with regard to buildability and CDM regulations within the available funding. As part of this package, some money was allocated specifically for amenity enhancement. An extensive public consultation ensued to develop a list of desirable enhancements from which, subsequently, interested parties would then select within the budget constraints.

How the team approached the delicate balance of practicality with enhancement is perfectly demonstrated by a problem encountered as the design developed. The full length of the frontage was very narrow and would prove problematical for access with construction plant. However, a crane platform would be required along the full length to place the precast concrete revetment units. Working from the beach would cause significant time delays through restricted tidal windows availability. The team could have implemented a temporary works solution, but this would prove costly. Therefore, the team incorporated into the design a second 'lower' promenade (*figure 2 refers*) – used specifically for crane access during construction, but which would remain as an added amenity in the permanent works.

The choice of a stepped, precast concrete revetment as the primary defence itself was also a carefully balanced decision of practicality versus amenity value. The rock solution could have been continued from the preceding Frontage B works. However, the precast 'steps' with a gently sloping 1:3 gradient allowed easy and much more user-friendly access directly onto the prized beach along the full length of the frontage. Access ramps also ensured those parents with pushchairs and wheelchair users were not excluded from access to the new lower promenade.

Fig 2: The new 'lower' promenade

Dymchurch town lies approximately 2m below MHWS and, therefore, the top of the old sea wall was at or above first floor level for most of the properties. Local residents were concerned that the new sea wall would further isolate the town from the sea, but were also in fear of past events of the sea wall being overtopped causing localised flooding and damage.

To address these concerns, the design included both a wave return wall and a rear primary defence wall (*see fig 3*). The aim was to use the wave wall to dissipate the wave energy and to capture any overtopping on the upper promenade between the wave wall and the primary defence wall. By using a double wall approach, the height of the new defences was kept lower than if it had been designed as a single wall.

Fig 3: Artists impression of the new design

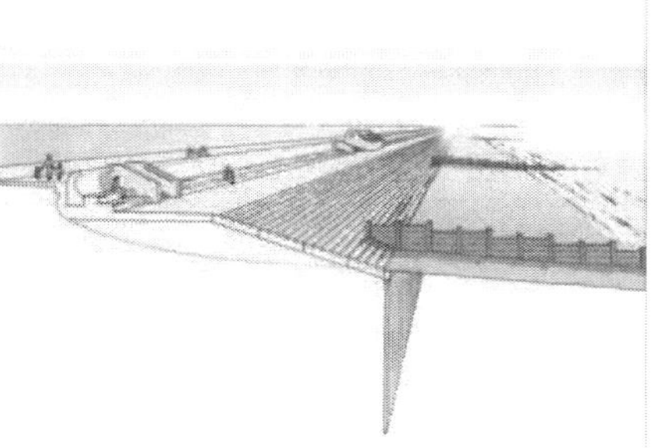

This concept was developed such that the top part of wave return wall also formed a seaward parapet to the upper promenade. This was a major safety improvement over the vertical unprotected drop of the old wall and effectively has transformed the upper promenade into a wider, safer, user-friendly area along the full length of the frontage for walkers, cyclists and tourists. This entire amenity was designed as part of the inherent structure without impinging on the enhancement budget.

Other than an old slipway, the town centre frontage lacked a focal point and was largely featureless. The first use of the enhancement budget was employed to restore the slipway and open up the central upper promenade into a large 70m x 9m platform as a centre for access to the beach, slipway, town centre, and local beach facilities and amusements. To achieve this, the wave wall was positioned further forward on the lower promenade and the additional cost arose only from the addition of curved end walls and increased fill and slab work behind the wall.

The design, which fulfilled the new flood defence criteria, was already a major improvement over the old frontage in terms of amenity without substantial use of the enhancement budget.

The origins of the sea wall at Dymchurch are believed to date from Roman times. Throughout the years, the wall has been improved, maintained and, in the 19th century, supplemented with a strategic line of Martello towers and a Redoubt structure as defence against possible invasion by Napoleonic forces. This rich heritage was a prime consideration in the design concept and was embraced fully by the public; for example, some of the desirable enhancements chosen for the scheme included two substantial, round viewing platforms, which were designed to reflect the Martello Towers.

Although not forming an essential part of the defence, these platforms offer satellite wind-break seated areas. They offer excellent views of the beach and feature step and ramp access between the upper and lower promenades in areas where there will be substantial tourist movement between the town amusements and the beach, further improving the amenity value of the frontage.

The beach, as Dymchurch's greatest asset, is a very large and safe (in coastal terms) area of sand at low tide and attracts tourists in vast numbers during the summer months. The depth of sand however is relatively shallow (less than a metre) and it was vital that the new defence design did not upset the sensitive balance of littoral drift along the frontage. The level of the beach is also an important, although lesser, factor in the stability of the toe of the new defences. Therefore a new groyne field consisting of 26 timber groynes (including 3 existing), each 80m long and at approximately 80m centres was constructed and was considered an essential part of the design to protect the amenity beach. Groynes have been a feature of this frontage throughout the last century and were in an advanced state of decay. Their replacement will help to maintain the shallow 1:50 gradient and wide beach plan that contributes to a safe amenity beach.

Other enhancement features selected included central area car parking and ramp facilities, offering DDA compliant access from the town to the upper promenade. These were built within the footprint of the central site compound, substantially reducing disturbance to the public. This car park employed some of the materials recycled from the old defences, saving further costs and providing environmental benefits.

In total, approximately 4% of the overall construction budget was expended on amenity enhancements which, together with relatively minor adjustments to the functional design, have created substantial improvements to the frontage in terms of amenity and public safety.

Challenging Convention: An Innovative Approach To Construction

The challenging construction of the Frontage A defences called for an innovative method to suit the environment and tidal conditions. This was a project in the very heart of a traditional seaside town that depended upon its visitors throughout the holiday periods. Placing over 25,000m^3 of in-situ concrete and forming a stepped revetment to the required quality by traditional means would have been virtually impossible. The precast concept was universally agreed to be the way forward to meet quality and programme requirements and to minimise disruption to the amenity areas.

It was also decided that significant advantages could be gained from utilising precast construction for both the wave return wall and the rear primary defence wall to be constructed on the town's promenade.

The use of a precast concrete revetment on a coastal scheme was not a new idea, Birse Coastal had utilised the process on a number of other schemes. However, the whole process had to be adapted to suit the existing (and somewhat problematical) environment. The placing of over 1,650 precast 'stepped' revetment units (*figure 4*) required detailed strategic planning, practicable temporary works and above all, a safe system of work which would fit in with the community and existing surroundings.

Fig 4: Precast concrete revetment units

One of the principal difficulties of construction faced by the team was access. The existing promenade was too narrow for large plant, hence traditional construction methodology would dictate works be undertaken from the beach. However, this would place severe tidal restrictions on the work, particularly with the time required for plant to travel along the beach from the limited number of access points. There were also concerns regarding plant stability due to soft ground, probably requiring temporary haul routes and working platforms on the beach, which would be expensive, and prone to washout and unwanted debris.

Some works in the tidal zone were inevitable to construct the 2.2km of sheet piled foundation toe for the revetment along the town's frontage and for works to the groyne field. However, placing of the precast units (some of which were in excess of 20 tonnes) required craneage from a safe, non-tidal range position. As demonstrated previously, this problem led to the innovative design of a platform and haul road to accommodate the crane during construction, which subsequently became a 'lower promenade' – an additional amenity in the permanent works.

The crane platform and haul road was constructed using gabion baskets to form the retaining edge, fill material was then used to form the foundation for the 120 tonne crane. Construction commenced by forming a uniform base utilising part of the existing failed revetment. The re-used revetment materials combined with imported granular material were levelled and then underwent a of series temporary works checks, including plate bearing tests carried out prior to crane mats being placed. All these works enabled the construction of the revetment foundation in preparation for placing the precast revetment.

The precast concrete units were cast in Belfast, after extensive research by the team into facilities which were capable of the manufacture of the very high quality product required. Each mould used was purpose designed and manufactured, which enabled the casting of 'special units' – slightly tapered to meet the specific coastal contours of the Dymchurch frontage. Shipments of approximately 200 units at a time were transported by vessel from Belfast into nearby Rye Harbour in East Sussex, where the units were unloaded and taken to a harbour yard for inspection and storage. Following a quality check, each unit was transported on specially adapted short trailers with angled frames to the site at Dymchurch. Logistics between Belfast and Dymchurch played a crucial role in the project and the team utilised numerous trailers to ensure an even and constant supply of units.

The description above of the manufacture and transportation of precast units is another prime example of the how the integration of the design with the construction methodology was both innovative and essential in securing the benefits of challenging conventional construction methodology. The size of the precast concrete revetment units were optimised to the size that could be carried on a standard trailer in the UK without a wide load or heavy load licence. This, in turn, defined the size of the crane required to lift the units and the positioning of the crane. These crucial decisions influenced the final design of the lower and upper promenade, saving significant costs in aggregate fill, concrete and revetment units and avoiding the need for a temporary piling platform on the existing beach.

Costs were also saved during the manufacture of the precast units through a reduction in the amount of steel reinforcement used. Over the past 6 years, Birse Coastal have perfected a technique of vacuum lifting, meaning that the precast concrete revetment units need no lifting eyes and therefore, have no points of additional reinforcement required. The even distribution of forces during the loading process also allows a reduction in amount of reinforcement in each unit. Lifting the revetment units was carried out using a purpose built vacuum lifter. The self-contained unit, lifted by the 120 tonne crane, was manufactured with five lifting pads and an engineered capacity of 25 tonnes to ensure it could handle the varied weight and size of the Dymchurch units.

Fig 5: Ro-Ro 'shunting' tractor

Before the units could be placed, the team faced a significant challenge with regards to delivery of almost 1700 revetment units and 900 wave wall units to their required location on site. Health & Safety became, as always a paramount factor, and the team had to develop a way to avoid a tractor and trailer reversing a distance of over 1km up to 10 times each day after delivering its load. Eventually, the team came up with a solution modified from many of the UK ports – a Ro-Ro unit was used (*see figure 5*). This is a 'shunting' tractor which enables the driver to either push or pull his load, depending on the direction of travel. This simple technology avoided all the hazards and potential pitfalls of reversing on a construction site and avoided a full time reversing banksman dedicated to the one vehicle.

Best Practice: Planned And Incidental!

A variety of innovative technology, methodology and general best practice was developed throughout the Dymchurch Frontage A project. Some was developed specifically to overcome the challenges the site team faced and some was an incidental result of the solutions developed to the challenges faced.

Recycling/Reuse of excavation and demolition arisings

During the design development it was known that during construction it would be necessary to demolish and excavate significant volumes of the original sea defence structure. This would generate large quantities of revetment stone and broken concrete which being inert was ideal for reuse in the new works. Approval was sought to do this and with careful planning and stockpiling this action reduced the importation of processed fill material by approximately 30%. This had multiple benefits of a significant reduction in the scheme carbon footprint by reduced haulage and associated noise and disturbance and a clear reduction in the use of natural resources.

Reuse of original precast concrete sea defence units

Redundant original precast concrete sea defence units were reused in the new works for construction of a retaining wall supporting the side of a new car park constructed behind the new defence near the town centre. This action involved no modification of the units which were simply stockpiled and then relocated in the works.

Use of precast concrete

There are many benefits in the use of precast concrete in the tidal zone, including:

- High quality, robust and consistent finishes.
- Potential very long life due to controlled conditions of manufacture.
- Manufacture of complex shapes not achievable in in-situ construction.
- Controlled delivery to site only when required.
- Beneficial health and safety risk reduction as a consequence of removing complex construction activities from the site working area to a more controlled environment.
- Precast concrete allowed the design to include complex shapes for some of the elements which could not be achieved by insitu concrete methods
- Initial high cost of precast units, offset by simplified on site construction; allowing a higher rate of progress and reduce weather dependency.
- Reduced exposure risk in the tidal zone.
- Reduced defect rectification requirements.

Multiple environmental benefits achieved, including:

- Reduced carbon footprint by use of pulverised fuel ash (pfa) in the pre-cast concrete design to reduce cement content.
- Delivery of pre-cast concrete by sea to nearby Rye Harbour for off site storage which reduced local land take required in a very constrained site area and minimised local disruption. Land availability is at a premium in this generally urban environment. This also reduced the project's carbon footprint.
- Use of pre-cast concrete reducing potential for in-situ concrete wash out from rising tides or wave action.
- Reduced number of vehicle deliveries into the local road system by optimising largest practicable pre-cast concrete loads.
- Use of large pre-cast concrete units maximised construction productivity and hence shortened the programme, minimising the impact on the local community and traders.
- The superior quality of pre-cast concrete reduced the amount of on site rework – saving programme time, noise and dust associated with breaking out defects and reduced surplus waste concrete. All waste concrete that was generated was recycled as beneficial granular material within the scheme.

Non- traditional groyne design/material selection brings sustainability benefits

The project team re-designed the 23 timber groynes on the project. Uniquely to the coastal environment, a fully FSC European Hardwood post and Douglas Fir softwood planking option was chosen, as opposed to the 'typical' tropical hardwood/greenheart timber commonly used. This is expected to prolong the serviceable life of the groyne by allowing the planking to be replaced once or twice with relative ease during the longer life of the hardwood posts. Serviceable existing groynes were also retained and incorporated into the new design.

Maximising the design of the precast revetment units reduces costs

The concrete revetment units are the main structural element of the project occupying an 8.5 to 14.5m width over a length of 2.2km. The design thickness of the revetment therefore has a big influence on the construction costs and so the design team placed emphasis on reducing this. This required consideration of a number of factors including strength, stability against hydrodynamic loading, lifting, placement and rate of progress. The predominant design criteria was disturbing forces from hydrostatic pressures behind the revetment units on a falling tide and also pressure differentials generated by wave run-up. The effectiveness of drainage was an important factor in determining thickness of the revetment units. *Figure 6* shows a cross section through the revetment and the drainage system.

Fig 6: Cross section through revetment and drainage system

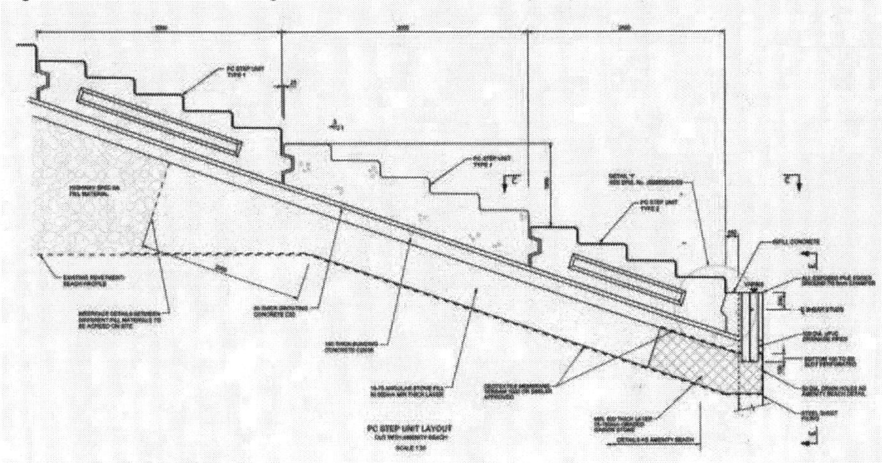

Wave return wall designed to maximise effectiveness and minimise visual impediment

The design team recognised that the wave return wall units could meet a number of scheme objectives, in addition to their primary function to reduce overtopping. The size and positioning was chosen so that the upper promenade could remain in use during all but the most extreme storm events and also so that the wall could act as a parapet at the edge of the upper promenade. The profile shape of the wave wall was designed with a curved' back face (*figure 7 refers*). This reduced the visual impact of the wall but also, most importantly, discouraged people from climbing onto the units. This resulted in a reduction in the height of the wall and the impediment to sea views.

Fig 7: Profile shape of the new wave wall

Keeping the beach and promenade open helps maintain goodwill within the local community

The previous Frontage B phase of the scheme closed sections of the entire beach and promenade to the public for up to three years. The team on the current phase programmed the construction works in such a way that the majority of the beach and promenade remained open. When closure is unavoidable during construction of the upper promenade, a phased closure plan was devised, ensuring access to the beach and lower promenade was always maintained.

Conclusion

Successful completion of the Dymchurch sea defences provides an important link in the chain of defences protecting the whole Dungeness peninsular and, in particular, the direct positive impact on those living and working in or visiting the town of Dymchurch.

References

Halcrow. (2008). 'Draft *Folkstone to Cliff End flood and erosion management strategy Edition*'. Environment Agency, Worthing, West Sussex.

Dymchurch Online Website (http://www.dymchurchonline.com) accessed 24 February 2011.

Innovative Coastal Zone Management
ISBN 978-0-7277-5749-4

ICE Publishing: All rights reserved
doi: 10.1680/iczm.57494.525

Working with Coastal Processes for Coastal Defence – A Case Study from Tywyn

Nicholas Elderfield, DHI, Denmark (Previously at Atkins Ltd, Taunton, UK)
Rob Morgan, Atkins Ltd, Swansea, UK
Derek Fenn, Atkins Ltd, Swansea, UK
Huw Davies, Gwynedd Council, UK

Introduction

Victorian coastal defences around the UK have left a legacy of structures that have artificially fixed the coastline in positions that today are unsustainable. With predictions of future climate change and forecasts of increasing budgetary constraints the challenge for managers of such structures is to ensure the most cost effective and suitable programme of upgrade, replacement and/or repair.

It is important therefore to develop a level of understanding of the dominant coastal processes, the existing problems at the defence and then to define a suitably effective solution. At Tywyn, Cardigan Bay, Wales (Figure 1), a range of options had been proposed over 15 years from rock revetments to offshore breakwaters, none of which received formal approval to proceed to construction due to either marginal cost benefit ratio or public ill-feeling over the solutions proposed.

The main problem at Tywyn has been loss of beach from the urban frontage, which has led to toe scour (Figure 3). With the beach in a denuded state the risk of overtopping, leading to flooding of the hinterland, increases (Figure 6). In addition wave impact on the exposed wall then exacerbates the erosion of the beach. This has led historically to toe scour protection being installed, which over time has led to a situation of "chasing the toe" down the seawall. Added to this was the ageing of the wider coast protection assets including the groyne system.

To make best use of the funds available for the Tywyn scheme a prioritised approach to the flooding and coastal erosion problems was proposed. This sought to achieve the greatest benefit with the smallest spatial extent of works and to minimise the requirements for costly direct maintenance and annual beach recharge schemes. Coastal process investigations and subsequent design had previously been done on schemes that provided whole frontage solutions with associated expense.

Consequently this study sought to build on the detailed work undertaken to develop the previous solutions, however at its core is an adaptive management approach that seeks to work with the dominant coastal processes.

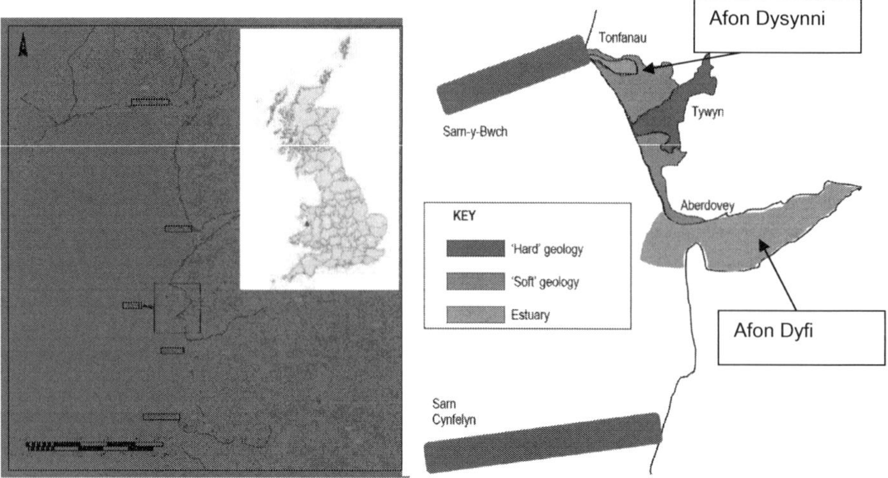

Figure 1. Location map of Tywyn

Figure 2. Geomorphological position of Tywyn on the Cardigan Bay shoreline (from ABPmer 2004a)

Figure 3. Tywyn 1908 © Tywyn Historical Society.

Figure 4. Tywyn 2009 © Atkins

Figure 5. Tywyn 1930's © Tywyn Historical Society.

Figure 6. Flooding of the promenade © Gwynedd Council

Where there is still considerable uncertainty in the ongoing natural processes and potentially the response of the system to management intervention, adaptive management is proposed as a method for managing coastal assets. This approach is advocated in the recently released Second Edition of the Beach Management Manual. At Tywyn there was a requirement to understand how the beach would respond to this "light touch" approach, to reduce the number of unknowns and to provide a quantification of the "known unknowns". To this end a range of studies were developed to provide further understanding of the proposed scheme to satisfy both coastal engineering questions such as optimal layout of structures and to understand questions with respect to the environmental impact assessment such as will the breakwater stop littoral transport along the coast.

Coastal processes operate on a range of spatial and temporal scales, therefore it is important to develop the understanding of coastal processes at a similarly suitable range of scales to resolve the processes of interest. In this situation, it is considered that adaptive management would be a more effective approach to managing the coastline. However this is subject to developing a suitable level of understanding of the controlling processes and a suitable conceptual approach to managing the beach into the future.

Developing the coastal process understanding
Historical Evolution

Historical understanding of the system is vital to both understand the evolution to the present day situation and to enable reasoned assessment of the likely future development.

The beach at Tywyn has largely developed in response to the transgression of sea levels that has taken place since the last ice age. Towards the end of the last ice age glaciers extended some distance into Cardigan Bay. Their retreat led to the development of a series of lateral moraines, seen on the present day seabed as the sarnau (Figure 8). In addition between these sarnau large quantities of gravel, cobble and sand material were deposited from upland erosion and dropout from the glaciers. Much of this material formed into a shore parallel cobble/shingle ridges, which over time have retreated across the seabed as sea levels have risen. This is seen in both the Tywyn and Borth coastlines, with the Borth coastline being the recipient of a small quantity of additional material from the cliffs to the South. The location of the Afon Dyfi in the centre of the bay is likely to have restricted the movement of this material to the Tywyn coastline, suggesting that much of the cobble/shingle material in the Tywyn ridge is relict, and over time will be lost from the system through abrasion, lock up into stores such as at the Afon Dysynni spit and anthropogenic removal. As this storm ridge provides much of the protection to the softer glacial material found along the coastal frontage. This has significant implications for coastal managers.

The ice age has left an additional legacy on the Tywyn coastline in the form of the impact of the sarnau on the wave propagation into the bay. This may be a beneficial impact as it is likely that, with the sarnau being at relatively shallow depths, they could control the bay shape (ABPmer 2004a) due to the diffractive effects of the sarnau leading to the formation of equilibrium bay shapes. This was the hypothesis for the historical evolution as defined in the previous works by ABPmer 2004a and provides a whole bay understanding of the historical evolution.

Further work (Atkins 2009) investigated direct evidence of the historical evolution through historical mapping since pre-1900 to the present day. This work suggested approximately 100m of erosion over 100 years, a figure similar to that seen around many of the UK coastlines. Tywyn was extensively developed by the Victorians with the construction of a promenade. This Marine Parade consisted of a seawall with slipway access. Photographic evidence (Figure 3) suggests that the beach was significantly higher at this stage and that the seawall was behind vegetated areas of shingle with a wide storm crest available.

Subsequent coastal erosion led to the installation of groynes to maintain a beach, however deterioration of the groyne field over time has led to the loss of the remaining beach material and subsequent lowering of the entire foreshore.

Present day system

From this larger scale historical morphological understanding, it is then important to understand the local effects at the beach scale. On this coastline, waves are the dominant control on the movement of sediment and the subsequent interaction of wave induced currents with any background currents. As such any study should seek to develop the understanding of the wave environment in both the nearshore and offshore environments and how this can impact the beach and any structures sited on the beach.

Understanding the wave conditions

Waves reaching the Tywyn coastline are from both locally generated sources in Cardigan Bay and also longer period swell waves from further afield in the Northern Atlantic. The latter are typically from the southwest as this is the primary longer fetch into Cardigan Bay, however locally generated waves across the Irish Sea can include significant waves from the northwest. Measurement of waves in Cardigan Bay is limited to the Aberporth wave buoy operated by the Met Office as part of its European wave model validation system. Whilst this gauge has a relatively long period of record critically it doesn't record direction. This allows an assessment to be made of the magnitude of waves in the deeper waters of Cardigan Bay but it doesn't provide any detail on the directional distribution and consequently any longer term changes in direction.

As part of this study an Acoustic Wave and Current Profiler (AWAC) was deployed for a month in November 2009 to provide a more detailed understanding of the wave and current environment immediately offshore of the Tywyn frontage. This gave an additional dimension to the data already held at Aberporth and crucially provided directional information as well as an understanding of the wave spectrum inshore.

The detailed data collection allowed numerical models of the wave conditions to be developed that could represent both the offshore and nearshore situation. The MIKE 21 Spectral Wave model by the Danish Hydraulic Institute (DHI) was used to represent the generation, growth and decay of waves in both the offshore and nearshore. The model was developed utilising Met Office European wave model data as boundary conditions and utilised the fully spectral capabilities of the MIKE package. Importantly these wave models could be calibrated to a level of detail previously impossible. Accurate calibration of the wave model to the prevailing conditions was vital prior to any further coastal process assessments as it provided increased confidence in the ongoing assessments.

Whilst data collection in the late Autumn/early winter period has operational issues, such as launch and recovery timings, the value to coastal process assessments is high. Capturing larger storm events provides invaluable data for assessment of beach response. During the data collection stages for this study a series of large storms were recorded with wave heights (Hm0) over 3m.

The predominant wave conditions immediately offshore of Tywyn consist of wind wave events for a large proportion of the time, and significantly some periods of swell dominated conditions. From the offshore Met Office data the swell waves are typically below 3m in height, whilst wind sea waves can reach up to 6m in the modelled period of record. This gives a feel for the distribution of wave energy in the offshore, however the transformation to the nearshore highlighted the dominant control of the sarnau to the North and South of the frontage. Figure 7 shows the typical annual distribution of wave heights as a wave rose for the offshore and nearshore conditions from the MIKE model. Northerly wave conditions in the offshore are significantly reduced or removed from the nearshore conditions, highlighting the protection afforded by the sarnau, however the effect on diffraction of the wave patterns is to spread the wave energy out around the Tywyn/Borth bay (Figure 8), with the subsequent swash aligned layout observed. This finding was important to the coastal process understanding as this provided the control on the dominant direction of transport and supported the original ABPmer hypothesis of control from the sarnau.

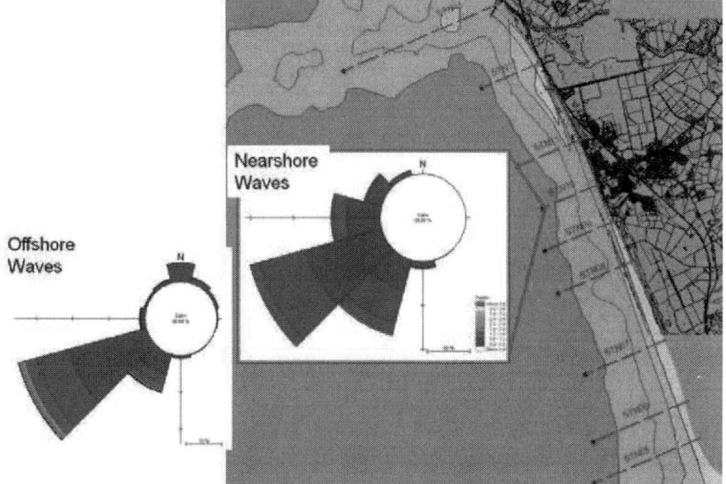

Figure 7. Typical annual wave conditions in the offshore (left) and nearshore (right).

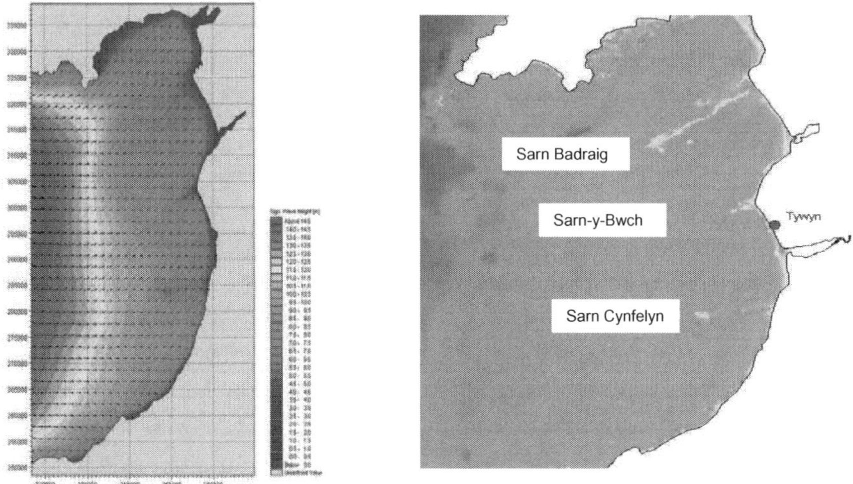

Figure 8. 1 in 100 year wave conditions (left) and bathymetry in Cardigan Bay (right)

Understanding the beach response and sediment transport environment

The response of the beach to wave forcing could be split into the two principal morphological components of the beach system, the relatively flat lower sand foreshore and the steeper shingle/cobble storm ridge at the rear of the beach. Along much of the urban frontage, the shingle/cobble ridge is non-existent due to wave impact interaction with the seawall. This leaves a shingle/sand mix in the upper beach on the urban frontage which is a combination of smaller sediment types.

For assessment of sediment transport mixed beaches often pose a problem to both empirical and numerical solutions. Importantly for this scheme extensive beach monitoring data had been collected since 2002 on both the sand and shingle parts of the beach. This data has proved invaluable in understanding the response of the beach volumetrically.

A volumetric assessment was undertaken which showed the gradual decline of the beach (Figure 9), however it also showed that the beach had responded positively to significant denusion since large storms in 2002. This suggested that there was a source of sediment (considered likely to be sand based on the knowledge of the frontage) that could respond following very large storms. The data also highlighted that typically there was a pattern of winter volume loss and summertime resupply of material. This supported the widely held observation that large winter storms lead to loss and draw down of material from the intertidal zone and summer time, swell dominated conditions, lead to a trend of beach building.

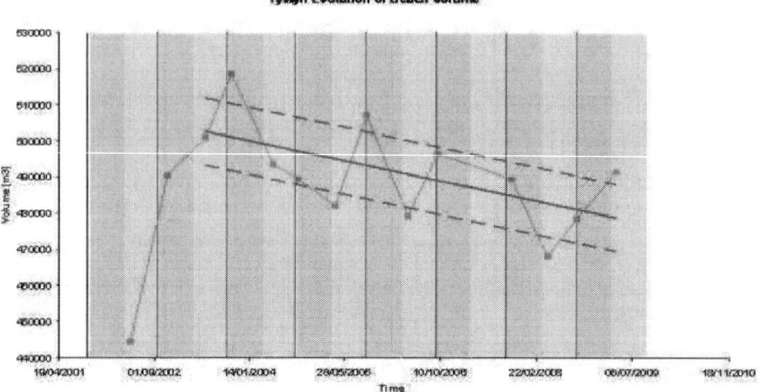

Figure 9. Volumetric change in the beach at Tywyn (Blue- Winter, Yellow – Summer)
In conjunction with the AWAC monitoring period for waves and currents, the beach surveys undertaken by Gwynedd Council were supplemented by surveys at the commencement and conclusion of the AWAC deployment. This provided a valuable assessment of the beach response to the series of storms that occurred in the period.

The swash aligned nature of the coastline immediately around Tywyn has led to a divergence of opinion on dominant local sediment transport processes (ABPmer, 2004a). Many studies are available (CGC, 2003; HR Wallingford, 1999; Pethick, 1996) with conclusions ranging from net northwards to net southwards movement. Evidence also suggested that over time the mixed sand/shingle nature of the beach on the main town frontages had altered to leave the flat sand lower and narrow shingle/cobble upper foreshore currently seen.

To understand the sediment transport processes numerical models were applied to the frontage to assess the cross-shore and longshore sediment transport. Initially these consisted of 1D models to develop a large scale understanding of how sediment transport varied along the entire coastline. 2D models then assessed the more complex processes on the urban frontage. The aim of both approaches was to develop models that could be considered representative of the key natural processes taking place on the beach.

The 1D models represent the longshore drift potential and cross shore transport with a relatively high level of confidence, however the inclusion of mixed sediment can lead to errors. Working with the model developers Atkins were able to account for the coarser sediment fraction, increasing the confidence in the 1D model results. The work highlighted that the coastline was very sensitive to small variations in offshore wave conditions. A five degree change in offshore wave climate could reverse the sediment drift from South to North. In assessing individual years with the 1D model it was apparent that the sequence, direction and strength of storms in any one year could dramatically alter the resultant sediment drift, highlighting again the sensitivity of the frontage. In addition the cross shore modelling identified the importance of the onshore/offshore mechanism for the transport of sand on the lower foreshore. This mechanism showed it was possible for sand to be removed from between the timber groynes and then added to a zone of longshore transport below low water, suggesting the groynes on the lower beach are ineffective.

This work highlighted an alternative way of considering sediment transport on the frontage. It suggested a gradual shuffling of the coarser material up and down the frontage in response to wave events and a strong offshore/onshore movement of sand. The modelling suggested that there were high gross rates but low net rates of transport, indicative of a system that is close to swash aligned. As a proportion sand movement consisted of between 66-75% of the total transport on the frontage, focused on the lower foreshore.

Detailed assessment of the existing sediment transport system was achieved using 2D morphological models that considered the wave field, tidal currents, the combination of these into resultant wave induced currents and their ultimate ability to transport sediment. In the MIKE 21 Coupled Wave, Flow and Sediment Transport model used for this study, these forcings and sediment transport were used to update the beach morphology prior to the next timestep wave event. Through this, a detailed appreciation of the development of peak transport during storm events was gained, in addition to the ultimate morphological response observed pre and post storm.

This provided a valuable insight, as it suggested that the zone of peak longshore transport being fed from sand transport between the groynes was further offshore below LAT, a factor that highlighted the need for considering the instantaneous transport at various stages of the tide rather than a tidally averaged assessment (as undertaken in the simpler 1D modelling).
The detailed understanding of sediment with respect to the coastal processes highlighted that the supply of sediment is critical to the ongoing effectiveness of any scheme that seeks to work with coastal processes. As described, the shingle/cobble fraction is relict and therefore has a finite supply. Transport onto the urban frontage is limited to movement of the small resources to the South and North.

Sand on the frontage however appears to be more abundant and moves up and down the coast in the lower bar area of the beach, typically below LAT but often moving up during the flood tide. The onshore/offshore transport of material from this zone onto the lower foreshore is significant, however the beach appears able to recover from periods of extreme denusion of sand (particularly during winter storms, when the fossil forest beds are exposed in the underlying peat). This is either due to an abundant supply of sand in this transport zone, which is transient in location between winter and summer, or that there is a significant onshore push of sand into the system from further offshore. ABPmer (2004a) hypothcsiscd that the latter is more likely given the relative speed of transport of sand.

Limited actual data exists to support/deny this hypothesis and in fact the bathymetry data collected as part of the survey for this study is some of the most extensive and recent publically available since the original surveys of this section of coast in the late 1800's. Initial assessment of the data have suggested that in the offshore zones (out to the 10m contour) there were elevation differences of approximately 2m (lower) between the original charts and the latest surveys. Care must be taken in interpreting this result as the sparseness of data temporally and the relative difference in data collection methods could be a large part of the differences observed.

Whilst extensive study had previously been undertaken during the previous 15 years of scheme development, the work developed for this study moved the understanding of coastal processes on by filling in the gap between the large scale geomorphological understanding outlined previously by ABPmer (2004a) and the scheme specific assessment of structure response gained from previous physical modelling of individual elements (ABPmer 2004b).

Being able to represent the narrow linear area of the urban frontage in a single model provided a whole area consideration of the coastal processes and provided valuable insight to aide in developing effective management solutions.

Developing effective management solutions

Concept for the Design

The concept for the full (2005) scheme at Tywyn was for 3 offshore shore parallel breakwaters with significant beach recharge, terminal groynes at each end and additional works to the walls. These would form a series of headlands and equilibrium bays along the Tywyn frontage, which would reduce the impact on the exposed vertical walls by creating a minimum width of beach in front. In doing so, both overtopping and potential collapse of the walls could be avoided. Whilst effective at reducing wave impact and overtopping, the cost/benefit ratio was not favourable, leading to a lack of funding through the traditional grant in aid routes. In addition it had a requirement for ongoing beach recharge for maintenance and possible capital recharge at points in the future to maintain the equilibrium beaches formed between the rock structures.

Following further assessment (Atkins 2008), the scheme was altered to focus on the priority southern end of the frontage which suffered the worst incidences of flooding from overtopping. In addition, alternative funding sources (EU Convergence Funding) were identified which were one off capital funds with no recourse for ongoing maintenance, hence an increased desire to work with the natural processes on the frontage.

The concept for the 2009 scheme was based on the understanding of coastal processes to date. The sections of the urban frontage at Bryn-y-Mor and Warwick Place were beginning to form headlands, with a slight embayment forming between the two (Pier Road). With the priority on the southern end at Warwick Place, it was proposed that there was a need to increase beach widths and reduce the energy of waves reaching the promenade to limit flooding of the hinterland.

As a headland was already forming in this location the most suitable method of formalising this to create a strong point in the coast was the placement of a shore parallel structure. However it was important that sediment transport between the new bays either side of this headland was not completely interrupted, particularly the sand fraction. The knowledge of the sediment transport processes developed for this study, suggested that whilst some onshore/offshore transport would be blocked by a shore parallel structure this was small in comparison to the longshore transport of sand that took place on the lower foreshore beyond the structure. The model testing also showed the formation of a salient with tendencies towards a tombolo behind this structure which suggested the need for additional beach material to avoid the scheme consuming material from the adjacent areas of coastline, in particular the coarser fractions.

At Bryn-y-Mor it was considered that whilst subject to deeper water at high tide, the length of wall exposed to exceptionally low beach levels was relatively short. In addition, the land rises as part of a series of old sand dunes that have since been built upon, reducing the number of properties at risk of flooding from overtopping. It was identified that there was a need to increase the resilience of the wall and toe to deal with overtopping,

To be effective, the coastal management solutions proposed for Tywyn were seeking to achieve a suitable balance between open and constrained coastal processes. Consequently certain aspects, such as allowing movement of sand on the lower foreshore, were considered important. To enable this, the ineffective longer timber groynes were to be removed, whilst the more critical coarser material could be controlled through the use of much shorter and steeper groynes on the upper foreshore in the areas away from rock structures.

Developing the detail

With a concept firmly rooted in the coastal processes the detailed design stage required the assessment of a series of key parameters to refine the schemes effectiveness. The detailed understanding of the propagation of waves in the surfzone from the numerical modelling allowed the optimisation of the rock required for the structure. Considering the requirements for minimum maintenance of the scheme by the client the stone was sized suitably to reduce the allowable damage factor. Importantly, consideration was given to the highly turbulent nature of breaking waves in the surf zone.

As noted above, the concept for the replacement timber groynes was to manage the shingle on the upper foreshore. Consequently the length and spacing of the groynes was optimised through use of the numerical tools and observation of the existing performance of the groynes. The resulting 40m long groynes limit their negative impact on the sand foreshore and seek to retain the steeper slopes associated with shingle. The long term monitoring data provided measurements within 1-2m of the groynes which allowed investigation of the historical variation in beach levels immediately adjacent to the structures. With typical beach profile measurement the arbitrary spacing would not have provided this additional detail. This allowed the top and bottom plank heights to be optimised to the envelope of measured conditions.

The height and position of the single shore parallel control structure at Warwick Place had been shown to allow larger storms to disrupt the salient that formed in its lee, preventing total blockage of sediment transport in this area. The angled alignment of the structure also provided an openness to the salient behind in the predominant southwesterly conditions that would aide in transporting material in the longshore direction.

The numerical assessment also showed that the performance of the structure in reducing overtopping at the seawall was enough to achieve the required standard of protection, whilst allowing relatively unconstrained sediment movement. Importantly too it highlighted zones of scour on the front face of the structure that could lead to undermining. Consequently the design was optimised to ensure sufficient toe protection and to consider the impact on the structure itself.

Scheme implementation, value engineering and ongoing monitoring

To achieve the convergence funding a rapid 2 year consenting, detailed design and construction period was adopted. This meant that decisions needed to be made rapidly on key factors such as the amount of timber required for the groynes and the overall requirement for rock, as these had relatively long lead times. Atkins worked closely with the client Gwynedd Council and contractor Jones Brothers, Ruthin, to enable this. This collaborative working allowed further value engineering savings to be made during construction.

Jones Bros. assisted Atkins in improving the amount of site investigation data available particularly under the rock structures. This removed a great deal of uncertainty in regard to founding depths of the rock structures and the amounts of underlying clay material that would have to be removed. With good detail on the underlying clay layer, Atkins was able to fine-tune the structure foundation depths to both reduce disposal quantities and reduce the quantities of rock imported. This resulted in significant cost and programme savings.

The works also specified timber sizes for the new timber groynes. Sourcing these exact sizes would have meant that hardwood timber would take 5 months to arrive on site, which could have delayed the start of piling for the groynes and adversely affected the project's overall programme. Instead the contractor found second-hand greenheart piles that had only been used on a temporary scheme and were available at short notice. As these were in very good condition, Atkins and the client agreed these timbers could be recycled and reused in the scheme.

The detailed coastal process assessment, undertaken as part of this study, was utilised to both provide input to the consenting and EIA process and the detailed design, an approach facilitated by the compressed nature of the programme. This allowed a much closer working relationship to be developed between the environmental and design team and elements of the work were beneficial to both parties. For example the quantity and composition of shingle/sand to be used in the recharge behind the breakwater was of interest to both parties though for different reasons. From an environmental point of view there needed to be enough material placed as beach nourishment to resist trapping of coarser material from either side of the scheme. From a construction point of view the quantity would influence the sources and methods of placement at site and from a client point of view there needed to be sufficient material to minimise requirements for future beach recharge and management.

Representation of groynes in the model allowed for their influence on sediment transport to be included. This provided additional confidence in the final level of planking and the led to the removal of timber sheeters from the groyne design.

Construction started on site on the 1st week of January 2009 and a very quiescent period through the early spring allowed rapid construction of the rock structures in the scheme almost two months ahead of schedule. This had the benefit of being able to hand over large parts of the beach before the busy summer holiday period.

The additional data collection process undertaken for the Tywyn scheme highlights the paucity of nearshore data on oceanographic and bathymetric conditions. Whilst the beach itself is monitored to a level of detail significantly greater than many coastal frontages, this forms only part of the picture in understanding the ongoing changes to coastal processes. The forcing conditions that create the observed changes on the beach are relatively poorly studied. Consequently whilst it is vital to continue to conduct regular surveys of the beach, it is also recommended that this is coupled with further collection of nearshore wave, water level and current data.

Following scheme implementation, monitoring of the beach was increased in temporal frequency to four times a year. In addition the survey was extended to cover the more natural beach areas to the North and South of the urban frontage. This ongoing monitoring is key to the adaptive management approach proposed for the Tywyn frontage as it not only provides warning of achievement of trigger levels for action but it also provides the scientific evidence base for improving decision making in the future. Monitoring is still critical, even in times of ever squeezed management and maintenance budgets, a fact borne out by the recent announcement of secured funding for a framework for the next 5 years of Regional coastal monitoring in England. Gwynedd Council have been tasked with development of a similar Coastal monitoring network in Wales to build on the extensive data sets already collected and to formalise this around the Welsh coastline.

Conclusions

The Tywyn project provides an example of the use of a combination of management options to provide coastal protection on a Victorian promenade that will continue to be at risk of flooding and erosion whilst a hold the line policy is implemented. Learning from history it can be seen that decisions made by Coastal Engineers in the present can have a considerable effect into the future. Consequently there is a need to minimise the impact of coast protection structures on the entire coastal system and focus on the areas of critical importance, in the case of Tywyn the management of suitable beach levels at the defence. Making coast protection schemes more efficient and effective relies on improving the coastal process understanding to a sufficient level of detail to allow a more harmonious solution.

In practice does the Tywyn scheme actually work with the coastal processes? A great deal of effort has gone into understanding and defining the dominant coastal processes from a detailed process level all the way up to a wider geomorphological level. Notwithstanding this, there will always be uncertainty due to the high level of natural variability in the system. This lends itself to adopting an adaptive management approach to the system as the scheme is focused on operating within one envelope of coastal process conditions. Changes to this envelope of conditions in the future will be related to a variety of functions, to numerous and interrelated to consider in detail. It poses the question: Do we ever really have enough knowledge of coastal processes?

In this case, the detailed coastal process understanding has allowed the development of a solution that seeks to work with coastal processes. In addition it has been possible to define the "known unknowns". These remaining uncertainties include but are not limited to:

- storm sequencing in any one year, with additional factors of strength and direction;
- sources of sediment for the frontage now and into the future; and
- longer term climate "change" such as shifting of net transport directions in response to forcing.

This study has highlighted that there are several stages to the implementation of a management scheme that seeks to work with coastal processes. These are :
- the understanding of the historical evolution is very important, with the past the key to the future;
- the understanding of the effectiveness of the present solutions and understanding what the shortfalls of the insitu defences in addition to the immediate engineering problem at the site, and
- the use of suitably detailed tools to enable effective coastal engineering.

From the initial observations at the site, the scheme is effective in its original objectives of reducing the scour and overtopping problem (see Figure 10), however its true effectiveness will only be measurable in the periods of years subsequent to its equilibration with the controlling coastal processes. To this end there is a critical need to develop suitable measurements for longer term monitoring, to both measure performance and to develop the adaptive management approach to ensure future interventions are suitable and timely. The key to the adaptive management approach to the frontage in the future is likely to be the ongoing monitoring.

Whilst monitoring of the response of the beach has been improving in the past decade, this does pose a problem when considering longer term design as it is widely believed that the past decade has been relatively quiescent in terms of wave events. Consequently alignment of monitoring on this coastal frontage with a wider national framework across England and Wales will bring benefit in terms of improved science and potentially a more level playing field when considering the business case for future intervention on a national scale.

Figure 10. Comparison images of Tywyn 2009 (left) and 2010 (right).

ABPmer. (March 2004a). Tywyn Coast Protection Review - Volume 1: Main Report. ABPmer R.1073.
ABPmer. (August 2004b). Tywyn Coast Protection Scheme: Physical Model Tests. ABPmer R.1125.
Atkins Ltd. (July 2008). Tywyn Coastal Strategy - Strategic Prioritisation of Preferred Scheme
Atkins Ltd. (2009). Tywyn Coastal Defence Improvements: Coastal Process Report.
Cygnor Gwynedd Council. (2003). North Cardigan Bay Shoreline Management Plan. Cygnor Gwynedd Council.
HR Wallingford. (1999). Afon Dysynni to Afon Dyffryn-Gwyn Coast Defence Scheme - Second Opinion Report . HR Wallingford Report EX 3491.
Pethick, J. (1996). Shoreline Intervention Proposals: Afon Dysynni to Aberdyfi. The Dyfi Estuary and the Aberdyfi Coast. Report to the Countryside Council for Wales. CCW.

Innovative Coastal Zone Management
ISBN 978-0-7277-5749-4

ICE Publishing: All rights reserved
doi: 10.1680/iczm.57494.537

Evaluating the Performance, Cost and Amenity Value of Timber & Rock Groyne Beach Management Structures: Deriving a Coastal Management Scheme for Central Felixstowe

Dr. Rosalind Turner, Mott MacDonald Ltd., Croydon, UK
Jonathan Kemp, HR Wallingford, Oxfordshire, UK
Dr. Alan Brampton, HR Wallingford, Oxfordshire, UK
Victoria Tonks, Mott MacDonald Ltd., Croydon, UK
Peter Phipps, Mott MacDonald Ltd., Croydon, UK
Terry Oakes, Terry Oakes Associates Ltd., Suffolk, UK

Introduction

Timber groynes have been used in UK beach management schemes for centuries whilst building groynes with quarried rock only began in the early 1980s. The capital costs of schemes using rock groynes are usually greater than for equivalent timber groynes. However, the current UK policy of appraising coastal protection schemes on the basis of whole life cost, i.e. including maintenance expenditure, means that rock groyne schemes can often be economically favourable over a 100 year period. As part of the Central Felixstowe Project Appraisal Report, a study to identify an appropriate long-term scheme to protect a 1.3km section of seafront, Suffolk Coastal District Council (SCDC) commissioned a review to compare timber and rock groynes. This examined the efficiency and performance of groynes constructed from timber, rock or hybrid materials alongside the cost of construction, maintenance expenditure, aesthetic value, sustainability and health & safety considerations.

Background

Groynes are shore-normal structures, i.e. oriented perpendicular to beach contours in the surf zone. Groynes are designed to alter longshore currents, reducing transport of beach sediments and changing beach plan-shape, which improves retention of beach sediment (Fleming 1990 & CIRIA 2010). The enhanced beach width can both create or improve an amenity function and can provide protection to backshore defences that may otherwise become exposed and vulnerable. A series of groynes along a coastline is known as a field of groynes.

The main design characteristics of a groyne scheme can be summarised by their:
- Length, i.e. distance from the crest of the beach to their seaward end;
- Spacing, i.e. distance apart along the shoreline; and
- Profile, i.e. the height of the crest of the groyne above the beach surface.

Note here that the length and especially the profile of any groyne, varies according to the changing profile and width of the beach. Generally in designing a groyne system, the length of groynes is measured relative to the intended beach crest position, e.g. after an initial recharge and subsequent beach reorientation.

Other groyne characteristics include:

- Permeability, i.e. the extent to which currents and sediment can travel through them;
- Orientation, i.e. some groynes are built at an angle to the beach; and
- Roughness or wave absorption, i.e. the extent to which they dissipate the energy of waves and currents they encounter, for example timber groynes may reflect wave energy, whilst rock groynes are more dissipative.

Once a groyne field is in place, beach material will collect against the updrift side of the groynes until the volumetric capacity of the bay is exceeded and the material collected spills into the adjacent bay. As beach material collects against the updrift side of the groynes, the beach orientation will tend towards the wave direction at breaking, forming a saw-tooth beach plan shape between the groynes (Figure 1), which will continuously realign according to prevailing wave conditions (Van Rijn, 2004).

Figure 1: A plan view of a typical timber groyne field showing a saw-tooth beach plan shape

As with the placement of any structure or obstruction on any shoreline within a zone of active littoral drift, a groyne scheme will cause impacts on the downdrift shoreline as a result of the reduced sediment input and these effects must be carefully considered. To reduce problems of beach erosion downdrift of the last groyne, the groyne bays should be recharged immediately after their construction. It may also be helpful to "taper off" groyne lengths in the downdrift direction where practicable.

Traditionally groynes in the UK have been constructed from a combination of timber, sheet piles and/or concrete. More recently rock, open stone asphalt, cribwork and plastic groynes have been trialled, with rock presently being the most prevalent construction material.

The use of timber groynes as beach control structures

Timber groynes have been used to manage beaches around the UK coastline for hundreds of years (Crossman & Simm 2004). Timber groynes are now more commonly constructed of tropical hardwoods such as Ekki and Greenheart rather than the more traditionally used woods such as Oak which have a shorter lifespan and are prone to cracking under the constant wetting and drying processes of the intertidal environment. One of the benefits of using timber is that it is relatively lightweight and has a good strength to weight ratio. Timber groynes also have a smaller overall footprint, when compared to rock groynes, which can prove more suitable for amenity beaches. However, timber can also contain natural flaws and the quantity of timber available in the required length is limited. Timber groynes can also be susceptible to damage by biological attack and abrasion, particularly on shingle beaches.

Timber above the beach level can be damaged by fungal decay and rot, whilst marine borers (molluscs or crustaceans which bore into and eat wood) can cause damage to timber below the beach level.

The design and detail of timber groynes affect their performance. Conventional straight timber groynes are constructed using piles and horizontal planking and occasionally vertical steel sheet piles are used to provide resistance to ground and wave forces. The spacing of timber groynes cannot easily be altered once they are constructed, since groynes need to be piled into the underlying substrate, which makes removal and adaptation of the structure more costly compared to rock groynes. The profile of most timber groynes can be modified by removing or adding planks, so improving their effects on beach levels. However, this requires frequent monitoring and intervention and since Coastal Management Authorities have limited resources for such maintenance this is no longer common practice. There are now many elderly, deteriorating and increasingly ineffectual timber groyne fields around the UK coastline

The use of rock groynes as beach control structures

In the early 1980s, the growing availability of durable rock and of plant large enough to transport and handle rock efficiently, led to an increased interest in building rock groynes. Although there are suitable UK sources, the two most popular materials used in modern rock groyne schemes have been Norwegian granite and French limestone which are very resistant to scouring by beach sediments under wave action and marine weathering.

Rock groynes are generally founded between 1.5m and 2m below the beach level (CIRIA, 2007) and their width increases the deeper they are founded, increasing the volume of rock required. The majority of the rock used should not move significantly under extreme storm conditions if the appropriate formula has been used to calculate rock size and the groyne has been constructed to industry standards.

Rock groynes have proved to be durable and, if built simply, their maintenance and any necessary adjustments to their profile should be less expensive than for timber groynes. If they eventually do need to be removed, this should also be easier, with fewer concerns about the dangers posed by any groyne remnants. Some of the rock should also be re-usable, increasing the sustainability of such groynes.

Relative performance of groyne fields

It is difficult to definitively compare the effectiveness of timber and rock groynes using field observations, since the groynes usually have different profiles and spacing even when they are installed on very similar beaches and subject to similar wave conditions.

For this reason, physical modelling was undertaken by Coates (1994) to compare the performance of rock and timber groynes on beach evolution. These tests were carried out for various groyne profiles using a variety of water levels and wave heights.

For each comparison, rock and timber groynes were built with virtually the same crest levels and lengths, only differing in that the end of the rock groynes sloped downward at 1:2 while the timber groynes ended vertically. The changes in beach widths and profiles under each wave and water level condition were then measured and compared to assess the effects of the different groyne types.

As an example of the results obtained, tests run with a (prototype) significant wave height of 2m showed that the beach profile just downdrift of the rock groynes retreated by 2m while for timber groynes the retreat was 7m. This difference was mainly attributed to the capacity of the rock groyne to absorb wave energy, so reducing turbulence and sediment transport close to it. In addition, the beach immediately updrift of each timber groyne was scoured by reflection of the large waves from its vertical face. These results supported the conclusions of an earlier modelling study (Coates and Lowe, 1993) in which it was observed that the difference in performance of rock and timber groynes became greater as wave heights increased.

Different beach plan-shapes between timber and rock groynes have also been noted during this study, though further work on comparable or adjacent frontages will be required to confirm these observations. Whilst timber groynes typically tend to produce the more classic saw-tooth beach plan shape seen in the example in Figure 1; rock groynes, perhaps due to their increased roughness, reduced reflectivity and their dissipation of wave energy, typically give rise to a different plan shape (as shown in the example in Figure 2). With rock groynes the minimum beach width tends to occur towards the middle of the groyne bay, as opposed to immediately downdrift of the groyne resulting in a curved or asymmetric bay-type beach.

Figure 2: A plan view of a typical rock groyne field showing a curved beach plan shape

Deriving suitable beach control structures to assist in long-term coastal management at Felixstowe, Suffolk.

Felixstowe is an attractive Victorian resort on the Suffolk coast with a lively urban seafront environment, which is a key tourist attraction for East Anglia (East of England Tourism, 2009). The existing defences consist of more than fifty timber/concrete groynes from the War Memorial to Cobbold's Point (see Figure 3), the original groynes were initially constructed over 100 years ago. Some have been repaired over the years by underpinning and/ or by placing rock on the beach adjacent to some of the groynes in order to provide lateral support. The backshore defence consists of a mass concrete wall topped with a promenade. In recent years, the dilapidated and increasingly ineffectual groynes have not prevented serious beach lowering; the subsequent loss of lateral support to the wall from the beach; and ultimately the undermining of the seawall. In some areas, as a short-term measure, rock has been placed in front of the seawall to provide extra lateral support.

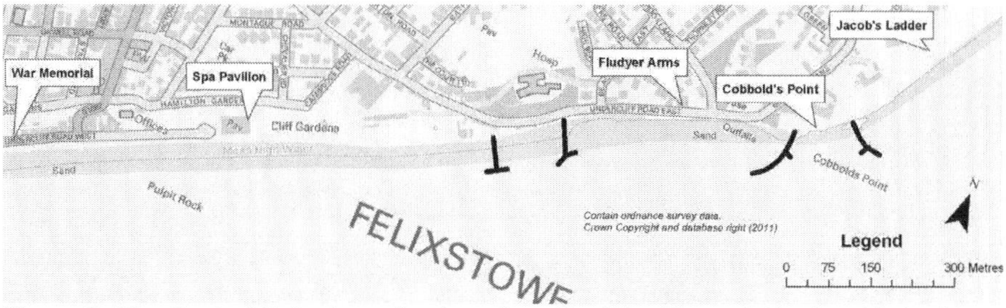

Figure 3: The Central Felixstowe frontage from the War Memorial to Jacob's Ladder

Central Felixstowe beach is a mixture of shingle and sand, with the majority of the shingle deposited on the upper beach. Owing to a drift divide around Cobbold's Point and a number of beach control structures little material now enters the site naturally. Through joint probability analyses design waves and water depths were derived: for example, a 1:100 year return period storm event has a nearshore water depth of 3.65m OD coupled with a significant wave height of 3.38m. The predominant wave direction approaches from an oblique angle, leaving a considerable angle between their crests and the seawall. To cope with this, groynes have been built close together with a typical spacing of 35m and typical length of 50m. Along the South Felixstowe frontage the beach has longer and more efficient groynes, including rock groynes only recently installed so that reducing the drift out of the Central Felixstowe frontage is likely to be less of a problem than in many cases where new groynes are being considered.

SCDC commissioned Mott MacDonald, with assistance from HR Wallingford, to produce a Project Appraisal Report identifying suitable coastal management methodologies for the next 100 years and to provide an application to the Environment Agency for funding of any initial capital works. Throughout the optioneering process for the Project Appraisal Report, Mott MacDonald undertook a costing exercise into timber groyne and rock groyne solutions for the frontage (Mott MacDonald, 2009).

The groyne design basis for the costing exercise assumes 50m long groynes at varying intervals (spacings) along the 1,300m long frontage of the existing groyne field from the War Memorial, Undercliff Road West to Cobbold's Point, Undercliff Road East.

Figure 4 shows the initial capital costs associated with timber and rock groyne fields at spacings of between 20m and 70m. Rock groynes require more expensive material, cover a larger footprint and require more labour for construction and when compared side-by-side the initial capital costs are greater. The assumptions made in preparing the costs were as follows:
- All costs are based on 2011 rates.
- The cost of the timber groynes was based on £1,714 per linear metre and for the rock groynes £3,384 per linear metre.

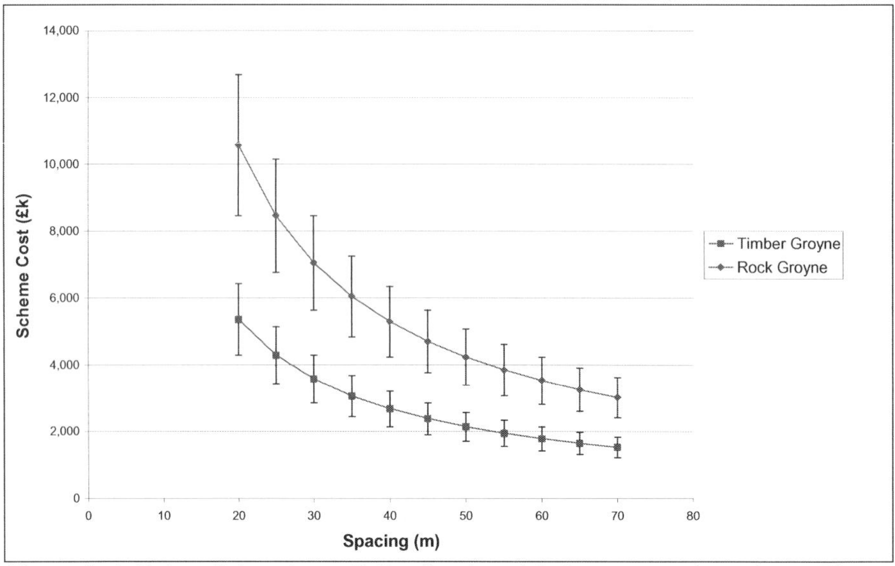

Figure 4 - Initial capital cost sensitivity of rock and timber groynes based on a 50m groyne length

Empirical review with Local Authority Coastal Managers to determine whole life costs for groyne schemes:

To improve the understanding of whole life costs of each structure type, Local Authority Coastal Managers responsible for the management of groyne fields across the south and east coasts of England were interviewed in order to obtain a first hand appreciation of timber or rock groynes on their frontages, as well as to understand typical issues, effective life and maintenance costs, which were comparable to issues likely to arise on the Central Felixstowe frontage.

Maintenance and expected lifetime

For timber groyne fields, Coastal Managers reported that annual maintenance is necessary to ensure continued effectiveness. Costs are heavily dependent upon the design of the timber groyne, the aspect of coastline, the type of beach material, meteo-oceanographic conditions and the position & spacing of the timber groynes along the beach. The average annual spend which Local Authorities reported for maintenance and repair of timber groynes varied between £500 and £2,000 per groyne. Two of the East Anglian Coastal Authorities interviewed reported that after 10 years in the intertidal environment the cost of groyne maintenance increased to between £7,000 and £8,000 per groyne, per annum. This increased cost relates to the groyne design which in both cases included sheet piling, the exposed nature of the sites and the abrasive and coarse shingle beach material at both of the frontages.

Few Authorities reported having undertaken any maintenance on rock groynes. In one instance, no maintenance had been necessary on a rock groyne field since initial construction 15 years ago. Another Authority reported undertaking repair works costing *ca.* £1,000 per groyne following a severe storm 8 years after construction.

The reported useful life of a timber groyne is significantly shorter than for a rock groyne for an exposed coastline. Rock groynes are assumed to have a 50-year design life and at the end of this period much of the material used in the initial installation can be re-used in a renovated scheme. Timber groynes for an aggressive wave environment and predominantly shingle beach are likely to have a maximum design life of 25 years, even when coupled with annual maintenance, to ensure continued effectiveness.

Sustainability

The predominant sustainability issue for both timber and rock groynes relates to the source of the materials. Whilst hardwood used for timber groynes can originate from certified plantations, Local Authority managers reported some uneasiness associated with the acceptability of using such a valuable and sensitive resource. The removal and disposal of obsolete timber groynes can prove problematic when the scheme reaches the end of its useful life. Though rock is not considered as scarce a material as a tropical hardwood, rock also has some sustainability issues associated with transport distances and extraction methods. However, in general, rock groynes were considered more sustainable than tropical hardwood timber groynes and the rocks have the added benefit of being more easily re-usable in any future re-configuration or renovation of the existing scheme, or for re-deployment in any future coastal protection scheme.

Attractiveness and amenity considerations

Both groyne types have different visual and amenity considerations, though such an appraisal can be rather subjective. The footprints of timber groynes are smaller, which means that more beach surface remains available for use by holiday makers (termed 'carrying capacity' in Spain). Timber groynes are often perceived as part of the traditional image associated with UK beaches. Rock groynes, however, have proved attractive to visitors, especially children investigating the marine plants and animals that the groynes host. Rock groynes have the added benefit of allowing the construction of a walkway along the top of the groyne thus allowing access along the groyne to less-abled people.

Health and Safety

Health and safety concerns have been reported for both rock and timber groynes. From the information provided by a number of Local Authorities, rock groynes appear to have a number of risks associated with them, but this is a general perception and is often anecdotal. However, appropriate signs and walkways along the top of rock groynes can reduce the risk of accidents. In one case study, particular care was taken with the construction of the rock groynes to reduce the number of large voids between individual large rocks which has proven successful in reducing health and safety concerns.

Costs

Whole life costs for a groyne field in Central Felixstowe were derived from the observations made by Local Authority Coastal Managers. The whole life cost of rock and timber groynes is presented in Figure 5. The assumptions made in preparing the costs were as follows:

- Timber groynes are considered to require full replacement every 25 years, whilst rock groynes have been considered with 25% replacement at year 50. These are both conservative estimates and with due maintenance both groyne fields could last longer.
- The costs include annual maintenance for timber groynes at a rate of £1,000 per groyne. This value was based on a low average maintenance cost of a variety of Local Authorities participating in the research.

- The maintenance for rock groynes was based on £5,000 per groyne every 10 years (i.e. £500 per groyne per annum).

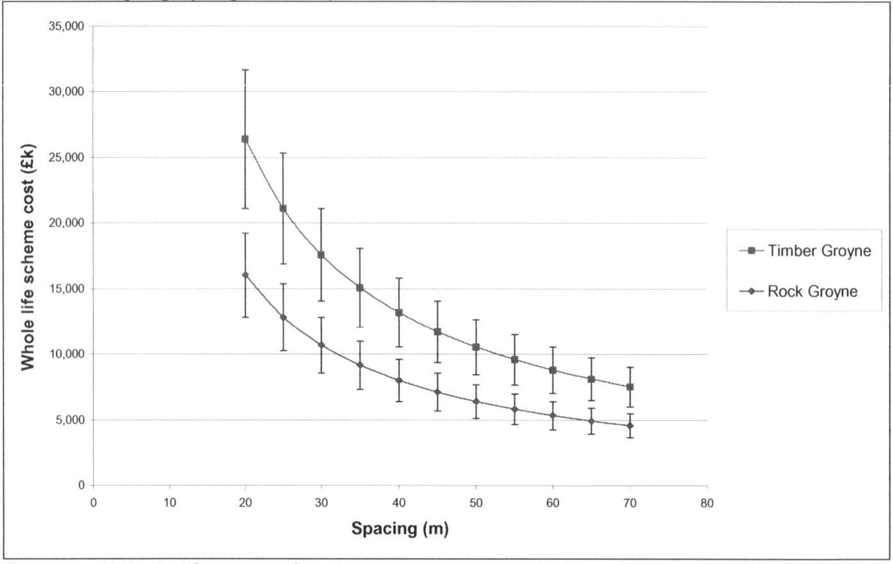

Figure 5 - Whole life costs of rock and timber groynes over 100 years for the Central Felixstowe frontage

Under these assumptions, Figure 5 clearly demonstrates that rock groynes have a reduced whole life cost when compared to equivalent timber structures at equivalent spacings. In addition to the whole life costing undertaken on a like-for-like basis, it was noted from the review that rock groynes are generally more widely spaced than timber groynes on the same frontage. The footprint of rock groynes reduces the size of the internal bay between groynes and additionally, anecdotal evidence has demonstrated that dissipative groyne structures such as rock groynes may be spaced further apart due to the lower reflection of wave energy within the groyne bays.

Areas identified for future research

Throughout this investigation and optioneering process it has become evident that there are significant gaps in scientific understanding of groyne behaviour, which could better inform the design, management and use of groynes. Two areas of further research were identified from this review:

- Empirical data collection and analysis of existing groyne systems and laboratory-based physical modelling observations, comparing different groyne construction materials. This will allow improvements in understanding behaviours/efficiencies obtained using different construction materials and to provide information for improving representation of groyne systems in numerical models.
- Anecdotal evidence has identified that dissipative groyne structures (such as rock groynes) can be more widely spaced when compared to timber groynes. However further scientific investigations and design guidance still need to be developed.

Dynamic systems require sustainable solutions to allow for adaptable management

During design of the beach management works for Central Felixstowe the Project Team used the data collated from the empirical review of Coastal Management Authorities to set out likely management actions required for the scheme over the next 100 years. However, there remains considerable uncertainty about future wave conditions e.g. as a result of climate change, particularly in mean wave directions and in the occurrence of storms. A changed wave climate will result in a change in beach behaviour. There is also considerable uncertainty regarding the amount and timing of maintenance that might be needed, relating not only to future wave and current climate, but also to future labour and material costs and likely future availability of funding.

The team concluded that a preferred scheme should be resilient to uncertainties in both the meteorological and financial climates in the coming decades.

Coastal protection based on the concept of maintaining a healthy beach is often the most economical, sustainable and environmentally acceptable way of reducing flooding and erosion risks to acceptable levels. A combination of groynes and beach recharge is therefore often preferable to building inflexible structures such as seawalls (and particularly so for a resort such as Felixstowe). In the longer term beach management through beach material recycling may be considered to counter the effects of long-term net drift or localised beach lowering problems, e.g. in individual groyne bays. Groyne scheme design, however carefully undertaken, may not prove optimal against climatic changes and could be improved by adjustments guided by long-term monitoring and experience, thus an adaptable solution was considered preferable.

Conclusions: the proposed coastal management solution for Felixstowe

On the basis of the whole life costing information, coupled with extensive public exhibitions and stakeholder engagement events, the Project Team put forward a Project Appraisal Report proposing a series of 18 straight rock groynes and 80,000m^3 beach recharge for the Felixstowe frontage. SCDC recognised that rock groynes would be more appropriate for the aggressive wave climate experienced here and valued the opportunity to be able to manage/alter the rock groynes, as required, over the 100-year life of the scheme. Through model testing and scheme refinement, the groynes designed for the frontage vary in length between 35m and 65m.

The Project Team at SCDC were keen to implement a groyne solution that physically met with the seawall at the back of the beach, as any gap between the end of the groyne and a hard backshore defence could result in outflanking and severe scour during extreme storms (as highlighted in CIRIA, 2010 & Coates, 1994). However, it was also necessary to ensure plant access between each groyne bay to allow for long-term management of the beach, since the promenade was not suitable for carrying heavy plant. In order to meet both of these objectives the Project Team derived a rock groyne design incorporating a 7m section of timber groyne towards the landward end of the structure. This allows for a 5m wide plant access bay when the timber planks are removed but when cross-planks are securely in place prevents outflanking and reduces potential for scour at the landward end of the groyne under storm conditions. An example sketch is presented in Figure 6.

The scheme will provide the people of Felixstowe with the ability to sustainably adapt and easily manage their coastline to 2110 and beyond.

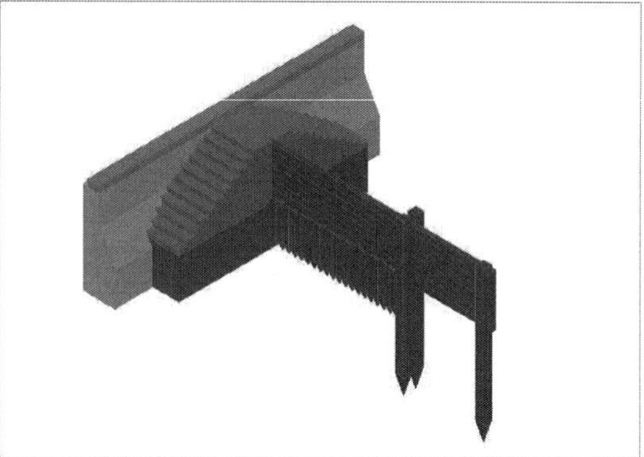

Figure 6 – An example visualisation showing a timber groyne root to close any gap between a rock groyne and the promenade. Steps are included to allow access.

Acknowledgements

The authors wish to thank Suffolk Coastal District Council and the Environment Agency for their continued support. The authors are also grateful for the input of Mr. Paul Patterson and Mr. David Wheeler of Waveney District Council who in conjunction with Suffolk Coastal District Council are leading the implementation of the Felixstowe Coast Protection Scheme.

The authors are also grateful to Mackley Construction, Dean & Dyball and to Van Oord for assistance in the costing exercise and to Coastal Managers at a number of Local Authorities who provided information on resilience and maintenance required for their *in situ* groyne fields.

References

CIRIA., (2010) *Beach Management Manual (2nd edition)*. C685, CIRIA, London.

CIRIA; CURl CETMEF., (2007) *The Rock Manual; The use of rock in hydraulic engineering (2nd edition)* C683, CIRIA, London.

Coates, T. T., (1994) *Effectiveness of control structures on shingle beaches, physical model studies.* Report SR 387. HR Wallingford, Oxfordshire, UK.

Coates, T. T., and Lowe J. P., (1993) *Three dimensional response of open and groyned shingle beaches.* Report SR 288. HR Wallingford, Oxfordshire, UK.

Crossman, M., and Simm, J., (2004) *Manual on the use of timber in coastal and river engineering.* Thomas Telford Publications. London.

East of England Tourism (2009) *Felixstowe Visitor Economy Report.* Bury St Edmunds, Suffolk.

Fleming, C. A., (1990) *Guide on the uses of groynes in coastal engineering.* CIRIA Report 119.

Mott MacDonald Ltd., (2009) *Performance review of rock and timber groynes.* Report 250927/010/A. Croydon, UK.

Van Rijn, L. C., (2004). *Principles of sedimentation and erosion engineering in rivers, estuaries and coastal seas.* Aqua Publications.

Innovative Coastal Zone Management
ISBN 978-0-7277-5749-4

ICE Publishing: All rights reserved
doi: 10.1680/iczm.57494.547

Probabilistic Design of Breakwaters in Shallow, Hurricane-prone Areas

Vana Tsimopoulou, HKV Consultants, Lelystad, The Netherlands / Hydraulic Engineering, Delft University of Technology, Delft, The Netherlands
Wim Kanning, Hydraulic Engineering, Delft University of Technology, Delft, The Netherlands
Hessel G. Voortman, ARCADIS, Amersfoort, The Netherlands
Henk Jan Verhagen, Hydraulic Engineering, Delft University of Technology, Delft, The Netherlands

Introduction

One of the failure mechanisms of a rubble mound breakwater is the failure of its armour layer. In order to determine the stability of an armour layer, the design load has to be defined, which is in fact the wave that attacks the structure. Being a highly stochastic phenomenon, the wave action is not easily defined, while there is always some uncertainty inherent to its definition. In a deterministic calculation this uncertainty is being left to engineering judgment, as the possible variations of the design wave height are not taken into account in a coherent way. In order to explicitly incorporate uncertainties into the design process, and therefore increase its reliability, probabilistic design methods should be applied. A commonly used approach is a semi-probabilistic computation, which introduces the application of partial safety coefficients. Nevertheless the indicated methods to derive and apply them do not clarify the uncertainties incorporated, adding an undefined degree of safety in the process, or end up with incorrect results under certain conditions. Another approach is a fully probabilistic computation. This type of design tackles explicitly a great deal of uncertainties, hence its results can be considered much more accurate. However it is not commonly used, due to the fact that there are not straightforward guidelines to support it, and therefore a number of critical decisions by the designers are required.

This paper focuses on the application of probabilistic methods for armour layer design of rubble mound breakwaters. The main objective is to indicate the weaknesses of the previously mentioned methods, and to suggest a probabilistic design approach that is both attractive to designers and sufficiently reliable. This can be achieved through elaboration of a design example with the various methods, followed by a critical evaluation of the results.

Case study description

The example application, through which a critical assessment of the design methods can be realized, needs to concentrate some particular characteristics that facilitate this process, and create a strong basis for the development of a new design approach. An interesting case for demonstration is the jetties at the entrance of Galveston Bay, which is a large estuary located along the upper coast of Texas in the Gulf of Mexico (figure 1). The main function of the structures is to stop siltation at the entrance of the estuary, but they must also be able to resist occurring waves. They are part of the network of structures that protect not only the port of

Galveston, but also the vital industrial shipping facilities in and around Galveston Bay, where a significant amount of America's oil and chemicals are produced.

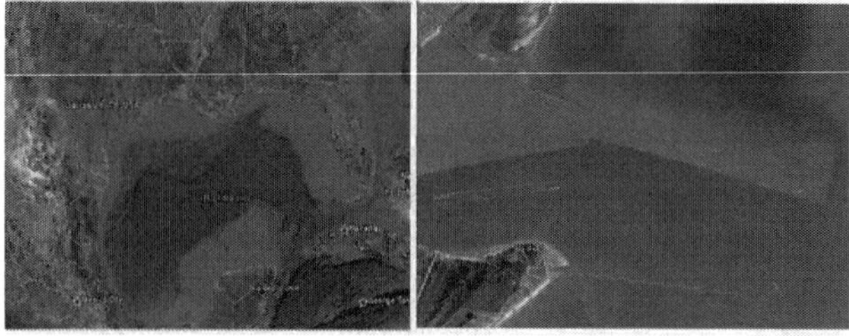

Figure 1: Location and footprint of Galveston jetties

The features of interest on which the choice of case has been based, are first the hurricane-dominated hydraulic climate of Galveston, and second the fact that the structure is located in relatively shallow waters. Both characteristics generate a number of drawbacks in the design process, which allow for a thorough overview of the way that each method deals with them. The existence of hurricanes imposes a lot of uncertainty in the definition of the design load and the design process, which has to be taken into account. The shallow water implies that the structure is attacked by depth-limited waves, which likewise, cannot be disregarded in a design.

Hydraulic climate

The hydraulic processes that take place along the coasts of the Gulf of Mexico are affected in a high rate by the occurrence of hurricanes. Depending on the bathymetry, hurricanes contribute in the occurrence of extreme storm surges and waves, which, in combination with other unfavourable conditions can contribute in the determination of the hydraulic boundary conditions in a particular area. The general hydraulic climate can be described with the following conceptual framework of hydraulic processes.

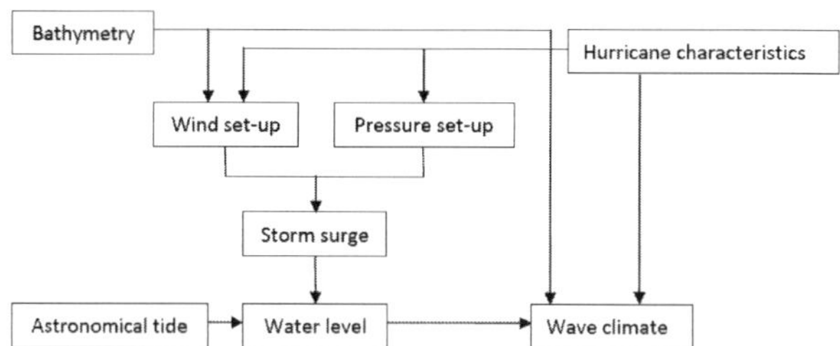

Figure 2: Framework of hydraulic climate (Maaskant, Van Vuren, Kallen, 2009)

Based on this framework, the design conditions can be derived, which are the design wave and water levels at Galveston. In order to determine the aforementioned parameters some local data need to be analysed. The data available for Galveston are some local bathymetric maps, hurricane records for the period 1850-2009, and storm wave data supplied by Argoss[1]

---

[1] www.waveclimate.com

database and Global Wave Statistics (Hogben et al, 1986). From hurricane records, information about hurricane-induced waves is extracted, while from Argoss and Global Wave Statistics information about normal storm waves can be gained. It is noted that except for the action of short waves, the action of swells is also taken into account, whose energy can have considerable effect to the hydraulic load. Based on the records, exceedance probability curves are produced for the various types of waves. The most unfavourable design conditions are indicated by hurricane-induced short waves.

Elaboration of design

The design presented below concerns a rubble mound breakwater, which is assumed to be the optimum solution for Galveston area. The design focuses on the stability of its armour layer. In total three designs are elaborated with three different methods; 1) a deterministic design, 2) a quasi-probabilistic design with a method indicated by PIANC (1992), and 3) a fully probabilistic design with a Monte Carlo simulation. As design equation the stability formula of Van der Meer for plunging waves in shallow waters is used (Verhagen, Mertens 2007). An analytic description of the parameters included in this equation can be found in the references. Apart from the significant wave height $H_{2\%}$, all parameters are consistent in the three designs. The determination of design wave height, which is the main load parameter, is not consistent in the presented methods. This is the key element of difference of the three designs.

Target probability of failure

An economic lifetime for a breakwater is in the order of 50 years; hence a lifetime of 50 years is chosen. During lifetime, a probability of failure equal to 20% is assumed to be acceptable for a structure functioning as outer breakwater, like Galveston jetties. This probability corresponds to a yearly probability of failure equal to 0.4%, and to a return period of design storm of 225 years. In reality the target probability of failure is chosen by means of economic optimization. This issue is out of the scope of this research, hence not elaborated. According to the above failure considerations and the available exceedance curves, the hydraulic boundary conditions can be determined.

Deterministic design

The classical method for designing a breakwater is application of all the common dimensioning rules with the use of deterministic values for all parameters. The wave characteristics derived from hurricane and wave records refer to deep-water conditions. As the jetties of Galveston are located in relatively shallow water, the design wave is limited due to breaking. For this reason the deep-water wave data cannot be used directly in the design, but they need to be transformed to shallow water. This is possible with the use of SwanOne software (Verhagen et.al, 2008). The local tide and storm-surge are included in the input parameters of SwanOne modelling. For these parameters average values are used based on local data. Substituting the correct parameter values into Van der Meer equation, an armour unit of 10-15 tones proves to be appropriate. The type of loading that determines this design is hurricane-induced short waves.

Quasi-probabilistic design (PIANC method)

An alternative method for design of a breakwater is the method of partial safety coefficients, which was worked out by PIANC (1992). This method introduces the use of safety factors for load and resistance in the armour stability formula, which is the equation of Van der Meer. According to the manual of PIANC the applied factors depend on the wave height distribution, which is supposed to be an extreme value distribution. This is in fact the case for deep-water conditions. Provided that no distinction is made concerning application of the method in deep and shallow water, an extreme value distribution of the wave height is assumed to be representative for Galveston as well. Based on this assumption, the safety factors are calculated with the formulae indicated by PIANC. For the parameters of Van der Meer equation, deterministic values are used as indicated for the deterministic design. The local tide and storm-surge are not needed for application of this method. The result is an armour unit of 60 tones. This design is also determined by hurricane-induced short waves.

Compared to the previously presented deterministic design, the method of PIANC results in an extremely larger armour unit. Although it is still not clear which of the two results is more appropriate, it can be already concluded that the latter result cannot be correct, due to the fact that the assumption of an extreme value distribution of the wave height is not correct in shallow waters, where wave breaking takes place. In this case the exceedance curve of the wave height cannot increase infinitely like in deep waters, but there is a point that it becomes constant (figure 2). This variation of the exceedance curve cannot be taken into account with the PIANC method.

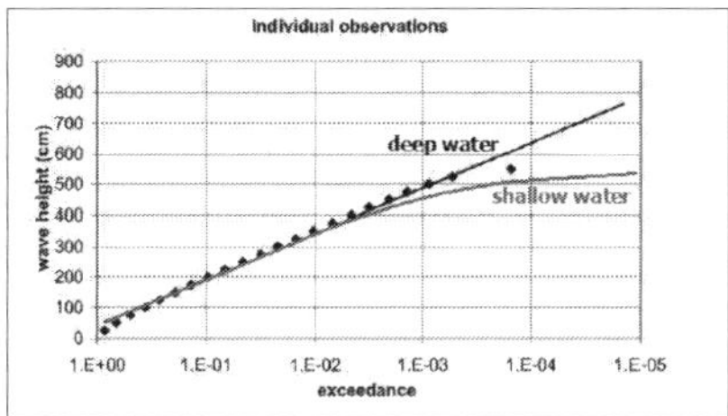

Figure 2: Exceedance curves of wave height in deep and shallow waters

Fully probabilistic design (Monte Carlo simulation)

The application of a fully probabilistic computation is the design method that deals explicitly with all uncertainties. This computation can be performed with a First Order Reliability method or Monte Carlo simulation. All simulations can be elaborated with appropriate MATLAB routines. A Monte Carlo simulation is chosen for Galveston jetties, as it is more accurate than a First Order Reliability method. According to it the results of the previously presented designs can be assessed.

The limit state function is as follows:

$$Z = c_{pl} \cdot P^{0.18} \left(S / \sqrt{N} \right)^{0.2} \left(s_{m-1.0} \right)^{0.25} \sqrt{\cot \alpha} \cdot \Delta d_{n50} - H_{2\%} \tag{1}$$

The wave height $H_{2\%}$ is the parameter with the highest degree of uncertainty in this function. In deep waters the wave height follows an extreme value distribution. In the case of Galveston that the water is shallow, the wave height can be approximated to a function of the water depth h. As water depth h the total depth is considered, i.e. the depth below mean sea level h_d, plus the rise of water level due to tide h_t and storm surge. The surge in shallow water is a function of different parameters depending on the hydraulic conditions that are examined each time. If a hurricane pass is the determining design condition, which is the case for Galveston, the surge is defined as the sum of wind set-up h_w and pressure set-up h_p (equation 2).

$$H_{2\%} = \gamma_b H_{surge} = \gamma_b (h_t + h_p + h_w)\tag{2}$$

The wind and pressure set-up are functions of many other parameters, among which some hurricane parameters exist. The hurricane proves to be sufficiently described by three uncorrelated variables; its speed u, the angle of its trajectory β, and the distance between its landfall and the design spot C. These three variables contribute the highest degree of uncertainty in the process. The wave height ends up being a function of 20 variables, while the limit state function ends up with 28 uncorrelated variables (Tsimopoulou, 2010).

In order to run a Monte Carlo simulation, a probabilistic determination of all variables is necessary. This means that their distributions have to be defined. After thorough investigation of each variable separately, their best-fitted distributions have been concluded. This information comprises the input for running the simulation in MATLAB. The output is a probability of failure for a certain armour unit size, which is represented in the stability formula by the nominal armour unit mass M_{50}. Following a trial and error procedure the armour size corresponding to a probability of failure equal to the target $P_{f,lifetime}=20\%$, can be derived. The appropriate unit size is 48 tones. This is a size that cannot be achieved with rock units, but artificial concrete units should be considered instead. Moreover this result is different from the results of the previously presented designs. By entering the unit masses of the deterministic and PIANC design, probabilities of failure other than 20% are extracted. The result of the trial and error procedure is shown in the graph below.

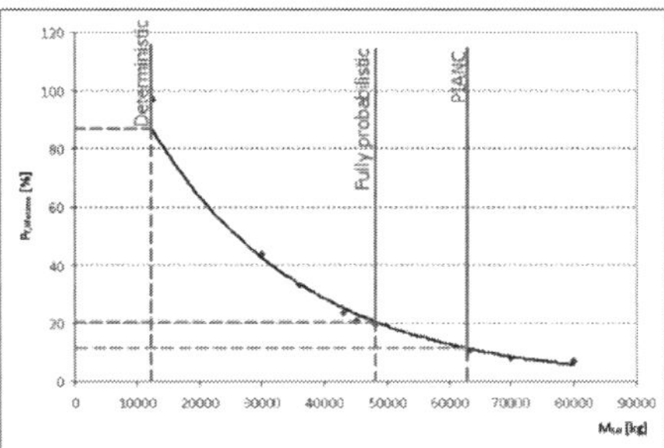

Figure 3: Design results

Assessment of results

In the above graph it is clearly shown that the results of the three elaborated designs are totally different. Considering the substantial differences of the three approaches, such a divergence of results was expected. These differences can be identified in three points: 1) the definition of safety level, 2) the determination of load, and 3) the deep and shallow water considerations. Given the fact that the fully probabilistic approach deals explicitly with the major uncertainties inherent to the design process, it is assumed to be the most accurate one. Based on this assumption the classical deterministic design is not sufficient, as it is almost certain that the structure will fail during lifetime. On the other hand, the design with PIANC method is quite conservative, since the resulted probability of failure during lifetime is about 50% lower than the target. This result was expected, since as mentioned before, the PIANC method gives incorrect results in cases that depth-limited waves determine the design load.

New design approach

In all the above presented design methods there is a contradiction between two equally important qualities; the reliability of the extracted results and the ease of use of the method itself. This contradiction can be the starting point for the development of a design approach, whose objective could be the optimum combination of the two qualities. Based on the elaborated designs, in areas like Galveston, where a high degree of uncertainty is present in the design process, a sufficiently reliable result is only feasible with a probabilistic computation. In order to have an easily applicable probabilistic method, a quasi-probabilistic computation is suggested, with the use of partial safety factors in the design equation. The concept of safety factors is a certainly effective method for designs that involve a lot of uncertainty like breakwater designs, while also very practical and easy to apply. It is commonly used in building codes, such as the Eurocodes. However, in many cases the degree of uncertainty incorporated by the factors is not well defined, leading to incorrect results. The method of PIANC applied in Galveston is an example of a quasi-probabilistic computation with incorrect results. For the case of Galveston a fully probabilistic design has already been presented, in which all uncertainties have been explicitly indicated. Although there are improvements to be made in the model used for the fully probabilistic calculation, it is assumed that its results are sufficiently reliable, and therefore they can be used as a basis for calibration of a new set of safety factors. Calibrating the new factors with respect to the results of Monte Carlo means that the outcome of the new approach will be the same as a fully probabilistic computation. The analytical steps for deriving the safety factors are presented below. It is important to note that the following steps are meant to be elaborated by code-makers rather than designers, while designers are expected to use the extracted safety factors in a proper way in order to achieve reliable designs.

Safety format

In order to derive the safety format, the scope of the new approach needs to be defined. This is to create a reliable and handy set of safety factors, which will cover an important degree of the uncertainty inherent to the physical problem and the design. In order to have a handy tool, the number of safety factors should be reduced to the minimum possible. Its reliability can be maximized if the maximum possible degree of uncertainty is incorporated. The least number of safety factors with which the maximum degree of uncertainty is incorporated is two: one factor for the total load, γ_S, and one for the resistance, γ_R. The stability formula takes the following form:

$$\gamma_R R - \gamma_S S = 0 \tag{3}$$

where R and S are the total load and resistance respectively, and can are defined as follows:

$$S = \gamma_b (h_t + h_d + h_p + h_w) \tag{4}$$

$$R = c_{pl} P^{0.18} \left(S/\sqrt{N} \right)^{0.2} (s_{m-1.0})^{0.25} \sqrt{\cot \alpha} \Delta d_{n50} \tag{5}$$

All the uncertainties covered by the load safety factor are connected with load parameters. There are still some uncertainties related to the load which are not dealt with the load safety factor, the ones inherent to the wave steepness and the number of waves. These parameters are included in the resistance term of the chosen limit state function, and therefore incorporated to the factor of resistance. Also uncertainties associated with the probabilistic model are not incorporated in the safety factors.

Calculation of load safety factor

The safety factor for load is defined as the ratio of the design load S^* over the characteristic load S^k:

$$\gamma_S = S^* / S^k \tag{6}$$

The design load can be calculated with the fully probabilistic model. In particular, based on information used as input in Monte Carlo, a new fully probabilistic calculation of the total load can be elaborated in MATLAB with a new Monte Carlo simulation. The outcome is an exceedance curve of the load, which is supposed to be the one with the highest possible accuracy, as all uncertainties have been incorporated in a satisfactory way. Therefore, they can be used as design values, while a set of satisfactory results of a semi-probabilistic calculation should converge to the outcome of this simulation.

The limit state function for the new simulation is the following:

$$Z = S - \gamma_b (h_t + h_d + h_p + h_w) \tag{7}$$

Where S = total load, while the rest of the parameters are already known. There are in total 12 variables in the above function. For every particular deterministic value of the total load, a probability of failure is extracted, which is in fact the probability that the Z-function becomes negative. By giving various deterministic values to the total load, a design exceedance curve is created.

If in the above simulation some of the variables are replaced by deterministic values, the outcome will be a different exceedance curve. This difference is indicative of the degree of uncertainty inherent to those particular variables, which is supposed to be incorporated by the safety factors. Using deterministic values for all the load variables, the total load becomes deterministic too, and the exceedance curve turns to be a straight line parallel to the x-axis. In order to come up with a line that represents the characteristic exceedance curve of the total load, all chosen deterministic values of the variables need to be their characteristic values. Since there is no standard rule for the choice of characteristic values, but they vary in different designs depending on the overall design approach, a choice for all the load variables is necessary, which can be reasonably substantiated. The most commonly used choices for characteristic values are either mean variable values or values with probability of non-exceedance equal to 95%. For this project mean values are chosen for the majority of variables, in accordance to PIANC (1987). It should be noted that the choice of characteristic values is not critical for the final design. A different set of characteristic values would result in a different set of safety factors; the same degree of uncertainty would always be incorporated though.

The design and characteristic exceedance curves are presented in figure 4. Using values of this graph, the load safety factor can be easily derived from equation 6. This factor accounts for the uncertainties that are neglected when the design parameters take deterministic values, and literally constitutes a measure for the divergence between the characteristic and design exceedance curves.

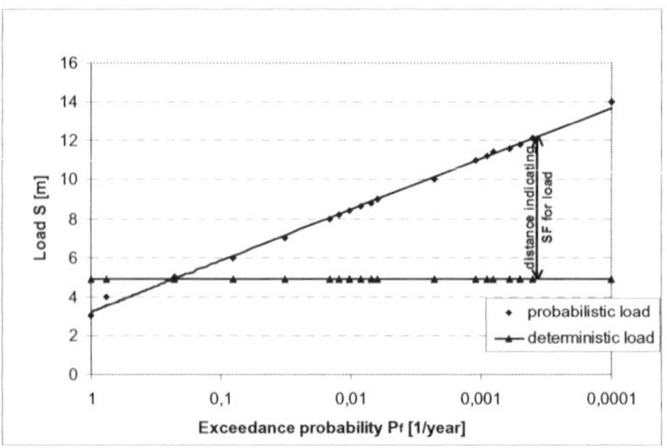

Figure 4: Design and characteristic exceedance curves

Calculation of resistance safety factor

The safety factor for resistance is meant to incorporate uncertainties inherent in the resistance parameters of the stability formula. In this case, the concept of exceedance curves makes no sense, as all resistance parameters are related to design and construction, and therefore their degree of randomness is a matter of choice. For this reason, the resistance factor is derived with an iterative process for certain discreet probabilities of failure, once the design hurricane and load factor are specified. This process is summarised in the following steps.

1. Calculation of characteristic load S^k, using the design hurricane values for all hurricane parameters, i.e. u, β, C, and mean values for the remaining load parameters.
2. Calculation of design load with equation 6, using the selected load safety factor γ_S.
3. Application of Van der Meer stability formula for calculation of characteristic nominal armour unit mass $M_{50}{}^k$, using mean values for all resistance parameters. The definition of mean values is the same as in step 1.
4. Validation of process through a Monte Carlo simulation. The limit state function is equation 1. All resistance input parameters are deterministic, $M_{50}{}^k$ for the nominal unit mass and mean values for the remaining parameters. The load parameters are inserted as random variables, with their own distributions, as determined before. The extracted probability of failure has to be equal to the target probability of failure.
5. Actual derivation of resistance safety factor with a trial and error procedure, applied in a Monte Carlo simulation, with limit state function corresponding to equation 3. All resistance and load parameters are inserted as random variables, except for the nominal mass, which takes the characteristic value calculated in step 3. The safety factor γ_S is inserted as constant and its value is already known, while the resistance safety factor γ_R is the trial parameter. Starting with $\gamma_R=1$, a probability of failure is calculated, which is higher than the target probability of failure. The simulation is repeated with gradual increase of γ_R until the target probability of failure is reached.

Once the resistance safety factor is derived, the design nominal mass is calculated as follows:

$$M_{50}{}^* = \gamma_R M_{50}{}^k \tag{8}$$

Sensitivity analysis

The final values of the safety factors depend to an important extend on the designers' choice of characteristic values for the various variables. Through a first order reliability method simulation in MATLAB, it is concluded that the probability of failure of the structure is determined in a high degree by two load variables, the hurricane speed u, and the distance between the hurricane landfall and the design spot C. As both parameters are connected with the hurricane, their variation is very high. For this reason a sensitivity analysis is performed and different values of load safety factors are extracted for the various values of the two parameters. Some indicative results of this analysis are presented in the graphs of figure 5.

According to the graphs the load factors increase as the target failure probability becomes lower. It is also perceptible that for higher hurricane speed the load factor decreases, meaning that when a higher design hurricane is used then a lower safety factor is needed. Both conclusions were expected. From the variations of parameter C, the most unfavourable hurricane landfall can be concluded, which is the one requiring the highest safety factor. All curves are maximized for C=-5000, which corresponds to a hurricane with landfall 5 kilometres to the south of Galveston jetties.

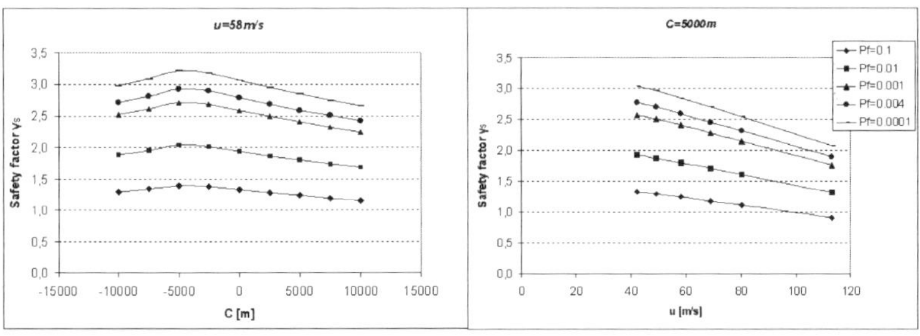

Figure 5: Load safety factor for variations of hurricane parameters

Concluding remarks

Based on this research a number of concluding remarks have been drawn and are summarized below with respect to the various issues that have been examined.

Design methods

- In a deterministic design uncertainties are only dealt implicitly, as the choice of all design parameters is un-prescribed and relies only on engineering judgement. Therefore the assigned safety level is not clearly defined; hence there is always a risk of insufficient designs results.
- The design of PIANC results in extremely high values for the armour unit mass. This is due to the fact that depth-limitations in the wave height are not taken into account in the method, and therefore incorrect results are extracted in case of shallow water designs, like Galveston jetties.
- A fully probabilistic computation deals explicitly with the majority of uncertainties inherent to the design, and therefore it can be considered the most reliable design method.

Suggested design approach

- The new approach is a quasi-probabilistic design method. The main accomplishment of the new development is that an easy to apply quasi-probabilistic method results to equally reliable designs with a fully probabilistic method, as it incorporates the same uncertainties.

- The procedure for deriving the new safety factors is generic and can be used as a guideline for code-makers.

Safety factor values

- The derived safety factors are site-specific, because a site-specific probabilistic model is used for their calibration, which is the model created for the fully probabilistic design of Galveston jetties. A generic set of safety factors should be appropriate for use in a greater area than just Galveston, e.g. the entire Gulf of Mexico. Hence a generic use of the safety factors requires incorporation of results of a number of different case studies.
- The load safety factor is basically a function of hurricane parameters. If a different design hurricane is used in a design, then the safety factors take different values.

Recommendations

- In all stages of the performed analysis a number of simplified assumptions have been made, many of which have not been validated. As a consequence, the reliability of the extracted results and the follow-up conclusions can be questioned. An optimization of the total analysis is therefore necessary, which can be achieved through reconsideration of all weak points that limit the value of the overall outcome.
- In order to derive safety factors that can be used for generic breakwater design more case studies need to be conducted. As different locations have different hurricane characteristics and bathymetry, the derived safety factors will have different values. The result of different case studies will be a scatter of safety factors. Subsequently, one safety factor suitable for a greater area than just Galveston could be chosen.
- A validation of the performed analysis is necessary. This can be done with its application in a new location with similar hydraulic features, such as the coast of Vietnam.
- Based on the new sets of safety factors, a guideline for future designers could be developed, containing indication of safety factor values in different locations.

References

CIRIA, (1977), "CIRIA report 63: Rationalisation of safety factors in structural codes", CIRIA, London, pp 18-20, 35-36

Hogben, H., Dacunha, N.M.C., Olliver, G.F. 91986) Global wave statistics, Urwin Bros, UK

Kanning W. (2005), Safety format and calculation methodology slope stability of dikes, MSc thesis, Delft University of Technology

Maaskant B., Van Vuren S., Kallen M.J. (2009), Typhoon-induced hydraulic boundary conditions in Vietnam, HKV

PIANC (1987), Risk analysis in breakwater design, report of MarCom subgroup 12c

PIANC, (1992), Analysis of rubble mound breakwaters", WG MarCom12, pp 21-28

Verhagen, H.J., Van Vledder, G., Eslami Arab, S. (2008) A practical method for design of coastal structures in shallow water, proc. ICCE 2008, Hamburg, pp. 2912-2922

Verhagen H.J., Mertens M. (2009), Riprap stability for deep water, shallow water and steep foreshores, proc. ICE breakwaters conference, Edinburgh

Tsimopoulou V. (2010), Probabilistic design of breakwaters in shallow hurricane-prone areas, MSc thesis, Delft University of Technology

Innovative Coastal Zone Management
ISBN 978-0-7277-5749-4

ICE Publishing: All rights reserved
doi: 10.1680/iczm.57494.557

The Use of Alternative Tropical Hardwood Timbers in the Marine Environment

Phil Welton, Environment Agency, Newcastle-upon-Tyne, UK
Melanie Meaden, Environment Agency, Cardiff, UK
John Williams, TRADA Technology, UK
Jonathan Simm, HR Wallingford, UK

Summary

A research project commissioned by the Environment Agency on lesser used species of hardwood timber has generated technical data which will allow engineers to use a wider range of timbers for marine construction, thereby supporting more sustainable forestry on a global scale.

Introduction

Within the UK coastal engineering industry there is a demand for strong, durable, cost-effective and environmentally acceptable construction materials. The marine environment is challenging for all construction materials but timber suffers remarkably little from the effects of the salt content of seawater compared to, for example, concrete and steel. In addition, the resilience of timber and its favourable strength to weight ratio, and the relative ease of fabrication and repair, make it an attractive construction material for engineers to design with, and specify. Timber is also a natural material and a sustainable option if sourced from a sustainably managed forest.

Dense tropical hardwoods are extensively used in maritime structures having the necessary strength, section sizes, lengths and durability, including resistance to abrasion and to attack by marine borers. Timber has the ability to withstand dynamic loads such as those caused by wind, waves or vessels better than most other structural materials and the natural resilience of timber has considerable benefits in coastal and river situations where impact loads often produce the highest stresses.

The two best known structural timbers are perhaps Greenheart or Ekki, although a few other dense tropical hardwoods such as Yellow Balau, Opepe and Jarrah have also been used successfully. These have the advantage over softwood species of much higher resistance to physical abrasion and damage, and to biological degradation by marine borers.

Specifiers and engineers have a key role in making informed decisions on the type of materials to be used in the schemes they design and construct. Technical requirements must be balanced with cost and environmental considerations, and in the case of timber these considerations must include where the timber comes from and the forestry practises adopted in the country of origin.

Illegal logging and unsustainable forest management are now recognised as a global problem. Forests are a precious natural resource, and their destruction has wide-ranging social, economic and environmental impacts. Since 1 April 2009 the UK Government's timber procurement policy requires central government departments, their executive agencies and non-departmental public bodies to procure timber and wood derived products originating from either independently verifiable legal and sustainable or FLEGT (Forest Law Enforcement, Governance and Trade) licensed or equivalent sources. The policy requires evidence of chain of custody (movement of the timber through the supply chain from the forest source to the end product or user) and that the source of the timber is legally and sustainably managed or FLEGT-licensed.

In relation to the long-list of candidate timbers assessed as part of this research project, all can be sourced with evidence of legality, sustainability and chain of custody (FSC certification).

The Problem

Tropical hardwood timbers used in the marine environment are generally selected from a restricted number of species with proven track records of in-service performance, particularly Greenheart and Ekki.

Over-reliance on this small number of timber species is exerting disproportionate environmental pressure on these species in their forests of origin. Potentially suitable timbers are often discarded during the sourcing of these commercially valuable timbers leading to environmental damage in the forest far greater than the loss of the timber being sought. This is not compatible with sustainable forest management in the longer term. Over-dependence on a few species is also likely to cause security of supply and inflationary price pressures in the future. Taking a holistic view of the timber trade, this makes profitable forestry and sustained yield management increasingly difficult to achieve.

Mr Herbert Reef from Transformation Reef Cameroon (TRC) commented (abridged):

"If you have an FSC forest you [can only] explore [one thirtieth of the forest] in [any] one year and than you have to wait 30 years to come back in that area. So it is very important to take out under the FSC rules as much timber as possible which can be sold in the market for reasonable prices [to ensure that the forest is as commercially viable as possible]. That's why it is very important that the users of wood take a wide range of species and not only one species."

In the marine construction industry there is a tendency to over-specify the technical properties of timber needed for a required end use, particularly if strength is critical. This means that timbers such as Greenheart and Ekki are often selected by default rather than on the basis of sound engineering judgement. The industry is typically conservative and there is a general reluctance to specify timber species without a proven track record. This is understandable as often the material cost of timber in a construction scheme is dwarfed by the overall construction cost. However, it means that this deadlock can only be overcome if reliable data on the performance of lesser used species of timber are obtained.

Research into Lesser Used Species (LUS)

The Environment Agency is the leading public body responsible for protecting and improving the environment in England and Wales. The Environment Agency is also a major consumer of timber for marine construction and has significant influence throughout this industry sector with its strategic overview role of coastal defence. The Environment Agency itself spends between £1 million and £3 million on timber in any year depending on the extent of flood risk management activity occurring. Concerns within the organisation regarding the wider environmental impacts from over-reliance on Greenheart and Ekki led to the desire to widen the inventory of suitable timber species for marine construction.

The Environment Agency, working in partnership with TRADA Technology and HR Wallingford commissioned a research project with the objective of identifying a range of lesser used timber species available from certified legal and sustainable sources that have the potential for use in marine construction. Other project partners included British Waterways, The Crown Estate, CETMEF (Centre d'Etudes Techniques Maritimes Et Fluviales) and VolkerStevin Ltd. Support, technical knowledge and supplies of timber were provided by Ecochoice UK Ltd and Aitken & Howard Ltd.

The key material attributes for hardwood timber for use in marine construction are resistance to attack by marine borers (shipworm and gribble), abrasion resistance and strength (Williams et al 2004A), so obtaining data on these attributes was the focus of the research project.

The research project was split into three stages:

Stage 1

A desk based study to identify lesser used species of timber which may merit further investigation (known as the 'long list') based on a review of previous research, existing literature and database reference sources. See Table 1 below.

Table 1: The long list of lesser-used candidate species

Commercial name	Region of Origin	Commercial name	Region of Origin
Angelim vermelho	S. America	Mukulungu	W. Africa
Basrolocus	S. America	Niove	W. Africa
Cloeziana	S. Africa	Okan (denya)	W. Africa
Cupiuba	S. America	Piquia	S. America
Dabema (dahoma)	W. Africa	Sapucaya	S. Amercia
Evuess (kruma)	W. Africa	Souge	W. Africa
Garapa	S. America	Tali	W. Africa
Massaranduba	S. America	Tatajuba	S. America
Mora	S. America	Timborana	S. America

Stage 2

Fast-track novel laboratory screening trials to assess the potential marine borer resistance and abrasion resistance of the long list of timbers, together with a marine exposure trial. Marine borer resistance focussed on the crustacean *Limnoria* spp. (gribble) and the mollusc *Teredo* spp. (shipworm).

Novel, fast track screening trials were chosen because guidelines in BS EN 275, which explains how species for use in marine construction can be evaluated, specifies a 5-year test period. This is too long a period for screening tests to be economically viable. Moreover, this standard does not provide a route by which abrasion resistance can be determined. Resistance to marine borers and abrasion was therefore established by using test methods developed by Borges *et al* (2003) Williams *et al* (2004B) and Sawyer and Williams (2005).

Stage 3

Testing to determine the strength properties of five timbers on the long list selected on the basis of the findings of Stage 2 and various other considerations (e.g. cost, availability within project timescales). The test material was selected from commercial stocks of timber and was visually strength graded to HS grade in accordance with BS 5756:2007.

Strength properties were determined by the requirements of BS EN 384: 2004. The programme of tests determined the bending strength, stiffness, density and moisture content of graded timber. Testing was conducted in accordance with BS EN 408: 2003, and the timber species were allocated to their appropriate strength classes as detailed in BS EN 338: 2003.

In each stage of the project, the performance of the lesser used species was compared to that of Greenheart and Ekki, which were used as benchmark timbers. In addition, comparisons were made with a number of reference timbers which also have a track record of being used in marine construction projects, e.g. Opepe.

Research results

Marine borers - gribble

The data in Figure 1 presents average faecal pellet production rate per day (*pellets/d*) over a 28 day period for candidate species compared to the Greenheart and Ekki benchmark timbers.

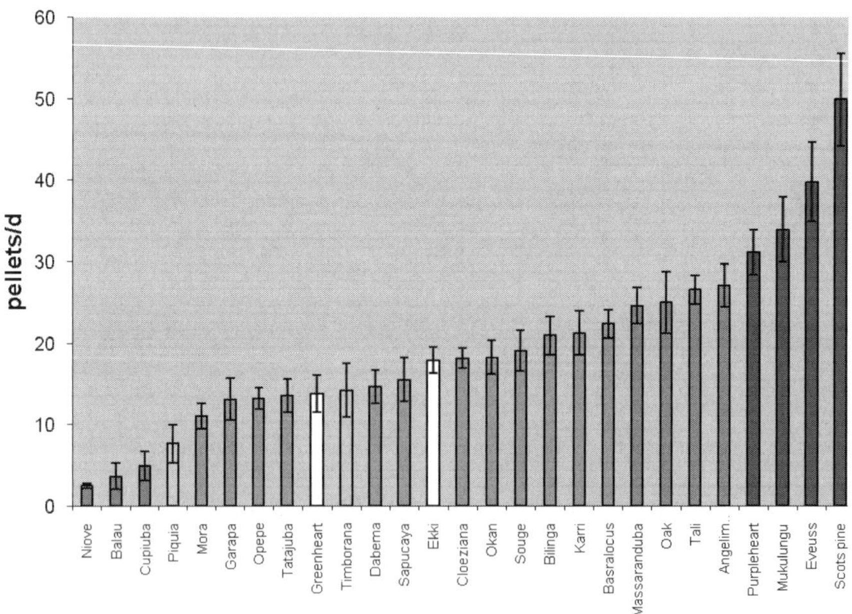

Figure 1: Comparison of daily feeding rates for compared to Greenheart and Ekki.

Higher faecal production represents a higher feeding rate and hence a lower resistance of the timber to attack. Significantly lower feeding rates than those recorded on both benchmark timbers (Greenheart and Ekki) are shown in dark green and feeding rates lower than Ekki alone are shown in pale green. Significantly higher feeding rates than both benchmark timbers are shown in red and higher than Greenheart alone are shown in pink.

Marine borer - shipworm

To test the resistance of the timbers to shipworm, a marine exposure test site was set up in Olhão harbour on the Ria Formosa lagoon, Portugal; where there is known habitat for aggressive marine borer populations. The marine exposure trial lasted for 18 months, from March 2008 to September 2009.

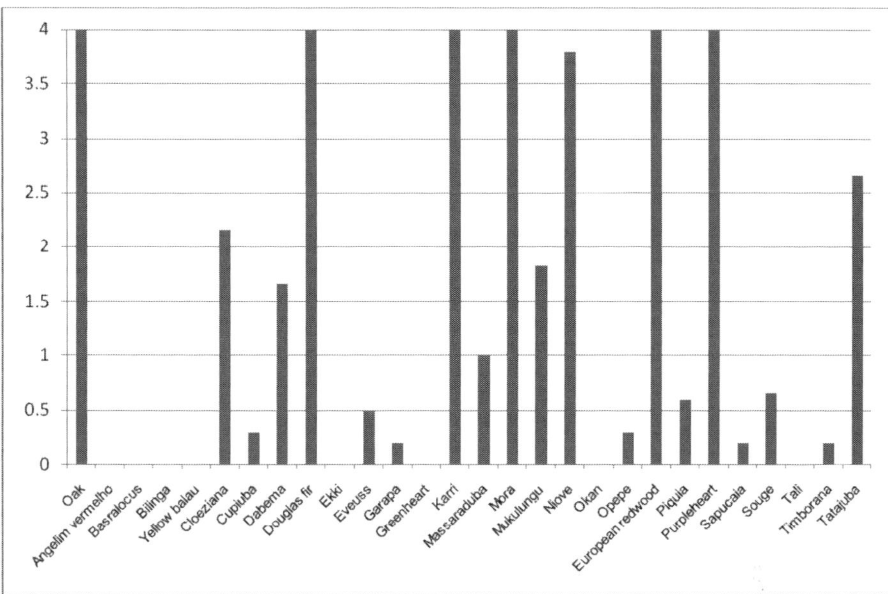

Figure 2: Performance of timbers on the long list after 18 months marine exposure, compared to Greenheart and Ekki.

The data in Figure 2 presents the comparative resistance to attack by shipworm in a marine exposure trial lasting 18 months where 0 equates to no attack and 4 equates to severe attack.

Abrasion Resistance

The effects of abrasion in marine trials are difficult to study owing to the extreme variability of weather and site conditions. The data in Figure 3 presents volume loss after 160,000 cycles, or around 1350 tides, for the candidate species compared to Ekki and Greenheart.

Significantly less volume loss (abrasion) than that recorded on Greenheart is shown in pale green. Significantly higher volume loss than both benchmark timbers are shown in red and higher than Ekki alone are shown in pink.

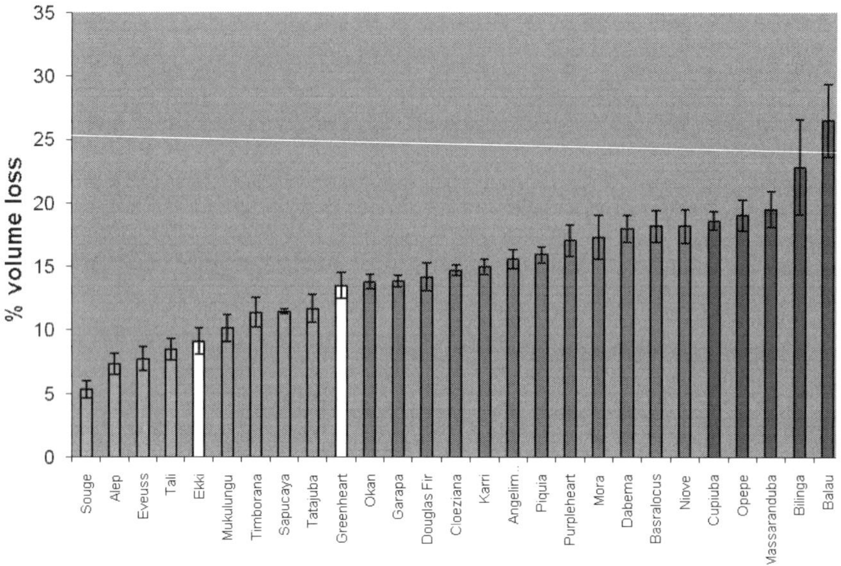

Figure 3: Average volume loss after 160,000 cycles for timbers on the long list compared to Ekki and Greenheart

Strength properties of the five selected species

Five timber species from the long list of candidate species were selected for strength testing on the basis of their performance in the initial trials and commercial factors.

These have been allocated to strength classes, as detailed in Table 2, on the basis of test results for bending strength, bending stiffness and density. The strength classes provide values that may be used in design for all the necessary mechanical properties. Figure 4 shows a section of a lesser used species being tested for strength.

Figure 4 – Strength testing

Table2 – Characteristic values as obtained from test data using BS EN 384 for five HS grade (BS 5756) lesser used species of timber

Species	Strength Class	Bending Strength	Modulus of Elasticity		Density	
		$f_{m,k}$ (N/mm²)	$E_{m,mean}$ (N/mm²)	$E_{m,05}$ (N/mm²)	ρ_{mean} (kg/m³)	$\rho_{k,05}$ (kg/m³)
Angelim Vermelho	D60	60.4	22084	18551	1082	1012
Cupiuba	D50	53.1	21414	17987	822	729
Eveuss	D50	51.0	20998	17638	1019	981
Okan	D40	47.3	19318	16227	998	898
Tali	D35	40.5	17200	14448	815	672

Discussion

The research has proven that there are at least five lesser used species of timber with suitable properties for use in critical marine construction applications, where a strength class for design purposes is required. These timbers, along with their properties, are shown in Table 3. For comparison purposes, data is also provided for Greenheart and Ekki and other reference timbers such as Opepe.

Table 3 – Performance of five HS grade (BS 5756) lesser used species of timber benchmarked against Greenheart and Ekki

Species	Resistance to fungal decay	Resistance to shipworm	Resistance to gribble	Bending Strength		Abrasion
		(benchmarked against Ekki/ Greenheart)	(benchmarked against Ekki)	$f_{m,k}$ (N/mm²)	Strength class	Benchmarked against Greenheart
Benchmark species						
Greenheart	Very durable	Resistant	Resistant	70.0	D70	Resistant
Ekki	Very durable	Resistant	Resistant	70.0	D70	Better
Lesser used species						
Angelim Vermelho	Very durable	Comparable	Comparable	60.0	D60	Comparable
Cupiuba	Durable	Comparable	Better	50.0	D50	Worse
Eveuss	Very durable	Comparable	Worse	50.0	D50	Better
Okan	Very durable	Comparable	Comparable	40.0	D40	Comparable
Tali	Very durable	Comparable	Comparable	35.0	D35	Better
Reference species						
Yellow Balau	Durable	Comparable	Better	70.0	D70	Worse
Opepe	Very durable	Comparable	Comparable	50.0	D50	Worse
Karri	Durable	Worse	Comparable	50.0	D50	Comparable
Oak	Durable	Worse	Comparable	30.0	D30	Better
Douglas Fir	Moderately durable	Worse	N/a	24.0	C24	Comparable
Purpleheart	Durable	Worse	Worse	N/a	N/a	Comparable

In addition to the five lesser used species where strength information has been obtained, there are a further 13 species where data on their resistance to marine borers and abrasion are known. This allows an even greater choice of timbers to be used where strength class is not critical but where marine borer and/or abrasion resistance is required. Data for these 13 timbers is shown in Table 4.

Table 4 – Performance of thirteen lesser used species of timber benchmarked against Greenheart and Ekki

Species	Resistance to fungal decay	Resistance to shipworm (benchmarked against Ekki/ Greenheart)	Resistance to gribble (benchmarked against Ekki)	Mean density (kg/m³)	Resistance to Abrasion Benchmarked against Greenheart
Benchmark Species					
Greenheart	Very durable	Resistant	Resistant	1080	Resistant
Ekki	Very durable	Resistant	Resistant	1080	Better
Newly tested species					
Basralocus	Very durable	Comparable	Comparable	720 - 840	Comparable
Cloeziana	Durable	Worse	Comparable	No data	Comparable
Dabema	Durable	Worse	Comparable	600 - 840	Comparable
Garapa	Durable	Comparable	Comparable	840 - 960	Comparable
Massaranduba	Very durable	Comparable	Comparable	960 - 1080	Worse
Mora	Durable	Worse	Comparable	960 – 1080	Comparable
Mukulungu	Durable	Worse	Worse	840 – 1080	Comparable
Niove	Durable	Worse	Better	840 - 1080	Comparable
Piquia	Durable	Comparable	Better	840 – 1080	Comparable
Sapucaia	No data	Comparable	Comparable	No data	Comparable
Souge	Moderately durable	Comparable	Comparable	840 - 1080	Better
Tatajuba	Durable	Worse	Comparable	720 – 840	Comparable
Timborana	Durable	Comparable	Comparable	960 -1080	Comparable

As a result of the research project, significantly more data is now available on the properties of a range of alternative tropical hardwood species that are suitable for use in marine and freshwater construction.

Specifying lesser used timbers

The promotion of lesser used species requires an approach to specifying timber based on key risk parameters (i.e. service requirements and service hazards) to ensure that these timbers are considered alongside more common tried and tested timbers such as Greenheart and Ekki. Timber should be specified on the basis of an understanding of the most significant hazard(s) encountered in service, i.e. resistance to marine borers, abrasion etc.

Where structural performance is critical, the five lesser used species identified in Tables 2 and 3 may be considered as alternatives to Greenheart and Ekki, provided their resistance to marine borer attack and abrasion, where relevant, and their strength properties meet project requirements. These timbers have been allocated to strength classes that may be used in design for all the necessary mechanical properties.

If structural performance is not critical, the longer list of thirteen lesser used timber species, as detailed in Table 4, can be considered.

Although it is recommended that the functional performance of a timber and its ability to withstand the most dominant site-specific hazards should drive the specification of timber species, it is recognised that other factors such as availability within project timeframes, cost and required section sizes may also influence the decision making process. Early consideration of the requirements is essential to avoid self-imposed barriers to specifying lesser used species. Ultimately, the specification of suitable timber requires consideration of a wide range of factors including technical requirements, environmental and commercial considerations.

Outcomes and Conclusions

Whilst none of the five lesser used species tested at Stage 3 (Angelim Vermelho, Cupiuba, Eveuss, Okan and Tali) timbers quite matched Greenheart or Ekki for strength, some came close with a strength class of D60 compared to the D70 strength class of Greenheart and Ekki. All reasonable testing on these five timbers has now been undertaken and their key material attributes are known so they can be used with confidence in marine and freshwater construction application where strength is critical.

In terms of the five lesser-used species' resistance to the marine environment, all were comparable with Greenheart and Ekki in their resistance to shipworm and all but one were comparable or better in their resistance to gribble. In addition, all but one were comparable or better in their resistance to abrasion and all but one were assessed as being very durable in their resistance to fungal decay (the other being durable).

In addition to the comprehensive data now available for Angelim Vermelho, Cupiuba, Eveuss, Okan and Tali, reliable data on marine borer resistance and abrasion for the other thirteen timbers on the long list is also available. Although their strength class has not been determined, they can still be used with confidence in applications where strength is not critical.

By generating technical data, this research has overcome the principal obstacle associated with specifying lesser used species, namely the absence of reliable data. A soon to be published User Guide will include a step-by-step methodology to help designers, engineers and specifiers identify lesser used species that are suitable for different project applications. By taking a risk-based approach to the specification of timber species, design and procurement decisions can help support more sustainable forestry on a global scale. At the project level, being able to specify a range of timbers gives engineers the flexibility to choose more available and potentially cheaper timbers while having confidence in their properties in the marine environment

References

Meaden, M, Williams, J and Simm, J. (2011) Alternative hardwood timbers for use in marine and freshwater construction – Science Report SC070083. Environment Agency, Rio House, Waterside Drive, Aztec West, Almondsbury, Bristol, BS32 4UD

Research summary Paper. (2010) Assessing the durability and engineering properties of lesser known hardwood timber species for use in marine and freshwater construction. TRADA Technology Ltd.

Williams *et al* (2010). Assessment of the durability and engineering properties of lesser known hardwood timber species for use in marine and freshwater construction

Borges, L. M. S., Cragg, S. M. and Williams, J.R. (2003) Comparing the resistance of a number of lesser known species of tropical hardwoods to the marine borer Limnoria using a short term laboratory assay. International Research Group on Wood Preservation IRG/WP 03.

Sawyer, G.S. and Williams, J.R. 2005. An investigation to assess the feasibility of developing an accelerated laboratory test to determine the abrasion resistance of lesser-used timber species for use in marine construction. International Research Group on Wood Preservation IRG/WP 05.

Williams, J. R., Sawyer, G.S., Cragg, S. M. And Simm, J: 2004A. A questionnaire survey to establish the perceptions of UK specifiers concerning the key material attributes of timber for use in marine and fresh water construction. International Research Group on Wood Preservation IRG/WP 04.

Williams, J.R., S.M. Cragg, L. M. S Borges and J D Icely: 2004B. Marine exposure assessment of the natural resistance of a number of lesser known species of tropical hardwoods to teredinid and limnoriid borers. International Research Group on Wood Preservation, Document No. IRG/WP 04.

Innovative Coastal Zone Management
ISBN 978-0-7277-5749-4

ICE Publishing: All rights reserved
doi: 10.1680/iczm.57494.567

Leigh Creek Realignment Feasibility Study

Louise Trim, Halcrow Group Ltd., Chichester, UK.
Nigel Pontee, Halcrow Group Ltd., Swindon, UK.
Richard Atkins, Southend-on-Sea Borough Council, Southend, UK.

Introduction

Two Tree Island lies to the south west of Southend within the Thames Estuary and is separated from the mainland by a stretch of water known as Leigh Creek. Two Tree Island is currently a Country Park and a Local Nature Reserve and lies within an area of great environmental importance, known for its populations of wildfowl and wading birds and commercial shellfisheries. The tidal mudflats surrounding Two Tree Island are designated as a Special Area of Protection (SPA), Ramsar site and National Nature Reserve. However, Two Tree Island has a complex past with a legacy of contamination that presents a current threat to human health, ecosystems and amenity.

In recent years, Southend Borough Council has become concerned regarding the condition of the embankments on the north-eastern frontage of Two Tree Island. Tidal flows through the southern branch of Leigh Creek have been exerting erosion pressures on this frontage resulting in a risk of undermining. In 2005, work was undertaken to install a proprietary revetment system to protect the embankment but continued pressure exerted from the meandering of the creek to the south is continuing to reduce the stability of the defence. Collapse of the revetments and the resulting erosion could potentially lead to the release of leachate and contaminated waste materials from the formerly landfilled areas of the island.

The leaching of harmful contaminants into the Thames Estuary may result in a local exceedance of water quality standards and would be detrimental to the condition of the SPA. Under the Habitats Regulations, the UK Government and its agencies, including SBC, have a legal obligation to maintain the conservation value of the site.

This paper outlines the approach and findings of a feasibility study completed to examine the viability and sustainability of realigning the position of the southern branch of Leigh Creek to reduce the risk of further damage to the defences protecting the contaminated lands of Two Tree Island.

Background information
Study Area

Leigh Creek separates Two Tree Island from the mainland at Leigh-on-Sea in the outer Thames Estuary (see Figure 1). At the eastern end the creek diverges into the two channels. The main southern branch of the creek which runs along the northern boundary of Two Tree Island and a northern branch of the creek which has been partially closed by siltation. To the south-west Two Tree Island is separated from Canvey Island by a broader stretch of water known as Hadleigh Ray.

Figure1: Photograph of Leigh Creek, near Southend-on-Sea. Source: Google Earth.

Two Tree Island was reclaimed from an area of saltmarsh in the 18th century when embankments were built around the perimeter. It was used for rough grazing until around 1910 when a sewage treatment works was built on the eastern tip. In 1936, Southend Borough Council (SBC) acquired the whole island and it was used as a rubbish tip until the late 1970s, when the activities ceased. The eastern half of the island, together with the adjoining salt marshes and a larger area of mudflats, was the first part of the island to be protected as an open space when, in 1974, SBC rejected proposals for commercial development. The western half was soon to follow. Two Tree Island has been enjoyed by the public and wildlife enthusiasts since it was partially restored in the 1990s.

Two Tree Island lies within the internationally designated Benfleet and Southend Marshes Special Protection Area (SPA), Ramsar site and Site of Special Scientific Interest (SSSI) and within the Leigh National Nature Reserve. The adjacent saltmarsh and mudflat habitats within the Benfleet and Southend Marshes SPA and Ramsar site represent one of the most important wildfowl and wading bird areas in Essex providing feeding and roosting opportunities for internationally important numbers of waterbirds in winter and during passage periods.

Leigh has a long history of boatbuilding and fishing. The fishing industry reached its peak in the 18th and early 19th centuries but significantly declined in the latter part of the 19th century when the deep water of Leigh Creek silted up. The silting up of Leigh Creek also meant that large ships could no longer get into Leigh. This led to a sharp decline in its fortunes, which was slightly abated by fishing, but not reversed until the arrival of the railway.

Oysters appear to have been the foundation of much of Leigh's early prosperity. The cockling that Leigh is known for today took over in the 20th century. There are a number of shellfisheries in the Thames estuary adjacent to and downstream of Two Tree Island. However, navigation access to the quay sides at Old Leigh has deteriorated over the last few decades due to siltation, which significantly restricts the extent of commercial fishing currently undertaken here.

The area is also important for recreational boating. There are several sailing and water sports clubs with a number of moorings within Leigh Creek and a number of others along the Southend coast of the Thames Estuary. There is also a boat repair yard at Leigh.

Study Objectives

The main objective of the creek diversion is to reduce erosion of the northern revetment protecting the landfill site on Two Tree Island to reduce the risk of contamination of Leigh Creek and the surrounding environment. SBC was also keen to determine whether additional benefits could be obtained through the scheme, namely:

- Improved access for both commercial and recreational boating interests along the Leigh frontage. Significant siltation of the northern branch of Leigh Creek, has constrained access to the boat repair yard and landing facilities for the local shell fish and fishing industry. Creek diversion, and possible deepening, could offer a significant benefit to navigation for fishermen using Leigh Quay that may bring benefits to the local fishing industry.

- Creating additional environmental benefits by transforming the area of the existing creek course to saltmarsh. Realigning Leigh Creek offers opportunities to establish rare habitats and to enhance the Benfleet and Southend Marshes SPA, Ramsar site and SSSI.

The purpose of the study was to determine whether permanent, stable and sustainable diversion of the creek and creation of neighbouring saltmarsh are technically achievable and what conditions would be imposed on the Council by the environmental interests in the area in order to achieve acceptance.

Scope of study

The scope of the feasibility study involved the investigation of a number of issues that were seen as possible show stoppers. The scope of the study included consultation with key parties including Natural England and the Environment Agency, local user groups and businesses; the completion of a contamination sampling survey of the Leigh Creek mudflats to inform disposal costs for dredged material; discussions with contractors regarding appropriate construction methods and costs; and the development of a Digital Terrain Model (DTM) of the site to inform dredging volume costs and habitat development maps.

Halcrow proposed to complete the technical feasibility assessment using information based on desk studies and a detailed review of the geomorphological history of the site. This approach allowed key factors that could represent 'show stoppers' (e.g. contamination, consents and approvals) to be investigated, and an enhanced view of the viability of the scheme based on existing data to be determined, ahead of undertaking costly numerical modelling. Furthermore, if the study indicates that the scheme is potentially feasible, any numerical modelling that is undertaken can be better focussed.

Geomorphological Setting

A comprehensive review of the environmental conditions that exist within Leigh Creek and the Thames Estuary, including a review of the hydrodynamic climate surrounding Leigh Creek and a summary of the geomorphological evolution of the area, was undertaken in order to provide an understanding of the processes at work in the vicinity of the study area to inform option development and design.

Thames Estuary

The Leigh to Southend frontage lies within the outer Thames estuary. In terms of hydrodynamic forcing the main features of the study area are:

- The predominant waves come from the north east sectors, but the inshore wave heights are limited both by the presence of a number of offshore sand banks and extensive intertidal flats. Waves tend therefore to be locally generated and relatively small, with 1 in 100 wave heights reaching 1.3 to 1.5m.
- The spring tidal range is 5.26m at Southend.
- North east of Two Tree Island maximum flow speeds on spring tides are typically between 0.5 and 0.8m/s.

The Outer Thames area is characterised by high suspended sediment concentrations derived primarily from the Thames. Additional sources include the erosion of the glacial drift deposits, London Clay and background suspended sediments within the southern North Sea. In terms of sediment transport, the Outer Thames estuary is a net sink with fine grained material being imported via a number of routes. Spring tide residuals show a clockwise movement into the Thames and in towards the Southend frontage.

Ongoing siltation is a known concern within the Thames Estuary. Discussions with the Environment Agency highlighted that increased siltation at outfalls and other structures in the estuary has been observed for a number of years resulting in the need for increased maintenance.

Historical Analysis

Historical admiralty charts and ordnance survey maps ranging from 1777 to the current day were examined. The maps showed that the main low water channel of Leigh Creek has flowed in a southerly path along the northerly shore of Two Tree Island since at least 1777. However, a secondary branch of Leigh Creek has followed a more northerly route along the shore of the mainland. It is the siltation of this branch that has reduced water depths in front of the working docks, cockle sheds and boat repair yards.

Analysis of historical maps indicated that two significant changes that occurred in the 1800s were the reclamation of Hadleigh Marsh and Leigh Marsh (now known as Two Tree Island). It is estimated that the reclamation of Hadleigh and Leigh Marshes occurred between 1805 and 1836. This was followed by rapid marsh growth in Leigh Creek over the next 50 years. The growth of marsh in this area would have reduced the capacity of tidal flow through the creek network and is likely to have increased siltation in the area of the northern and southern branches of the creek. It was noted that the period of growth of Hadleigh and Leigh marshes was coincident with the closure of two docks in the upper part of Leigh Creek.

The historical maps show that from 1898 to 1946, the overall configuration of the Leigh Creek Channel remained essentially the same.

The maps and charts also show that the offshore part of Leigh Creek where it joins Hadleigh Ray has changed over time. Between 1805 and 1898 there were changes in the configuration of this area and a narrowing of Leigh Creek north of this point by some 20 to 50m.

Halcrow (2003) made site-specific estimates of saltmarsh change at Two Tree Island using georectified aerial photographs from 1964 and 2001. The main losses of saltmarsh occurred on the south side of Two Tree Island although there was some loss on the north west side of the island. Since 2003, Southend Borough Council has undertaken work to restore lost saltmarsh to the south-east of Two Tree Island.

Siltation Estimates

Despite the losses of saltmarsh around Two Tree Island indicating erosion, the Leigh Creek area has undergone continued siltation. Recent anecdotal evidence from fishermen and boat owners reports the continued accretion of Leigh Creek to the north of Two Tree Island. The cockle fishermen also report a significant reduction in depth at the quays.

Discussions with local boat users indicated that they had 'lost' about 30 minutes of tidal access over the last 30-40 years. From local Admiralty tidal prediction curves for mean spring tide it was determined that this could relate to an elevation difference of around 0.5m. The implication of this is that there has been an increase in bed levels due to siltation. During consultation the boat users also reported that there had been approximately 1m of accretion on the foreshore at Belton Way Small Craft Club.

The siltation that has been reported at Leigh relates predominately to the northern branch of Leigh Creek. There have been no reports during consultation with boat users of significant siltation in the deeper southern branch. Given the predominance of siltation in the Outer Thames and Leigh Creek area it is unlikely that the southern channel has escaped this. It is more likely that the amount of siltation has been insufficient to affect the navigation due to the deeper water channel which exists.

Further analysis of the changes in depths between the 1926/27 chart and the present day bathymetry data collected for this study was undertaken. This analysis was subject to a number of uncertainties which limit application of the information. However, the analysis showed that the general trend across the majority of the Leigh Creek area, including both the northern and southern branches of Leigh Creek, has been for accretion. Across the areas analysed, average accretion during the period ranged from 1.2m to 2.3m. This equates to an overall average of 1.7m accretion in 83 years which corresponds to around 2cm per year.

Key Process Drivers

From the review of wider changes in the Thames Estuary and local changes, it is determined that the siltation within Leigh Creek has been due to:

- A wider tendency for the Thames estuary to act as a sink for fine grained sediment, particularly where wave energy is lowered by offshore banks or extensive intertidal flats.
- The possible decrease in flow speeds and sediment carrying competency within Leigh Creek due to the reclamation of Hadleigh (by 1836) and Leigh Marshes (by 1836) which used to drain into Leigh Creek upstream of the present day bridge to the mainland.
- Rapid expansion of marshes along Leigh Creek and the mainland between 1805 and 1898.
- Changes at the confluence of Leigh Creek and Ray Gut between 1836 and 1898, which may have contributed to the siltation occurring in the upper regions of Leigh Creek by reducing flows there.

Even allowing for a rise in sea level in the future it is unlikely that Leigh Creek will undergo significant deepening. It is likely that there will be a continued import of fine grained material into the area in the future until a new equilibrium is reached between the flows in the creek and the overall dimensions. Relative to the deeper more southern branch of Leigh Creek which flows between Two Tree Island and the mainland, the northern arm of Leigh Creek is likely to show higher rates of siltation since the discharge through this area is lower.

Due to the ongoing siltation in the study area, it was concluded that over the longer term ongoing dredging will be needed to maintain any implemented option in the study area.

Option Development

The preferred route for a diversion of the southern branch of the creek is to the northern creek which flows past a large boat repair yard and the landing facilities of the local shellfish and fishing industry.

Figure 2: Fully excavate 'New Creek' along northern branch and completely infill the southern branch

As the ongoing siltation is linked to long term siltation throughout the entire Thames Estuary, it is unlikely that siltation in the creek area can be reduced by solely changing the dimensions of the creeks to increase tidal flushing. Therefore, as a starting point for assessing the feasibility of a creek diversion it was determined that the form of the new northern creek should be of the same cross sectional area and depth as the existing southern and northern creeks combined to maintain tidal flushing through the system. A single efficient scheme could potentially be provided by infilling the southern branch with material excavated from the northern branch, as shown in Figure 2. Opening up the northern creek and not closing the southern creek would not guarantee a 'switch' in the system to predominate the northern creek. Therefore, the southern creek had to be fully closed to reduce the risk of this reopening.

The new channel will be positioned to avoid existing defences, but the downstream margins of the northern channel will lie against the quay walls at Leigh in order to provide deeper water access.

At this stage it has been assumed that if the channels are kept the same total size in the upper part of Leigh creek, then the flows at Ray Gut and Hadleigh Ray should also remain unchanged.

Technical Issues

A contamination sampling survey and laboratory testing of Leigh Creek sediments was carried out in February 2011. This confirmed that the sediments of Leigh Creek were dominated by alluvial clays and silts. The presence of alluvium materials in the proposed dredged areas raises a number of issues of technical concern with regards to the scheme:

- The structure of alluvial clays and silts is weak, and such material collapses and slumps naturally. To ensure the design depth of the channel is achieved and maintained, the channel side slopes required need to be significantly shallow or slope protection will be required.
- The disturbance of alluvium materials can result in high concentrations of suspended sediments and can generate releases of ground borne gases. The release of high concentrations of suspended sediments can cause changes to the dissolved oxygen levels in the surrounding waters, which can have detrimental effects on marine life.

Option appraisal and assessment has included the examination of a range of techniques to address these issues including the use of slope protection, containment bunds, trailer suction hopper dredging, and the use of silt curtains.

Contamination

Initial analysis of the contamination sampling laboratory testing results also highlighted that sediments tested within the Leigh Creek study area do not wholly conform to the contaminant restrictions of Cefas Guideline Action Levels for the Disposal of Dredged Material or to the Environment Canada's TEL/PELs (Threshold Effects Level and Predicted Effects Level) guidelines which have been adopted by the Environment Agency's Water Quality group in response to a need for guidance on how to assess the impact of chemicals in sediments for the Habitats Directive Review of Consents (RoC).

The initial analysis indicates that approval for direct reuse of material dredged within Leigh Creek for the development of mudflat and saltmarsh habitat would not be given without further investigation. However, the analysis indicated that the contaminant levels were found to be below the highest Action Level where materials are considered wholly unsuitable for sea disposal. Contaminant levels lie in the ambiguous area between the two action levels. Further investigation is therefore required to further delineate areas where elevated concentrations have been detected, investigate historical and current information regarding potential sources and type of discharge into the area of interest and obtain possible biological data information, from which the planning authority can make a decision on the suitability of material for dredging. Further qualitative analysis would also be required to statistically analyse the risk of contamination from these contaminants and it is likely that further testing at sites of the planning authorities choosing would be required once the exact dredge area has been confirmed.

For option appraisal, a range of options including direct reuse of dredged material for the closure of the southern branch of Leigh Creek and development of mudflat and saltmarsh, and the disposal of dredged material and use of alternative material for closure of the southern branch of Leigh Creek, have been examined.

Additional Environmental Benefits

Dredged material from the northern branch of Leigh Creek, or infill sourced from elsewhere, would be placed to a level suitable for saltmarsh growth (mean high water neap to mean high water spring tides) to close the southern branch of Leigh Creek. The provision of refill material will aim to provide a slope between the western saltmarsh level and the old southern creek depth to encourage the formation of protective outer mud flats and to reduce the likelihood of the collapse of the recharged surface.

Experience in managed realignment schemes has shown that providing there is a suitable supply of propagules and seeds from existing marshes, the colonisation of new areas is fairly rapid (approximately 5 years) provided the sedimentological and geomorphological conditions are suitable. Overall, elevation within the tidal frame has been shown to be the most crucial factor in successful vegetation development. It is therefore proposed that artificially reseeding the mudflat would not be necessary and that the marsh is left to revegetate itself from the surrounding marshes.

Access for Fishermen

Consultation with the fisherman outlined the aspiration to dredge the creek to the east by a depth of 1 metre at the moorings out towards Hadleigh Ray. This would provide an extra hour on the flood tide and an extra three quarters of an hour on the ebb tide. This would allow them to:

- travel to the Maplin sands;
- catch their allowed quotas of shellfish; and,
- return to the quayside in Leigh with the catch still fresh.

At present the fishermen go to sea to get their quota, but by the time they return to Leigh there is insufficient water for them to return to the quaysides. The fishermen therefore have to wait in Hadleigh Ray for the next tide before being able to return to the quayside. In summer months it is undesirable to have the cockles laying on the boat for 13 hours before landing. This is worse for the boats that land and send the live catch away by lorry to Kings Lynn or Wales to be processed. In these circumstances the quality and yield of the catch decreases.

Consultation with recreational boat users also indicated that a deepening of the northern branch of Leigh Creek would be supported by them.

To improve access for the fishermen along Leigh Creek (and in order to assist in providing a transition from the eastern end of the excavated northern channel) additional dredging works would be required to the east of Two Tree Island. Additional dredging would be undertaken between the newly excavated northern branch of Leigh Creek and the confluence with Hadleigh Ray running in front of fishermen's huts at Leigh. Volumes have been derived based on deepening the channel in front of the fishermen's huts by 1m over the depths at present. This additional dredging requires a significant element of work over and above that required to divert the creek to achieve the key objective.

The impact of increasing the lower part of Leigh Creek for fisherman's access on the wider estuary is felt to be more significant than the works to realign the southern branch of Leigh Creek to the north. Dredging of the lower part of Leigh Creek has the potential to change flows in this area and possibly around the confluence with Hadleigh Ray. However the relatively small increase in channel cross sectional area suggests that impacts would be relatively small. Such aspects would need to be investigated with a more detailed numerical modelling approach at the next stage.

Discussion – feasible or unfeasible?

The proposed realignment of Leigh Creek offers a number of benefits but there remain a number of constraints that will need to be overcome before the scheme can go ahead.

The proposed option offers benefits in terms of removing the erosion risk to the landfill site on Two Tree Island and providing deeper water access to the quays. Although there will be a loss of saltmarsh along the route of the new expanded northern creek, this will be matched by the creation of new marsh in the blocked southern creek. Additionally, there is potential to use the additional material dredged from the seaward portions of Leigh Creek (east of Two Tree Island) to create and restore additional areas of marsh if further contamination testing and investigation determines that the sum of impacts are below the required thresholds. The possibility to increase saltmarsh will be explored through further project development.

The option layout proposed in Figure 2 offers the best chance of gaining deeper water in the northern creek since all the tidal flows will be channelled through the northern creek. Nevertheless, the long term siltation across the entire Thames Estuary, suggests it is likely that the study area will continue to silt up and maintenance dredging will be required over the longer term to maintain the northern creek.

There is also a risk that the realigned and widened north creek will meander causing undermining and structural instability of defences along the Leigh frontage. Investigations of the existing quay walls will be needed to ensure that the structures are not undermined. The final alignment of the creek will require fine tuning at design stage to reduce the risk of undermining. However, it is understood that a landfill site also lies along the north bank and the risk of this failing due to erosion pressure from a meandering of creek remains a risk of implementing this option.

There are also a number of environmental considerations that need to be taken into account. Key potential impacts on the flora and fauna of the international, European and nationally designated sites include:
1. The loss of mudflat habitat and gain of saltmarsh habitat with no net loss of intertidal habitat. Loss of saltmarsh due to coastal squeeze is a major concern in Essex, and the provision of new saltmarsh to compensate for these losses is seen as a conservation objective in the Essex Biodiversity Action Plan and to comply with Regulation 3(4) of the Habitats Regulations (2010). Initial consultation with Natural England has indicated that a gain in saltmarsh may be seen as a benefit to the scheme.
2. Affects on feeding and roosting patterns of the birds of the SPA and Ramsar site.
3. The risk of suspended sediment released during construction contaminating local commercial cockle beds and eel grass habitats.

As the proposed creek realignment is within a European Site (a Special Protected Area (SPA)) it is subject to the Conservation (Natural Habitat &c) Regulations 2010, which implement the Habitats Directive into UK law. Under the EU Habitats Directive, an assessment of the scheme will be required to determine whether it will have a *significant effect* on designated sites. If this is deemed to be the case, than an Appropriate Assessment will be required to determine whether the scheme will *adversely affect the integrity* of the site. If a scheme is deemed to adversely affect the integrity of the site it will only proceed if (i) there are no available alternatives; (ii) it is for reasons of over-riding public interest; and (iii) the provision of compensation is made where required.

If the scheme is deemed necessary to protect the SPA and Ramsar interests, then no Appropriate Assessment will be required. If however, the scheme is not deemed necessary, and the scheme is deemed to have a *significant effect* then an Appropriate Assessment will be required and the scheme will have to continue through the route of Habitat Regulations procedure.

Conclusion

The realignment of Leigh Creek is a bold and innovative solution to two problems. As with any innovative approach there are a number of risks. The work that has been completed has demonstrated the constraints that need to be overcome in order to progress the scheme. Final consideration of whether the scheme can be progressed under the Habitats Regulations requires the completion of further studies to determine whether any impacts can be satisfactorily mitigated against to determine the level of impact and to inform Natural England's and other consultees consents. Further work is also being undertaken on the costs of the various scheme options. Once these are confirmed, then the scheme may be proceeded under the rationale that the diversion is for maintenance of the international conservation sites. As the dredging works required to improve access for the fishermen at the wharfs is over and above that required to address the erosion issue at Two Tree Island, it is likely that this portion of the work will need to apply separately for consent and gain private funding.

References

Halcrow Group Ltd. (2003). *Two Tree Island Management Strategy. Final Study Report.* Report produced by Halcrow Group Ltd for Southend Borough Council, August 2003. 63pp.

Essex County Council. (2008). *Leigh Old Town. Conservation Area Appraisal and Management Plan.* The Essex Design Initiative. Southend on Sea Borough Council.

Crossbow Partnership and Southend-on-Sea Borough Council. (2002). *Southend Fisheries Strategy - Fisheries Consultation Study.* Volume 1. Vantage Point.

Innovative Coastal Zone Management
ISBN 978-0-7277-5749-4

ICE Publishing: All rights reserved
doi: 10.1680/iczm.57494.577

Physical Modelling of the New Coastal Defence Scheme at Borth

Charlotte Obhrai, HR Wallingford, Wallingford, UK
Tom Rigden, HR Wallingford, Wallingford, UK
Keith Powell, HR Wallingford, Wallingford, UK
Alice Johnson, Royal Haskoning, Peterborough, UK

Introduction

The village of Borth, located on the west coast of Wales in Cardigan Bay (Figure 1), was originally a small fishing hamlet which has grown into a holiday resort due to its wide beaches, sand dunes, good bathing water and visitor attractions. The prevailing south westerly weather and gently sloping beach also provides good conditions for surfing and the main industry within Borth is now tourism. The beach is currently (2011) protected by an extensive, but ageing, groyne system with a timber breastwork to strengthen the cobble (a coarse gravel-sized rock material) ridge at the beach crest. The standard of the defence provided by the beach and breastworks is low, allowing frequent overtopping during winter storms. During more severe storms, overtopping can lead to flooding and shingle being washed into the village. As the defences are near the end of their life the risk of the shingle ridge breaching is also increasing, which could lead to significant damage and widespread flooding of the village. A strategic appraisal report, carried out in 2006, recommended that the frontage should be protected by the phased replacement of the breastwork, renourishment of the beaches, and building a series of beach control structures incorporating a multi-purpose reef at the southern end of the frontage. The first phase of the scheme described by Johnson et al (2011) comprises two rock groynes, two nearshore detached breakwaters, a multi-purpose reef and beach nourishment. The cobble ridge and particularly its response to storm conditions (longshore transport and beach drawdown) will be integral to the maintenance of an acceptable standard of defence.

The main objective of the physical modelling was to optimise the layout of the defence structures to provide sufficient cobble beach widths and thus reduce wave overtopping at Borth. Determining the impact of the proposed coastal structures on sediment transport to further north of the scheme was also a key aim of the study. The surfing performance of the offshore multi-purpose reefs was also assessed during the physical modelling tests. This paper describes the design and use of a 1:45 3-dimensional (3D) mobile bed physical model to investigate the performance of the proposed coastal defence scheme.

Figure 1   Location of Borth, UK

Design of the Physical Model

The physical model looked at a 1.66km (prototype) stretch of the Borth coastline that covered Phase 1 of the scheme. The main processes to be modelled were the morphological beach response, overtopping, stability of the proposed groynes and the surfability of the multipurpose reef. Three dominant wave directions and associated representative conditions were determined to be replicated in the model. The morphological condition (WL = 1.81mOD, H_s = 3.7m, T_p = 10.2s) was run from three directions (255°N, 270°N, 285°N) and tests with these conditions were used to determine the stable beach plan shape. The extreme conditions (WL = 2.0-4.5mOD, H_s = 3.8-5.9m, T_p = 10.2-12.4s) were run from 255°N and looked at overtopping and stability performance.

Test Facility

The model was constructed in one of HR Wallingford's larger wave basins with a working area of up to 1000m². An outline sketch of the basin layout is shown in Figure 2. A geometric scale of 1:45 was used and the model beach was approximately 37m long (1.66km prototype). Detailed measurements of beach response were, however, confined to a 30m long (1.5km prototype) test section in the central zone of the beach to ensure that edge effects do not influence study findings. The nearshore bathymetry was formed using cement mortar over compacted fill to survey data. The mobile bed sediment was laid to at least 0.45m deep (prototype).

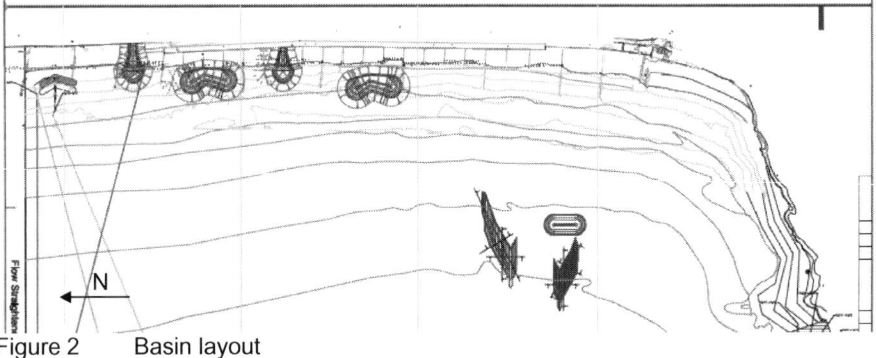

Figure 2   Basin layout

Model Scaling

Two distinct types of prototype beach material were represented in the physical model: fine sand and cobbles. Sediment samples were analysed at HR Wallingford to obtain sediment grading curves.

The sediment grading curve for the constituent parts of the cobble ridge sample C is shown in Figure 3 and the sediment size analysis is summarised in Table 1. The significant amounts of sand contained in the cobble ridge samples tend to skew the value of the median grain size. The resulting bimodal distribution has a pronounced gap grading (Figure 4) with a 20% sand fraction lying between 0.125mm and 0.5mm and around 75% shingle lying between 8mm and 35mm. The distribution was therefore re-sampled to include only the coarse fraction, which primarily dictates the hydrodynamic response of the upper cobble ridge, (>8mm) and the two distributions are shown in Figure 3. The size analysis of the shingle only fraction is also given in Table 1 and it is these values that were used to scale the cobble material for the physical model.

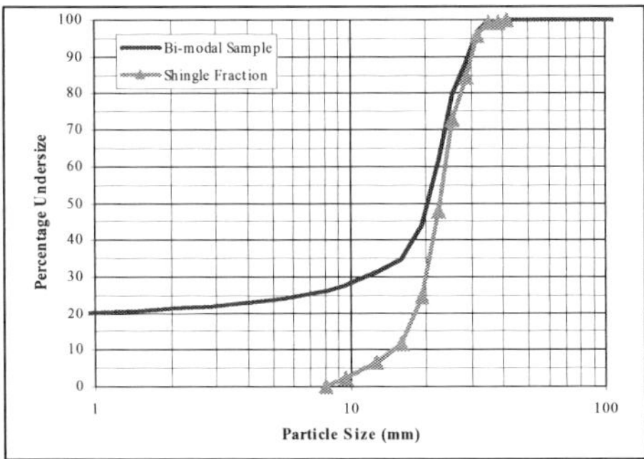

Figure 3   Sediment grading curve for the "Borth C" sample

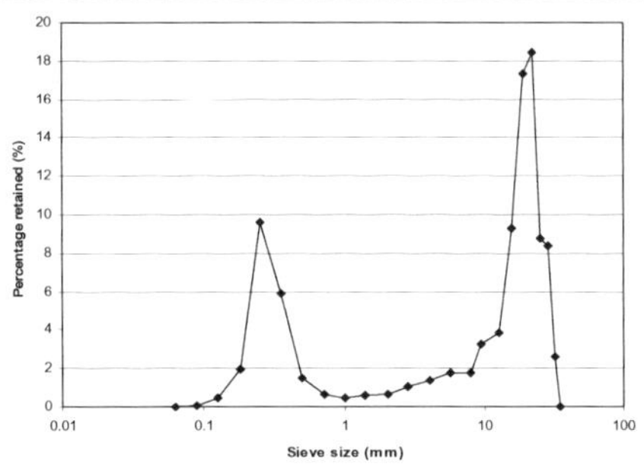

Figure 4   Bimodal distribution of the "Borth C" sample

Table 1   Sediment size analysis of Borth beach

Sample	D_{10} (mm)	D_{50} (mm)	D_{90} (mm)
Borth A	0.38	9.2	47.5
Borth B	0.39	10.8	43.6
Borth C	0.33	20.4	29.0
Borth E	0.14	0.23	0.43
Borth C Shingle Fraction	14.9	23.5	30.3

To produce the correct beach response in a mobile bed model, the model should satisfy three main criteria:

- The permeability of the beach (hence the correct beach slope).
- The relative magnitudes of the onshore and offshore motion (hence determining whether the beach erodes or accretes).
- The threshold of motion (hence the bed shear stress at which the point of erosion occurs).

In the selection of model sediment, in practise there are only two variables: sediment size and specific gravity. Thus, for most studies one of the modelling criteria will need to be relaxed. The threshold of motion criterion did not need to be satisfied for this study as the hydrodynamic conditions being tested were well in excess of the threshold of motion. Details of the sediment scaling for the cobbles and the fine sand for permeability and onshore / offshore criteria are given in Table 2. The results suggest that it is not possible to scale the cobbles based on both theoretical scaling criteria (permeability & onshore / offshore motion) using sediment of the same density (2650 kg/m^3). The scaling of permeability is important for the correct reproduction of the beach slope and hence overtopping performance while the scaling of the onshore / offshore motions determines beach profile development during storm conditions and hence also potentially influences wave overtopping. Previous work at HR Wallingford by Russo et al (2010) had included physical measurements of flow velocities through a range of very uniform model sediments. These particular model sediments were highly permeable relative to their median grain size due to their narrow grading. For this study the option was taken to utilise a model sediment size to meet the onshore / offshore motion criterion and to then compensate for the reduced permeability by narrowing the grading.

Table 2 Sediment scaling of the cobbles and fine sand using sand (density=2650kg/m^3)

Cobbles – D_{50} = 23.5mm D_{10} = 14.9mm	Sediment Scale	Model D_{50} (mm)
permeability:	5.50	4.18
onshore / offshore motion (Soulsby 1997):	26.4	0.87
Fine sand - D_{50} = 0.22mm		
onshore / offshore motion (Soulsby 1997):	2.83	0.08

Beach profile data indicates that the natural slope of the cobble ridge at Borth varies between 1:6 and 1:8. The model sediment as described in Table 3 was selected as this satisfied the onshore / offshore criterion and it was also known from the previous work (see Russo et al, 2010) to sit at a slope similar to the cobble ridge of 1:6.

Table 3   Sediment characteristics of the model sediment used in this study

Model Sediment	D_{90}	D_{50}	D_{10}	Porosity	Permeability
	mm	mm	mm		m/s
Cobble Ridge	1.10	0.87	0.70	0.40	0.019
Fine Sand					
Tracer	0.17	0.11	0.08	-	-

If the fine foreshore sand were to be scaled directly for the onshore / offshore criterion, this would result in cohesive model sediment that would not be representative of the prototype beach. Very fine sand (D_{50} = 0.11mm) was therefore selected to act as a tracer material to indicate areas of increased erosion or accretion (see Table 3).

The material selected to model the cobbles reproduced the plan shape of the upper beach, and its ridge, and the interaction between the various structural elements, with a good reproduction of the locations of erosion and deposition. Quantitative estimates of depths of cross-shore erosion or deposition could also be provided which was a key requirement to determine the overtopping performance of the cobble ridge. The tracer material was used to model the lower foreshore sand and reproduce patterns of seabed erosion and accretion expected to occur under the prevailing wave and current conditions. Only qualitative estimates of depths of erosion or deposition could be obtained with the fine sand.

Beach and Cobble Ridge Morphology

The main beach morphology measurements were taken using a 3D laser scanner to recreate the beach surface before and after a test, thus indicating where accretion and erosion of beach material has occurred. Overtopping and armour movement measurements were taken and the surfability analysis was carried out using overhead video analysis.

Measurements

During the morphological tests, the beach was surveyed using a 3D laser scanner (Figure 5). The survey data (coordinates and spot heights) was analysed to produce regular grids, provisionally specified as 1m x 1m in prototype terms, describing the beach surface. The grids were used to produce contour plots and cross-shore profiles of the model beach. The initial and final scans were then differenced to provide areas of erosion or deposition.

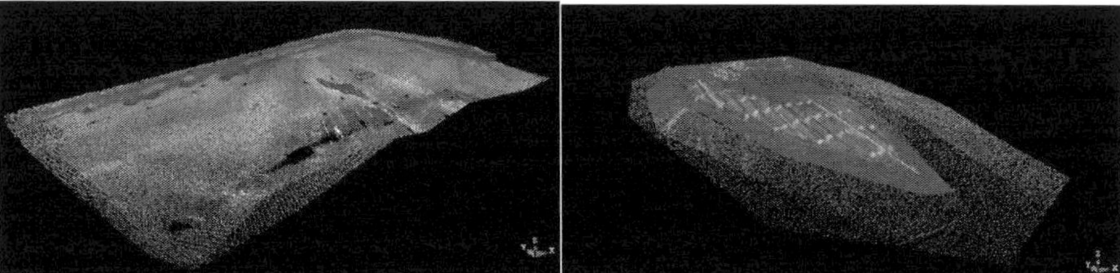

Figure 5   3D Laser data from the Borth physical model

Cobble Ridge

The main aim of the tests was to determine the quasi-equilibrium beach plan under morphologically averaged conditions and the resultant overtopping performance.   A morphological condition (from 255°N), representative of the typical annually averaged wave climate, was used to record the shoreline position at key locations at 12 hour intervals until a quasi-equilibrium plan shape was reached (Figure 6).

Figure 6   Final shoreline position with an incoming wave direction of 255°N

The difference plot of the final and initial laser scans for this series of tests is shown in Figure 7. This clearly shows that there is accretion of the cobble ridge on the southern side of each of the nearshore structures and behind the tip of the northern offshore reef.   The fine tracer material on the lower sandy foreshore indicates that there is accretion between the two arms of the offshore multi-purpose reef structures.   It is important to note that the thick black line represents the line of the upper breastwork and thus the edge of the physical model. The accretion at the lower left hand corner of Figure 7 is a model effect caused by build up of sand against the wave guides at the edge of the model.   In reality the sediment would be free to migrate further north and offshore.

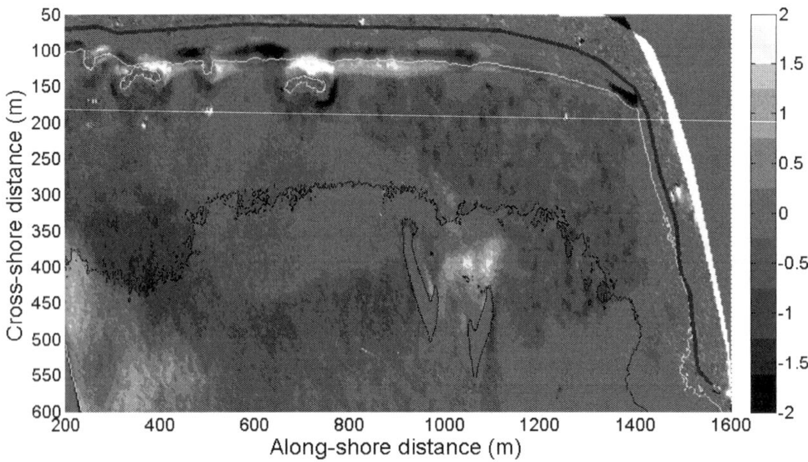

Figure 7   Difference plot for wave direction of 255°N after 108 hours

Observations made during these tests indicate that once the sediment bays become "full" sediment starts to bypass both the groynes and the nearshore breakwaters. The photographs in Figure 8 show sediment bypass observed during one of the tests.

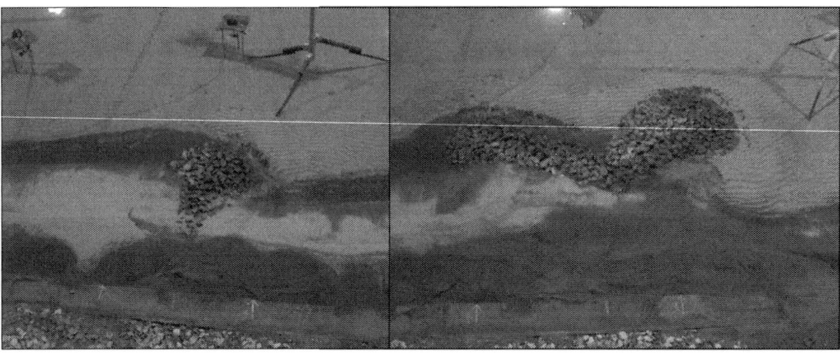

Figure 8   Sediment bypass of groyne and nearshore breakwater

Comparisons between the final beach plan shapes for two different layouts of the nearshore are shown in Figures 9 & 10. These are for incoming wave directions of 255°N & 285°N respectively. In both figures the +1.81mODN contour is indicated by the dark line for the first layout and the light line for the second layout.

The final beach plan shape is an important factor when considering the overtopping performance of the cobble ridge. Observations from the overtopping tests indicate that overtopping was highest at the positions where the cobble ridge was most narrow. It is therefore important to ascertain the likely extremes in beach plan shape to ensure that overtopping performance of the cobble ridge is not compromised.

582

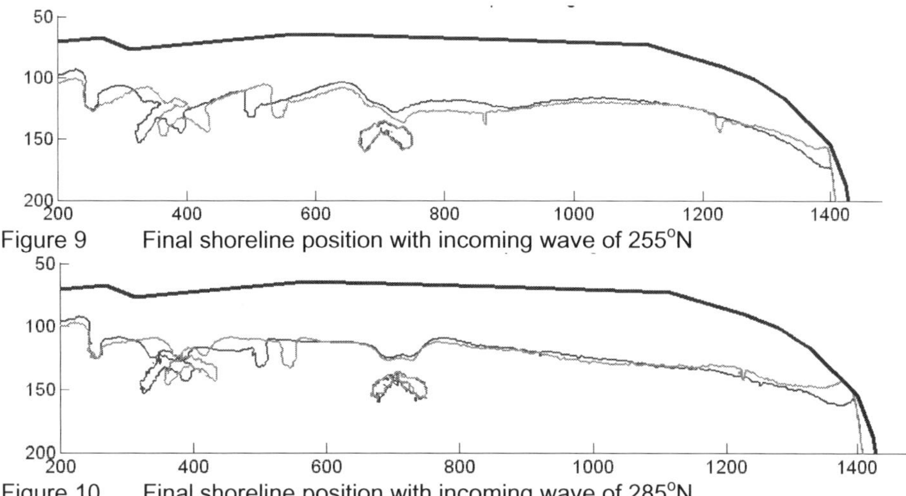

Figure 9   Final shoreline position with incoming wave of 255°N

Figure 10   Final shoreline position with incoming wave of 285°N

Sandy Foreshore

Anthracite tracer material was placed at several locations around the multipurpose reef to illustrate the sand transport around it (Figure 11). Photographs were then taken at time intervals to determine the sediment transport pathways around the offshore structures. At locations A & C in Figure 11 the transport of sediment is to the north and then offshore. At locations B & D the transport of sediment is initially onshore to the toe of the cobble ridge, then continues alongshore, remaining at the toe of the cobble ridge.

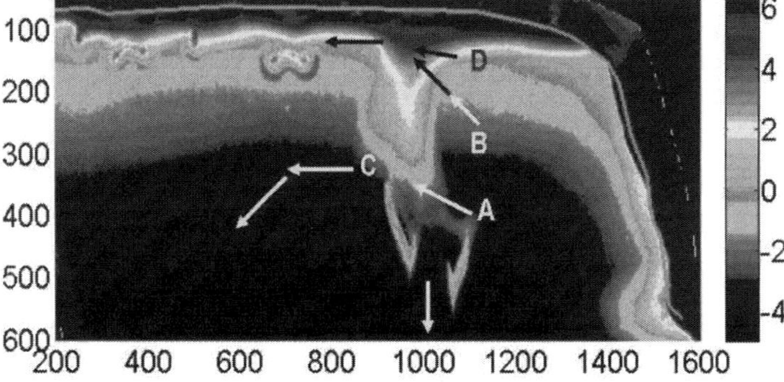

Figure 11   Laser scan for Test 121 showing movement of anthracite tracer

Surfability Assessment

The original scheme design included a multi-structure configuration with two truncated delta shaped reefs separated by approximately 70m. This multipurpose reef, developed by ASR Ltd (2010), was intended to provide both protection to the cobble ridge at the southern end of Borth frontage, and to enhance the surfing potential of the area. The northern reef was larger with a longer northern arm designed to provide a left-hand surfing break. The southern structure was somewhat smaller and set further offshore and was designed to provide a right-hand surfing wave.

Three different reef configurations were tested using 16 monochromatic waves (H_s = 1.0m-2.0m, T_p = 8s-12s WL = 0.86-2.36mOD) and 4 random wave conditions (H_s = 1.0m, T_p = 8s-12s WL =1.36-1.86mOD). The first two configurations were variations of the original 2-reef layout. For these cases, the model reefs were constructed from mortar to create a smooth, impermeable reef shape. The second case included a widening of the reef crest and the inclusions of grooves to represent geo-container construction. For the third case, the northern reef was built from rock armour and the southern reef was removed and replaced with a shore parallel breakwater. The three plan configurations of the northern reef are shown in Figure 12, the profile is shown Figure 13.

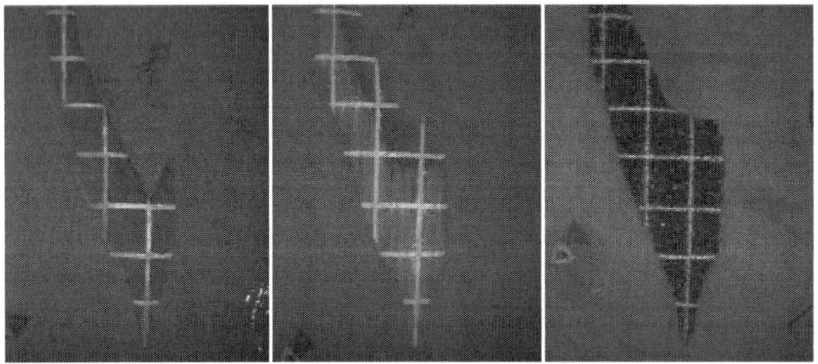

Figure 12   The three configurations of the northern reef

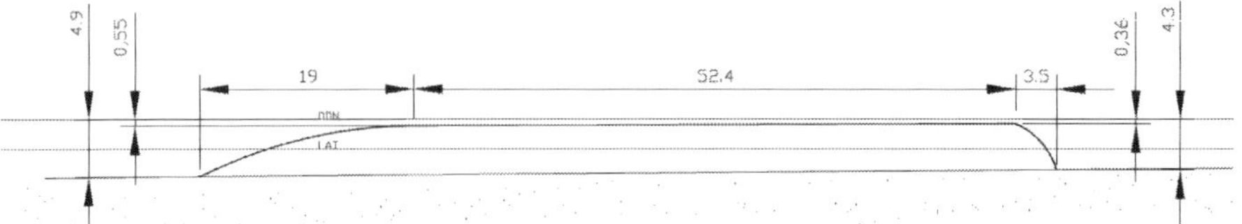

Figure 13   Profile of northern reef

The key parameters required to determine surfability are usually taken to be the wave 'break rate' and 'peel angle' (Hutt et al, 2001) (Figure 14). These are best estimated from short overhead video recordings (<100 waves) which allow the necessary parameters to be determined relative to a grid marked on the reef. Overhead video cameras installed above each of the reef structures were used to record surfability parameters during specific tests for surfing (Figure 14). An additional camera with a side view was installed to give information on the quality of the breaking wave and the wave shape.

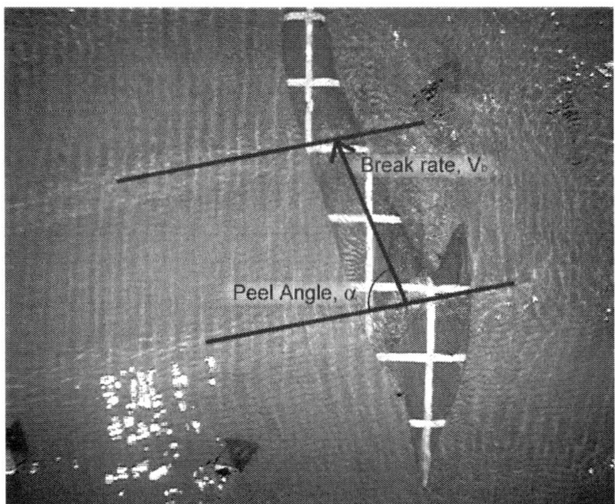

Figure 14   Overhead video shot of the northern reef

The offshore reef structures were optimised for both surfing amenity and coastal protection. Surfability assessments of three potential designs were made by ASR over a range of wave conditions and water levels. Overhead video and detailed wave measurements were taken which allow parameters such as the break rate and the peel angle to be extracted.

Observations from the tests, suggested that the southern reef in the initial design tended to interfere with the breaking wave process on the northern reef. As a result the southern reef was replaced by a simplified structure and the surfing assessment was then focussed on the northern reef. The northern reef was extended to increase the ride time of the waves and provide a more gradual inshore slope. Visual observations suggested that this did indeed improve the wave breaking over the reef. Two methods of construction were tested in the physical model: a smooth solid impermeable construction and a rock construction. The form of construction was found to have a significant influence on the wave breaking. The solid construction tended to prevent the waves breaking at lower water levels, perhaps due to the reduced roughness of the structure. In general the northern reef produced consistent breaking waves when tested from WSW and the waves tended to be spilling breakers.

Discussion on study results

Morphology of the Cobble Ridge:
Initial runs for the present layout confirmed that the model provided a realistic picture of the beach development. The mobile material used in the physical model was selected to reproduce the morphological response of the prototype cobble beach, and its overtopping performance. For both structure layouts, the cobble ridge tended to form tombolos behind each of the breakwaters and, depending on the incoming wave direction, material would build up on either side of the groynes. By applying the morphological wave conditions for a significant period of time a stable plan shape was obtained. This was then used to investigate the extreme response of the cobble ridge. In all the cases tested, once a stable beach plan shape was obtained, sediment started to bypass the groynes and the breakwaters. These results suggest that the scheme, when the beaches are fully developed, does not block

transport of shingle material to the north provided there is sufficient nourishment material available.

Morphology of the Sandy Foreshore:
The mobile material selected to represent the behaviour of the lower sandy foreshore was fine sand with anthracite tracers. These materials were able to provide only qualitative measurements of the sand transport due to limitations of model scaling responses of the beach material. Under waves from 255°N (WSW), sand tended to move onshore at the southern end of the frontage, propagate alongshore behind the offshore structures and then continue offshore. These observations suggest that sand transport will continue to the north with the scheme in place.

Surfability observations:
The offshore multipurpose reef structures were optimised for surfing amenity and coastal protection by changing the design of the Southern reef to a shore parallel breakwater. The Northern reef produced the most consistent breaking waves and was extended to increase the ride time of the wave. It was found that constructing the reef out of a solid structure reduced its performance at lower water levels.

Optimisation of Nearshore Rock Structures:
Two main layouts were tested during the physical model testing. The proposed initial layout resulted in narrow cobble beach widths and increased overtopping between the central groyne and breakwater. It was then proposed to move the two central structures 40m south which resulted in a more equidistant spacing between the nearshore rock structures. This layout was subsequently tested and found to increase the cobble beach widths in the lee of the nearshore breakwaters.

Acknowledgments

The authors are grateful for support of this work by Ceredigion County Council and ASR marine consultants.

References

1) Johnson, E.A., Ohbrai C., Llwyd R., Newman M., and Sarker M.A. (2011), Development of Borth Coastal Defence Scheme, ICE conference on Coastal Management 2011, ICE
2) Russo, V., Allsop, W., Sutherland, J., Obhrai, C., Arena, F., (2010) Physical modelling of mobile beach material to study short-term beach response, Proc. 3rd Int. Conf. on Application of Physical Modelling to Port and Coastal Protection, CoastLab, June 2010
3) Soulsby R. (1997), Dynamics of Marine Sands, Thomas Telford, London
4) Hutt, J.A., Black, K.P., and Mead, S.T. (2001), Classification of Surf Breaks in Relation to Surfing Skill, Journal of Coastal Research, Special Issue No. 29
5) ASR Ltd (2010), Borth Reef: Surfing Assessment Study, ASR, Raglan

Innovative Coastal Zone Management
ISBN 978-0-7277-5749-4

ICE Publishing: All rights reserved
doi: 10.1680/iczm.57494.587

Application of One-line Numerical Models to Beach Management on the Central Felixstowe Frontage

Jonathan Kemp, HR Wallingford, Wallingford, UK
Dr. Rosalind Turner, Mott MacDonald Ltd., Croydon, UK
Terry Oakes, Terry Oakes Associates Ltd., Lowestoft, UK

Introduction

The urban frontage at Felixstowe between the War Memorial in Undercliff Road West and Jacob's Ladder has been affected by long term beach lowering (Figure 1). Levels have fallen so dramatically in recent years that some of the groynes have been undermined and the frontage is at risk of flooding from overtopping and failure of the seawall (Plate 1).

To counter the ongoing erosion Suffolk Coastal District Council commissioned a Project Appraisal Report (PAR) to review the long term management of the frontage and to apply for funds to support the scheme.

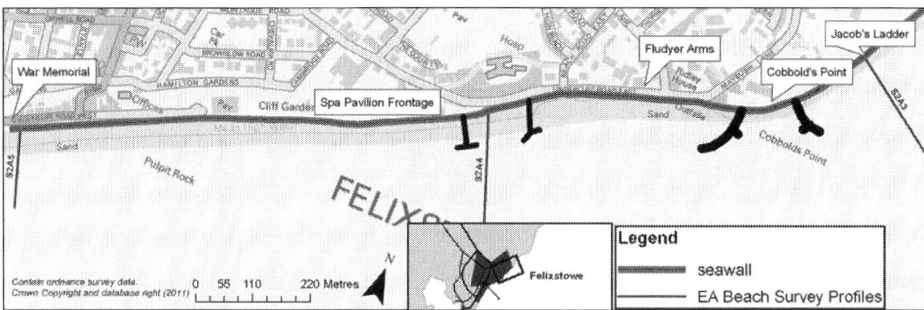

Figure 1.   Location of study area and beach surveys

Plate 1.   Low beach levels and undermined groynes

The main purpose of the planned improvements to the Central Felixstowe frontage is to improve the standard of coastal defence through a combination of beach recharge and the installation of structures designed to control the beach plan-shape and obtain the best value from the recharge operations i.e. maintain uniform beach widths along the whole frontage.

A review of coastal processes, a condition survey of the existing defences and a multi-criteria assessment were undertaken. This work led to the identification of seven viable coastal defence options for the Central Felixstowe frontage. Following a series of public exhibitions and further investigations of the cost, benefits and coastal processes, a preferred coastal risk management scheme was developed. This scheme consists of 50m long groynes at 50m spacing between the War Memorial and the T-head groynes; a 45m groyne between the two T-head groynes; and four 60-70m groynes and a beach recharge of 16,000m³ between the T-head groynes and Cobbold's Point. The groynes are sloped at around 1:9 gradient and are 1.5m above the recharged beach level.

As part of this study the evolution of the beach plan-shape has been numerically modelled, using HR Wallingford's one-line numerical BEACHPLAN model, to assist in the appraisal of the preferred coastal management scheme. Modelling of the beach plan-shape along the frontage was used to provide guidance on the refinement of this beach management scheme and then taken forward to final design and construction.

one-Line numerical beach plan-shape models are designed as a tool for understanding the behaviour of a beach and the impact of engineering works upon it and are ideally suited to assessing the shoreline response to beach control structures. One-line numerical beach plan-shape models, such as the one used in this study, are better applied to the short stretches of coastline generally covered in PARs rather than the larger stretches of coastline covered in Shoreline Management Plans and Strategy Studies (CIRIA, 2010).

This paper describes the modelling process with reference to the Felixstowe beach frontage study and goes on to explain and demonstrate the calibration of the model and presents results for the preferred coast defence scheme, including sensitivity testing under a hypothetical increased drift scenario.

Model input information

Initialisation and set-up of the model for the frontage required analysis of the in-situ sediment size, beach profiles and tidal levels, incorporation of any existing beach control structures and a beach contour line extracted from the latest topographic survey, and an assessment of shoreline orientation.

Seawall location

A seawall behind a beach affects the evolution of its plan shape in two ways. First, it limits the landward recession of the beach contours and prevents material being added to the beach as a result of the erosion of the hinterland. Second, if obliquely incident waves reflect off the face of the seawall, this alters the capacity of those waves to transport sand in front of the seawall along the coastline. To ensure that these effects are realistically reproduced, the position of the seawall along the whole frontage was obtained from surveys and input into the model.

Definition of nearshore wave conditions

The most significant long-term changes in beach morphology are likely to be caused by variations in longshore drift, i.e. the rate of transport of beach sediments along this shoreline. The rate of drift, at any moment in time, depends on the wave direction, wave height and wave period. Consequently, a good knowledge of wave conditions along the coastline is required for the calculation of sediment transport rates and beach plan-shape evolution.

Inshore wave sequences were produced at several locations along the shoreline of Felixstowe during an earlier study (HR Wallingford, 1997). As part of this study the wave model was re-commissioned and wave sequences over a 10 year period were predicted at two nearshore locations within the study area.

The majority of waves approach the coast from south-easterly and southerly directions although larger waves approach the coast from north-easterly and easterly directions. Thus, beach sediments available for transport are likely to travel both northwards and southwards along the frontage at different times. The exact extent to which this occurs depends strongly upon the prevailing wave conditions.

Beach morphology

Bi-annual beach surveys have been undertaken at three locations along the Central Felixstowe frontage by the Environment Agency since 1992 (Figure 1). These beach surveys have been used to define the representation of the average beach profile and to establish rates of erosion along the frontage for calibration of the BEACHPLAN model. Based on the analysis of the survey data between 1992 and 2008, a rate of beach erosion of between 0.4 and 1.0 metres per annum was calculated at the three locations along the Central Felixstowe frontage.

Data from a topographic survey undertaken by Anglia Land Surveys in March 2009 was used to provide input for the initial beach position for the calibration model runs. The model represents the changing plan-shape of the beach by predicting the evolution of a single contour, and in this study the Mean High Water contour (MHW = 1.5m ODN) has been selected. Cross-sectional profiles of the Cobbold's Point groynes and the T-head groynes were obtained from the "as built" drawings provided by Suffolk Coastal District Council and entered in to the model to ensure that their performance was reproduced as accurately as possible.

Calibration of model

In the calibration process it is necessary to demonstrate that the model can represent the main features of the beaches along the Central Felixstowe frontage and their evolution as measured in previous years. This requires the model predicting the beach alignments correctly and reproducing the measured erosion rates of between 0.4 and 1.0 metres per annum along the frontage as well as the observed loss of beach material from Cobbold's Point. In order to allow the model to demonstrate the measured amount of erosion along the frontage, including Cobbold's Point, a nominal beach width of 15m from the seawall to the MHW contour was selected as the starting shoreline position in the calibration runs.

The main purpose of the numerical modelling was to examine how beach widths would alter in the future with or without the preferred beach control structure schemes. It is not possible to predict future wave conditions or their sequence of occurrence precisely. The standard modelling technique is to use a long sequence of wave conditions that have occurred previously. Provided all the options are studied using the same set of wave conditions, this produces a reasonable basis for comparing their likely performance. Sensitivity testing to check that the preferred option is not unduly sensitive to a change in wave conditions (i.e. sequence or mean direction or mean height) is also carried out.

Following a considerable number of model runs, as is normal in such studies, a satisfactory comparison with existing beach characteristics (i.e. its orientation and measured rates of retreat), was attained and this is presented in Figures 2 and 3. Figure 2 presents the ten "end-of-year" shoreline positions. While these outputs are only instantaneous "snapshots" in time of a constantly changing beach (MHW) position, they provide a good guide to the general plan-shape and orientation of the beach.

Of particular interest from a coastal protection viewpoint is the situation where beach widths become very narrow. Figure 3 presents the predicted mean, minimum and maximum MHW positions during the ten years of this calibration run. The points showing the mean, minimum and maximum beach widths are not joined to form a continuous line in this, or in similar subsequent figures, to emphasise the fact that these positions do not occur concurrently. This form of presentation is considered the most useful way of summarising the important results from the modelling for this and all other scenarios considered.

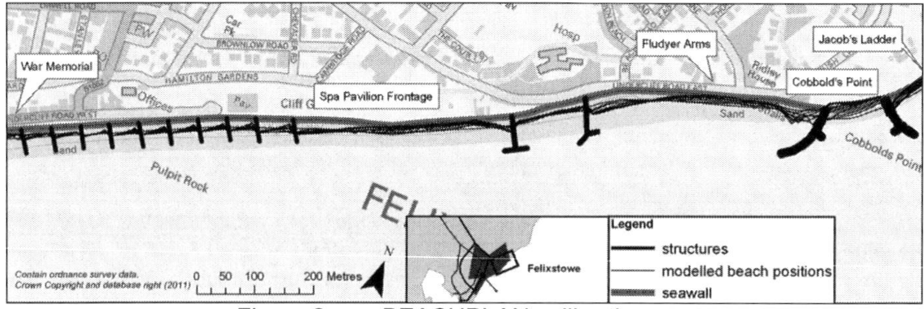

Figure 2.   BEACHPLAN calibration run

The minimum predicted beach widths shown in Figure 3 indicate a particularly narrow beach at Cobbold's Point and along Spa Pavilion where the seawall 'bulges' further seaward. At these locations, the beach can become very narrow and low in front of the seawall, and at times there would be no visible beach at MHW, with the tide reaching the seawall. A narrow beach at these two locations was also observed during the site visit. The model continues to calculate the changes in the beach, predicting how low its level will drop at the toe of the seawall. To show this graphically, the model then uses the beach level and the beach gradient to indicate an "equivalent" position of the MHW contour on the beach face in the absence of the seawall. As a result, the minimum beach position is then shown as being landward of the seawall; the lower the beach levels fall, the further landward this shoreline position is plotted. There is therefore the potential for problems with overtopping/scour and even undermining of the seawall during these periods.

The model has predicted that the minimum beach positions at Cobbold's Point are up to 15m behind the seawall, indicating a beach level in front of the wall of around 0mODN. This correlates with the measured beach level in 2009. The main concerns raised from this calibration run, therefore, are the erosion leading to a very narrow beach around Cobbold's Point and at Spa Pavilion. The main target in the design of different schemes is to reduce these problems through the introduction of beach recharge and beach control structures, namely groynes.

It was considered that the model demonstrated a satisfactory comparison with existing beach characteristics and could therefore be used to form a reasonable basis on which to examine the beach recharge and the effects of beach control structures. The calibration of the model was agreed with the local council and the general public during a public meeting. The 10-year period of wave conditions chosen to predict changes in beach widths in the existing scenario, produces an average longshore drift rate, and thus provides a reasonable view of the long-term evolution of the beach, identifying the main challenges that need to be met. This same sequence of wave conditions was used for the subsequent testing of the coastal management schemes.

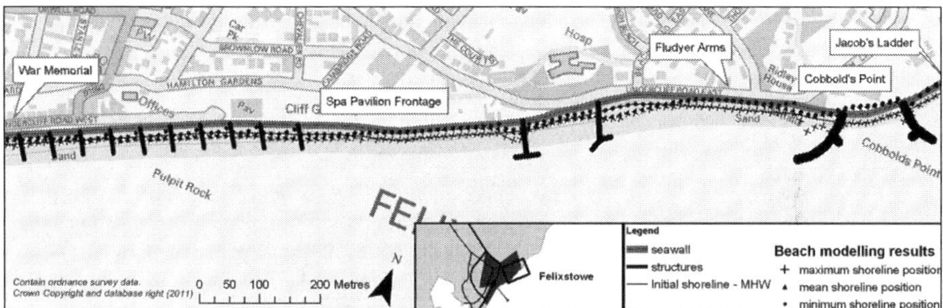

Figure 3.   Predicted mean, minimum and maximum beach positions from the BEACHPLAN calibration run

Testing the preferred scheme

Method for assessing modelling results

When undertaking numerical modelling to assess or compare different coastal defence schemes, it is important to decide and clearly define the output parameters that are to be used as a basis for any comparisons.

The primary purpose of the beach control structures is to maintain an overall wider and uniform beach width along the Central Felixstowe frontage, so that it is the combination of the seawall and the beach, which provides an appropriate standard of defence against coastal erosion.

The beach widths, measured out to the MHW contour (1.5mODN), and the area above the MHW contour, are used as a basis on which the results from the beach plan-shape modelling have been assessed. Of particular interest are those areas where the beach width is narrowest, since it is in these areas that the greatest flooding and erosion risks will occur.

From a coastal protection viewpoint, the ideal scenario is to maintain an equal beach width along the entire frontage, so that none of the sediment imported by recharge operations is wasted in providing, at some locations, a greater beach width than that required to protect the seawall. Thus a uniform narrowing of the beaches along the entire Central Felixstowe frontage would occur, rather than localised reductions in beach width at critical locations.

Scheme 0: Beach recharge

Scheme 0 consists of the proposed beach recharge only, which increases the beach crest to a width of 5m along the Fludyer Arms frontage; this corresponds to an increase in beach volume of approximately 16,000m³.

Figure 4 shows the mean MHW position during the final year, together with the minimum and maximum positions during the whole period.

In the first instance, it is useful to examine how the beach recharge along the Fludyer Arms frontage responds without any beach control structures in place. A general re-orientation of the recharge material occurs, which reflects the predominant wave direction and the resulting net southwards longshore drift. This results in a general decrease in beach width at the northern end of the frontage, apart from some localised accretion in the lee of the Cobbold's Point breakwater, and a slight increase in beach width at the southern end. There is a loss however, in beach area above the MHW line of **480m²/year** and the beach retreats back to the seawall at the northern end within 5 years. This erosion could result in continued undermining of the seawall and increased overtopping. This modelling result demonstrates a requirement for beach control structures along this frontage.

South of the Undercliff T-head rock groynes, the seawall has a convex plan alignment so it will be more difficult to maintain healthy beach levels here without installing any beach control structures. Indeed the BEACHPLAN modelling results indicate that the MHW shoreline position is against the seawall for much of this section of the frontage.

The main target in the design and modelling of different schemes is to reduce the loss of the planned shingle recharge through the introduction of beach control structures, and maintain a generally wider beach along the Central Felixstowe frontage.

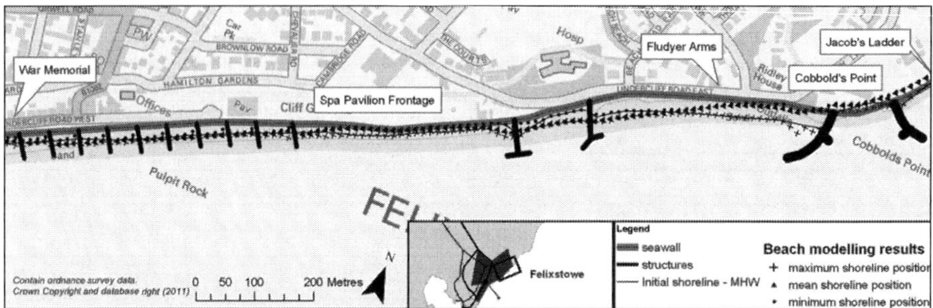

Figure 4.   BEACHPLAN results for scheme 0

Scheme 1: Beach recharge and 17 groynes

The preferred scheme for the Central Felixstowe frontage consists of a series of groynes, a beach recharge (same as Scheme 0) and a rock revetment. From the War Memorial to the Southern T-head Groyne, a series of 50m long groynes spaced at 50m intervals will replace the existing scheme of timber groynes. Between the two T-head groynes a shorter 45m groyne will be installed to help maintain a wider beach in the centre of the bay. Between the northern T-head groyne and Cobbolds Point, four straight groynes of varying lengths between 60m and 70m will be constructed combined with a shingle recharge. The mean MHW position during the final year, together with the minimum and maximum positions during this period are presented in Figure 5.

The modelling results show that the beach responds quickly to the introduction of the groyne scheme, as shown by the saw-tooth pattern of the shoreline in Figure 5. The results are generally encouraging. The beach along the whole frontage remains fairly wide throughout, with only a limited number of locations where the annual shoreline positions are landwards of the initial shoreline before the scheme was implemented. Comparing the shoreline position in Figure 5 with that predicted in Figure 4 (i.e. without groynes), the beach width along the Spa Pavilion area is greater than without the groynes, indicating the groynes are reducing longshore transport southwards. Furthermore, the calculation of change in beach area above the MHW contour indicates a reduced amount of loss to only **160m²/year**.

In places, Figure 5 shows the minimum beach position as being landward of the seawall. This is done deliberately to emphasise the prediction of low beach levels along the seawall in a situation where, at high tide, there would be no beach in front of it. The actual beach level at the base of the seawall and the assumed beach slope is used to predict how far landward the MHW contour would be if the seawall was not present. So the further landward the beach position is shown, the lower the predicted beach level just in front of the seawall. There is the potential for problems with overtopping, scour and even seawall undermining at these locations where the beach is very narrow.

These problems are worst, as would be expected, to the south of the bulge in the seawall along the Spa Pavillion frontage and the groyne spacing and lengths could be subtly altered to at least provide a more uniform beach width. The beach also becomes narrow to the south of the 45m groyne between the existing two T-head groynes. The aim of this shorter groyne is to help maintain a wider beach in the centre of the bay. The existing beach width however, is already narrow at this location (i.e. at high tide there would be no beach in front of the seawall) and there is erosion on the south side of the shorter groyne. This area may benefit from some shingle renourishment and the groyne could be made a bit shorter to allow some more bypassing.

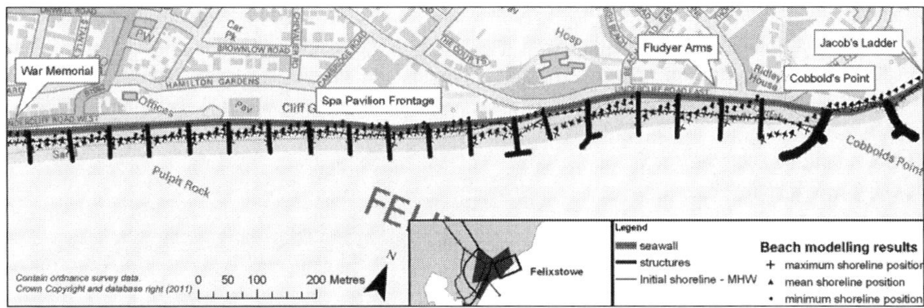

Figure 5.   BEACHPLAN results for Scheme 1

Scheme 2: Scheme refinement

Based on the analysis and assessment of the modelling results of Scheme 1 and the results of further modelling, a refined scheme was developed. This scheme consists of a larger beach recharge, which increases the beach crest to a width of 5m from Cobbold's Point to the southern end of the Spa Pavilion, and a series 18 groynes of various lengths along the frontage, see Figure 6. Table 1 provides further details of the groyne lengths. This scheme was then tested in the model.

Table 1   Details of groyne dimensions

Groyne Number	Groyne Length	Groyne Number	Groyne Length
1 to 9	50m	14	35m
10 and 18	45m	15 and 16	65m
11 to 13	40m	17	60m

The modelling results show a general improvement, from Scheme 1, in beach widths along the whole frontage and none of the annual shoreline positions south of Cobbold's Point are landwards of the seawall (Figure 6). The saw-tooth pattern of the shoreline is still evident in places but the downdrift erosion is reduced and the shoreline does not cut-back as much as seen in Scheme 1. Comparing the shoreline positions in Figure 6 for the refined scheme with that predicted in Figure 5, the beach width is generally greater along the whole frontage; largely because of the greater amount of recharge being placed along the beach. Adjustments made (as part of this refinement phase) in groyne numbers, spacing and length, mainly seek to make beach widths more uniform along the frontage; therefore reducing problems of narrow beach widths in some areas by reducing excess amounts of sediment retained in other areas. The overall improvement to beach widths is primarily down to the increased volume of recharge material being applied to the frontage. There could also be opportunities for further refinements/adjustments to the groynes, if deemed necessary, following the analysis of beach survey data, which should be undertaken as part of a long-term monitoring programme.

The long-term erosion problem, however, can only be solved by providing more shingle into the system than there is leaving the system. The installation of the groyne scheme helps to reduce the loss of shingle to the south but not prevent it entirely, and hence cannot do more than reduce the rate of beach erosion i.e. there will always be downdrift erosion, so the problem of narrow beach can not be totally eliminated. In order to maintain a healthy beach in the long-term there will be a need for observation and monitoring of beach levels coupled with regular beach recharge in the future. The model can also be used to indicate the likely maintenance requirements for the beach by calculating the difference between the volume of shingle entering the system through longshore drift and that leaving the system. For the refined scheme, Scheme 2, the modelling shows an average volume of around 1,500m³ would be required each year to replace the amount lost south to longshore drift. There is also the possibility of recycling shingle from some of the healthier beach sections further south and leaving any periodic recharge to be an operation undertaken at longer intervals to make the mobilisation costs more acceptable. The best timing and placement of this extra beach sediment should be informed by the analysis of the data collected during the proposed monitoring programme.

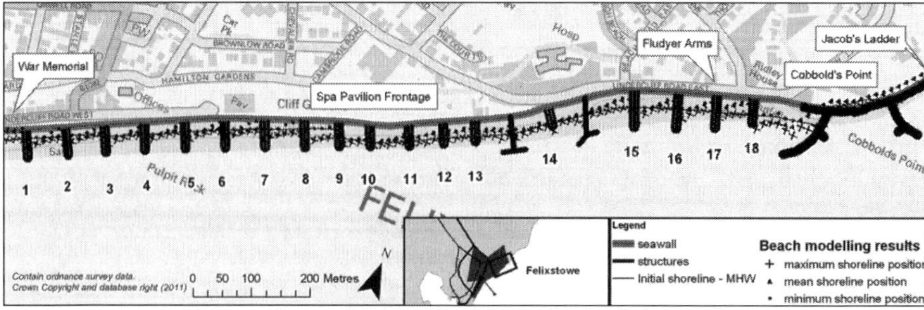

Figure 6.   BEACHPLAN results for Scheme 2

Sensitivity testing

The long-term evolution of the beaches along this part of the Felixstowe frontage is dominated by the likelihood that beach sediment, including that placed during the recharge operation, will migrate southwards, thus reducing beach widths. So far the various schemes designed to reduce the loss of beach sediment, have been tested under wave conditions that result in approximately average alongshore drift conditions.

It is possible however, that this process may occur more rapidly in the future, for example should there be an increase in the percentage or energy of waves approaching the site from the east. It is therefore sensible to assess the likely response of the beach against such a possible change. For this purpose, the model was re-run for a hypothetical five degree change in wave direction, leading to a greater longshore drift rate (mimicking the potential impacts of climate change).

In comparison with the results from the wave conditions giving average drift conditions, it was apparent that there was little difference in beach widths apart from a localised decrease in beach width to the south of Groyne 18. The implication is that the optimised scheme generally performs well under both the average drift conditions and the increased drift conditions.

Conclusions

The main purpose of the planned improvements to the Central Felixstowe frontage is to improve the standard of coastal defence through a combination of beach recharge and the installation of structures designed to control the beach plan-shape and obtain the best value from the recharge operations i.e. maintain uniform beach widths along the whole frontage. A numerical model was used to assist in the appraisal of the preferred coastal management scheme.

During the calibration process the model demonstrated, to councillors and the general public, a satisfactory comparison with existing beach characteristics. It was therefore agreed that it could be used to form a reasonable basis on which to examine the beach recharge and the effects of beach control structures

Through the use of the model it was demonstrated that a combination of a beach recharge from the southern groyne at Cobbold's Point to Spa Pavilion, a rock revetment and a groyne scheme (Scheme 2) generally performs well. The beach along the whole frontage remains fairly wide throughout, with only a small number of locations where the shoreline positions are landwards of the initial shoreline before the scheme was implemented. These erosion problems are generally worst to the south of the groynes, as a result of the net southwards longshore drift, and cannot be totally eliminated in the long–term.

In order to maintain the beach in the long-term, there will be a need for further beach recharge at some point in the future. For the refined scheme, Scheme 2, the modelling shows an average volume of around $1,500m^3$ will be required each year to replace the amount lost south to longshore drift. To minimise mobilisation costs any periodic recharge operation should be undertaken at a longer interval. It may also be possible to manage beach levels through recycling shingle from some of the healthier beach sections and it is recommended this is investigated during the detailed design process. The best timing and placement of this extra beach sediment should also be informed by the analysis of the data collected during the monitoring programme.

Sensitivity testing of Scheme 2, under a hypothetical increased drift rate scenario indicates that this scheme still maintains a fairly wide beach along the whole frontage with only localised areas of increased erosion. Overall, the model was a useful and appropriate tool for aiding the appraisal of the coastal management scheme.

Recommendations

It was recommended that a regular monitoring programme is established in order to assess shoreline evolution along the Central Felixstowe frontage, and performance of the new scheme. This should involve monitoring of beach widths particularly within the new groyne bays, and should include plan-shape surveys as well as profile surveys.

Acknowledgements

The authors would like to thank Dr. Alan Brampton of HR Wallingford and Peter Phipps of Mott MacDonald for all their technical advice and support throughout the project. The authors would also like to thank Suffolk Coastal District Council and the Environment Agency for their support and data supplied.

References

CIRIA. 2010. *Beach Management Manual. C685.* London: CIRIA.

HR Wallingford (1997). "Harwich harbour approach channel deepening – Impact on wave and coastal conditions", HR report EX3703, September 1997.

Innovative Coastal Zone Management
ISBN 978-0-7277-5749-4

ICE Publishing: All rights reserved
doi: 10.1680/iczm.57494.597

COST EFFECTIVE IMPLEMENTATION OF A SHORELINE MANAGEMENT PLAN IN AN ENVIRONMENTALLY SENSITIVE AREA: SEASALTER TO GRAVENEY SEA DEFENCES.

Daniel Glasson[1], Geert Meijers[2], Alexandra Schofield[1] and Kevin Talbot[3]
[1]Black & Veatch Ltd, UK, [2]Team Van Oord, UK and [3]ncpms - Environment Agency, UK (Southern Region)

Introduction

The Environment Agency has recently completed a £1.1m coastal defence project on the North Kent coast.

The 6.7km frontage between Faversham Creek and Seasalter had been identified in the regional plan as a key contributor towards the successful delivery of the region's targets for flood protection. The resulting solution was designed to prolong the life of the existing defences in accordance with the approved Shoreline Management Plan (SMP) and included the construction and refurbishment of a number of strategically placed timber groynes and the importation of 7,500m^3 of beach recharge material.

The main challenge faced by the project team was overcoming the numerous environmental constraints, particularly in terms of the programme, whilst implementing this adaptive solution and helping to achieve regional targets.

Background

The existing sea defences between Faversham Creek in the west and Seasalter in the east protect the 1,700ha of flood risk area behind, known as the Graveney floodplain. This floodplain is bounded to the south and east by high ground and to the west by Faversham Creek. The area is generally rural in nature with the predominant land use being agricultural. Much of the hinterland is marshland below Mean High Water Springs (MHWS) that has been progressively drained over the past few centuries.

The existing coastal defences along the frontage predominantly consist of a concrete seawall on top of a clay embankment. The seaward face is largely being protected by either a concrete blockwork or a stone pitching revetment, both of which are fronted by a small sand/shingle beach. The crest level of the defence is high and currently provides a minimum standard of protection of 1 in 50 years (annual probability of flooding of 2%). Whilst the seawall is generally in fair condition it has been vulnerable to undermining through beach losses at isolated locations. Historically beach recharge and groyne maintenance was routinely undertaken along the frontage, however the latest record of this was in 1996. The maintenance of existing defences and beach management is carried out by the Environment Agency using its permissive powers under the Land Drainage Act (1991).

The majority of the 871 residential and 517 non-residential properties (including caravans) at risk within the floodplain are located at either Faversham or Seasalter. At Seasalter, many of the properties are below MHWS, which would represent a significant risk to life should the defences fail. Other assets at risk include the Faversham to Thanet main line railway and high voltage trunk power line (on pylons) which both cross the floodplain. There are also two Scheduled Monuments and several listed buildings within the flood risk area.

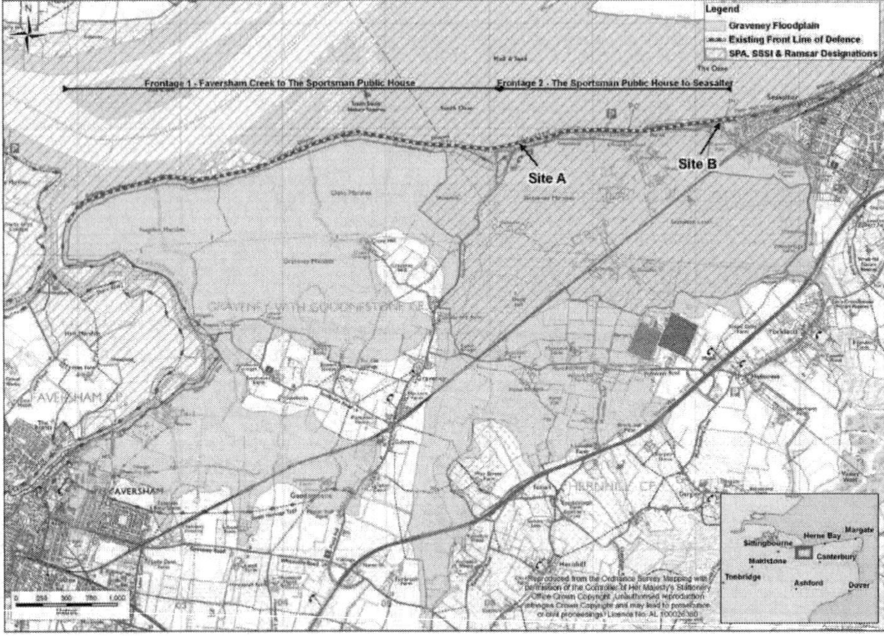

Figure 1 – Graveney Floodplain

In addition, the eastern part of this floodplain along with both the beach and inter-tidal mud flats in front of the seawall are internationally designated as a Special Protection Area (SPA) and as a Ramsar wetland for its rare breeding and over-wintering birds. It is also nationally designated as a Site of Special Scientific Interest (SSSI).

Strategic Context

The Isle of Grain to South Foreland SMP was adopted in 1996, proposing a preferred management policy of 'Hold the Line' for this section of the coastline protecting the Graveney floodplain. The SMP noted, however, that future strategies for the Graveney area may involve managed realignment.

In 2004 Canterbury City Council (CCC) in partnership with the Environment Agency completed a strategy for the 10.7km frontage from Faversham Creek to Whitstable Harbour that was approved by Defra in January 2005. The strategy recommended a preferred option of a "minimum beach" for the Graveney frontage to reduce the risk of undermining and thereby prolong the life of the existing flood defences. In July 2007 the Draft SMP Review (SMP2) was issued for consultation. This proposed a 'Hold the Line' policy for the Graveney frontage in the short term (0 – 20 years), with managed realignment in the medium and long term (20 – 100 years). However, significant local pressure has led to the Graveney frontage being split into two differing policy units, and the subsequently revised SMP2 for the area outlines the following policies for the frontage:

1. The defences from Faversham Creek to the eastern end of the Sportsman Public House has policy of 'Hold the Line' in the short term (0-20 years) and a managed realignment policy over the medium to long term (20-100 years).
2. The defences between the Sportsman Public House and Seasalter has a policy of 'Hold the Line' in the medium term (0-50 years) and a managed realignment policy in the long term (50-100 years).

Delivering Protection Targets

The Environment Agency is committed to the justifiable protection of properties against the risk of flooding. Outcome Measures have been developed by Defra to help operating authorities such as the Environment Agency design programmes of work to meet the targets set out in the latest government Comprehensive Spending Review (CSR 2007) for the period April 2008 to March 2011.

It was identified that by undertaking works to protect the Graveney flood risk area by March 2011 a relatively large contribution could be made to Defra's regional Outcome Measure targets for protecting both households and deprived communities, which in turn would contribute towards the CSR 2007 target relating to protecting properties from flooding. For the project to qualify the works needed to be completed by the end of the 2010-11 financial year.

To comply with the strict environmental constraints (outlined below), it was essential that the construction works were undertaken between mid-August and October 2010. However, funding was only secured following the completion of the appraisal phase in April 2010, therefore resulting in an extremely tight programme for the delivery phase.

Adaptive Approach

The most recent SMP2 policy states that 4.4km of the frontage is to be realigned in 20 years and the remaining 2.3km will be realigned in 50 years. Consequently, the main challenge faced by the project team was how to cost effectively continue maintaining the entire frontage and achieve Defra's Outcome Measures until such time as the frontage is realigned.

The adaptive approach taken by the project team was to firstly assess the entire frontage and identify the areas that were most at risk of failure, and secondly consider each of these sections independently with a view to prolong the life of the existing defences until such time as the frontages are realigned. This involved a thorough condition survey of the existing frontages, where all the defences were scrutinised for weaknesses that would prevent the defence lasting the required time and an analysis of the local coastal processes that affected the shoreline.

Using this adaptive approach a number of possible options were considered at the project appraisal phase that focussed on implementing the approved SMP2 policy, i.e. extend the life of the sea defence until such time as the defences are realigned.

Existing Defence Issues

Following a condition assessment of the existing defences the following issues were identified:

Frontage 1 - Faversham Creek to the Sportsman Public House
In recent years, only minor maintenance has been carried out along this 4.4km frontage, yet the existing defences were generally found to be in reasonable condition, with the following exceptions:

- At various points along the revetment the blockwork has started to break away. In time this will expose the clay embankment to erosion and eventually lead to the failure of the seawall on top.
- The seawall itself appears to have suffered some isolated differential settlement and the majority of the seals between the sections are no longer functioning.

However, given the existing high standard of protection and the SMP2 policy for this frontage of managed realignment in 20 years, these issues were not considered to be critical and going forward should only be addressed through any available maintenance budgets.

Frontage 2 - Sportsman Public House to Seasalter
The SMP2 policy for this 2.3km frontage is to 'Hold the Line' for the next 50 years. However, several significant issues with the defences along this frontage were identified:

- Western end of the frontage (East of the Sportsman Public House - Site A): several of the timber groynes were in a particularly poor state of repair and localised erosion had resulted. Beach levels here were lower than elsewhere and the stone pitching at the base of the seawall was exposed. The foundations to the nearby access steps were also exposed.
- Eastern end of the frontage (Seasalter - Site B): beach erosion was occurring due to the limited supply of material from the east (the beach to the east is retained by the CCC groynes). In addition, the sea wall changes alignment at this location and moves more seaward to join with the CCC seawall to the east. Consequently the beach levels were extremely low and the sheet piling foundations to the seawall were exposed and at risk of failure.

Since the issues identified along Frontage 2 would potentially have prevented the SMP2 policy from being implemented, it was clear that any proposed works would need to address the deficiencies in the defences at both the East of the Sportsman (Site A) and Seasalter (Site B) locations.

Figure 2 – Before Works Photographs

Coastal Processes

Following the condition survey a coastal processes desk study was undertaken. The purpose of this was twofold: 1) to gain an understanding of the existing coastal processes along the frontage and any potential detrimental impacts they have had on the existing defences; 2) to further investigate how these processes may be impacted by any subsequent coastal defence works, especially since the local coastal processes are a key feature of the SPA/Ramsar designations.

Following the completion of this analysis it was demonstrated that a strong net transfer of beach material from east to west took place along the frontage and the following conclusions were drawn on the impact of coastal process at the proposed working locations:

Site A - East of the Sportsman Public House Frontage

In general the existing frontage is relatively stable. However, over the past 5-10 years some localised erosion has taken place in areas where there are some dilapidated groynes that are failing to hold the beach material in place. Therefore the proposed works at this frontage are essentially mitigating for the deterioration of the existing groynes due to the lack of maintenance over the years and returning the beach to the levels that were present before the groynes failed.

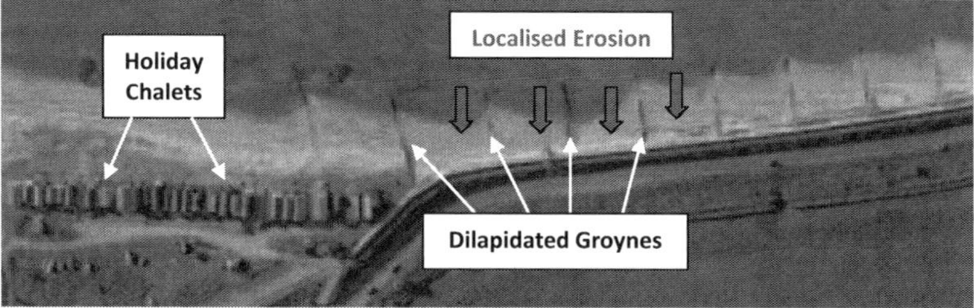

Figure 3 – Site A Beach Dynamics

The proposed works at this frontage are very localised. The drift of material from the frontage to the immediate west is likely to be reduced, but this will be counteracted by the importation of new material to the system. The proposed works will effectively restore the frontage to its previous condition prior to the deterioration of the groynes. The overall impact on the local coastal processes is expected to be negligible.

Site B - Seasalter Frontage

The net analysis of longshore sediment transport at this particular frontage shows that this localised loss of beach material is in spite of a continuous drift of material from the CCC frontage. The concrete slipway acts as a terminal groyne that restricts the direct flow of material to the west of the frontage, consequently material builds up against the slipway until it can eventually pass around it; as a result the western side of the frontage is stable.

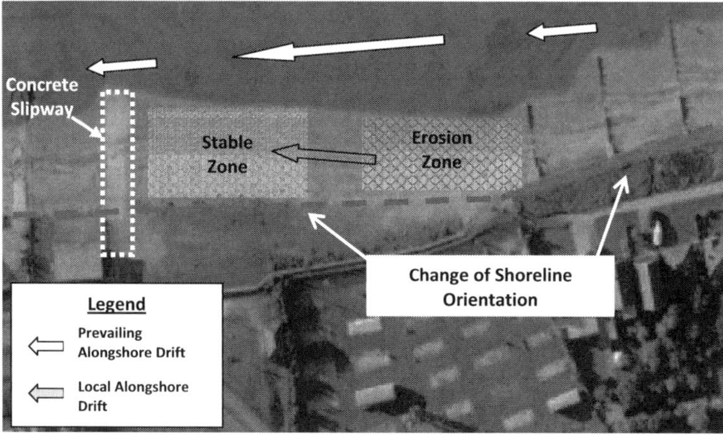

Figure 4 – Site B Beach Dynamics

The analysis has shown that the end of the neighbouring CCC groyne field combined with the presence of the slipway acting as a terminal groyne had effectively created an over sized groyne bay at this location. The natural effect of longshore drift on this bay had starved the eastern section of the frontage of beach material. This, coupled with a change in shoreline orientation and seawall alignment, had left the seawall on the eastern side of the frontage more exposed and therefore vulnerable to failure.

The proposed works of some localised beach recharge are required to provide the vulnerable section of the seawall with some additional protection. The two proposed new groynes will act to hold this new material in place. This will effectively extend the CCC groyne field and reduce the size of the Seasalter bay (between the concrete slipway and the CCC groynes) and smooth the transition in shoreline orientation. Therefore, by maintaining this current drift regime the impact of the works on the local coastal process is expected to be minimal.

Delivery Team

Following the completion of the appraisal phase and the funding award in April 2010 the delivery phase of the project commenced. The organisation of the project's delivery team is shown below in Figure 5.

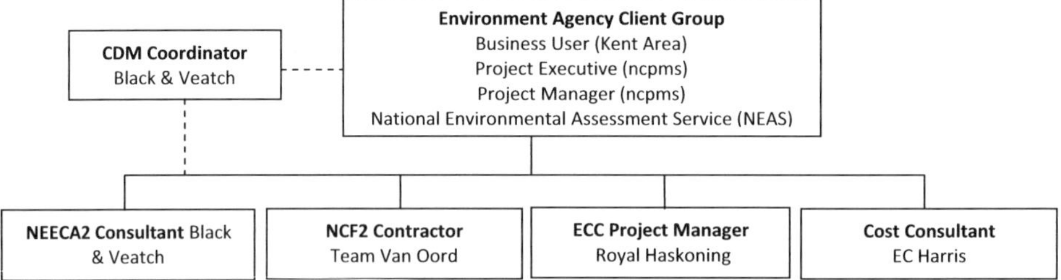

Figure 5 – Team Organagram

The Implemented Solution

Following the completion of an extensive appraisal process and the funding award, the resulting preferred option and ultimately implemented solution has addressed the deficiencies with the existing defences so that they can continue to protect the floodplain in the short to medium term. The works included the following:

East of the Sportsman Public House – Site A
- Replacement or refurbishment of four existing timber groynes that were in a particularly poor state of repair and were no longer effective in holding the beach at this location.
- Beach recharge of approximately 3,000m³ of imported material to reinstate the locally depleted beach levels.

Seasalter – Site B
- Beach recharge of approximately 4,500m³ of imported material to reinstate the locally depleted beach levels.
- Construction of two new timber groynes to address a localised focus of erosion through beach loss, by helping to retain this material in place and thereby reducing the amount of "repeat" recharge that would be required.
- Install a Navigational Marker at the request of Trinity House.

However, despite the completion of these works continued beach recharge will be required at regular intervals to counteract the continued westwards net drift of material and to mitigate the impact of storm events. It is predicted that approximate quantities of $1,000m^3$ at Site A and $2,000m^3$ at Site B will be required every ten years. In addition, the continued maintenance of both the new and existing timber groynes will also be required.

This continued beach management would be sufficient to prolong the life of the seawall at Seasalter for the next 50 years. In addition, through the natural processes, these works have the potential to provide increased protection to the neighbouring western frontages from the longshore westward net drift of material.

Figure 6 – Photographs of Construction Activities

Health, Safety and Environmental Management

Managing the health, safety and environmental risks in an efficient and effective manner has been essential in achieving a successful project delivery. Some of the key aspects managed through the delivery of this project are detailed below:

Habitat Regulations

The beach, inter-tidal mudflats and part of the area protected by the sea defences are all designated internationally as both SPA and Ramsar wetland for breeding and over-wintering birds. This meant that works could only take place from mid-August to end of October and/ or February to March.

Given the additional need to deliver in the 2010-11 financial year to help meet Defra's Outcome Measure targets and in order to maximise cost efficiencies it was preferable for the works not be split between the two working windows. However, the estimated construction period was 11 weeks, therefore it was essential that the construction works commenced on schedule in mid-August 2010 to enable completion by the end of October the same year.

This programme constraint was identified at the earlier appraisal stage of the project and therefore on commencement of the delivery phase, all of the project team were aware of the challenging time scales. Team members were therefore aware that deliverables, particularly submissions for consents for the works, needed to be right first time as any delay would jeopardise the project programme.

In addition to the constrained working window, proximity to a sensitive SPA meant that tidal working restrictions were also required for environmental purposes as well as the usual health and safety restrictions. Following negotiations with Natural England it was established that works could only be undertaken during daylight hours, and in October, no work could take place 1.5 hours before and 0.5 hours after high tide, due to roosting birds. Careful planning of the construction activities was therefore required.

The habitat regulations assessment (HRA) was undertaken with a close integration between the engineering (designer and contractor) and environmental teams so that the environmental, technical and buildability issues were well understood by all. In addition, early engagement and close working with Natural England was also undertaken so that the environmental constraints were clear at an early stage. Both of these meant that a detailed method of working was planned at an early stage, incorporating the required constraints and mitigation measures, allowing for Natural England to issue their approval upon their first review.

Legislative Consent

During the environmental screening process, it was identified that the works would require a statutory Environmental Impact Assessment (EIA) under the Marine Works EIA Regulations (2007). Being a relatively new piece of legislation, the project team had only limited experience of these regulations, consequently the Marine Management Organisation (MMO) was engaged early to discuss the legislative requirements.

Due to the impact of other environmental constraints on the project programme it was also agreed with the MMO that we would submit the EIA consent request to them at the same time as the required marine consents applications. This streamlined the process so that we only needed to undertake one period of consultation. In addition, by liaising with the MMO and their statutory consultees in advance, the project team were able to ensure that they had captured the requirements and key issues within the submitted application. This was supplemented throughout the determination process where communication with the MMO was maintained, enabling the project team to respond to consultee queries and comments as and when they were received by the MMO rather than at the end of the determination period. This proactive approach by the team meant that the process went very smoothly and all the consents were received on time enabling the project to proceed on programme.

Traffic Management

Due to the flat nature of the internationally designated foreshore, and the relatively low quantities of material required it was neither cost effective nor environmentally desirable to import beach material from an off-shore source. Consequently all materials had to be delivered via the local road network.

However, during the public exhibition in January 2010 it became apparent that access to the site using the local road network was a key issue, particularly as some of the local roads would be used at the same time by another large construction project (London Array). As a result early engagement with Kent Highways and London Array commenced without delay. This led to the early identification and development of traffic mitigation measures that were recorded in a pre-contract traffic management plan (TMP). These traffic management mitigation measures included:

- Implementing a temporary speed limit on the local roads to reduce the differential between local traffic and site vehicles.
- Ensuring that the restricted access at each site from the main road was very carefully managed by the contractor and each delivery was supervised by two members of the site

team. This management extended to proactively keeping the road surface clear of excessive mud that may have presented a hazard to other road users.

- Laying a temporary road surface along the access track to Site A to improve the ground conditions. On completion of the works, this temporary surface was reused on another Environment Agency site to avoid its disposal to landfill.
- Implementing a controlled crossing of the Public Right of Way (PRoW) and public parking area at Site A that utilised Banksmen to effectively marshal both site traffic and members of the public. Fencing also provided a barrier between the site vehicle's beach access route and pedestrians whilst still allowing the public and beach chalet owners to access the beach.
- Consultation with both Natural England and the CDM-Coordinator to develop a safe and environmentally preferred route across the beach that negotiated both the existing timber groynes and protected vegetated shingle.

Protected Flora and Fauna

Rare vegetated shingle, eel grass and saltmarsh were all located along this frontage, the protection of which meant that there was restricted access to the working areas. Access routes across Sites A and B were discussed with the contractor and were clearly marked on the drawings for construction, and protected areas were segregated from site activities.

Where the works were required to encroach on vegetated shingle, the top layer of shingle and vegetation was stripped and stockpiled for replacement on completion. Stripping and storing locations were agreed in advance and clearly shown on the construction drawings.

A survey undertaken by an expert identified that there were several dock plants with eggs of the rare Fiery Clearwing moth within the area of the works. This moth is now restricted to a small number of coastal sites in Kent and it is therefore not possible to disturb these plants without a special licence. By working closely with the contractor and Natural England we were able to agree a revised working method that avoided disturbing the identified plants. However, to avoid any potential delays during the works, the required licence was obtained in case the relocation of these dock plants did prove to be necessary.

Supervision of all of the works associated with the protection, stripping and replacement of vegetated shingle was coordinated by the NEAS Officer and the Environmental Clerk of Works. This required close communication with the contractor to ensure availability of key personnel at the required time.

Pacific Oyster Herpes Virus

There are some oyster beds located just offshore from Seasalter. Only weeks before works were due to commence on site the Centre for Environment, Fisheries and Aquaculture Science (CEFAS) declared the coastal waters at Seasalter a contamination zone due to the discovery of the Pacific Oyster herpes virus; the first known case in UK coastal waters. The project team therefore had to develop bio-security measures at short notice to mitigate the risk of spreading this disease to other sites. These measures were developed through close liaison with CEFAS and the contractor and the methodology has now been communicated internally within NEAS.

Active Environmental Management

A detailed Environmental Action Plan (EAP) was developed to identify the outstanding actions to be undertaken. The works were audited against the EAP on a monthly basis, with an updated EAP and audit report being sent to the team ahead of the progress meeting, where any identified issues were raised and discussed. In this way the EAP was successfully used to monitor the achievement of actions. Prior to the commencement of construction activities on site the NEAS officer worked with the contractor to develop the site induction so that it adequately covered the environmental challenges of the site. In addition, a toolbox talk was also undertaken by the NEAS Officer therefore ensuring that the site team understood the restrictions that were in place, and why it was so important that they worked within them.

Public Engagement
The works were undertaken in close proximity to a number of properties. Therefore noise and vibration monitoring was carried out to provide evidence of our compliance with the agreed limits. To mitigate possible complaints an extensive newsletter drop was undertaken, which included details of the access route and provided information on the works and its anticipated impacts as well as contact details for more information.

Conclusion
A cost effective solution was delivered both on time and under budget enabling it to contribute to the region's delivery targets. The real success of this project was implementing a coastal defence scheme in an extremely environmentally sensitive area, working under strict constraints with minimal lasting impacts and no incidents throughout construction.

The need to deliver to a very challenging programme, yet under strict environmental constraints required the project team to proactively manage safety, health and environmental risks. As a result a number of best practices were developed throughout the delivery phase of this project that have been paramount to its success.

The key success factor is good communication and planning with an emphasis on:
- **Engaging consenting bodies early** to ensure we meet their expectations and pass through the consenting process as smoothly as possible.
- **Good team work** to ensure all understand the project and the challenges/risks to be managed.

Figure 7 – Completion Photographs

Innovative Coastal Zone Management
ISBN 978-0-7277-5749-4

ICE Publishing: All rights reserved
doi: 10.1680/iczm.57494.607

'Stitching up' the Past – The Strengthening of Three Heritage Marine Structures in Jersey with Needling Technology

Ray Hine, Technical Services Manager, Port of Jersey, States of Jersey
Steve Hold, Arup, Cardiff, United Kingdom
Ana Ulanovsky, Arup, London, United Kingdom

Introduction

Figure 1 - St Aubin's North Pier and St. Aubin's Fort

There are a considerable number of difficulties in understanding the behaviour of old mass gravity marine structures of dressed granite blocks with secondary rock core fill. The original form of these structures was often constructed over earlier breakwaters or rock outcrops and sited at geographically or historically important locations. Frequently, they have great historical and heritage value as well as local population appreciation as is the case with the three structures discussed in this paper; the three of them located in St. Aubin's Bay on the south coast of the Channel Island of Jersey (Figure 1). The three structures are the North Pier and South Pier of St. Aubin's Harbour (1 & 2) and the St. Aubin's Fort Breakwater (3). In the interest of clarity, this paper will focus on the North Pier and the Fort Breakwater. Both structures are exposed to waves from the south west and act as breakwaters. Frequently, heritage maritime structures (360 years old in the case of the St. Aubin's Fort Breakwater), provide the custodians with little income but unfortunately they do require indefinite maintenance expenditure.

Figure 2 - North Pier

This paper describes the development of a cost effective solution to keep the integrity of these historic structures that have suffered accelerated structural deterioration. The paper explains in some detail the levels of investigation that took place, including the use of innovative survey technology and the design evolution, together with the construction methodologies that were adopted for the site works are described in some detail.

Figure 3 - St. Aubin's Fort Breakwater

The practical evolution of the 'stitching' concept was developed with input from the Jersey civil and marine Contractor, Geomarine Ltd and the manufacturer of the stitching anchors, Cintec.

The problem

Both structures have suffered ongoing failures or unacceptable movements during the past decades, on the inner sheltered side of the breakwaters.

In 2001 the North Pier was assessed as being in a state of instability due to movement of its internal wall and apparent settlement of the interior of the pier. The deck level of the North Pier is surfaced in asphalt which waterproofs the deck but also masks (to a certain extent) masonry movements. Large cracks however had developed and could be seen in the asphalt indicating a threat of impending wedge failure of a section of the inner wall (Figure 4).

Figure 4 - North Pier - cracks at deck level

St Aubin's Fort Breakwater has in turn suffered a re-occurrence of masonry movements and loss of pointing at its inner, lower deck wall (Figure 5). In 2008 a bulge was also identified in part of this inner wall which prompted concerns about this part of the breakwater's structural integrity.

The two structures, at first sight, appear to be constructed of similar large pink dressed granite blocks that are similar to those found in other marine structures around the Jersey coast. Our studies of both the construction and behaviour of these piers revealed that both their integrity and performances were very different. The causes of their recent accelerating deterioration, their structural problems and the loads they are subjected to

Figure 5 - Fort Breakwater - area of 'bulging' and rebuild

vary significantly. Therefore, the solutions discussed in this paper address two different failure mechanisms. The main challenge for both however, was to develop a semi-flexible design solution through holistic investigation, which would aid the structures in behaving as they have during centuries without imposing current engineering canons.

Investigations and Analysis

Historical Desk Study

As the two marine structures were built centuries ago and have 'evolved' in terms of geometry and their constituent materials, meaningful detailed drawing information was not available.

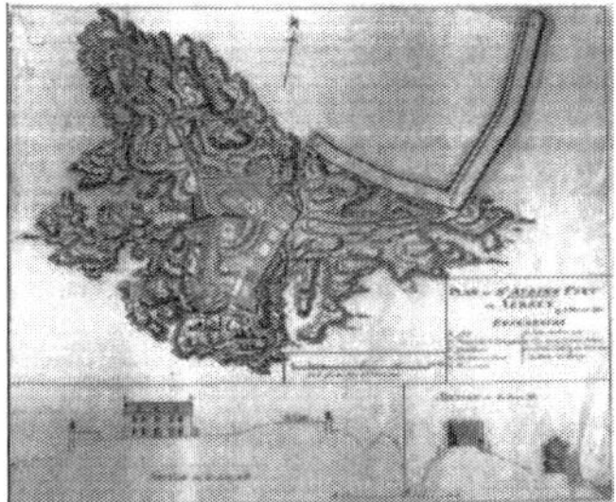
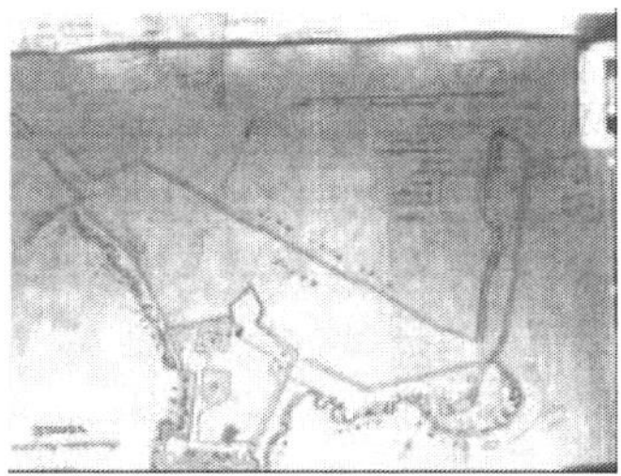

Figure 6 - 1775 Map Fort & Wharf and then 1876 Plan of Fort Breakwater enlargement

Literature and web searches were undertaken to determine the history of the sites. The records for St. Aubin's area revealed that a structure first appeared in the 1540s when St. Aubin's Fort was built to protect a boat building wharf. A plan of St. Aubin's Fort with a wharf and breakwater in 1775 shows the initial construction of the Fort pier and a wharf. A plan known as the 'perambulation plan' of 1876 appears to show a new breakwater on a different alignment, which includes an inner deck at a lower level and a full wharf for boatbuilding at the Fort. Visible construction joints suggest that the inner deck was an addition to the original Breakwater. This area appears to have an inferior quality of design and construction.

Investigations also revealed that the inside lower deck had failed and had collapsed four times

during the previous 40 years. The largest of these collapses was in 1972 when the inner wall of the Breakwater was re-built and pressure grouted (Figure 7). The extent of pressure grouting was such that most of the final leg of the Fort Breakwater can now be considered as a 'solid' structure.

Figure 7 - St. Aubin's Fort Breakwater - 1972 failure of inner wall

This is significant in as much that a permeable core more readily absorbs wave energy from waves striking the outer wall reducing wave loads on to the inner face of the stones forming the inner wall.

The original construction of the Fort Breakwater was intended to act as a mass gravity free draining structure. It had external walls of dress stone with secondary core material that over time disintegrated and was washed out. Throughout its earlier life it would have had intensive labour care and maintenance i.e. patched and repaired as necessary and void loss topped up.

Investigations at the North Pier identified the cause or 'trigger' of the inner wall movement to possibly be due to over dredging in the harbour in the late 1990s exacerbating poor soils, and removing toe stability from the base of the inner wall.

Intrusive Site Investigations

Intrusive site investigations took place on both structures. To identify the reason for the settlement and rotation of the inner face of the North Pier, boreholes and trial pits were dug in 2003. Investigations showed that the inner wall was founded on runny sands. The monitoring enabled preliminary conclusions to be drawn in 2005 that the inner wall of the North Pier required rebuilding and a stable foundation. Conventional surveying and levelling enabled cross sections to be produced that were used to monitor movements of the North Pier.

Environmental considerations

As part of the planning application, the first stage of an environmental impact assessment was initially carried out. In the case of St. Aubin's North Pier, Société Jersiaise responded as part of the consultation that they had knowledge of two types of endangered mollusc species that had been found in the voids and openings of the upper levels of the North Pier [Truncatella Subclindirca and Palunderiell Littorina].

LiDAR Surveys

The verticality of the inner face of the North Pier wall and the instability of the St. Aubin's Fort Breakwater required accurate measurements to assist in engineering analysis. Conventional surveying of both of these marine structures was impractical because they are of varying vertical and horizontal geometry and access to their vertical faces in exposed tidal conditions was not feasible. To overcome this and to provide accurate dimensions in polar coordinates, LiDAR techniques were used to survey both structures. This allowed 'virtual',

Figure 8 - LiDAR Survey of Fort Breakwater

accurate scale models of both piers and their surroundings to be produced to an accuracy of 2 to 3mm. The models also enabled accurate two and three dimensional drawings to be created so that the structures could be studied and remedial works designed and tendered.

Figure 9 shows the importance of this data in assessing the verticality of the interior wall at the North Pier. The risk of wall collapse was evaluated with the aid of this technology. Notional cross sections of the structures were also produced when the LiDAR superficial

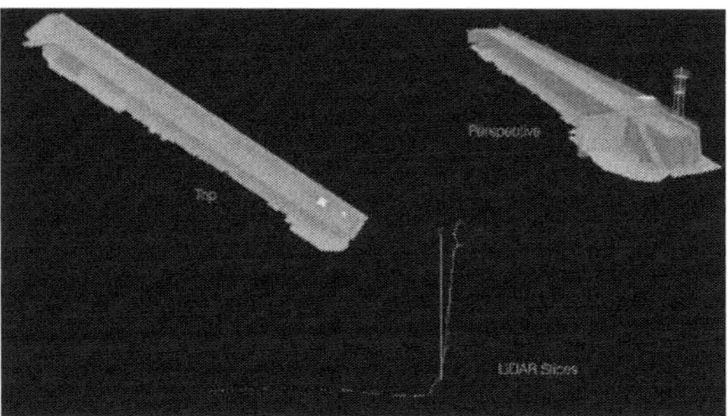

Figure 9 - Graphical model demonstrates movement of North Pier

profile was integrated with the intrusive site investigation data. The production of the cross sections enabled a better understanding of the structures and therefore the reasons for their types of failure.

Wave climate

The prevailing wave climate at the site is predominantly generated by wind from between the southerly and south-easterly directions. Jersey is well known for its very large tidal range, 9.5m at spring tide. Although both structures are high and dry at low tides, water depths of approximately 10m at high tides, allow deep water waves to reach them. For the purpose of design and in the absence of size specific wave model outputs, an estimate was made on the extreme wave climate to impact upon the North Pier and the Fort Breakwater. For this effect, results from modelling by HR Wallingford were made available by the States of Jersey Engineers Department. Information on water levels, tidal currents and prevailing wind directions were used to calibrate the wave estimates.

Figure 10 - Example modelling summary of wave climate at St. Aubin's

It was recognised that the wave climate at St. Aubin's North Pier would be reduced due to the sheltering effects of the St. Aubin's Fort Breakwater and the orientation of the North Pier. Therefore a simple MIKE21 Model was set up to assess the impact of the sheltering effect (Figure 10). The results suggest that a 50 year return period wave height at the North Pier is approximately the same as a 1 year return period wave height in the unsheltered St. Aubin's Bay. In this manner, design wave parameters were adopted for the assessment of the North Pier.

Wave pressure propagation

The intrusive site investigation showed that the Fort Breakwater's core near to the roundhead consists of grouted boulders and cobbles in a matrix of cementious grout with non-grouted boulders at the base of the Breakwater, founded on sands. Overall, it can be assumed that the mortar between the faceblocks has disintegrated locally and that the grouted core provides a rigid structure behind the blockwork walls. Wave pressure propagation mechanisms on the fort roundhead and waves arriving at the roundhead produce slowly-varying pulsating wave pressures, treated as quasi-static for structural design purposes below. Short duration or impulsive wave-loads may also occur at the structure during breaking wave scenarios. These are of greater intensity than the pulsating waveloads, but are also of shorter duration. A number of load propagation routes across the Fort Breakwater structure exist that provide a pressure path to the roundhead area in disrepair. The main mechanisms are summarised below:

- Impermeable areas will reflect wave energy away from the structure. However, the material forming the core in the bulging area is fractured and the mortar is deteriorated, this produces weak spots and fracture planes. Pulsating pressures penetrating through the front face blockwork can produce sliding forces on the segments which could be the reason for the bulging of the structure at the inner face.

- Water jets by incoming waves will travel across masonry joints and propagate impulsive pressures to the back of the masonry. Wave impulsive pressures can also propagate through fractures across the core and travel along the back of the masonry if voids exist between the blockwork and the grouted core (considered likely).

- Pulsating pressures from the bottom layers of the breakwater which consists of un-grouted core.

- Downfall pressures produced by overtopping. Also a source for washout of loosened or deteriorating grouted core materials.

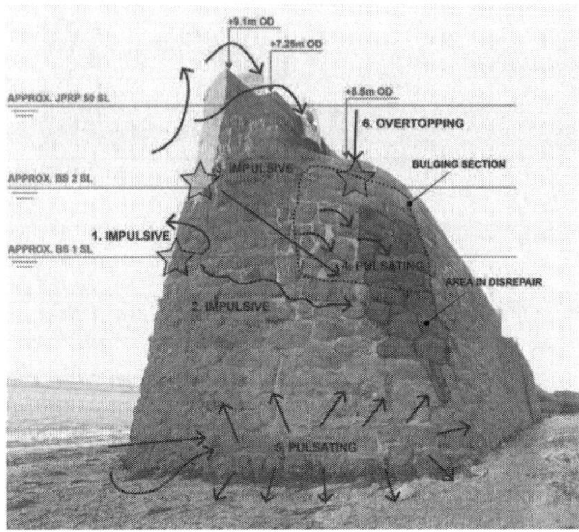

Figure 11 - Combination of loads in Fort Breakwater

In the same way, the major mechanisms that can impose loads on the deteriorated inner harbour pier wall at St. Aubin's North Pier were analysed for use in design. The capacity of the rock core material to dampen the transfer of loads to the inner walls is the subject of a follow up research programme by Port of Jersey and Arup.

The pressure paths above together with more readily identifiable dead-load forces, such as superimposed deck loads, allowed an engineering analysis of the current structural capacity of the North Pier to be made. The imposed loading due to vehicles, boat storage or lifting on to the North Pier was also considered.

Conclusions on Mechanisms

The verticality assessments of the walls, other data obtained from the LiDAR and findings from previous works lead us to conclude that the St. Aubin's North Pier had technically 'failed' and that structural rehabilitation works to the inner wall were vital. The monitoring of the North Pier movement indicated that approximately 2/3 of the length from the landward end was seen to be prone to a catastrophic wedge failure. The three recognized modes of distress on the inner wall were: vertical settlement because of poor substrata and no foundations; horizontal sliding; and overturning towards the vertical at the toe; with the combined effect of increasing the risk of a wedge failure. A triangular section of inner pier from the centre line of the deck to the toe of the inner wall would be vulnerable to collapse (similar to that at the St. Aubin's Fort Breakwater in 1997, see Figure 7).

The situation at the St. Aubin's Fort Breakwater was different. The sea state conditions are more severe as positioned further out in the bay, the wave energy impacting on the outside of the Fort outer wall results in significant transfer of energy to the inner walls causing distress at the inner lower deck. The loss of pointing material at deck level and at the inner face around the 'bulge' had allowed overtopping and tidal water to wash unbounded materials out of its core. It was also identified from the site investigations that the inner wall is situated over an area of softer, more compressible silt. The engineering problems that needed resolution at the Fort Breakwater's inner wall can therefore be summarised as follows:

- Cracks and fissures across the grouted core are providing paths for impulsive wave pressure propagation;
- Settlement of the harbour wall is causing the loosening of blockwork in the deteriorated areas;
- Hydrostatic pressure is building up in the core of the breakwater;
- The construction of the inner lower deck and wall is of a lesser quality than that of the original seaward section of the breakwater and is not toothed into it.

Strategy for Repairs

Many options were considered in 2005 to protect and strengthen the North Pier. The basic engineering solution of constructing a foundation on the outside of the existing alignment and then rebuilding the inner wall on this new foundation was the initial solution favoured. However, the rebuild proposals were not endorsed by the States of Jersey Planning and Heritage Departments, who asked for a 'tying in place' solution to be investigated for stabilising the inner wall. Environmental concerns regarding the endangered mollusc species further reinforced this argument.

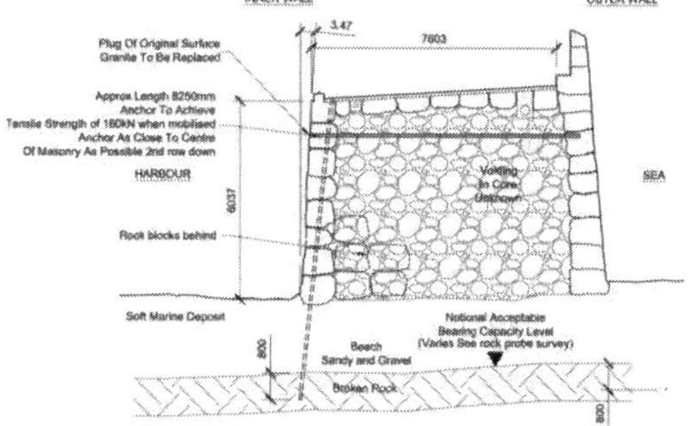

Figure 12 - Cross section of North Pier with stitching anchors

The chosen 'stitching' solution used anchors consisting of a number of solid stainless bars contained within a fabric sleeve injected with cementitous grout after positioning. The bond of the grout that seeps through the fabric sock used in the anchor between the masonry and the reinforced anchor is very high tying areas of loose masonry blocks together. This 'stitching' methodology was also able to provide a foundation solution; the numerous small diameter vertical anchors become mini piles when drilled down through the granite wall stones, through the beach deposits and into the Jersey rock shale beneath. The natural arching of the masonry blocks between mini piles at beach level then provides significant underpinning support to the existing mass granite masonry wall over and above the sea bed level materials.

To resist the inner wall overturning, after pinning the base of the wall with vertical mini piles, requires a form of top restraint in the outward direction (Figure 12). The final design consisted of horizontal 'passive' ties from the inner wall to the outer face although inclined raking anchors through the core of the Pier tied into the Jersey rock shale were also considered.

Several options were also considered for repairing and strengthening the Fort Breakwater inner wall but due to financial constraints, the options for structural integrity repairs and long-term protection were limited. The budget was focused upon dealing with the bulging section only. The engineering options considered were:

- Partial rebuild and removal of cementitious grout;
- Take down and partial rebuild in sections;
- Mini-piling to provide an effective inner foundation;
- Reduction of wave energy forces on the outside face of the breakwater – rock armour;
- The 'stitching' option repairs and strengthening.

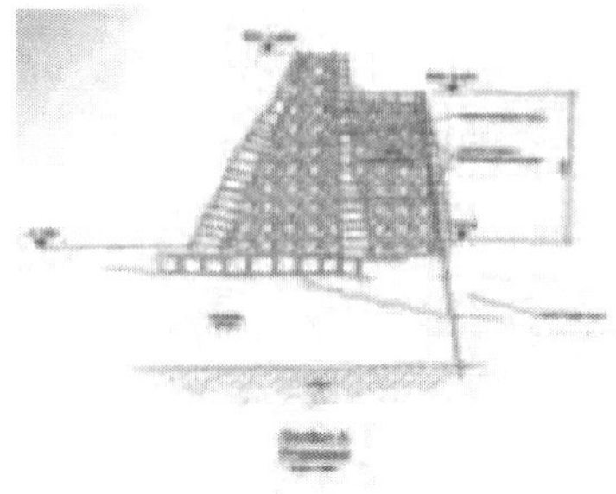

Figure 13 - St. Aubin's Fort inner wall ties and mini piles

The quality of the cementitious grout that was pumped into the end of the structure in 1972 was found (in 2009) to now be of deteriorating variable quality and consistency. However, it was not thought pragmatic or cost effective to dismantle this end of the lower deck structure and rebuild it in its original form, whilst the 'stitching' anchor solution provided clear benefits allowing the structure to be anchored and stitched together without changing its character and with minimal impact. In order to optimise the number of anchors, an iterative process was used, finding the optimum spacing to satisfy the onerous load criteria imposed by the sea conditions in such an exposed location. Horizontal ties were inserted through the inner wall of the main breakwater so that the inside face wall of the lower deck was not only tied to the inner bulk of the original wall but also had additional cantilevered 'beam' support.

Another design philosophy applied in the design of both structures repair works was that within the weaker strata such as sand, this grout injected anchor/pile system expands, reducing the potential for buckling and increasing skin friction.

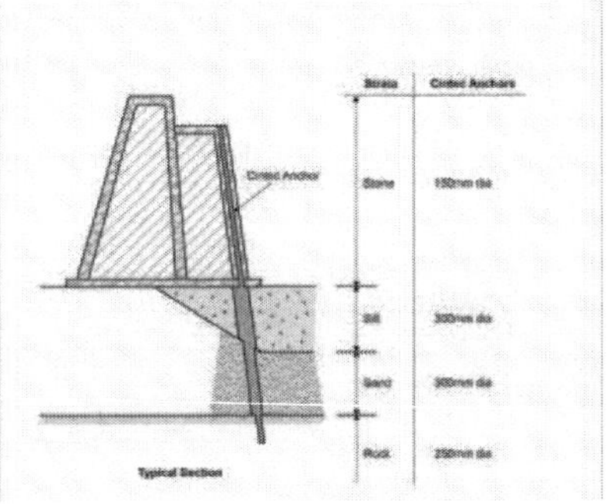

Figure 14 - Cintect anchor design philosophy

Implementation

Not only did the final design solution comply with the planning authority requirements, it

Figure 15 - Anchors with coupler

Figure 16 - Pre-injection with masking tape

Figure 17 - Anchor expanded

furthermore mitigated pollution control concerns by the 'sock' principle of the anchor system where the grout is contained within a small radius of the sock diameter. Only a small surface area of grout oozes through the sock material to bond to the adjacent substrates and when cured, forms a concrete skin over the 'sock' material (Figure 16 before grouting injection/Figure 17 after grouting injection).

During trial anchor works at the North Pier, Jersey Ports refocused the contractor's programme to include similar work at the South Pier and the St. Aubin's Fort Breakwater roundhead, where further significant masonry movements had been observed and stone loss had taken place. This way the repair works to the three structures were wrapped into one project, realising benefits in terms of cost and programme and lessons learnt from the trials.

Figure 19 - North Pier vertical drilling

At the North Pier the vertical drilling works for the anchors took place through the inner wall concrete up-stand used for vessel mooring. Plugs in the masonry were reinserted to enable invisible fix (such as that shown in Figure 20) once the anchor had been installed. The supplier utilised standard 2.5m anchor lengths for both structures, coupled together to achieve vertical and horizontal lengths as required (Fig 15).

Another valuable aspect of the installation process was that each individual anchor drilling provided its own borehole information. This allowed the length of the anchors to be reduced where, for example, the rock outcrop was found to be at a shallower depth.

From a large machine platform aided by localised, demountable scaffolding the horizontal anchors were drilled and fixed (Figure 19). The number of verticals and horizontals at the inner wall of the Fort Breakwater required careful setting out to avoid conflicts and also to allow for flexibility with respect to deck positioning of the rig on the structure. In-situ vertical load tests took place on the anchors to confirm design assumptions,

Figure 19 - Fort horizontal drilling

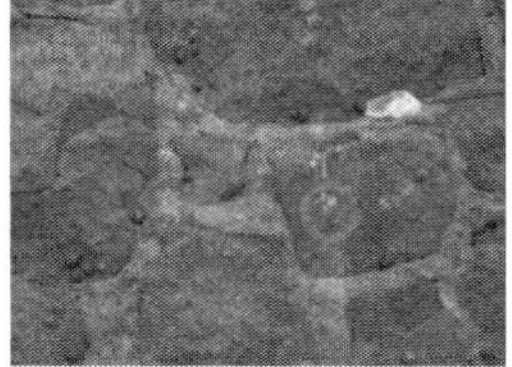

Figure 20 - Anchor with plug cap replaced, 'Secret Fix'

configuration and spacing of the anchors. The storm events at St. Aubin's Fort in December 2010 coincided with high tides, so the contractor re-focused upon the St Aubin's Harbour North Pier work, another advantage of having wrapped both works into one. In February 2011 they remobilised back out at the Fort Breakwater.

The site works were completed in May 2011, including some localised horizontal 'stitching' necessary in one section of the St. Aubin's Harbour South Pier to prevent further horizontal sliding of the inner wall masonry there.

Conclusion

The 'stitching' method used to stabilise and strengthen these two marine heritage structures has proved to be effective on a number of fronts. The 'secret' or 'hidden' fix of the structure means that the heritage planning aspects of the strengthening works are achieved. The predicted wave pressure paths and loadings were analysed in an empirical way to maximise the effectiveness of the solution in areas of local maximum distress. Environmentally, the impact on the endangered mollusc species is now negligible and the risk of grout spillage is low. Economically, the costs budgeted for the original rebuilding of the inner wall of the North Pier on a new foundation, were of a similar magnitude to that for the 'stitching' techniques. There is a certainty with respect to the capacity of each anchor or mini pile as the drilling technique means that every element's bearing capacity is known and recorded. The technique therefore proved itself adaptable to the engineering judgements so necessary in this type of work; effective in terms of providing strengthening and repairs to threatened parts of heritage structures; and cost effective.

We can conclude that combining methods of historical research, intrusive site investigations and virtual modelling through LiDAR surveys, enabled even these unique and geometrically difficult structures to be secretly and effectively strengthened and preserved for future generations.

Figure 21- The Fort Breakwater Heritage Monument restored

Thanks to the training of the contractor Geomarine Ltd by the anchor manufacturer Cintec, Jersey now has a resident 'stitching' practitioner that can stitch up the future.

The authors acknowledge and thanks for contributions from: Iain Barclay – Geomarine Ltd, John Brookes– Cintec Ltd, and Clon Ulrick – Arup.

Innovative Coastal Zone Management
ISBN 978-0-7277-5749-4

ICE Publishing: All rights reserved
doi: 10.1680/iczm.57494.617

Training of Coastal Engineers to Work in a Non-Engineering Environment

Madelon Burgmeijer, Delft University of Technology, Delft, Netherlands
Henk Jan Verhagen, Delft University of Technology, Delft, Netherlands
Mick van der Wegen, Unesco-IHE, Delft Netherlands

Introduction

Integrated Coastal Zone Management (ICZM) is by definition an interdisciplinary process. Individual actors in CZM need to cooperate and communicate to come to an optimum management strategy for the coastal zone. Because engineers focus on finding the "best" solution in engineering terms, they are often quite surprised to find that at the end of the day the solution they think is best is not considered to be the best solution by the other participants in the process. In 1990 Delft University of Technology and UNESCO-IHEstarted a course on ICZM, which evolved into a training course for engineers to make them aware of this point and to train them in communication with other professionals [Verhagen, 1995]. This paper will give an overview of the experience gained with this course over the last 20 years.

Present practice

Often engineers consider ICZM as finding the best engineering option to solve coastal problems. In this respect "best" means that for example the coastline is maintained at a given place at the lowest cost. However, other considerations are often of equal importance. Although ecological values are usually considered as important (also by engineers), social and political aspects as well as associated decision making processes are often completely neglected. In the education of engineers not much time is devoted to the interaction with other professions, especially not to the interaction with non-professional stakeholders. In this respect the word "non-professional" means groups which main profession is not the management of the coastal zone. Examples of non-professional stakeholders are inhabitants of the coastal zone, fishermen, but also local politicians.

It is not the primary task of coastal engineers to manage the political decision making process. However, engineers should be aware of this process and their specific role in it. They should write reports, etc. in such a way that they can be used effectively in the decision making process. Due to the weight to non-technical arguments in the political discussion, engineers should not be surprised when the opposite decisions are made instead of the engineering recommendations. .

> **Intermezzo 1**
>
> *An example of a well prepared, but completely failed decision in coastal zone management is the "de-poldering" of some areas around the Westerscheldt estuary in southwestern Netherlands. To improve the access to the port of Antwerp extensive dredging works have to be carried out in the Westerscheldt, being a Natura2000 area. As compensation the creation of more "nature" was needed. Ecological studies indicated that the main problem was the decrease of intertidal area. Morphological studies revealed that inundating a few polders along this estuary (by giving them an open connection to the tide of the estuary) would increase the intertidal area. Economically this is a feasible project, because the area to be inundated is mainly agricultural land. An engineering proposal was made for this de-poldering project, supported by ecological and economical studies. All experts agreed that this was the best option for the problem.*
>
> *However, during the compulsory EIA-hearings with the public it became clear that de-poldering in this area was completely unacceptable for the local population. Mainly for emotional reasons. This land had been reclaimed by their ancestors at great cost. It had been defended against extreme storm surges, at great cost as well. It was simply not possible for the local population to even think about inundating this area. This in spite of all good quality background research. This strong reaction of the public came as a complete surprise to the engineers.*

Definitions of "Coastal Zone Management"

Coastal zone management has many definitions. One of them is "to planhuman activities in the coastal zone in such a way that the best strategy is followed for management of existing activities and the development of future activities". In order to elaborate this, one first has to define what the coastal zone is. In order to clarify the topic of discussion we must distinguish the following "types of ICZM":

- Coastline management
  A mainly technical type of management to keep the coastline in good shape. This can be done by construction of revetments, beach nourishment, or measurements to allow a controlled retreat. This is the realm of the engineer.
- Coastal strip management
  The legal and institutional management of the coastal area. Building permits, etc. belong to this domain. This is the realm of the lawyer and the administrator.
- Coastal Zone Management
  The integrated planning of the complete coastal zone, including its effect on the abiotic and biotic systems. This is the realm of the political decision maker.

In some cases people call Coastline Management or Coastal Strip Management "Coastal Zone Management", but this is in fact a misnomer. Especially engineers (and mainly involved in Coastline Management) often refer to Coastline Management as "Coastal Zone Management". This is often the starting point of quite some misunderstanding between engineers and other stakeholders in the coastal zone. The optimum solution for coastline management is often not the optimum solution for coastal zone management. A standard training program for coastal engineers focuses on coastline management, but engineers should realise that this is only one aspect in the decision making regarding coastal zone management.

What is best?

In the above definition, the term "best strategy" is used. However, there is no objective definition of "best". People have different point of views on this term. Therefore the decision on what is "best" will always remain related to a political decision, which can only be taken by what is called in this course, the "decision maker". This decision maker can be an elected official, but it can also be a non-elected influential person (a village elder for example). In any case this person is not an expert in some kind of field, and certainly not an external advisor. Engineers are used to compute the best technical alternative, and therefore it is difficult for them to accept that sometimes a sub-optimal solution (in their view) is selected as the best option by the decision makers.

The Delft course in Integrated Coastal Zone Management

In 1990 a number of institutions in the Netherlands started the development of a Master course in ICZM. As a start-up a short course was given, trying to accommodate students and practitioners with various backgrounds. The course focused on experts, and consisted of participants from various fields of expertise (See figure 1, upper left block). It proved that this did not work at all, because nearly all groups considered the depth of the course as not sufficient in their own field of expertise. Also we concluded that in fact ICZM is not really a science, but merely a matter of good cooperation in planning and decision making, requiring skills to join forces with scientists and practitioners from different backgrounds. So we shifted the focus to the communication between the experts (represented by the arrows in the upper left block) instead of the interactions between the stakeholders (lower right block).

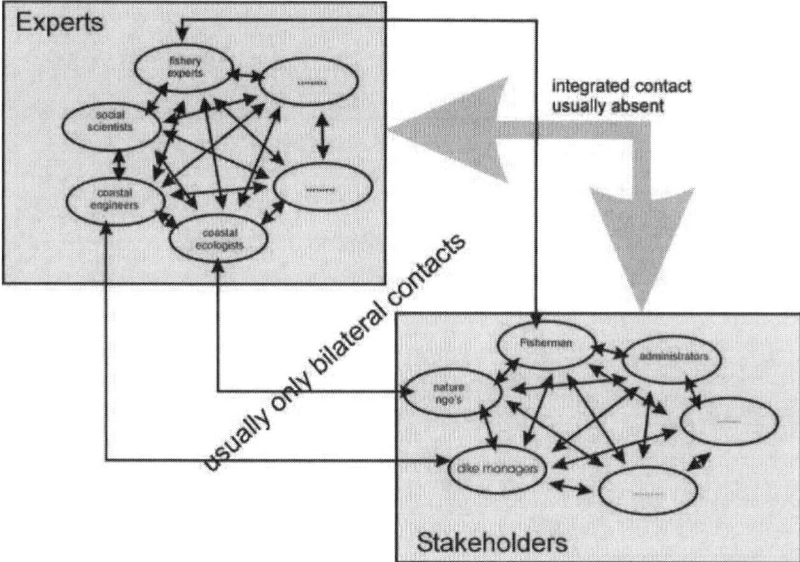

Figure 1. Communication between experts, between stakeholders and between groups

From 1992 onwards we focussed our course on students with a background in the exact sciences (coastal engineers and physical geographers, from Delft University of Technology, Utrecht University and from UNESCO IHE in Delft). The course is given in the final year of the master programme. This means that all students have a good knowledge of the physical processes in the coastal zone. Therefore there is no need to pay much attention to these processes. The course also includes some short introductions in ecology of the coast and in the sociological problems of the coastal zone.

However, the main line of the course is the process of decision making in the coastal zone. In order to make this practical, a case study is made of a fictitious coastal region in a developing country: Pesisir Tropicana. The coast of Pesisir Tropicana has all standard problems of a coastal zone. There is an eroding coastline, a rising sea level and an increasing and poor population migrating to the main city. There is some mining activity (Cassiterite Dredging), artisanal fishery, agriculture and tourism.

The students challenge is simple: make the best strategy for the development of the coastal zone given the amount of money available.

In the first phase of the exercise the students are divided into small groups, playing the role of a Policy Consultant Company. Each group basically gets the same assignment (as defined above), but has a different client. All clients are stakeholders from society (e.g. local government, fishermen association, Cassiterite Dredging Company). A short list of the primary and secondary objectives of the clients is given to the groups. They are requested to prepare a report for their client to use in a decision making meeting.

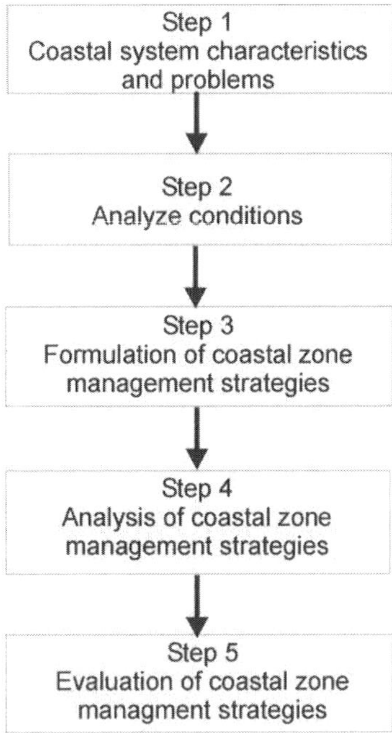

Figure 2.Steps for setting up a CZM strategy (following WCC'93)

For this exercise we follow the steps of WCC'93 (see Figure 2). For step 4, a resource allocation model is used for the calculation of the effect of the various strategies. This is a spreadsheet where controlled parameters like investments can be entered, as well as scenario values (like the expected sea level rise). The spreadsheet then calculates the effect of the strategy on employment, income, pollution, ecological quality, etc. Finally the strategies are evaluated using a simple comparison programme with weight factors. For this purpose we use Jesew, a computer program for 'Joint Ecological and Socio-economic Evaluation of Water resources development. The program is originally written in Basic (Torno, 1988), but for the purpose of this course it is rewritten as a spreadsheet application (Zitman, 2003).

At the end of this phase the students have made a nice report and are convinced that they have found the most optimum solution for their client. At this stage they usually do not realise that the weight factors included in Jesew are not objective at all, but in fact reflect their own preferences.

In the second phase of the exercise each student gets a different role. They will now act as stakeholders (like the mayor of the town, the owner of the mining company, etc) and will participate in a decision making meeting on the way to plan the future of the region. They also receive an "expert report", which is in fact one of the reports written by another group in phase 1. During the discussion in this meeting they are confronted with the strengths and weaknesses of their own reports and with the experience on how background material for such a decision making meeting has to be prepared.

Lessons learned from the Delft course on Integrated Coastal Zone Management

One of our first experiences is that it is not very useful to make a course in ICZM with the aim to train people to become Coastal Zone Manager. Coastal Zone Management is more skill than knowledge and more process than discipline. In this process information is needed from several disciplines and the information has to be presented in such a way that it helps the decision maker to come to the right conclusion. Decision makers can only make good decisions when they are aware of the alternatives and of the effects of their choices. This means that in the preparation phase alternatives (management options) have to be generated before final decisions can be made. The effect of the alternatives has to be predicted (preferably for different scenarios)as well. In fact these are steps 2 and 3 of the scheme of Figure 2.

The person in charge of facilitating this process could be called the Coastal Zone Manager, but in fact that is not right. There is no Coastal Zone Manager. The process of ICZM develops by stakeholder interaction. There may be facilitators for this process but they will never have a directive, leading role. Therefore it was not useful to make a course to train people in becoming Coastal Zone Managers. It also means that the focus of the course is more on communication and less on analysis.

The conclusion is that for good ICZM one needs well educated experts in their own field of expertise, augmented with some skills in managing decision making processes. The ICZM course trains experts in these skills, so that they are aware of what kind of input they should provide to the ICZM process.

For good Coastal Zone Management one needs three groups: the decision makers, the process facilitators and experts in various disciplines. In the next paragraphs these three groups are described in relation to the course.

The decision makers

One cannot set-up a course in becoming decision maker (however one could train decision makers in making better decisions). Making a training program to illustrate decision makers how the process works is therefore a useful process, but it is not typically a Master course. It has to be a short course and may include simple simulation games. In fact the computer program Cosmo, developed for the WCC 1993 functioned as a training tool for CZM decision makers.

The strong point of Cosmo was that it focused quite well on the decision making process and illustrated on what basis one could make a decision. We did apply Cosmo and similar programs at the WCC and at other short courses for decision makers all over the world. In some cases it was successful in showing decision makers how the process works. In some cases decision makers got the idea that it was possible to put the decision making process into a computer and then let the computer calculate the best solution. We found that it is very difficult to tell decision makers that this cannot be done, because it depends on what aspects they consider as most relevant.

Intermezzo 2:

For some computer tools decision makers have to give weight factors to various aspects of the problem. Then the computer calculates what the best alternative is according to the viewpoint of this decision maker. Some other type of programs ask decision makers to choose between two options, and after a list of questions the best option for this decision maker is presented.

However, we found that decision makers usually answer these questions on the basis of their preferred choice. The thinking pattern of the decision maker is: "I want alternative 3, so I should answer the questions in such a way that alternative 3 appears in the end as the best option".

Because of this behaviour it is not very useful to apply these tools in this way. It is better to give a list of alternatives with their quantified effects to the decision makers (a scorecard format works quite well).

The process facilitators

Process facilitation is a typical process that has to be done by people with a good feeling for inter-human relations. In fact no knowledge of the coast is necessary; knowledge is needed on how people interact and how one can come to well supported decisions. Also the facilitator should act as the interpreter between the experts and the decision makers. The process facilitator has to guard that only relevant information is presented to the decision makers and in such a way that it is understandable for non-experts. We concluded during our courses that, although process facilitators are essential, it is not our task (i.e the task of a University of Technology) to train such facilitators. It seems to be more of a task for Universities with humanities departments.

The experts

In the ICZM-process many disciplines are needed, and consequently many experts with different backgrounds. Communication between the experts is important, even more so is communication between them and the decision makers.. Because all groups have their own jargon, such communication is not easy. Therefore one could think of a course for experts with various backgrounds with introductions to adjacent fields in order to help them to understand the other disciplines.

We found that it is impossible to make one course for experts with various backgrounds in order to give them more background knowledge on adjacent disciplines. In the initial stages of the course we did try so, but that was a failure. The biologists were bored during the introductory lectures on ecology, while the engineers did not like the introductions in how to manage the coastline. The general opinion of all participants at the end of these courses was that the content was to general and lacked sufficient depth. Therefore we decided to focus the course on how coastal engineers and physical geographers should act within the ICZM-process. This focus became quite successful.

From the course we learned that engineering students take it for granted that they can compute the best solution. They do not automatically acknowledge that many parameters are in fact unknown and that some scenario parameters are in fact highly uncertain (such as sea level rise and its effects). Furthermore, engineers and other experts are usually able to explain their findings to their "own" stakeholders (coastal engineers communicate with the ministry of public works and environmental scientist with the ministry of the environment) However, they insufficiently realize that there are many different stakeholders, each with their own agenda. Students are only little aware that decisions regarding large investments in the coastal zone often have a dominant political dimension.

We found that it is of no use to explain these aspects by pure lecturing. However, when students experience this in a simulation game in which they are confronted with the limited value of their own engineering reports, they start to understand their role in the coastal zone management process. This kind of course made that engineering students and engineering professionals were more able to put their findings on paper in such a way that this becomes a valuable contribution to the decision making process.

Figure 3. Participants from various countries active in the Simulation Game

Cultural aspects

The past 20 years about 30 students per year enrolled in the ICZM course ending up in about 400 trained people. About 25% of these students were from non-Western nationality (as Figure 3 shows). Apart from the courses in Delft we have given the same course in many overseas countries. We experienced several discussions on the government style in different (non-Western) countries in relation to the decision making processes and stakeholder participation advocated in our course. We believe that there are no fundamental cultural differences in the decision making process. Most governments have an interest of serving the needs of people in an optimal way, although stakeholder participation and procedures may differ considerably over different countries in terms of management (even already within the group of 'Western' countries).

Conclusions and recommendations

It is important to train students (working in a given discipline) in communication with other disciplines. For example, for engineering students focus has to be on political decision making. They have to be trained in communication with stakeholders as well. Furthermore they should realize that "communication" differs from "providing information". We found that a mixture of lectures and simulation games works quite well to create awareness with students on the matter that the "best" solution to solve a coastal zone management problem is not always the obvious one..

Coastal engineers should be educated in coastal engineering and trained in the interaction with decision makers.

References

Torno, H.C. [1988] Training guidance for the integrated environmental evaluation of water resources development projects, UNESCO report FP/5201-85-01

Verhagen, H.J. [1995]. *The requirements for Coastal Engineers in Integrated Coastal Zone Management*. Proc. Copedec-4, Rio de Janeiro, Brazil

World Coast Conference [1993] Preparing to meet the coastal challenges of the 21st century, Noordwijk, The Netherlands, 1-5 November 1993

Zitman, T.J. [2003] Quick manual for Jesew, TU Delft, section Hydraulic Engineering, course document

Innovative Coastal Zone Management
ISBN 978-0-7277-5749-4

ICE Publishing: All rights reserved
doi: 10.1680/iczm.57494.625

The Benefits of Managing Spit Evolution: a Case Study in the Exe Estuary, UK

Paul Canning PhD CEng MICE, Atkins Ltd, Bristol, UK.
Dan Fox PhD, Halcrow, Exeter, UK.
Nigel Pontee PhD MCIWEM, Halcrow, Swindon, UK.

Introduction

A significant proportion of the UK population and industry is located within estuary flood plains. Assessing the flood and erosion risks to these areas is therefore of significant interest. Spits located at estuary mouths are one factor in controlling wave and water levels within estuaries and as such, their management is important as regards managing flood and coastal risk. However the integrity of many of these features under a scenario of limited sediment supply and rising sea level is uncertain (Hartley and Pontee, 2008).

The Exe Estuary Flood and Coastal Risk Management (FCRM) Strategy sets out how flood and coastal risks around the Exe Estuary, in Devon, UK, will be managed over the next 100 years. The Exe Estuary is a 'spit enclosed drowned river valley' (Defra, 2007), which has been subjected to marine inundation caused by a rise in sea level at the end of the most recent glaciation (which ended c. 12,000 years ago). The Exe Estuary has a shoreline length of 40km, channel length of 16km, valley width 2km and a mouth width of 380m. It is classified as macrotidal with a range of 4m, whilst the River Exe (the main tributary) has a mean flow of $23m^3/s$, and a maximum of $371m^3/s$. At the mouth of the Exe Estuary, the sand spit of Dawlish Warren covers approximately three quarters of the estuary mouth width, and consequently potentially shelters the estuary from the coastal swell wave climate, as well as influencing the propagation of extreme tide levels. The in-estuary extreme wave climate is consequently limited to significant wave heights of less than 1.1m, with extreme tide levels between 3-4mAOD. The estuary exhibits high tidal velocities through its mouth, with flood/ebb tidal deltas and offshore banks present. The management of Dawlish Warren sand spit is therefore of particular importance to the strategy, as it potentially reduces flood risk to up to 7000 properties in the wider Exe Estuary.

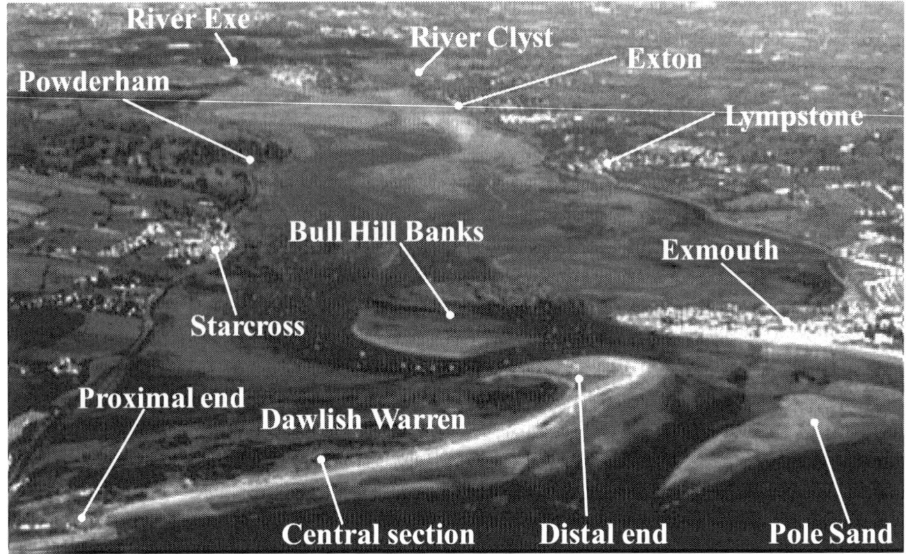

Figure 1. The Exe Estuary and main features (sourced from the Environment Agency).

Strategic context

The FCRM Appraisal Guidance (Environment Agency, 2010) notes that baseline scenarios should be defined, to support assessment of potential options for management of flood and coastal risks. The Do Nothing and Do Minimum scenarios represent progressively diverging situations for estuaries (Canning *et al*, 2010), where FCRM assets are allowed to deteriorate, or are structurally maintained, respectively. In the Exe Estuary, this divergence starts occurring by 2030 when weaker assets would have failed, and is fully realised by 2060 when all standard FCRM assets would have failed under the Do Nothing scenario. However, for maintained FCRM assets (the Do Minimum scenario), by 2110 the magnitude of climate change would likely still result in loss of their FCRM function (i.e. they would experience breach or allow flooding on at least an annual basis). This would result in the Do Nothing and Do Minimum scenarios converging to the same situation.

In the Exe Estuary a complicating factor is that there are two key issues that could influence the baseline scenarios. The first is the future evolution of Dawlish Warren sand spit, whilst the second is the future management of the railway embankments. The former is of particular interest herein. If a significant amount of damage is allowed to occur at Dawlish Warren, flood risk within the estuary could potentially be increased due to increased wave and tide climate. This issue is less important in the Do Nothing scenario, as the majority of FCRM assets could have already failed by the time Dawlish Warren was seriously damaged. This means that there would be limited FCRM or socio-economic impact over and above the wider Do Nothing assumptions. However, it is of greater importance to the Do Minimum scenario, as the maintained FCRM assets around the Exe Estuary shoreline could experience an increased extreme wave and tide climate. These issues are shown conceptually in Figure 2.

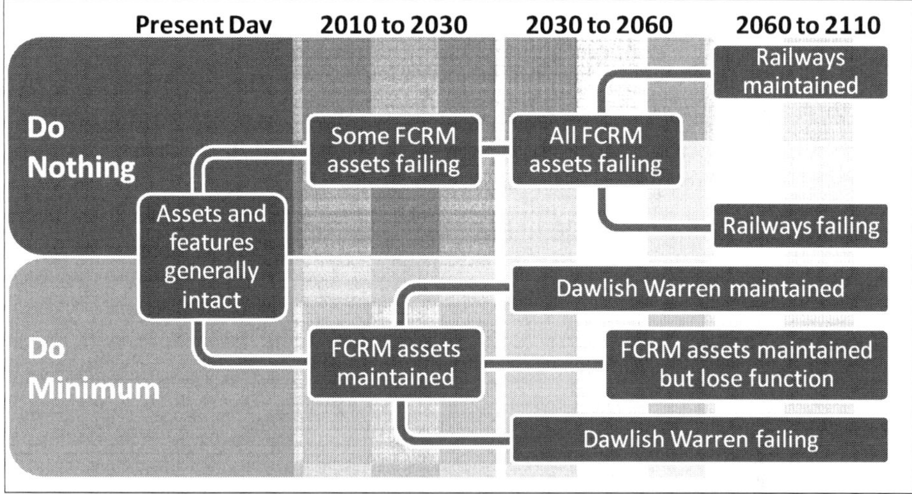

Figure 2. Overview of potential future baseline scenarios.

Geomorphological Assessment

A number of studies have already been completed on the coastal and estuarine processes surrounding the Exe Estuary, particularly including Posford Duvivier (1998), FutureCoast (Defra, 2002), the Lyme Bay and South East Devon Sediment Transport Study (SCOPAC, 2004), the Exe Estuary Coastal Management Study, EECMS (Halcrow, 2008; Fox *et al.*, 2008), and the South Devon and Dorset Shoreline Management Plan 2 (Halcrow, 2010). The approach taken to further develop understanding of the estuarine and coastal processes was considered in light of Defra (2007), with the intention to build a robust evidence base for future predictions of geomorphological change. The approach included:

- Human and natural impacts reviews, which provided a timeline of human impacts and storm events, affects on the evolution of the estuary, and qualitative confirmation of key areas of risk.
- Historic and modern spatial and volumetric analyses, which provided a long-term (centuries) assessment of the spatial and volumetric changes to the intertidal and supratidal features in the Exe Estuary around the estuary mouth.
- Breach analyses, which assessed the likelihood of damage to Dawlish Warren.
- Modeling of waves and water levels in the estuary with different configurations of Dawlish Warren.

A data search resulted in the collation of Ordnance Survey maps for the years 1890, 1900, 1930 and 2000; Admiralty Charts for the years 1830, 1890, 1930 and 2000; vertical aerial photography (focused primarily around the estuary mouth) for the years 1941, 1946, 1950, 1960, 1969, 1979, 1988 and 1999; and LiDAR for the years 1998 and 2007. The above assessments were synthesised in a conceptual model, developing an improved understanding of how the estuary system works, how it has evolved historically, and how it could evolve into the future. In particular, the spatial and volumetric datasets were used to confirm or provide new quantitative estimates of sediment transport rates within the coastal and estuarine system.

Historic spatial and volumetric evolution

Geo-referencing to well defined, long-term unchanging features resulted in an estimated horizontal and vertical accuracy of the order of 1-10m and 0.1-1m respectively, improving for the post 1960 datasets. Spatial mapping of Low Water Mean Ordinary Tide (LWMOT) and High Water Mean Ordinary Tide (HWMOT) from the Ordnance Survey maps was used as an indicator of movement of intertidal and supratidal features, specifically Dawlish Warren, Pole Sand (the ebb delta) and Little Bull Hill and Great Bull Hill Banks (forming the flood delta, subsequently referred to as Bull Hill Banks), as shown in Figure 3.

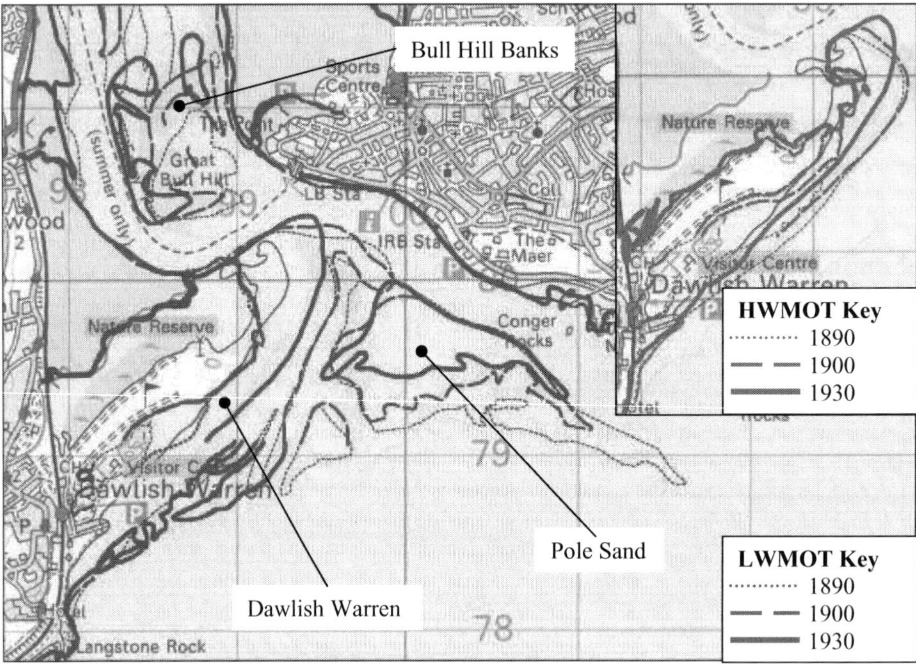

Figure 3. Spatial mapping of intertidal and supratidal features.

Interpretation of the changes in LWMOT and HWMOT over time are also shown in Figure 4, with the following commentary. Between 1890 and 1930, at Dawlish Warren it is apparent that the central and distal sections moved up-estuary, with formation and expansion of a sub-tidal channel between itself and Pole Sands. At the same time, the planform and volume of Pole Sand increased, whilst Bull Hill Banks varied in planform, drifted westwards and increased in volume. Between 1930 and 2007, the trend for Dawlish Warren to move in-estuary decelerated, with evidence of beach steepening and significant erosion along the central section. The distal section accreted and experienced variable expansion/retreat across the estuary mouth, whilst the subtidal channel between Dawlish Warren and Pole Sands also became less coherent, with the planform and volume of Pole Sands showing variable trends (inferring an expansion and flattening). These observations are generally corroborated by Estevez and Williams (2010), with the suggestion that the distal end movement was related to the Pole Sands expansion. Further in-estuary, Bull Hill Banks continued to vary in planform, reversed its drift westwards and decelerated its increase in volume.

Estuary-wide analysis of intertidal bathymetry change between 1998 and 2007 further indicated an absolute average increase in volume an order of magnitude less than the accuracy of the LiDAR dataset. Nevertheless the whole estuary appears to have shown a net increase in bed elevation of 1mm/year, which is exceeded by the estimated rise in sea level for the area (3.5mm/year).

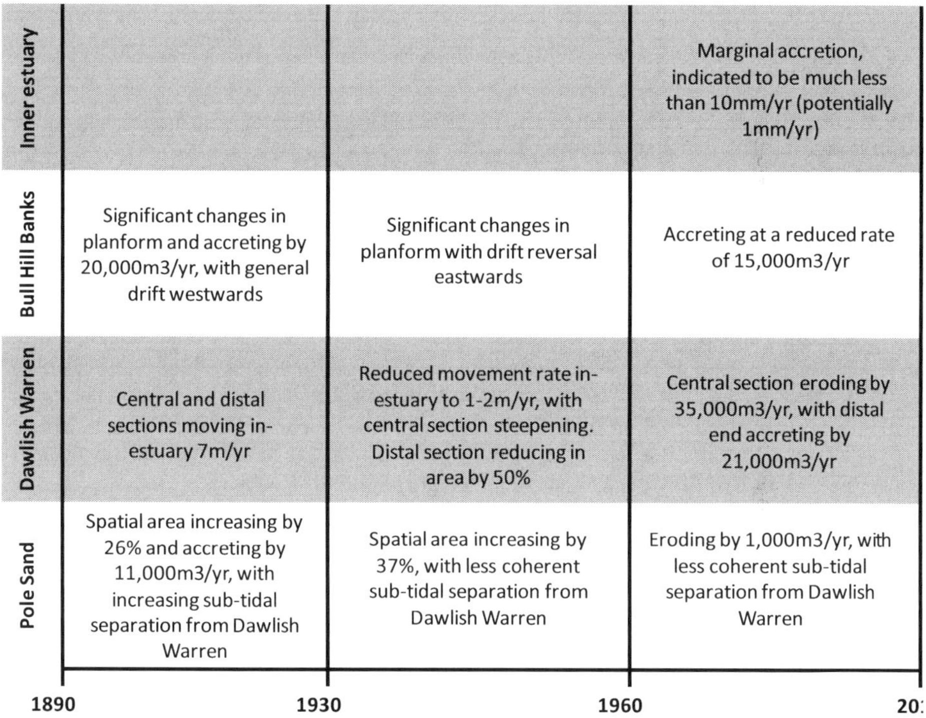

	1890	1930	1960	20:
Inner estuary			Marginal accretion, indicated to be much less than 10mm/yr (potentially 1mm/yr)	
Bull Hill Banks	Significant changes in planform and accreting by 20,000m3/yr, with general drift westwards	Significant changes in planform with drift reversal eastwards	Accreting at a reduced rate of 15,000m3/yr	
Dawlish Warren	Central and distal sections moving in-estuary 7m/yr	Reduced movement rate in-estuary to 1-2m/yr, with central section steepening. Distal section reducing in area by 50%	Central section eroding by 35,000m3/yr, with distal end accreting by 21,000m3/yr	
Pole Sand	Spatial area increasing by 26% and accreting by 11,000m3/yr, with increasing sub-tidal separation from Dawlish Warren	Spatial area increasing by 37%, with less coherent sub-tidal separation from Dawlish Warren	Eroding by 1,000m3/yr, with less coherent sub-tidal separation from Dawlish Warren	

Figure 4. Timeline of spatial feature changes.

Present day sediment transport trends

Consideration of the spatial and volumetric analyses described above and in more detail in Atkins and Halcrow (2010), in conjunction with previous studies (SCOPAC, 2004; and Posford Duvivier, 1998), enabled quantification of sediment transport pathways to build towards a sediment budget model as summarized in Figure 5 (volumetric rates of exchange are estimated to have an accuracy of the order of 10%, and hence are quoted to the nearest 1,000m^3/year). This particularly informed the strategic understanding of erosion of Dawlish Warren sand spit, and potential accretion in the inner estuary.

SCOPAC (2004) indicated sediment transport towards the estuary mouth from both the eastward and westward coastlines, which has been corroborated by the spatial and volumetric analyses. In detail, sediment input to Dawlish Beach is sourced from erosion of Holcombe cliffs (1,000m^3/year) whilst Langstone Rock and groyne, under non-extreme conditions, stop further eastward transport through to Dawlish Warren. The proximal and central sections of Dawlish Warren are eroding (35,000m^3/year) with sediment transported alongshore to both the distal end and offshore to the near-shore banks, and the distal end of Dawlish Warren accreting (21,000m^3/year). Similarly, sediment input (10,000m^3/year) to Sandy Bay is sourced from erosion of Orcombe Cliffs, and potentially can be completely transported westward to Exmouth beach. However, Exmouth beach experiences sediment transport (6,000m^3/yr) from the upper beach to the lower beach, and subsequently to the tidal channel and presumably to the estuary mouth sub-tidal.

Both Posford Duvivier (1998) and SCOPAC (2004) suggested that the nearshore banks along Dawlish Warren and Exmouth beach may have a sediment transport link to Pole Sand, with sediment entering the ebb delta system becoming subject to tidal processes which redistribute sediment between the flood and ebb deltas. The spatial and volumetric analyses indicate that the total estimated input to the estuary mouth sub-tidal is 31,000m^3/year, drawn from Exmouth Beach, Pole Sand and the distal end of Dawlish Warren.

Moving up-estuary, SCOPAC (2004) notes that the transport of sediment from the estuary mouth sub-tidal to Bull Hill Banks is 17,000m^3/year, This has close agreement with our volumetric analysis, with the remaining balance (14,000m^3/year) corresponding well to the averaged historic rate of dredged removal (15,000m^3/year). SCOPAC (2004) makes the further judgement that a total 3,000m^3/year is passed from Bull Hill Banks and the River Exe to the inner estuary: whilst there is large uncertainty, this has order of magnitude agreement with our volumetric analysis which suggests net transfer into the estuary of 4,000m^3/year from all sources.

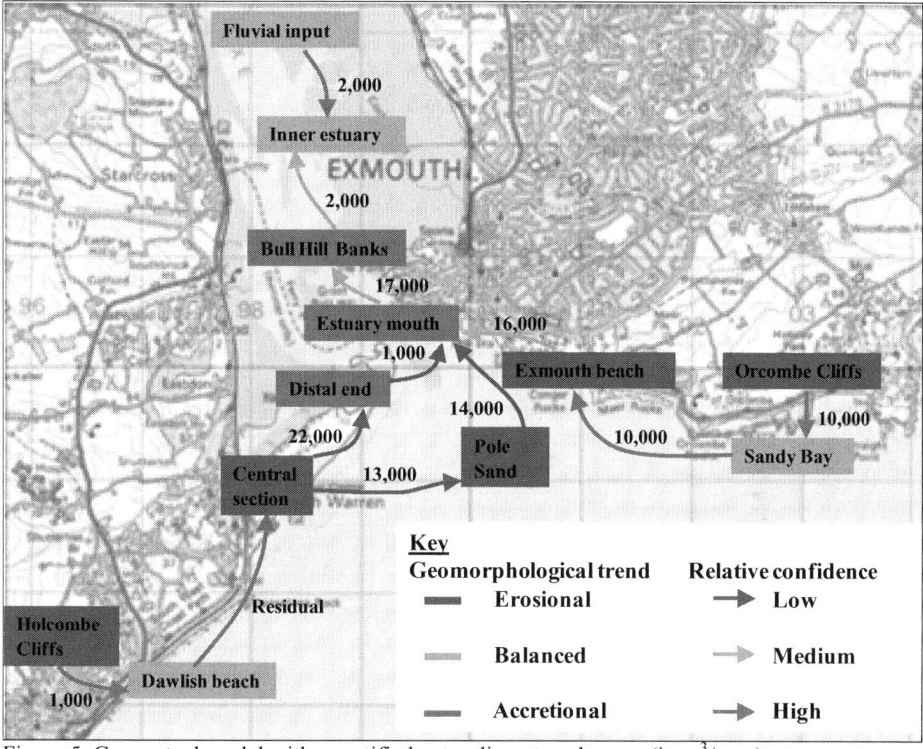

Figure 5. Conceptual model with quantified net sediment exchanges (in m^3/year).

Historic and future potential for breaching of Dawlish Warren

Anecdotal records since the 18th century have noted that Dawlish Warren has periodically experienced large scale change in geomorphological form, eight times in the 19th century, four times in the pre-1940 20th century, and eight times since then. This broadly indicates an annual probability of serious events of 5-15%, increasing towards the modern day. The historic vertical aerial photography provided a more detailed source of visual information relating to damage to Dawlish Warren. In particular, the effect of a well documented (Fox *et al.*, 2008) series of storms in the early 1940s was captured. The 1941 and April 1946 photography showed Dawlish Warren with its distal end increasingly damaged (with remains of buildings on the distal end present in 1941) and partially detached (dependent on tidal state) as a relatively small island. However, by November 1946, just 7 months later, the island had reattached, possibly lowered (due to visual lack of vegetation) and become more active, with evidence of sand recurves. Subsequent to this period, the distal end grew significantly in the 1950s, although still un-vegetated, and there is evidence of protection works (linear structures apparent with north-south orientation). By 1969 revegetation of the distal end had occurred, expanding consistently through to 1999. This process indicates that whilst the distal end may be prone to damage and separation from the main sand spit, it can recover substantially within a decade (assuming no further significant events in the interim period).

Whilst quantified analysis of damage and breach to natural features such as Dawlish Warren is the subject of ongoing research as part of the strategy development, in order to build upon the anecdotal evidence a range of quantified analyses relating to the likelihood of damage to Dawlish Warren were carried out, with the intention of building an evidence base from a variety of methods:

- Overwashing, as identified in Bradbury (2000). It is noted that this method is specific to shingle beaches.
- Erosion of cover of inner slope by overflow, as identified in EC (2007a).
- Erosion of seaward face of sand by waves, as identified in EC (2007a).

Breach as discussed herein is identified with a basal level around mean low water tides, whether infilling, temporary or deepening, as identified in Hartley and Pontee (2008). Visual inspection, LiDAR analysis, and known points of weakness were used to identify a range of points along Dawlish Warren to apply the various analyses. These were applied for the strategic time horizons of 2010, 2030, 2060 and 2110, using climate change guidance (Defra, 2006) with the assumption that Dawlish Warren does not change form. A summary of the results from the various analyses is given in Figure 6, along with the weak point locations. Overall consideration of the findings indicates that serious erosion of the foredunes at Dawlish Warren, and potentially temporary breaching, could occur during 4% Annual Exceedance Probability (AEP) storm events in the present day, although temporary breaching would have a greater likelihood of occurring during a 1% AEP event. The locations of P2 and P4 are particularly found to be susceptible, and these are close to the previously documented breach occurrences in 1962 and 1946 respectively. Further to this, the increase in breach probability between 2030 and 2060 indicates that repeated breaching could over-ride the ability of the distal end to recover naturally.

Potential future evolution of Dawlish Warren

The future 100 year evolution of Dawlish Warren is uncertain. However, it appears that over the last 100 years the central and distal sections of Dawlish Warren have been rotating anti-clockwise into the estuary hinged around its proximal end, and in the last decade the proximal and central areas have been eroding, whilst the distal end is generally accreting. Consideration of the cumulative volumetric impact of present day sediment transport trends, and various breach analyses, indicates that without continued maintenance regular breaching could occur between 2030 and 2060, with repeated breaching leading to a separation of the distal end of the spit from its central section. Previous occurrences of spit breaching have resulted in the distal end reducing in height as it re-attaches to the central section. The detailed morphological form of the distal end of the spit is subject to much higher levels of uncertainty. The broad predicted future evolution of the Exe Estuary is summarised in Table 1 for the Do Nothing and Do Minimum scenarios.

<table>
<thead>
<tr><th rowspan="2">Method</th><th rowspan="2">Weakest point</th><th colspan="4">Annual exceedance probability (%)</th></tr>
<tr><th>2010</th><th>2030</th><th>2060</th><th>2110</th></tr>
</thead>
<tbody>
<tr><td>Over-washing, Bradbury (2000)</td><td>P2: low spot along central section and site of 1960s breach</td><td colspan="2">1</td><td>2</td><td>100</td></tr>
<tr><td>Erosion of cover of inners slope by overflow</td><td rowspan="2">P4: neck of distal end and site of 1940s breach</td><td rowspan="2">4</td><td>10</td><td colspan="2">100</td></tr>
<tr><td>Erosion of seaward face of sand by waves</td><td colspan="3">10-100</td></tr>
</tbody>
</table>

Figure 6. Points of interest along Dawlish Warren, with summarised breach analysis results.

Table 1. Timeline of future changes.

Epoch	Do Nothing	Do Minimum
Present Day	Dawlish Warren coast facing length has breach SoP of 4% AEP, more generally 1% AEP. Estuary facing length has breach SoP of 10% AEP.	
2010 to 2030	Dunes, beach, groynes and gabions deteriorating rapidly. Central Dawlish Warren seriously damaged, with the distal end separated on a temporary basis, but would continue to provide shelter from offshore waves.	Dunes, beach, groynes and gabions would require regular maintenance. Central Dawlish Warren intermittently damaged, with the distal end separated on a temporary basis. However, Dawlish Warren would continue to provide shelter from offshore waves.
2030 to 2060	Central Dawlish Warren permanently damaged and open to in-estuary movement of up to 60m. The distal end would be regularly separated from the main body and lost in supratidal form. This would allow some penetration of offshore waves into the inner estuary.	Central Dawlish Warren protection would require extensive regular maintenance to remain fixed in place. The distal end would be regularly separated but still exist as an island feature. This would provide shelter from offshore waves for parts of the inner estuary.
2060 to 2110	Dawlish Warren would likely move in-estuary and become an intermittent intertidal sandbank feature, but would continue to provide some shelter from offshore waves.	

Flood and coastal risk, and socio-economic benefits

The impact that the predicted evolution of Dawlish Warren could have on flood risk and related socio-economic consequences was based on the comparison of the predicted future geomorphological change, and no future change, at Dawlish Warren. These two scenarios can be considered to provide likely extreme bounds for testing flood risk and socio-economic response (Environment Agency, 2010), and the consequent benefits of future management of Dawlish

Warren.

The prediction of geomorphological change at Dawlish Warren was used to inform wave and tide extremes modelling within the estuary using MIKE21SW, ISIS and DAWN. These models were based on:

- present day bathymetry with Dawlish Warren intact; and also,
- an amended bathymetry for the distal end of Dawlish Warren, based on the well-recorded large scale geomorphological changes during the 1940s, showing the distal end reduced to a sub-tidal morphological form.

These two bathymetries were used to define the change in extreme wave and tide climate around the estuary shoreline. Assessment of the corresponding performance (applying EC, 2007a and b) of the FCRM assets, and inundation, around the Exe Estuary for the two scenarios highlighted that flood and coastal risk could be significantly increased with the partial breakdown of Dawlish Warren spit. Along the east and west banks of the estuary the standard of protection against breach, and consequent flooding, provided by assets (to the settlements of Lympstone, Exton and Powderham) was found to decrease by up to 100%. However, at Starcross which is more directly sheltered by Dawlish Warren, the increase was significantly greater with annual breaching of assets predicted to occur. Further application of these findings to the baseline socio-economic analyses (Atkins and Halcrow, 2010) indicated that the comparative Present Value benefit that the distal end of Dawlish Warren could provide to the wider estuary is in the range of £5 million to £60 million, heavily weighted towards Starcross. The Present Value benefit range is dependent on the particular timing of large scale, permanent geomorphological change at the distal end.

Conclusions

Assessment of the historic evolution of the Exe Estuary, and Dawlish Warren, has enabled future predictions to be made to the extent that the benefit of Dawlish Warren sand spit can be quantified. Analysis of the sheltering function of Dawlish Warren sand spit to the inner estuary has identified that flood risk benefits are material for the settlements and infrastructure adjacent to the estuary shoreline, and are quantified at between £5-60 million Present Value benefit for the Do Minimum (maintaining the structural integrity of FCRM assets) scenario, as compared to the Do Nothing (allowing FCRM assets to naturally deteriorate) scenario. This finding provides a clear basis for informing engineering decisions to manage this feature.

The geomorphological assessment of the Exe Estuary has identified the value of building a wide evidence base, by using a broad range of qualitative and quantitative analyses. The value of historic Ordnance Survey and Admiralty Chart mapping is confirmed in quantifying historic and recent sediment transport trends, providing a key dataset for framing the prediction of future trends. Historic vertical aerial photography is particularly valuable in providing visual documentary evidence of events that have temporarily but significantly affected spit features. Analysis and prediction of overwashing and erosion, leading to potential breach, of sand spit features benefits from considering range of failure modes, in conjunction with anecdotal records of historic storm events and their related impacts. The overall approach adopted in this study is applicable to other parts of the world where spits or coastal barriers potentially influence flood and coastal erosion risks.

References

Atkins and Halcrow, 2011. Baseline Flood and Coastal Risk Assessment, Draft Report. Exe Estuary FCRM Strategy. Report to the Environment Agency, 236p.

Canning, P., Hills, K., Samphier, R., Quarrier, G. and Dunsford, D., 2010. Climate change adaptation at an estuary-wide scale. Flood and Coastal Risk Management Conference, Telford, UK. 30th June – 1st July 2010.

Defra, 2002. Futurecoast. Set of 4 CDs (including security patch update) produced by Halcrow Group Ltd, Swindon, UK, as part of the Futurecoast study for DEFRA 2002.

Defra, 2007. Project FD2117: Development and Demonstration of Systems Based Estuary Simulators. Website: www.discoverysoftware.co.uk/estsim/EstSim.html.

Esteves, L. and Williams, J., 2010. Dawlish Warren: an assessment of coastal erosion and mitigation options. School of Marine Science and Engineering, University of Plymouth.

Environment Agency, 2010. Flood and Coastal Erosion Risk Management appraisal guidance. European Community, 2007a. Failure Mechanisms for Flood Defence Structures, report T04-06-01.

European Community, 2007b. Wave Overtopping of Sea Defences and Related Structures: Assessment Manual.

Fox, D., Pontee, N.I., Fisher, E., Box, S., Rogers, J.R., Reeve, D.E., Chadwick, A.J. and Sims, P., 2008. Spits and Flood Risk: The Exe Estuary. Paper 02a-3, Flood and Coastal Risk Management Conference, University of Manchester, UK, 1st to 3rd July 2008.

Halcrow, 2008. Technical Findings Report. Exe Estuary Coastal Management Study.

Halcrow/Environment Agency, 2010. South Devon and Dorset Shoreline Management Plan Review. Available to view at http://www.sdadcag.org/.

Hartley, L., and Pontee, N.I., 2008. Assessing Breaching Risk in Coastal barriers: Lessons from Suffolk, UK. *Proceedings of the Institution of Civil Engineers, Maritime Engineering Journal.* 161, MA4, 143-152.

Posford Duvivier, 1998. Lyme Bay and South Devon Shoreline Management Plan.
SCOPAC, 2004. Lyme Bay and South East Devon Sediment Transport Study.

Innovative Coastal Zone Management
ISBN 978-0-7277-5749-4

ICE Publishing: All rights reserved
doi: 10.1680/iczm.57494.637

Author Index